D0072793

IEE ELECTROMAGNETIC WAVES SERIES 15
SERIES EDITORS: PROFESSOR P. J. B. CLARRICOATS.
E. D. R. SHEARMAN
AND J. R. WAIT

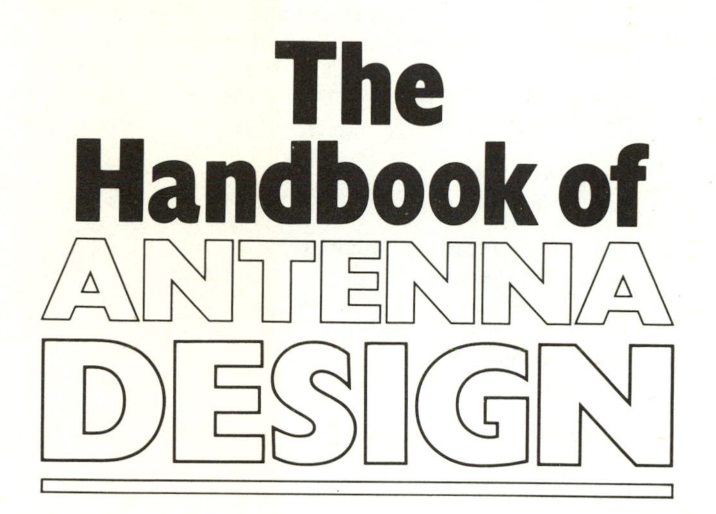

Previous volumes in this series

Volume 1	Geometrical theory of diffraction for electromagnetic waves Graeme L. James
Volume 2	Electromagnetic waves and curved structures Leonard Lewin, David C. Chang and Edward F. Kuester
Volume 3	Microwave homodyne systems Ray J. King
Volume 4	Radio direction-finding P. J. D. Gething
Volume 5	ELF communications antennas Michael L. Burrows
Volume 6	Waveguide tapers, transitions and couplers F. Sporleder and H. G. Unger
Volume 7	Reflector antenna analysis and design P. J. Wood
Volume 8	Effects of the troposphere on radio communications Martin P. M. Hall
Volume 9	Schuman resonances in the earth-ionosphere cavity P. V. Bliokh, A. P. Nikolaenko and Y.F. Filippov
Volume 10	Aperture antennas and diffraction theory E. V. Jull
Volume 11	Adaptive array principles J. E. Hudson
Volume 12	Microstrip antenna theory and design J. R. James, P. S. Hall and C. Wood
Volume 13	Energy in Electromagnetism H. G. Booker
Volume 14	Leaky feeders and subsurface radio communications P. Delogne

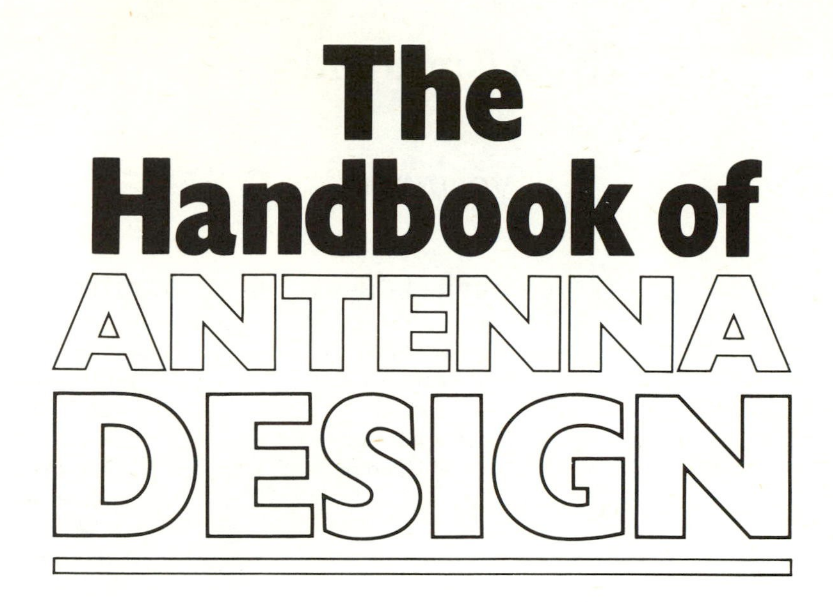

Editors

A. W. Rudge K. Milne
A. D. Olver P. Knight

Peter Peregrinus Ltd on behalf of the Institution of Electrical Engineers

List of Contributors

VOLUME 1

EDITORS

A.W. Rudge
K. Milne
A.D. Olver
P. Knight

CHAPTER LEADERS

W.V.T. Rusch
R.J. Mailloux
L.J. Ricardi
J. Bach Anderson
J. Appel-Hansen

CONTRIBUTORS

T.S. Chu
A.R. Dion
J.D. Dyson
E.S. Gillespie
T.G. Hickman
P.A. Jensen
A.W. Love
A.C. Ludwig
W.C. Wong

Published by Peter Peregrinus Ltd., London, UK.

British Library Cataloguing in Publication Data

The Handbook of antenna design.
Vol.1.—(Electromagnetic waves; 15)
1. Antennas (Electronics)—Design and construction
I. Rudge, A. W. II. Series
621.3841'35 TK7871.6

ISBN 0-906048-82-6

Printed in England by Short Run Press Ltd., Exeter

Contents

Preface

The Handbook of Antenna Design was conceived at a discussion between two of the editors following a meeting of the Professional Group on Antennas and Propagation at the Institution of Electrical Engineering in London during 1977. The original concept was to produce a text which dealt with the principles and applications of antenna design with particular emphasis upon developments which had occurred in the previous decade. The authors of the text were to be invited from a select list of internationally respected experts in the field, to provide a comprehensive treatise on the design of antennas in the frequency bands from LF to microwaves.

In the event the sheer magnitude of the task has imposed its own limitations. Many topics which are worthy of treatment have been omitted and although the coverage of this two-volume text is undoubtedly broad, the editors are fully conscious of the fact that additional volumes would not be difficult to fill. To condense the material to the greatest extent possible the contributing authors were asked to include detailed theoretical developments only where these were fundamental to a proper understanding of the design principles, and were not readily available in existing well-established texts. Although much emphasis has been placed upon the provision of formuli and data which is applicable to the design process, it was recognised from the outset that an adequate understanding of the basic principles involved is essential for good design practice and in encouraging innovation.

The extensive efforts which have been devoted to the development of antennas and the associated design and measurement techniques over the past few decades is, perhaps, not surprising. The electromagnetic spectrum is a limited resource and with the very rapid growth of system applications in communications, navigation and radar, there has been a continuous pressure for more effective and efficient transmission and reception of electromagnetic radiation. The crowded spectrum has led to more demanding specifications upon the performance of radiating devices and has thereby created an underlying need for improved design techniques.

The need for more accurate design and the availability of powerful digital computers have combined to guide antenna design along a path which is providing a continuous transition from empirical art to mathematical science. This trend has been nowhere more evident that at the shorter wavelengths, where the combination of digital computer and mathematical method has found numerous applications in the design of quasi-optical antennas. Computer codes based upon geometric optics, physical optics, the geometric theory of diffraction, Fourier transform theory, and spherical-wave analysis now provide essential tools in modern antenna design. Thoughtful application of these analytical methods has lead, not only to improved computations of antenna radiation characteristics, but also to improved insight into the underlying physical phenomena.

The first volume of this text provides the mathematical background and a large number of examples of the use of mathematical methods in the design of reflector and lens antennas, including their primary-feeds. Quasi-optical antennas, including hybrid configurations, where the reflector or lens is used in conjunction with a complex array feed, are particularly well suited to the use of computer-aided design methods and major performance improvements can be realised. These topics are dealt with in Chapters 2-6 of the first volume.

Low and medium gain microwave antennas have many applications, of which serving as a primary-feed is only one. Chapter 7 provides design data for a diverse range of radiators including helices and spirals, slots, microstrip antennas, backfire antennas, dielectric rods and horns.

Advances in antenna performance must be matched by improved electrical measurements and there have been a number of significant developments in antenna metrology during the past decade. The key developments in this important area are described in the last chapter of the first volume.

The fundamental principles and design of antenna arrays has been the focus of a great deal of attention during the past decade. Although cost still remains a barrier in the wide scale deployment of large phased-arrays a continued growth in the applications of antenna arrays can be anticipated. The importance of array technology has been recognised here in the dedication of the first five chapters of the second volume to this theme. The subject matter covered includes linear, planar, conformal and circular geometries in addition to array signal processing. It is evident that the combination of antenna arrays with integrated-circuit technology will have much to offer in the future in terms of both performance flexibility and lower cost.

In practice many antennas are operated either behind or within a radome cover. In such cases the electromagnetic characteristics of the radome must be considered as a factor in arriving at the overall system performance. In Chapter 14 this important topic is considered and design data are provided for a variety of radome types.

The last few chapters of the second volume are concerned with the design of antennas and coaxial components in the LF to UHF frequency bands. In general design data for these bands are less well documented than their counterparts in the microwave region, and the opportunity has been taken here to bring together design information from a number of sources including a rather sparse and widely dispersed literature.

In preparing these volumes the editors are indebted to the authors, all of whom are well known experts in their fields, who have somehow found time in their very busy schedules to organise and prepare their contributions. The authors have exhibited a remarkable degree of patience in seeing through a project which has taken several years to complete. The work described here has been performed by a great many engineers and scientists over a period of many years and although a sensible effort has been made to ensure that the references and credits are accurate, the editors offer apologies for any errors or omissions in this respect.

For our part as editors we have sought to bring together the contributions into a cohesive whole and have on occasions made modifications or deletions to achieve this end. Any resulting omissions, misprints or errors in the text remains our responsibility and we shall be glad to have notice of them to correct any future editions. Although some effort has been made to standardise the notations employed throughout the text, the two volumes represent the efforts and individual styles of 28 contributing authors and, in view of the excessive work entailed, a comprehensive cross-correlation between chapters has not been undertaken. Vector notations have been indicated by the use of either bold type as in $\boldsymbol{E}, \boldsymbol{H}$, or by use of bars as $\bar{\mathrm{E}}$, $\bar{\mathrm{H}}$. References, equations and figures are numbered in sequence on a chapter basis, and the references are listed at the end of each chapter. A comprehensive index is provided at the end of each volume.

It is the editors hope that this handbook will be of value both to practising design engineers and to students of antenna theory. In the preparation of the text it has been assumed that the reader has at least a working knowledge of electromagnetic theory and antenna technology, however, the material provided should be sufficiently detailed to guide the reader from this stage. The review of the basic properties of antennas contained in Chapter 1 provides an indication of the technical level required and includes appropriate references to further background reading.

Finally the editors wish to acknowledge with thanks the many organisations and institutions for their kind permission to use figures, photographs and other information.

Chapter 1

Basic properties of antennas

A.D. Olver

1.1 Introduction

The fundamental principles of antenna design are based on classical electromagnetic field theory which were established during the second half of the 19th and the beginning of the 20th Centuries when the first radio experiments were being performed. Since that time the applications of antenna design have undergone a steady expansion which accelerated during and after the Second World War with the explosive growth in communications, radar, remote sensing and broadcasting. Antennas are a necessary, and often critical, part of any system which employs radio propagation as the means of transmitting information, and this has been recognised in the very considerable effort devoted to the study of their properties and the engineering of practical radiating systems.

The material contained within this book represents the work of many antenna engineers over a long period of time, but is particularly intended to cover the ideas and designs which have emerged during the last 20 years. This period has seen the development of new technologies such as space and satellite communications and solid-state devices which have stimulated new requirements and new possibilities in communications. There has also been a phenomenal growth in the absolute number of radiating systems, all of which must share the available radio frequency bands. As a direct consequence of the pressure upon the available frequency spectrum and the problems arising from interference and electromagnetic compatibility, modern antennas are required to satisfy much tighter specifications, operate at higher frequencies, perhaps with dual polarisation, and possibly provide variable parameters to take advantage of the power of digital processing.

Compared to an earlier period, the external appearance of many modern antennas may appear to be similar to their predecessors but the electrical performance is likely to be completely different. High-performance antennas are now required to satisfy specifications which were not previously considered attainable.

These performance demands have led to an increasing degree of analytical modelling and computer-aided methods in the design of antenna systems. The theoretical methods and mathematical models used to study antennas have been forced to become increasingly more precise and sophisticated and extensive use is made of numerical techniques to derive optimised designs.

The antenna design process typically commences with the selection of a general type of antenna which is known from experience to be broadly capable of satisfying the specifications. The specifications may include some, or all, of the following electrical performance parameters: *frequency of operation, radiation pattern, gain and efficiency, impedance, bandwidth, polarisation, noise temperature.* In addition there may be the major practical limitations such as dimensional and structural constraints, weight, materials, environmental factors and, of course, cost. It is the range and variety of the combination of demanding electrical and non-electrical specifications which makes the whole process of antenna design both challenging and interesting.

High-performance antenna design is a combination of theoretical and experimental techniques. Wherever possible a theoretical model will be developed to provide predictions of the radiated electromagnetic fields and the input impedance of the antenna in terms of its principal parameters. The validity and accuracy of the theoretical model, which may involve advanced mathematical and computational techniques, must be well understood if it is to be used to investigate and optimise a high-performance antenna system. Occasionally closed-form mathematical expressions can be found for an antenna's radiation or impedance characteristics but it is increasingly common that the theoretical complexity of the system will demand the use of numerical techniques and a digital computer.

Having established the design employing theoretical techniques it becomes necessary to fabricate and evaluate an experimental model. The experimental model may be a full-scale model or alternatively to reduce costs, or to improve the accuracy or convenience of the experimental evaluation, frequency scaling may be employed. This technique can be very effective with many types of antennas but it must be remembered that the scaling must be correctly applied to all parts of the antenna structure which influence its electromagnetic properties. For example, some antenna systems contain frequency-dependent dielectrics while in others their electromagnetic characteristics are strongly dependent upon their immediate environment, the electrical properties of which must be similarly scaled with respect to frequency.

To meet the more demanding performance specifications, electrical measurements on the experimental antenna must be performed to a known degree of accuracy. A well instrumented high-performance antenna range thus becomes an essential requirement for the antenna designer. Considerable progress has been made in antenna measurements during the last decade and this important topic warrants special attention. Important advances have been made in measurement concepts, in precision instrumentation and automation, and in the storage, digital processing and display of measured data.

General classification of antenna types

Antennas can be very broadly classified either by the frequency spectrum in which they are commonly applied or by their basic mode of radiation. For example, employing the 'mode of radiation' basis four groups could be defined as follows: elemental electric and magnetic currents, travelling-wave antennas, array antennas and radiating aperture antennas. Although these definitions are somewhat arbitrary the four groups can be broadly distinguished by the size of the antenna measured in wavelengths, which in turn can be related to the region of the spectrum in which they are commonly applied. To illustrate this division Fig. 1.1 shows these groupings against typical sizes and frequencies and Table 1.1 lists some examples of antennas which can be allocated to each group.

The classification of antenna types into these groupings is obviously only approximate and there are numerous exceptions to these simple categories. Nevertheless, this classification provides a convenient general basis for organising the

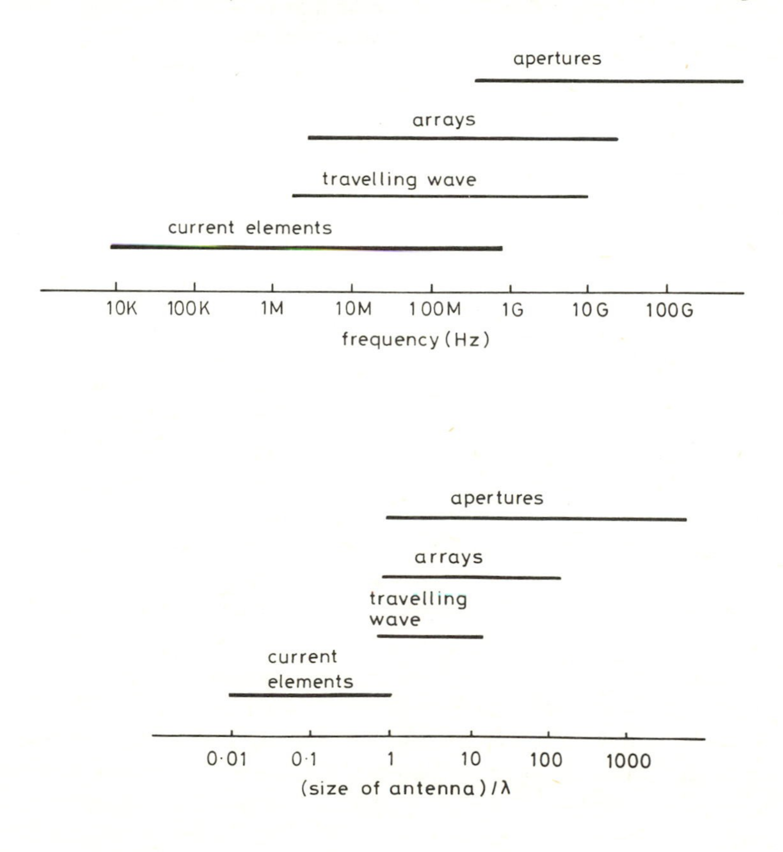

Fig. 1.1 *Classification of antennas*

Table 1.1

Types of Antennas

Elements	Travelling wave	Arrays	Apertures
Monopoles	Line source	Broadside	Reflectors
Doublets	Long wire	Endfire	Single
Dipoles	Rhombic	Linear	Dual
Loops	Slotted Waveguide	Dipole	Paraboloidal
Slots	Spirals	Slots	Non-paraboloidal
Stubs	Helices	Microstrip	Spherical
Rods	Log periodic	Planar	Cylindircal
Vertical radiators	Slow wave	Circular	Off-set
Whip	Fast wave	Multiplicative	Multibeam
Biconical	Leaky wave	Beacon	Contoured
Wire	Surface wave	Conformal	Hybrid
Notch	Long dielectric rod	Commutative	Tracking
Spheroidal		Signal processing	Corner
Disc		Adaptive	Horns
Microstrip radiators		Null steering	Pyramidal
Patch		Synthetic	Sectoral
Active		Log periodic	Conical
			Biconical
			Box
			Hybrid mode
			Multimode
			Ridged
			Lenses
			Metal
			Dielectric
			Luneberg
			Backfire
			Short dielectric rod
			Parabolic horn

material presented in these volumes and it has been adopted for this purpose. Information relating to radiating-aperture type antennas and some travelling-wave antennas is largely contained within this volume while data on current element and array antennas are considered in Volume II. The division is not absolute since primary feed antennas for example, which are dealt with in Volume I, employ a mixture of both current element and radiating aperture techniques. Similarly array antennas, which appear in Volume II, include arrays of small radiating apertures.

The overall performance of a radiating system may be very significantly influenced by the performance of the passive transmission-line components immediately behind the antenna and, in many cases, the radome within which it is contained. Radomes and coaxial components which form a critical part of many radiating systems, are treated in Volume II.

Coordinate systems and unit vector rotation

In the mathematical analysis of antennas the correct choice of coordinate system is often an important contributory factor in simplifying the expressions for the unknown parameters. These parameters are the electromagnetic fields and currents associated with the antenna system, either on the structure itself or at some distant field point. Depending upon the geometry involved, it is common practice to make use of the conventional cartesian, polar, cylindrical and spherical coordinate systems. It is often convenient to employ combinations of these coordinate systems, particularly where calculations of electromagnetic fields radiated from the antenna are required. For example, a point on or close to the antenna structure may be defined in terms of cartesian coordinates (x, y, z) while the distant point at which the field is to be calculated is expressed in terms of spherical coordinates (r, θ, ϕ). This very common hybrid arrangement is illustrated in Fig. 1.2.

Conventional vector notation has been used throughout this text with unit vectors being given either the symbol a or $\hat{a}$ and their direction indicated by an appropriate

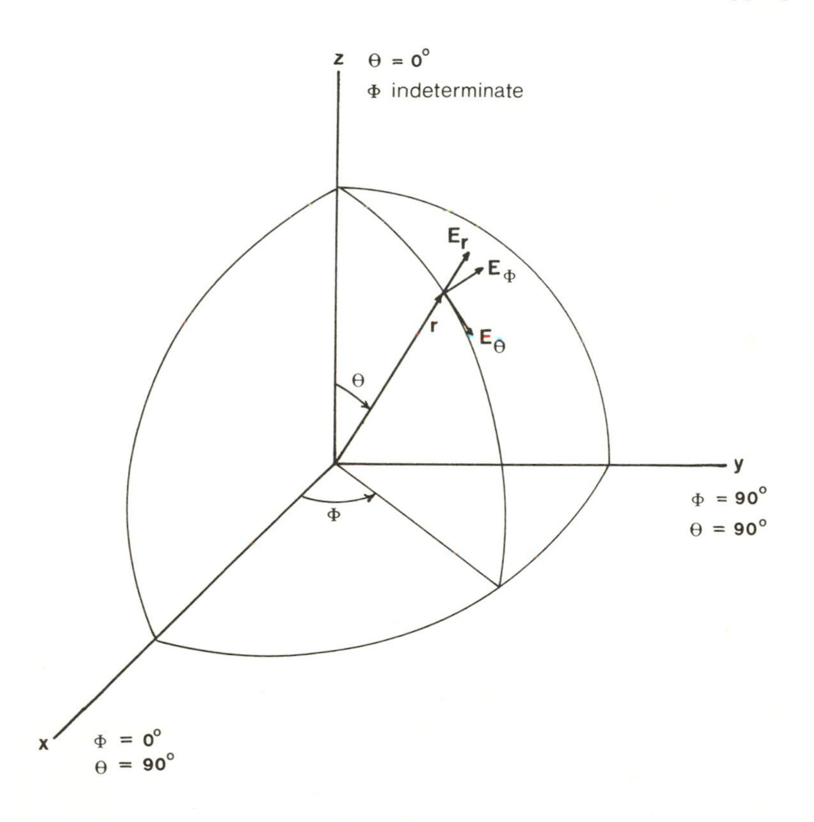

Fig. 1.2 *Spherical coordinate system used with antennas*

coordinate subscript. Referring to Fig. 1.2 we have $\boldsymbol{a}_r$, $\boldsymbol{a}_\theta$, $\boldsymbol{a}_\phi$ for the three orthogonal unit vectors in spherical coordinates or $\boldsymbol{a}_x, \boldsymbol{a}_y, \boldsymbol{a}_z$ for the cartesian case. The recommended orientation of an antenna with respect to the coordinate system is shown in Fig. 1.3 for current elements, small current loops and apertures. The current element (electric dipole) is placed along the z axis. The axis of the small current loop (magnetic dipole) is also placed along the z axis. The aperture is in the xy plane with the axis of the structure behind the aperture oriented along the z axis. In complex antennas, the orientations of the elements may be in different

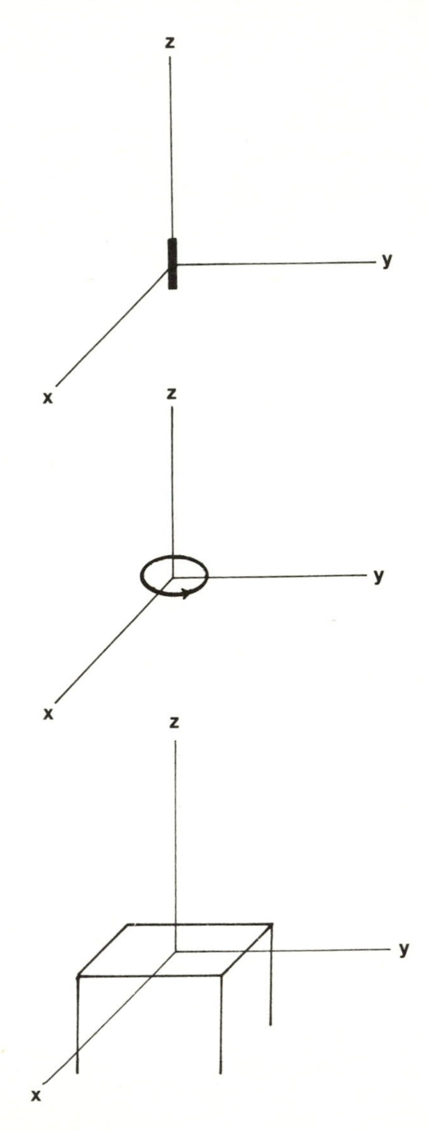

Fig. 1.3 *Orientation of current element, current loop, and aperture*

directions to those shown and it is usually necessary to perform some co-ordinate transformations to be able to evaluate the fields.

1.2 Electromagnetic fields

There are many excellent texts dealing with the principles of electromagnetic fields as applied to antenna theory. For example Collin and Zucker[1], Silver[2], Jordan and Balmain[3], Ramo *et al.*[4], Schelkumoff[5], Stratton[6], Plonsey and Collin[7]. In the theoretical sections of this book the reader is assumed to be familiar with basic electromagnetic field and wave theory. For completeness and by way of definition the main equations used in antenna theory are summarised here. Their derivation can be found in any of the above books.

Maxwells equations

Maxwells equations for a homogeneous isotropic media, including source currents $(\boldsymbol{J}, \boldsymbol{J}_m)$ and with an assumed time dependence of the form $\epsilon^{j\omega t}$, can be expressed in a generalised form as

$$\nabla \times \boldsymbol{E} = -j\omega\mu \boldsymbol{H} - \boldsymbol{J}_m$$

$$\nabla \times \boldsymbol{H} = (\sigma + j\omega\epsilon)\boldsymbol{E} + \boldsymbol{J} \qquad (1.1)$$

$$\mu\nabla \cdot \boldsymbol{H} = \rho_m$$

$$\epsilon\nabla \cdot \boldsymbol{E} = \rho$$

where $\boldsymbol{E}$ and $\boldsymbol{H}$ are the electric and magnetic fields respectively,

$\boldsymbol{J}$ = the electric current density

$\boldsymbol{J}_m$ = a fictitious magnetic current

ρ = the electric charge density

ρ_m = the magnetic charge density associated with σ

σ = the conductivity of the media

ϵ = the electric permittivity of the media

μ = the magnetic permeability of the media

ω = angular frequency (2π times frequency)

$$\nabla = \frac{\partial}{\partial_x} a_x + \frac{\partial}{\partial_y} a_y + \frac{\partial}{\partial_z} a_z \text{ in cartesian coordinates}$$

The fictitious magnetic current J_m is a formalism introduced to symmetrise the equations and thereby simplify certain mathematic developments. A more complete discussion can be found in the references.

Wave equations

The wave equation, which can be derived directly from Maxwell's equations, describe the propagation of an electromagnetic wave in non-dispersive homogeneous isotropic media in the form,

$$\begin{aligned} \nabla \times \nabla \times \boldsymbol{E} &= k^2 \boldsymbol{E} - j\omega\mu \boldsymbol{J} + \triangle \times \boldsymbol{J}_m \\ \nabla \times \nabla \times \boldsymbol{H} &= k^2 \boldsymbol{H} + \triangle \times \boldsymbol{J} - j\omega\mu\epsilon \boldsymbol{J}_m \end{aligned} \tag{1.2}$$

where the wave number is given by $k = \sqrt{\omega^2 \mu\epsilon} = 2\pi/\lambda$

In a source-free region the wave equation can be expressed in a simplified form as

$$\begin{aligned} \nabla^2 \boldsymbol{E} + k^2 \boldsymbol{E} &= 0 \\ \nabla^2 \boldsymbol{H} + k^2 \boldsymbol{H} &= 0 \end{aligned} \tag{1.3}$$

These equations are also termed the Helmholtz equations.

Power

The time average power transmitted across a closed surface s is given by the integral of the real part of one half the normal component of the complex Poynting vector $\boldsymbol{E} \times \boldsymbol{H}^*$

$$P = \text{Re}\, \tfrac{1}{2} \oint_s \boldsymbol{E} \times \boldsymbol{H}^* \,.\, d\boldsymbol{s} \tag{1.4}$$

where $\boldsymbol{E}$ and $\boldsymbol{H}$ are the peak values of the fields (if r.m.s. values are used the ½ is omitted).

Boundary conditions

The boundary conditions describe the relationship between electric or magnetic fields and currents on either side of an interface between two isotropic, homogeneous media.

(a) Two different media with parameters ϵ_1, μ_1 and $\epsilon_2\ \mu_2$

$$\begin{aligned} \boldsymbol{a_n} \cdot \epsilon_1 \boldsymbol{E_1} &= \boldsymbol{a_n} \cdot \epsilon_2 \boldsymbol{E_2} \\ \boldsymbol{a_n} \cdot \mu_1 \boldsymbol{H_1} &= \boldsymbol{a_n} \cdot \mu_2 \boldsymbol{H_2} \\ \boldsymbol{a_n} \times \boldsymbol{E_1} &= \boldsymbol{a_n} \times \boldsymbol{E_2} \\ \boldsymbol{a_n} \times \boldsymbol{H_1} &= \boldsymbol{a_n} \times \boldsymbol{H_2} \end{aligned} \tag{1.5}$$

$\boldsymbol{a_n}$ is the unit vector normal to the surface.

(b) One media a conductor

$$\begin{aligned} \boldsymbol{a_n} \times \boldsymbol{E} &= 0 \\ \boldsymbol{a_n} \times \boldsymbol{H} &= \boldsymbol{J_s} \\ \boldsymbol{a_n} \cdot \epsilon \boldsymbol{E} &= \rho_s \\ \boldsymbol{a_n} \cdot \mu \boldsymbol{H} &= 0 \end{aligned} \tag{1.6}$$

$\boldsymbol{J_s}$ and ρ_s are the current and charge densities on the surface.

Vector potential

It is convenient when working with some antennas to introduce the concept of an electric vector potential, $\boldsymbol{A}$. Many problems can be solved by use of the vector potential because the expressions for the vector potential are much simpler than those for the electric and magnetic fields. The fields can always be derived from the vector potential.

$$\boldsymbol{B} = \nabla \times \boldsymbol{A}; \tag{1.7}$$

where $\boldsymbol{B} = \mu \boldsymbol{H}$. Then

$$\boldsymbol{E} = -\mathrm{j}\omega \boldsymbol{A} + \frac{\nabla \nabla \cdot \boldsymbol{A}}{\mathrm{j}\omega\mu\epsilon} \tag{1.8}$$

$$\nabla^2 \boldsymbol{A} + k^2 \boldsymbol{A} = -\mu \boldsymbol{J} \tag{1.9}$$

Assuming an infinitesimal current source located in a homogeneous isotropic media the equation for $\boldsymbol{A}$ can be solved to give

$$A(r) = \frac{\mu}{4\pi} \frac{e^{\pm jk|r-r'|}}{|r-r'|} a \quad (1.10)$$

$|r-r'|$ is the magnitude of the distance from the source point to the field point at which A is evaluated and a is the unit vector describing the direction of the current element source, Fig. 1.4. The term e^{jkR}/R is often called a Green's function since a

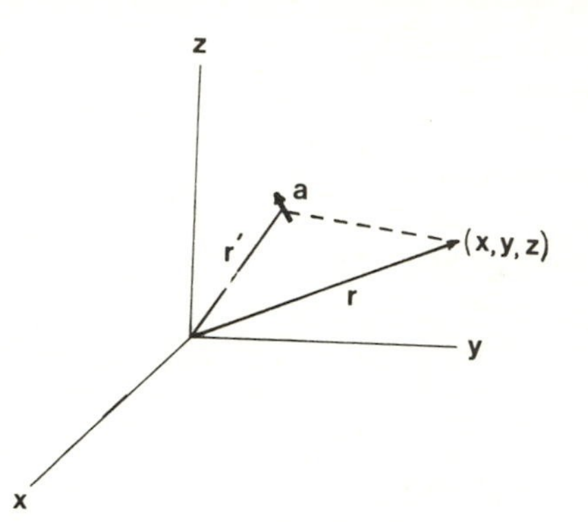

Fig. 1.4 *Vector potential from a current element*

Green's function is, by definition, the solution of the differential equation of a unit source. The vector potential for a general current distribution $J(r')$ enclosed in a finite volume v is given by the integration of the effects of the elemental currents forming the general distribution. Hence

$$A(r) = \frac{\mu}{4\pi} \int_V \frac{J(r')\, e^{\pm jk|r-r'|}}{|r-r'|} dv \quad (1.11)$$

Note that if J describes a surface current then the integration is performed over the closed surface. The electromagnetic fields (E, H) in the region external to the closed surface can then be obtained without further integration from

$$E = -j/\omega\mu\epsilon\,(\nabla \times \nabla \times A)$$

$$H = 1/\mu(\nabla \times A)$$

Similarly a magnetic vector potential (F) can be defined in integral form as

$$F = \frac{\epsilon}{4\pi} \int_V J_m(r') \frac{e^{\pm jk|r-r'|}}{|r-r'|} dv$$

and the associated fields arising from the magnetic current J_m are given by

$$E = -1/\epsilon\,(\nabla \times F)$$

$$H = -j/\omega\mu\epsilon\,(\nabla \times \nabla \times F)$$

1.3 Reciprocity

The principle of reciprocity is of fundamental importance in antenna theory and practice because, for a reciprocal antenna, the properties may be explained (or determined) either by analysis (or measurement) with the antenna as a transmitter or with the antenna as a receiver. An antenna is said to be reciprocal if the constituent parameters of the transmitting media through the antenna can be characterised by symmetric tensors. This generally means no ferrite or plasma devices within the antenna or in the transmission media, and that the media is linear, passive and isotropic. The antenna in this instance includes the transmitting or load impedance because reciprocity applies to the interchange of energy sources.

The reciprocity theorem, as applied to antennas, states that if a (zero-impedance) voltage source is applied to the terminals of an antenna A and the current measured at the terminals of a separate antenna B, then an equal current (in both amplitude and phase) will be obtained at the terminals of antenna A if the

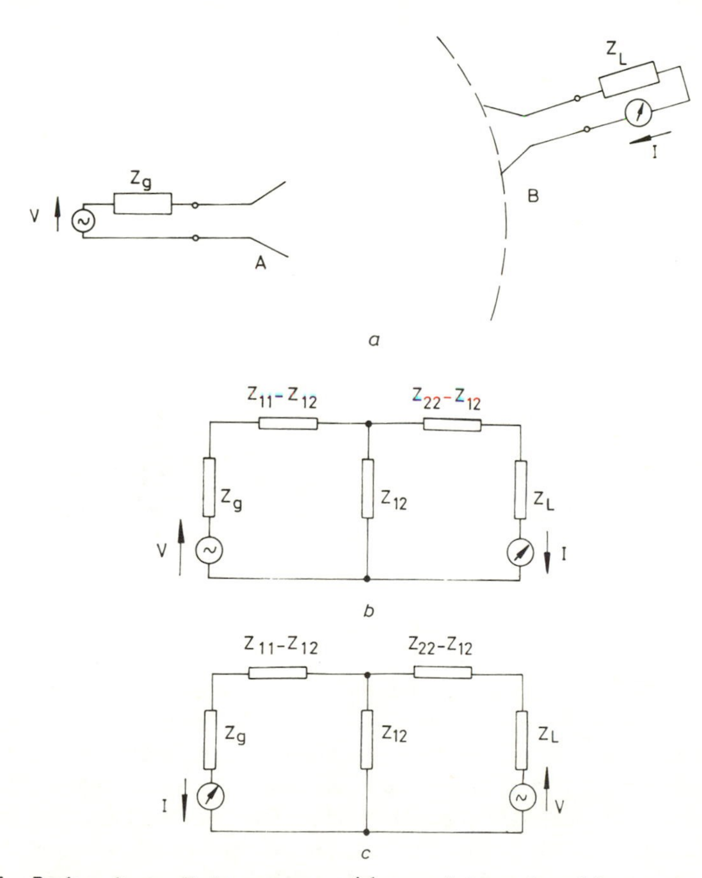

Fig. 1.5 ***Reciprocity applied to antennas:*** *(a)* transmitting and receiving antennas; *(b)* antenna A excited; *(c)* antenna B excited

same voltage source is applied to the terminals of antenna B and the permeability and permittivity tensors of the media between the antennas are transposed.

The general reciprocity relations for electromagnetic fields have been derived by Collin and Zucker Chap. 1[1] and will not be considered here. To show that the radiation patterns of an antenna transmitting or receiving are identical we can treat the communication system as an impedance network.

In circuit-theory terms, the free-space 'connection' between the two antennas in Figure 1.5*a* can be thought of as a 4-terminal network, where the parameters of the network depend upon the separation and orientations of the antennas. For fixed separation and orientations, the network can be reduced to the simple equivalent circuit shown in Fig. 1.5*b*, where Z_g represents the internal impedance of the generator feeding antenna A, Z_{11} and Z_{22} are the self-impedances of the antennas, Z_L the load impedance on antenna B and Z_{12} the 'mutual coupling' impedance. (Under far-field conditions, Z_{12} is very much smaller than Z_{11} or Z_{22}). Straightforward application of circuit theory shows that the current induced in antenna B is

$$I = \frac{V Z_{12}}{(Z_g+Z_{11})(Z_L+Z_{22}) - Z_{12}^2} \tag{1.12}$$

If the voltage source and current meter are interchanged, as in Figure 1.5*c*, the current flowing through the load impedance Z_g on antenna A is given by the same equation, thus demonstrating the reciprocity relationship.

If the separation or orientation of the antennas is altered, the value of Z_{12} is modified, but the reciprocity relationship still applies and hence the variation of received signal with rotation of one antenna is the same whether the antenna is transmitting or receiving.

The load and source impedances and the antenna impedances do not have to be the same, i.e. the relationship applies whether or not the antennas are matched. However, it should be noted that the actual power received is $I^2.\mathrm{Re}(Z_L)$ when the voltage generator is connected to antenna A and $I^2.\mathrm{Re}(Z_g)$ when it is connected to antenna B, i.e. the received powers are only identical when $Z_L = Z_g$.

1.4 Field regions surrounding a radiating source

The electromagnetic fields set up by the oscillating charges of any radiation vary characteristically with distance from their source. Although these changes are continuous with distance it is often useful to identify three distinct regions around a radiating antenna located in free-space. The zones, or regions, are termed the *reactive field region,* the *radiating near-field region* and the *radiating far-field region*, respectively. Although the boundaries of the zones are not defined precisely, the general characteristics of the field distributions in each region can be established.

The reactive near-field region exists very close to the antenna where the reactive components of the electromagnetic fields are very large with respect to the radiating fields. The reactive fields arise from the electromagnetic charges on the structure, they do not radiate but form an essential part of the radiating mechanism. These field components decay with the square or cube of the distance from the source and are generally considered to be negligible relative to the radiating fields at distances of greater than a wavelength from the source. An infinitesimal magnetic or electric current element will produce reactive and radiation fields which have equal magnitudes at a distance of $\lambda/2\pi$ from the source. For any other source current distribution this crossover point will occur at a lesser distance. An infinite plane wave has no reactive field region for example. Large, but finite, aperture antennas will thus have a negligible reactive zone in their centre region but reactive fields will exist at the aperture edges.

Beyond the reactive region the field is radiating, and is divided into two sub-regions: the radiating near-field region and the far-field region. Not all antennas have a radiating near-field region (e.g. electrically small antennas) but all antennas possess a far-field region. In the radiating near-field region the radiation pattern is dependent on the distance from the antenna as well as the observation angle. This is because the distance from different parts of the antenna to the observation point varies considerably and consequently the phase and amplitude of field contributions from the different parts changes proportionally.

The radiating near-field region is sometimes referred to as the *Fresnel* region, by analogy with optical terminology. However, this analogy is only strictly valid for situations where the antenna is focussed at infinity.

As the distance from the antenna is increased the point is reached where the relative amplitude and phase of components from different parts of the antenna become essentially independent of distance. The field then decays monotonically with an inverse dependence on distance. In the far-field region the angular radiation characteristics are independent of distance. The phase term in the expression for the radiation field is approximated accurately by a linear exponential of the form $e^{j(\omega t-kr)}/r$ and only transverse components of the electric and magnetic field appear. This considerably simplifies the mathematical analysis of the angular radiation characteristics.

It is normal to take the $e^{j(\omega t-kr)}/r$ dependence as implicitly stated in many antenna analysis problems. For an antenna focused at infinity the radiating far-field region is sometimes referred to as the *Fraunhofer* region, again by analogy with optical terminology.

Antennas are commonly employed to transmit energy to their far-field region. Since it is not practical to measure the radiation characteristics at infinite distances, it is important to know where the transition can be said to occur between the radiating near-field and the far-field regions. This distance varies from one type of antenna to another. For electrically large antennas which have a well defined cophased radiating aperture a commonly used criteria is to state that the far-field region starts when

$$R = \frac{2D^2}{\lambda} \tag{1.13}$$

where D is the largest dimension of the aperture. At this distance the difference in path length between the centre of the aperture and the edge of the aperture is $\lambda/16$ corresponding to a phase difference of 22½ degrees. If an aperture antenna is measured at this distance it is found that the recorded patterns do not deviate by more than a small factor from the value which would be measured at an infinite distance. Non-aperture antennas behave in a different way and no simple formulae can be derived to define the far-field distance. However, in practice the $2D^2/\lambda$ distance tends to give a conservative estimate of distance and so this is widely used as a criteria for all antennas.

For a large antenna with a constant phase across its aperture D, most of the energy in the far-field region occurs within an angular interval of $\pm\lambda/D$ radians. Close to the antenna, the energy is largely confined to a tube of width D. As illustrated in Fig. 1.6, the radiation can be regarded as being substantially parallel in the first part of the near-field region, diverging into a cone of semi-angle λ/D at a transition range $R_0 = D^2/2\lambda$, where the path difference between centre and edge rays is $\lambda/4$. As noted earlier, the field becomes a good approximation to its value at infinite range for $R > 2D^2/\lambda$, where the path difference is $\lambda/16$.

The exact detail of the field variation within the near-field region is complicated,

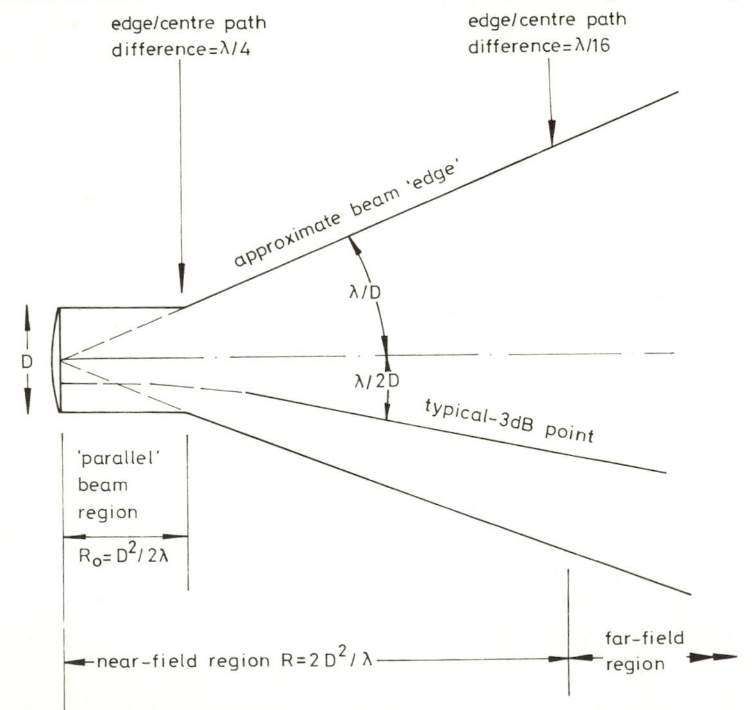

Fig. 1.6 *Radiation from a large antenna*

depending upon the particular amplitude distribution across the aperture, but the power flow across any cross-section of the near-field 'tube' is equal to the total radiated power. The power density gradually settles down to a $1/R^2$ law as the far-field region is approached.

1.5 Radiation patterns

An antenna which is being used as a transmitter does not radiate uniformly in all directions, or conversely, an antenna being used as a receiver does not detect energy uniformly from all directions. This directional selectivity of an antenna is characterised in terms of its radiation pattern, which is a plot of the relative strength of radiated field as a function of the angular parameters θ and ϕ, for a constant radius r, Fig. 1.2.

To completely define the antennas' radiation characteristics it would be necessary to measure or calculate the absolute amplitude, phase and polarisation of the radiated fields over the surface of a sphere and to do this at every frequency of operation. This is rarely feasible so that in practice relative amplitude and phase is sampled at a number of points and the absolute amplitude level calibrated by a measurement of power gain in the direction of the peak radiated field.

For example, with the frequency of operation and polarisation fixed, a value of angle ϕ could be chosen and the relative field measured for increments of angle θ. The angle ϕ is then incremented and the process repeated. The calibration of this relative distribution would then be performed by a single measurement of power gain typically in the $\theta = \phi = 0$ direction. Each two dimensional radiation pattern taken at an increment of ϕ is called a "pattern cut", referred to the test antenna. If the radiation pattern is taken along the $\phi = 0°$ or $\phi = 90°$ planes it is described as a "principal-plane cut". For antennas radiating a linearly polarised field, the principal planes are usually those chosen to be parallel and perpendicular to the electric field vector of the antenna. A radiation pattern described as an electric plane or E-plane pattern records the relative field in a plane perpendicular to the magnetic vector. Fig. 1.7 shows a waveguide aperture with the electric vector along the x axis. The E-plane pattern is recorded in the $\phi = 0°$ plane and the magnetic plane or H-plane pattern is recorded in the $\phi = 90°$ plane.

The radiation pattern may be recorded in a variety of formats. The most common are two-dimensional patterns with either polar or cartesian coordinates. The relative amplitude of the radiated energy may be recorded as a relative power pattern, a relative field pattern or a logarithmic decibel pattern. Each has advantages and the choice is determined by the antenna and its application. Polar plots provide a 'bird's eye' view of the radiation pattern. They are particularly useful for general explanation. Figure 1.8 shows polar plots of the above three types. The relative power pattern, Fig. 1.8*a*, illustrates the variation of power density between 0% and 100%. These are most useful in assessing power variations between 10% and 100% of the peak value. As can be seen from the Figure, small variations are difficult to resolve, so the plot would not be useful in applications where low sidelobe levels and high gains were required. The relative field pattern

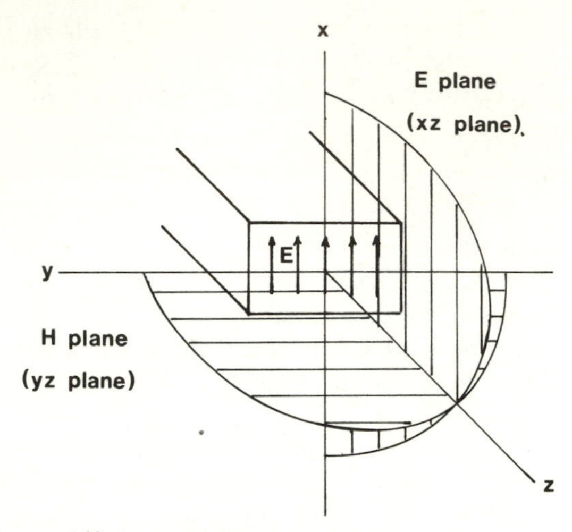

Fig. 1.7 *E-plane and H-plane pattern cuts*

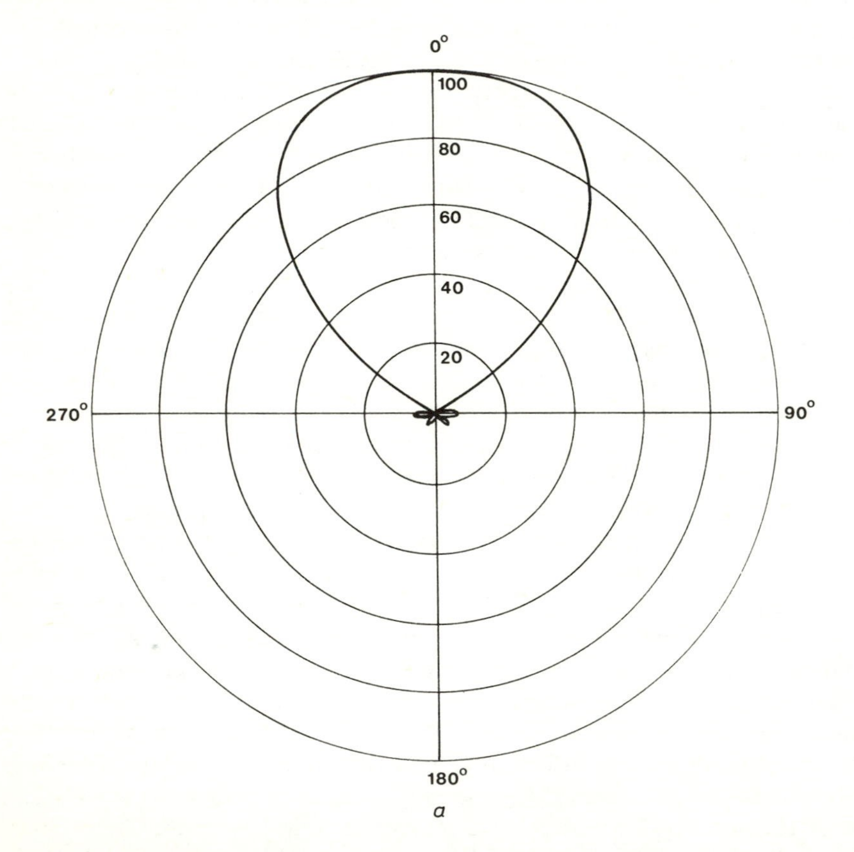

Fig. 1.8 *Polar radiation patterns; (a)* relative power pattern; *(b)* relative field pattern; *(c)* decibel pattern

shows the variation of electric field intensity at a fixed distance from the antenna as a function of the angular coordinates, Fig. 1.8*b*. By comparison with the relative power pattern the sidelobe structure is enhanced but the main beam shape remains clear. It is important when drawing linear plots to record whether the figure represents a relative power or a relative field pattern. This possible confusion

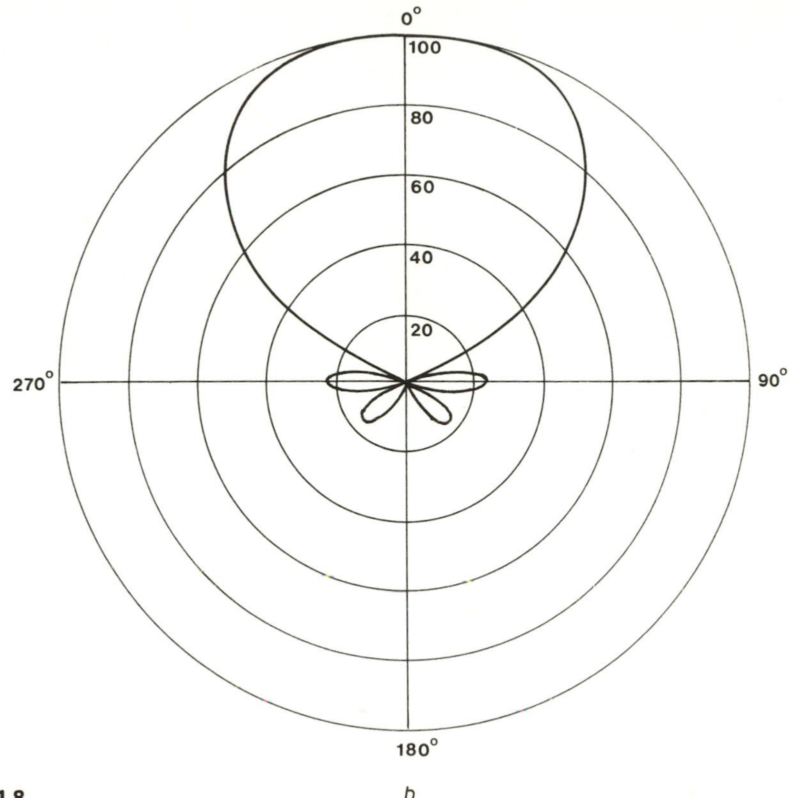

Fig. 1.8 *b*

is avoided by the logarithmic decibel relative gain pattern which records 10 $\log_{10}$ (relative power) or 20 $\log_{10}$ (relative field). The decibel pattern, Fig. 1.8*c*, provides a constant resolution over the entire display range and a wider dynamic range than can be shown on a linear plot. A recording range of 40 dB is often used because this provides good resolution of the main beam and indicates the principal sidelobe structure.

The polar type of display is limited to recording patterns of low and medium gain antennas. The polar scale is such that insufficient detail of the main-beam region of high-gain antennas is provided by this form of presentation. The cartesian plot of amplitude (or phase) against pointing angle can be readily expanded to examine the fine structure of the radiation pattern to any required degree of detail and the expanded scales need only extend over the range of angles and amplitudes of most interest for the particular application. Fig. 1.9*a* and *b* show relative field and logarithmic cartesian plots, respectively. The relative field is again most useful for low and medium gain antennas while the decibel cartesian plot is the most common method of recording radiation patterns.

The radiation patterns can be used to obtain the 'beamwidth' of the main beam. This is normally specified as the angular difference between the points on the radi-

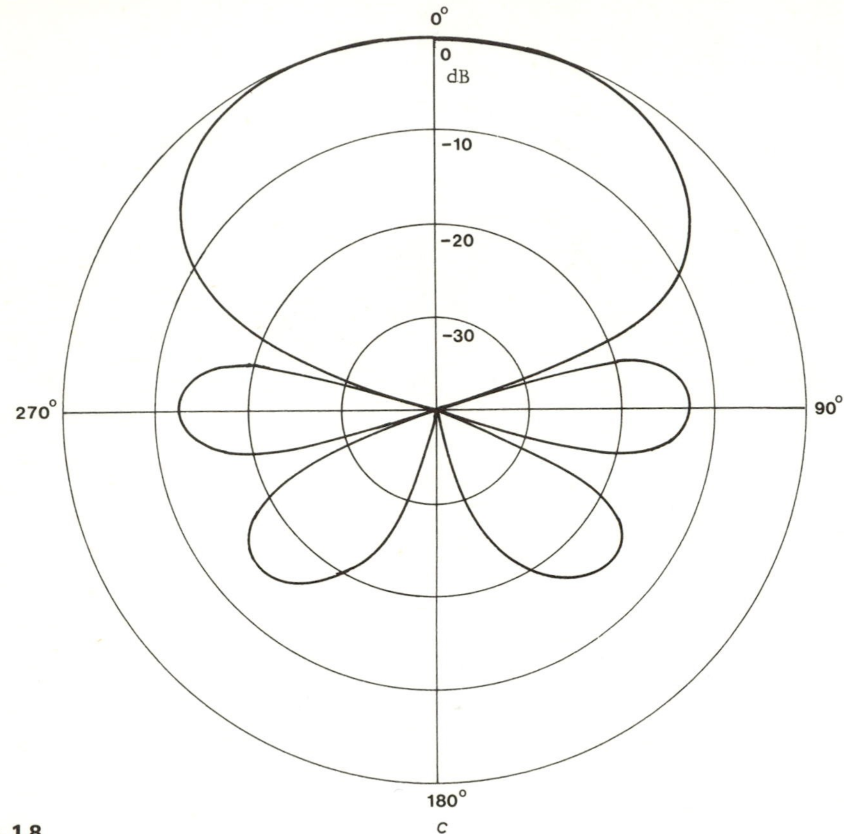

Fig. 1.8
c

ation pattern where the power has fallen to one half of the peak value or -3dB on the decibel scale.

The need for more complete data on the radiation characteristics of high performance antennas, combined with the ability of computers to process radiation pattern data and the ability of computer graphics to display the data has resulted in the growing use of other forms of radiation-pattern presentation. Fig. 1.10 shows a three-dimensional radiation pattern which gives a pictorial representation. This is not commonly employed for detailed design purposes but is useful for demonstration and for a qualitative assessment of the total radiation patterns for a given polarisation and frequency. A more practical design aid is the contour plot of Fig. 1.11 which displays lines of constant amplitude; it gives a relative power contour map of the antenna's radiation characteristics and provides a very effective means of presenting the large quantity of data associated with a complete radiation pattern on a single sheet of paper. Variations on this technique can be used to

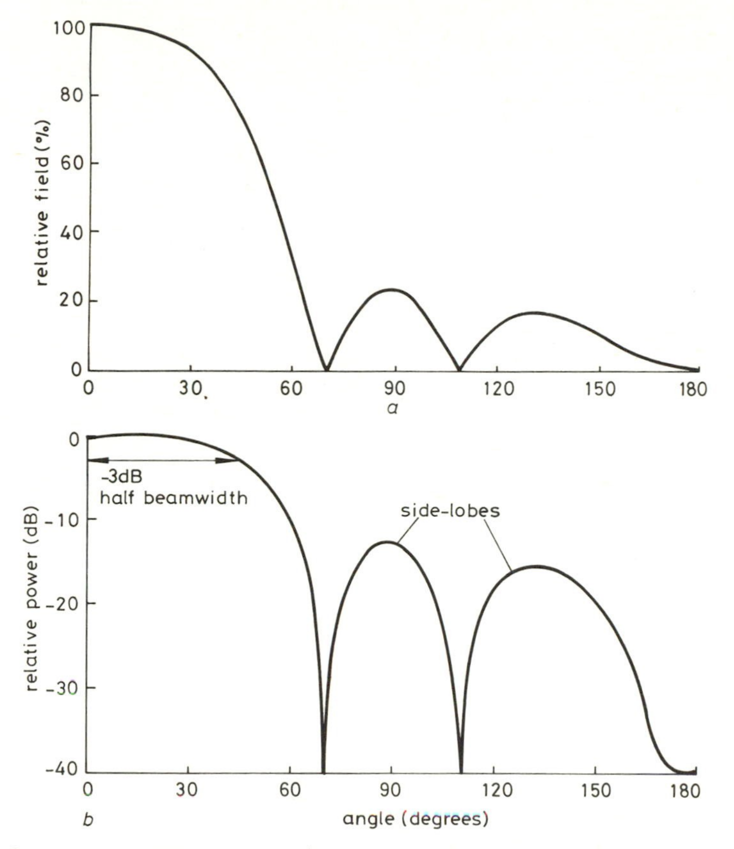

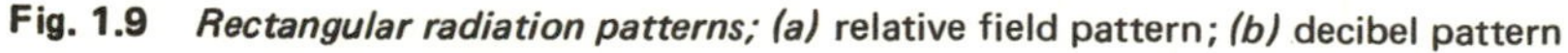

Fig. 1.9 ***Rectangular radiation patterns; (a)*** **relative field pattern;** ***(b)*** **decibel pattern**

provide colour TV graphical representations of radiation patterns. These are particularly useful as a diagnostic aid in evaluating an antenna design.

The discussion above has been largely concerned with radiation patterns depicting the relative amplitude of the radiated field. Patterns of the relative phase, polarisation, frequency or other secondary parameters are also commonly displayed in cartesian format.

1.6 Gain, directivity and efficiency

The *power gain* and *directivity* are quantities which define the ability of an antenna to concentrate energy in a particular direction. The power *gain* of an antenna in a specified direction (θ,ϕ) is

$$G(\theta,\phi) = \frac{4\pi \text{ power radiated per unit solid angle in direction } \theta,\phi}{\text{total power accepted from source}} \qquad (1.14)$$

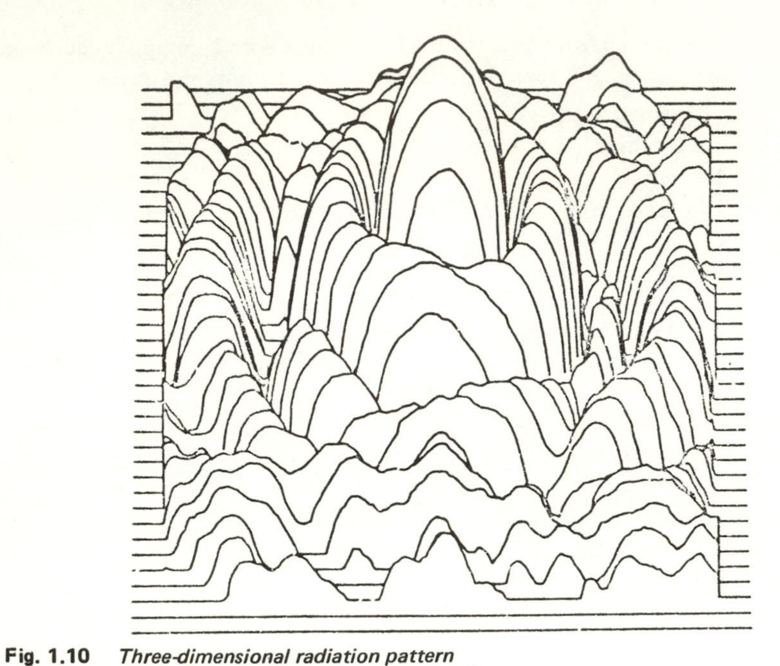

Fig. 1.10 *Three-dimensional radiation pattern*

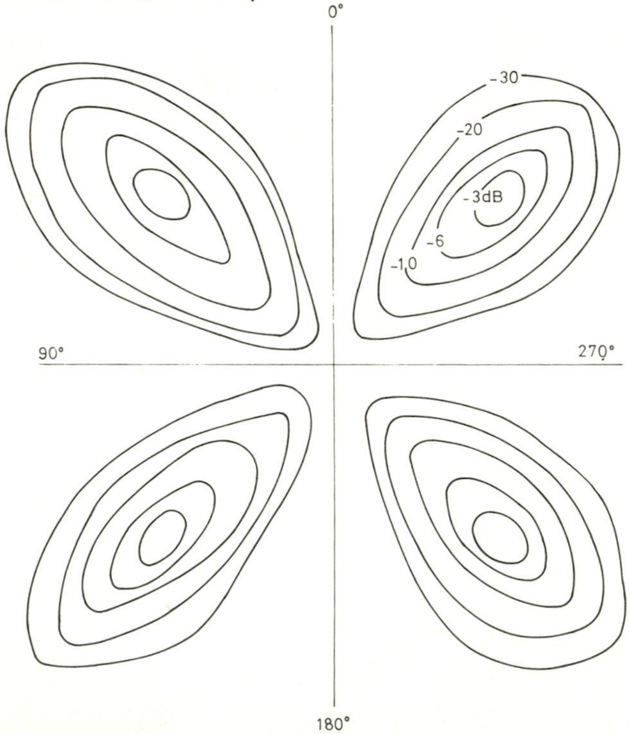

0°

-30
-20
-3dB
-6
-10

90°

270°

180°

Fig. 1.11 *Contour plot showing lines of constant amplitude (typical plot showing cross-polarisation pattern of a horn). Parameter: amplitude relative to peak cross polarisation*

This quantity is an inherent property of the antenna and does not include losses which may arise from mismatch of impedance between the antenna and power source or mismatch of polarisation of the receiving antenna. It does, however, include the ohmic or dissipative losses arising from the conductivity of metal and dielectric loss. The *directivity* of an antenna does not include the dissipative losses and is defined in a specified direction (θ,ϕ) as

$$D(\theta,\phi) = \frac{4\pi \text{ power radiated per unit solid angle in direction } \theta,\phi}{\text{total power radiated by antenna}} \tag{1.15}$$

In the literature the term 'gain' is often used synonymously for either directivity or power gain. 'Gain' is often used when the writer has strictly calculated the directivity, that is, the dissipative losses which are often difficult to quantify, have been excluded.

Losses in the antenna structure are also difficult to quantify experimentally unless a complete integration of all of the power emitted by the antenna is performed. This is a relatively complicated and costly procedure and conventional calibrated antenna ranges provide a measure of power gain rather than directivity. The indiscriminate use of the term gain for either power gain or directivity is commonplace and the implications should be fully understood to avoid errors, particularly when comparing different systems or theory with practice. Clearly the two parameters are only equal when the antenna is lossless.

The ratio of power gain to directivity is termed the radiation efficiency (η) of the antenna.

$$\text{radiation efficiency} = \frac{\text{power gain in direction } \theta,\phi}{\text{directivity in direction } \theta,\phi} \tag{1.16}$$

or

$$\eta = G(\theta,\phi)/D(\theta,\phi)$$

Although power gain or directivity can be specified in any direction it is usual to refer to the peak value which coincides with the direction of the principal lobe or main beam radiated by the antenna. In any reference to the gain or directivity of an antenna the peak figure is normally implied. The peak gain or directivity of an antenna is usually expressed in decibels relative to either a short current element, a thin lossless half wavelength long dipole, or an ideal isotropic radiator. The isotropic radiator is a hypothetical antenna which radiates uniformly in all directions (i.e. for one watt input power it radiates $1/(4\pi)$ watts per steradian). Although this is not realisble in practice with vector fields, the isotropic radiator provides a very convenient point of reference. For example, the short current element has a peak gain of 1·5 times relative to the isotrope (i.e. 1·76dB$_i$) while the lossless half-wave dipole peak gain is 1·64 times, or 2·15dB$_i$.

The peak directivity of an antenna with reference to an isotope can also be expressed as

$$\text{Peak directivity } (D) = \frac{\text{peak power received}}{\text{average power received}}$$

If E is the field strength of the antenna at a point in the far field, and since the directivity will be the same in either transmission or reception, the peak directivity can be expressed in spherical coordinates as

$$D(\theta,\phi) = \frac{\left|E(\theta_0\phi_0)\right|^2}{\frac{1}{4\pi}\int_0^{\pi}\int_0^{2\pi}\left|E(\theta,\phi)\right|^2 \sin\theta \, d\theta d\phi} \tag{1.17}$$

where θ_0, ϕ_0 correspond to the direction from which the peak power is received.

For an antenna which radiates a pencil beam which is essentially axis-symmetric about $\theta=\phi=0$, the peak directivity can be obtained approximately by replacing $E^2(\theta,\phi)$ with the average of the radiation patterns measured in the principal E– and H– planes of the antenna, i.e.

$$D \triangleq \frac{4\,E^2\,(0,0)}{\int_0^{\pi}(E^2(\theta,\phi) + E^2(\theta,\pi/2))\sin\theta \, d\theta} \tag{1.18}$$

This approximation has been found to be accurate within about 10% for pencil-beam antennas in which not too large a portion of the radiated energy is contained in the sidelobe structure.

While reciprocity ensures that the calculated values of gain will apply equally well to either a transmitting or receiving antenna, the performance of the latter can also be usefully expressed in terms of a receiving cross-section or an effective area. A receiving antenna will collect energy from an incident plane wave and, if properly matched, will transfer this power to a load. The proportion of the incident energy which will find its way to the load is a function of the polarisation properties of the antenna and its gain in the direction of the incident plane wave. The effective aperture of an antenna can be defined as the area of an ideal antenna which would absorb the same power from an incident plane wave as the antenna in question. For our purposes here the antenna polarisation will be assumed to be such that maximum power is absorbed and the ideal antenna will, by definition, absorb all of the power incident upon its aperture. The effective area which is a function of the angle of arrival of the incident wave can be related to the power gain by[1]

$$A_{eff}(\theta,\phi) = \frac{\lambda^2 G(\theta,\phi)}{4\pi} \tag{1.19}$$

from which it is apparent that the effective area of an isotropic radiator is just $\lambda^2/4\pi$.

The effective area can also be expressed as a proportion of the physical area of the antenna. This is particularly meaningful for antennas which have a well defined collecting aperture, where the ratio is termed the *aperture efficiency*.

$$\text{aperture efficiency} = \frac{\text{effective area of antenna aperture}}{\text{physical area of antenna aperture}} \tag{1.20}$$

The aperture efficiency of an optimised Cassegrainian reflector antenna can be as much as 90% although a more usual figure for reflector antennas is in the range 50 - 70%.

Fundamental-mode pyramidal horns have aperture efficiencies between 50% and 80% depending on how optimum are the design parameters. The relation between effective area and gain allows the calculation of equivalent areas for antennas which do not have an aperture. For instance, the effective area of a thin half wavelength dipole is 0·13 square wavelengths which approximates to an area of

$$\frac{\lambda}{4} \times \frac{\lambda}{2}$$

However, for rod-type antennas designed to radiate at right angles to the rod, it is more useful to use the term *effective length*. This can be defined for both transmitting and receiving antennas, Jordan[3]. The effective length of a transmitting antenna is the length of an equivalent linear antenna that has a constant current at all points along its length equal to the input current of the rod and that radiates the same field as the actual antenna in the direction perpendicular to its length. If $I(0)$ is the constant current and L is the length, then

$$l_{eff}\,(\text{transmitting}) = \frac{1}{I(0)} \int_{-L/2}^{+L/2} I\, dl \tag{1.21}$$

Thus for the case of a thin half-wave dipole, assuming a sinusoidal current distribution,

$$I = I_0 \cos \frac{2\pi l}{\lambda} \tag{1.22}$$

then $l_{eff} = \lambda/\pi$ (1.23)

For a receiving antenna the effective length can be expressed in terms of the terminal voltage and field intensity (Jordan[3]).

$$l_{eff}\,(\text{receiving}) = \frac{\text{open circuit voltage (V)}}{\text{electrical field intensity (V/m)}} \tag{1.24}$$

The two expressions for effective length are equivalent, as can be shown by reciprocity.

Krauss[8] has derived an expression relating the effective length and the effective area of a rod type antenna,

$$A_{eff} = \frac{l_{eff}^2 Z_0}{4R_r} \tag{1.25}$$

where R_r is the radiation resistance and Z_0 is the intrinsic impedance of free space ($120\pi\Omega$). For a thin half wavelength dipole,

$A_{eff} = 0{\cdot}13\,\lambda^2$ and $l_{eff} = \dfrac{\lambda}{\pi}$ so the radiation resistance is $R_r = 73\Omega$.

1.7 Polarisation

The polarisation of an electromagnetic wave at a single frequency describes the shape and orientation of the locus of the extremities of the field vectors as a function of time. In antenna practice the electromagnetic waves are either plane waves or may be considered as locally plane waves so that the electric (E) and magnetic (H) fields are related by a constant, the intrinsic admittance of the media of propagation, $H/E = \sqrt{\epsilon_0/\mu_0}$. In these circumstances it is sufficient when describing the polarisation of a wave to specify the polarisation of the electric field vector $\boldsymbol{E}$. The magnetic field vector $\boldsymbol{H}$ can be obtained by a 90° rotation about the vector defining the direction of propagation followed by multiplication by the intrinsic admittance of the media.

A wave may be described as linearly polarised, circularly polarised or elliptical polarised. A single current element oriented along the a_θ axis will radiate a linearly polarised field with an electric field vector always in the θ direction. A more complicated antenna may radiate a field with components of electric field in both θ and ϕ directions. If these components differ in phase the direction of the resultant electric field vector at a given point in space will rotate at an angular rate ω and the extremity of the total electric field vector will in general trace out an ellipse. The field is then said to be elliptically polarised. When the components have equal magnitudes the ellipse degenerates into a circle and the field is described as circularly polarised.

In some circumstances, the wave received by an antenna may be randomly or partially polarised; for instance, a signal received from a celestial radio source. This case will not be considered here; the reader is referred to works by Deschamp[19] and Cohen[20] for treatment of the subject.

An elliptically polarised wave may be regarded as either the resultant of two linearly polarised waves of the same frequency or as the resultant of two circularly polarised waves of the same frequency but with opposite directions. Consider the case of two linearly polarised fields at a given point in space and with phase difference δ, oriented along the x and y axis in a plane transverse to the direction of propagation.

$$\begin{aligned} E_x &= E_1 \cos\omega t \\ E_y &= E_2 \cos(\omega t + \delta) \end{aligned} \tag{1.26}$$

E_1 and E_2 are real constants. The 'instantaneous' resultant vector has an amplitude

$$\surd(E_x^2 + E_y^2)$$

and is inclined to the z axis at an angle $\tan^{-1}(E_y/E_x)$. The locus traced out by this resultant vector may be determined by eliminating time t. Expanding the equation for E_y, then squaring and substituting the equation for E_x leads to

$$\left(\frac{E_x}{E_1}\right)^2 + \left(\frac{E_y}{E_2}\right)^2 - \frac{2E_xE_y}{E_1E_2}\cos\delta = \sin^2\delta \tag{1.27}$$

This is the equation of an ellipse. In the case when $E_1 = E_2$ and $\delta = \pm\pi/2$ the equation represents a circle and when $\delta = 0$ the ellipse degenerates into a straight line inclined at an angle $\tan^{-1}(E_2/E_1)$. For $\delta = \pm\pi$ the straight line is inclined in the opposite direction. The direction of rotation of the tip of the electric field vector of an elliptical or circular polarised wave, is termed the *sense* or *hand* of the polarisation. The sense is called righthand (lefthand) if the direction of rotation of the electric field in a transverse plane of polarisation is clockwise (counterclockwise) when viewed by an observer, looking from the transmitter to the receiver, along the direction of propagation. This is the IEEE Standard Definition of Polarisation[12]. Confusion has often arisen in the past partly because classical physics used a different definition and partly because of the distinction between the rotation of an electric vector in a static transverse plane (xy plane if propagation in the z direction) and the helical path corresponding to the locus of the instantaneous field vector. A field vector rotating counter-clockwise in a transverse plane as time advances appears as a right hand helix in space when viewed at fixed time (Fig. 1.12). The IEEE definition is in terms of the vector rotating in the transverse plane and therefore a left hand polarised wave is produced by a right hand helix.

Elliptical polarisation is characterised by three parameters: *the axial ratio* (ratio of major axis to minor axis); the *tilt angle* (angle between the reference direction

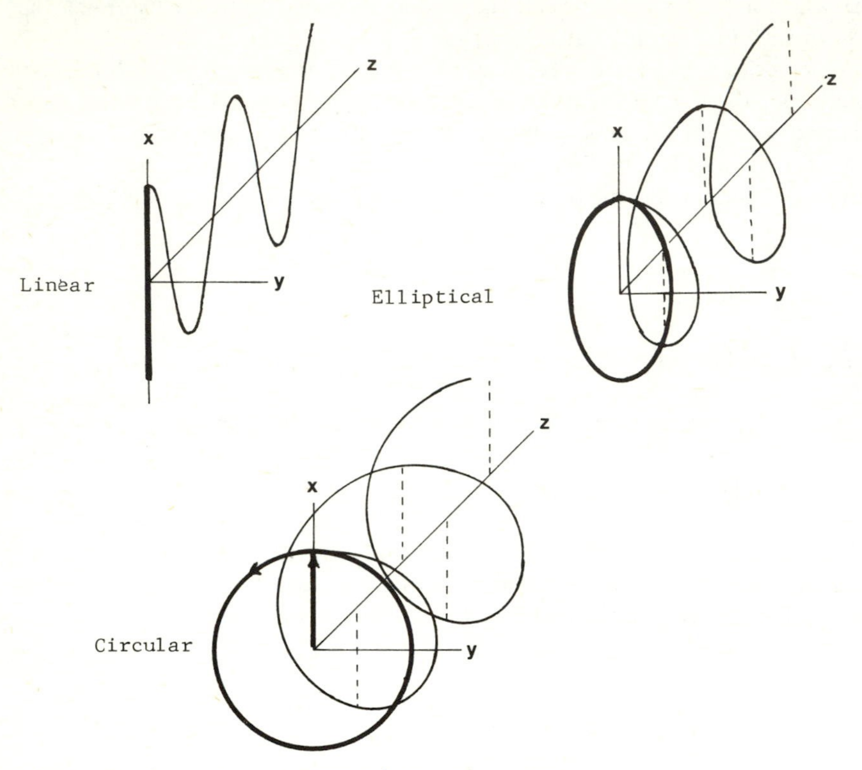

Fig. 1.12 *Linear, circular and elliptical (lefthand) polarised waves*

and major axis of ellipse, clockwise when viewed in direction of propagation); and the *sense* of rotation. Polarisation can be illustrated with a standard spherical coordinate system, Fig. 1.13. The reference direction for establishing the orienta-

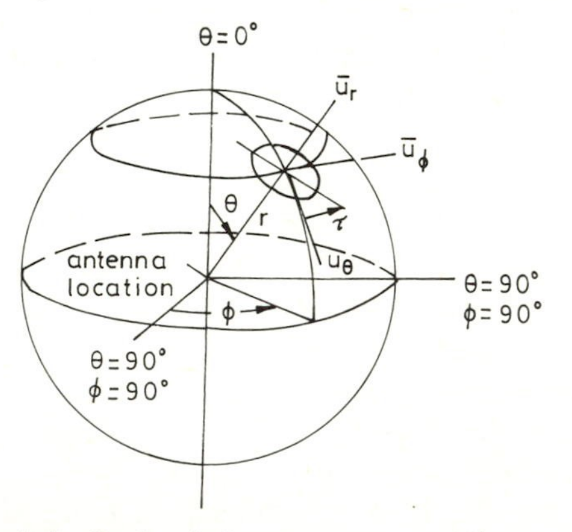

Fig. 1.13 *Elliptical polarisation in relation to antenna coordinate system*

tion of the polarisation ellipse is arbitrary but it is common practice to use the a_θ axis as the reference direction.

An antenna receiving a plane wave from a given direction will pick up the maximum signal when the polarisation ellipse of the incident electric field has the same axial ratio, the same tilt angle and the same sense of polarisation as that of the receiving antenna in the given direction. If the polarisation of the incident wave does not match the polarisation of the receiving antenna then a polarisation loss occurs due to the mismatch. The amount of mismatch is given by the *polarisation efficiency* which is defined as the ratio of the power actually received by the antenna divided by the power that would be received if a wave came from the same direction and with the correct polarisation were incident on the antenna.

Various graphical aids have been devised to aid the description of polarisation. The reader is referred to the following texts for descriptions Krauss[9], IEEE Standard[12], Born and Wolf[21], Descamps[19], Hollis *et al.*[22].

The *cross-polarisation* of a source is becoming of increasing interest to antenna designers. This is defined as the polarisation orthogonal to a reference polarisation. The *co-polar* field of a source is the component of the field which is parallel to the field of the reference source and the *cross-polar* field is the orthogonal component.

This definition is however not unique for linear and elliptically polarised fields because it does not define the direction of the reference polarisation for all observation angles. The choice of definition has been considered by Ludwig[23] who identified three possibilities for the case of a linearly-polarised source (Fig. 1.14).

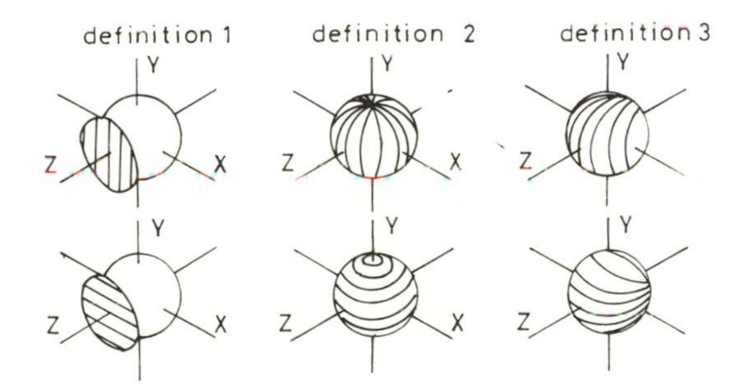

Fig. 1.14 *Definitions of cross-polarisation*

These have become known as the Ludwig 1st, 2nd and 3rd definitions according to whether:-

(1) the reference field is a plane wave in a rectangular coordinate system.

(2) the reference polarisation is that of an electric dipole lying in the aperture plane. The cross-polarisation is then the polarisation of a magnetic dipole with the same axes as the electric dipole.

(3) the reference polarisation is that of a Huygens source (electric and magnetic dipoles with orthogonal axes lying in the aperture plane and radiating equal fields

in phase along the z axis). The cross-polarisation is then the polarisation of a similar source rotated 90° in the aperture plane.

The co-polar $E_p\ (\theta,\phi)$ and cross-polar $E_q\ (\theta,\phi)$ fields using definition 3 are given by:

$$E_p(\theta,\phi) = E_\theta(\theta,\phi)\sin\phi\,\boldsymbol{a}_\theta + E_\phi(\theta,\phi)\cos\phi\,\boldsymbol{a}_\phi$$
$$E_q(\theta,\phi) = E_\theta(\theta,\phi)\cos\phi\,\boldsymbol{a}_\theta - E_\phi(\theta,\phi)\sin\phi\,\boldsymbol{a}_\phi \tag{1.28}$$

with the principal vector aligned with the y axis.

Ludwig recommended the use of definition (3) because it most closely corresponds to what is normally measured when an antenna radiation pattern is recorded and interchanging the co- and cross-polar fields as measured in any direction corresponds to a 90° rotation of the reference source. This definition has become widely accepted. It is the only one which does not lead to confusion when an elliptically polarised source is considered. Note that in the case of an elliptical wave it is not sufficient to rotate the reference source by 90° about the direction of propagation. The rotation must be accompanied by a change in the sense of polarisation.

1.8 Impedance

The antenna is essentially a transducer between the radio or radar system on the one hand and the propagation medium on the other. The antenna designer must be concerned, therefore, with the characteristics of the electromagnetic fields which the device transmits or receives from the medium and the load characteristics which the antenna presents to the system. Typically the antenna will be coupled to the transmitter or receiver by a transmission line, which may take the form of wires, coaxial cable, dielectric or metallic waveguide, or one of the newer forms of transmission line such as stripline. It is generally desirable to achieve the maximum power transfer from the transmission line to the antenna, and vice versa, without distortion of the information which is being conveyed.

The impedance concept can be very useful for certain classes of antenna in defining the required characteristics at the input terminals of the radiating device. Providing a unique value for impedance can be defined for the transmission line then the design objective for the antenna impedance will be to provide a matching value, thereby ensuring a maximum power transfer on the basis of the power transfer theorem.

The impedance concept can be particularly useful for lower frequency antennas where a pair of input terminals can be readily defined and the impedance is single valued and relatively easy to measure. The concept does not fail at higher frequencies but there may be practical difficulties in defining and measuring this quantity which mitigate against its use.

For example, at microwave frequencies the antenna may well be connected to a waveguide transmission line. The impedance of a waveguide is not single valued since, unlike coaxial cable, the electric and magnetic fields inside the waveguide are not purely transverse. An impedance can be defined based upon the transverse components of the electromagnetic fields of the fundamental mode in the waveguide, but in practice the input terminal to the antenna is often a waveguide flange and it is more convenient to employ waveguide matching techniques coupled with direct measurements of the voltage-standing-wave ratio (vswr) or return loss. These measurements can be converted into an effective input impedance and displayed on a Smith chart if required.

Where it is applicable the impedance concept is helpful as an aid to the understanding and design of antenna systems. In these cases the antenna input impedance can be considered as a two terminal network terminating the physical antenna (see Fig. 1.15). In general this impedance can be considered to be comprised of two parts; a self-impedance and a mutual impedance such that,

input impedance = self impedance + mutual impedance

The *self impedance* is the impedance which would be measured at the input terminals of the antenna in free space, that is in the absence of any other antennas or reflecting obstacles. The *mutual impedance* accounts for the influence of coupling to the antenna from any outside source. Clearly nearby objects are potentially greater sources of coupling and for many antennas the mutual impedance is effectively zero, either because the antenna is sufficiently isolated in space or because the influence of nearby objects is much less than the self impedance.

On the other hand some antennas rely on the mutual coupling between elements to produce the desired specifications. A classic example is the Yagi antenna where all but one of the elements are passive and unconnected to any other elements but have currents induced in them by mutual coupling from the one driven element. In array antennas the mutual coupling may not be desired but will often be a significant factor in the total radiation characteristics. The calculation of mutual impedance is usually theoretically complicated because the coupled antennas are in their reactive near-field regions and the geometry of the antennas is often difficult to model analytically.

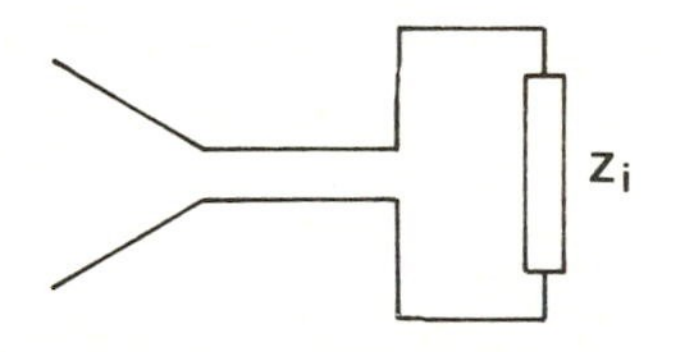

Fig. 1.15 *Aperture input impedance*

The self impedance of an antenna has both resistive and reactive components, i.e. employing complex algebra it has the form

self impedance = (antenna resistance) + j (self reactance)

The *self reactance* arises from the reactive energy which is stored in the near-field region surrounding the antenna, while the *antenna resistance* accounts for all the power absorbed by the antenna as seen at its input terminals. The power absorbed includes that which is ultimately readiated by the antenna and thus the antenna resistance comprises a so-called *radiation resistance* R_r and a *loss resistance* R_L which accounts for the dissipative and ohmic losses in the antenna structure. Hence

$$\text{antenna resistance} = R_r + R_L$$

where the radiation resistance is defined as the equivalent resistance which would dissipate a power equal to that radiated, P_r, when carrying the current flowing at the input terminals I_o, i.e.

$$R_r = P_r/I_o^2 \tag{1.29}$$

For an efficient antenna it is necessary that the radiation resistance be much greater than the loss resistance. For a practical thin half-wave dipole for example, the radiation resistance may have a value of approximately 73Ω, with a loss resistance of perhaps 2Ω. On the other hand an h.f. notch antenna on an aircraft may have a radiation resistance of 0·01Ω with a loss resistance of several ohms. Many standard textbooks have dealt with the calculations of the radiation resistance for cylindrical rods and wires; see, for example, Kraus[8] and Collin and Zucker[1].

Measurements of input impedance can be performed using conventional impedance bridge techniques and this is common practice at the lower frequencies. At higher frequencies measurements of reflection coefficients or voltage-standing-wave-ratios (vswr) are favoured. Providing the characteristics of the transmission line are known these measurements can be converted to an impedance if required.

For example, if an antenna of impedance Z_L terminates a transmission line of characteristic impedance Z_0 and an impressed sinusoidal voltage at the input to the the antenna (E_1) gives rise to a reflected voltage (E_2) then the reflection coefficient of the antenna (Γ) is simply given by

$$\Gamma = (|E_1|/|E_2|)\, e^{j\theta} \tag{1.30}$$

where θ is the phase difference between the transmitted and reflected voltages.

The voltage standing-wave ratio is a measure of the ratio of the maximum and minimum voltages set up on the transmission line. In terms of the reflection coefficient this will have the value

$$\text{vswr} = (1 + \Gamma)/(1 - \Gamma) \tag{1.31}$$

and will take values in the range unity (i.e. perfect match) to infinity. Typically microwave horn antennas have vswr values in the range 1·01 to 1·5 while a broadband antenna may operate with vswr values of the order of 1·5 to 2·5. A vswr of 5·8 implies that one half of the incident power is reflected from the antenna.

The antenna impedance (Z_L) is given by

$$Z_L = Z_0 (1 + \Gamma)/(1 - \Gamma) \quad (1.32)$$

and it must be noted that this is a complex quantity which requires that the phase angle of the reflection coefficient be established. The return loss, which is often specified as a performance parameter in microwave applications, is given by

$$\text{return loss} = 20\log_{10}|\Gamma| \text{ decibels} \quad (1.33)$$

Return losses of −15dB are typical for many antennas but values of the order of −30dB or more may be demanded for a high performance satellite communications ground station.

1.9 Bandwidth

The receiving antenna serves both as a spatially selective and a frequency selective collector of energy. Electromagnetic waves at frequencies which may range over the entire radio spectrum, may be incident upon the antenna from all directions. The antenna is a spatially selective filter in that it accepts only a portion of the total incident energy by transforming free-space electromagnetic waves from specific directions into guided waves which are then detected in the receiving system. The antenna and the radio frequency components behind it are frequency selective in that they are designed to perform effectively over an operating band of frequencies. Outside of this band the electrical performance of the antenna and the radio frequency system will be non-optimum and will tend to reject the out of band signals.

The term *bandwidth* is used to describe the frequency range over which the antenna will operate satisfactorily. There is no unique definition for satisfactory performance and this will differ from application to application. Operating bandwidths are commonly defined in terms of both radiation pattern characteristics and input impedance or vswr requirements. Some typical examples are indicated below.

Gain or directivity: The frequency range over which the gain or directivity is within specified limits.
Beamwidth: The frequency-range over which the main beamwidth (specified at the half power point −3dB or at the −10dB or −20dB levels) is within some specified limit.
Polarisation or cross-polarisation: The frequency range over which the polarised field components are within some specified limits.
Impedance v.s.w.r. or return loss: The frequency range over which any or all of these parameters are within some specified limits

In practice it is usually found that one or more of the above parameters is more sensitive to frequency change than the others and this then determines the bandwidth of the antenna. Note, however, that this will be very dependent upon the particular performance requirements which are specified. As a consequence the terms broadband and narrowband can mean very different things in different applications.

Bandwidth is expressed either as an absolute frequency range in hertz, or as a percentage of the centre frequency. A wideband antenna could imply anything from several octaves down to perhaps a 40% band, although in some applications a 15% band would be considered broadband operation. A narrowband antenna may have a bandwidth of a few percent or less.

The physical design factors limiting the bandwidth vary from antenna to antenna. In monopoles, dipoles, slots and microstrip elements the structures are resonant at particular frequencies and the bandwidth is well defined by the impedance characteristics at the antenna terminals. Many types of waveguide horn antenna are potentially broadband but most are band limited by the modal nature of the propagation in the horn which has a low frequency cut-off for the fundamental mode and a high frequency value where higher order modes which will distort the radiation pattern are free to propagate. Dielectric rod and surface wave antennas suffer from similar limitations. In some senses large aperture-type antennas such as paraboloidal reflectors are inherently broadband because the reflecting properties of the surface do not depend on frequency, being limited only by the diffraction at the edge when the reflector is small in wavelengths. However, in practice the reflector must have a primary feed at the paraboloidal focus and this will limit the bandwidth. In addition the beamwidth of any fixed aperture antenna tends to vary with frequency and this may constitute a bandwidth limitation in some applications.

The bandwidth of linear and planar array antennas are limited by the distance apart of the array elements measured in wavelengths, partly by the phase of the driving currents, but also by the method of feeding. For example, in the 'series-fed' array the elements are placed along a transmission line with the result that the beam angle changes with frequency. In a 'parallel-fed' array the elements are fed from a branching network which provides equal path lengths to all elements: the main-beam angle does not change with frequency, but the beamwidth and sidelobe structure is still frequency-dependent. Genuine broad-band operation is only achieved where the elements radiating are made to change with frequency in such a way as to keep the radiation characteristics constant; an example is the log-periodic antenna.

1.10 Antenna noise temperature

The minimum signal power which can be detected by a receiving system is limited by the inherent noise in the system. Part of this noise originates with the receivers connected to the antenna terminals, but this is not generally the concern of the antenna engineers. The remainder of the noise power arises as a consequence of electromagnetic energy which is received by the antenna along with, or even in the absence of the desired signal. The noise power introduced into the system via the antenna can be conveniently expressed as the temperature of an equivalent matched resistance which, when connected to the receiver input in place of the antenna, would produce an identical noise power. This resistance is the radiation resistance of the antenna and the associated temperature is termed the *antenna noise temperature.* The available noise power arising from the radiation resistance is given by:

$$P = kT\Delta f \tag{1.34}$$

where k is Boltzmanns constant, T is the effective absolute temperature of the radiation resistance and Δf is the bandwidth. This is Nyquist's formula. For detailed discussion of the noise temperature of antennas the reader is referred to Collin and Zucker[1] or Krauss[9].

The temperature of the antenna structure itself does not determine the antenna noise temperature. The noise temperature arises as a consequence of the electromagnetic energy entering the antenna and, as such, is determined by the temperature of the emitting regions which the antenna 'sees' through its directional pattern. It is a weighted average of the temperature from all external sources, where the weighting depends on the antenna gain in the various source directions. These sources can be: manmade noise; emission from the ground and objects located near to the antenna; emission from the atmosphere; and celestial or extraterrestrial emission. Manmade noise is different in character and it is usually assumed that it can be avoided or controlled. Emission from the ground near to the antenna can be reduced by designing the antenna to minimise spillover radiation and other sidelobes in the direction of the ground. The actual observed temperature of the sky is dependent on many factors, including frequency and direction. It also exhibits daily and seasonal variations. The average noise temperature of celestial sources, $\overline{T}_s$, is found to lie in the range[1]

$$\frac{\lambda^2}{5} \leqslant \frac{T_s}{290} \leqslant 5\lambda^2$$

where λ is in metres. In the h.f. band, at 18 MHz, a typical value is 140 000°K. At microwave frequencies, 1-100 GHz, the cosmic noise is relatively low.

A rough estimate of antenna noise temperature for a beam directed at an elevation of θ, neglecting ohmic losses in the antenna structure, can be made as follows:

Let T_s = sky temperature, P_s = fraction of power in sidelobes intercepting the ground and α = fractional power absorbed by the total atmospheric path at θ. Assuming that the ground temperature is 290°K and that the effective temperature of the attenuating mechanism in the atmosphere (which is partially scattering and partially true absorption) is the same, the effective antenna noise temperature becomes

$$T_A = (1-P_s)[T_s(1-\alpha) + 290\,\alpha] + 290P_s \tag{1.35}$$

At 5GHz, T_s may be 10°K and α is typically 0·4 at zero elevation, falling rapidly to 0·05 at 10° elevation and to 0·008 at 90°. In a well-designed antenna, P_s may be less than 0·05, so that at this frequency T_A is typically 130°K near zero elevation, 37°K at 10° and 26°K at 90° elevation. Rainfall in the path increases α and degrades the antenna noise temperature.

At frequencies above 20 GHz, atmospheric attenuation is much greater and the antenna temperature can be assumed to be 290°K for most practical purposes.

1.11 Field equivalence principles

To determine the radiated fields from an electromagnetic source *field equivalence principles* are a useful aid. Although they do not make an exact solution any easier to obtain, they often suggest approximate methods which are of value in antenna problems. Field equivalence principles are treated at length in the literature and it will not be appropriate to consider the many variants here. Schelkenoff[5] and Collin and Zucker[1] are useful sources of reference in this respect. The basic concept is illustrated in Fig. 1.16. The electromagnetic source region is enclosed by a surface S which is sometimes referred to as the Huygen surface.

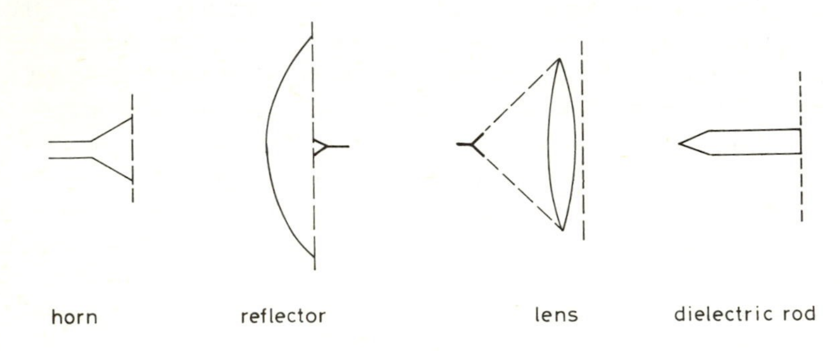

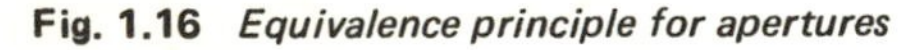

Fig. 1.16 *Equivalence principle for apertures*

Without changing the fields $\boldsymbol{E}, \boldsymbol{H}$ external to S the electromagnetic source region can be replaced by a zero field region with appropriate distributions of electric and magnetic currents $\boldsymbol{J}, \boldsymbol{J}_m$ on the Huygen surface. This example is in fact overly restrictive and we could specify any field within S with a suitable adjustment to $\boldsymbol{J}, \boldsymbol{J}_m$. However, the zero internal field approach is particularly useful when the tangential electric fields over a surface enclosing the antenna are either known or can be evaluated, or approximated. In this case the surface currents can be obtained directly from the tangential fields, and the external field can be determined by means of vector potential theory (see Chapter 4, for example).

In fact a knowledge of both $\boldsymbol{E}$ and $\boldsymbol{H}$ over the surface is not necessary since from uniqueness concepts it can be shown that only $\boldsymbol{E}$ or $\boldsymbol{H}$, or some combination of $\boldsymbol{E}$ over part of S and $\boldsymbol{H}$ over the rest, need be known to determine the fields outside of S uniquely. When we are concerned only with the fields outside of S we are left with a degree of freedom in postulating the configuration within S. If, for example, we assume a sheet of perfect electric conductor just inside, and infinitesimally close to S, then the currents on S can be considered to be short-circuited and the radiated fields can be obtained in terms of only a magnetic current J_m over the surface. However, the presence of this conducting surface implies that there can be no inwardly directed radiation and by image theory the effective magnetic current (J'_m) must be given a value of $2E \times a_n$. It follows that while maintaining the same electromagnetic field distribution E, H outside the surface S we could postulate any of the following

(i) J and J_m on S
(ii) J'_m and $2J_m$ on S (for a perfect electric conductor within S)
(iii) J' and $2J$ on S (for a perfect magnetic conductor within S)

In case (i), with a zero internal field postulated, the electromagnetic sources inside S can be considered to be removed and the radiated fields outside of S is then determined from the electric and magnetic surface current distributions alone. For cases (ii) and (iii) however, the radiated fields must be determined in the presence of the internal conducting surfaces and, in general, the problems are no easier to deal with than the original source geometry.

There is one special configuration, however, which offers significant advantages. In this case the Huygen surface is chosen to be an infinite plane which separates space into two hemispherical regions, with the sources contained on one side of the plane. If either the electric or magnetic fields arising from these sources can be determined (albeit approximately) over the Huygen surface then the radiated fields on the far side of the plane can be calculated. The introduction of an infinite conducting sheet just inside the Huygen surface in this case will not complicate the calculations of the radiated fields in the other half space. This model is useful for antennas whose radiation is largely directed into the forward hemisphere and has found wide application in dealing with aperture antennas.

When the limitations of the half-space model are acceptable it offers the important advantage that only either the electric *or* magnetic currents need be specified. It must be emphasised that any of the above methods will produce exact results providing J or J_m are known exactly over the Huygen surface. When these current distributions are known only approximately the methods will not, in general, agree. The preferred method then rests upon the accuracy of the approximations involved in each case. It is sometimes difficult to establish the relative accuracies on a theoretical basis and the optimum choice is determined by experience based upon comparison with experimental data.

1.12 Radiation from apertures

There are a large number of antenna types for which the radiated electromagnetic fields can be considered to emanate from a physical opening or aperture. Identifying this general class of antennas is advantageous in that it provides a very convenient basis for analysis and permits a number of well established mathematical techniques to be applied to provide expressions for the distant radiation fields. Antennas which fall into this category include several types of reflector, planar arrays, lenses, horns and some surface-wave antennas. The aperture surface comprises a finite opening in an infinite plane. The electric and/or magnetic fields in the aperture region are generally determined by approximate methods, while the fields elsewhere are assumed to be zero. As a first approximation the distant radiated fields are then obtained from the fields on the aperture surface alone.

Aperture techniques are especially useful for parabolic reflector antennas, where the aperture plane can be defined immediately in front of the reflector (see Fig. 1.17). Parabolic reflectors are generally large electrically with diameters typically varying from tens to thousands of wavelengths. Perhaps more surprisingly aperture techniques can be successfully applied to small aperture waveguide horns. However, for very small horns with aperture dimensions of less than about one wavelength, the assumption of zero fields outside the aperture will fail unless the horn is surrounded by a planar conducting flange.

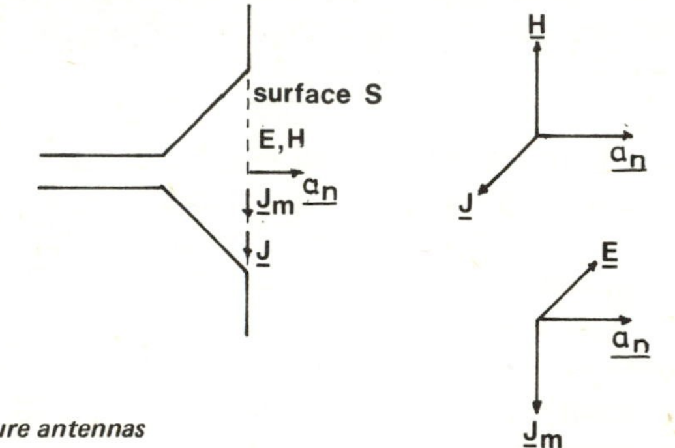

Fig. 1.17 *Aperture antennas*

The steps involved in the analysis of aperture antennas are first to calculate the electromagnetic fields over the aperture due to the sources on the rearward side of the infinite plane and to use these field distributions as the basis for the prediction of the distant fields in the forward half-space. The electromagnetic fields in the aperture plane can rarely be determined exactly but approximate distributions can be found by a variety of methods, which are dependent upon the antenna.

For example, geometric optics techniques are commonly applied in reflector antennas to establish the fields in the reflector aperture plane. This procedure is referred to as the aperture field method and it is employed as an alternative to the induced current method, which is based upon an approximation for the electric current distribution on the reflector surface. Both of these techniques are dealt with elsewhere in this text (Chapters 2 and 3) and are treated extensively in the literature.

In applying either the aperture field or the induced current methods to reflector antennas it is necessary to establish a model for the radiation fields of the primary-feed horn. Aperture methods can be employed successfully to predict the radiation from small waveguide horns. In this case solutions of Maxwell's equations inside rectangular and cylindrical waveguides can be employed to obtain approximate field distributions in the aperture planes of pyramidal and conical horns.

For any antenna for which the aperture concept is valid, and for which an approximate aperture field distribution can be established the expressions for the distant radiated fields can be formulated because the aperture and radiation fields are the Fourier transformation of each other.

The Fourier transform relationship

The Fourier transform is a mathematical relationship which can be expressed in two dimensional form as[24]

$$f(k_x, k_y) = \int_{-\infty}^{\infty}\int_{-\infty}^{\infty} u(\epsilon,\eta) \exp\left[j(k_x\epsilon + k_y\eta)\right] d\epsilon\, d\eta \tag{1.36}$$

with its inverse transformation given by

$$u(\epsilon,\eta) = \frac{1}{4\pi^2}\iint f(k_x, k_y) \exp\left[-j(k_x\epsilon + k_y\eta)\right] dk_x dk_y \tag{1.37}$$

In this formulation the functions f and u are such as to be Fourier transform pairs and this relationship can be expressed symbolically as $f = FT < u >$ and $u = FT^{-1} < f >$

The Fourier transformation has been studied extensively and its mathematical and implied physical characteristics for engineering purposes are well understood.[10,24]

Either by application of the plane-wave spectrum method due to Booker and Clemmow[24] or by use of vector potential theory as described in Chapter 4, the co-polarised radiated fields (Ludwigs third definition) from a linearly polarised aperture in the x, y, o plane (Fig. 1.18) can be expressed as

$$E_p(\theta,\phi) = \cos^2\theta/2\,(1 - \tan^2\theta/2\,\cos 2\phi)\, f(\theta,\phi) \tag{1.38}$$

where

$$f(\theta,\phi) = \int_{-\infty}^{\infty}\int_{-\infty}^{\infty} E_a(x,y,o) \exp\left[jk(x\sin\theta\,\cos\phi + y\sin\theta\,\sin\phi)\right] dx\,dy \tag{1.39}$$

and E_a is the tangential electric field distribution of the principle component in the aperture plane.

Comparison of eqns. (1.36) and (1.39) i.e. with $k_x = k \sin\theta \cos\phi$, $k_y = k \sin\theta \sin\phi$ and $\epsilon = x, \eta = y$ confirms that f and E_a are transform pairs. For high gain antennas the radiation pattern is largely focused into a small range of θ angles and since for these cases $\cos^2\theta/2 \approx 1{\cdot}0$ and $\tan^2\theta/2 \approx 0$ it can be seen that the Fourier transform relationship exists between the far fields E_p and the tangential aperture field E_a. It is also evident that even for moderate gain antennas the Fourier transform relationship will be the dominant characteristic in the expression for the radiated fields.

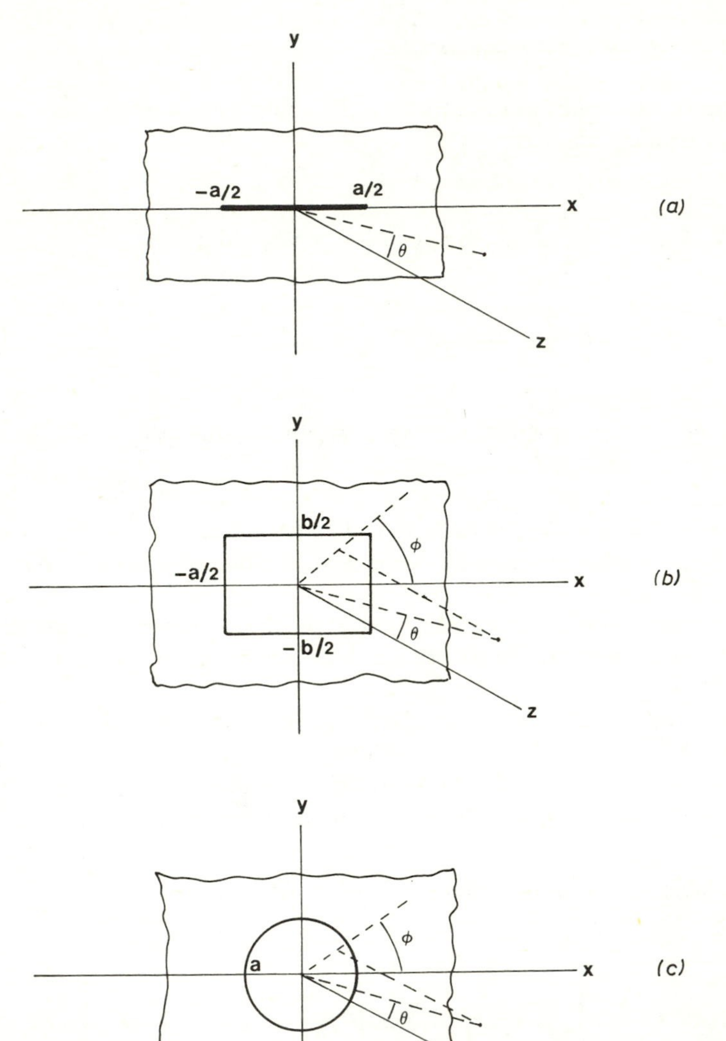

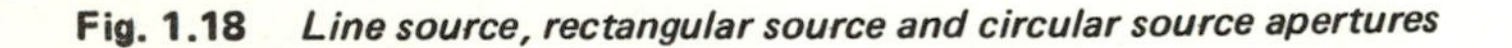

Fig. 1.18 *Line source, rectangular source and circular source apertures*

The existence of this dominant Fourier transform relationship is extremely important since it makes available all of the operational properties of Fourier transform theory for the analysis and synthesis of aperture antennas. Analytical solutions for eqn. 1.39 can be obtained for many simple aperture distributions to provide general guidance in the design of aperture antennas. A considerable literature has been established on transformable functions and some of the basic ones are considered in the following Sections.

More complex aperture distributions, which do not lend themselves to analytical solution, can be solved numerically. The increasing power and lower costs of the digital computer have resulted in its acceptance as a conventional tool of the antenna designer. The Fourier transform integral is generally well behaved and does not present fundamental computational problems. The main practical limitation is the computing effort required to evaluate the argument at each field point. In the region corresponding to the antenna main beam the exponential term in eqn. 1.39 varies slowly and comparatively few integration intervals are required for satisfactory convergence. At field points located at large angles from the antenna main beam, the exponential term varies rapidly, requiring large numbers of small integration intervals and hence a significant increase in the computational effort.

Line source distribution

The line source is a one dimensional aperture where the field radiates from a source, length a, along the x axis. Fig. 1.18a. The source is positioned in a ground plane of infinite extent. This model is simple and yet the analysis gives results which illustrate the main features of the more practical two dimensional apertures. The line source distribution does have a practical realisation, namely in a long one dimensional array which has sufficient elements to enable it to be approximated to a continuous distribution.

For the geometry shown in Fig. 1.18a, the Fourier transform relation becomes:

$$F(u) = \frac{1}{2\pi} \int_{-\pi}^{\pi} E_x(p) e^{jpu} dp \tag{1.40}$$

where $p = \dfrac{2\pi x}{a}$ and $u = \dfrac{a \sin\theta}{\lambda}$. The far field amplitude pattern is given by the magnitude $|F(u)|$.

The simplest distribution of field across the aperture is when the electric field $E_x(p)$ is constant. Inserting $E_x(p) = 1$ in the above equation and integrating gives

$$F(u) = \frac{\sin \pi u}{\pi u} \tag{1.41}$$

The sin $(x)/x$ distribution is very important in antenna theory and is the basis for many antenna designs. It is plotted in Fig. 1.19. The level of the first sidelobe is at −13·2 dB below the peak level of the main beam. This is relatively high and is the main reason why a uniform aperture distribution is unacceptable in a large number of antenna situations. The field has a null, or zero at a $\sin\theta/\lambda = \pm 1, \pm 2, \ldots$

The width of the main beam is another factor usually specified in antenna design. In this case the half power, or 3dB, beamwidth is $0{\cdot}88\lambda/a$ (radians), so the beamwidth is inversely proportional to the aperture size. This is generally true for all apertures operating at radio or optical frequencies. As the aperture increases in size the main beamwidth narrows and there are an increasing number of zeros

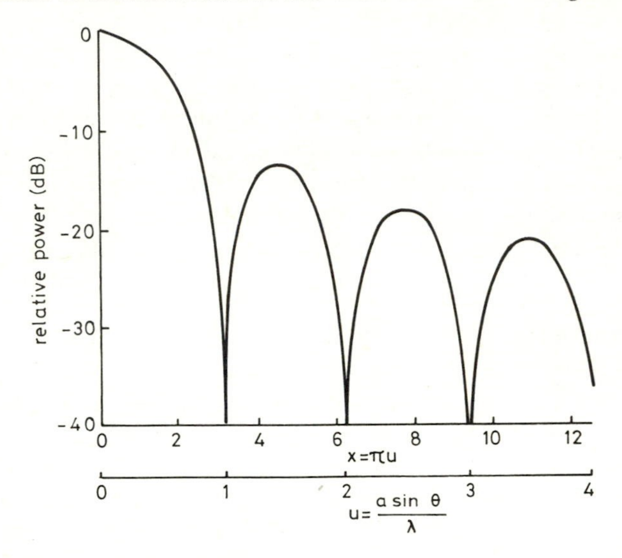

Fig. 1.19 *sin (x)/x amplitude radiation pattern*

in any given angular distance. The sin $(\pi u)/\pi u$ distribution has a phase pattern which changes sign by π every time a zero is encountered.

As mentioned above, the high sidelobe level is a disadvantage of a uniform source distribution. This can be reduced considerably by a tapered aperture distribution where the field is greatest at the centre and reduces to a low level at the edge of the aperture. An example of this is a cosine distribution where $E_x(p) = \cos^n (p/2)$. For n=1 the first sidelobe level drops to −23 dB at the expense of a broadening of the main beamwidth to $1{\cdot}2\lambda/a$. The characteristics of some aperture distributions are summarised in Table 1.2. The Taylor distribution (see Hansen[10]) is included as an example of an optimised distribution where the sidelobes are low and the main beam is still acceptably narrow.

Rectangular apertures

The radiation pattern of rectangular distributions is calculated in the same manner as line source distributions except that the integration is carried out over two dimensions rather than one dimension. For many types of antenna, such as the rectangular horn, the x and y functions are separable and the pattern in the principal xz plane (Fig. 1.18b) can be determined from a line source distribution $f(x)$ while the pattern in the yz plane can be determined from a line source distribution $f(y)$. The total pattern is then given by

$$f(x,y) = f(x)\,f(y) \tag{1.42}$$

The results quoted for line source distributions thus apply for the principal planes of separable rectangular apertures. For cases where the variables are not separable, numerical integration is probably the best solution.

Table 1.2 Line-source distributions

Distribution	Aperture field	3 dB Beamwidth	Level of 1st Sidelobe	Angular position of 1st zero
Uniform	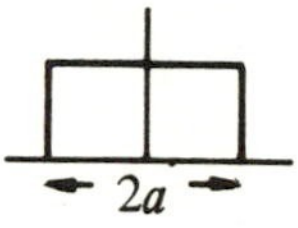 (2a)	$0{\cdot}88\frac{\lambda}{a}$	$-13{\cdot}2$ dB	$\frac{\lambda}{a}$
Cosine $\cos(\frac{\pi x}{2a})$		$1{\cdot}2\frac{\lambda}{a}$	$-23{\cdot}0$ dB	$1{\cdot}5\frac{\lambda}{a}$
Cosine square $\cos^2(\frac{\pi x}{2a})$	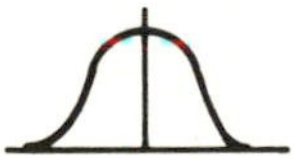	$1{\cdot}45\frac{\lambda}{a}$	$-32{\cdot}0$ dB	$2{\cdot}0\frac{\lambda}{a}$
Pedestal $1-(1-0{\cdot}5)(\frac{x}{a})^2$	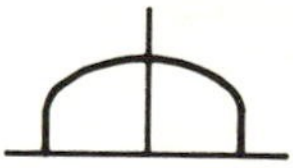	$0{\cdot}97\frac{\lambda}{a}$	$-17{\cdot}1$ dB	$1{\cdot}14\frac{\lambda}{a}$
Taylor n=3 −9 dB edge	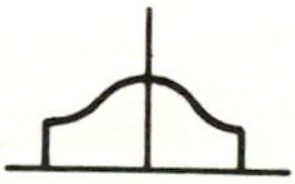	$1{\cdot}07\frac{\lambda}{a}$	$-25{\cdot}0$ dB	

Circular apertures

Circular apertures form the largest single class of aperture antennas. The symmetric paraboloidal reflector is used extensively for microwave communications and is often fed with a conical horn. The simplest form of aperture distribution is where the field does not vary with ϕ, that is, it is rotationally symmetric. This is not always true in practice because many horns produce an unsymmetric radiation pattern. However it is an ideal and effort is expended trying to realise symmetric patterns. It will be assumed here in order to illustrate the main features of circular apertures. The radiation pattern for the geometry shown in Figure 1.18*c*, written in normalised form, is

$$F(u) = \frac{1}{\pi^3} \int_{\phi'=0}^{2\pi} \int_{p=0}^{\pi} E_x(p) \exp j\,[pu \cos(\phi - \phi')]\, p\,\mathrm{d}p\,\mathrm{d}\phi' \qquad (1.43)$$

where $p = \frac{\pi r}{a}$ is the normalised radius and ϕ' the angle of a point in the aperture;

$$u = \frac{2a \sin\theta}{\lambda} .$$

Upon integrating with respect to ϕ'. the result is

$$F(u) = \frac{2}{\pi^2} \int_0^{\pi} E_x(p) J_0(pu) p\,\mathrm{d}p \qquad (1.44)$$

where J_o is the Bessel function of the first kind and order zero.

A useful aperture distribution is $(1-r^2/a^2)^n$ on a pedestal. The pedestal determines the level of the field at the edge of the aperture, the so-called edge taper. If the pedestal height is b then eqn. 1.44 becomes

$$F(u) = \int_0^{\pi} [b+(1-p^2/\pi^2)^n]\, J_0(pu) p\,dp \qquad (1.45)$$

After integration this becomes

$$F(u) = 2b\frac{J_1(\pi u)}{\pi u} + \frac{n! J_{n+1}(\pi u)}{(\frac{\pi u}{2})^{n+1}} \qquad (1.46)$$

By analogy with the line source the simplest case is uniform illumination, when b=0 and n=0. Then $F(u) = 2J_1(\pi u)/\pi u$ which can be compared to $\sin(\pi u)/\pi u$. The function is plotted in Fig. 1.20. The sidelobe level is −17·6 dB, compared to −13·2 dB for the line source, but the beamwidth is slightly broader at $1{\cdot}02\lambda/D$ (D=2a). Table 1.3 shows a number of circular aperture distributions and the corresponding radiation pattern properties. The table shows the results for a Taylor distribution (Hansen[10]) giving very low sidelobes. The addition of an amplitude taper reduces the

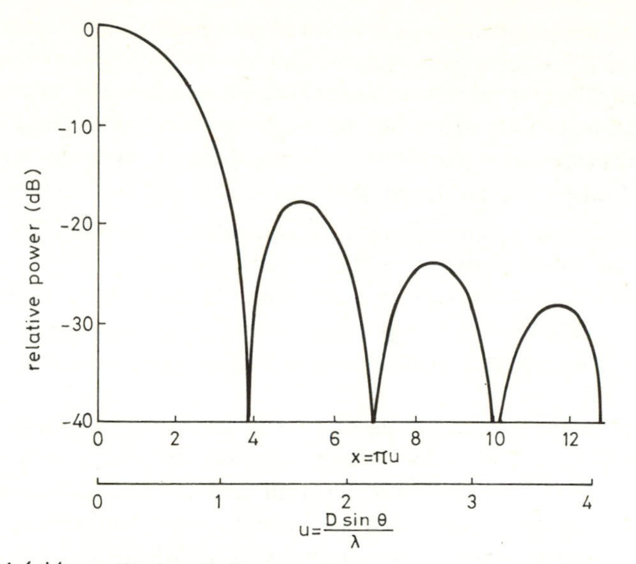

Fig. 1.20 *$2J_1(x)/x$ amplitude radiation pattern*

Table 1.3 Circular aperture distributions

Distribution	Aperture field	3 dB Beamwidth	Level of 1st Sidelobe	Angular Position of 1st zero
$0 \leqslant r \leqslant 1$				
Uniform	← D →	$1{\cdot}02\frac{\lambda}{D}$	-17·6 dB	$1{\cdot}22\frac{\lambda}{D}$
Tapered to zero at edge $(1-r^2)$		$1{\cdot}27\frac{\lambda}{D}$	-24·6 dB	$1{\cdot}63\frac{\lambda}{D}$
Tapered to zero at edge $(1-r^2)^2$		$1{\cdot}47\frac{\lambda}{D}$	-30·6 dB	$2{\cdot}03\frac{\lambda}{D}$
Tapered to 0·5 at edge $0{\cdot}5+(1-r^2)^2$		$1{\cdot}16\frac{\lambda}{D}$	-26·5 dB	$1{\cdot}51\frac{\lambda}{D}$
Taylor distribution		$1{\cdot}31\frac{\lambda}{D}$	-40·0 dB	

level of the sidelobes and increases the width of the main beam.

Sometimes a phase taper is present in addition to the amplitude taper. This is usually the case when the pattern of an aperture antenna is measured on an antenna range. The incident field across the aperture has a spherical phase front. The patterns for this case can be found by replacing the exponential term in eqn. 1.40 by $e^{j(pu+\psi(p))}$ where $\psi(p)$ is the phase variation.

Numerical integration is usually needed to obtain the solution. The effect of a symmetric phase taper is always to fill in the nulls of the amplitude radiation pattern so that a zero is replaced by a minimum. In addition the sidelobe level is raised. The simplest example of an asymmetric phase taper is a linear phase variation across the aperture, which produces a beam shift, without any change in sidelobe structure.

An important type of circular aperture distribution not listed in the Table is a Gaussian distribution. This is of interest because the Fourier transform of a Gaussian distribution is itself of Gaussian form, and the radiation pattern contains no sidelobes, falling off gradually as angle increases away from boresight. However, a Gaussian distribution implies an infinite aperture which in practice must be truncated. A horn aperture antenna producing an approximation to a Gaussian beam is often used as a feed with a reflector. The main practical example is a hybrid mode horn or corrugated horn, the properties of which are described in Chapters 4 and 7. To satisfy the Gaussian criteria the field at the edge of the horn must be essentially zero, say less than 20 dB. This implies that the horn aperture is being inefficiently illuminated so that the feed size required is much greater than that of simple feed horns.

Aperture synthesis

The Fourier transform (1.40) can be found for any aperture illumination $E_x(p)$ by expressing the latter as a complex Fourier series

$$E_x(p) = \sum_{-\infty}^{\infty} C_k \exp(jkp) \qquad (-\pi < \rho < \pi) \tag{1.47}$$

where the coefficients C_k are found in the usual way from the given illumination, specifically

$$C_k = \frac{1}{2\pi} \int_{-\pi}^{\pi} E_x(p) \exp(-jkp) \mathrm{d}p \tag{1.48}$$

The Fourier transform can then be written as

$$F(u) = \frac{1}{2\pi}\int_{-\pi}^{\pi} \sum_{-\infty}^{\infty} C_k \exp [j(k+u)p]\, dp \tag{1.49}$$

yielding

$$F(u) = \sum_{-\infty}^{\infty} C_k \frac{\sin [\pi (u+k)]}{\pi (u+k)} \tag{1.50}$$

Thus each Fourier coefficient C_k is responsible for a $(\sin x)/x$ type of beam. These beam components are spaced by unit increments in u, i.e. by increments of λ/a in $\sin \theta$ space.

The above equations form the basis for Woodward's aperture synthesis technique (Chapter 9, Volume 2), which enables the aperture illumination $E_x(p)$ required to produce a given beam shape to be approximated. The desired pattern $F(u)$ is sketched (in amplitude and phase) and ordinates erected at unit intervals of u. The values of these ordinates are used directly as trial coefficients C_k in the Fourier series representation of the aperture illumination (1.47). The actual pattern obtained (1.50) is correct at the ordinates, but differs marginally from the desired pattern between the ordinates. The technique is illustrated in Fig. 1.21 where it is

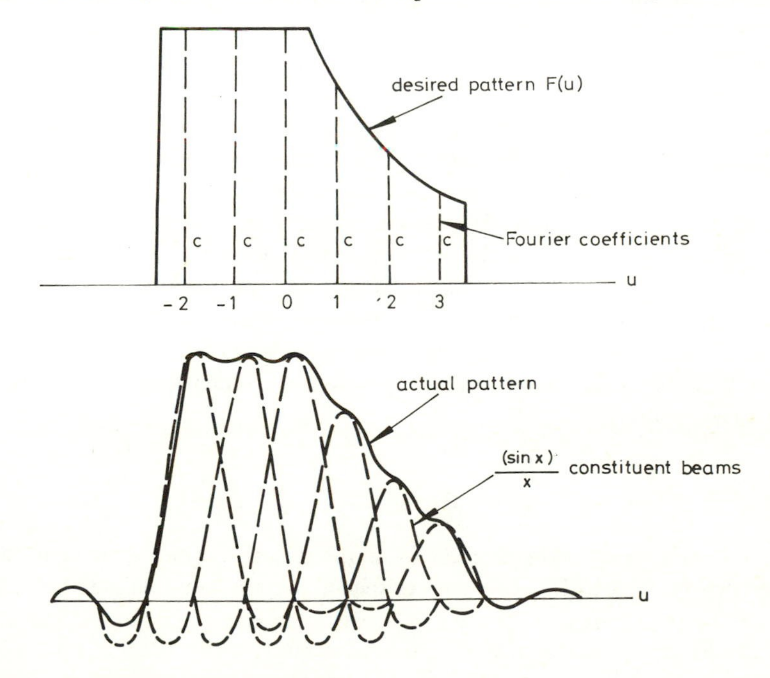

Fig. 1.21 *Aperture synthesis technique*

assumed, for simplicity, that $F(u)$ is of constant phase. Adjustments to the actual pattern may be made by varying the values of the coefficients (C_k).

1.13 Radiation from arrays

Array antennas are made up of a matrix of discrete sources. The discrete sources radiate individually but the pattern of the array is also determined by the relative amplitude and phase of the excitation fields on each source and the geometric spacing of sources. The total radiation pattern is the multiplication of the pattern due to an individual source, or element, the *element factor* and the pattern due to an array of omnidirectional sources, the *array factor.* Array theory is largely concerned with the array factor and the synthesising of an array to generate a particular array factor. Usually the individual sources will be low directivity elements so the total pattern depends very little on the element factor, though sometimes high directivity elements are used in which case the two factors may be equally important. Typical elements in arrays are dipoles, monopoles, slots in waveguides, open-ended waveguides and microstrip radiators. The choice of the type of element depends on the operating frequency and other factors such as the power handling capability, the polarisation desired, the feeding arrangements and the mechanical constraints. The array may be one dimensional with the elements in a line, forming a *linear array* or it may be two dimensional with the elements arranged in or around a rectangle or a circle. Usually the elements are equally spaced in order to reduce the number of variables in the total design. The basic concept has been implemented in a wide variety of different forms. The array is particularly attractive for electronic scanning where the direction of the main beam is controlled by altering, electronically, the phase of the signals applied to the individual elements. These arrays are termed *phased arrays*. The idea of combining the power of electronics with the array antenna has led to the ability to use signal processing techniques to guide the electronically steered beams. The result are *signal processing arrays* and *adaptive arrays.* In the latter the array radiation pattern adapts itself, under computer control to a particular situation. An example is a *null steering array* where the null of the radiation pattern is steered so that it points in the direction of an unwanted, interfering, signal. The study of arrays is dealt with in detail in Volume II. Here we will illustrate the basic design features by considering simple linear arrays.

Two-element arrays

The two-element array is the simplest form but indicates some of the main features of array theory. We consider two sources spaced d apart and with radiated fields E_1 and $E_2\, e^{j\delta}$ where δ indicates the phase difference between the fields and the element, Fig. 1.22. We are not concerned with the radiation characteristics of the individual elements so the elements are taken as point sources having omnidirectional patterns. The general expression for the far-field in the plane of the array is

$$E(\phi) = E_1 + E_2 e^{j(\delta + kd\cos\phi)} \tag{1.51}$$

($k = 2\pi/\lambda$). The amplitude radiation pattern is given by the magnitude of the above equation.

The case of most interest is when the amplitude of the individual elements is equal, $E_1 = E_2$. In this case the radiation pattern is determined by the element spacing and the phase differential. Three cases illustrate some basic properties. In the first the phase differential between elements is zero and the amplitudes are equal.

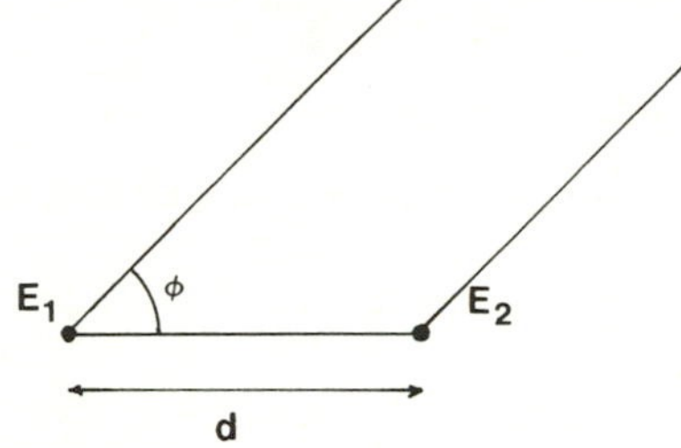

Fig. 1.22 *Two point source array*

The pattern for very small spacings will be almost circular, and as the spacing is increased the pattern has a maxima perpendicular to the axis of the array. At $\lambda/2$ spacing a null appears along the array axis, Figure 1.23*a*. This is termed a *broadside array* because energy is transmitted or received broadside to the array.

The second case is when the amplitudes are equal but the phase differential between the elements is 180° or π radians. The pattern for $\lambda/2$ spacing is shown in Fig. 1.23*b*. Now the main beam direction is along the axis of the elements and the

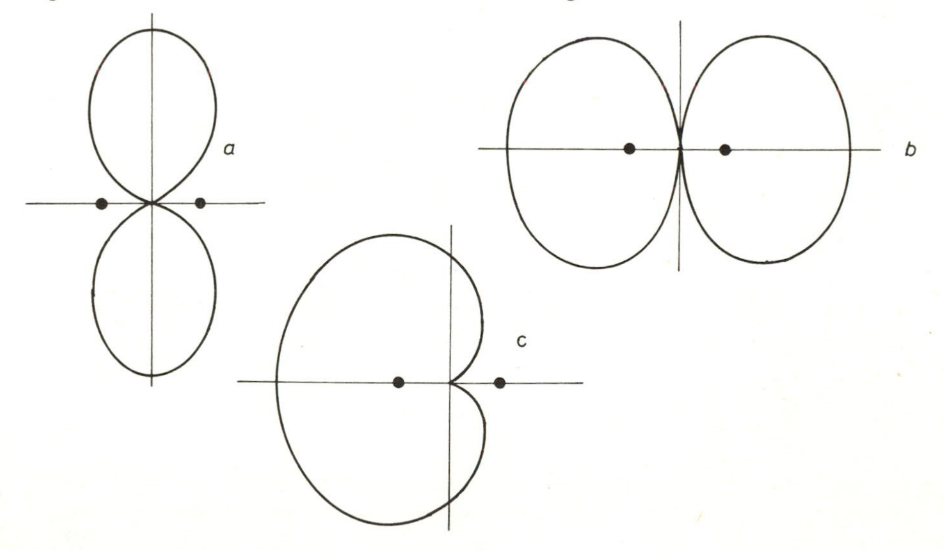

Fig. 1.23 *Array patterns for two point sources (a) $d = \lambda/2$, $E_1 = E_2$, $\delta = 0$, (b) $d = \lambda/2$, $E_1 = -E_2$, $\delta = 0$, (c) $d = \lambda/4$, $E_1 \doteq E_2$, $\delta = -kd$*

array is called an *end-fire array*. Reducing the spacing will maintain the same shape but increasing the spacing will cause the shape to change and more lobes to appear. When $d = \lambda$ there are four lobes with nulls at $\phi = 0, 90°, 180°, 270°$ and maxima at $\phi = 60°, 120°, 240°, 300°$. Since this pattern is not particularly useful, the element spacing is usually kept less than one wavelength apart.

The third case is when the array elements have equal amplitudes and a phase differential $\delta = \pi - kd$. The pattern for a spacing of λ/d is shown in Fig. 1.23*c*. This type of pattern is called a cardiod and has no radiation in the backward direction. It is widely used as the basis for broadcast receiving antennas.

Multielement linear array

If the linear array has *n* elements equally spaced and with equal amplitudes, Fig. 1.24, the array factor is given by

$$E = E_1\,(1 + e^{j\psi} + e^{j2\psi} + \ldots e^{j(n-1)\psi}) \tag{1.52}$$

where $\psi = \delta + kd\cos\phi$, d is the spacing between elements, δ is the impressed phase shift between elements, and E_1 is amplitude of the radiated field of an element

Multiplying eqn. 1.52 by $e^{j\psi}$ and then subtracting (1.52) gives

$$E = E_1 \frac{1 - e^{jn\psi}}{1 - e^{j\psi}} = E_1 e^{\frac{j(n-1)}{2}\psi} \cdot \frac{\sin(n\psi/2)}{\sin(\psi/2)} \tag{1.53}$$

If the centre of the array is chosen as the reference for the phase this equation becomes

$$E = E_1 \frac{\sin(n\psi/2)}{\sin(\psi/2)} \tag{1.54}$$

For a broadside array with the maximum at $\phi = 90°$ the elements must be in phase with $\delta = 0$. In this case the nulls occur when

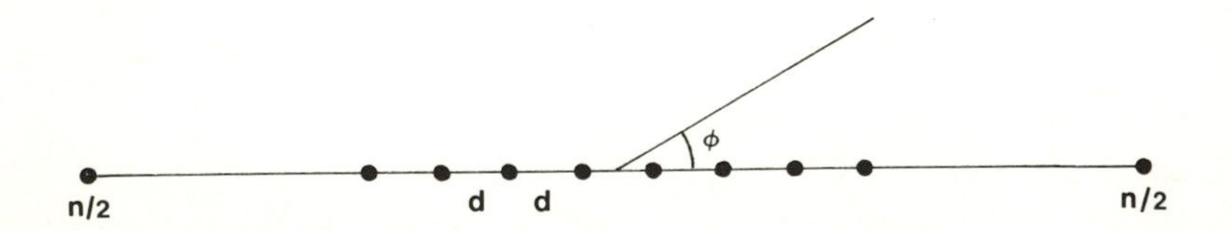

Fig. 1.24 *Multisource linear array*

$$\phi = \cos^{-1} \left[\pm \frac{N\lambda}{nd}\right] \text{ where } N = 1, 2, 3 \text{ (but } N \neq mn \text{ where } m = 1, 2, 3 \ldots).$$

Notice that the cosine of the angle is inversely proportional to the length of the array in wavelengths. If the array is large and ϕ is restricted to angles around the maximum at 90° then $\sin(\psi/2) \simeq \psi/2$ and eqn. 1.54 can be written

$$E = nE_1 \frac{\sin(n\psi/2)}{(n\psi/2)} \tag{1.55}$$

This is now a $\sin x/x$ pattern and shows that a large broadside linear array behaves in the same way as the continuous line source aperture discussed in Section 1.12. The deductions concerning the line source radiation pattern therefore apply to long arrays. In particular the first sidelobe level can be reduced from −13·2 dB in the case of equal amplitude (uniform source) elements by tapering the amplitude distribution across the array. There is more control available with an array of discrete elements than with a continuous aperture distribution and optimum distributions have been devised based on polynomial distributions. These are optimum in the sense that the narrowest beamwidth is obtained for a given sidelobe level, or vice versa. Tchebyscheff polynomial distributions give one of the optimum designs, (Chapter 9, Volume II).

The basic concept of electronic scanning can be illustrated by considering the case when there is a progressive phase delay along the array. Putting δ negative in eqn. 1.52, the direction of the main beam is given by

$\phi = \cos^{-1}(\delta/kd)$. Thus changing δ from 0 to kd will scan the main beam from $\phi = 90°$ to $\phi = 0$. The array then changes from a broadside array to an endfire array.

1.14 Radiation from current elements

At frequencies below the microwave region the most common basic radiating element is the electric current element. Physically this is realised by a very short thin dipole or doublet. There are few antennas which actually use a doublet but a large number of antennas which are derivatives of the basic element. Examples are the vertical radiators used at low and medium frequencies, cylindrical rod antennas, linear wire and monopoles. The electric current element is also of interest historically as being the first widely used form of antenna. The practical realisation is simple and relatively cheap and is probably responsible for the common public belief that a short piece of wet string makes an adequate antenna.

We will illustrate the basic radiating properties by stating the fields arising from the electric current element and the short dipole. The exact radiation pattern of a linear antenna can be computed if the exact current distribution on the dipole is known, and considerable effort has been devoted to the solution of the current

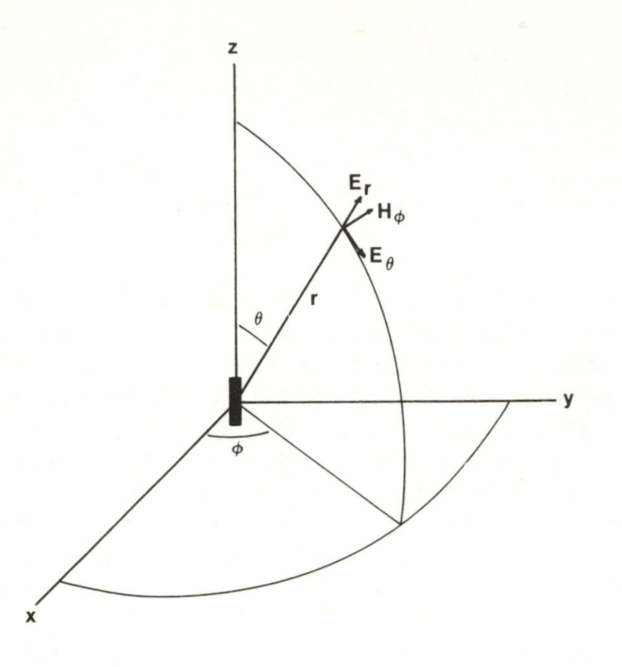

Fig. 1.25 *Electric current element*

distribution on antennas such as the cylindrical rod antenna, King[16]. The problem is simpler if the linear antenna is assumed to be infinitely thin so that the circumferential current flow can be neglected. The solution is then found by knowing the radiated fields of a differential element and integrating over the length of the antenna. The current distribution along the antenna is either known or can be approximated, allowing for end effects, to a sufficient accuracy for most practical purposes.

The differential electric current element will be assumed to be positioned along the z axis, Fig. 1.25. The derivation of the radiated fields is done with the help of the vector potential A which in this case has only a z component, A_z, given by

$$A_z = \frac{\mu_o e^{-jkr}}{4\pi r} \tag{1.56}$$

The details of the derivation are contained in many texts, for instance Collin and Zucker[1], Jordan and Balmain[3]. The electric and magnetic components of the radiated field are

$$E_r = \frac{\eta_0}{2\pi k} Idz \left[\frac{k}{r^2} - \frac{j}{r^3}\right] e^{-jkr} \cos\theta$$

$$E_\theta = \frac{j\eta_0}{4\pi k} Idz \left[\frac{k^2}{r} - \frac{jk}{r^2} - \frac{1}{r^3}\right] e^{-jkr} \sin\theta$$

(1.57)

$$H_\phi = \frac{1}{4\pi k} Idz \left[\frac{jk}{r} + \frac{1}{r^2}\right] e^{-jkr} \sin\theta$$

$$E_\phi = H_r = H_\theta = 0$$

where Idz is the current in the differential current element, η_0 is the intrinsic impedance of free space

$$= \sqrt{\frac{\mu_0}{\epsilon_0}} = 120\pi \text{ and } k = 2\pi/\lambda = \omega\sqrt{(\epsilon_0\,\mu_0)}$$

The expressions give the field at all values of distance r from the antenna. The terms in $1/r^3$ and $1/r^2$ predominate in the near-field region ($kr < 1$), and the terms in $1/r$ predominate in the radiating far-field region ($kr > 1$). In this case, because the element is short, the reactive and radiating near-field regions are the same.

Normally only the far-field region is of interest, in which case the equations simplify to

$$E_\theta = j\,60\pi \frac{Idz}{r\lambda} e^{-jkr} \sin\theta \qquad (1.58)$$

$$H_\phi = E_\theta/\eta_0$$

All other field components are zero. The field consists of a transverse electromagnetic wave propagating away from the source. The field amplitude at any angle is proportional to the 'projected length' of the element, ($\sin\theta\, dz$).

These equations are only valid for short elements where the current distribution along the antenna is constant. This can be approximately realised in practice by 'top loading' a thin wire. To calculate the field for an element of any length, eqn. 1.55 is integrated over the length l of the element:

$$E_\theta = j\frac{60\pi \sin\theta}{r\lambda}\int_{\frac{-l}{2}}^{\frac{l}{2}} I(z)\; e^{-jkr(z)}\, dz \tag{1.59}$$

The most common type of linear element is a half wave dipole. The current distribution is assumed to be $I(z) = I_0 \cos kz$, a form which is not precisely correct but is a good approximation. Then from eqn. 1.59 using the far-field approximation $r(z) = r - z\cos\theta$, the field after integration becomes

$$E_\theta = j60 I_0 \frac{e^{-jkr}}{r}\;\frac{\cos((\pi/2)\cos\theta)}{\sin\theta} \tag{1.60}$$

The pattern for a half wavelength dipole is shown in Fig. 1.26. This is shown in the plane which contains the axis of the antenna. In the orthogonal plane, xy plane in Fig. 1.25, the radiation is omnidirectional and the pattern a circle.

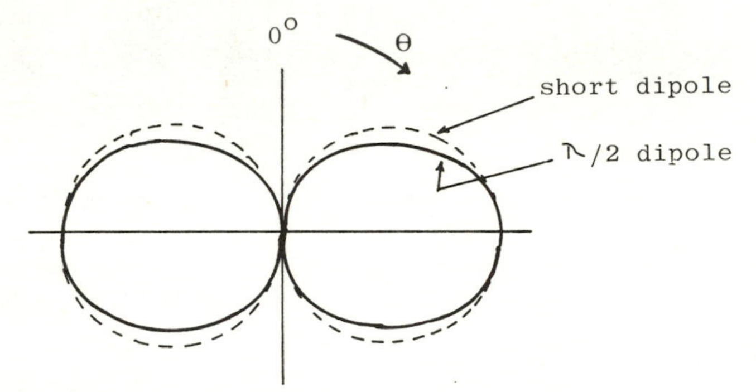

Fig. 1.26 *Horizontal radiation pattern of a half wavelength vertical dipole*

Impedance properties of electric current elements

The self impedance of a linear antenna is an important property which has a significant bearing on the usefulness of the antenna. The antenna resistance or radiation resistance R (assuming no ohmic losses) can be calculated by evaluating the radiated power P and using $P = I^2 R$, where I is the rms current. The radiated power is found by integrating the Poynting vector $\mathrm{Re}(E_\theta H_\phi^*)$ over a sphere surrounding the antenna. (E_θ and H_ϕ^* are the rms values of the field produced by I).

$$P = \int_0^\pi \mathrm{Re}(E_\theta H_\phi^*)\, 2\pi r^2 \sin\theta\, d\theta \tag{1.61}$$

Substituting E_θ for an electric current element, eqn. 1.58 and integrating gives

$$R = 80\pi^2 \frac{dz^2}{\lambda^2} \text{ ohms} \tag{1.62}$$

The radiation resistance is proportional to the square of the normalised length. An infinitely short element has zero resistance. A linear element with uniform current and one tenth of a wavelength in length has a resistance of 7·9 Ω. The radiation resistance of a thin half wave dipole can be found by the same method as above and has a value of 73·1 Ω. The problem of computing the radiation resistance of linear antennas with finite diameter has been considered by many writers and is dealt with in Volume II.

The reactive component of the input impedance cannot be found by the far-field method because it is determined by the fields in the reactive near-field region immediately surrounding the antenna. For a thin half wavelength dipole the reactance is nearly zero, but for a very short element it has a very large negative value. This indicates that the half wave dipole is easy to match into a standard transmission line but that the short antenna will have a bad mismatch and hence be a very inefficient radiator.

Attempts have been made to overcome this disadvantage by devising impedance transforming networks. The most successful is the *active antenna*, devised by Meinke[25] where an active element, usually a transistor, is placed at the terminals of the short antenna. The impedance of one side of the active network can then be matched to the antenna and the other side to the transmission line.

1.15 Radiation from travelling wave sources

A travelling wave antenna is one where the fields and currents which produce the radiation can be represented by one or more travelling waves. The antenna structure supports propagating waves and it is from these propagating waves that the radiation occurs. There are two parts to the analysis and design of travelling wave antennas. Firstly the prediction of the propagation coefficients along the structure due to the propagating wave and secondly the calculation of the radiation characteristics, which will be a function of the propagation coefficients.

In many cases the waves are travelling only in one direction. That is, the structure is properly terminated so that the reflections are small. Examples of this type of antenna are the long wire, the rhombic, dielectric rod, long slots in waveguide and helical antennas. If the antenna structure is not properly terminated a reflected wave will exist and as a result a standing wave caused by two oppositely travelling waves will occur. Thus a standing wave, or resonant, antenna can be considered as a travelling wave antenna. An example is a resonant slot in a waveguide. A dipole can also be analysed as a travelling wave antenna. Some long arrays such as the Yagi-Uda array or slots in waveguides can also be considered from the viewpoint of having a continuous travelling wave along the array. It is thus clear that there are few antennas which fall only into the category of travelling wave antenna. It is however a useful method of studying an antenna if the physical or mathematical representation aids in the analysis and design process. A study of travelling wave antennas has been done by Walter[11] and the information in this section is taken from his work.

The travelling wave can be represented by a current flowing along the z axis:

$$I(z) = \sum B_n e^{-\gamma nz} \tag{1.63}$$

where B_n are coefficients and γ_n are complex propagation constants, one for each of the n travelling waves. The propogation constant γ is given by

$$\gamma = \alpha + j\beta \tag{1.64}$$

where γ is the attenuation constant and β is the propagation constant. α and β describe the characteristics of the propagating waves. Also $\beta = 2\pi/\lambda_g$ where λ_g is the guide wavelength.

The travelling wave has a phase velocity $v = \omega/\beta$ where ω is the angular frequency. v may be greater than or less than c, the velocity of light. If $v > c$ then the travelling waves are described as *fast waves*, whilst if $v < c$ the travelling waves are described as *slow waves*.

A travelling wave that loses energy continually as it propagates along the guiding structure is termed a *leaky wave*. This means that the energy remaining in the travelling wave decays along the structure. Most leaky wave antennas are fast wave structures. An example of a leaky wave antenna is a slotted waveguide propagating the dominant mode.

A *surface wave* is one where the energy is bound to an interface between two different media and radiation occurs at the end of the structure or discontinuities, curvatures and non-uniformities. The surface wave differs from the leaky wave because the latter radiates continually whereas the former does not radiate until some physical change in the guiding structure creates a means for the energy to radiate. A surface wave is a slow wave structure. An example is a dielectric rod in free space. The field is bound to the rod and decays exponentially away from the rod in a radial direction.

As an example of a typical travelling wave antenna consider the line source shown in Fig. 1.27, which has a current propagating along it of the form

$$I(z) = I_0\, e^{-(\alpha + j\beta) z} = I_0\, e^{-\gamma z} \tag{1.65}$$

The radiating field is found from the vector potential for the source

$$A_z = \frac{\mu I_0}{4\pi} \int_{-l/2}^{+l/2} \frac{e^{-\gamma z - jkr'}}{r'}\, dz \tag{1.66}$$

The electric field is given by $E_\theta = -j\,\omega \sin\theta A_z$ which after integration gives

$$E_\theta = \frac{j\omega\mu I_0 l \sin\theta\, e^{-jkr}}{4\pi r} \quad \frac{\sin X_0}{X_0} \tag{1.67}$$

where $X_0 = \frac{l}{2}(k\cos\theta + j\,\Upsilon)$. For the case of uniform amplitude $(\alpha = 0$ and $\Upsilon = j\,\beta)$, X_0 becomes $\frac{l}{2}(k\cos\theta - \beta)$.

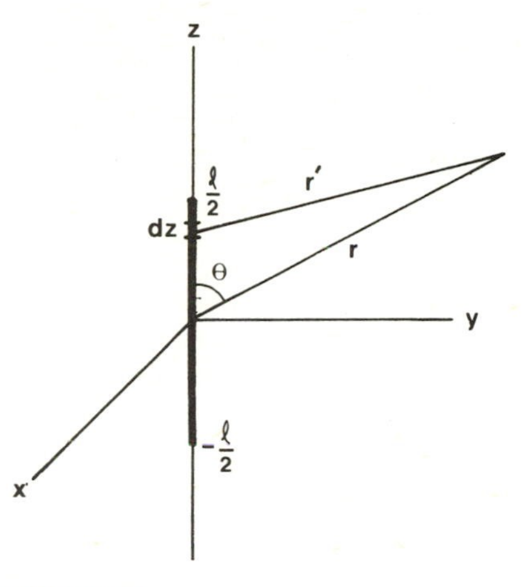

Fig. 1.27 *Travelling wave line source*

The sin (x)/x pattern has already been studied with reference to line source apertures and long linear arrays. The direction of maximum radiation occurs when $\cos\theta = \beta/k = c/v$. Thus the cosine of the angle of radiation is proportional to the propagation constant, and for a travelling wave just above cut-off ($\beta = 0$), the radiation is broadside whilst when the phase velocity approaches the velocity of light the radiation is endfire.

1.16 Frequency independent antennas

Antennas which have radiation and impedance characteristics which do not change over a very wide range of frequencies form a class of antenna on their own. No antenna is completely frequency independent because any physical structure has a limited size which constrains the lower and upper frequency limits. However some antennas have an inherently frequency independent characteristic and can operate over as much as a 20:1 range of frequencies. It is important to emphasise that we are not dealing here with broad-band antennas such as the parabolic reflector where the pattern characteristics depend on the normalised dimensions expressed in wavelengths.

A frequency independent antenna is one where the part of the structure radiating changes as frequency changes in order to keep the dimensions of the radiating part of the antenna constant in wavelengths. For true frequency independence this can be achieved by using a design for which geometrically similar forms are obtained by rotation. The radiating section then depends on angle and not on distance. A change in wavelength leads only to a rotation of the pattern. The basic idea was proposed by Rumsey[26] who suggested using equiangular or logarithmic spirals given by

$$r = c\, e^{a(\phi - \delta)} \tag{1.68}$$

where r and ϕ are the usual spherical coordinates and c, a and δ are constants. A physical realisation is shown in Fig. 1.28. To obey the equation exactly the spiral would have to start at $r = 0$ and extend to $r = \infty$. In practice the feeding point at the centre, and the terminations at the outer edge, determine the frequency limits.

The mechanism of radiation can be understood by considering the two spirals in Fig. 1.28 as a transmission line. A wave travelling out from the feed point will propagate along the spirals without radiating until it reaches a point where resonance occurs, and then strong radiation occurs. This loses almost all the energy so that there is very little to be reflected by the outer limits of the spiral. The region of radiation is near to the centre for high frequencies and near to the outer edge for low frequencies. No radiation must take place in the plane of the spiral, for in this direction waves would be reflected from the finite diameter of the structure. If radiation is desired in one direction only, this can be achieved by winding the spiral onto a conical surface to give a conical spiral.

Fig. 1.28 *Equiangular spiral antenna*

A second type of frequency independent antenna is the log-periodic antenna, shown in Fig. 1.29. This can be considered as a development of the equiangular conical spiral. If a plane is inserted along the axis of the cone the parts of the spiral which cut the plane will radiate when they resonate. The resonance can be helped by adding a half wavelength dipole as in the Figure. The feed lines must cross over between resonant points as in the spiral. In the case of the log-periodic dipole array the characteristics repeat periodically as a function of frequency. The change in

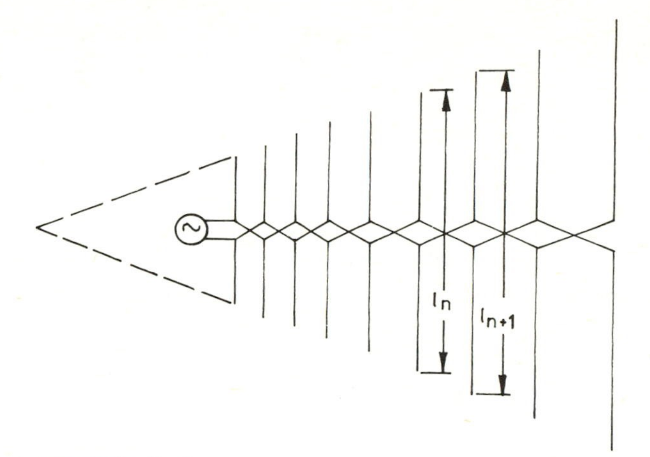

Fig. 1.29 *Log periodic dipole antenna*

frequency from that at which l_n resonates to the frequency at which l_{n+1} resonates should be small to obtain nearly constant characteristics. Thus the log-periodic antenna is not a truly frequency independent antenna because the characteristics repeat periodically rather than depend on the frequency independent parameter of angle.

The lengths of the dipoles increase in the fixed ratio

$$l_{n+1} = r\, l_n \tag{1.69}$$

The distance to successive dipoles from the apex increases in the same ratio, so that the whole structure is scaled by r for every periodic increase in the logarithm of frequency. The frequency limits are determined by the lengths of the shortest and longest dipoles.

1.17 Re-radiation from antennas

When an antenna is receiving a signal from a direction which is away from its main beam peak, only a fraction of the energy incident on the aperture is delivered to the receiver. Sometimes it is necessary to know what happens to the rest of the incident energy, e.g. a radar antenna may itself form a target which is being illuminated by another radar.

In the case of a paraboloid antenna, it will be apparent that the reflected energy is focused to a 'spot' which lies outside the feed area and this energy is therefore re-radiated as a diverging beam which has roughly the same beamwidth as the feed horn.

In an array, the nature of the re-radiated energy depends not only on the properties of the elements, but also on the feeding network. As a simple example, Fig. 1.30 shows an array of two isotropic elements which are assumed to be small enough for mutual coupling effects to be ignored. The elements are connected together by either a simple matched T-junction or by a 3 dB hybrid coupler. The symbols on the diagrams show incident and reflected voltage coefficients at various points in the circuits.

For reception on boresight, equal amplitude signals arrive in-phase at the two elements and simply add together at the T-junction or at the output of the hybrid coupler. There are no reflections within the feeding networks. Off boresight, signals arrive with a relative phase of $\pm\,\phi$, where $\phi = (\pi s \sin\theta_1)/\lambda$. This causes reflections at the T-junction and the reflected waves (which are of equal amplitude but anti-phase) are re-radiated by the elements in another direction. In the case of the

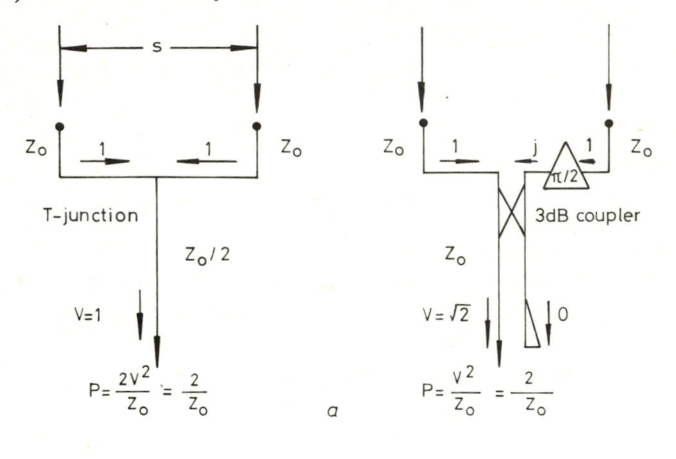

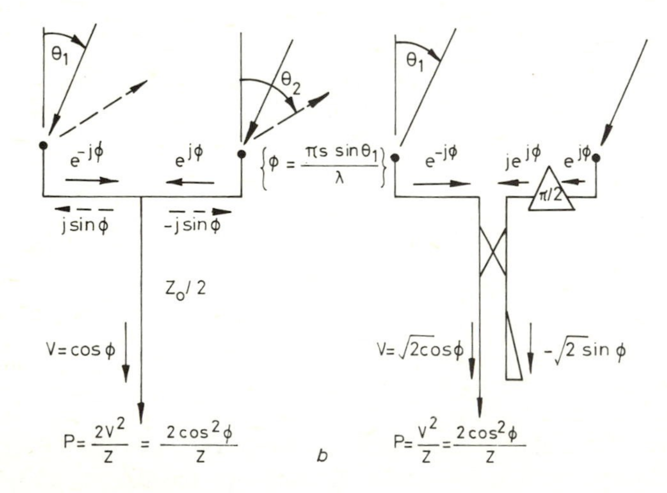

Fig. 1.30 *Voltage relationships in two-element arrays:* (*a*) reception on boresight; (*b*) reception off boresight, $\sin\theta_2 = (\lambda/Z_s + m\lambda/s)$

hybrid coupler, the surplus energy is directed into the auxiliary load and is not re-radiated. Both arrays are identical when considered as transmitters or receivers, but their scattering properties are completely different. It should be noted that this does not contravene the reciprocity theorem discussed earlier.

More detailed treatments of the scattering properties of antennas can be found in references 1 and 2.

1.18 References

1 COLLIN, R.E. and ZUCKER, F.J.: 'Antenna theory', McGraw Hill, NY, 1969
2 SILVER, S.: 'Microwave antenna theory and design', McGraw Hill, NY, 1949
3 JORDAN, E.C., and BALMAIN, K.G.: 'Electromagnetic waves and radiating systems', Prentice Hall, NY,
4 RAMO, S., WHINNERY, J.R. and VAN DUZER, T.: 'Fields and waves in communication electronics', Wiley, NY, 1965
5 SCHELKUNOFF, S.A.: 'Electromagnetic waves', VAN NOSTRAND, Princeton, N.J., 1943
6 STRATTON, J.A.: 'Electromagnetic theory', McGraw Hill, NY, 1941
7 PLONSEY, R., and COLLIN, R.E.: 'Principles and applications of electromagnetic fields', McGraw Hill, NY, 1961
8 KRAUSS, J.D.: 'Antennas', McGraw Hill, NY, 1950
9 KRAUSS, J.D.: 'Radio astronomy', Ch. 6, McGraw Hill, NY, 1966
10 HANSEN, R.C.: 'Microwave scanning antennas', Vol I, Academic Press, NY, 1964
11 WALTER, C.M.: 'Travelling wave antennas', McGraw Hill, NY, 1965
12 IEEE: 'Standard Test Procedures for Antennas', IEEE, 1979
13 SCHELKUNOFF, S.A. and FRIIS, H.T.: 'Antenna theory and practise', Wiley, NY, 1952
14 SCHELKUNOFF, S.A.: 'Advanced antenna theory', Wiley, NY, 1952
15 WILLIAMS, H.P.: 'Antenna theory and design', Pitman, London, 1950
16 KING, R.W.P.: 'The theory of linear antennas', Harvard Univ. Press, Cambridge, Mass., 1956
17 KING, R.W.P., MACK, R.B., and SANDLER, S.S.: 'Arrays of cylindrical dipoles'. Cambridge Univ. Press, New York, 1968
18 JASIK, H. (ed): 'Antenna engineering handbook', McGraw Hill, NY, 1961
19 DESCHAMP, G.A., and MAST, P.E.: 'Poincare sphere representation of partially polarised fields', *IEEE Trans.*, **AP-21**, July 1973, pp. 474-478
20 COHEN, M.H.: 'Radio astronomy polarization measurements', *Proc. IRE*, **46**, Jan. 1958, pp.172-183
21 BORN, M., and WOLF, E.: 'Principles of Optics', Macmillan, 1964
22 HOLLIS, J.S., LYON, T.J., and CLAYTON, L.: 'Microwave antenna measurements', Scientific Atlanta, Atlanta, Georgia, 1970
23 LUDWIG, A.C.: 'The definition of cross-polarisation', *IEEE Trans.*, **AP-21**, Jan. 1973, pp.116-119
24 BOOKER, H.G., and CLEMMOW, P.C.: 'The concept of an angular spectrum of plane waves, and its relation to that of polar diagram and aperture distribution', *JIEE*, **97**, 1950, pp.11-17
25 MEINKE, H.A.: 'Aktive antennen', *Nachrichtentech.Z.*, **26**, H.8, 1973, pp.361-362
26 RUMSEY, V.H.: 'Frequency independent antennas', Academic Press, New York, 1966

Chapter 2

Analytical techniques for quasi-optical antennas

W.V.T.Rusch, A.C.Ludwig and W.C.Wong

2.1 Introduction

The earliest radio antennas and other microwave components were based on ray-optical principles because it was known that electromagnetic and optical phenomena were closely related. The radio antennas of Hertz, Righi and Marconi were parabolic cylinders, doubtless because of the ease with which they could be constructed. Paraffin refracting lenses, dual-prism directional couplers, and wire-grid polarisers were other types of early radio equipment using ray-optics for their concepts and designs. Indeed, the Hertzian era gave birth to the term 'quasi-optics' which, along with 'microwave optics' continues to this very day. Most modern microwave engineers, in fact, spend as much time studying optical texts as those dealing with Maxwell's equations.

This chapter and Chapter 3 will deal with the two principal types of quasi-optical antenna: reflectors and lenses. It will be seen that the principles of ray-optics, as opposed to guided-wave or constrained-wave theory, will be instrumental in their design. The term optics, however, is considered in its most general sense, and diffraction (e.g. physical optics, geometric optics and the geometric theory of diffraction (GTD)) and aberrations are also included. In fact the commonest microwave antenna, the collimating reflector, is a diffraction-limited device which is incapable of being analysed using ray optics alone. Other material in these chapters will be seen to go significantly beyond the normal realm of optical principles: e.g. spherical wave theory, tolerance theory, aperture blocking, frequency-selective surfaces, contoured beams, and low-noise antennas. Yet these subjects are held together by the common thread of their direct applications to quasi-optical antennas.

2.2 Basic scattering theory for reflector analysis (Contribution by W.V.T.Rusch)

2.2.1 Differential geometry of surfaces

A surface can be presented by a parametric equation with parameters (u, v)[1-3]:

$$\bar{r} = \bar{r}(u, v) = x(u, v)\hat{a}_x + y(u, v)\hat{a}_y + z(u, v)\hat{a}_z \tag{2.1}$$

where $u_1 < u < u_2$ and $v_1 < v < v_2$. The unit surface normal at (u, v) is defined by

$$\hat{n}(u, v) = \mu \frac{\bar{r}_u \times \bar{r}_v}{|\bar{r}_u \times \bar{r}_v|} \tag{2.2}$$

where $\bar{r}_u$ and $\bar{r}_v$ are, respectively, the partial derivatives of $\bar{r}$ with respect to u and v. Lee[2] has pointed out that in electromagnetic diffraction problems it is convenient to choose $\mu = \pm 1$ such that $\hat{n}$ always points toward the source, whether the surface is a wavefront or a mirror.

The parameters E, F, G, e, f, and g may be used to determine the curvature properties of the surface described in eqn. 2.1:

$$E = \bar{r}_u \cdot \bar{r}_u, \quad F = \bar{r}_u \cdot \bar{r}_v, \quad G = \bar{r}_v \cdot \bar{r}_v \tag{2.3a}$$

$$e = \mu \frac{\bar{r}_{uu} \cdot (\bar{r}_u \times \bar{r}_v)}{\sqrt{EG - F^2}}, \quad f = \mu \frac{\bar{r}_{uv} \cdot (\bar{r}_u \times \bar{r}_v)}{\sqrt{EG - F^2}}, \quad g = \mu \frac{\bar{r}_{vv} \cdot (\bar{r}_u \times \bar{r}_v)}{\sqrt{EG - F^2}}. \tag{2.3b}$$

The principal curvatures at a point (u, v) on the surface described by eqn. (2.1) are defined by:

$$\kappa_1, \kappa_2 = \kappa_M \pm \sqrt{\kappa_M^2 - \kappa_G} \tag{2.4}$$

where

$$\kappa_M = \frac{\kappa_1 + \kappa_2}{2} = \frac{Eg - 2fF + eG}{2(EG - F^2)} \tag{2.5a}$$

$$\kappa_G = \kappa_1 \kappa_2 = \frac{eg - f^2}{EG - F^2}. \tag{2.5b}$$

If the normal section of the surface bends toward $\hat{n}$, the sign of κ computed in eqn. (2.4) is positive. Conversely, if the bending is away from $\hat{n}$, the sign is negative.

The two principal directions are given by

$$\hat{e}_1 = \frac{1}{\gamma_1} [1\bar{r}_u + \alpha \bar{r}_v] \tag{2.6a}$$

$$\hat{e}_2 = \frac{1}{\gamma_2} [\beta \bar{r}_u + 1\bar{r}_v] \tag{2.6b}$$

where

$$\alpha = \frac{e - \kappa_1 E}{\kappa_1 F - f} = \frac{f - \kappa_1 F}{\kappa_1 G - g} \tag{2.7a}$$

$$\beta = \frac{f - \kappa_2 F}{\kappa_2 E - e} = \frac{g - \kappa_2 G}{\kappa_2 F - f} \tag{2.7b}$$

$$\gamma_1 = (E + 2\alpha F + \alpha^2 G)^{1/2} \tag{2.7c}$$

$$\gamma_2 = (\beta^2 E + 2\beta F + G)^{1/2} \tag{2.7d}$$

The elements of a 2 x 2 curvature matrix $\bar{\bar{Q}}$ may be defined from the formulas

$$Q_{11} = \frac{eG - fF}{EG - F^2} \qquad Q_{12} = \frac{fE - eF}{EG - F^2} \tag{2.8a}$$

$$Q_{21} = \frac{fG - gF}{EG - F^2} \qquad Q_{22} = \frac{gE - fF}{EG - F^2}. \tag{2.8b}$$

To first-order in transverse dimensions, adjacent surface normals on a line of curvature intersect at a point along the extension of the normals (Fig. 2.1). The signed distance from the surface to this point of intersection is the principal radius of curvature. The locus of these points is a caustic surface.

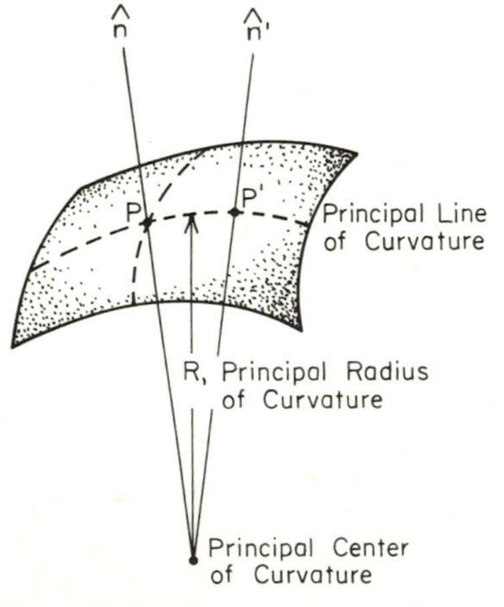

Fig. 2.1 *Geometry of normal intersection and radius of curvature*

Consider the surface S shown in Fig. 2.2. It can represent either a reflector or a wavefront. Define a Cartesian coordinate system at a point 0 on the surface such that the z-axis points (*a*) toward the incident field region if S is a reflector, or (*b*) in the direction of propagation if S is a wavefront. Alternatively, the z-axis points (*a*) toward the sources ($\hat{n}$) if S is a reflector, or (*b*) away from the sources ($-\hat{n}$) if S is a wavefront. Thus, using Lee's convention, the surface normal $\hat{n}$ always points torwards the sources. The x_1- and x_2-axes lie in the tangent plane perpendicular to $\hat{n}$. Then at surface points P in the vicinity of S the surface can be approximated by

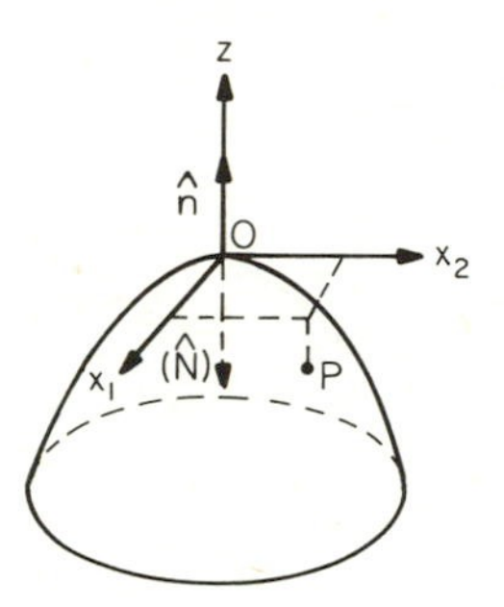

Fig. 2.2 *Geometry of principal coordinate system*

$$z \cong \pm\tfrac{1}{2}\,[x_1\ x_2]\,\bar{\bar{Q}}\begin{bmatrix} x_2 \\ x_2 \end{bmatrix} . \tag{2.9}$$

If the x_1- and x_2-axes are aligned with the principal directions, then the curvature matrix $\bar{\bar{Q}}$ is diagonalised:

$$\bar{\bar{Q}} = \begin{bmatrix} \kappa_1 & 0 \\ 0 & \kappa_2 \end{bmatrix} \tag{2.10}$$

where, as before, the signs of κ_1 and κ_2 are determined by the bending of the respective normal sections toward or away from n̂. If the x_1- and x_2-axes make an angle ψ with the principal directions:

$$\bar{\bar{Q}} = \begin{bmatrix} \cos\psi & -\sin\psi \\ \sin\psi & \cos\psi \end{bmatrix} \begin{bmatrix} \kappa_1 & 0 \\ 0 & \kappa_2 \end{bmatrix} \begin{bmatrix} \cos\psi & -\sin\psi \\ \sin\psi & \cos\psi \end{bmatrix} . \tag{2.11}$$

Example: Surface of revolution

For the surface of revolution shown in Fig. 2.3 the defining equation is

$$\bar{r} = \rho\cos\varphi\,\hat{a}_x + \rho\sin\varphi\,\hat{a}_y + f(\rho)\hat{a}_z\ . \tag{2.12}$$

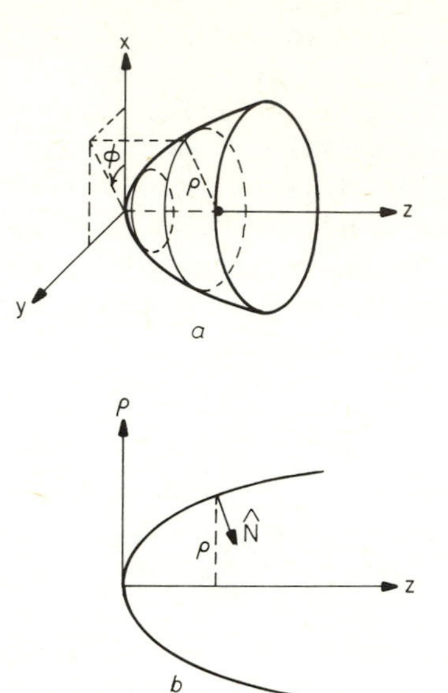

Fig. 2.3 *Geometry of surface of revolution*
a oblique view
b cross-sectional view

From eqn. 2.2 one obtains

$$\hat{n} = \frac{-f'(\rho)\,\hat{a}_\rho + \hat{a}_z}{g(\rho)} \tag{2.13}$$

where $u = \rho$, $v = \varphi$, $g = \sqrt{1 + [f'(\rho)]^2}$, and $\mu = +1$ for the normal direction shown in Fig. 2.3 when $f'(\rho) > 0$. The mean curvature is

$$\kappa_M = \frac{f'(\rho)g(\rho) + \rho f''(\rho)}{2\rho\,[g(\rho)]^3} \tag{2.14a}$$

and the Gaussian curvature is

$$\kappa_G = \frac{f'(\rho)\,f''(\rho)}{\rho\,[g(\rho)]^4} \tag{2.14b}$$

from which

$$\kappa_1 = \frac{f''(\rho)}{[g(\rho)]^3} = 1/R_1 \tag{2.15a}$$

$$\kappa_2 = \frac{f'(\rho)}{\rho g(\rho)} = 1/R_2 . \tag{2.15b}$$

Notice that for $f'(\rho) > 0$ (already assumed) and $f''(\rho) > 0$ (as shown in Fig. 2.3) the surface bends toward $\hat{n}$. In this case the centres of curvature are on the same side of the surface as n. For this axially symmetric geometry it can be shown that $|R_2|$ is the distance from the surface point to the axis measured along the normal direction (or its backward extension). In addition

$$\hat{e}_1 = \frac{\hat{a}_\rho + f'(\rho)\hat{a}_z}{g(\rho)} \tag{2.16a}$$

$$\hat{e}_2 = \hat{a}_\varphi . \tag{2.16b}$$

Thus the lines of curvature are the meridians ($\hat{e}_1$) and the parallels ($\hat{e}_2$), and the principal radii of curvature are R_1 for the meridians and R_2 for the parallels.

2.2.2 Asymptotic solutions: Wavefronts and rays

The electromagnetic field may be postulated in the form of an asymptotic series in inverse powers of $(-jk)$:

$$\bar{E}(\bar{r}) = e^{-jks(\bar{r})} \sum_{m=0}^{\infty} (-jk)^{-m} \bar{e}_m (\bar{r}), k \to \infty \tag{2.17a}$$

$$\bar{H}(\bar{r}) = e^{-jks(\bar{r})} \sum_{m=0}^{\infty} (-jk)^{-m} \bar{h}_m (\bar{r}), k \to \infty \tag{2.17b}$$

For eqn. 2.17 to satisfy the source-free field equations for like orders of $(-jk)$, it is subject to the following constraints:

(eikonal equation): $(\nabla s)^2 = 1$ (2.18a)

(transport equation): $2(\nabla s \cdot \nabla)\bar{e}_m + \nabla^2 s\, \bar{e}_m = -\nabla^2 \bar{e}_{m-1}$ (2.18b)

(Gauss's law): $\nabla s \cdot \bar{e}_m = -\nabla \cdot \bar{e}_{m-1}$ (2.18c)

(Faraday's law): $\bar{h}_m = \sqrt{\frac{\epsilon}{\mu}}\, [\nabla s \times \bar{e}_m + \nabla \times \bar{e}_{m-1}]$ (2.18d)

where $m = 0, 1, 2, \ldots$ and $\bar{e}_{-1} = 0$.

The surfaces of constant phase, $s(\bar{r})$ = constant, are called wavefronts. The curves everywhere orthogonal to wavefronts are called rays. The equation of a ray in free space may be obtained from eqn. 12.18*a*;

$$\bar{r} = \sigma \nabla s + \bar{b} \tag{2.19}$$

where ∇s and $\bar{b}$ are constant vectors, independent of $\bar{r}$, and σ is the arc length along the ray. Thus the rays in free space are straight lines. The phase at a point $\bar{r}$ on a ray is related to the phase at any other point $\bar{r}_0$ on the ray by

$$s(\bar{r}) = s(\bar{r}_0) + \sigma(\bar{r}) - \sigma(\bar{r}_0). \tag{2.20}$$

A pencil of rays is a small tube of (paraxial) rays surrounding a central (axial) ray. With reference to Fig. 2.2, select the z-axis to lie along the axial ray passing through the point 0 on the wavefront. Then in terms of the coordinates (x_1, x_2, z) for a point in the pencil, the phase function $s(\bar{r})$ can be approximated by

$$s(x_1, x_2, z) \cong s(0,0,0) + z + \tfrac{1}{2}\begin{bmatrix} x_1 & x_2 \end{bmatrix} \bar{\bar{Q}}(z) \begin{bmatrix} x_1 \\ x_2 \end{bmatrix} \tag{2.21}$$

where

$$\left[\bar{\bar{Q}}(z)\right]^{-1} = \left[\bar{\bar{Q}}(0)\right]^{-1} + \begin{bmatrix} z & 0 \\ 0 & z \end{bmatrix} \tag{2.22}$$

and $\bar{\bar{Q}}(0)$ is obtained from eqns. 2.10 or 2.11.

If the cross-sectional area of the pencil at z_0 is $da(z_0)$, then the transport eqn. 2.18*b* yields:

$$\bar{e}_0(z) = \bar{e}_0(z_0) \left[\frac{da(z_0)}{da(z)}\right]^{1/2} \tag{2.23}$$

which is a statement of the conservation of power along the pencil. Furthermore, from eqns. 2.18*c* and 2.18*d*:

$$\nabla s \cdot \bar{e}_0(z) = 0 \tag{2.24a}$$

$$\bar{h}_0(z) = \sqrt{\frac{\epsilon}{\mu}}\, \nabla s \times \bar{e}_0(z) \tag{2.24b}$$

which indicate that the zeroth-order asymptotic field is TEM. It should be pointed out that the cross-sectional area ratio is a signed quantity:

$$\frac{da(z_0)}{da(z)} = \frac{(R_1 + z_0)(R_2 + z_0)}{(R_1 + z)(R_2 + z)} \tag{2.25}$$

so that the square root in eqn. 2.23 should take positive real, positive imaginary, or zero values.

The higher-order field vectors propagate according to[2]

$$\bar{e}_m(z) = \bar{e}_m(z_0)\left[\frac{da(z_0)}{da(z)}\right]^{1/2} - \frac{1}{2}\int_{z_0}^{z}\left[\frac{da(z')}{da(z)}\right]^{1/2} \nabla^2 \bar{e}_{m-1}(z')dz' \tag{2.26}$$

where $m = 0, 1, 2, \ldots$.

2.2.3 Geometrical optics for reflectors

Fermat's Principle

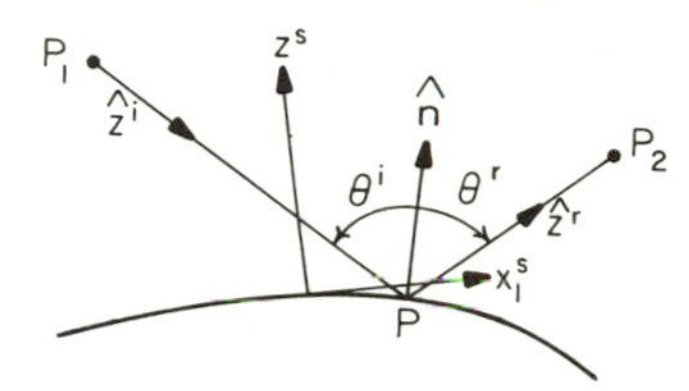

Fig. 2.4 *Geometry for Fermat's principle*

In free space the ray or rays from a source P_1 to an observer at P_2 is the curve along which the path length is stationary with respect to infinitesimal variations in the path. Thus, in free space the ray paths are straight lines. For a reflector (Fig. 2.4) where $\hat{n}$ is determined from eqn. 2.2 and z^S from eqn. 2.9, Fermat's principle yields:

$$\frac{\partial \overline{P_1PP_2}}{\partial x_1^S} = 0 \qquad \frac{\partial \overline{P_1PP_2}}{\partial x_2^S} = 0. \tag{2.27}$$

Consequently, $\hat{z}^i$, $\hat{z}^r$, and $\hat{n}$ are coplanar, and $\theta^i = \theta^r$ (Snell's law). Thus

$$\hat{z}^r = \hat{z}^i - 2(\hat{n}\cdot\hat{z}^i)\hat{n}. \tag{2.28}$$

Solution of eqn. 2.27 may be difficult, time consuming, or ambiguous. Consequently, it is useful to determine beforehand whether or not a solution exists. This may be done for a convex surface of revolution as shown in Fig. 2.5.[4] If R_0 is a specular point on the surface, the quantity

$$d = |\overline{OR}| + |\overline{RP}| \tag{2.29}$$

is a global minimum at R_0 so that at most one specular point may exist on the surface. Next define $\overline{\nu}$

$$\overline{\nu} = \nabla_R d - (\nabla_R \cdot \hat{n})\,\hat{n} \tag{2.30}$$

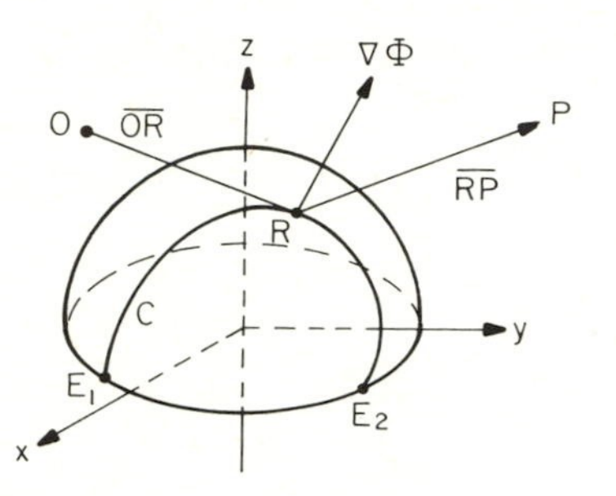

Fig. 2.5 *Geometry for specular-point determination*

where $\nabla_R d$ is the gradient of d with respect to the coordinates of R. Thus $\overline{\nu}$ is the projection of $\nabla_R d$ on a plane tangent to the surface at R, and $\overline{\nu}$ also lies in the plane containing O, R, and P. For R in the vicinity of R_0, $\overline{\nu}$ will be directed away from R_0 since d is a minimum at R_0.

Define the axially symmetric surface by:

$$\Phi(\rho, z) = \rho^2 - f(z) = 0 \tag{2.31a}$$

$$\nabla\Phi = 2x\hat{a}_x + 2y\hat{a}_y - f'(z)\hat{a}_z \tag{2.31b}$$

Fermat's principle requires that

$$\nabla_R \Phi \cdot (\overline{OR} \times \overline{RP}) = 0 \tag{2.32}$$

Expanding eqn. 2.32 yields

$$Ax_R + By_R + C = 0 \tag{2.33}$$

where

$$A = -2[(y_O - y_P)(z_R - z_O) + y_O(z_O - z_P)] - f'(z_R)(y_O - y_P) \tag{2.34a}$$

$$B = +2[(x_O - x_P)(z_R - z_O) + x_O(z_O - z_P)] + f'(z_R)(x_O - x_P) \tag{2.34b}$$

$$C = f'(z_R)(x_P y_O - y_P x_O). \tag{2.34c}$$

These equations represent a three-dimensional surface which interesects the axially symmetric reflector in a curve C. The specular point, R_0, if it exists, lies along C. At every point R along C $\bar{\nu}$ is tangent to C. Consequently, the problem reduces to finding the two points E_1 and E_2 where C intersects the reflector rim and to evaluate $\bar{\nu}$ at those points. A specular point will exist if and only if $\bar{\nu}$ points away from the surface at both E_1 and E_2.

The two edge points may be found by substituting

$$x_R^2 + y_R^2 = f(z_{RIM}) \tag{2.35}$$

into eqn. 2.33. Thus

$$y_{RIM} = -\frac{A}{B}x_{RIM} - \frac{C}{B} \tag{2.36a}$$

$$x_{RIM}^2\left[1 + \left(\frac{A}{B}\right)^2\right] + \left[\frac{2AC}{B^2}\right]_{RIM} + \left[\frac{C^2}{B^2} - f(z_{RIM})\right] = 0. \tag{2.36b}$$

The descriminant of eqn. 2.36*b* is never negative so that two real roots always exist.

Reflected field

If a geometrical-optics field illuminates a perfect reflector:

$$\bar{E}^i(\bar{r}) = e^{-jks^i(\bar{r})} \sum_{m=0} (-jk)^{-m}\bar{e}_m^i(\bar{r}) \tag{2.37a}$$

$$\bar{H}^i(\bar{r}) = e^{-jks^i(\bar{r})} \sum_{m=0} (-jk)^{-m}\bar{h}_m^i(\bar{r}). \tag{2.37b}$$

Then, if a specular point exists, the reflected field may also be expanded in asymptotic form:

$$\bar{E}^r(\bar{r}) = e^{-jks^r(\bar{r})} \sum_{m=0} (-jk)^{-m}\bar{e}_m^r(\bar{r}) \tag{2.38a}$$

$$\bar{H}^r(\bar{r}) = e^{-jks^r(\bar{r})} \sum_{m=0} (-jk)^{-m}\bar{h}_m^r(\bar{r}). \tag{2.38b}$$

At a point of reflection, 0, the initial values of the zeroth-order amplitude coefficients are given by:

$$\bar{e}_0^r = -\bar{e}_0^i + 2(\hat{n} \cdot \bar{e}_0^i)\hat{n} \tag{2.39a}$$

$$\bar{h}_0^r = \bar{h}_0^i - 2(\hat{n} \cdot \bar{h}_0^i)\hat{n} \tag{2.39b}$$

where $\bar{e}_0^{ir}$, $\bar{h}_0^{ir}$, and n̂ are evaluated at 0. Along a reflected ray passing through 0, the zeroth-order reflected field is given by

$$\bar{E}^r(z^r) \cong e^{-jks^r(z^r)} \sqrt{\frac{R_1^r R_2^r}{(z^r + R_1^r)(z^r + R_2^r)}}\, \bar{e}_0^r(z^r = 0) \tag{2.40}$$

where $\bar{e}_0^r(z^r = 0)$ is defined in eqn. 2.39*a* and $s^r(z^r)$ is:

$$s^r(z^r) = s^i(z^i) + z^r \tag{2.41}$$

where $z^r = z^i = 0$ at the reflection point. The two radii of curvature R_1^r and R_2^r will be calculated in the following Section. An expression similar to eqn. 2.40 applies to the zeroth-order reflected H-field. The zeroth-order field is locally TEM, i.e.

$$\nabla s^r \cdot \bar{e}_0^r(z^r) = \nabla s^r \cdot \bar{h}_0^r(z^r) = 0 \tag{2.42a}$$

$$\bar{h}_0^r(z^r) = \sqrt{\frac{\epsilon}{\mu}}\, \nabla s^r \times \bar{e}_0^r(z^r) \tag{2.42b}$$

Determination of the higher-order field coefficients is straightforward but considerably more complicated[5-7]. The surface-current density at the specular point is given to zeroth order by

$$\bar{J}_S = 2e^{-jks^i}\, \hat{n} \times \bar{h}_0^i \tag{2.43}$$

Determination of principal curvatures of the reflected wavefront

Define the following matrix elements:

$$p_{mn}^i = \hat{x}_m^i \cdot \hat{x}_n^S, \qquad m,n = 1,2 \tag{2.44a}$$

$$p_{mn}^r = \hat{x}_m^r \cdot \hat{x}_n^S, \qquad m,n = 1,2 \tag{2.44b}$$

$$p_{33}^i = \hat{z}^i \cdot \hat{n} \quad , \qquad p_{33}^r = \hat{z}^r \cdot \hat{n} \tag{2.44c}$$

where $(x_1^i, \hat{x}_2^i)$ and $(\hat{x}_1^S, \hat{x}_2^S)$ are the principal directions of, respectively, the incident wavefront and the reflector at the specular point; $\hat{z}^i$, $\hat{z}^r$, and $\hat{n}$ are, respectively, the incident wave-normal, the reflected wave-normal, and the outward surface normal; and $(\hat{x}_1^r, \hat{x}_2^r)$ are two mutually orthogonal directions each of which is perpendicular to $\hat{z}^r$. Then, in terms of two matrices $\bar{\bar{P}}^i$ and $\bar{\bar{P}}^r$ defined

$$\bar{\bar{P}}^i = \begin{bmatrix} p_{11}^i & p_{12}^i \\ p_{21}^i & p_{22}^i \end{bmatrix} \qquad \bar{\bar{P}}^r = \begin{bmatrix} p_{11}^r & p_{12}^r \\ p_{21}^r & p_{22}^r \end{bmatrix} \tag{2.45}$$

the curvature matrix of the reflected wavefront at the specular point, $\bar{\bar{Q}}^r$, can be determined from:

$$(\bar{\bar{P}}^i)^T \bar{\bar{Q}}^i \bar{\bar{P}}^i + p_{33}^i \bar{\bar{Q}}^S = (\bar{\bar{P}}^r)^T \bar{\bar{Q}}^r \bar{\bar{P}}^r + p_{33}^r \bar{\bar{Q}}^S \tag{2.46}$$

where $\bar{\bar{Q}}^i$ and $\bar{\bar{Q}}^S$ are the diagonalised curvature matrices of, respectively, the incident wavefront and the reflector at the specular point. If $(\hat{x}_1^r, \hat{x}_2^r)$ are the principal directions of the reflected wavefront, then $\bar{\bar{Q}}^r$ will be diagonal, leading immediately to the principal radii of curvature of the reflected wavefront. Otherwise it will be necessary to diagonalise $\bar{\bar{Q}}r$ if the principal directions and curvatures are to be found. If $(\hat{x}_1^r, \hat{x}_2^r)$ are chosen so that

$$\hat{x}_n^r = \hat{x}_n^i - 2(\hat{n} \cdot x_n^i)\hat{n}, \; n = 1,2 \tag{2.47}$$

then eqn. 2.46 can be solved directly for $\bar{\bar{Q}}^r$:

$$\bar{\bar{Q}}^r = \bar{\bar{Q}}^i + 2p_{33}^i \, ((\bar{\bar{P}}^i)^{-1})^T \bar{\bar{Q}}^S (\bar{\bar{P}}^i)^{-1} \tag{2.48}$$

The solution for eqn. 2.48 is:

$$Q_{11}^r = 1/R_1^i + 2p_{33}^i \left[(p_{22}^i)^2/R_1^S + (p_{21}^i)^2/R_2^S\right] / |\bar{\bar{P}}^i|^2 \tag{2.49a}$$

$$Q_{22}^r = 1/R_2^i + 2p_{33}^i \left[(p_{12}^i)^2/R_1^S + (p_{11}^i)^2/R_2^S\right] / |\bar{\bar{P}}^i|^2 \tag{2.49b}$$

$$Q_{12}^r = Q_{21}^r = -2p_{33}^i \left[p_{12}^i p_{22}^i/R_1^S + p_{11}^i p_{21}^i/R_2^S\right] / |\bar{\bar{P}}^i|^2 \tag{2.49c}$$

Note that the sign conventions are such that R_1^i and R_2^i are positive if the wavefront bends away from the direction of propagation whereas R_1^S and R_1^S are negative if the reflector bends away from the space containing the incident and reflected waves.

By diagonalising eqn. 2.49 the principal curvatures of the reflected wavefront can be obtained:

$$1/R^r_{1,2} = 1/2(1/R^i_1 + 1/R^i_2) + p^i_{33}\left[\frac{(p^i_{12})^2 + (p^i_{22})^2}{R^S_1} + \frac{(p^i_{21})^2 + (p^i_{11})^2}{R^2_2}\right]/|\bar{\bar{P}}^i|^2$$

$$\pm\left\{(1/R^i_1 - 1/R^i_2)^2 + 4(1/R^i_1 - 1/R^i_2)p^i_{33}\left[\frac{(p^i_{22})^2 - (p^i_{12})^2}{R^S_1} + \frac{(p^i_{21})^2 - (p^i_{11})^2}{R^S_2}\right]\right.$$

$$/|\bar{\bar{P}}^i|^2$$

$$\left. + 4(p^i_{33})^2\left[\left(\frac{(p^i_{22})^2 + (p^i_{12})^2}{R^S_1} + \frac{(p^i_{21}) + (p^i_{11})^2}{R^S_2}\right)^2 - 4|\bar{\bar{P}}^i|^2/R^S_1R^S_2\right]\right\}^{1/2} / |\bar{\bar{P}}^i|^4 \tag{2.50}$$

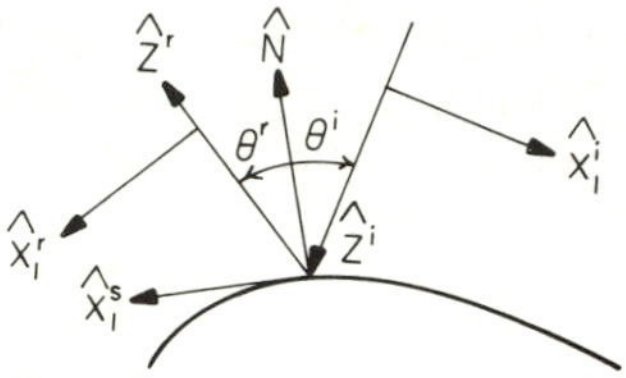

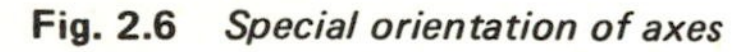

Fig. 2.6 *Special orientation of axes*

In the special but commonly encountered case that $\hat{x}^i_1$, $\hat{x}^S_1$, and $\hat{x}^r_1$ lie in the plane of incidence while $\hat{x}^i_2 = \hat{x}^S_2 = \hat{x}^r_2$, perpendicular to the plane of incidence (Fig. 2.6), then eqn. 2.50 reduces to

$$1/R^r_1 = 1/R^i_1 + 2/R^S_1\, p^i_{33} \tag{2.51a}$$

$$1/R^r_2 = 1/R^i_2 + 2p^i_{33}/R^S_2. \tag{2.51b}$$

As an example, consider a paraboloid coaxial with the z-axis for which $f(\rho)$ is $\rho^2/4F$, from which eqn. 2.50 yields:

$$R^S_1 = 2F[1 + (\rho/2F)^2]^{3/2} \tag{2.52a}$$

$$R^S_2 = 2F[1 + (\rho/2F)^2]^{1/2}. \tag{2.52b}$$

Then, for a spherical wave emanating from the focus at $z = F$:

$$R^i_1 = R^i_2 = [(F-z)^2 + \rho^2]^{1/2} = F[1 + (\rho/2F)^2] \tag{2.53a}$$

$$p^i_{33} = -\cos\theta^i = -1/[1 + (\rho/2F)^2]^{1/2}\ . \tag{2.53b}$$

Then eqn. 2.51 yields the principal curvatures of the reflected wavefront:

$$1/R_1^r = 1/F[1+(\rho/2F)^2] - 2/2F[1+(\rho/2F)^2] = 0 \tag{2.54a}$$

$$1/R_2^r = 1/F[1+(\rho/2F)^2] - 2/2F[1+(\rho/2F)^2] = 0\,. \tag{2.54b}$$

Thus the wavefront reflected from a paraboloid with a focused spherical wave source is planar, as expected.

An alternative technique for determining the caustic distances of the reflected wavefront exists if $\bar{R}$, the position vector describing the surface, and $\hat{z}^r$, the reflected wave normal, can be expressed in terms of two independent variables, u and v:[1,3]

$$\bar{R} = x(u, v)\,\hat{a}_x + y(u, v)\hat{a}_y + z(u, v)\hat{a} \tag{2.55a}$$

$$\hat{z}^r = z_x^r(u, v)\hat{a}_x + z_y^r(u, v)\hat{a}_y + z_z^r(u, v)\hat{a}_z\,. \tag{2.55b}$$

Then, compute the following six quantities:

$$E = \frac{\partial \hat{z}^r}{\partial u}\cdot\frac{\partial \hat{z}^r}{\partial u},\; F = \frac{\partial \hat{z}^r}{\partial u}\cdot\frac{\partial \hat{z}^r}{\partial v},\; G = \frac{\partial \hat{z}^r}{\partial v}\cdot\frac{\partial \hat{z}^r}{\partial v} \tag{2.56a}$$

$$e = \frac{\partial \bar{R}}{\partial u}\cdot\frac{\partial \hat{z}^r}{\partial u},\; f = \frac{\partial \bar{R}}{\partial v}\cdot\frac{\partial \hat{z}^r}{\partial u},\; g = \frac{\partial \bar{R}}{\partial v}\cdot\frac{\partial \hat{z}^r}{\partial v}\,. \tag{2.56b}$$

The two roots, β_1 and β_2, of the quadratic equation

$$(EG - F^2)\beta^2 + (Eg - 2Ff + Ge)\beta + (eg - f^2) = 0 \tag{2.57}$$

then yield the principal radii of curvature of the reflected wavefront

$$R_1^r = -\beta_1 \tag{2.58a}$$

$$R_2^r = -\beta_2\,. \tag{2.58b}$$

For a problem with axial symmetry, when circular coordinates (ρ, φ) are used to describe both R and $\hat{z}^r$, the solution reduces to

$$R_1^r = e/E \tag{2.59a}$$

$$R_2^r = g/G\,. \tag{2.59b}$$

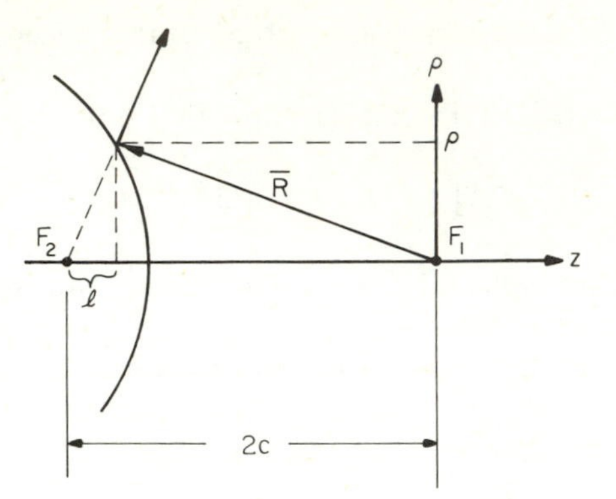

Fig. 2.7 *Hyperboloid geometry*

As an example, consider the hyperboloid geometry in Fig. 2.7, for which the ray incident at (ρ, φ, z) from the external focus F_1 reflects as if it were coming from the internal focus F_2:

$$\bar{R} = \rho \cos\varphi \hat{a}_x + \rho \sin\varphi \hat{a}_y - [c + (a/b)\sqrt{\rho^2 + b^2}\,]\hat{a}_z \tag{2.60a}$$

$$\hat{z}^r = \sin\gamma \cos\varphi \hat{a}_x + \sin\gamma \sin\varphi \hat{a}_y + \cos\gamma \hat{a}_z \tag{2.60b}$$

where $\sin\gamma = \rho/\sqrt{l^2 + \rho^2}$, $\cos\gamma = l/\sqrt{l^2 + \rho^2}$, and $l = c - (a/b)\sqrt{\rho^2 + b^2}$. Then eqns. 2.56-2.59 yield:

$$R_1^r = e/E = R_2^r = g/G = (c/b)\sqrt{\rho^2 + b^2} - a = \sqrt{l^2 + \rho^2} \tag{2.61}$$

and the reflected wavefront appears to be a spherical wave from F_2.

2.2.4 Physical optics

The fields radiated by a known surface-current distribution flowing on a perfectly

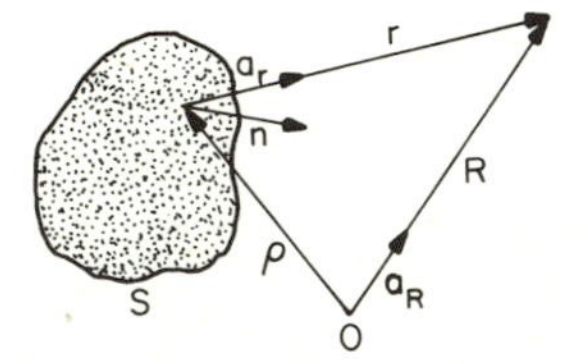

Fig. 2.8 *Geometry of vectors for surface integrals*

conducting surface are, in terms of the geometry of Fig. 2.8:

$$\bar{E}(P) = -\frac{j\omega\mu_0}{4\pi}\int_S [\bar{J}_S\psi + \frac{1}{k^2}(\bar{J}_S\cdot\nabla)\nabla\psi]\,dS \tag{2.62a}$$

$$\bar{H}(P) = \frac{1}{4\pi}\int_S [\bar{J}_S\times\nabla\psi]\,dS \tag{2.62b}$$

where $\psi = \exp(-jkr)/r$. The operator ∇ within the field integrals operates only on the coordinates of the source point $\bar{\rho}$. Consequently

$$\nabla\psi = (jk + 1/r)(\exp(-jkr)/r)\hat{a}_r, \tag{2.63a}$$

$$(\bar{J}_S\cdot\nabla)\nabla\psi = \Big[-k^2(\bar{J}_S\cdot\hat{a}_r)\hat{a}_r + \frac{3}{r}(jk + \frac{1}{r})(\bar{J}_S\cdot\hat{a}_r)\hat{a}_r.$$

$$-\bar{J}_S(j\frac{k}{r} + \frac{1}{r^2})\Big](\exp(-jkr)/r). \tag{2.63b}$$

The integrand of eqn. 2.62 is frequently written in terms of a linear operator $\underset{\sim}{\Gamma}$ known as the free-space dyadic Green's function

$$\bar{E}(P) = (-j\omega\mu)\int_S \underset{\sim}{\Gamma}\cdot\bar{J}_S\,dS \tag{2.64}$$

where the components of the dyadic $\underset{\sim}{\Gamma}$ are found in eqn. 2.63*b*. Near-field calculations, e.g. focal-region calculations, require that this most general form of $\underset{\sim}{\Gamma}$ be used.

When $r > \lambda$, the leading term dominates eqn. 2.63*b* and the fields become

$$\bar{E}(P) = -\frac{j\omega\mu_0}{4\pi}\int_S \Big[\bar{J}_S - (\bar{J}_S\cdot\hat{a}_r)\hat{a}_r\Big]\psi\,dS \tag{2.65a}$$

$$\bar{H}(P) = \frac{1}{4\pi}\int \Big[\bar{J}_S\times\nabla\psi\Big]\,dS \tag{2.65b}$$

These somewhat simpler expressions are used in Fresnel-region calculations. When the field point P is sufficiently far from the sources that the Fraunhofer approximations can be made, the fields at such distant points become

$$\bar{E}(P) = -\frac{j\omega\mu_0}{4\pi}\frac{\exp(-jkR)}{R}\int_S \left[\bar{J}_S - (\bar{J}_S \cdot \hat{a}_R)\hat{a}_R\right] \exp(jk\bar{\rho}' \cdot \hat{a}_R)\, dS \quad (2.66a)$$

$$\bar{H}(P) = \frac{jk}{4\pi}\frac{\exp(-jkR)}{R}\int_S \left[\bar{J}_S \times \hat{a}_R\right] \exp(jk\bar{\rho}' \cdot \hat{a}_R)\, dS. \quad (2.66b)$$

Eqns. 2.66, considerably simplified by the fact that $\hat{a}_R$ is constant with respect to the integration variables, are used for determining such far-field properties as gain and radiation pattern.

If the induced surface-current distributions are known, evaluation of the fields becomes straightforward, although the computations may be lengthy and laborious. However, only in a few special cases, e.g. sphere, ellipsoid, etc., can the currents be determined rigorously. The method-of-moments can be used to determine induced currents on small or moderately sized scatterers; however, the applicability of this technique to large focusing reflectors may be limited by computation cost and accuracy.

In the event that the integrals have stationary points, the fields can be evaluated in terms of one or more simple, closed-form expressions which frequently lend themselves to greatly simplified geometrical ray-tracing interpretations. Examples of these geometrical techniques are presented in a subsequent section.

In the event that focused or nearly focused conditions obtain, however, the entire reflector is part of a single Fresnel zone. An obvious example is the main beam of the radiation pattern of a paraboloid reflector. Simple, localised stationary points do not exist in the field integrals. It is then necessary to evaluate contributions from all parts of the reflector. A technique that has found wide acceptance under those conditions is known in scattering theory as *physical optics*. Physical optics (PO) simply approximates the currents on the reflector by the currents calculated from the theory of *geometrical optics* (GO) [cf. (2.43)]. These approximate current distributions are then used in eqns. 2.62-2.66 to determine the scattered field.

No rigorous justification for the PO approximations has been established. On the contrary, it can be shown that PO in general fails to satisfy the reciprocity theorem everywhere except in the direction of a specular return. In spite of this and other shortcomings, PO is an approximation technique that has proven very successful in the analysis of large reflector antennas, particularly under focused or nearly focused conditions.

Aperture formulation

In the event that the antenna configuration has a well-defined aperture, e.g.

focusing reflectors, horns, and arrays, an equivalent aperture formulation may be used to compute the scattered field

$$\bar{E}(P) = -\frac{j\omega\mu_0}{4\pi}\left(\frac{\exp(-jkR)}{R}\right)\int_{aperture}\left\{-\frac{1}{\eta}(\hat{n}\times\bar{E})\times\hat{a}_R\right.$$

$$\left.+[\hat{n}\times\bar{H}-(\hat{n}\times\bar{H})\cdot\hat{a}_R)\hat{a}_R]\right\}\exp(jk\bar{\rho}\cdot\hat{a}_R)\,dS. \qquad (2.67)$$

The aperture fields must be approximated using, for example, geometrical optics, known unperturbed waveguide fields, etc. In many circumstances eqn. 2.67 is more appropriate for solution than the surface integrals. In the case of the paraboloid, at boresight the PO surface integral of eqn. 2.66*a* and the aperture integral of eqn. 2.67 are equivalent. In directions far from the reflector axis, however, the aperture formulation introduces a significant path-length discrepancy.

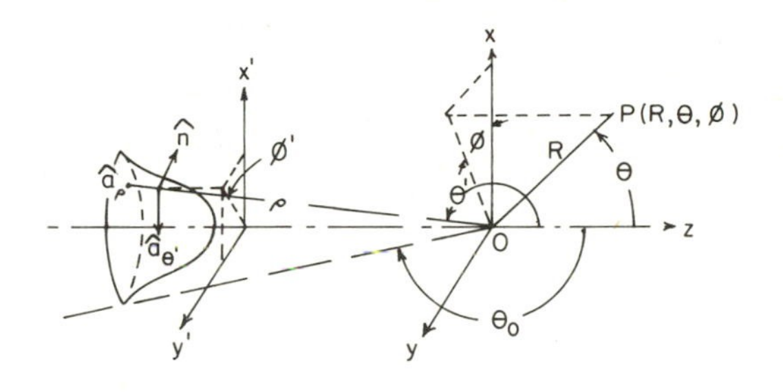

Fig. 2.9 *Geometry for figure of revolution*

Orthogonal source expansions

Under certain conditions of geometrical symmetry (Fig. 2.9) it is convenient to expand the surface-current/aperture-field distribution in a complex Fourier series, thus enabling the azimuthal part of the two-dimensional physical-optics integrals to be evaluated analytically.[8-9] These analytical manipulations introduce more complicated functions into the integrand of the remaining numerical integrals, which still must be evaluated numerically for each field point, but the net computer time is reduced typically to be a factor of 50 or more from the time required for an equivalent two-dimensional integration for the same reflector.

The incident electric field due to localised sources in the vicinity of O may be described by vector spherical waves (cf. Section 2.3). In the limit of 'point source' feeds, the spherical-wave expressions reduce to Fourier series. The Fourier coefficients are, in general, complex, thus encompassing linearly, circularly, or elliptically polarised feeds, feed-system phase errors, etc. In the special case of a linearly-polarised circular aperture (yz-plane = E-plane):

$$\bar{E}_{inc} = \frac{\exp(-jk\rho)}{\rho}\left\{a_1(\theta')\cos\varphi'\hat{a}_{\theta'} + d_1(\theta')\sin\varphi'\hat{a}_{\varphi'}\right\} \tag{2.68a}$$

$$a_1(\pi) = -d_1(\pi). \tag{2.68b}$$

The resulting transverse components of the far-zone field scattered from a paraboloid immersed in the field of eqn. 2.68 are:

$$E_\theta(P) = jkF\sin\varphi\,\frac{\exp(-jkR)}{R}\int_{\theta_0}^{\pi}\frac{\exp[-jk\rho(1-\cos\theta\cos\theta')]}{(1-\cos\theta')}$$

$$\left\{a_1\cos\theta\,[J_0(\beta)-J_2(\beta)] - d_1\cos\theta\,[J_0(\beta)+J_2(\beta)]\right.$$

$$\left.-2j\sin\theta\,\mathrm{ctn}\frac{\theta'}{2}\,J_1(\beta)a_1\right\}\sin\theta' d\theta', \tag{2.69a}$$

$$E_\varphi(P) = jkF\cos\varphi\,\frac{\exp(-jkR)}{R}\int_{\theta_0}^{\pi}\frac{\exp[-jk\rho(1-\cos\theta\cos\theta')]}{(1-\cos\theta')}$$

$$\left\{a_1[J_0(\beta)+J_2(\beta)] - d_1[J_0(\beta)-J_2(\beta)]\right\}\sin\theta' d\theta', \tag{2.69b}$$

where F = the paraboloid's focal length, $\beta = k\rho\sin\theta\sin\theta'$, and $J_n(\beta)$ is the Bessel function of nth order.

Galindo-Israel and Mittra[10] have extended the usefulness of the orthogonal-source-expansion approach by removing the requirements for axial symmetry and demonstrating how the source distribution in the radiation integral can be expanded in a double series of trigonometric functions in azimuth and modified Jacobi polynomials in radius. The orthogonality of these functions permits the field integrals to be reduced to a rapidly converging Fourier-Bessel series, the first term of which is the well-shaped Airy function. The coefficients of this double series are independent of the observation angles. Hence, once they are computed, the field may be determined very rapidly at large numbers of observation points. This technique is particularly efficient for geometries with lateral components of feed defocusing.

Numerical integration

In general, the physical-optics field integrals are of the form

$$\bar{E}(R,\theta,\varphi) = \int\int \bar{F}[R,\theta,\rho(\theta',\varphi'),\theta',\varphi'] \times \exp\left\{ jk\gamma\,[R,\theta,\varphi,\rho(\theta',\varphi'),\theta',\varphi']\right\} d\theta'\,d\varphi' \tag{2.70a}$$

or alternatively

$$\bar{E}(x,y,z) = \int\int \bar{F}[x,y,z,x',y',z'(x',y')] \times \exp\left\{ jk\gamma[x,y,z,x',y',z'(x',y')]\right\} dx'\,dy' \tag{2.70b}$$

where (R, θ, φ) or (x, y, z) are the coordinates of the field point; (ρ, θ', φ) or (x', y', z') are the coordinates of the integration (source) point; $\bar{E}$ and $\bar{F}$ are complex vector functions and γ is a real scalar function.

Only in rare instances is it possible to evaluate complex integrals of the form of eqns. 2.70 in closed form. Frequently, various factors of the integrand take the form of tabular data or empirical data. When the integrations are carried out over large (in terms of wavelengths) surfaces, the integrals become proportionately more difficult to evaluate. Consequently, the material to follow will deal with numerical techniques by which satisfactorily accurate approximations can be obtained.[11-14]

Before discussing specific integration techniques, the problem of numerically evaluating this type of integral will be illustrated with a specific example. Suppose the field $\bar{E}(R, \theta, \varphi)$ is to be evaluated on a set of points (the output grid) consisting of 91 θ values at each of 2 φ values. Also suppose that the integrand must be specified at a set of points (the integration grid) consisting of 400 θ' values and 500 φ' values. Then the functions $\bar{F}$ and γ must be evaluated at $91 \times 2 \times 400 \times 500 = 3.64 \times 10^7$ points, and the processing involved in the numerical integration must occur at each of these points. Several hours of computer time could be required to evaluate the field of this example.

Two basic methods can be employed to reduce excessive computer time: (i) to devise efficient methods for numerically evaluating the integral, or (ii) to analytically transform the integral to another form which may be numerically evaluated more efficiently than the integral. In both of these cases the problem is considerably easier if the function $\bar{F}$ is dependent only on the integration grid coordinates and not dependent on the output grid coordinates R, θ, φ. The reason that this is important for the first case is that if $\bar{F}$ is a function of the integration grid coordinates θ', φ' only, then it need be computed only once for each of the grid coordinate points. It is then stored and used for each output point. In the example given above this reduces the number of times $\bar{F}$ must be computed from 3.64×10^7 to 2×10^4. For these reasons it is relatively easy to evaluate far-field fields, where $\bar{F}$ can be made dependent only on the integration grid coordinates, and difficult to evaluate near-field fields, for which $\bar{F}$ is dependent on the output coordinates as well.

Consider, for example, the field of eqn. 2.66*a* for Fig. 2.8. Since the unit vector $\hat{a}_R$ is a function of the output grid coordinates, the integral is not yet in the desired form. This dependency can be removed by converting the integrand to rectangular components, since rectangular unit vectors are constants, independent of the coordinate system. The dependency on output grid coordinates can then be factored *outside of* the integral:

$$\begin{bmatrix} E_\theta(\theta,\varphi) \\ E_\varphi(\theta,\varphi) \end{bmatrix} = \begin{bmatrix} \cos\theta\cos\varphi & \cos\theta\sin\varphi & -\sin\theta \\ -\sin\varphi & \cos\varphi & 0 \end{bmatrix} \begin{bmatrix} I_x(\theta,\varphi) \\ I_y(\theta,\varphi) \\ I_z(\theta,\varphi) \end{bmatrix} \tag{2.71a}$$

where

$$\bar{I}(\theta,\varphi) = -\frac{j\omega\mu_0}{4\pi}\frac{\exp(-jkR)}{R}\int_S \bar{J}_S(\theta',\varphi')\exp(jk\bar{\rho}\cdot\hat{a}_R)\,dS\,. \tag{2.71b}$$

The physical-optics integrals are fundamentally two-dimensional. These forms are frequently evaluated by consecutive application of one-dimensional integration algorithms over each independent variable. Certain geometries are amenable to analytic reduction of the original forms to one-dimensional integrals with slightly more complicated argument functions. In either case, the basic integration formula used is

$$\int_a^b f(x)\,dx = \sum_{i=1}^{n}\left[W_i f(x_i)\right] + E_n \tag{2.72}$$

where x_i is the point at which the integrand is evaluated, W_i is the weighting coefficient at x_i, and E_n is the truncation error. Extensive literature is available for dealing with the theory and application of numerical quadrature integration techniques, primarily of the Gaussian and Newton-Cotes varieties[15,16] and their coefficients are extensively tabulated.[17] Richardson's extrapolation or Romberg integration[29] is available to reduce the truncation error at minimum cost in terms of integrand evaluations.[18,19] Numerous studies have been made to compare the effectiveness of the various techniques in the context of radiation integrals for reflector antennas.[20–23] Computer listings of efficient algorithms are widely available.[24–26]

Very few numerical integration algorithms are known which can deal efficiently with the two-dimensional integrals directly. For less than three variables, the application of Monte Carlo multi-dimensional integration is not economical in the number of integrand evaluations.[27] No satisfactory theory of multivariable orthogonal polynomials exists which is appropriate for the physical-optics integrals. A two-dimensional version of Romberg integration has been applied to reflectors.[28] The most successful direct two-dimensional integration technique for reflector fields is the Ludwig method,[29] which achieves a significant reduction in computer time by breaking down the field integrals into a series of wavelength-sized patches. The functions $\bar{F}$ and γ are expanded in truncated Taylor series, which can then be integrated in closed form. In general, the Ludwig algorithm is most advantageous when the phase of the integrand varies rapidly as a function of the integration coordinates.

Stationary points of the physical-optics integrals

The physical-optics field integrals generally possess stationary points.[30–32] In the limit of large k, major contributions to the field come from the immediate vicinity of these stationary points, provided that the magnitude of the integrand does not vanish at these points. Standard saddle-point techniques can be used to evaluate these asymptotic contributions.[33]

Stationary points of the first kind satisfy the condition:

$$\frac{\partial \gamma}{\partial x'} = \frac{\partial \gamma}{\partial \gamma'} = 0 \cdot \qquad (2.73)$$

This condition is satisfied at points on the reflector intersected by the straight line from the source to the field point. Thus these points exist only when P lies within the shadow 'cast' by the reflector when it is illuminated by the source. In the limit of large k, the field due to the induced currents at these points exactly cancels the source field behind the reflector, producing a perfect geometrical shadow there. Stationary points of the first kind are also found at specular points on the reflector (cf. Section 2.2.3).

Stationary points of the second kind are defined by the condition

$$\frac{\partial G}{\partial x'}\,\frac{\partial \gamma}{\partial y'} - \frac{\partial G}{\partial \gamma'}\,\frac{\partial \gamma}{\partial x'} = 0 \qquad (2.74)$$

where the reflector rim is specified by the condition $G(x', y') = 0$. These stationary points correspond to points on the reflector rim. The field contributions from these points can also be evaluated from saddle-point theory. For example, for any axially symmetric reflector with an on-axis feed, two stationary points of the second kind lie where the plane containing the reflector axis and the field point intersects the rim of the reflector. These ray-like contributions which appear to emerge from each stationary point may be expressed as the product of:

(*a*) the incident-field component
(*b*) a phase factor corresponding to the path length from the stationary point to the field point
(*c*) a caustic divergence factor
(*d*) a diffraction coefficient.

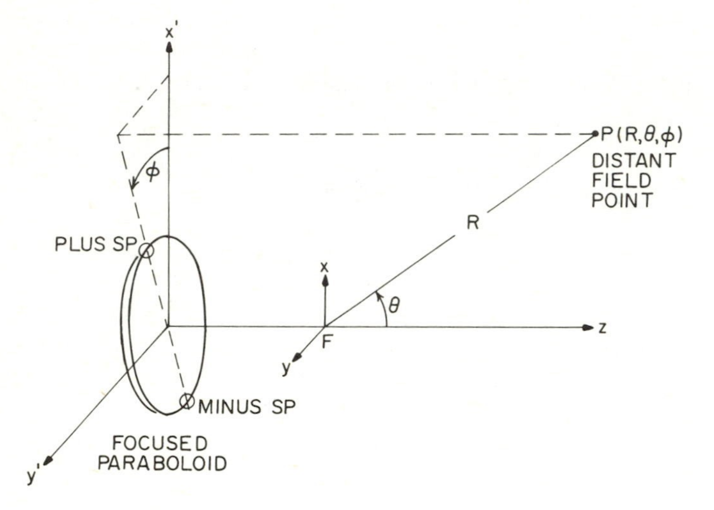

Fig. 2.10 *Focused paraboloid immersed in field of vector, shperical-wave source*

Consider an infinitely thin paraboloid immersed in the field of a vector spherical-wave point source located at its focus[34] (Fig. 2.10). The illumination fields are assumed to be:

$$E_f = [E_{f\theta}(\theta, \varphi)\hat{a}_\theta + E_{f\varphi}(\theta, \varphi)\hat{a}_\varphi] \frac{\exp(-jkR)}{R} \,. \tag{2.75}$$

For the focused paraboloid, and for any axially symmetric reflector with an on-axis spherical-wave feed, two stationary points of the second kind lie where the plane containing the reflector axis and the field point intersects the rim of reflector. The stationary point on the same side of the axis as the field point is designated the 'plus' stationary point in Fig. 2.10 and the point on the opposite side of the axis is designated the 'minus' stationary point. Application of standard saddle-point procedures then yields the total scattered field associated with thise two stationary points:

$$\bar{E}_S(P) = \left\{ E^{+}_{S\theta}(P) + E^{-}_{S\theta}(P) \right\} \hat{a}_\theta + \left\{ E^{+}_{S\varphi}(P) + E^{-}_{S\varphi}(P) \right\} \hat{a}_\varphi \qquad (2.76)$$

where

$$E^{+}_{S\varphi}(P) = E^{+}_{i\varphi}\,\mathrm{PF}^{+}\mathrm{CDF}^{+}D^{+}_{SPO} \qquad (2.77)$$

$$E^{+}_{i\varphi} = E_{f\varphi}(\theta'_0, \varphi)\,\frac{\exp(-jk\rho_0)}{\rho_0}\;, \qquad (2.78a)$$

the φ component (parallel to the edge) of the E-field incident at the plus stationary point,

$$\mathrm{PF}^{+} = \exp(-jkR)\exp\left\{ jk\left(\frac{D}{2}\sin\theta + Z_{SP}\cos\theta\right)\right\}\;, \qquad (2.78b)$$

the phase-factor pathlength from the plus stationary point to P,

$$\mathrm{CDF}^{+} = \tfrac{1}{2}\sqrt{\frac{D/2}{\sin\theta}}\;, \qquad (2.78c)$$

the caustic divergence factor at great distances from the plus stationary point, and

$$D^{+}_{SPO} = -\frac{\exp\left(-j\frac{\pi}{4}\right)}{2\sqrt{2\pi k}}\left[\frac{2}{\sin\theta - \left(\dfrac{\sin\theta'_0}{1-\cos\theta'_0}\right)(1-\cos\theta)}\right], \qquad (2.78d)$$

the physical-optics edge-diffraction coefficient for the plus stationary point, parallel (soft) incidence, and where ρ_0 and θ'_0 are the polar coordinates of the edge, D is the reflector diameter, and Z_{SP} = Z-coordinate of the rim. Similar expressions can be derived for the other edge and the other linear polarisation.

The diffraction coefficient (only the quantity in square brackets) is compared in Fig. 2.11 with the comparable edge-diffraction coefficient from the geometrical theory of diffraction (θ_0 = 116°). The differences are seen to be second-order. Knop[35,36] has applied more rigorous saddle-point theory to yield a continuous, finite expression through the shadow boundary which is shown as a singularity in Fig. 2.11.

In the event that the source illumination changes rapidly in the vicinity of the stationary points, modified expressions to describe slope-diffraction effects are used in place of eqns. 2.77-2.78.

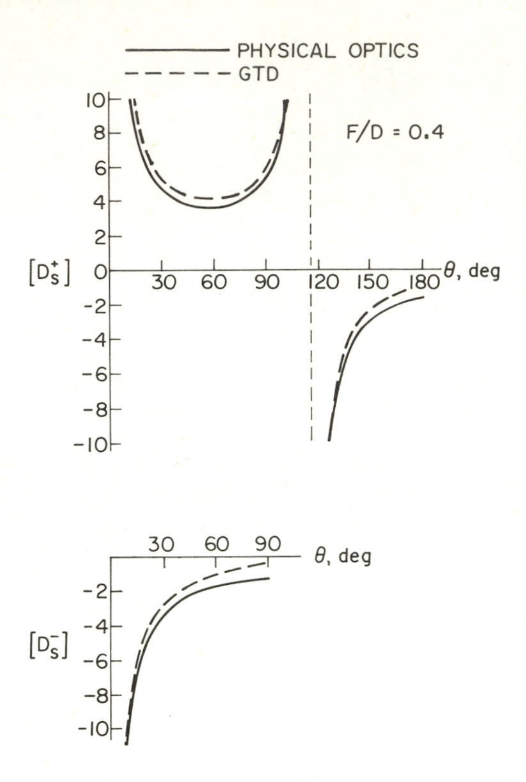

Fig. 2.11 *Comparison of GTD and asymptotic PO edge-diffraction coefficient*

2.2.5 Scalar physical optics (Contribution by W.C.Wong)

High gain spacecraft antennas frequently require apertures too large to be deployed before launch. One solution to this problem exists in the utilisation of a deployable umbrella reflector. While there may exist numerous deployable configurations, the material in this Section treats a particular type of deployable antenna in which the supporting ribs are parabolic in shape and the gore surface between any two adjacent parabolic ribs is the surface of a parabolic cylinder.[37] The quasiparabolic reflector surface has the effect of spreading the focal point into a focal region, resulting in an antenna gain loss. The range of the focal region for practical low-gore-loss antennas can be shown to be expressable in terms of antenna parameters. In addition, the relationship between the number of ribs and reflector sizes is such that scaling of the gore-loss curve for any gore reflector is possible.

The gore reflector is assumed to be perfectly conducting and illuminated by a linearly polarised, time-harmonic point source with no back lobes. The resulting radiated Fraunhofer field can be expressed in terms of the physical-optics integral [cf. (eqn. 2.66)]. In terms of the geometry of Fig. 2.12, $\bar{r}$ and $\bar{r}'$ are, respectively, the field vector and the source vector with respect to a coordinate system defined by the reflector boresight as its z-axis; the feed polarisation (assumed linear) is parallel to the x-axis; and the feed phase centre is chosen as the coordinate origin.

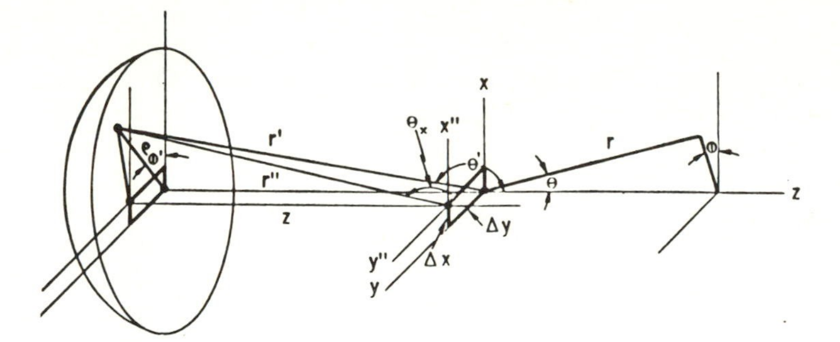

Fig. 2.12 *Geometry of problem*

The geometrical-optics surface current density for the nth gore surface, $\bar{J}_n$, has the following x and y components:

$$\bar{J}_n = 2f(\theta', \varphi')\sqrt{\epsilon_0/\mu_0}\ \hat{a} \exp(-jkr')/r' \tag{2.79a}$$

$$\hat{a} = \begin{bmatrix} \dfrac{-\cos\dfrac{\alpha}{2}\sin\theta'\sin\varphi'\sin\varphi_n + \sin\dfrac{\alpha}{2}\cos\theta'}{\sqrt{1-\sin^2\theta'\cos^2\varphi'}} \\ \\ \dfrac{\sin\theta'\sin\varphi'\cos\varphi_n\cos\dfrac{\alpha}{2}}{\sqrt{1-\sin^2\theta'\cos^2\varphi'}} \end{bmatrix} \tag{2.79b}$$

where r', θ' and φ' are the spherical coordinates defining the source vector $\bar{r}'$; μ_0 and ϵ_0 are the free-space constitutive parameters; $\varphi_n = 2\pi(n-1)/N_g$ and

$$\cos\alpha = \frac{\rho_g^2 - 4f_r^2\cos^4(\pi/N_g)}{\rho_g^2 + 4f_r^2\cos^4(\pi/N_g)} \tag{2.80}$$

where f_r is the focal length of the parabolic supporting ribs; N_g is the number of such ribs; and ρ_g is a cylindrical coordinate defined in Fig. 2.13. $f(\theta', \varphi')$ is the primary illumination function expressable in a finite Fourier series:

$$f(\theta', \varphi') = \sum_{m=-M}^{M} a_m(\theta') \exp(jm\varphi'). \tag{2.81}$$

It can be shown that the intersection of any plane containing the reflector axis with any gore panel is a parabola whose focal length is a periodic function of the azimuthal angle ψ:

$$fc = f_r \cos^2 (\pi/N_g)/\cos^2 \psi. \tag{2.82}$$

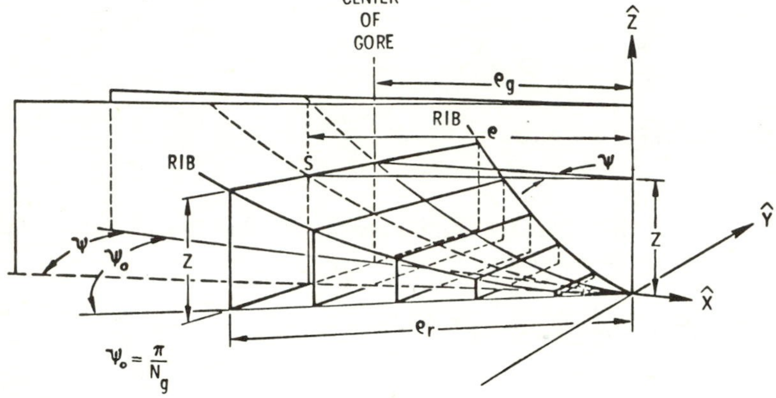

Fig. 2.13 *Geometrical relations of a gore*

f_c is therefore bounded between the focal length of the gore panels (ψ = 0) and the focal length of the ribs ($\psi = \pi/N_g$). Physically, the entire gore reflector can be thought of as made up of an infinite set of parabolas with a continuous distribution of foci. Thus, if a point source is in focus for one parabola, it is out of focus for the rest of the set. It will be shown in Chapter 3 that the best feed position can be expressed in terms of f_r and N_g.

Cylindrical coordinates (ρ, ψ) shown in Fig. 2.13 are a convenient set of integration variables to evaluate the radiation integral. These variables are related to (r', θ', φ'), defining the source vector $\bar{r}'$, by the following transformations:

$$r' = \frac{2f_c}{1 - \cos\theta'} - (f_s - f_c)\cos\theta' \tag{2.83a}$$

$$\theta' = \cos^{-1}\left[\frac{\rho^2 - 4f_c^2}{\rho^2 + 4f_c^2}\right] \tag{2.83b}$$

$$\varphi' = \varphi_n + \psi \tag{2.83c}$$

where f_c is given by eqn. 2.82 and f_s is the distance of the point source from the reflector vertex in the absence of lateral defocusing.

The vector radiation integral can be considerably simplified if the following approximations are valid:

(i) The secondary field of interest is restricted to a small angular region around the boresight.
(ii) The lateral feed displacement, if any, is small so that the parallel-ray approximation is valid.
(iii) The angular spectrum of the primary field is relatively smooth in the azimuthal direction, so that M in eqn. 2.81 is not too large.
(iv) The gore loss over the frequency band of interest is no more than a few dB.

Under these conditions, the vector radiation integral reduces to a far simpler scalar integral:

$$E_x(\bar{r}) = -4jk\frac{e^{-jkr}}{r}N_g\sum_{m=-M}^{M}\int_0^{\pi/N_g} d\psi\int_0^{\rho_0} \rho d\rho a_m(\theta')j^{-m}\,{-}_m(k\rho A)$$

$$\cdot \exp(jm\xi)\exp[jkr_0(\cos\theta' - 1)]\exp[jk\Delta f(1-\cos\theta')] \qquad (2.84a)$$

where

$$A = \sqrt{(\sin\theta\cos\varphi + \Delta x/r')^2 + (\sin\theta\sin\varphi + \Delta y/r')^2} \qquad (2.84b)$$

$$\xi = \tan^{-1}\left[\frac{\sin\theta\sin\varphi + \Delta y/r'}{\sin\theta\cos\varphi + \Delta x/r'}\right] \qquad (2.84c)$$

$$\Delta f = f_s - f_c \quad . \qquad (2.84d)$$

In the limit as N_g approaches infinity, integration with respect to ψ reduces to π/N_g, exp $[jkr_0$ $(\cos\theta' - 1)]$ reduces to a constant, and the resulting expression is that for a paraboloidal reflector with axial defocusing described by Δf and lateral defocusing described by A. The following expression for antenna gain can then be used to obtain data for gore loss curves, defocusing curves and antenna patterns:

$$G_A(\bar{r}) = \frac{4\pi r^2\,|E_x(\bar{r})|^2}{\int |f(\theta',\varphi')|^2\,d\Omega'} \qquad (2.85)$$

The radiation field of a 503.289 λ reflector was evaluated using the scalar formula given by eqn. 2.84. The results are shown as triangles in Fig. 2.14 for 28 gores, 12%

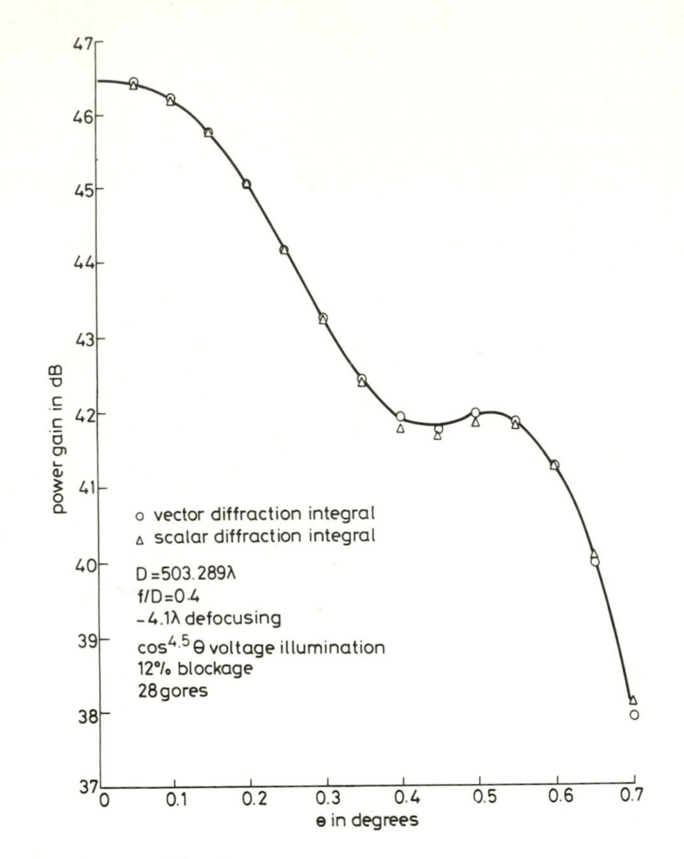

Fig. 2.14 *Scalar and vector diffraction patterns*

blockage, $\cos^{4.5}(\pi-\theta')$ voltage illumination, defocusing = -4.1λ. The circles correspond to the results for the same reflector obtained using the exact vector diffraction integral. It should be noted that, for this particular reflector, the loss in accuracy in using the scalar integral is seen to be negligible within the polar angle of interest.

Best-fit point of a gored reflector

The best focal point of a gored reflector can be determined if the reflector size and N_g are such that the radiated field can be accurately described by eqn. 2.84. This integral is maximised (approximately) by reducing the mean of the phase term $k\Delta f(1-\cos\theta')$ to zero as ψ varies across the gore. Thus

$$\frac{1}{\pi/N_g}\int_0^{\pi/N_g}\left[f_s - f_r\,\frac{\cos^2(\pi/N_g)}{\cos^2\psi}\right]d\psi = 0. \tag{2.86}$$

In arriving at eqn. 2.86 θ' has been assumed independent of ψ, which is justified when the gore loss is not excessive. The quantity f_s is the distance of the feed from the reflector vertex, and f_r is the rib focal length. From eqn. 2.86 it is found that

$$f_s = f_r \frac{N_g}{\pi} \sin(\pi/N_g)\cos(\pi/N_g) \cong f_r(1 - 2\pi^2/3N_g^2). \tag{2.87}$$

In the practical range of design parameters, this formula is highly accurate.

Reflective properties of wire mesh

The reflecting surface of deployable antennas for space applications is usually made of wire mesh due to its light weight. Electrodynamic properties of plane wire meshes have been studied by several researchers using different methods. Kontorovich[38] first formulated the square mesh problem using the so-called averaged boundary condition method in which he replaced the periodic variations of the fields and currents over the mesh surface with quantities averaged over one mesh cell. The method was later extended[39] to solve more general mesh problems involving rectangular mesh cells with non-perfect contacts between the wires at cell nodes. Chen[40] treated the problem with an entirely different approach using Floquet modes to describe the fields over the periodic mesh surface and obtain exact expressions for the reflection and transmission coefficients.

The last method is very general and accurate. It can be used to solve either rectangular (square) or circular meshes with the size of meshes ranging from less than one tenth of a wavelength to one wavelength. Frequently these meshes are used as bandpass structures instead of as reflecting surfaces.

The reflecting properties of a plane wire mesh are frequently characterised by four reflection coefficients: R_{MM}, R_{ME}, R_{EM} and R_{EE}. These coefficients are functions of mesh parameters such as the mesh geometry and wire diameter, and a set of angles (θ_M, φ_M). θ_M is the angle of incidence, while φ_M is the angle between some reference direction of the wire mesh and the plane of incidence spanned by the incident k vector and the unit vector normal to the plane mesh at the point of incidence. Given θ_M and φ_M, both the incident field $\bar{E}_i$ and the scattered field $\bar{E}_r$ can be resolved into components parallel to the plane of incidence $(\hat{U}_M)$ and normal to the plane of incidence $(\hat{U}_E)$.

$$\bar{E}_i = (\bar{E}_i \cdot \hat{U}_m)\,\hat{U}_m + (\bar{E}_i \cdot \hat{U}_E)\,\hat{U}_E \tag{2.88a}$$

$$\bar{E}_r = \begin{pmatrix} R_{MM} & R_{ME} \\ R_{EM} & R_{EE} \end{pmatrix} \begin{pmatrix} \bar{E}_i \cdot \hat{U}_m \\ \bar{E}_i \cdot \hat{U}_E \end{pmatrix}. \tag{2.88b}$$

Total field at the point of incidence is therefore

$$\bar{E}_t = \hat{U}_E\ [(1 + R_{EE})(\bar{E}_i \cdot \hat{U}_E) + R_{EM}(\bar{E}_i \cdot \hat{U}_M)]$$

$$+ \hat{U}_M [(1 + R_{MM})(\bar{E}_i \cdot \hat{U}_M) + R_{ME}(\bar{E}_i \cdot \hat{U}_E)] \tag{2.89a}$$

$$\bar{H}_t = \hat{U}_E [(1 - R_{MM})(\bar{E}_i \cdot \hat{U}_M) - R_{ME}(\bar{E}_i \cdot \hat{U}_E)] Y_M$$

$$- \hat{U}_m [(1 - R_{EE})(\bar{E}_i \cdot \hat{U}_E) - R_{EM}(\bar{E}_i \cdot \hat{U}_M)] Y_E \tag{2.89b}$$

where

$$Y_M = \frac{1}{\cos\theta_M} \sqrt{\frac{\epsilon_0}{\mu_0}} \tag{2.89c}$$

$$Y_E = \cos\theta_M \sqrt{\frac{\epsilon_0}{\mu_0}} \quad . \tag{2.89d}$$

The radiation integral in terms of tangential field components is then given by

$$\bar{E}(\bar{r}) = jk\,\hat{a}_r \times \hat{a}_r \times \iint_S \sqrt{\frac{\mu_0}{\epsilon_0}}\, \hat{n} \times \bar{H}_t \frac{e^{-jk|\bar{r} - \bar{r}'|}}{4\pi|\bar{r} - \bar{r}'|} dS'$$

$$-jk\hat{a}_r \times \iint_S \hat{n} \times \bar{E}_t \frac{e^{-jk|\bar{r} - \bar{r}'|}}{4\pi|\bar{r} - \bar{r}'|} \tag{2.90}$$

The analysis therefore assumes that at the point of incidence on the parabolic reflecting surface, the mesh surface is locally an infinite plane and both the incident and scattered fields are locally plane waves. These assumptions are consistent with the physical-optics approximations.

2.2.6 Geometrical theory of edge-diffraction

The geometrical theory of diffraction (GTD) postulates that the total field, $\bar{E}^t(\bar{r})$, can be expressed as

$$\bar{E}^t(\bar{r}) = \bar{E}^g(\bar{r}) + \bar{E}^d(\bar{r}) \tag{2.91}$$

where $\bar{E}^g$ is the classical geometrical-optics field [cf. (2.17)], and $\bar{E}^d$ the 'diffracted' field can be expressed as an asymptotic series in inverse powers of $(-jk)$:

$$\bar{E}^d(\bar{r}) = e^{-jks(\bar{r})}(k)^{-1/2} \sum_{m=0}^{\infty} (-jk)^{-m}\, \bar{e}^d_m(\bar{r}), k \to \infty \tag{2.92}$$

GTD represents a systematic procedure [41–43] whereby the properties and behaviour of the leading term of eqn. 2.92 can be described on an essentially 'ray' basis, much in the same way that the leading term of the GO series is handled. The material of this section will deal with the behaviour of the leading term, which, for simplicity, will be designated the GTD ray or field, in the context of diffraction from the thin edge of a reflector.* Before considering specific details of the GTD, some general principles should be enumerated:

(i) The GTD field is asymptotically of order $k^{-1/2}$, whereas the GO field is of order k^0.
(ii) The GTD field does not possess shadow regions.
(iii) The diffracted field propagates along rays which are determined by a generalisation of Fermat's principle to include points on edges, vertices and smooth surfaces in the ray trajectory.
(iv) Diffraction, like reflection and transmission, is a local phenomenon at high frequencies, i.e. it depends only on the nature of the boundary surface and the incident field in the immediate neighbourhood of the point of diffraction.
(v) The diffracted wave propagates along its rays so that (*a*) power is conserved in a tube of rays, and (*b*) the phase delay along the ray path equals the product of the wave number of the medium and the distances.

As a result of these basic postulates, edge-diffraction effects can be accurately described in terms of diffracted rays with the following properties, many of which are also common to the GO field:

(*a*) Diffracted rays emerge radially from edges.
(*b*) At the place of origin the intensity and polarisation of the diffracted field are linearly related to the incident field by dyadic diffraction coefficients.
(*c*) In homogeneous media they travel in straight lines.
(*d*) The associated fields are TEM.
(*e*) In a homogeneous medium polarisation is constant along a ray.
(*f*) The field strength is inversely proportional to the square-root of the cross-sectional area of a tube of rays.
(*g*) At places where the cross-section of a tube of rays converges to a line (caustic line or focal line), there is a phase shift of 90°, and to a point (focus), there is a phase shift of 180°.
(*h*) An edge is always a caustic.
(*i*) Special correction techniques must be applied to determine the intensity at caustics or foci.

Fermat's principle for an edge

Fig. 2.15 shows a 'point source' O at (x_O, y_O, z_O) and an observation point P at (x_P, y_P, z_P). The z-axis lies along the edge of a conducting, semi-infinite plane.

*So-called 'surface' and 'tip' diffraction have not yet achieved a state of general usefulness for three-dimensional reflectors.

GTD postulates that a possible ray from O to P is by means of a point Q_E on the edge:

$$d = \overline{OQ_E} + \overline{Q_EP} = \sqrt{x_O^2 + y_O^2 + (z_Q - z_O)^2} + \sqrt{x_P^2 + y_P^2 + (z_P - z_Q)^2} \quad (2.93)$$

Applying Fermat's principle, $\partial d/\partial z_Q = 0$, yields

$$\beta_0' = \beta_0. \quad (2.94)$$

Thus, for Q_E to be a possible point for edge diffraction, P must lie along a cone with its vertex at Q_E and half-angle equal to β_0', the angle of incidence between the ray $\overline{OQ_E}$ and the edge. This condition is known as the *law of edge diffraction.*

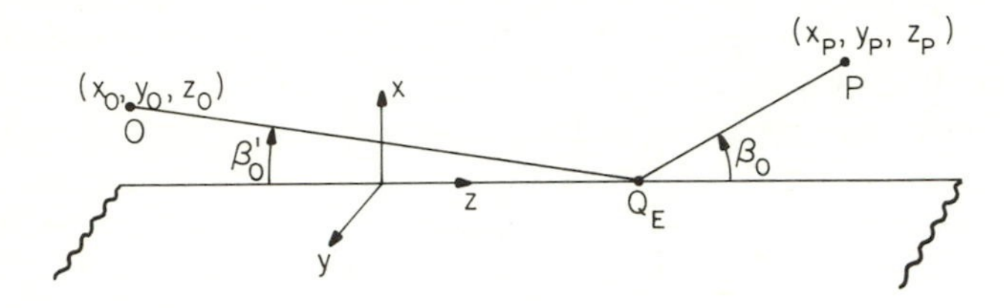

Fig. 2.15 *Geometry of Fermat's principle for an edge*

Fig. 2.16 shows a cross-sectional view of the same phenomenon. According to Fermat's principle, three possible ray trajectories from O to P are: $\overline{OP}$, the direct ray; $\overline{OQP}$, the geometrically reflected ray subject to Snell's law; and $\overline{OQ_EP}$, the 'edge-diffracted' ray discussed above. The position of the diffracted ray on the 'cone of diffraction' is specified by φ, while the position of the incident ray on the 'cone of incidence' is specified by φ'. Thus Fig. 2.16 can be regarded as the projection of the rays in the three dimensional problem onto a plane perpendicular to the edge at the point of diffraction Q_E. In the case that the incident ray is *normally* incident on the edge, the cone of diffracted rays degenerates into a planar disc.

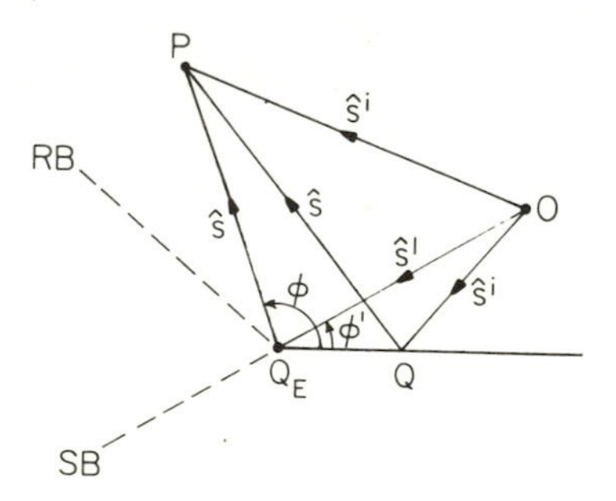

Fig. 2.16 *Reflection and diffraction by a straight edge on a conducting half-plane*

Coordinate systems

Kouyoumjian[42] has defined two ray-fixed coordinate systems which considerably simplify the dyadic diffraction coefficients in the case of a curved edge as shown in

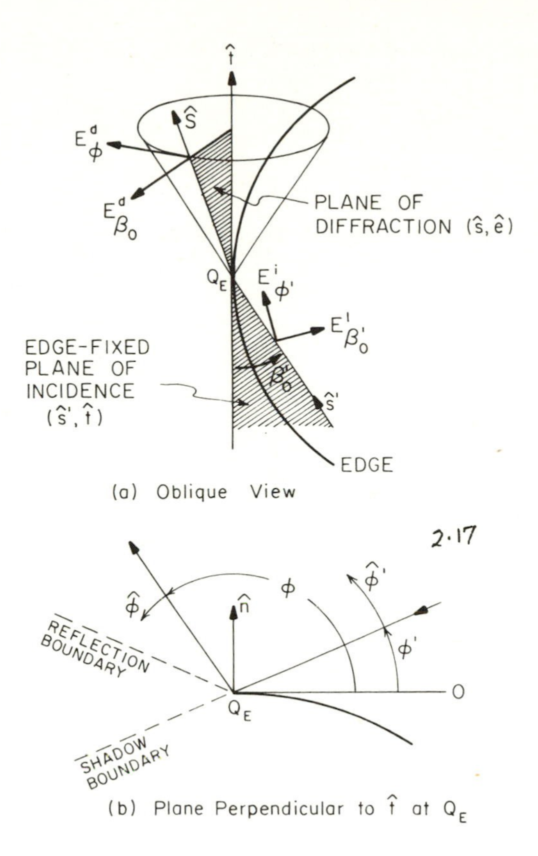

Fig. 2.17 *Geometry of edge-fixed coordinate systems*

Fig. 2.17. The edge-fixed plane of incidence is the plane containing the vector $\hat{t}$, tangent to the edge at Q_E, and $\hat{s}'$, the incident wave direction (Fig. 2.17*a*). The unit vector $\hat{\varphi}'$, in the direction of increasing φ' (Fig. 2.17*b*), is perpendicular to the edge-fixed plane of incidence. It may be computed by

$$\hat{\varphi}' = \frac{\hat{t} \times \hat{s}'}{|\hat{t} \times \hat{s}'|} \tag{2.95a}$$

where it is assumed that grazing incidence will not occur. A third unit vector $\hat{\beta}_0'$, in the direction of increasing β_0', is defined:

$$\hat{\beta}_0' = \hat{s} \times \hat{\varphi}'. \tag{2.95b}$$

Similarly, an edge-fixed plane of diffraction can be defined to contain both $\hat{t}$ and $\hat{s}$, the diffracted wave normal (Fig. 2.17*a*). The unit vector $\hat{\varphi}$, in the direction of increasing φ, is perpendicular to the edge-fixed plane of diffraction

$$\hat{\varphi} = \frac{\hat{s} \times \hat{t}}{|\hat{s} \times \hat{t}|} . \tag{2.96a}$$

The third member of the triplet of unit vectors is then, correspondingly,

$$\hat{\beta}_0 = \hat{s} \times \hat{\varphi} . \tag{2.96b}$$

Thus $\hat{s}'$, $\hat{\beta}'_0$, $\hat{\varphi}'$ are unit vectors in an 'incident-ray-fixed' spherical coordinate system and $\hat{s}$, $\hat{\beta}_0$, $\hat{\varphi}$ are unit vectors in a 'diffracted-ray-fixed' spherical coordinate system.

GTD field equation

The electric field of the edge-diffracted ray at a distance s from Q_E is;

$$\bar{E}^d(P) = \bar{E}^i(Q_E) \cdot \bar{\bar{D}}(\varphi, \varphi'; \beta'_0) \sqrt{\frac{\rho}{s(\rho + s)}}\, e^{-jks} \tag{2.97}$$

where $\bar{E}^i(Q_E)$ is the incident field at Q_E;

$\bar{\bar{D}}(\varphi, \varphi'; \beta'_0)$ is a matrix (dyadic) diffraction coefficient;

$\sqrt{\dfrac{\rho}{s(\rho + s)}}$ is the *caustic divergence factor* and accounts for changes in the cross- section of the tube of rays diffracted from Q_E;

ρ is the 'caustic' distance, i.e. the distance from Q_E to the principal centre of curvature of the diffracted wavefront.

The edge is always one caustic of the diffracted rays. Therefore, at a distance s from the point Q_E on the edge the principal radii of curvature of the diffracted wavefront are s and $\rho + s$. The quantity ρ is negative if the caustic lies along the positive direction of the diffracted ray. Under these conditions, the quantity under the radical will be negative if the caustic lies between Q_E and P and a 90° phase shift will be experienced. Thus, if $\rho < 0$ and $s > |\rho|$,

$$\sqrt{\frac{\rho}{s(\rho + s)}} = e^{j\pi/2} \sqrt{\frac{|\rho|}{s\,|\rho + s|}} . \tag{2.98}$$

The caustic distance ρ is given by:

$$\frac{1}{\rho} = \frac{1}{\rho_e^i} - \frac{\hat{n}_e \cdot (\hat{s}' - \hat{s})}{a \sin^2 \beta_0} \tag{2.99}$$

where ρ_e^i is the distance from O to Q; a is the radius of curvature of the edge at Q; $\hat{n}_e$ is the unit normal to the edge at Q (pointing *away* from the centre of curvature). For a circular edge with an on-axis point source, the caustic lies where the extension of the diffracted ray intersects the axis (Fig. 2.18).

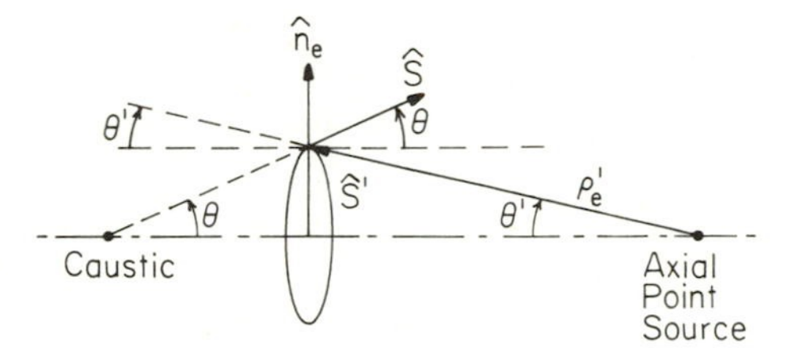

Fig. 2.18 *Caustic distance for a circular edge with an axial point source*

The dyadic edge-diffraction coefficient $\bar{\bar{D}}(\varphi, \varphi'; \beta_0')$ indicates that a linear relationship exists between $\bar{E}^d(P)$ and $\bar{E}^i(Q_E)$. The components of $\bar{\bar{D}}(\varphi, \varphi'; \beta_0')$ are dependent upon choice of a particular coordinate system. In an edge-fixed coordinate system $\bar{\bar{D}}(\varphi, \varphi'; \beta_0')$ is the sum of seven dyads, but in the ray-fixed coordinate system defined in this section, it is the sum of only two dyads:

$$\bar{\bar{D}}(\varphi, \varphi'; \beta_0') = -\hat{\beta}_0'\hat{\beta}_0 D_s(\varphi, \varphi'; \beta_0') - \hat{\varphi}'\,\hat{\varphi} D_h(\varphi, \varphi; \beta_0') \tag{2.100}$$

where D_s is the scalar diffraction coefficient for the acoustically soft (Dirichlet) boundary condition at the edge and D_h is the scalar diffraction coefficient of the acoustically hard (Neumann) boundary condition. In matrix notation, $\bar{\bar{D}}(\varphi,\varphi'; \beta_0')$ is expressed as a 3 x 3 matrix with seven nonvanishing components in an edge-fixed coordinate system, while in the ray-fixed systems it becomes a 2 x 2 diagonal matrix, enabling eqn. 2.97 to be written

$$\begin{bmatrix} E_{\beta 0}^d \\ E_\varphi^d \end{bmatrix} = \begin{bmatrix} -D_s & 0 \\ 0 & -D_h \end{bmatrix} \begin{bmatrix} E_{\beta_0'}^i \\ E_{\varphi'}^i \end{bmatrix} \sqrt{\frac{\rho}{s(\rho + s)}}\, e^{-jks} \cdot \tag{2.101}$$

Keller's diffraction coefficients

J.B.Keller[44] has formulated the scalar diffraction coefficients for the infinitely thin, curved edge to be

$$D_{\substack{s\\h}} = \frac{-e^{-j\frac{\pi}{4}}}{2\sqrt{2\pi k}\,\sin\beta_0'}\left\{\frac{1}{\cos\left(\dfrac{\varphi-\varphi'}{2}\right)} \mp \frac{1}{\cos\left(\dfrac{\varphi+\varphi'}{2}\right)}\right\}. \qquad (2.102)$$

The minus sign in the brackets is associated with the soft (s) coefficient and the plus sign is associated with the hard (h) coefficient. The coefficients are very simple trigonometric functions of β', φ', and φ, and consequently are readily amenable to calculation.

Trouble occurs, however, in regions where the denominator of the first term is excessively small because $\varphi \to \pi + \varphi'$ (shadow boundary) or the denominator of the second term becomes excessively small because $\varphi \to \pi - \varphi'$ (reflection boundary). Consequently, GTD cannot be used in these so-called transition regions unless the original formulations are modified in some manner.

Kouyoumjian's treatment of the transition region

Kouyoumjian has eliminated the transition-region singularities in Keller's diffraction coefficients by introducing factors in the numerator of each term in eqn. 2.102 which go to zero when the denominators go to zero. Thus:

$$D_{\substack{s\\h}} = \frac{-e^{-j\frac{\pi}{4}}}{2\sqrt{2\pi k}\,\sin\beta_0'}\left\{\frac{F[kL^i\,a(\varphi-\varphi')]}{\cos\left(\dfrac{\varphi-\varphi'}{2}\right)} \mp \frac{F[kL^r\,a(\varphi+\varphi')}{\cos\left(\dfrac{\varphi+\varphi'}{2}\right)}\right\}. \qquad (2.103)$$

The common function $a(\varphi \pm \varphi')$ is defined simply

$$a(\varphi \pm \varphi') \equiv 2\cos^2\left(\frac{\varphi \pm \varphi'}{2}\right). \qquad (2.104)$$

From the first term in the bracketed expression

$$L^i = \frac{s(\rho+s)\,\rho_1^i\,\rho_2^i\,\sin^2\beta_0'}{\rho(\rho_1^i+s)(\rho_2^i+s)} \xrightarrow[s\to\infty]{} \frac{\rho_1^i\,\rho_2^i\,\sin^2\beta_0'}{\rho} \qquad (2.105)$$

where s is the distance from Q_E to P, ρ_1^i and ρ_2^i are the principal radii of curvature of the incident wavefront at Q_E, and ρ, the caustic distance, is evaluated at the shadow boundary so that $\hat{s}' = \hat{s}$ and, from eqn. 2.99, $\rho = \rho_e^i$. For a spherical-wave source $\rho_1^i = \rho_2^i = \rho_e^i = \rho_0$, so that for this special (but very common case)

$$L^i = \rho_0 \sin^2 \beta_0' \tag{2.106}$$

where ρ_0 is the distance from the source to Q_E.

The transition function $F(kL^i a)$ is a transition function of argument $kL^i a$, a measure of the angular distance between P and the shadow boundary. For example, exactly at the shadow boundary $\varphi = \varphi' + \pi$ so both $a(\varphi - \varphi') = 0$ and $kL^i a(\varphi - \varphi') = 0$. The transition function is given by

$$F(kLa) = 2j\sqrt{kLa}\, e^{jkLa} \int_{\sqrt{kLa}}^{\infty} e^{-j\tau^2}\, d\tau \,. \tag{2.107}$$

At the shadow boundary, the factor $\cos\dfrac{\varphi - \varphi'}{2}$ in the denominator of the first term in the brackets of eqn. 2.103 becomes zero. However, $F[kL^i a(\varphi - \varphi')]$ simultaneously becomes zero, and the quotient remains finite:

$$\frac{F(kL^i a(\varphi - \varphi')]}{\cos\left(\dfrac{\varphi - \varphi'}{2}\right)} \xrightarrow[\varphi \to \pi + \varphi']{} 2\pi \sqrt{\frac{L^i}{\lambda}}\, e^{j\pi/4} \,. \tag{2.108}$$

From the second term in the brackets of eqn. 2.103

$$L^r = \frac{s(\rho + s)\, \rho_1^r \rho_2^r \sin^2 \beta_0}{\rho(\rho_1^r + s)\,(\rho_2^r + s)} \tag{2.109}$$

where ρ_1^r and ρ_2^r are the principal radii of curvature of the geometrical-optics ray reflected from the surface at Q_E and ρ is evaluated in the direction of the reflection boundary, i.e. $\varphi + \varphi' = \pi$.

At the reflection boundary the factor $\cos(\frac{\varphi+\varphi'}{2})$ in the denominator of the second term in the brackets of (2.103) becomes zero. However, $F[kL^r\, a(\varphi + \varphi')]$ simultaneously vanishes, yielding

$$\frac{F[kL^r\, a(\varphi + \varphi')]}{\cos\left(\dfrac{\varphi + \varphi'}{2}\right)} \xrightarrow[\varphi \to \pi - \varphi]{} 2\pi \sqrt{\frac{L^r}{\lambda}}\, e^{j\pi/4} \,. \tag{2.110}$$

Lee's treatment of the transition region

Lee has postulated a somewhat different treatment of the transition regions near geometrical shadow- or reflection-boundaries, entitled the *uniform asymptotic theory* (UAT) of edge-diffraction[45]. As in eqn. 2.91 the total field is again expressed as the sum of specular and diffracted components

$$\bar{E}^t(\bar{r}) = \bar{E}^G(\bar{r}) + \sum_n \bar{E}_n^d(\bar{r}) \tag{2.111}$$

where the summation is carried out over all of the edge-diffracted GTD rays (using the Keller diffraction coefficients). However, when the observation point lies in the vicinity of a reflection boundary (RB),* the diffraction coefficient of one of the edge rays (e.g. $n = p$) becomes unphysically large, due to the denominator of the second term in eqn. 2.102. In that case, $\bar{E}^G(\bar{r})$ is modified as follows:

$$\bar{E}^G(\bar{r}) = \left[F(\xi_p^r) - \hat{F}(\xi_p^r)\right] \bar{E}^r(\bar{r}) \tag{2.112}$$

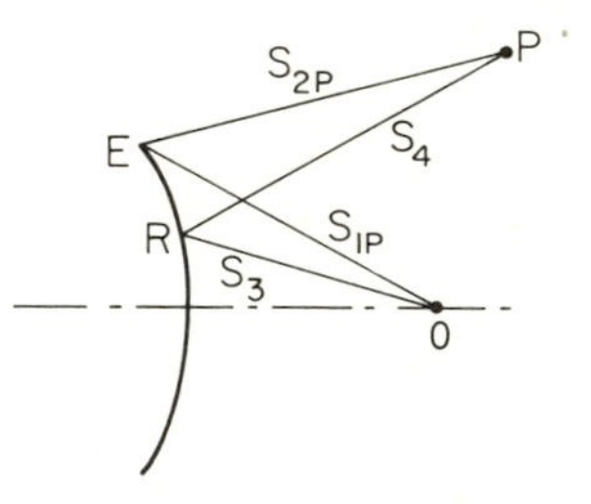

Fig. 2.19 *Geometry for UAT detour parameter*

The specular ray ORP and the pth edge ray OEP are shown in Fig. 2.19. The pth detour parameter ξ_p^r is defined by

$$\xi_p^r = \epsilon^r(\bar{r})\sqrt{k\,|(s_{1p} + s_{2p}) - (s_3 + s_4)|} \tag{2.113}$$

and the 'shadow indicator' $\epsilon^r(\bar{r})$ is

$$\epsilon^r(\bar{r}) = \begin{cases} -1, \bar{r} \text{ on 'lit' side of RB} \\ +1, \bar{r} \text{ on 'shadow' side of RB} \end{cases} \tag{2.114}$$

*Similar relationships are used near transmission shadow boundaries

The Fresnel functions are defined in UAT as

$$F(x) = \frac{e^{+j\pi/4}}{\sqrt{\pi}} \int_x^\infty e^{-jt^2}\, dt \qquad (2.115a)$$

$$\hat{F}(x) = \frac{1}{2x\sqrt{\pi}}\, e^{-j(x^2 + \pi/4)} \qquad (2.115b)$$

and $\bar{E}^r(\bar{r})$ is the conventional GO reflected field. Near the reflection boundary, both $\bar{E}_p^d(\bar{r})$ and $\bar{E}^G(\bar{r})$ will become singular but will cancel each other in such a way as to provide a smooth transition from $\bar{E}_p^d(\bar{r})$ on the shadow side of RB to $\frac{1}{2}\bar{E}^r(\bar{r})$ at RB to $\bar{E}^r(\bar{r}) + \bar{E}_p^d(\bar{r})$ on the lit side of RB. Note that a fictitious specular path length $S_3 + S_4$ and a fictitious $\bar{E}^r(\bar{r})$ are required in order to extend eqn. 2.112 into the shadow.

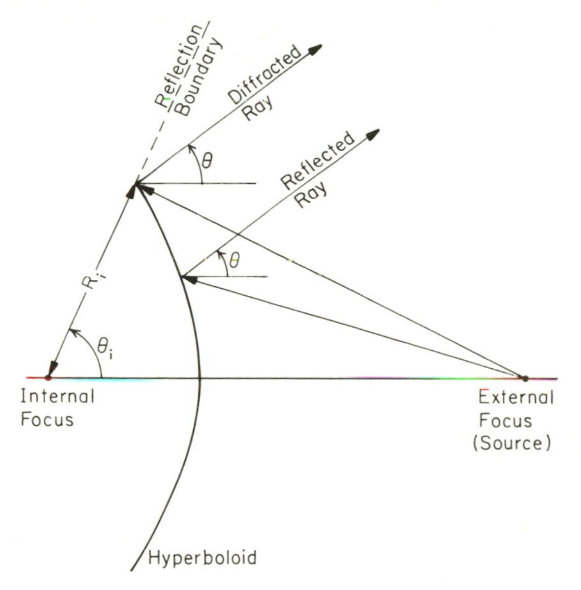

Fig. 2.20 *Hyperboloid geometry*

For the hyperboloid scattering geometry shown in Fig. 2.20, the argument of the Kouyoumjian Fresnel-integral transition function is $kL^r a$, which, for the hyperboloid:

$$kL^r a = 2\pi(R_i/\lambda)\,[1 - \cos(\theta_i - \theta)] \qquad (2.116)$$

which vanishes at the reflection boundary $\theta = \theta_i$. Similarly, the argument of the Lee

Fresnel-integral transition function is the detour parameter, ξ^r, which, for the hyperboloid:

$$\xi^r = \epsilon^r(\theta_i - \theta)\sqrt{2\pi(R_i/\lambda)\,[1 - \cos(\theta_i - \theta)]}\,. \tag{2.117}$$

Thus the magnitude of one transition-function argument equals the square root of the other. However, the Fresnel integrals are formulated in such a way as to yield identical total fields at the reflection boundary and very nearly equal total fields throughout the transition region. For example, for an isotropic source and a 25-wavelength hyperboloid with an eccentricity of 2.295, the differences are less than 1–2% in magnitude and 1–2° in phase. Application of the two techniques to other problems, however, may yield substantially greater differences.

Slope diffraction

Under some conditions the illumination of the edge is changing very rapidly, so that the derivative of the illumination is comparable in significance to the illumination level itself. A prime example of this occurs when a null in the illumination pattern is directed toward the edge. Inclusion of a slope diffraction term in eqn. 2.97 then yields[46]

$$\bar{E}^d(P) = \left[\bar{E}^i(Q_E)\cdot\bar{\bar{D}} + \frac{\partial\bar{E}^i}{\partial n}(Q_E)\cdot\bar{\bar{d}}\right]\sqrt{\frac{\rho}{s(\rho+s)}}\;e^{-jks} \tag{2.118}$$

where $\partial/\partial n$ is with respect to the normal to the edge. The dyadic slope diffraction coefficient is given by

$$\bar{\bar{d}} = \begin{bmatrix} -d_s & 0 \\ 0 & -d_h \end{bmatrix} \tag{2.119}$$

and the elements are related to the elements of the diffraction coefficient defined in eqn. 2.103 by

$$d_s = \frac{1}{jk\sin\beta_0}\,\frac{\partial}{\partial\varphi'}\,D_s \tag{2.120a}$$

$$d_h = \frac{1}{jk\sin\beta_0}\,\frac{\partial}{\partial\varphi'}\,D_h \tag{2.120b}$$

The UAT description of slope-diffraction effects is contained implicitly in the basic UAT equations 2.111-2.114 without requiring additional derivative terms.

2.3 Spherical-wave theory[47] (Contribution by A.C. Ludwig)

Spherical wave theory can be applied to a wide range of antenna problems of which quasi-optical applications represents only one special group. Nevertheless it is finding an increasing number of uses in the analysis of quasi-optical antennas and their primary feeds and as such cannot be omitted from the range of fundamental analytical techniques.

An electromagnetic field can in general be expressed in terms of either an integral over the source currents and/or boundary fields, or in terms of a series expansion in which the expansion coefficients are defined by integrals over the source currents and/or boundary fields. When the integral formulation is used, the integral must be evaluated for each field point; for the series formulation an integration is performed only once to determine the coefficients, and thereafter the series is summed for each field point. When there are many field points, the best technique will then depend on the relative ease of evaluating an integral or summing a series. It is frequently, particularly in problems involving the near field, far easier to sum the series.[48,49]

Solutions of Maxwell's equations may be obtained using the separation of variables technique in rectangular, cylindrical, or spherical coordinates. In general, the rectangular and cylindrical coordinate systems lead to a continuous spectrum of solutions (e.g., a plane-wave spectrum), and a discrete set of modal solutions arises only within certain bounded regions, such as a waveguide. In contrast, the spherical

Table 2.1 Spherical-wave applications

	Problem	*References*
1	Boundary value solution of scattering problems	Mie;[83] Kennaugh and Ott;[84] Liang and Lo;[78] Hizal and Marincic;[85] Bruning and Lo;[76] King and Harrison;[86] Hizal and Yaza[79]
2	Basic physical limitations on antennas	Chu;[54] Dicke;[87] Taylor;[53] Harrington;[55] Collin and Rothschild;[57] Ludwig;[88] Potter;[58] Wasyilkiwskyj and Kahn;[89] Ricardi[90]
3	Prediction of far-field patterns from near-field data	Brown and Jull;[91] James and Longdon;[92] Jensen;[93,94] Hansen and Jensen;[95] Wacker;[67] Jensen and Larsen;[69,72] Wood[70] Larsen[96]
4	Calculation of focal region fields or fields near a source	Kennaugh and Ott;[84] Clarricoats and Olver;[97] Ricardi;[90] Wood[98]
5	Reflector antenna synthesis; feed synthesis	Potter;[58] Ricardi;[90] Wood;[99] Ludwig and Brunstein[100]
6	Prediction of near-field patterns from far-field data	Ludwig[59,101]
7	An alternative to Kirchoff aperture integration	Clarricoats and Saha[81]
8	Near-field coupling between two antennas	Ludwig and Norman[102]

9	Modified physical optics using near-zone magnetic fields	Ludwig;[101] Potter and Ludwig[103]
10	An alternative to physical optics current integration	Wood[48,49,98]
11	Inverse scattering	Boerner and Vandenberghe[104]

coordinate system leads to a discrete set of modal solutions even for an unbounded region, so spherical waves can be considered to be 'modes' of free space.

Spherical wave expansions are particularly useful for problems involving fields arising from a finite distribution of sources radiating into an unbounded region, since, in addition to satisfying the radiation conditions, spherical waves exhibit a cutoff phenomenon, described in detail later in this Section, which results in an upper limit on the number of terms in the expansion. The limit is related to the physical size of the source, and provides the basis for a realisability criterion for synthesis problems, for determining basic physical limitations on antenna properties such as gain or aperture efficiency, or for filtering experimental data. (Cylindrical waves fill an analogous role for two-dimensional problems.) There is also a limit on terms needed for expansions which are not required to satisfy the radiation conditions. Other uses of spherical waves range from expansion of a plane wave of infinite extent (see Section 2.3.2), to the expression of modes inside a spherical cavity. It should also be emphasized that applications are not restricted to problems involving spherical boundaries; for example, the case of a plane wave incident on a paraboloidal reflector is handled readily by the technique (see Section 2.3.2).

The primary difficulty of using spherical wave analysis is the complexity of the formulation; the technique is generally not suited for calculator analysis and requires computer implementation. On the other hand, the method is well suited to computers, and has some unique features suitable for computer program validation, as discussed later. As Table 2.1 shows, sperical-wave analysis is extremely versatile, and the same basic computer software can typically be applied to a broad range of problems.

2.3.1 Fundamental properties

Spherical waves are a solution of Maxwell's equations with the sources equal to zero everywhere except the origin, and *any* physically realisable electromagnetic field in an isotropic homogeneous source-free region may be represented as a spherical-wave expansion.[50,51] An expansion is typically valid only in a particular region of space and will not be equal to the specified field outside of this region. If the origin of the expansion is shifted, the wave coefficients will change. Therefore one cannot really say that certain spherical waves are physically generated by a given antenna. If only outgoing waves (i.e. satisfying the radiation conditions) are present, wave coefficients are uniquely determined for a given origin if the tangential $\bar{E}$ (or $\bar{H}$) field is specified on a surface enclosing the origin. If both incoming and outgoing waves may be present, both the tangential $\bar{E}$ and $\bar{H}$ fields are required.

Definition of spherical waves

Unfortunately, notation for spherical waves is complex and not standardised; the notation used here will generally follow Stratton[52] except for using the engineering time dependence $e^{j\omega t}$, and the $\hat{a}_R, \hat{a}_\theta, \hat{a}_\phi$ notation for the unit vectors of the standard R, θ, ϕ spherical coordinate system.

A spherical wave expansion of an electromagnetic field may be written:

$$\bar{E} = -\sum_{n=1}^{N} \sum_{\substack{m=0 \\ e,o}}^{n} \left(a^{(i)}_{\substack{e \\ o}mn} \bar{m}^{(i)}_{\substack{e \\ o}mn} + b^{(i)}_{\substack{e \\ o}mn} \bar{n}^{(i)}_{\substack{e \\ o}mn} \right) e^{j\omega t} \tag{2.121a}$$

$$\bar{H} = \frac{k}{j\omega\mu} \sum_{n=1}^{N} \sum_{\substack{m=0 \\ e,o}}^{n} \left(a^{(i)}_{\substack{e \\ o}mn} \bar{n}^{(i)}_{\substack{e \\ o}mn} + b^{(i)}_{\substack{e \\ o}mn} \bar{m}^{(i)}_{\substack{e \\ o}mn} \right) e^{j\omega t} \tag{2.121b}$$

where $\bar{m}^{(i)}_{\substack{e \\ o}mn}$ and $\bar{n}^{(i)}_{\substack{e \\ o}mn}$ are the spherical vector wave functions*

$$\bar{m}^{(i)}_{\substack{e \\ o}mn} = \mp \frac{m}{\sin\theta} z^{(i)}_n (kR) P^m_n (\cos\theta) \begin{matrix}\sin \\ \cos\end{matrix} m\phi \hat{a}_\theta$$

$$- z^{(i)}_n (kR) \frac{\partial P^m_n (\cos\theta)}{\partial\theta} \begin{matrix}\cos \\ \sin\end{matrix} m\phi \hat{a}_\phi \tag{2.122a}$$

$$\bar{n}^{(i)}_{\substack{e \\ o}mn} = \frac{n(n+1)}{kR} z^{(i)}_n (kR) P^m_n (\cos\theta) \begin{matrix}\cos \\ \sin\end{matrix} m\phi \hat{a}_R$$

$$+ \frac{1}{kR} \frac{\partial}{\partial R} \left[R z^{(i)}_n (kR) \right] \frac{\partial}{\partial\theta} P^m_n (\cos\theta) \begin{matrix}\cos \\ \sin\end{matrix} m\phi \hat{a}_\theta \tag{2.122b}$$

$$\mp \frac{m}{kR \sin\theta} \frac{\partial}{\partial R} \left[R z^{(i)}_n (kR) \right] P^m_n (\cos\theta) \begin{matrix}\sin \\ \cos\end{matrix} m\phi \hat{a}_\phi$$

where $k = \omega\sqrt{\epsilon\mu} = 2\pi/\lambda$, and ω, ϵ, and μ are specified to be real constants.

The spherical wave coefficients $a^{(i)}_{\substack{e \\ o}mn}$ and $b^{(i)}_{\substack{e \\ o}mn}$ have the following interpretation:

$a^{(i)}_{\substack{e \\ o}mn}$ for TE waves, $b^{(i)}_{\substack{e \\ o}mn}$ for TM waves;

*As defined here $\bar{m}^{(i)}_{\substack{e \\ o}mn}$ and $\bar{n}^{(i)}_{\substack{e \\ o}mn}$ are dimensionless; therefore for $\bar{E}$ and $\bar{H}$ to have the proper dimensions, the dimensions of $a^{(i)}_{\substack{e \\ o}mn}$ and $b^{(i)}_{\substack{e \\ o}mn}$ must be volts·m^{-1}.

e, o subscripts denote even, odd azimuthal (ϕ coordinate) dependence;

n represents polar (θ coordinate) wave order;

m represents azimuthal wave order;

(*i*) superscript represents the type of Bessel function governing the radial dependence, $z_n^{(1)}(kR) \equiv j_n(kR), z_n^{(2)}(kR) \equiv y_n(kR), z_n^{(3)}(kR) \equiv h_n^{(1)}(kR)$, and $z_n^{(4)}(kR) \equiv h_n^{(2)}(kR)$.

The m index is restricted to integer values by the requirement of field continuity between $\phi = 0$ and $\phi = 2\pi$; the n index is restricted to integer values by the requirement that $P_n^m(\cos\theta)$ be finite at $\theta = 0$ and $\theta = \pi$ (in regions bounded by a cone and/or wedge, these requirements do not apply). Therefore, in free space, the spherical wave solutions are fundamentally modal, as mentioned previously.

Explicit forms for the polar dependence of the lowest-order spherical waves are given in Table 2.2. The associated Legendre function $P_n^m(\cos\theta)$ is zero for $m > n$, so the summation in the m index of eqn. 2.121 terminates at $m = n$. For $m = 0$ the odd waves do not exist, so there are $2(2n + 1)$ TE and TM, odd and even, azimuthal waves for a given polar order n. Therefore, the total number of waves in eqn. 2.121 is $2N(N + 2)$, for a single value of (i).

In general, two independent solutions for the spherical Bessel function $z_n^{(i)}(kR)$ are required (any two of the four solutions given above); it is always valid to begin any problem with two solutions, and equations are given later to determine the coefficients for this case (see Section 2.3.2). However, for certain conditions, one solution is sufficient and the problem may be simplified by choosing it at the outset.

For a specified field arising from a finite source distribution and satisfying the radiation conditions, it is sufficient to use the outgoing wave solution $z_n^{(4)}(kR) = h_n^{(2)}(kR)$, which satisfies

$$\lim_{R\to\infty} h_n^{(2)}(kR) = j^{n+1}\frac{e^{-jkR}}{kR} \tag{2.123}$$

$$\lim_{R\to\infty} \frac{1}{kR}\frac{\partial}{\partial R}\left[Rh_n^{(2)}(kR)\right] = j^n \frac{e^{-jkR}}{kR} \tag{2.124}$$

the first three of these functions are shown in Table 2.3(*a*). These functions have a singularity at $R = 0$, and therefore the region of validity for the expansion must exclude the origin; the region of validity must also exclude the sources of the specified field, so the expansion is valid everywhere outside a sphere, centred at the origin, containing the sources of the specified field. Normally, the origin is located

Table 2.2 Polar functions

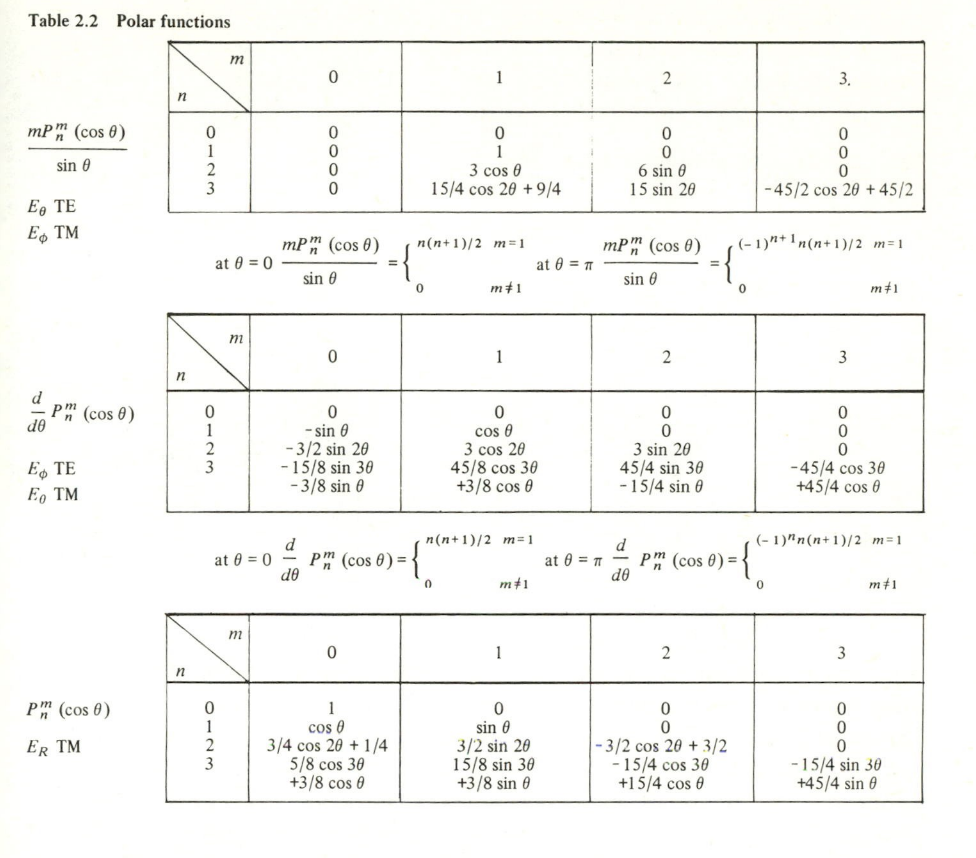

$\dfrac{mP_n^m(\cos\theta)}{\sin\theta}$ — E_θ TE, E_ϕ TM

n \ m	0	1	2	3
0	0	0	0	0
1	0	1	0	0
2	0	$3\cos\theta$	$6\sin\theta$	0
3	0	$15/4\cos 2\theta + 9/4$	$15\sin 2\theta$	$-45/2\cos 2\theta + 45/2$

$$\text{at } \theta = 0 \quad \frac{mP_n^m(\cos\theta)}{\sin\theta} = \begin{cases} n(n+1)/2 & m=1 \\ 0 & m \neq 1 \end{cases} \qquad \text{at } \theta = \pi \quad \frac{mP_n^m(\cos\theta)}{\sin\theta} = \begin{cases} (-1)^{n+1} n(n+1)/2 & m=1 \\ 0 & m \neq 1 \end{cases}$$

$\dfrac{d}{d\theta}P_n^m(\cos\theta)$ — E_ϕ TE, E_θ TM

n \ m	0	1	2	3
0	0	0	0	0
1	$-\sin\theta$	$\cos\theta$	0	0
2	$-3/2\sin 2\theta$	$3\cos 2\theta$	$3\sin 2\theta$	0
3	$-15/8\sin 3\theta$ $-3/8\sin\theta$	$45/8\cos 3\theta$ $+3/8\cos\theta$	$45/4\sin 3\theta$ $-15/4\sin\theta$	$-45/4\cos 3\theta$ $+45/4\cos\theta$

$$\text{at } \theta = 0 \quad \frac{d}{d\theta}P_n^m(\cos\theta) = \begin{cases} n(n+1)/2 & m=1 \\ 0 & m \neq 1 \end{cases} \qquad \text{at } \theta = \pi \quad \frac{d}{d\theta}P_n^m(\cos\theta) = \begin{cases} (-1)^n n(n+1)/2 & m=1 \\ 0 & m \neq 1 \end{cases}$$

$P_n^m(\cos\theta)$ — E_R TM

n \ m	0	1	2	3
0	1	0	0	0
1	$\cos\theta$	$\sin\theta$	0	0
2	$3/4\cos 2\theta + 1/4$	$3/2\sin 2\theta$	$-3/2\cos 2\theta + 3/2$	0
3	$5/8\cos 3\theta$ $+3/8\cos\theta$	$15/8\sin 3\theta$ $+3/8\sin\theta$	$-15/4\cos 3\theta$ $+15/4\cos\theta$	$-15/4\sin 3\theta$ $+45/4\sin\theta$

within the source region to minimize the size of this sphere.

For the reciprocal of the above situation, consisting of a field propagating inwards and being totally absorbed by a load, the incoming wave solution $z_n^{(3)} \equiv h_n^{(1)}(kR)$ is sufficient. Again, there is a singularity at $R = 0$ and the expansion will be valid outside a sphere centred at the origin and containing the load currents.

For a field arising from sources which are all exterior to a sphere centred at the origin, the solution $z_n^{(1)}(kR) = j_n(kR)$ is sufficient. This is the only Bessel function solution without a singularity at the origin. The expansion will be valid everywhere inside the sphere.* (Case 6 of Table 2.6 is an example of this situation.) This sphere can, in fact, approach infinite radius, leading to an expansion valid everywhere; a plane wave is an example of this limiting case (Case 1 of Table 2.6).

The singularities of $h_n^{(2)}(kR)$ and $h_n^{(1)}(kR)$ at $R = 0$ represent current sources

*Electromagnetic fields are analogous to analytic functions in complex variable theory in several ways; this expansion corresponds to a Taylor series, since it converges in the interior of a region around the origin, whereas the previous series correspond to a Laurent series; a vector field integral corresponds to Cauchy's integral formula.

(or loads) for the spherical waves; these are similar to the 'equivalent sources' of a surface integral. The superposition of the sources for an expansion of a specified field radiate a field identical to the specified field within the region of validity, but may radiate a very different field elsewhere.

Table 2.3(a) Radial functions $z_n^{(4)}(kR)$

n	$h_n^{(2)}(kR)$	$\frac{1}{kR}\frac{d}{dR}\left[Rh_n^{(2)}(kR)\right]$
0	$j\frac{e^{-jkR}}{kR}$	$\frac{e^{-jkR}}{kR}$
1	$-\frac{e^{-jkR}}{kR}\left[1-\frac{j}{kR}\right]$	$\frac{e^{-jkR}}{kR}\left[j+\frac{1}{kR}-\frac{j}{(kR)^2}\right]$
2	$-j\frac{e^{-jkR}}{kR}\left[1-\frac{3j}{kR}-\frac{3}{(kR)^2}\right]$	$-\frac{e^{-jkR}}{kR}\left[1-\frac{3j}{kR}-\frac{6}{(kR)^2}+\frac{6j}{(kR)^3}\right]$

Current sources of the lowest-order spherical waves are shown in Fig. 2.21. The $n = 1$ waves represent simple infinitesimal dipoles and loop radiation, and the higher order waves represent increasingly complex multiple sources. These current sources are not unique (for example, rotating the TM even $n = 2, m = 2$ source 90° around the z-axis and changing the phase by 180° results in the same wave), so the sources of spherical waves are not uniquely specified even if the origin is fixed. All of these sources indicate a limiting case in which the dimensions approach zero, and the current magnitudes approach infinity. These sources are more complex than electrostatic multipoles;[52] for example, the $m = 0$, $n = 3$ source includes a component of the $m = 0, n = 1$ source, and is not generated simply by displacing two $m = 0, n = 2$ sources, as for the electrostatic case.

Orthogonality

The spherical vector wave functions* have orthogonality properties that can be expressed in various ways. A convenient form is

$$\int_0^{\pi}\int_0^{2\pi}\left(\bar{m}^{(i)}_{\substack{e\\o}mn} \times \bar{n}^{(j)}_{\substack{e\\o}lk} \cdot \hat{a}_R\right) R^2 \sin\theta \, d\theta \, d\phi =$$

*'Spherical vector wave functions' refer to the functions $\bar{m}^{(i)}_{\substack{e\\o}mn}$ and $\bar{n}^{(i)}_{\substack{e\\o}mn}$ defined by eqns. 2.122*a* and 2.122*b*; 'spherical waves' refer to an electromagnetic field constructed from two of these functions.

Table 2.3(*b*) Radial functions $z_n^{(1)}(kR)$

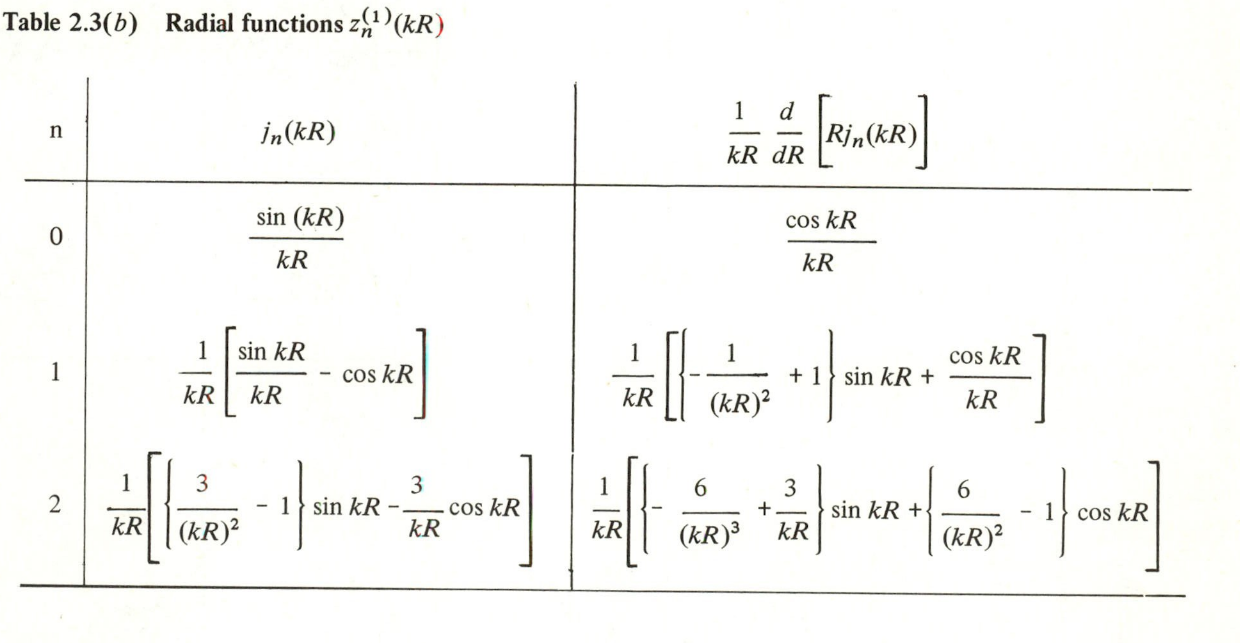

n	$j_n(kR)$	$\frac{1}{kR}\frac{d}{dR}\left[Rj_n(kR)\right]$
0	$\frac{\sin(kR)}{kR}$	$\frac{\cos kR}{kR}$
1	$\frac{1}{kR}\left[\frac{\sin kR}{kR} - \cos kR\right]$	$\frac{1}{kR}\left[\left\{-\frac{1}{(kR)^2} + 1\right\}\sin kR + \frac{\cos kR}{kR}\right]$
2	$\frac{1}{kR}\left[\left\{\frac{3}{(kR)^2} - 1\right\}\sin kR - \frac{3}{kR}\cos kR\right]$	$\frac{1}{kR}\left[\left\{-\frac{6}{(kR)^3} + \frac{3}{kR}\right\}\sin kR + \left\{\frac{6}{(kR)^2} - 1\right\}\cos kR\right]$

In general

$$\frac{1}{kR}\frac{d}{dR}\left[Rz_n^{(i)}(kR)\right] = Z_{n-1}^{(i)'}(kR) - \frac{n}{kR}z_n^{(i)}(kR).$$

$$
= \begin{cases} \gamma_{mn} R^2 z_n^{(i)}(kR) \dfrac{1}{kR} \dfrac{\partial}{\partial R} \left[R z_n^{(j)}(kR) \right] & \text{for } l = m,\ k = n, \\ \quad \text{and both functions even or both odd;} & \\ 0 \quad \text{if any of the three conditions above are not met} & \end{cases} \tag{2.125}
$$

where R must be constant, and

$$
\gamma_{mn} = 2\pi \frac{n(n+1)(n+m)!}{(2n+1)(n-m)!} \qquad m \neq 0 \tag{2.126a}
$$

$$
\gamma_{0n} = 4\pi \frac{n(n+1)}{(2n+1)} \quad \text{for} \qquad m = 0. \tag{2.126b}
$$

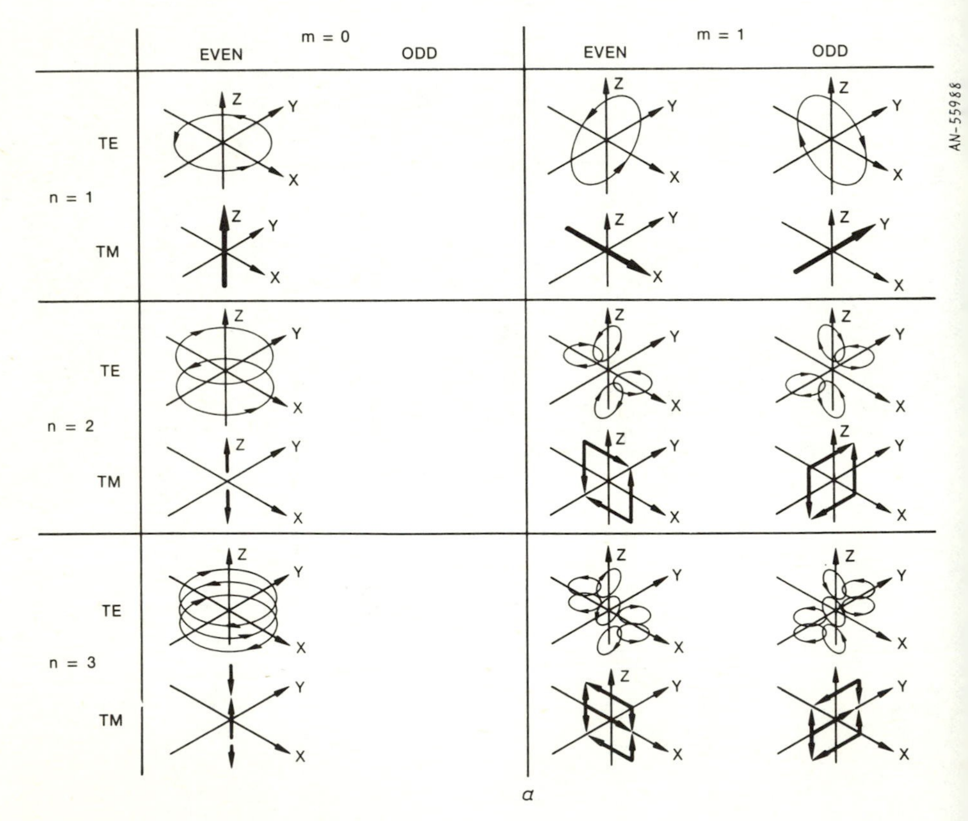

Fig. 2.21 *Current sources of spherical waves*

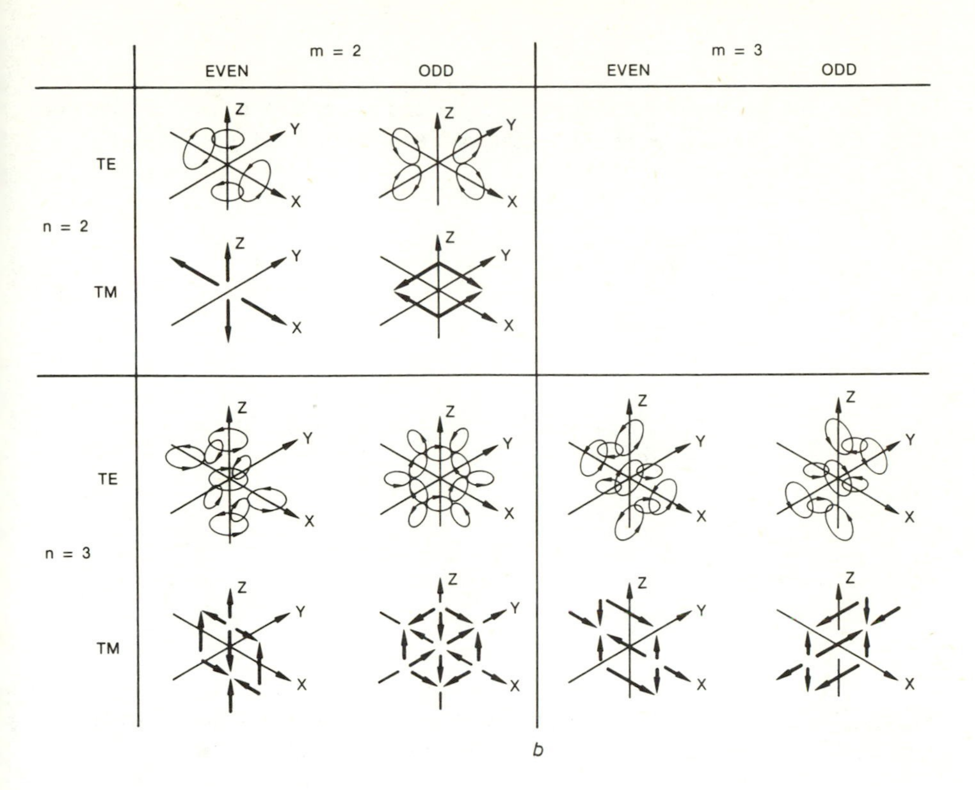

Fig. 2.21 *(continued)*

For cross products involving two $\bar{m}^{(i)}_{e_{mn}}$ type waves or two $\bar{n}^{(i)}_{e_{mn}}$ type waves the integral is zero

$$\int_0^{\pi}\int_0^{2\pi}\left(\bar{m}^{(i)}_{\substack{e\\o}mn} \times \bar{m}^{(j)}_{\substack{e\\o}lk} \cdot \hat{a}_R\right) R^2 \sin\theta\, d\theta\, d\phi = 0 \tag{2.127a}$$

$$\int_0^{\pi}\int_0^{2\pi}\left(\bar{n}^{(i)}_{\substack{e\\o}mn} \times \bar{n}^{(j)}_{\substack{e\\o}lk} \cdot \hat{a}_R\right) R^2 \sin\theta\, d\theta\, d\phi = 0\,. \tag{2.127b}$$

Eqns. 2.125-2.127 follow directly from the mathematical properties of the trigonomentric functions and associated Legendre functions.[52] Further orthogonality properties of spherical waves are described under the topic of calculating spherical wave coefficients.

Power flow

Using the orthogonality equations 2.125-2.127, the power flow through a sphere of radius R of a field expressed as a spherical wave expansion 2.121 is found to be

$$P = \frac{1}{2} \text{Re} \int_0^{\pi} \int_0^{2\pi} (\bar{E} \times \bar{H}^* \cdot \hat{a}_R) R^2 \sin\theta \, d\theta \, d\phi$$

$$= \frac{R^2}{2} \sqrt{\frac{\epsilon}{\mu}} \sum_{n=1}^{N} \sum_{\substack{m=0 \\ e,\, o}}^{n} \left(\left| a^{(i)}_{\substack{e \\ o}mn} \right|^2 + \left| b^{(i)}_{\substack{e \\ o}mn} \right|^2 \right) \gamma_{mn} \text{Re} \left\{ j \left[z_n^{(i)}(kR) \right]^* \frac{1}{kR} \frac{\partial}{\partial R} \left[R z_n^{(i)}(kR) \right] \right\} \tag{2.128}$$

When $z_n^{(i)}(kR)$ is a real linear combination of the Bessel functions $j_n(kR)$ and $y_n(kR)$, there is zero net power flow. If $z_n^{(i)}(kR) = h_n^{(2)}(kR)$,

$$\text{Re} \left\{ j h_n^{(1)}(kR) \frac{1}{kR} \frac{\partial}{\partial R} \left[R h_n^{(2)}(kR) \right] \right\} = \frac{1}{(kR)^2} . \tag{2.129}$$

The waves are outgoing and the power flow is positive and independent of R. If $z_n^{(i)}(kR) = h_n^{(1)}(kR)$, the waves are incoming and the power flow is negative.

Eqns. 2.128 and 2.129 show that for an expansion of out-going or incoming waves, each spherical wave maintains a constant total propagation power through space, and that this power is independent of all other waves. It can be easily verified that this is still true when both outgoing and incoming waves are present.

Propagation for $kR < n$: Wave cutoff

One of the most important properties of spherical waves is the cut-off behaviour they exhibit. We will consider outward propagating waves, where $z_n^{(i)}(kR) = h_n^{(2)}(kR)$, but effects for the other Bessel functions may be deduced from the behaviour of the real and imaginary parts of this function. For values of $kR < n$, a spherical wave exhibits behaviour similar to a waveguide mode beyond cutoff. Unlike waveguide modes, the cutoff point is not precisely defined, but the transition occurs within a rather short interval. For $kR \to 0$, the asymptotic forms for the Hankel functions are[17]

$$h_n^{(2)}(kR) \to \frac{(kR)^n}{1\cdot3\cdot5\cdots(2n+1)} + j\,\frac{1\cdot3\cdot5\cdots(2n-1)}{(kR)^{n+1}} \quad (2.130a)$$

$$\frac{1}{kR}\frac{\partial}{\partial R}\left[Rh_n^{(2)}(kR)\right] \to \frac{(n+1)(kR)^{n-1}}{1\cdot3\cdot5\cdots(2n+1)} - jn\,\frac{1\cdot3\cdot5\cdots(2n-1)}{(kR)^{n+2}}\,. \quad (2.130b)$$

At exactly $kR = n$, the Hankel function is given approximately by

$$h_n^{(2)}(n) \cong \frac{1}{n}\left\{[n^{1/6}.561 - n^{-1/6}.257] + j\,[n^{1/6}.971 + n^{-1/6}.446]\right\} \quad (2.130c)$$

Therefore, the amplitude at $kR = n$, relative to $1/R$ behaviour, is quite stable with increasing n; at $n = 1000$ it is only a factor of 3.6 higher. For $kR < n$, the asymptotic terms given by eqns. 2.130*a* and 2.130*b* begin to dominate, creating a sharp transition to the following behaviour:

(*a*) Amplitude grows extremely rapidly.
(*b*) Both functions of eqns. 2.130*a* and 2.130*b* become dominantly imaginary, so the rate of change of phase with respect to R becomes very small; thus the 'wavelength' of the spherical wave becomes very large.
(*c*) The relative phase of the $\bar{E}$ and $\bar{H}$ fields approaches 90° and the ratio of the imaginary and real parts of the Poynting vector becomes very large; this implies a large value for the field Q.

Typical amplitude behaviour of $h_n^{(2)}(kR)$ is shown in Fig. 2.22 relative to $1/R$ behaviour. The rapid transition in the vicinity of $kR = n$ is evident in this Figure.

Consider an example of an expansion of a field arising from sources enclosed in a sphere of radius ρ_0, where $k\rho_0 = 20$. Suppose that on the surface of the sphere at $R = \rho_0$, the $n = 20$ spherical wave and the $n = 30$ spherical wave have roughly the same amplitude; then as R increases from $kR = 20$ to $kR = 30$ it is seen from Fig. 2.22 that the amplitude of the $n = 30$ wave will be attenuated by roughly a factor of 1000 (60 dB) relative to the $n = 20$ wave. Therefore, in this example, for $kR > 30$ the higher order wave will make a negligible contribution to the field.

It can be shown that if the root-mean-square value of the field on the sphere of radius ρ_0 is E_0, then the value of the nth spherical wave coefficient is bounded by

$E_0\sqrt{4\pi/\gamma_{mn}}/h_n^{(2)}(k\rho_0)$ for TE waves, with a similar relationship for TM waves. Therefore, the cutoff phenomenon is normally manifested by a very rapid decline in values of the expansion coefficients for $n > k\rho_0$. The catch in this argument is that it concerns *maximum* values of the coefficients; if it happens that all of the coefficients up to $n = k\rho_0$ are identically zero, then even though the higher order coefficients are small, they are still dominant. This pathological situation is intimately related to the case of supergain (e.g. see case 2, Table 2.6).

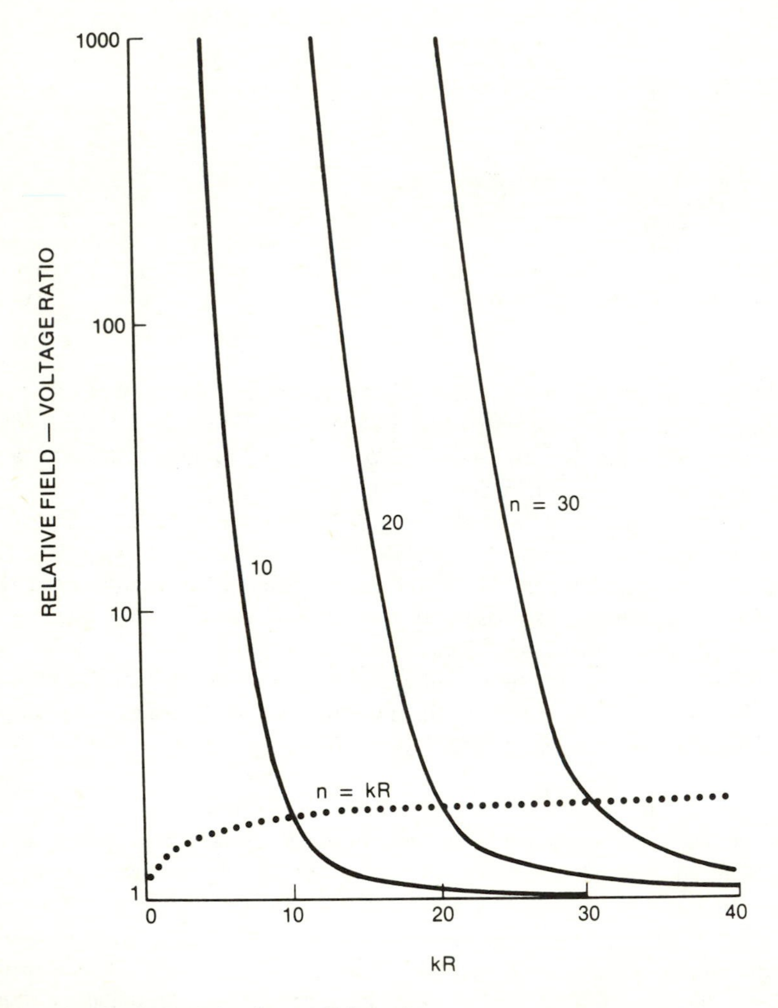

Fig. 2.22 *Amplitude deviation from 1/R behaviour*

As shown by many authors,[53–58] the Q of the nth spherical wave rises sharply for $n > kR$. This implies that if there is any loss in the medium, in the region near

ρ_0 waves with $n > k\rho_0$ will lose a very large amount of energy. In practice this also acts to eliminate these higher order waves.

Although the above arguments are not conclusive, they provide good reason to expect that higher order waves can normally be neglected. Experimental results have been obtained which confirm this expectation.[59] Fig. 2.23 illustrates the experimental source, consisting of the aperture of a horn antenna. The size of the sphere enclosing this source is dependent on the location of the coordinate system origin, as shown. Patterns of this source were measured and spherical wave expansions representing 90% to 99.9% of the total radiated power were computed from the data. This was done for several aperture sizes, and several locations of the origin; as shown in the Figure, 99.9% of the power was always contained in modes for which $n \leqslant k\rho_0$.

The cutoff phenomenon has an interesting interpretation in terms of the spacing of nulls of the field on a sphere of radius ρ_0. The function $P_n^m(\cos\theta)$ has $n - m$ zeros roughly equally spaced in the interval $0 < \theta < \pi$; for $m = 0$ and $n = k\rho_0$ the zeros are therefore spaced about $\lambda/2$ apart in θ, for any cut ϕ = constant; for $m > 0$ the spacing is larger. Since $m \leqslant n$, the functions $\sin m\phi$, $\cos m\phi$ have a

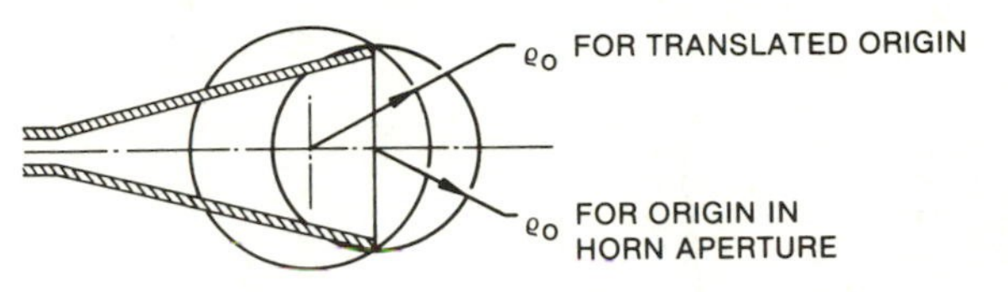

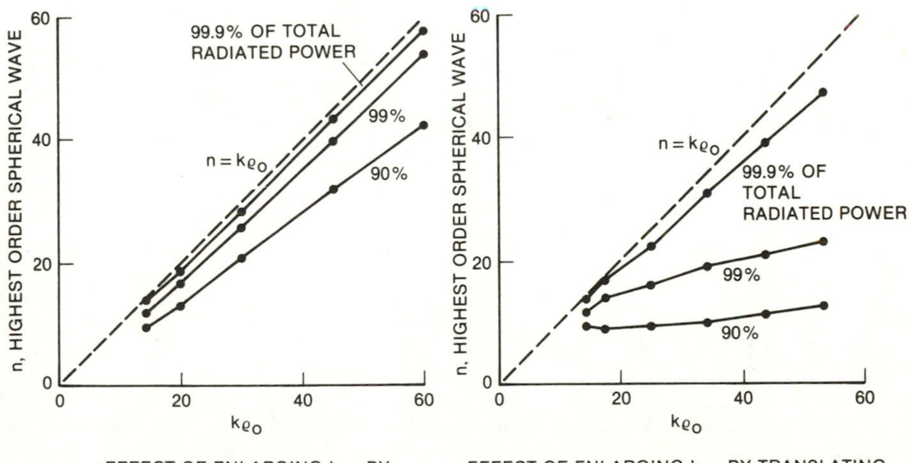

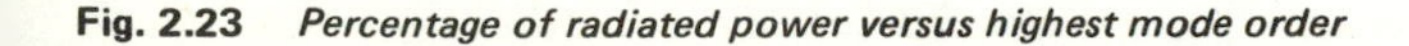

Fig. 2.23 *Percentage of radiated power versus highest mode order*

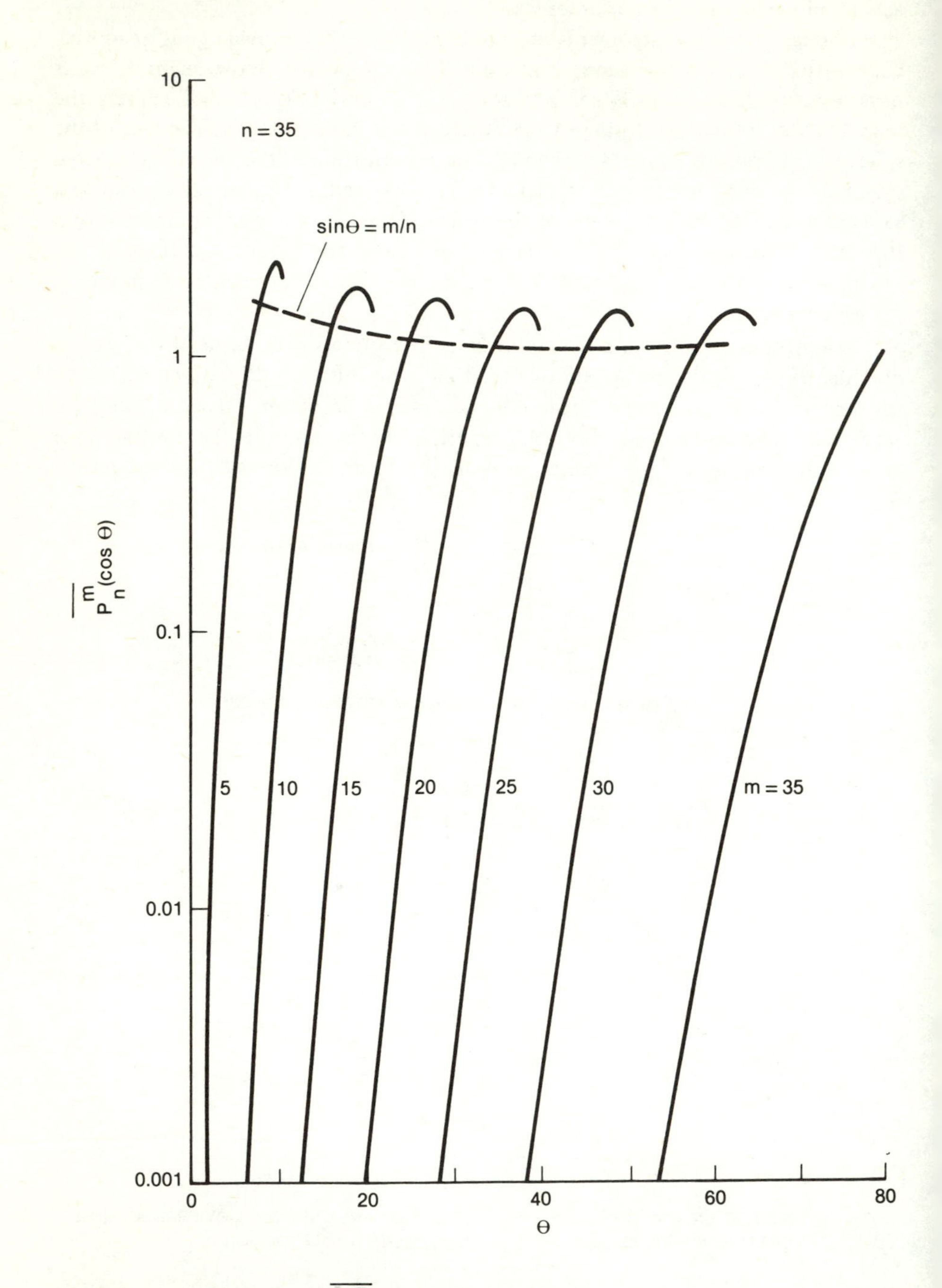

Fig. 2.24 *Cutoff property of $\overline{P_n^m}$ (cos θ) (Normalised)*

maximum of $2n$ zeros equally spaced on the interval $0 \leqslant \phi \leqslant 2\pi$. Therefore at $\theta = \pi/2$ these zeros are also $\lambda/2$ apart for $n = k\rho_0$. For $\theta < \pi/2$ it would appear that the zeros would get closer together; however, as shown in Fig. 2.24 the functions $P_n^m(\cos\theta)$ become small for $\sin\theta < m/n$, which eliminates terms with null spacing $< \lambda/2$. Therefore the cutoff phenomena eliminates waves which could lead to a null spacing of less than 1/2 wavelength on the sphere of radius ρ_0.

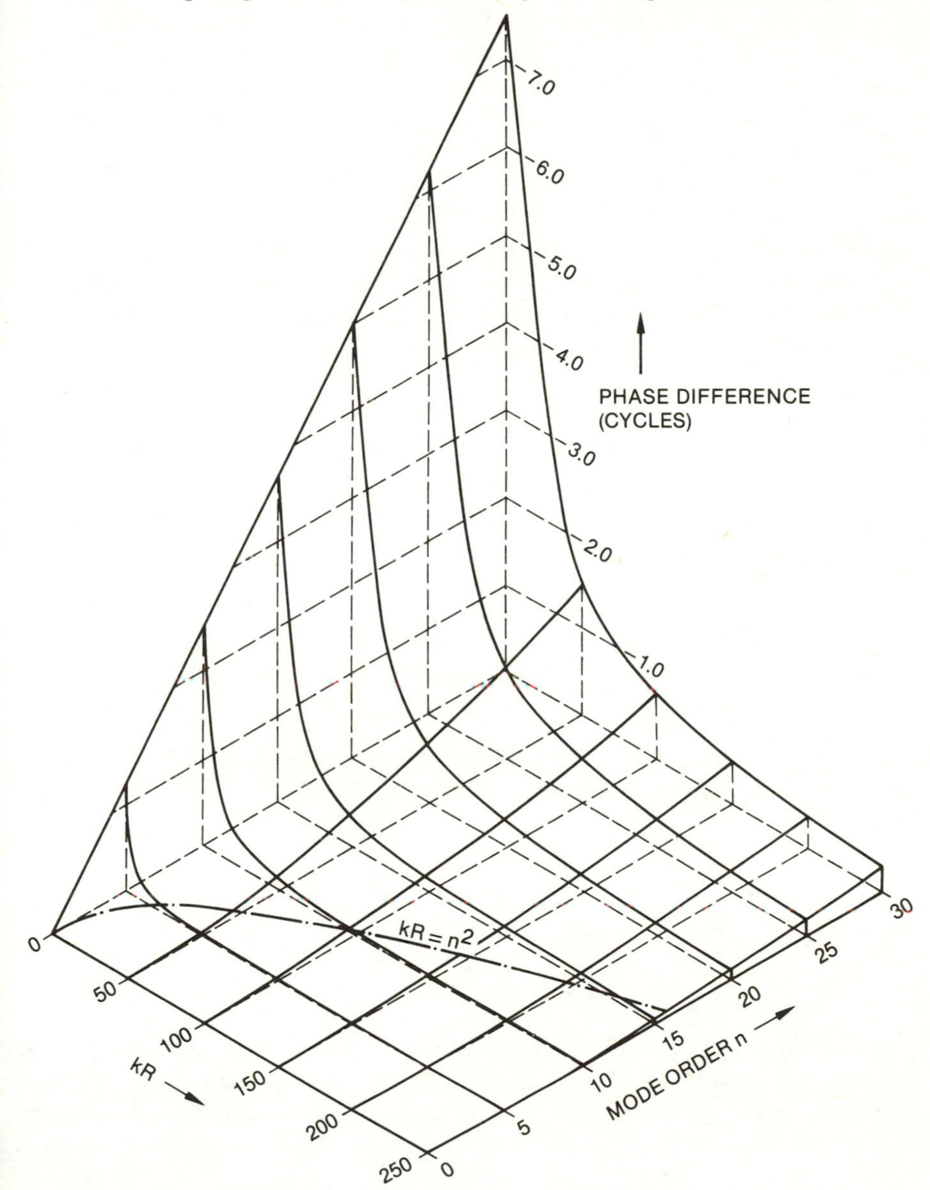

Fig. 2.25 *Phase deviation from e^{-jkR} behaviour*

In summary, several viewpoints lead to the conclusion that if a field arises from a source that can be enclosed in a sphere of radius ρ_0, then it can be expressed as a spherical wave expansion truncated at $N = k\rho_0$. This is not a precise rule, and other limits have been suggested as $k\rho_0 + 10$, or $7.7(2\rho_0/\lambda)^{78}$, (ref 60). It is also clear that when the expansion is evaluated close to $R = \rho_0$, more terms may be needed than when the expansion is evaluated for $R >> \rho_0$.* However, care must be taken when the series is evaluated for $R < N/k$ since the extremely fast amplitude growth of the higher-order waves will greatly magnify any errors due to numerical or experimental noise; for this reason errors introduced by using too many modes in an expansion can be far more severe than using too few modes, if the series is evaluated for small values of R.

Propagation for $n < kR < n^2$: Fresnel region

For $kR > n$ the amplitude of the Bessel functions $h_n^{(2)}(kR)$ approximates to $1/R$ behaviour, but the phase deviates significantly from e^{-jkR} behaviour, as shown in Fig. 2.25. This region will of course be different for each wave in an expansion, but the Fresnel region of the overall expansion may be considered to be $N < kR < N^2$.

Propagation for $kR > n^2$: Fraunhofer region

It can be shown that the phase deviation from e^{-jkR} behaviour for the nth spherical wave is approximated quite accurately by $n(n + 1)/2kR$ radians, for $n > 1$ and $kR \geqslant n^2$. At $kR = n^2$ this is about 28.6°, which corresponds well with the conventional 22.5° phase-error far-field criterion. Therefore for $kR > n^2$ the function $h_n^{(2)}(kR)$ may be approximated by its asymptotic form, eqn. 2.123. The relative magnitude of the $\hat{a}_R$ component of eqn. 2.122*b* is approximately proportional to n^2/kR, and can be neglected for $KR >> n^2$. The far-field form for the outward propagating spherical vector wave functions is then

$$\lim_{kR\to\infty} \bar{m}^{(4)}_{\substack{e\\o}mn} = j^{n+1}\left[\mp \frac{mP_n^m(\cos\theta)}{\sin\theta} \begin{matrix}\sin\\ \cos\end{matrix} m\phi\hat{a}_\theta - \frac{\partial}{\partial\theta} P_n^m(\cos\theta) \begin{matrix}\cos\\ \sin\end{matrix} m\phi\hat{a}_\phi\right] \frac{e^{-jkR}}{kR} \tag{2.131a}$$

$$\lim_{kR\to\infty} \bar{n}^{(4)}_{\substack{e\\o}mn} = j^{n}\left[\frac{\partial}{\partial\theta} P_n^m(\cos\theta) \begin{matrix}\cos\\ \sin\end{matrix} m\phi\hat{a}_\theta \mp \frac{mP_n^m(\cos\theta)}{\sin\theta} \begin{matrix}\sin\\ \cos\end{matrix} m\phi\hat{a}_\phi\right] \frac{e^{-jkR}}{kR} \tag{2.131b}$$

The resulting $\bar{E}$ and $\bar{H}$ fields of eqn. 2.121 are in-phase, orthogonal in space, in the ratio of the impedance of free space, and all waves propagate with e^{-jkR}/R amplitude and phase dependence. The Fraunhofer region of an overall expansion may be considered to be $kR > N^2$. For a source enclosed in a sphere of radius ρ_0,

*A paper by Ojeba and Walter[61] provides interesting numerical data on this point, as well as the general truncation question.

and for $N = k\rho_0$, the criterion $kR > N^2$ yields a value of R approximately equal to 0.8 times the conventional $2D^2/\lambda$ far-field criterion, with $D = 2\rho_0$.

Table 2.4 Coefficient equations; Incoming and outgoing waves present

$$a^{(3)}_{\substack{e\\o}mn} = -\frac{k\omega\mu}{2\gamma_{mn}}\int_0^{\pi}\int_0^{2\pi}\left(-\bar{m}^{(4)}_{\substack{e\\o}mn} \times \bar{H} + \frac{k}{j\omega\mu}\bar{n}^{(4)}_{\substack{e\\o}mn} \times \bar{E}\right)\cdot \hat{a}_R R^2 \sin\theta \, d\theta \, d\phi$$

$$a^{(4)}_{\substack{e\\o}mn} = \frac{k\omega\mu}{2\gamma_{mn}}\int_0^{\pi}\int_0^{2\pi}\left(-\bar{m}^{(3)}_{\substack{e\\o}mn} \times \bar{H} + \frac{k}{j\omega\mu}\bar{n}^{(3)}_{\substack{e\\o}mn} \times \bar{E}\right)\cdot \hat{a}_R R^2 \sin\theta \, d\theta \, d\phi$$

$$b^{(3)}_{\substack{e\\o}mn} = -\frac{k\omega\mu}{2\gamma_{mn}}\int_0^{\pi}\int_0^{2\pi}\left(-\bar{n}^{(4)}_{\substack{e\\o}mn} \times \bar{H} + \frac{k}{j\omega\mu}\bar{m}^{(4)}_{\substack{e\\o}mn} \times \bar{E}\right)\cdot \hat{a}_R R^2 \sin\theta \, d\theta \, d\phi$$

$$b^{(4)}_{\substack{e\\o}mn} = \frac{k\omega\mu}{2\gamma_{mn}}\int_0^{\pi}\int_0^{2\pi}\left(-\bar{n}^{(3)}_{\substack{e\\o}mn} \times \bar{H} + \frac{k}{j\omega\mu}\vec{m}^{(3)}_{\substack{e\\o}mn} \times \bar{E}\right)\cdot \hat{a}_R R^2 \sin\theta \, d\theta \, d\phi$$

where

$$\gamma_{on} = 4\pi\frac{n(n+1)}{2n+1}$$

$$\gamma_{mn} = 2\pi\frac{n(n+1)}{2n+1}\frac{(n+m)!}{(n-m)!} \qquad m \neq 0.$$

Superscripts (3) and (4) denote incoming and outgoing waves, respectively.

2.3.2 Calculating spherical-wave coefficients

Theoretical relationships

Explicit equations for the coefficients can be obtained by directly applying the orthogonality equations for a spherical surface of radius R (eqns. 2.125-2.127) to the expansion equations 2.121. In general, two solutions for $z_n^{(i)}(kR)$ will be present; considering the case where the solutions are represented as incoming and outgoing waves, the equations given in Table 2.4 may be derived. If only outgoing

Table 2.5 Coefficient equations; Outgoing waves only

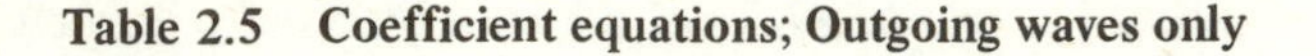

$$a^{(4)}_{\substack{e\\o}mn} = \frac{1}{\gamma_{mn}\, V_n^*(kR)} \int_0^{\pi}\int_0^{2\pi} \bar{n}^{(3)}_{\substack{e\\o}mn} \times \bar{E} \cdot \hat{a}_R \sin\theta \, d\theta \, d\phi$$

$$b^{(4)}_{\substack{e\\o}mn} = \frac{-1}{\gamma_{mn}\, V_n(kR)} \int_0^{\pi}\int_0^{2\pi} \bar{m}^{(3)}_{\substack{e\\o}mn} \times \bar{E} \cdot \hat{a}_R \sin\theta \, d\theta \, d\phi$$

where

$$V_n(kR) \equiv h_n^{(1)}(kR) \frac{1}{kR} \frac{\partial}{\partial R}\left[R h_n^{(2)}(kR)\right]$$

waves are present, the equations simplify to those given in Table 2.5.* It is seen that when two solutions for $z_n^{(i)}(kR)$ are present, it is necessary to know the tangential components of the $\bar{E}$ and $\bar{H}$ fields on the surface; for one solution the $\bar{E}$ field alone is sufficient. Equations similar to those in Table 2.5 can be derived for incoming waves, or in terms of the $\bar{H}$ field alone.

The derivation of the integrals of Table 2.4 requires the surface of integration to be a sphere, but the integrals are in the form of the surface integral of the Lorentz theorem[62] which enables the results to be easily generalised.[49] Since both the spherical waves and the specified field must satisfy Maxwell's equations, if S is an arbitrary closed surface containing a volume V, and $\hat{n}$ is a unit vector normal to S pointing out of V, then

$$\int_S (\bar{E}_1 \times \bar{H}_2 + \bar{H}_1 \times \bar{E}_2) \cdot \hat{n} \, dS = \int_V (-\bar{E}_1 \cdot \bar{J}_2 + \bar{E}_2 \cdot \bar{J}_1) \, dV. \tag{2.132}$$

To correspond with Table 2.4, the subscripts 1 and 2 represent the spherical waves and specified fields, respectively. The current sources of the spherical waves are $\bar{J}_1$ and the sources of the specified field are $\bar{J}_2$. (Eqn. 2.132 could be formulated to also include equivalent magnetic current sources.)

*For expansions that are finite at $R = 0$ it is possible to obtain an equation for

$a^{(1)}_{\substack{e\\o}mn}$ (or $b^{(1)}_{\substack{e\\o}mn}$)

similar to that of Table 2.5, except that

$h_n^{(1)}(kR)$ (or $h_n^{(2)}(kR)$)

is replaced with $j_n(kR)$ in the function $V_n(kR)$. Near roots of $j_n(kR)$ the equation then becomes ill-conditioned and it may be prefereable for this reason to use the equations of Table 2.4

The integrals in Table 2.4 are over a spherical surface which we will call S_1; if S_2 is any other closed surface such that the volume V enclosed by S_1 and S_2 is source free, then eqn. 2.132 shows that

$$\int_{S_1} (\bar{E}_1 \times \bar{H}_2 + \bar{H}_1 \times \bar{E}_2) \cdot \hat{a}_R \, dS = \pm \int_{S_2} (\bar{E}_1 \times \bar{H}_2 + \bar{H}_1 \times \bar{E}_2) \cdot \hat{n} \, dS \tag{2.133}$$

the ± sign depending on whether $\hat{a}_R$ points into or out of the volume enclosed by S_1 and S_2. This means that spherical waves are not only orthogonal over spherical surfaces, but are actually orthogonal over surfaces of general shape; this fact can be very useful in some applications.[49] * Note that the spherical wave sources are at the origin which must therefore be excluded from the volume enclosed by S_1 and S_2.

Eqn. 2.132 can also be used to transform the surface integrals of Table 2.4 into volume integrals, which then give the spherical wave coefficients in terms of the source currents of the specified field. If the surface encloses the origin, in order to eliminate the source currents of the spherical waves from the equations, use is made of the fact that the source currents of corresponding incoming and outgoing waves are identical except for a change in sign. This can be demonstrated by noting that the sum of an incoming and outgoing wave with equal coefficients results in the Bessel function $j_n\ (kR)$, which is finite at the origin; therefore, the resultant source currents must be zero. (The source of each outgoing wave generates a positive outward power flow, but the 'source' of an incoming wave absorbs power and should perhaps be called a 'sink' or load instead). Then it is found that

$$a^{(4)}_{\substack{e\\o}mn} - a^{(3)}_{\substack{e\\o}mn} = \frac{k\omega\mu}{\gamma_{mn}} \int_V \bar{m}^{(1)}_{\substack{e\\o}mn} \cdot \bar{J} \, dV \tag{2.134a}$$

$$b^{(4)}_{\substack{e\\o}mn} - b^{(3)}_{\substack{e\\o}mn} = \frac{k\omega\mu}{\gamma_{mn}} \int_V \bar{n}^{(1)}_{\substack{e\\o}mn} \cdot \bar{J} \, dV \tag{2.134b}$$

where $\bar{J}$ is the source current distribution of the specified field.

Eqn. 2.134 does not enable one to resolve the individual values of outgoing and incoming wave coefficients, but only their difference. This can be explained as follows: a generator could produce $\bar{J}$, or $\bar{J}$ could be currents in a load, and one must have additional prior information to resolve the ambiguity. For example, if one can argue that only outgoing waves are present, then one can set

*Note that this argument applies only to the integrals of Table 2.4, which therefore may be evaluated over an arbitrary closed surface, but not to the integrals of Table 2.5, which must be evaluated over a spherical surface.

$$a^{(3)}_{\substack{e\\o}mn} = 0 \text{ and } b^{(3)}_{\substack{e\\o}mn} = 0$$

in eqn. 2.134. The expansion resulting from this derivation will be valid outside of the volume V which contains the sources of the specified field and the coordinate origin. If it is desired to have an expansion that is valid inside a volume V^1 containing the origin but no sources, which is enclosed in a surface S^1, one can choose the surface of integration of eqn. 2.132 to be S^1 plus a sphere of radius $R \to \infty$. Assuming the specified field satisfies the radiation condition, by selecting outgoing spherical waves, which also satisfy the radiation condition, then the integral over the sphere of radius $R \to \infty$ will be zero. Then the equations for

$$a^{(3)}_{\substack{e\\o}mn} \text{ and } b^{(3)}_{\substack{e\\o}mn}$$

of Table 12.4 can be transformed into integrals over the volume exterior to V^1, enclosed by S^1 and S_∞. Finally, using the fact that

$$a^{(3)}_{\substack{e\\o}mn} = a^{(4)}_{\substack{e\\o}mn} \text{ and } b^{(3)}_{\substack{e\\o}mn} = b^{(4)}_{\substack{e\\o}mn}$$

for this case, the result is equations identical to eqn. 2.134 except that

$$\bar{m}^{(4)}_{\substack{e\\o}mn} \text{ and } \bar{n}^{(4)}_{\substack{e\\o}mn}$$

appear in the integral, and the integral yields the sum of the coefficients instead of the difference, which gives

$$a^{(1)}_{\substack{e\\o}mn} = a^{(3)}_{\substack{e\\o}mn} + a^{(4)}_{\substack{e\\o}mn}, \text{ and } b^{(1)}_{\substack{e\\o}mn} = b^{(3)}_{\substack{e\\o}mn} + b^{(4)}_{\substack{e\\o}mn}.$$

These two types of expansions, one valid outside a finite surface enclosing the sources, and the other valid inside a finite closed surface, have different truncation and convergence behaviour.[49] If the surface is a sphere of radius ρ_o, the first type converges due to a rapid decline in the value of the wave *coefficients* for $n > k\rho_o$, as discussed above. The second type converges due to a decline in the *spherical wave function terms* $j_n\ (kR)$ for $n > kR$, and this depends on the value of R at which the expansion is evaluated rather than the dimensions of the source.

It is important to note that the mode cutoff behaviour discussed previously is true for physically realisable fields only, and that if one attempts to expand non-realisable fields containing discontinuities in amplitude or phase, then the computed coefficients may not converge and serious errors can result.

Numerical considerations

Spherical wave coefficients can be numerically calculated for a source field either by (i) using the integral relationships of eqn. 2.134 or of Tables 2.4 or 2.5 or

(ii) by a point-matching technique. To solve for $K = 2N(N + 2)$ unknown coefficients, K equations can be established by requiring the expansion to equal a given source field at K points in space. (More equations can also be utilized and solved in a least-squares sense.) Although these two approaches appear to be distinct, it has been shown that they are in fact intimately related both theoretically and numerically.[63] The minimum of K data points required to determine the coefficients, if distributed uniformly over the surface of a sphere of radius ρ_0, would each represent an area $\Delta A = 4\pi\rho_0^2/K$. For a maximum mode order of $N = k\rho_0$, $\Delta A \cong \lambda^2/2\pi$; therefore the points are spaced roughly 0.4λ apart.* This corresponds closely to the minimum null spacing discussed previously. If the points were not distributed uniformly (e.g. no points in a region where the source field is very small) it would be possible to have more than one null or one maximum between data points, which intuitively seems bad. In fact, it can be shown for certain special cases that a non-uniform point spacing leads to severe numerical instabilities when the point-matching matrix is inverted.[63]

Numerical integration techniques such as Simpson's rule generally require a spacing of $\lambda/6$ between data points, so numerically integrating the equations in Table 2.4 or 2.5 would require roughly six times as many data points as the point-matching technique. †Several authors have considered this problem in depth, and efficient techniques have been derived which are a great improvement over both numerical integration, or inverting a matrix to obtain the point-matching solution.[63,65–68] Comparisons between various techniques are discussed in the literature.[69,70] Minimising the effects of noise in the specified field data and statistical tests for determining the series truncation part have also been studied;[63,71] this can be very important in case of transforming far field data to the near field, where errors tend to be greatly magnified. The transformation from the near field to the far field is well-conditioned and the effect of errors is less significant, but still may be important.

Change of coordinates

As stated previously, a spherical wave expansion of a specified field changes if the coordinate origin of the expansion is moved. The transformation of the spherical wave coefficients under a translation and/or rotation of coordinates is an important problem in applications such as the near-field measurement problem.[72] When the specified field is given in the far field, a shift in the coordinate origin represents a simple change in the phase pattern of the specified field, and the field may simply be expanded with the altered phase data to obtain the translated coefficients. It is important to note that the size of the minimum sphere enclosing the source changes when the origin moves, and this has an effect on the number of waves needed, as

*For an interesting discussion on the related question of sampling on a planar surface, see Joy and Paris.[64]

†In practice it has been found that as few as 1.5 times as many points can yield good results (WOOD, P.: Private communication, June 1978).

shown in Fig. 2.23. In general, however, it is possible to change coordinates by operating directly on the coefficients, and various aspects of this problem are well described in the literature.[68,73–77]

Special cases

Coefficients of spherical wave expansions for several important special cases are available in the literature as shown in Table 2.6. It is seen that cases 1 and 2 are complementary, the coefficients for the maximum directivity pattern being the first N coefficients of the plane wave expansion.* Cases 3 and 4 are complementary in a similar manner.† The results for cases 5 and 6 may be obtained directly from eqn. 2.134; these cases may in turn be used to obtain the coefficients for an arbitrary plane wave by locating the infinitesimal dipole at $R_o = \infty$ with the desired polarisation and direction of propagation (also see Liang and Lo,[78] and Hizal and Yaza[79] for this case). If the limit is taken such that both A and R_o approach infinity while maintaining

$$-\frac{k^2 A}{4\pi\epsilon}\,\frac{e^{-jkR_o}}{R_o} = 1 \tag{2.135}$$

then the resulting expansion will have unity amplitude and zero phase at $R = 0$, like the case 1 expansion.

Computer program validation

There are several techniques for checking a computer program based on spherical waves. Numerical coefficient calculations can be checked for the special cases given in Table 2.6 by comparing with the exact analytical results. If the power in the specified field can be obtained by direct numerical integration of Poynting's vector, this can be compared with the power represented by the coefficient values; i.e. one numerically evaluates and compares both sides of eqn. 2.128. The power in the coefficients before and after a shift in coordinates can also be compared.

A unique test for spherical wave expansions can be made by comparing the field values of two expansions, which have two different coordinate origins, but are evaluated on the same surface.[80] Direct comparisons with other analytical techniques such as Kirchoff integration or physical optics can provide a rigorous test of both techniques.[48,49,81,82]

*The plane wave expansion may be considered an infinitely directive outgoing wave in the $+z$ direction plus an infinitely directive incoming wave in the $-z$ direction.
†Neglecting near-field effects and assuming optical reflection, in the limit $N \to \infty$ the case 4 coefficients produce a field

$$\bar{E} = 2\hat{a}_x\, e^{-jkf\sec^2\psi/2}$$

in the aperture. The factor of 2 difference between the aperture fields in cases 3 and 4 is due to the use of different Bessel functions.

Table 2.6 Spherical wave coefficients for special cases

Case	Notes	Region of validity	$z_n(kR)$	$a_{\substack{e\\o}mn}$	$b_{\substack{e\\o}mn}$
1 Plane wave propagation in +z direction, $\hat{a}_x$ polarisation (Stratton[52])	$\bar{E} = \hat{a}_x e^{-jkz}$	everywhere	$j_n(kR)$	$(-j)^n \frac{2n+1}{n(n+1)}$ $m = 1$, odd only	$(-j)^{n-1} \frac{2n+1}{n(n+1)}$ $m = 1$, even only
2 Pattern with maximum directivity in +z direction, $\hat{a}_x$ polarisation, $n \leqslant N$ (Harrington[55])	Gain $G = N^2 + 2N$; For $N = kR$ $G = \left(\frac{2\pi R}{\lambda}\right)^2 \frac{4\pi R}{\lambda}$	$R > 0$	$h_n^{(2)}(kR)$	Same as case 1	Same as case 1
3 Plane wave with $\hat{a}_x$ polarisation incident on paraboloid and reflected toward focus at origin (Kennaugh and Ott,[84] Wood[49])	$\bar{E} = \hat{a}_x e^{jkf \sec^2 \psi/2}$ in aperture ψ = paraboloid edge angle f = paraboloid focal length	$R < f$	$j_n(kR)$	Multiply case 1 coefficients by $2jkf \tan\frac{\psi}{2} \frac{P_n^1(\cos\psi)}{n(n+1)}$	Multiply case 1 coefficients by $2jkf \tan\frac{\psi}{2} \frac{P_n^1(\cos\psi)}{n(n+1)}$
4 Paraboloid feed for maximum gain, $n \leqslant N$ (Potter[58])	Aperture efficiency $\eta = \sum_{n=1}^{N} \frac{2n+1}{n^2(n+1)^2}[P_n^1(\cos\psi)]^2$	$R > 0$	$h_n^{(2)}(kR)$	Same as case 3	Same as case 3
5 Infinitesimal dipole at $\bar{R} = \bar{R}_o$ with dipole moment $A\hat{p}_o$ (Jones[51])	$\bar{J} = j\omega A\hat{p}_o\, \delta(\bar{R} - \bar{R}_o)$	$R > R_o$	$h_n^{(2)}(kR)$	$\frac{jk^3 A}{\epsilon\gamma_{mn}} \hat{p}_o \cdot \bar{m}^{(1)}_{\substack{e\\o}mn}(\bar{R}_o)$	$\frac{jk^3 A}{\epsilon\gamma_{mn}} \hat{p}_o \cdot \bar{n}^{(1)}_{\substack{e\\o}mn}(\bar{R}_o)$
6 Same as case 5 (Jones[51])		$R < R_o$	$j_n(kR)$	Same as case 5 except use $\bar{m}^{(4)}_{\substack{e\\o}mn}(\bar{R}_o)$	Same as case 5 except use $\bar{n}^{(4)}_{\substack{e\\o}mn}(\bar{R}_o)$

2.4 References

1 EISENHART, L.P.: 'A treatise on the differential geometry of curves and surfaces'. Dover Publications, Chapter 4, 1909

2 LEE, S.W.: 'Differential geometry for GTD applications.' EM Lab Report No. 77-21, University of Illinois, Urbana, Illinois, Oct. 1977

3 YOUNG, F.A.: 'Toroidal scanning antennas with Gregorian correctors,' Ph.D. Dissertation, University of Southern California, Los Angeles, California, Sept. 1978

4 RUSCH,,W.V.T. and SORENSEN, O.: 'On determining if a specular point exists,' *IEEE Trans.* AP-27, No. 1, Jan 1979, pp.99-101

5 DESCHAMPS, G.A.: 'Ray techniques in electromagnetics', *Proc. IEEE,* **60,** Sept. 1972, pp.1022-1035

6 LEE, S.W. 'Electromagnetic reflection from a conducting surface: geometrical optics solution,' *IEEE Trans.* **AP-23,** No. 2, March 1975, pp.184-191

7 KOUYOUMJIAN, R.G. and PATHAK, P.H.: 'A uniform geometrical theory of diffraction for an edge in a perfectly conducting surface,' *Proc. IEEE,* **62,** Nov. 1974, pp.1448-1461

8 RUSCH, W.V.T. and POTTER, P.D.: 'Analysis of reflector antennas.' Academic Press, March 1970, pp. 131-137

9 RUSCH, W.V.T.: 'Reflector antennas', *in* 'Topics in applied physics; Vol. 3: Numerical and asymptotic techniques in electromagnetics', Springer Verlag, Heidelberg, 1975, pp. 217-256

10 GALINDO, V., and MITTRA, R.: 'A new series representation for the radiation integral with applications to reflector antennas', *IEEE Trans.,* **AP-25,** Sept. 1977

11 ALLEN, C.C.: 'Numerical integration methods for antenna pattern calculations', *IRE Trans.* Spec. Suppl. **AP-7,** 1959, pp.5387-5401

12 RICHMOND, J.H.: 'The numerical evaluation of radiation integrals', *IRE Trans.,* **AP-9,** 1961, p.358

13 COOLEY, J.W., and TUKEY, J.W.: 'An algorithm for the machine calculation of complex Fourier series', *Math. Comp.,* 1965, pp.297-301

14 MILLER, E.K., and BURKE, G.J.: 'Numerical integration methods', *IEEE Trans.,* **AP-17,** Sept. 1969, pp.669-672

15 RALSTON, A.: 'A first course in numerical analysis', McGraw-Hill, 1965

16 RALSTON, A. and WILF, S. (ed.): 'Mathematical methods for digital computers', Vol. II. Wiley, 1968

17 ABRAMOWITZ, M., and STEGUM, I.A.: 'Handbook of mathematical functions, Dover, 1965

18 FOX, L.: 'Romberg integration for a class of singular integrands', *Computer J.,* **10,** 1976, pp.87-93

19 SCHJAER-JACOBSEN, H.: 'Computer programs for one- and two-dimensional Romberg integration of complex functions', Laboratory of Electromagnetic Theory, Rpt. D187, Technical University of Denmark, Denmark, 1973

20 SHIRAZI, M., and RUDGE, A.W.: 'Numerical evaluation of two-dimensional integrals in antenna problems', DM No. 452, Dept. of Electronic and Electrical Engineering, University of Birmingham, England, 1974

21 RUSCH, W.V.T.: 'Antenna notes, Vol. II', NB84b, Electromagnetics Institute, Technical University of Denmark, August 1974, Appendix

22 LESSOW, H.A., RUSCH, W.V.T., and SCHJAER-JACOBSEN, H.: 'On numerical evaluation of two-dimensional phase integrals', *IEEE Trans.,* **AP-23,** Sept. 1975, pp.714-718

23 CHUGH, R.K., and SHAFAI, L.: 'Comparison of Romberg and Gauss methods for numerical evaluation of two-dimensional phase integrals', *IEEE Trans.* **AP-25,** July 1977, pp.581-583

24 IBM System/360 Scientific Subroutine Package, Version 3, Programmers Manual, 5th Edition (GH20-0205-4), Aug. 1970
25 SHIRAZI, M, and RUDGE, A.W.: *Op. Cit.*
26 SCHJAER-JACOBSEN, H.: *Op. Cit.*
27 SHIRAZI, M., and RUDGE, A.W.: *Op. Cit.*
28 SCHJAER-JACOBSEN, H.: *Op. Cit.*
29 LUDWIG, A.C.: 'Computation of radiation patterns involving numerical data integration', *IEEE Trans.* **AP-16,** 1968, pp.767-769
30 ALLEN, C.C.: 'Final Report on the Study of Gain-to-Noise Temperature Improvement for Cassegrain Antennas', General Electric Co., Schenectady, New York, 1967
31 RUSCH, W.V.T.: *Op. Cit.*
32 RUSCH, W.V.T.: 'Antennas notes', NB84b, Vol. II, Electromagnetics Institute, Technical University of Denmark, Aug. 1974, pp.32-58
33 FELSEN, L.B., and MARCUVITZ, N.: 'Radiation and scattering of waves', Prentice Hall, 1973, Chapter 4
34 RUSCH, W.V.T.: 'Physical-optics diffraction coefficients for a paraboloid', *Electron. Lett.,* **10,** No. 17, 22 Aug. 1974
35 KNOP, C.M.: 'An extension of Rusch's asymptotic physical optics diffraction theory of a paraboloid antenna', *IEEE Trans.,* **AP-23,** Sept. 1974, pp.741-743
36 KNOP, C.M., and OSTERTAG, E.L.: 'A note on the asymptotic physical optics solution to the scattered fields from a paraboloidal reflector', *IEEE Trans.,* **AP-25,** July 1977, pp.531-534
37 INGERSON, P.G., and WONG, W.C.: 'The analysis of deployable umbrella parabolic reflectors', *IEEE Trans.* **AP-20,** July 1972
38 KONTOROVICH, M.I.: 'Average boundary conditions at the surface of a grating with square mesh', *Radiotekhnika i Electronika,* 1963
39 ASTRAKHAN, M.I.: 'Reflecting and screening properties of plane wire grids', *Telecommun. and Radio Enging.,* **23,** (1), 1968
40 CHEN, C.C.: Private Communication
41 KELLER, J.B.: 'Diffraction by an aperture', *J. Opt. Soc. Am.,* **52,** 1962, pp.116-130
42 KOUYOUMJIAN, R.G., and PATHAK, P.H.: 'A uniform geometrical theory of diffraction for an edge in a perfectly conducting surface', *IEEE Proc.* **62,** Nov. 1974, pp. 1448-1461
43 LEE, S.W., and DESCHAMPS, G.A.: 'A uniform theory of electromagnetic diffraction by a curved wedge', *IEEE Trans.,* **AP-24,** 1967, pp.25-34
44 KELLER, J.B.: 'Diffraction by an aperture', *J. Appl. Phys.,* **28,** 1957, pp.426-444
45 LEE, S.W.: 'Uniform asymptotic theory of electromagnetic edge diffraction: A Review', in *Electromagnetic scattering,* P.L.E. USLENGHI, (ed.). Academic Press, 1978, pp.87-119
46 HWANG, Y.M., and KOUYOUMJIAN, R.G.: 'Dyadic deffraction coefficient for an Electromagnetic wave which is rapidly-varying at an edge', URSI 1974 Annual Meeting, Boulder, Colorado
47 LUDWIG, A.C.: 'Spherical-wave theory', Technical Report NB120, Technical University of Denmark, Lyngby, Denmark, December 1978
48 WOOD, P.J.: 'Spherical harmonic expansions in near-field aerial problems', *Electron. Lett.,* **6,** 1970, pp.535-536
49 WOOD, P.J.: 'Spherical waves in antenna problems, *Marconi Rev.,* **34,** 1971, pp.149-172
50 CALDERON, A.P.: 'The multipole expansion of radiation fields', *J. Rational Mechanics and Analysis,* **3,** Sept. 1954, pp.523-536
51 JONES, D.S.: 'The theory of electromagnetism', MacMillan, New York, Sec. 8.16, 1964
52 STRATTON, J.A.: 'Electromagnetic theory', McGraw-Hill, New York, 1941, Chap VII
53 TAYLOR, T.T.: 'A discussion of the maximum directivity of an antenna', *IRE Proc.,* **36,** Sept. 1948, p.1135

54 CHU, L.J.: 'Physical limitations on antennas', *J. Appl. Phys.,* **19,** Dec. 1948, pp.1163-1175

55 HARRINGTON, R.F.: 'Effect of antenna size on gain bandwidth and efficiency', *J. Res. NBS,* Sec. B, **64D,** June 1960, pp.1-12

56 HARRINGTON, R.F.: 'Time-harmonic electromagnetic fields', McGraw Hill, 1961

57 COLLIN, R.E., and ROTHSCHILD, S.: 'Evaluation of antenna Q', *IEEE Trans.,* **AP,** Jan. 1964, pp.23-27

58 POTTER, P.D.: 'Application of spherical wave thoery to Cassegrainian-fed paraboloids', *IEEE Trans.* **AP-15,** Nov. 1967, pp.727-736

59 LUDWIG, A.C.: 'Near-field far-field transformations using spherical wave expansions', *IEEE Trans.,* **Ap-19,** 1971, pp.214-220

60 WOOD, P.J.: "Comment on 'Numerical check on the accuracy of spherical wave expansions'," *Electron. Lett.,* **8,** May 4, 1972, pp.227-228

61 OJEBA, E.B., and WALTER, C.H.: 'On the cylindrical and spherical wave spectral content of radiated electromagnetic fields', *IEEE Trans.,* **AP-27,** Sept. 1979, pp.634-639

62 RUMSEY, V.H.: 'Reaction concept in electromagnetic theory', *Phys. Rev.,* **94,** June 1954, pp.1483-91

63 LUDWIG, A.C.: 'Calculation of orthogonal function expansions from imperfect data', Technical Univ. of Denmark, Lab. of Electromagnetic Theory, Report R-102, 1972

64 JOY, E.B., and PARIS, D.T.: 'Spatial sampling and filtering in near field measurements', *IEE Trans.,* **AP-20,** May 1972, pp.253-261

65 LUDWIG, A.C.: 'Computation of radiation patterns involving numerical double integration', *IEE Trans.,* **AP-16,** Nov. 1968, pp.767-769

66 RICARDI, L.J., and BURROWS, M.L.: 'A recurrence technique for expanding a function in spherical harmonics', *IEEE Trans.,* **C-21,** June 1972, pp.583-585

67 WACKER, P.F.: 'Non-planar near-field measurements: Spherical scanning', Report NBSIR 75-809, Electromagnetics Division, National Bureau of Standards, Boulder, Colorado, June 1975

68 LEWIS, R.L.: 'Highly efficient processing for near-field spherical scanning data reduction', AP-S International Symposium, October 1976, pp.251-254

69 JENSEN, F., and LARSEN, F.H.; 'Spherical near-field far-field techniques', 6th European Microwave Conference Proceedings, Sept. 1976, pp.98-102

70 WOOD, P.J.: 'The prediction of antenna characteristics from spherical near field measurements, Part I, Theory', *Marconi Rev.,* **XL,** 1977, pp.42-68

71 PARKINSON, R.G., and KHARADLY, M.M.Z.: 'Optimum determination of the scattered-field coefficients from near-field measurements', *Proc. IEEE,* Sept. 1970, pp.1396-97

72 JENSEN, R., and LARSEN, F.H.: 'Spherical near-field technique', Digest, International Symposium, Antennas and Propagation Society, June 1977, pp.377-381

73 CRUZAN, O.R.: 'Translational addition theorems for spherical vector wave functions', *J. Quart. Appl. Math.,* **20,** 1962, pp.33-40

74 STEIN, S.: 'Addition theorems for spherical wave functions', *J. Quart. Appl. Math.,* **19,** 1961, pp.15-24

75 WILCOX, C.H.: 'An expansion theorem for electromagnetic fields', *Comm. Pure Appl. Math.,* **9,** 1956, pp.115-34

76 BRUNING, J.H., and LO, Y.T.: 'Multiple scattering of EM waves by spheres, Pt. 1: Multipole expansion and ray-optical solutions', *IEEE Trans.,* **AP-19,** 1971, pp.378-390

77 EDMUNDS, A.R.: 'Angular momentum in quantum mechanics', Princeton Univ. Press, 1957

78 LIANG, C., and LO, Y.T.: 'Scattering by two spheres', *Radio Sci.* **2** (New Series), Dec. 1976, pp.1481-95

79 HIZAL, A., and YAZA, Z.: 'Scattering by perfectly conducting rotational bodies of arbitrary form excited by an obliquely incident plane wave or by a linear antenna',

Proc. IEE, **120,** Feb. 1973
80 LUDWIG, A.C.: 'Numerical check on the accuracy of spherical-wave expansions, *Electron. Lett.,* **8,** March 1972, pp.202-203
81 CLARRICOATS, P.J.B., and SAHA, P.K.: 'Radiation from wide-flare angle scalar horns',. *Electron. Lett.,* **5,** 1969, pp.376-378
82 RUSCH, W.V.T., and POTTER, P.D.: 'Analysis of reflector antennas', Academic Press, 1970
83 MIE, *Ann. Physik,* **25,** 1908, p.377
84 KENNAUGH, E.M., and OTT, R.H.: 'Fields in the focal region of a parabolic receiving antenna', *IEEE Trans.,* **AP-12,** May 1964, pp.376-377
85 HIZAL, A., and MARINCIC, A.: 'New rigorous formulation of electromagnetic scattering from perfectly conducting bodies of arbitrary shape', *Proc. IEEE,* **117,** 1970, pp. 1639-1647
86 KING, R.W.P., and HARRISON, C.W.: 'Scattering by imperfectly conducting spheres', *IEEE Trans.,* **AP-19,** March 1971, pp.197-207
87 MONTGOMERY, C.G., DICKE, R.H., and PURCELL, E.M.: 'Principles of microwave circuits', McGraw Hill, New York 1948
88 LUDWIG, A.C.: 'Radiation pattern synthesis for circular aperture horn antennas', *IEEE Trans.,* **AP-14,** July 1966, pp.434-440
89 WASYLKIWSKYJ, W., and KAHN, W.K.: 'Scattering properties and mutual coupling of antennas with prescribed radiation pattern', *IEEE Trans.,* **AP-18,** 1970, pp.741-752
90 RICARDI, L.J.: 'Synthesis of the fields of a transverse feed for a spherical reflector', *IEEE Trans.,* **AP-19,** 1971, pp.310-320
91 BROWN, J.D., and JULL, E.V.: 'The prediction of aerial radiation patterns from near-field measurements', *Proc. IEE,* **108,** Pt. B, 1961, 635-644
92 JAMES, J.R., and LONGDON, L.: 'Prediction of arbitrary electromagnetic fields from measured data', *Alta Freq.,* **38,** Speciale, May 1969, pp.286-290
93 JENSEN, F.: 'Electromagnetic near-field far-field correlations', Ph.D. dissertation LD 15, Lab. of Electromagnetic Theory, Tech. Univ. of Denmark, Lyngby, July 1970
94 JENSEN, F.: 'On the probe compensation for near-field measurements on a sphere', *Archiv. für Elektronik und Übertragungstechnik,* **29,** July/August 1975, pp.305-308
95 HANSEN, J.E., and JENSEN, F.: 'Near-field measurements using directive antennas', IEEE G-AP Int. Symp., 1970, pp.284-287
96 LARSEN, F.H.: 'Probe correction of spherical near-field measurements', *Elec. Lett.,* **13,** July 1977, pp.393-395
97 CLARRICOATS, P.J.B., and OLVER, A.D.: 'Near-field radiation characteristics of corrugated horns', *Electron. Lett.,* **7,** 1971, pp.446-448
98 WOOD, P.J.: 'Field correlation diffraction theory of the symmetrical Cassegrain antenna', *IEEE Trans.,* **AP-19,** March 1971, pp.191-197
99 WOOD, P.J.: 'Reflector profiles for the pencil-beam Cassegrain antenna', *Marconi Rev.,* 1972, pp.121-138
100 LUDWIG, A.C., and BRUNSTEIN, S.A.: 'A transform-pair relationship between incident and scattered fields from an arbitrary reflector', *Radio Sci.,* Sept/Oct. 1978, pp.785-788
101 LUDWIG, A.C.: 'Calculation of scattered patterns from asymmetrical reflectors', Ph.D. dissertation, Univ. of Southern California, Los Angeles, 1969. Also Jet Propulsion Lab., California Inst. of Tech., Pasadena, Tech. Rep. 32-1430, February 1970
102 LUDWIG, A.C., and NORMAN, R.A.: 'A new method for calculating correction feactors for near-field gain measurements', *IEEE Trans.,* **AP-21,** Sept. 1973, pp.623-628
103 POTTER, P.D., and LUDWIG, A.C.: 'Applications of spherical wave expansions to near-field problems', 1975 International IEEE/AP-S Symposium Digest, pp. 158-160
104 BOERNER, W.M., and VANDENBERGHE, F.H.: 'Determination of the electrical radius *ka* of a spherical scatterer from the scattered field', *Can. J. Phys.,* **49,** 1971, pp.1507-1535

Chapter 3

Quasi-Optical antenna design and application

W.V.T. Rusch, T.S. Chu, A.R. Dion, P.A. Jensen, A.W. Rudge, W.C. Wong

3.1 Prime-focus paraboloidal reflectors (Contribution by W.V.T. Rusch)

The symmetrical, prime-focus-fed paraboloid is the most commonly used reflector for medium and high-gain, pencil-beam applications. This reflector is relatively straightforward to analyse, design and fabricate. An extensive literature exists to describe its properties.[1-5]

3.1.1 Primary radiation characteristics

As indicated in Section 2.2.4, the surface-current and aperture-field formulations of physical optics are nearly identical for a focused paraboloid when the observation point is near the the boresight axis. Consequently, consider an aperture field of the form

$$\bar{E} = E_0 \hat{a}_x \, [B + (1-B)(1-\rho^2/a^2)^p] \tag{3.1a}$$

$$\bar{H} = (E_0/\eta)\hat{a}_y \, [B + (1-B)(1-\rho^2/a^2)^p] \tag{3.1b}$$

where a is the radius of the circular aperture, ρ is the radial aperture coordinate, and B is the 'edge taper' of this axially symmetric 'polynomial-on-a-pedestal' distribution. The total power passing through this aperture is

$$P_T = (\pi a^2/2\eta)|E_0|^2 \, [B^2 + 2B(1-B)/(p+1) + (1-B)^2/(2p+1)] \tag{3.2}$$

and, from eqn. 2.67 the normalised pattern of the radiated field is

$$F(\theta) = BJ_1\,(ka \sin\theta)/(ka \sin\theta)$$
$$+ \, [(1-B)/2(p+1)] \, [2^{p+1}\,(p+1)!J_{p+1}\,(ka \sin\theta)/(ka \sin\theta)^{p+1}]. \tag{3.3}$$

The aperture illumination efficiency from eqns. 3.2 and 3.3 is then

$$\eta_{AP} = [B+(1-B)/(p+1]^2/[B^2+2B(1-B)/(p+1)+(1-B)^2/(2p+1)] \qquad (3.4)$$

This value is plotted in Fig. 3.1 for $p = 0, 1, 2$ and B varying from 0 to 30 dB.

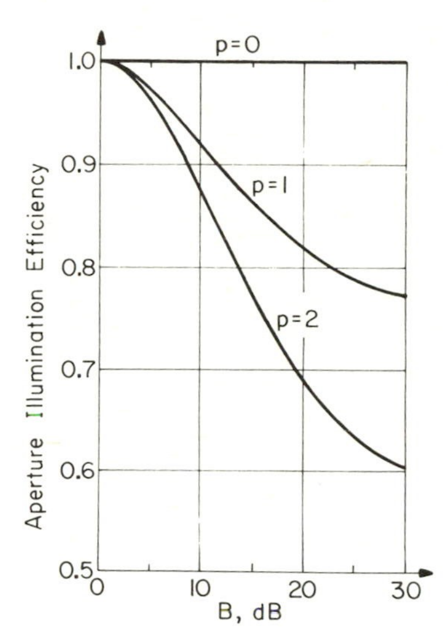

Fig. 3.1 *Aperture illumination efficiency of circular aperture*

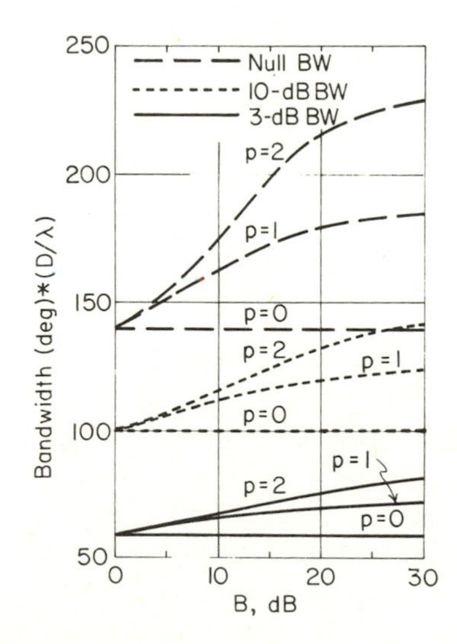

Fig. 3.2 *Beamwidths of circular aperture*

The beamwidths of the field radiated by the circular aperture are plotted in Fig. 3.2. The three sets of curves correspond to the angular separation between the 3-dB points on the main beam (solid curve), between the 10-dB points (short dashes), and between the first nulls on either side of the main beam (long dashes). The numerical values have approximated sin θ by θ. To avoid this approximation, the following transformation may be used

$$\text{Actual BW (degrees)} = (360/\pi)\sin^{-1}\ [(\text{value plotted})\ \pi\lambda/360D]. \tag{3.5}$$

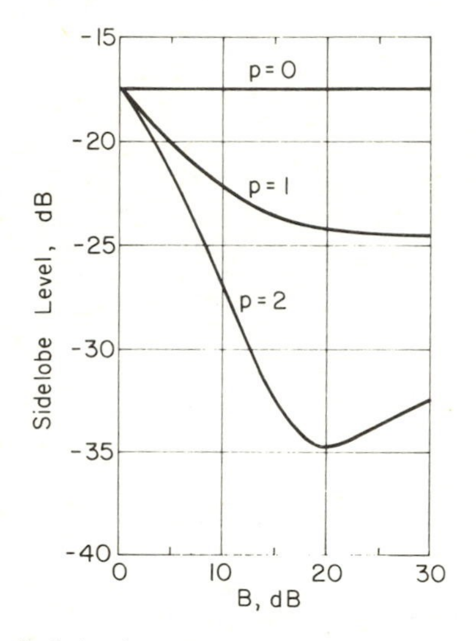

Fig. 3.3 *Sidelobe level of circular aperture*

The levels of the first sidelobe adjacent to the main beam relative to the peak of the main beam are plotted in Fig. 3.3. When $p = 0$ or $B = 1$ (i.e. 0 dB), the pattern of eqn. 3.3 becomes the classical Airy function and the first sidelobe has a level of $-17{\cdot}57$ dB.

3.1.2 Effects of random surface errors

Random deviations of the reflector profile from the ideal paraboloidal shape will cause its radiation pattern to deteriorate, primarily by a decrease in the antenna's aperture efficiency, and an increase in its sidelobe envelope. The antenna community is indebted to J. Ruze for his pioneering work in the area of aperture phase errors,[6,7] based on the aperture formulation of physical optics (eqn. 2.67). This work has subsequently been extended to random amplitude errors by Schanda.[8]

The statistical model used by Ruze assumed that phase errors at a point $\bar{r}$ have mean of zero and belong to a Gaussian population of rms deviation δ about this mean. Furthermore, the phase errors of neighbouring points are assumed to be correlated with a degree of correlation that decreases with increasing separation according to $\exp(-|\bar{r} - \bar{r}'|^2/c^2)$, where c is the 'correlation radius', such that

(i) c is constant over the aperture
(ii) c is much less than the aperture diameter
(iii) the aperture illumination function is substantially constant over distances of the order of c.

For a shallow reflector

$$\delta \cong 2(2\pi\epsilon/\lambda) \text{ radians} \tag{3.6}$$

where ϵ is the RMS surface error, and the additional factor of 2 accounts for the two-way path incurred by the reflected ray. As a rough rule of thumb, the RMS surface error has been found to be approximately one-third of the peak error for large structures.[7] For reflectors with substantial curvature (smaller F/D values) the effective ϵ along a ray is related to Δz (or Δn), the actual surface distortion measured in the axial (or normal) directions by

$$\epsilon = \Delta z/[1 + (r/2F)^2] \tag{3.7}$$

$$\epsilon = \Delta n/\sqrt{1 + (r/2F)^2} \tag{3.8}$$

where r is the radial coordinate in the aperture. These formulas can be used to generate correction factors to account for reflector curvature.[7]

The gain function resulting from the above assumptions can be shown to be[7]

$$G(\theta,\varphi) = G_0(\theta,\varphi)e^{-\delta^2} + \left(\frac{2\pi c}{\lambda}\right)^2 e^{-\delta^2} \sum_{n=1}^{\infty} \frac{(\delta^2)^n}{n!n} e^{-(\pi c \sin\theta/\lambda)^2/n} \tag{3.9}$$

where $G_0(\theta,\varphi)$ is the no-error gain function and (θ,φ) are the conventional polar angles. Thus the no-error gain is reduced by an exponential tolerance factor to account for power scattered into the sidelobes by the phase errors. The complete radiation pattern is the power sum of a reduced pencil-beam diffraction pattern, and a more diffuse scattering pattern containing the 'lost' power.

The reduction of axial gain ($\theta = 0$) thus becomes:

$$G(0,0) = \eta\left(\frac{\pi D}{\lambda}\right)^2 e^{-\delta^2} \left\{1 + \frac{1}{\eta}\left(\frac{2c}{D}\right)^2 \sum_{n=1}^{\infty} \frac{(\delta^2)^n}{n!n}\right\} \tag{3.10}$$

where η is the aperture efficiency. For small correlation regions and reasonable tolerance losses, the second term can be neglected, leaving the standard equation for loss:

$$G(0,0) \cong \eta \left(\frac{\pi D}{\lambda}\right)^2 e^{-(4\pi\epsilon/\lambda)^2}. \tag{3.11}$$

Contours of constant value for the exponential tolerance-loss factor in eqn. 3.11 are plotted in Fig. 3.4 as a function of frequency versus ϵ. Thus, for an assumed

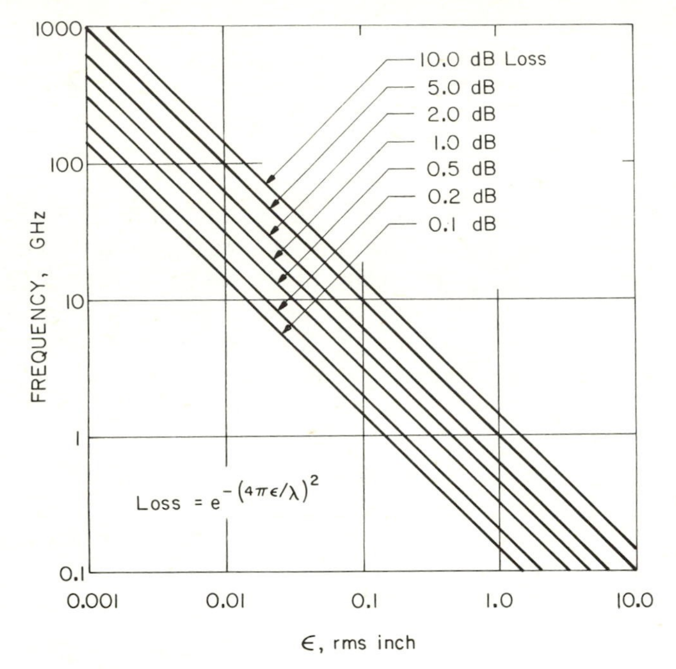

Fig. 3.4 *Cross-plot of tolerance-loss factor*

constant aperture efficiency, η, the axial gain increases with the square of the frequency until the tolerance factor becomes significant. Maximum gain is thus achieved when

$$\lambda_{max} = 4\pi\epsilon \tag{3.12}$$

which is 4·3 dB below the error-free gain at that wavelength:

$$G_{max} \cong \frac{\eta}{43}\left(\frac{D}{\epsilon}\right)^2 \tag{3.13}$$

Ruze has also pointed out that eqn. 3.9 is also valid in the focal plane of a receiving antenna and can thus be used to determine the focal-spot size due to surface errors or atmospheric inhomogeneities.

Eqn. 3.9 can be written

$$G(\theta,\varphi) = G_0(\theta,\varphi)e^{-\delta^2} + \left(\frac{\pi D}{\lambda}\right)^2 e^{-\delta^2}\left(\frac{2c}{D}\right)^2 f(\delta, c \sin\theta) \tag{3.14}$$

where the pattern factor of the diffuse term is

$$f(\delta, c\sin\theta) = \sum_{n=1}^{\infty} \frac{(\delta^2)^n}{n!n} e^{-(\pi c \sin\theta/\lambda)^2/n}. \tag{3.15}$$

This diffuse pattern factor is plotted in Fig. 3.5. The resultant radiation pattern based on eqn. 3.14 for a circular aperture with a 12 dB taper is plotted in Fig. 3.6

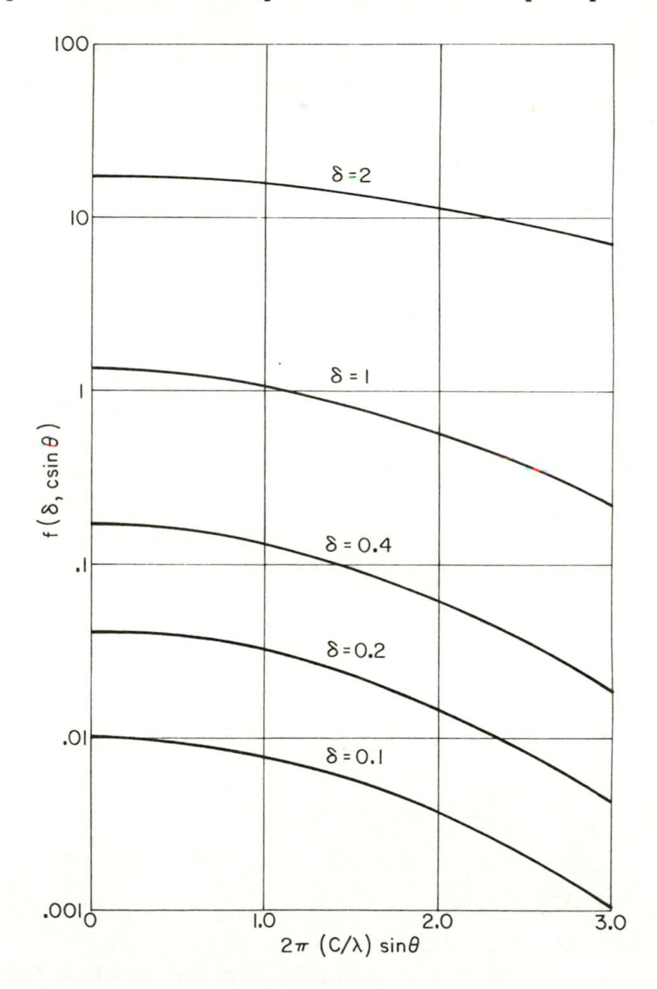

Fig. 3.5 *Diffuse pattern factor due to tolerance loss*

for $\delta = 0{\cdot}0$, $0{\cdot}4$ and $1{\cdot}0$ radians. As the error increases, the axial peak is degraded by the exponential tolerance-loss factor: the nulls are filled; and the sidelobe levels approach the asymptotic value dictated by the second term of eqn. 3.14.

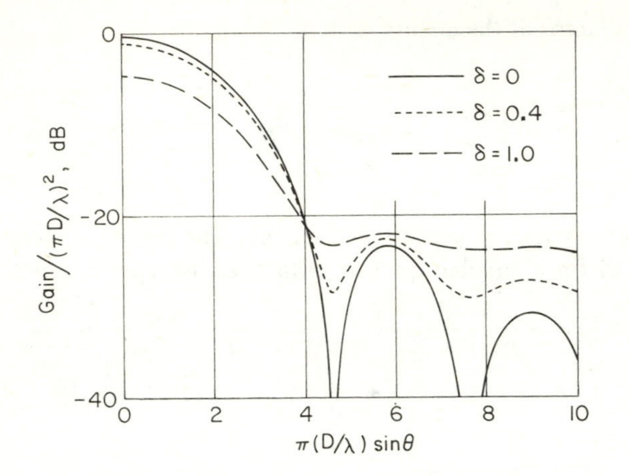

Fig. 3.6 *Radiation pattern as RMS phase-error for B = 12 dB; p=1*

3.1.3 Feed defocusing

Axial defocusing

If a y-directed, infinitesimal electric dipole is placed on the symmetry axis of a paraboloid, a distance d in front of the focus (Fig. 3.7), for $d \ll f$ the boresight field is[9]

$$\bar{E}_s(R,0.0) = -j\frac{k^2 p}{4\pi\epsilon_0}(2kf)\,\hat{a}_y\,\frac{\exp(-jkR)}{R}\sin^2(\theta_0/2)$$

$$\cdot\ \exp(-j2kf)\left\{\exp(-jkd\cos^2\theta_0/2\right\}\left\{\sin X/X\right\} \tag{3.16}$$

where

$$X = \frac{2\pi\,(d/\lambda)}{1+(4F/D)^2}. \tag{3.17}$$

The primary effect of displacing the source axially from the focus is to create a quadratic phase error across the aperture.

The magnitude of the field in eqn. 3.16 is plotted in Fig. 3.8 together with corresponding curves for more accurate approximations of the phase in deriving eqn. 3.16, and the curve obtained by generating eqn. 3.16 numerically without any approximations assuming $d \ll f$. The sin X/X natures of the curves are evident.

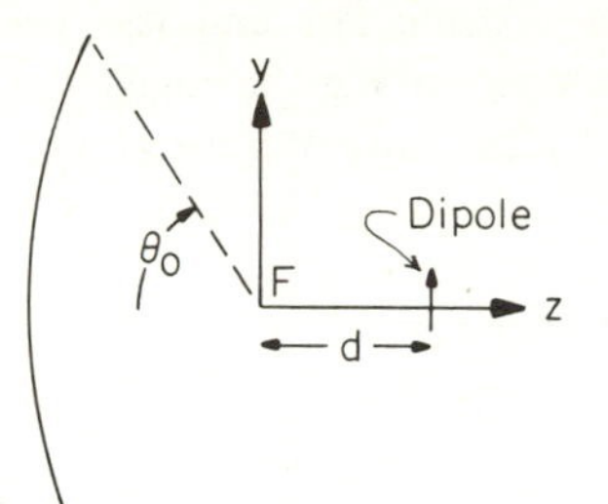

Fig. 3.7 *Defocused Dipole Feed*

For $X = \pm\pi, \pm 2\pi$, etc., the radiated field on axis becomes virtually zero. At these points

$$\frac{d}{\lambda} = \pm \frac{m}{2}\left[1 + \left(\frac{4F}{D}\right)^2\right], m = 1,2,3\ldots \tag{3.18}$$

and the main beam has widened and, in any azimuthal plane, the field on either side of the axis is greater than the value on axis so that the beam appears bifurcated.

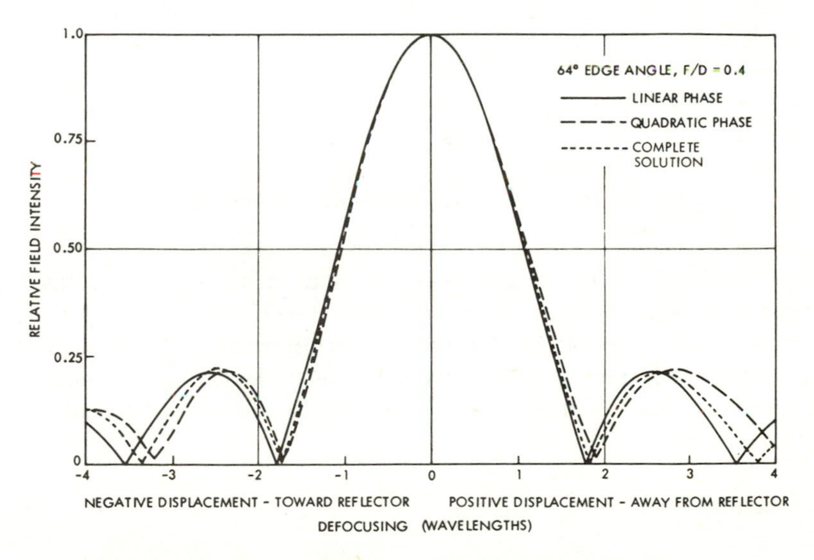

Fig. 3.8 *Relative boresight field for defocused infinitesimal electric dipole*

Defocusing curves for an entire family of feed functions are plotted in Fig. 3.9. Since a linearisation was made involving a square-root factor in the phase function, the curves are identical for positive and negative defocusing. The curves are normalised by the total feed power radiated so that spillover is included. Consequently, for example, cos θ is optimum for $F/D = 0{\cdot}4$, etc. The less tapered feeds exhibit deep minima. The dipole curve (not shown) has exact nulls at the same positions. As the taper is increased, the minima become less and less pronounced. At the deep minima in the less tapered defocusing curves, the angular beam exhibits bifurcation. Consequently, these less-tapered feeds are not generally suited to beam-broadening applications.

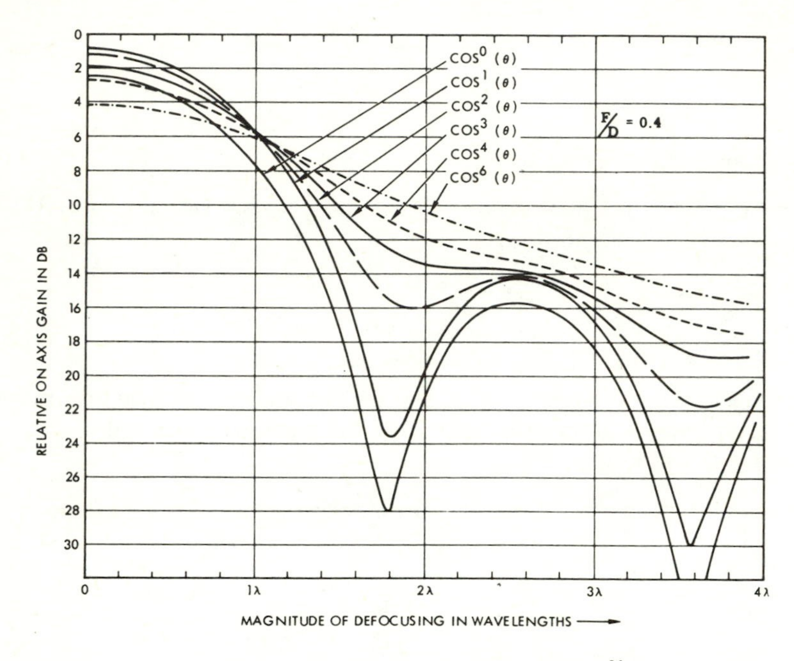

Fig. 3.9 *Relative boresight gain as a function of defocusing for $\cos^N\theta$ illumination*

A family of angular patterns for a cos θ feed is plotted in Fig. 3.10 for different values of defocusing. The beam breaks up and becomes poorly defined with increased defocusing. At a defocusing distance of $1{\cdot}78\lambda$, where the deep null occurs in the defocusing curves of Fig. 3.9, the beam is bifurcated by 10 dB or so on axis. On the other hand, the angular patterns for a $\cos^4\theta$ feed are plotted in Fig. 3.11. Bifurcation does not occur, and the beam remains well defined for all values of defocusing shown.

Lateral defocusing

Lateral displacement of the feed of a paraboloidal antenna causes the pencil beam to scan on the opposite side of the reflector axis. The ratio of beam scan angle to feed scan angle (feed squint) is defined as the *beam deviation factor,* generally of the order of 0·7 to 0·9. The *Petzval surface,* a term from classical optics, is the surface of best focus in an optical system in the absence of astigmatism. For a single mirror the radius of curvature of the Petzval surface is one-half the radius of curva-

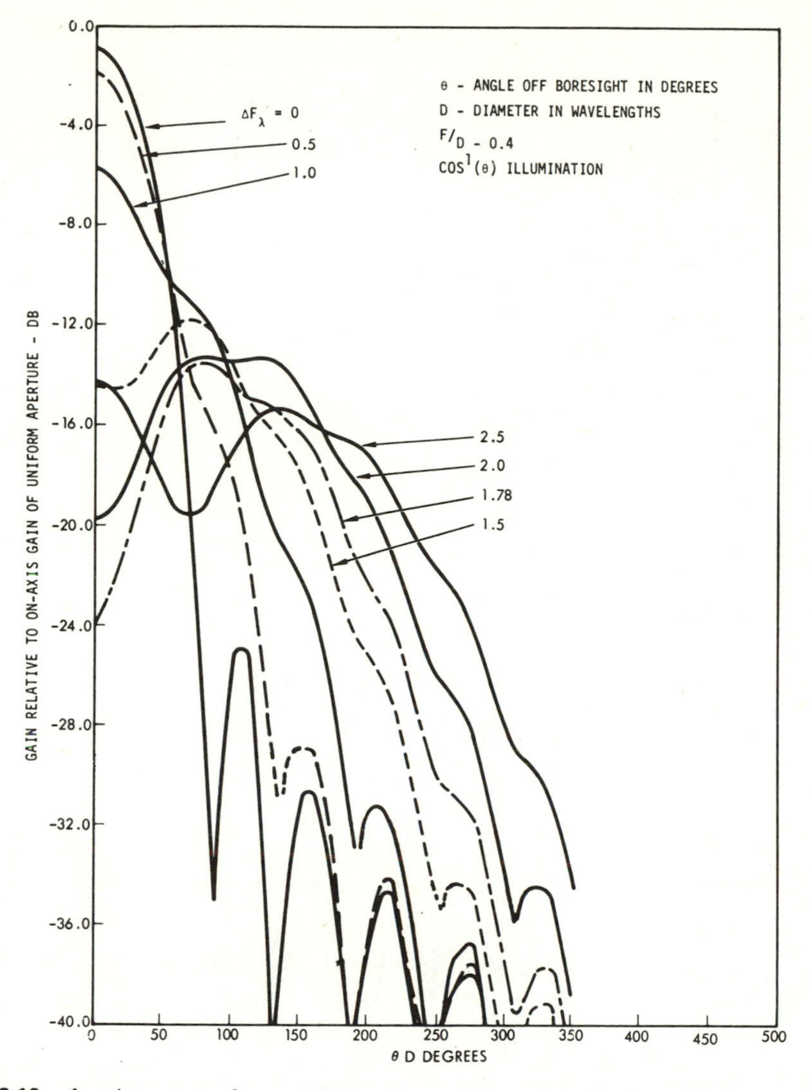

Fig. 3.10 *Angular patterns for cos θ illumination with variable axial defocusing*

ture of the mirror. In the primary microwave reference to the subject, the Petval surface of a paraboloidal mirror is derived by Ruze[10] to be another paraboloid of half the focal length, tangent to the focal plane at the focus, and described by the equation

$$\rho_{Petz}^2 = 2Fz_{Petz} \tag{3.19}$$

where ρ_{Petz} is the distance from the reflector axis and z_{Petz} is the axial distance beyond the focus.

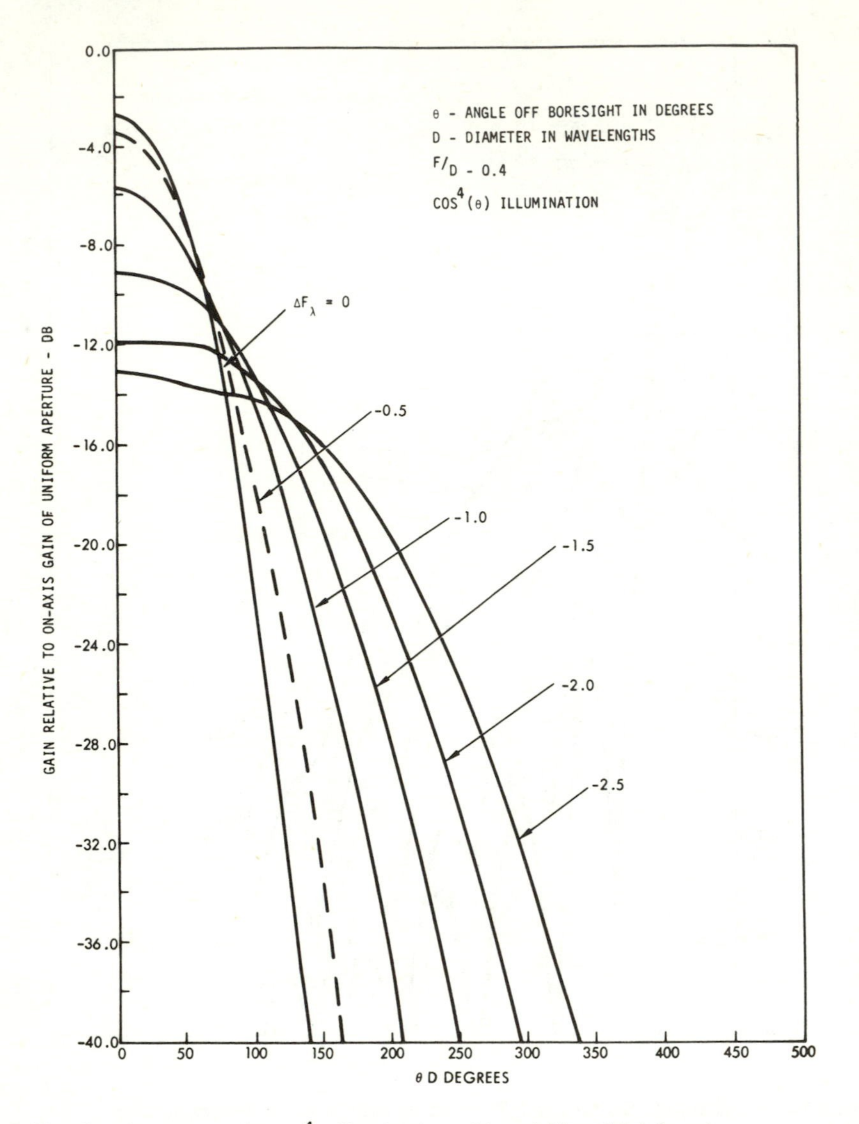

Fig. 3.11 *Angular patterns for $\cos^4\theta$ illumination with variable axial defocusing*

To illustrate the scanning properties of a microwave scanning reflector, a half-wavelength radius TE_{11} open-ended circular waveguide feed was scanned in the focal region of a 25-wavelength reflector with F/D = 0·4.[11] The feed was translated in such a manner that its beam peak remained parallel to the reflector axis. By trial and error, the contour to achieve maximum scan gain for a series of scan angles to 10 HPBWs was determined. This surface was plotted in Fig. 3.12 (solid curve) together with the corresponding Petzval surface (dashed curve). The scan plane patterns are plotted in Fig. 3.13 for focused conditions (A), one-HPBW scan (B), two-HPBW scan (C), four-HPBW scan (D), six-HPBW scan (E), eight-HPBW scan (F), and ten-HPBW scan (G). It is seen that:

(*a*) the gain drops with scan (scan loss)
(*b*) the beam broadens with scan
(*c*) the beam scan is a decreasing fraction of the feed squint
(*d*) the sidelobe on the axis side (coma lobe) increases
(*e*) the sidelobe on the other side decreases, changes sign, and merges with the main beam and second sidelobe causing additional beam broadening.

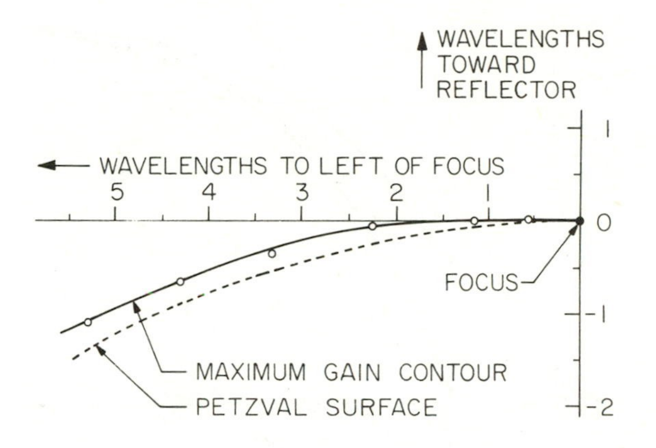

Fig. 3.12 *Maximum gain contour of paraboloid illuminated by laterally defocused TE_{11} feed for an H-plane scan where D = 25 wavelengths, F/D = 0.4, and a/λ = 0.5*

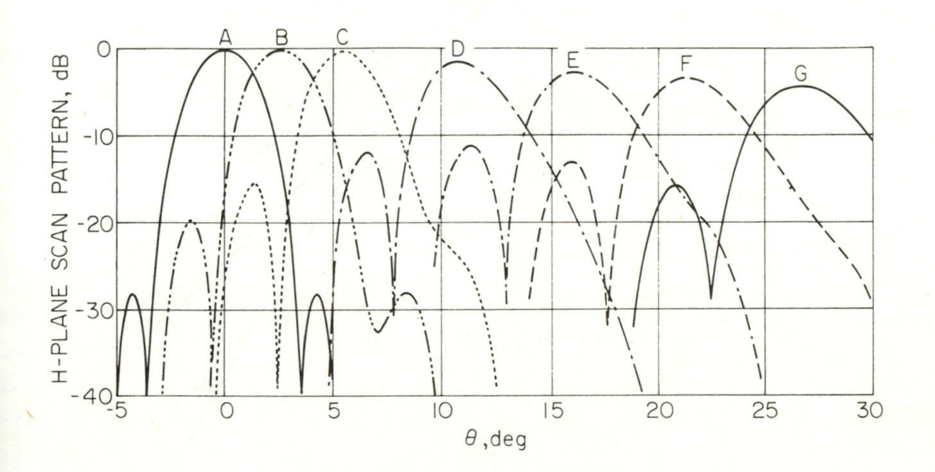

Fig. 3.13 *H-plane pattern of paraboloid illuminated by laterally defocused TE_{11} feed*

Under the condition of small feed-displacements, Ruze developed an expression for the beam deviation factor:

$$\mathrm{BDF} = \frac{\displaystyle\int_0^a \frac{f(r)r^3}{M(r)}\,dr}{\displaystyle\int_0^a f(r)r^3\,dr} \tag{3.20}$$

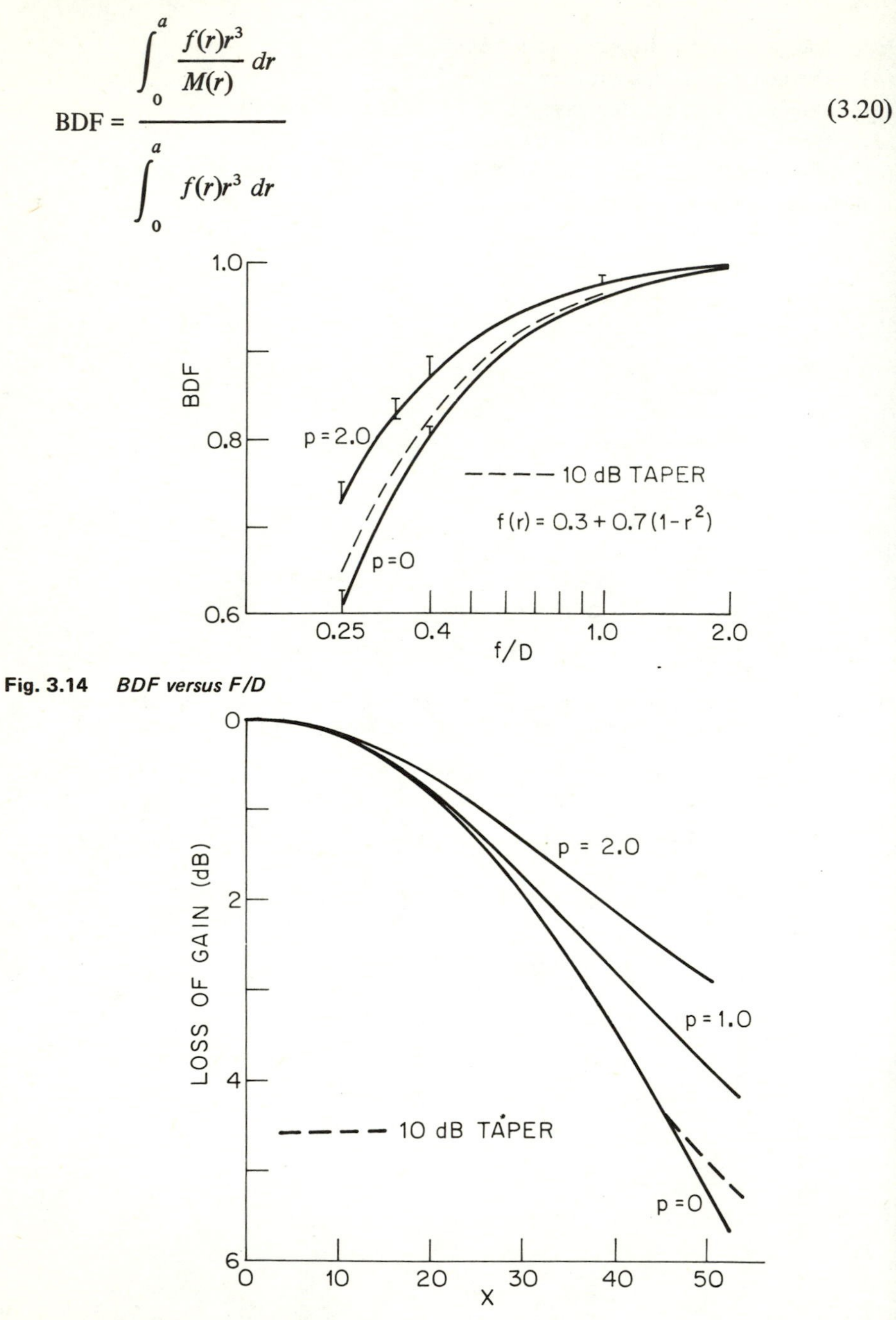

Fig. 3.14 *BDF versus F/D*

Fig. 3.15 *Scan Loss versus Scan Angle*

where $f(r)$ is the aperture illumination function and

$$M(r) = 1 + (r/2F)^2. \tag{3.21}$$

Eqn. 3.20 is plotted in Fig. 3.14 for

(*a*) $f(r) = 1{\cdot}0$

(*b*) $f(r) = \left[1 - \dfrac{r^2}{a^2}\right]^2$

(*c*) $f(r) = 0{\cdot}3 + 0{\cdot}7\left[1 - \dfrac{r^2}{a^2}\right]$

To evaluate scan loss, Ruze used the aperture-field radiation integral

$$E(\theta,0) = 2\pi \int_0^a f(r)\, J_0\left[kc\left(\sin\theta - \frac{s/F}{M(r)}\right) r\right] r\,dr \tag{3.22}$$

where s is the lateral component of defocusing. Results are plotted in Fig. 3.15 as a function of the normalized scan parameter

$$X = \frac{\dfrac{w_m}{2w_0}(D/F)^2}{1 + 0{\cdot}02\,(D/F)^2} \tag{3.23}$$

where $w_m/2w_0$ is the number of half-power beamwidths scanned.

Highly accurate determinations of scan-properties, or evaluations at larger scan

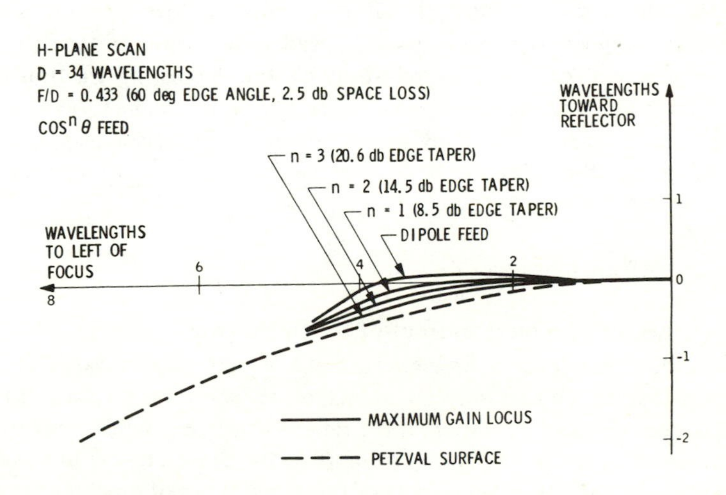

Fig. 3.16 *Comparison of maximum transmit-gain contours and Petzval surface; F/D = 0.433*

angles, must be carried out using the physical-optics surface integrals. For example,[12] Fig. 3.16 shows the maximum scan-gain contours for $F/D = 0{\cdot}433$ and the four feed functions n = 1,2,3 and infinitesimal dipole. The Petzval surface is also plotted in the Figure. It is evident that the maximum-gain contours are slightly on the focal-plane side of the Petzval surface, although the difference is a relatively small fraction of a wavelength. As the scan angle is increased, or the edge taper is increased, the maximum-gain contours approach the Petzval surface.

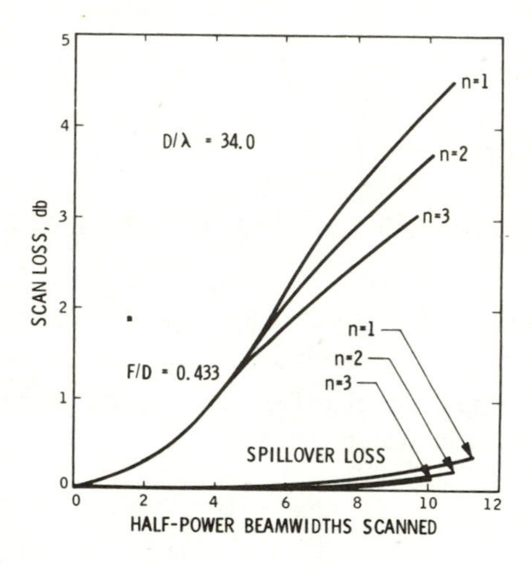

Fig. 3.17 *Beam-scan losses; F/D = 0.433*

Fig. 3.17 presents the scan losses along the maximum-gain contours. The abscissa is half-power beamwidths (HPBWs) scanned for the 34-wavelength aperture. The scan-loss curves *do* include spillover loss, since spillover occurs as the feed is scanned laterally while pointing straight ahead. Also plotted in the Figure, however, is the component of loss due exclusively to the decreasing fraction of the total power intercepted by the reflector as the feed is scanned. This spillover loss amounts to a fraction of a dB, while the remaining scan loss is due to phase defocusing and other effects.

3.1.4 Aperture blocking

The presence of an object in front of a reflector will cause significant changes in its radiation characteristics. In general, these objects may be classified as: (*a*) large, centrally located objects such as a smaller reflector or a feed horn, which are essential to the proper operation of the reflector system, and (*b*) long, thin cylindrical structures used for mechanical support of the central object but which serve no direct RF purpose. Different aspects of these two types of blocking are discussed in the following sections.

Central blocking of a focused paraboloid

For small polar angles, the field of an unblocked, uniformly illuminated paraboloid is

$$E_\phi \cong jkE_0 \cos\phi \frac{e^{-jkR}}{R} e^{-j2kF} \int_0^{D/2} J_0(kr \sin\theta)\, r dr. \tag{3.24}$$

For the paraboloid with an axially symmetric central object of diameter d it is physically reasonable to assume that the central, blocked section of the aperture does not contribute to the radiated field so that, with blocking, eqn. 3.24 becomes

$$E_\phi = jkE_0 \cos\phi \frac{e^{-jkR}}{R} e^{-j2kF} \int_{d/2}^{D/2} J_0(kr \sin\theta) r dr \tag{3.25}$$

which yields

$$E_\phi = jkE_0 \cos\phi \frac{e^{-jkR}}{R} e^{-j2kF} \left(\frac{D}{2}\right)^2 \left\{ \frac{J_1(u)}{u} - \beta^2 \frac{J_1(\beta u)}{(\beta u)} \right\} \tag{3.26a}$$

where $\beta = d/D$ and $u = \pi(D/\lambda)\sin\theta$. Similarly, the theta component is

$$E_\theta = jkE_0 \sin\phi \frac{e^{-jkR}}{R} e^{-j2kF} \left(\frac{D}{2}\right)^2 \left\{ \frac{J_1(u)}{u} - \beta^2 \frac{J_1(\beta u)}{(\beta u)} \right\}. \tag{3.26b}$$

The loss in axial gain is thus

$$\text{Loss (dB)} = -20\log_{10}(1-\beta^2) \cong 8{\cdot}7\beta^2 = 8{\cdot}7\left(\frac{d}{D}\right)^2 \text{ for} \left(\frac{d}{D}\right)^2 \ll 1 \tag{3.27}$$

The field of the blocked aperture expressed in (3.26) can be interpreted as a superposition of the field of the unblocked aperture with the field of a smaller, out-of-phase aperture. Since the phase of the unblocked aperture fields alternates between consecutive lobes of the radiation pattern, the 1st, 3rd, etc., sidelobes are raised while the 2nd, 4th, etc., sidelobes are lowered. This effect is illustrated in Figure 3.18.

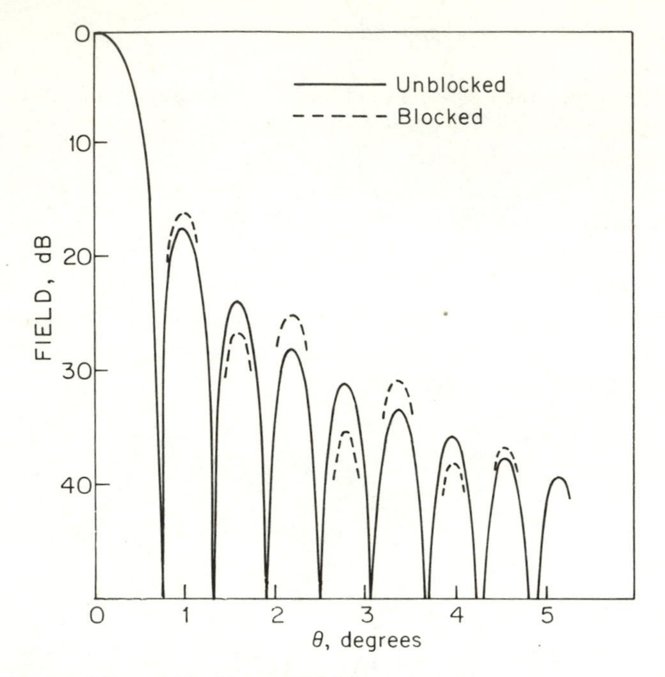

Fig. 3.18 *Effects of Central Blocking on Sidelobes*

For a non-uniform aperture distribution, the loss in boresight gain is approximately.

$$\text{Loss (dB)} = -20\log_{10}\left|1 - \frac{\int_0^{2\pi}\int_0^{\beta} F(t,\phi')\,t\,dt\,d\phi'}{\int_0^{2\pi}\int_0^{1} F(t,\phi')\,t\,dt\,d\phi'}\right| \tag{3.28}$$

where t = normalized radial aperture coordinate, ϕ' = azimuthal aperture coordinate, $F(t,\phi')$ = aperture distribution function, and $\beta = d/D$. As a 'rule of thumb', this generally becomes

$$\text{Loss (dB)} = -20\log_{10}(1 - a\beta^2) \cong 8{\cdot}7 a\beta^2 \tag{3.29}$$

where the parameter a is typically about 2. Potter[13] has estimated that for a prime-focus paraboloid of focal-length F

$$a \cong G_0/32(F/D)^2 \tag{3.30}$$

where G_0 is the gain of the feed.

Strut blocking

The presence of long, thin mechanical support structures in front of a radiating aperture has generally defied rigorous attempts to analyse their effects upon the RF performance of the antenna. Early[14-16] analyses of the effects of strut blocking have been based on the null-field hypothesis, i.e. that the currents/fields on the shadowed portsions of the surface/aperture are non-radiative. Rather elaborate

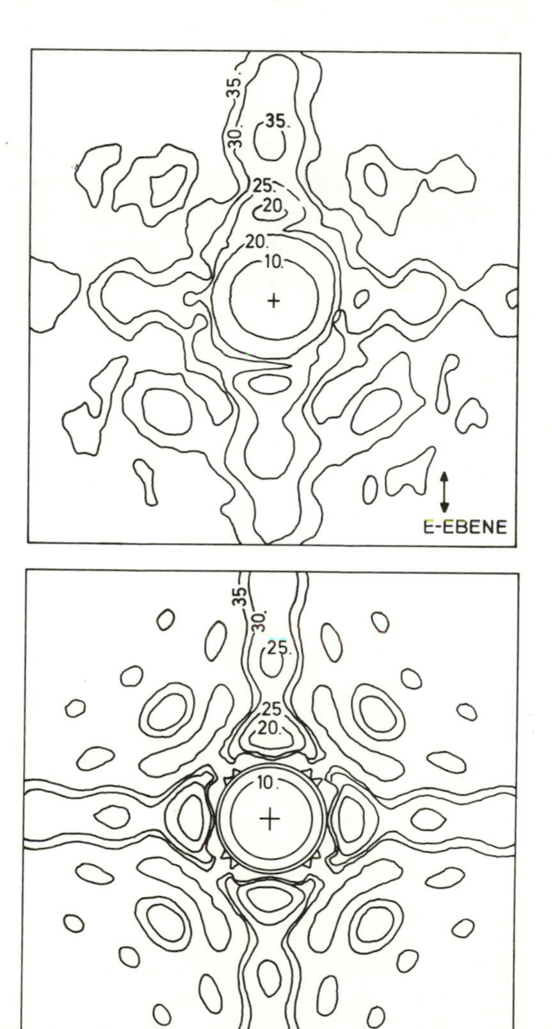

Fig. 3.19 *a Measured contours of the radiation field from a circular aperture with quadripod supports*
b Computed pattern based on null-field hypothesis

geometrical constructions have been made to determine the shape of the shadows cast by the quasi-planar and quasi-spherical wavefronts in the vicinity of the struts[17-18]. The null-field analysis has yielded results close to the measured values (Fig. 3.19) in cases where the strut width was much greater than a wavelength and the observation angle was not far from the boresight axis. However, this approach fails to take into account the depth, cross-section, or tilt of the struts, nor does it provide differences for frequency or polarisation effects. Furthermore, the struts frequently have widths of the order of a wavelength, so that no basis exists for the expectation that deep, clearly defined optical shadows will be cast by the wavefronts impinging on the struts.

A physically more-reasonable assumption[20-28] is that the strut currents due to the plane-wave component of focal-region field are the same currents that would flow on an infinite cylindrical structure of the same cross-section in free space immersed in an infinite plane wave with the same polarisation and direction of incidence as the local geometrical ray incident upon that part of the strut as it emerges from the aperture (transmit mode). This approximation is known as the 'IFR hypothesis'.

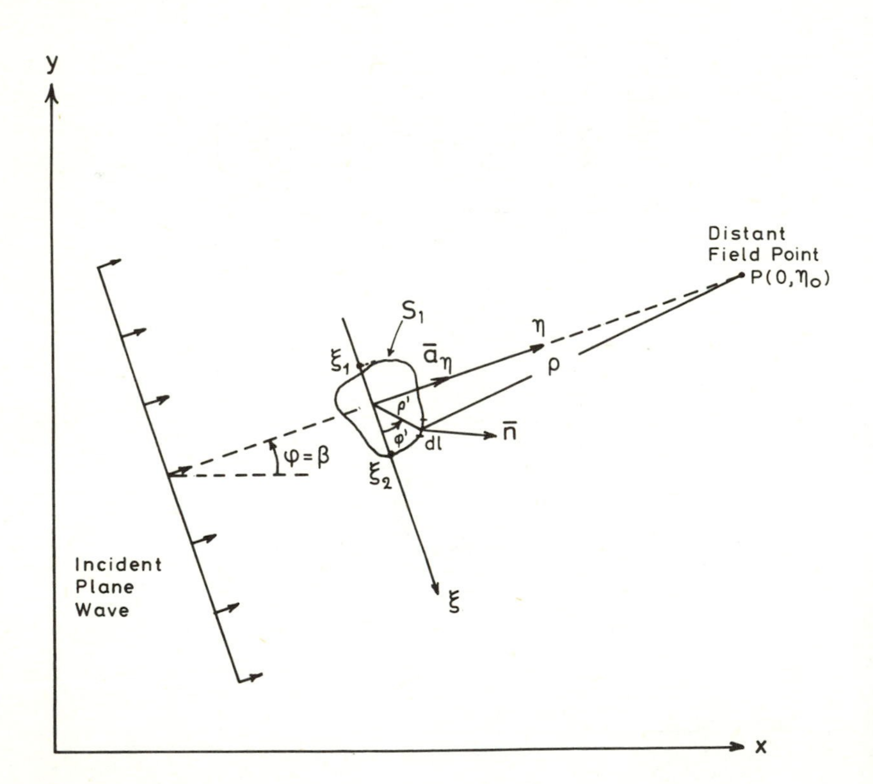

Fig. 3.20 *IFR geometry*

When an infinite cylinder is immersed in an incident plane wave (Fig. 3.20), its induced field ratio (IFR) is defined as the ratio of the forward-scattered field to the hypothetical field radiated in the forward direction by the plane wave in the reference aperture of width equal to the shadow of the geometrical cross section of the cylinder on the incident wavefront[20]. Thus for the E-vector of the incident plane wave parallel to the cylinder axis

$$IFR_E = -\frac{Z_0}{2(\xi_2 - \xi_1)E_0} \int_{S_1} J_{Sz}\, e^{jk\rho' \sin\phi'} dl \tag{3.31a}$$

where E_0 is the electric intensity of the plane wave, Z_0 is the intrinsic impedance of free space, J_{Sz} is the axial component of surface-current density, and S_1 is the line contour defining the cylinder's periphery. Similarly, for the H-vector parallel to the cylinder axis,

$$IFR_H = \frac{1}{2(\xi_2 - \xi_1)H_0} \int_{S_1} H_z\, (\hat{a}_\eta \cdot \hat{n})\, e^{jk\rho' \sin\varphi'}\, dl \tag{3.31b}$$

where H_0 is the magnetic field of the incident wave, H_z is the total magnetic field in the axial direction at the surface of the cylinder, and the unit vectors $\hat{a}_\eta$ and $\hat{n}$ are defined in the figure. The IFR_E and IFR_H for a right-circular cylinder of radius a are given by

$$IFR_E = -\frac{1}{ka\cos\alpha} \sum_{n=-\infty}^{+\infty} J'(ka\cos\alpha)/H_n^{(2)\prime}(ka\cos\alpha) \tag{3.32a}$$

$$IFR_H = -\frac{1}{ka\cos\alpha} \sum_{n=-\infty}^{+\infty} J_n'(ka\cos\alpha)/H_n^{(2)\prime}(ka\cos\alpha) \tag{3.32b}$$

where J_n and $H_n^{(2)}$ are, respectively, the Bessel and outgoing Hankel function, and the prime indicates the derivative with respect to the argument. These formulas have also included the effects of a tilt angle α between the incident wavefront and the cylinder axis. Thus for non-normal incidence the IFR is determined for an equivalent cylinder with linear dimensions of the cross section reduced by the factor $\cos\alpha$. This principle applies equally well to other cross sections in general. The IFRs for the right-circular cylinder are plotted on the complex plane in Fig. 3.21. In general, the IFR_E is larger in magnitude than the IFR_H and has a positive phase angle compared to a negative phase angle for the H polarisation. Both IFRs approach the value $-1{\cdot}0 + j\,0{\cdot}0$ as the radius increases, one from below and the other from above. The IFRs for a square cylinder are plotted on the complex plane in Fig. 3.22. This Figure resembles Fig. 3.21 for the circular cylinder except for its more complicated structure and enhanced magnitude for the H-polarization.

Using the IFR hypothesis in a practical application, the loss in boresight gain due to central blockage and the plane-wave component of strut current for a linearly polarised aperture is

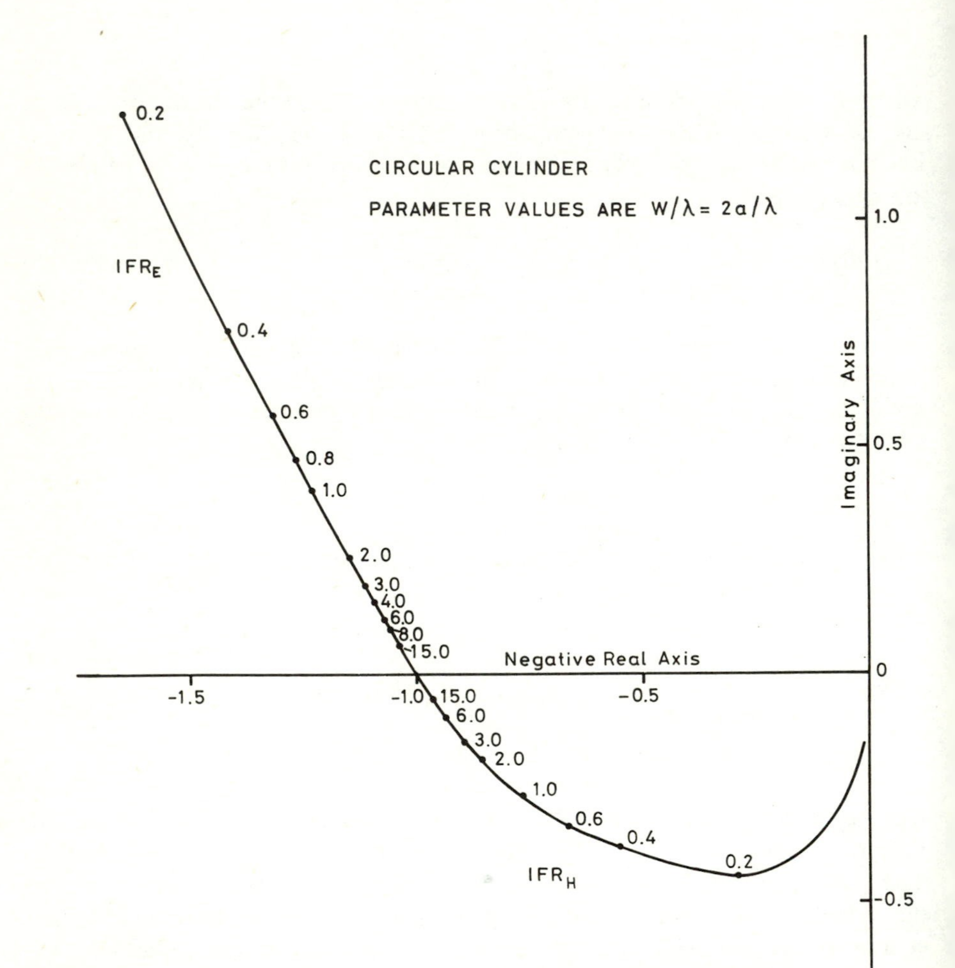

Fig. 3.21 *Complex IFR_E and IFR_H for circular cylinder*

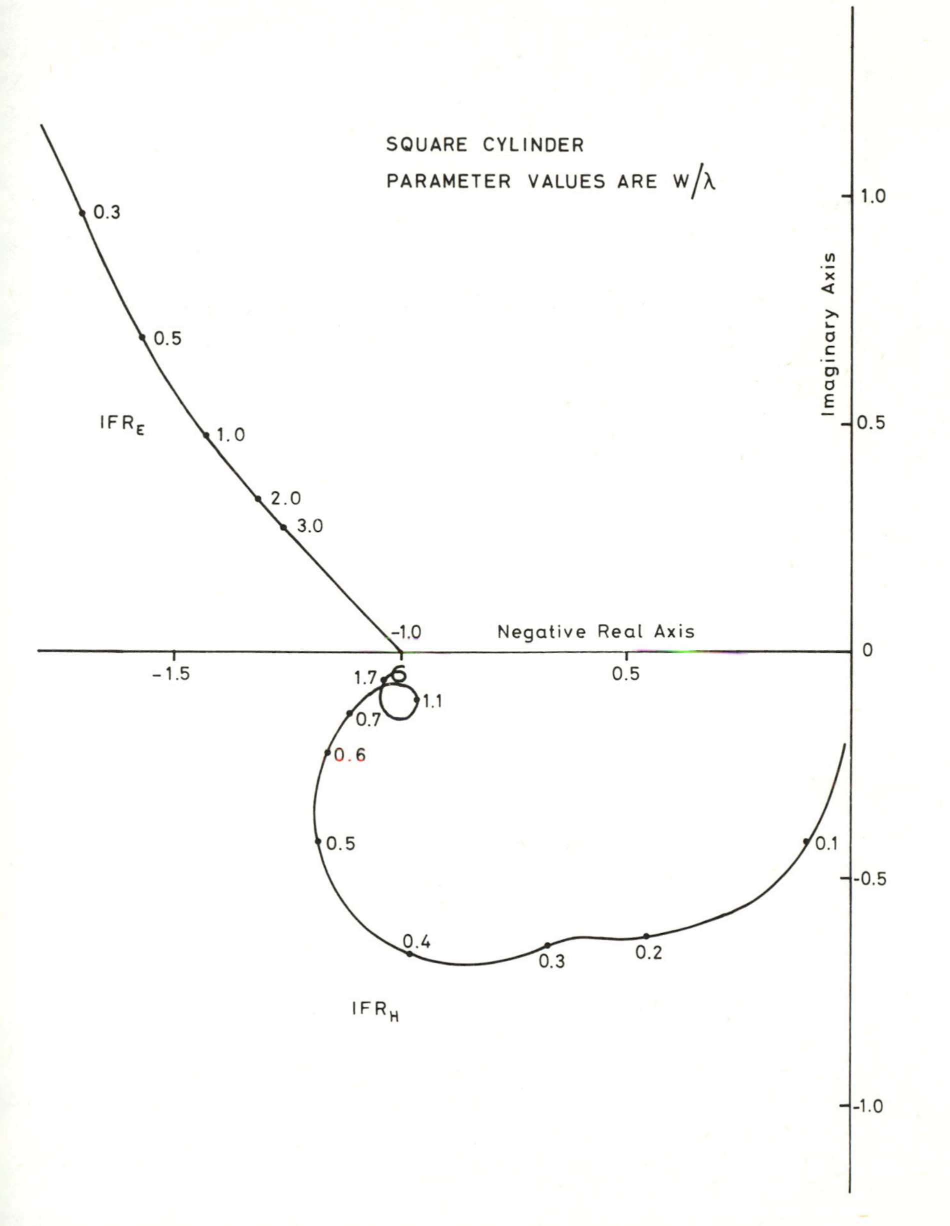

Fig. 3.22 *Complex IFR_E and IFR_H for square cylinder*

$$\text{Loss (dB)} = -20\log_{10}\left| 1 - \frac{\int_0^{2\pi}\int_0^{\beta} F(t,\phi)t\,dt\,d\phi}{\int_0^{2\pi}\int_0^{1} F(t,\phi)t\,dt\,d\phi} + \sum_{i=1}^{\text{number of struts}} \left\{ \left(w_i \frac{D}{2}\right)(IFR_{Ei}'\cos^2\gamma_i + IFR_{Hi}'\sin^2\gamma_i) \cdot \frac{\int_{\beta}^{\beta_i} F(t,\phi_i)\,dt}{\left(\frac{D}{2}\right)^2 \int_0^{2\pi}\int_0^{1} F(t,\phi_i)t\,dt\,\phi} \right\} \right| \qquad (3.33)$$

where

$\bar{t}$	=	normalised radial aperture coordinate
ϕ	=	azimuthal aperture coordinate
$F(t,\phi)$	=	aperture distribution function
β	=	fractional diameter blocking by central blockage
w_i	=	width of ith strut
D	=	aperture diameter
β_i	=	fractional radius blocking by ith strut
γ_i	=	angle between electric vector and ith strut
α_i	=	angle between ith strut and aperture plane
IFR_{Hi}'	=	IFR_H for ith strut cross section with linear dimensions reduced by $\cos\alpha_i$.

The second term represents the 'optical' approximation to axially symmetric blockage. The third term represents a summation over the plane-wave component of strut currents on each strut. For a uniformly illuminated aperture eqn. 3.33 reduces to

$$\text{Loss (dB)} = -20 \log_{10} \left| 1 - \beta^2 + \sum_i \left\{ \frac{A_i}{\left(\pi \frac{D}{2}\right)^2} (IFR_{Ei}{}' \cos^2 \gamma_i + IFR_{Hi}{}' \sin^2 \gamma_i) \right\} \right| \tag{3.34}$$

where A_i is the projected area of the ith strut on the aperture plane.

In addition to degrading boresight gain, a feed-support strut will generate boresight cross polarization if it is not aligned parallel to or perpendicular to the electric field in the aperture. The boresight cross polarization level due to the plane-wave component of strut currents is

$$CP\ (\text{dB}) = 20 \log_{10} \left| \sum_i \left\{ \left(w_i \frac{D}{2}\right) (IFR_{Hi'} - IFR_{Ei'}) \sin \gamma_i \cos \gamma_i \cdot \frac{\int_{\beta}^{\beta_i} F(t,\phi_i)\,dt}{\left(\frac{D}{2}\right)^2 \int_0^{2\pi} \int_0^1 F(t,\phi) t\, dt\, d\phi} \right\} \right| . \tag{3.35}$$

Clearly, if $\sin \gamma_i \cos \gamma_i$ is zero (e.g. principal-plane struts), cross-polarisation is not generated by the ith strut. Other strut configurations, e.g. an equiangular tripod, will also not generate boresight cross-polarisation because of cancellation properties of the total geometry, although each strut individually may generate a cross-polarised component.

Strut fields for non-axial field points

The geometry of a single, perfectly conducting strut is shown in Fig. 3.23, where the strut axis lies in the plane $\phi' = \phi_0$. The $x' - y' - z'$ coordinate system is centered at the paraboloid prime focus, with the z'-axis directed away from the reflector along its axis of symmetry. The strut lies at an angle α $(-90 < \alpha < 90)$ with respect to the r'-axis, which is perpendicular to z' in the plane $\phi' = \phi_0$.*

The right-handed $x''-y''-z''$ coordinate system is shown in Fig. 3.24, with its origin located at the nearer end of the strut (r_1', z_1'), the z''-axis lying along the strut, and the x''-axis lying in the $r'-z'$ plane, i.e. in the plane $\phi' = \phi_0$. Also shown in the Figure is the incident plane wave emerging from the reflector (transit mode) with $\bar{k} = k\hat{a}_{z'}$, i.e. in the positive z'-direction.

* For normal strut configurations $-90 < \alpha \leq 0$

E-Polarisation

In this Section it will be assumed that the $\bar{E}$-vector of the incident plane-wave emerging from the reflector lies in the plane $\phi' = \phi_0$ as indicated in Fig. 3.24. Under these conditions the currents induced on the cylinder (neglecting end effects because $L \gg \lambda$) will flow entirely in the z''-direction. Then the scattered field due to the plane-wave-induced component of strut current is

$$\bar{E}\,(P) = -\frac{j\omega\mu}{4\pi}\left(\frac{e^{-jkR}}{R}\right)\oint dl''\int_0^L dz''\left\{J_{Sz''}\,[\hat{a}_{z''}]_{trans}\,e^{jk\bar{\rho}\cdot\hat{a}_R}\right\} \tag{3.36}$$

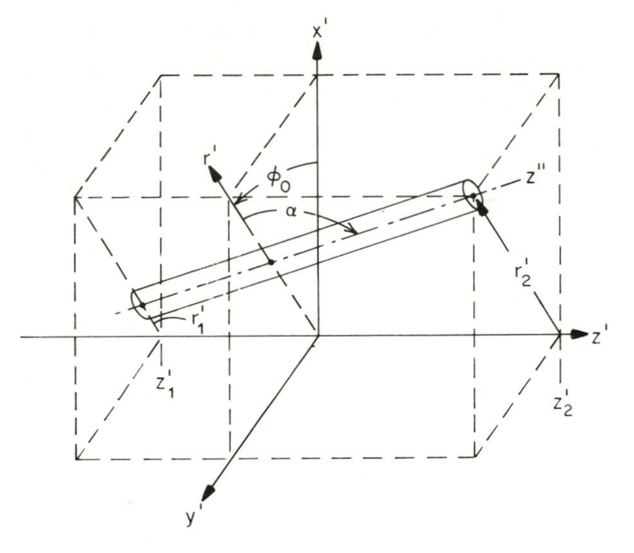

Fig. 3.23 *Strut Geometry — Oblique View*

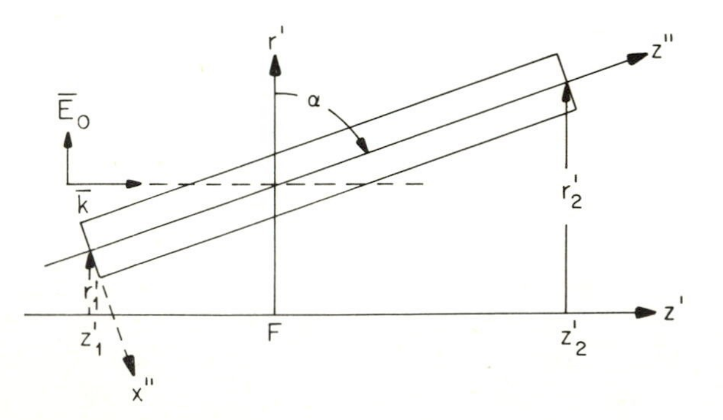

Fig. 3.24 *Strut Geometry — Cross-section*

where P is a field point with coordinates (R, θ, ϕ), R is the distance from the origin 0 to the field point, $\overline{\rho}$ is the vector from 0 to the integration point, $\hat{a}_R$ is a unit vector from 0 to P, and only the transverse (to $\hat{a}_R$) components of the current contribute to the scattered field.

Since the IFR hypothesis will be invoked to determine the strut currents, they will be expressed as

$$J_{Sz''} = e^{-jk\sin\alpha z''} J'_{Sz''}(l'') \tag{3.37}$$

where l'' is the path length around the cylinder periphery in the $x''-y''$ plane, as shown in Fig. 3.25. In the special case of a circular cross-section $l'' = a\cos\phi''$. Insertion of eqn. 3.37 into eqn. 3.36 yields

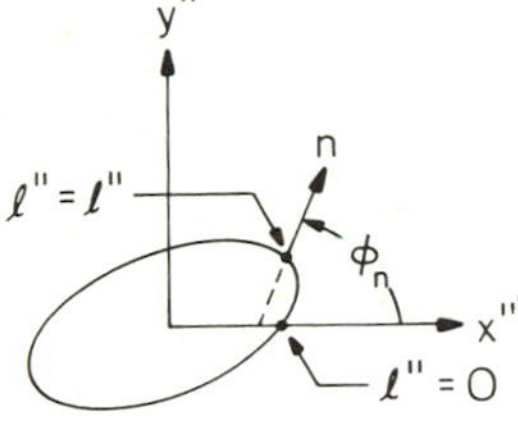

Fig. 3.25 *Strut Cross-section*

$$\overline{E}(P) = \left(\frac{jk}{2\pi}\right)\left(\frac{e^{-jkR}}{R}\right) e^{jP_0} \left\{ \overline{e}_0 \, IFRE(D, \delta, \alpha) \right\} w \int_{r'_1}^{r'_2} E_A(r') e^{jkr'A_0} dr' \tag{3.38}$$

where

$$\overline{e}_0 = [\cos\theta\cos(\phi - \phi_0) - \tan a \sin\theta]\, \hat{a}_\theta - \sin(\phi - \phi_0)\hat{a}_\phi \tag{3.39a}$$

$$P_0 = (kz'_1 - kr'_1 \tan\alpha)(\cos\theta - 1) \tag{3.39b}$$

$$A_0 = \sin\theta\cos(\phi - \phi_0) + \tan\alpha(\cos\theta - 1) \tag{3.39c}$$

$$B = \sin\alpha\sin\theta\cos(\phi - \phi_0) - \cos\alpha\cos\theta \tag{3.39d}$$

$$C = \sin\theta\sin(\phi - \phi_0) \tag{3.39e}$$

$$D = \sqrt{B^2 + C^2} \tag{3.39f}$$

$$\delta = \tan^{-1}\left[\frac{-C}{-B}\right]. \tag{3.39g}$$

$IFRE\,(D, \delta, \alpha)$ in eqn. (3.38) is obtained from the integral

$$IFRE = -\frac{\eta}{2\left(\frac{w}{\lambda}\right)} \oint \frac{1}{E_0} J'_{Sz''}(l'')\, e^{-j2\pi\left(\frac{x''}{\lambda} B + \frac{y''}{\lambda} C\right)} d\left(\frac{l''}{\lambda}\right) \tag{3.40}$$

where

$$\eta = 120\pi \text{ ohms} \tag{3.41}$$

w/λ = projected width of cylinder (in wavelengths) on incident wavefront. (3.42)

The function $J'_{Sz}{}''(l'')$ in eqn. (3.40) is generally evaluated in tabular form from the method of moments[22,24]. However, for a circular cylinder

$$\frac{1}{E_0} J'_{Sz}{}'' = \left(\frac{2}{\eta\pi ka \cos\alpha}\right) \sum_{n=-\infty}^{+\infty} (j)^{-n} \frac{e^{jn\phi''}}{H_n^{(2)}(ka \cos\alpha)} \tag{3.43}$$

and the corresponding generalised IFR function for the circular cylinder is

$$IFRE\,(D, \delta, \alpha) = -\frac{1}{ka\cos\alpha} \sum_{n=-\infty}^{+\infty} e^{jn\delta} \frac{J_n(kaD)}{H_n^{(2)}(ka\cos\alpha)}. \tag{3.44}$$

The z'' integral in eqn. (3.36) has been transformed in eqn. (3.38) to an integral of $E_A(r')$, the *focal-plane* E-field in the r'-direction, times the phase factor $e^{jkr' A_0}$.

When $\theta = 0$, i.e. P lies on the z'-axis (boresight axis), eqn. (3.38) reduces to

$$\bar{E}(R, 0, 0) = \left(\frac{jk}{2\pi}\right)\left(\frac{e^{-jkR}}{R}\right) \hat{a}_{r'}\, IFR_E\, w \int_{r'_1}^{r'_2} E_A(r')\, dr' \tag{3.45}$$

where IFR_E is the E-polarisation *IFR* of a cylinder defined in eqn. (3.31*a*). Except for the IFR_E factor, eqn. (3.45) is the field of the rectangular aperture defined by the projection of the strut on the $x' - y'$ plane as shown in Fig. 3.26. Thus, in the optical limit of $IFR_E \rightarrow -1$, eqn. (3.45) reduces to the classical null-field result. However, for typical strut widths of the order of a wavelength, IFR_E is far from -1, and a considerably different result is obtained.[26]

Eqn. (3.38) also provides more precise angular information about the strut tilt than does the conventional 'flat' shadow representation. For example, since $E_A(r')$ is constant-phase in the focal plane, the integral in eqn. (3.38) is maximum in those directions (θ, ϕ) for which $A_0 = 0$. Thus, from eqn. (3.39*c*), it is shown that local maxima of the scattered field occur along the contour

$$\cos(\phi_{MAX} - \phi_0) = \tan(\theta_{MAX}/2)\tan\alpha. \tag{3.46}$$

These are contours which are evident in measured wide-angle contour plots.[29,30]

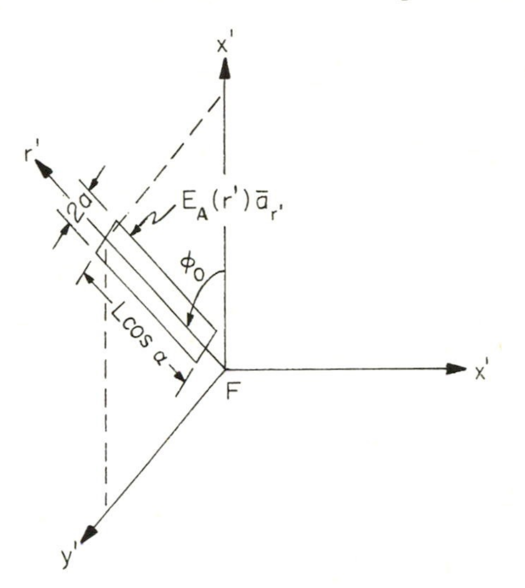

Fig. 3.26 *Projection of Strut on Aperture Plane*

H-polarisation

When the *H*-vector of the incident plane-wave lies in the plane of the strut, the scattered fields are considerably complicated by the existence of both axial and peripheral currents on the strut. Thus, if the total *H*-field at the surface of the strut is

$$H_{z''}(l'') = e^{-jk\sin\alpha z''}\, H'_{z''}(l'') \tag{3.47}$$

then the scattered field is

$$\bar{E}(P) = \left(\frac{jk}{2\pi}\right)\left(\frac{e^{-jkR}}{R}\right) e^{jP_0}\left\{\bar{e}_c JFRHN + \bar{e}_s\left(\frac{1}{\pi\frac{w}{\lambda}D}\right)JFRH1\right.$$

$$\left. - (\hat{a}_\theta e_\theta + \hat{a}_\phi e_\phi)\left(\frac{\sin\alpha}{\pi\frac{w}{\lambda}\cos^2\alpha}\right)JFRH2\right\} w\int_{r'_1}^{r'_2} \eta H_A(r')e^{jkr'A_0}\,dr' \tag{3.48}$$

where

$$\bar{e}_c = \hat{a}_\theta \; [\sin\delta \cos\alpha \sin\theta + \sin\delta \sin\alpha \cos\theta \cos(\phi-\phi_0) - \cos\delta \cos\theta \sin(\phi-\phi_0)] + \hat{a}_\phi \; [-\sin\delta \sin\alpha \sin(\phi-\phi_0) - \cos\delta \cos(\phi-\phi)] \tag{3.49a}$$

$$\bar{e}_s = \hat{a}_\theta \; [\cos\delta \cos\alpha \sin\theta + \cos\delta \sin\alpha \cos\theta \cos(\phi-\phi_0) + \sin\delta \cos\theta \sin(\phi-\phi_0)] + \hat{a}_\phi \; [-\cos\delta \sin\alpha \sin(\phi-\phi_0) + \sin\delta \cos(\phi-\phi_0)] \tag{3.49b}$$

$$e_\theta = \cos\alpha \cos\theta \cos(\phi-\phi_0) - \sin a \sin\theta \tag{3.49c}$$

$$e_\phi = -\cos\alpha \sin(\phi-\phi_0) \tag{3.49d}$$

and

$$IFRHN = \frac{1}{2\left(\frac{w}{\lambda}\right)\cos\alpha} \oint \cos(\phi_n - \delta) \left[\frac{H'_{z''}(l'')}{H_0}\right] e^{-j2\pi(x''B + y''C)/\lambda} \, d\left(\frac{l''}{\lambda}\right) \tag{3.50a}$$

$$JFRH1 = \frac{\pi D}{2\cos\alpha} \oint \sin(\phi_n - \alpha) \left[\frac{H'_{z''}(l'')}{H_0}\right] e^{-j2\pi(x''B + y''C)/\lambda} \, d\left(\frac{l''}{\lambda}\right) \tag{3.50b}$$

$$JFRH2 = \frac{-j}{4\cos\alpha} \oint \frac{\partial}{\partial\left(\frac{l''}{\lambda}\right)} \left[\frac{H'_{z''}(l'')}{H_0}\right] e^{-j2\pi(x''B + y''C)/\lambda} \, d\left(\frac{l''}{\lambda}\right) \tag{3.50c}$$

and ϕ_n is the angle between the surface normal at l'' and the x''-axis as shown in Fig. 3.25. For a circular cylinder

$$\frac{H'_{z''}}{H_0} = -\frac{2j}{\pi ka} \sum_{n=-\infty}^{+\infty} (j)^{-n} \frac{e^{jn\phi''}}{H_n^{(2)\prime}(ka\cos\alpha)} \tag{3.51}$$

and *IFRHN* reduces to *IFRH*(*D*, δ, α) and both *JFRH*1 and *JFRH*2 reduce to *JFRH*(*D*, δ, α) shown below:

$$IFRH(D, \delta, \alpha) = -\frac{1}{ka\cos\alpha} \sum_{n=-\infty}^{+\infty} e^{jn\delta} \frac{J'_n(kaD)}{H_n^{(2)\prime}(ka\cos\alpha)} \tag{3.52a}$$

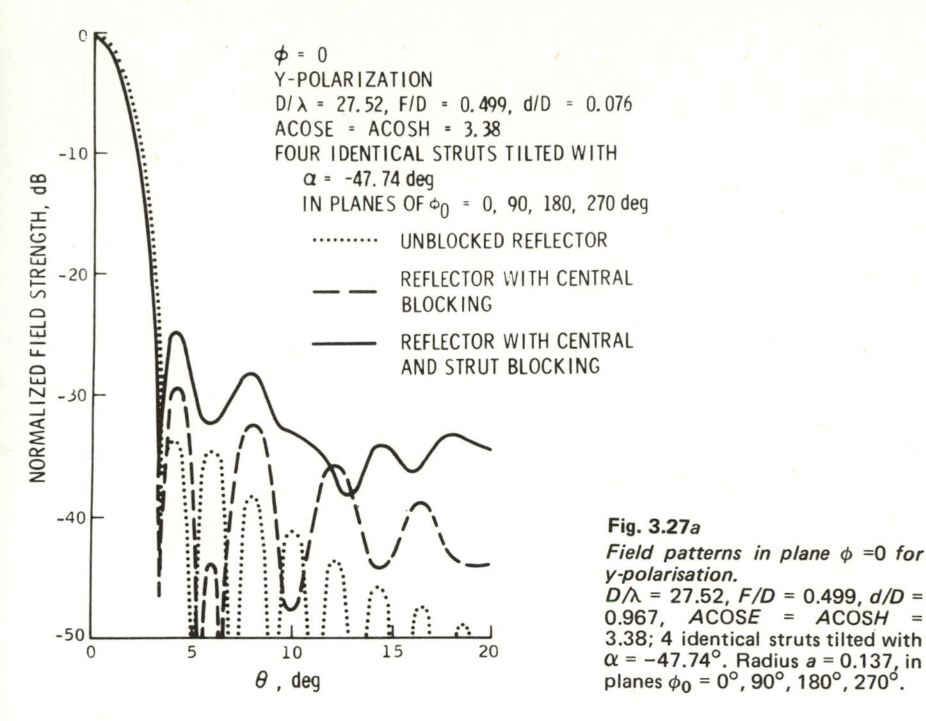

Fig. 3.27*a*

Field patterns in plane ϕ =0 for y-polarisation.
D/λ = 27.52, F/D = 0.499, d/D = 0.967, *ACOSE* = *ACOSH* = 3.38; 4 identical struts tilted with $\alpha = -47.74°$. Radius a = 0.137, in planes ϕ_0 = 0°, 90°, 180°, 270°.

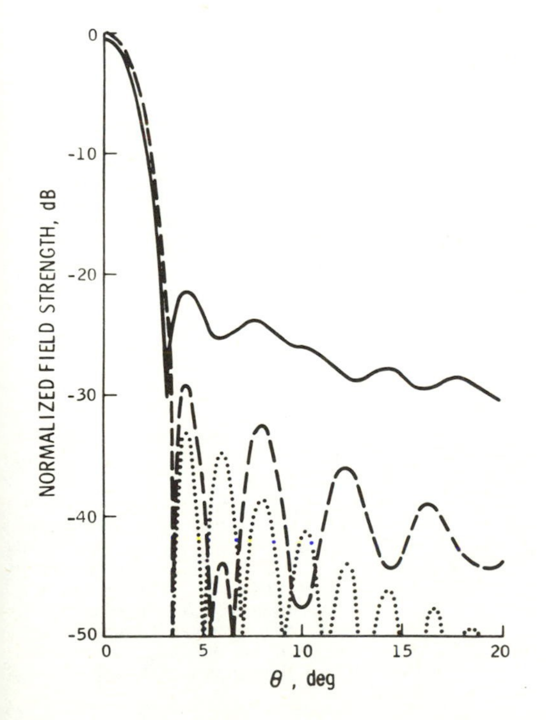

Fig. 3.27*b*

Co-polar and cross-polar field patterns in plane ϕ = 0 for circular polarisation.
D/λ = 27.52, F/D = 0.499, d/D = 0.076, *ACOSE* = *ACOSH* = 3.38, four identical struts tilted with $\alpha = -47.74°$, radius a = 0.137λ, in planes ϕ_0 = 0°, 90°, 180°, 270°.

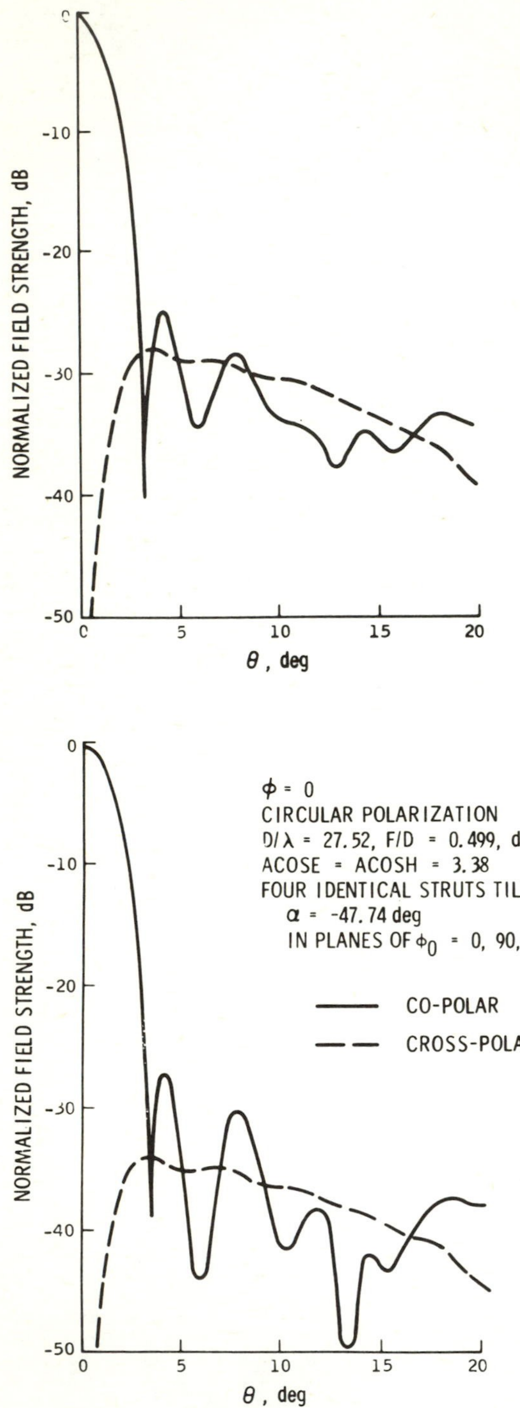

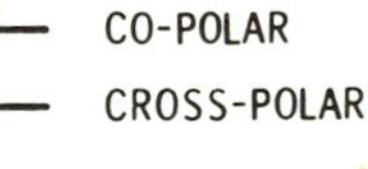

Fig. 3.28*a*
Field patterns in plane $\phi = 0$ for y-polarisation.
D/λ - 27.52, *F/D* = 0.499, *d/D* = 0.076, *ACOSE* = *ACOSH* = 3.38, four identical struts tilted with $\alpha = -47.74°$, rectangular cross-section 0.275λ by 1.603λ, narrow-end-first, in planes ϕ_0 = 0, 90°, 180°, 270°.

Fig. 3.28*b*
Co-polar and cross-polar field patterns in plane $\phi = 0$ for circular polarisation.
D/λ = 27.52, *F/D* = 0.499, *d/D* = 0.976, *ACOSE* = *ACOSH* = 3.38, four identical struts tilted with $\alpha = -47.74°$, rectangular cross-section 0.275λ by 1.603λ, narrow-end-first, in planes ϕ_0 = 0°, 90°, 180°, 270°.

$$JFRH(D, \delta, \alpha) = -\frac{1}{ka \cos \alpha} \sum_{n=-\infty}^{+\infty} jn e^{jn\delta} \frac{J_n(kaD)}{H_n^{(2)\prime}(ka \cos \alpha)}. \tag{3.52b}$$

In general, however, $H'_{z''}/H_0$ must be obtained numerically in tabular form using the method of moments. The final factor in eqn. 3.48 is an integral of $H_A(r')$, the *focal-plane* H-field in the r'-direction.

When $\theta = 0$, *JFRH* is identically zero, and eqn. 3.48 becomes

$$\bar{E}(R, 0\,0) = -\left(\frac{jk}{2\pi}\right)\left(\frac{e^{-jkR}}{R}\right)\hat{a}_{\phi_0}\, IFR_H\, w \int_{r_1'}^{r_2'} \eta H_A(r')dr' \tag{3.53}$$

where IFR_H is the $\acute{H}$-polarisation *IFR* of a cylinder defined in eqn. 3.31*b*. Except for the IFR_H factor, eqn. 3.53 is the field of the rectangular aperture shown in Fig. 3.26.

Figs. 3.27 and 3.28 are the normalised radiation patterns of a focused paraboloid with both central and strut blocking. The geometrical parameters are stated in the Figure captions. The paraboloid has a quadripod with 4 identical struts of width equal to 0.275λ. However, in one case the strut cross-section is circular, and in the other it is rectangular. It is clear that the rectangular struts give higher sidelobes for linear polarisation and higher cross-polarisation for circular polarisation.

3.1.5 Umbrella reflectors (Contribution by W.C.Wong)

The geometry of gored umbrella-like paraboloid reflectors was discussed in Section 2.2.5. The performance data presented below will correspond to the special case that eqn. 2.81 reduces to

$$f(\theta', \varphi') = \cos^N(\pi - \theta'), \quad \pi/2 \leqslant \theta' \leqslant \pi. \tag{3.54}$$

Frequency scaling

In the ranges of reflector diameters and gore numbers where eqn. 2.84*a* applies, a simple observation can be made on the relationship between the gore loss, the reflector diameter, and the number of gores. Thus from eqn. 2.82 it may be shown that the range of a focal region is

$$S_f = f_r - f_g = f_r \sin^2 \frac{\pi}{N_g}. \tag{3.55}$$

Furthermore, if, in eqn. 2.84*a*, the slight dependence on θ' is neglected, the ρ integration becomes independent of ψ and hence the magnitude of the integral only depends on the integration of S_f. Furthermore, when $\sin(\pi/N_g) \approx (\pi/N_g)$, two reflectors will have the same loss when

$$\frac{k_1 f_{r1}}{N_{g_1}^2} = \frac{k_2 f_{r2}}{N_{g_2}^2}. \tag{3.56}$$

This scaling was found to hold over the practical range of gore loss and can be verified numerically.

Performance characteristics

Fig. 3.29 plots gore number versus the deviation of the best feed position from the rib focal point. The solid lines are predicted results obtained by using the simple equation given by eqn. 2.87. The broken lines are the correct results extracted from actual defocusing curves obtained by plotting the on-axis gain versus axial defocusing as shown in Fig. 3.30. Clearly the differences between the two sets of curves in Fig. 3.29 are only significant for small gore number or extremely large values of D/λ.

Fig. 3.31 shows the effect of frequency on gore loss for three different gore numbers: 16, 28, and ∞. ($N_g = \infty$ corresponds to a smooth paraboloid.) For a 500λ reflector with 28 gores, the gore loss is seen to be more than 5 dB. Despite the enormous gore loss, Fig. 3.29 shows that the prediction of the best feed position using eqn. 2.87 is surprisingly accurate.

Fig. 3.32 shows the radiation patterns of a gore reflector and a comparable smooth reflector. The gore reflector has lower on-axis gain and wider beamwidth than the smooth reflector.

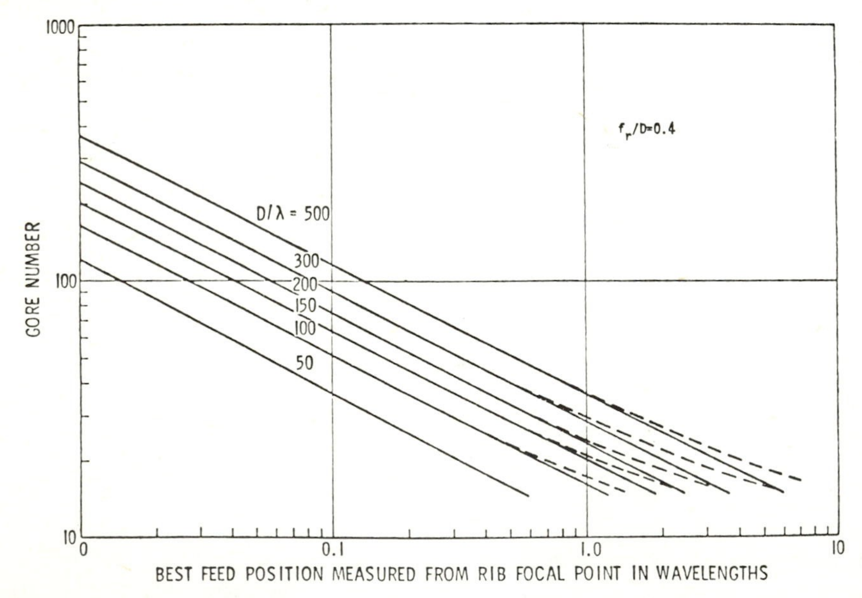

Fig. 3.29 *Best feed position*

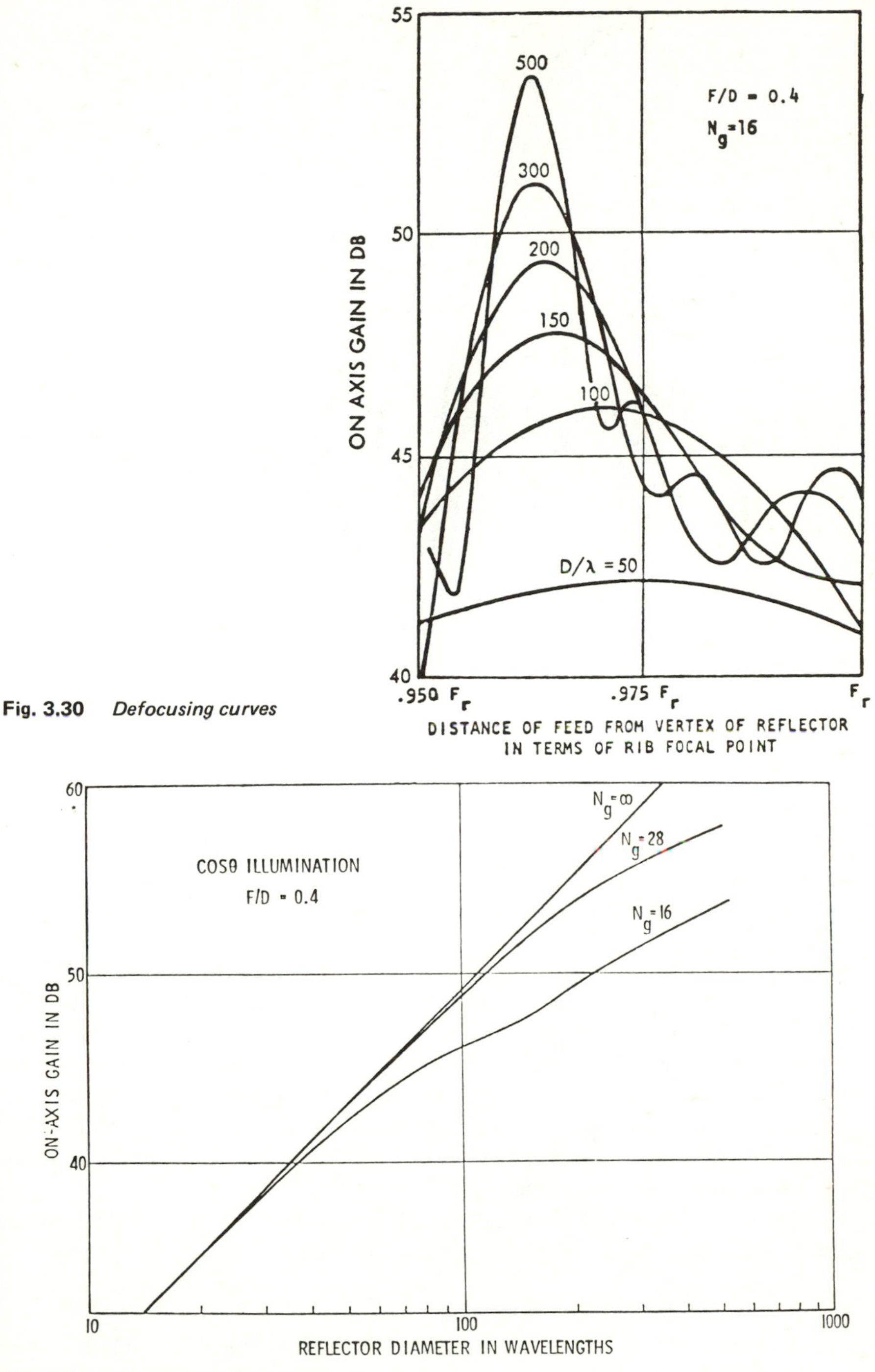

Fig. 3.30 *Defocusing curves*

Fig. 3.31 *On-axis gain frequency*

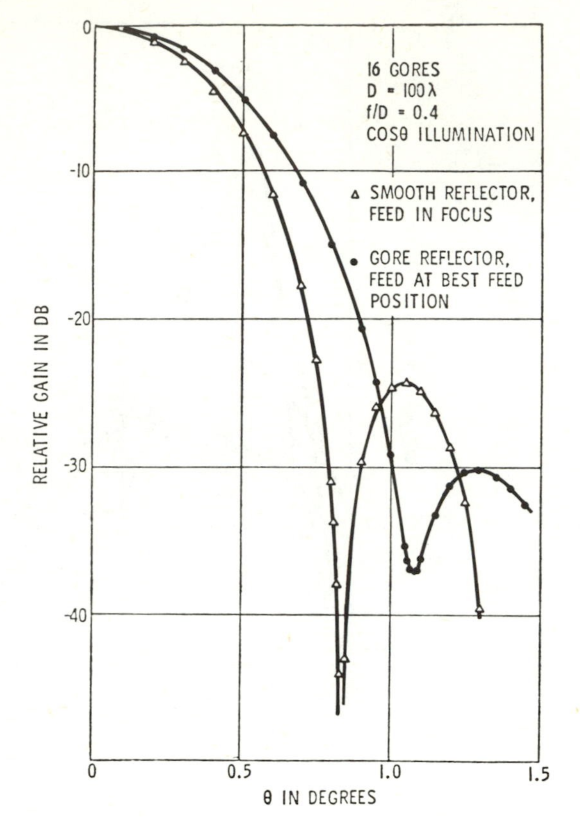

Fig. 3.32 *Radiation patterns*

3.2 **Cassegrain systems** (Contibution by P.A. Jensen)

In recent years increased use of the Cassegrain antenna system has been apparent, particularly since the advent of satellite communications. The feed is located near the vertex of the paraboloid and a hyperboloidal subreflector is placed with one focus coincident with the paraboloid focus (Fig. 3.33). Parallel rays from a distant source are reflected by the paraboloid and converge toward its focus. They are then intercepted by the hyperboloid and reflected to the primary feed at the second focal point. A family of hyperboloids can be used, and the choice must be made based on a variety of design criteria.

The Cassegrain antenna has several advantages, particularly for satellite ground terminals including:

- elimination of long transmission lines
- more flexibility in design of primary feeds than possible with front-fed paraboloids

- amplitude and phase control of the aperture illumination by subreflector shaping
- spillover past the subreflector is directed toward a cold sky
- low spillover past the paraboloid toward the ground for high elevation angles
- large depth of focus and field of view.

Some disadvantages are also apparent:

- greater blockage, particularly in small antennas (less than 100 wavelength apertures)
- higher sidelobes near the main beam
- not readily adaptable to use of poorly directional, frequency independent feeds feeds

Given $D =$
$F/D =$
Freq. $=$

(1) Hyperboloid Diameter

For Minimum Blockage

$d = \sqrt{k \lambda F} \approx$

$k =$

use $d =$

(2) Hyperboloid Focal Length

$$f = \frac{d}{2} (\text{ctn } \alpha + \text{ctn } \phi_o)$$

$$\phi_o = 2 \tan^{-1} \frac{D}{4F} =$$

$\alpha = \;^o$ (From Feed Design)

$f =$

(3) Cassegrain Magnification

$$M = \frac{D}{4F} \text{ ctn } \frac{\alpha}{2} =$$

(4) Hyperboloid Eccentricity

$$e = \frac{M + 1}{M - 1} =$$

(5) Hyperboloid Location

$$p = \frac{f}{2} \left(\frac{e - 1}{e} \right) =$$

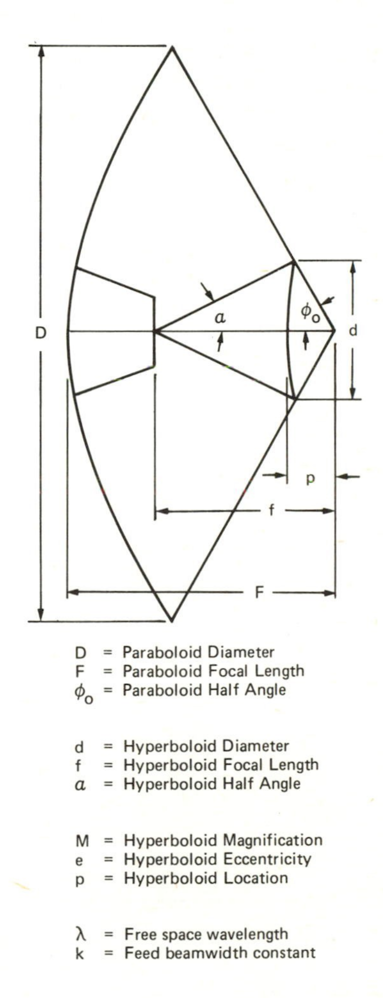

Fig. 3.33 *Determination of Cassegrain geometry*

- larger physical size of feed system
- higher cost.

3.2.1 Geometrical design considerations

Determination of the geometry of a Cassegrain antenna system has been reduced to a simple set of arithmetic calculations[31,32] (Fig. 3.33). This approach, however, sometimes leads to difficulties in feed design or excessive fabrication costs of the

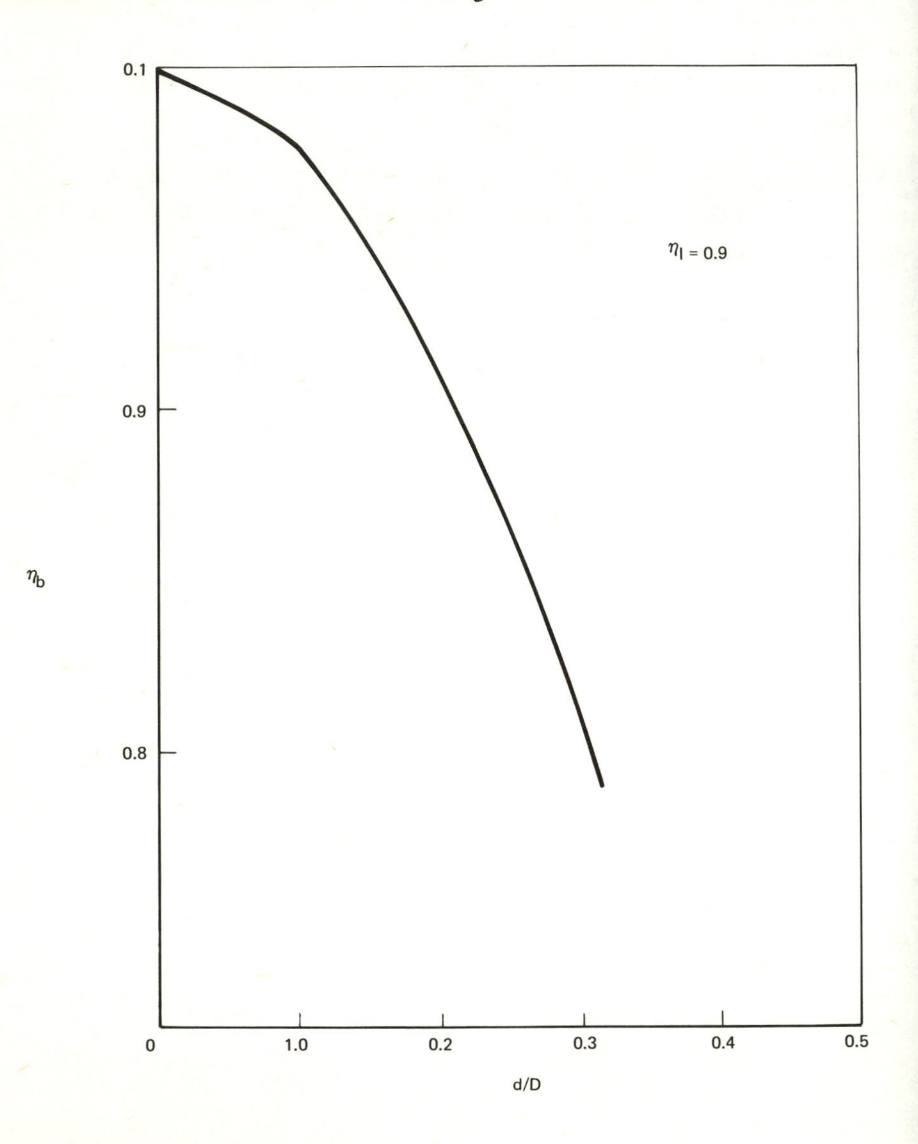

Fig. 3.34 *Blockage efficiency vs. blockage ratio*

feed, feed support, and subreflector. The principal factors in the design procedure which require engineering judgement to result in an optimum design are the choice of subreflector size and the location of the primary feed. The material in this Section represents an attempt to provide guidance in the choice of these parameters.

(i) *Cassegrain geometry*

Calculate hyperboloid diameter for minimum blockage

$$d_m = \sqrt{k\lambda F} \tag{3.57}$$

where d = hyperboloid diameter, F = paraboloid focal length, and the beamwidth constant k = 2 for an average feed with 10 dB taper. If $d_m/D > 0.2$, the hyperboloid diameter should be d_m, unless it is too small (in terms of wavelengths) for efficient operation. If $d_m/D < 0.1$, the value $d = 0.1\ D$ should be considered (unless the physical size becomes too large). This permits use of a smaller feed aperture.

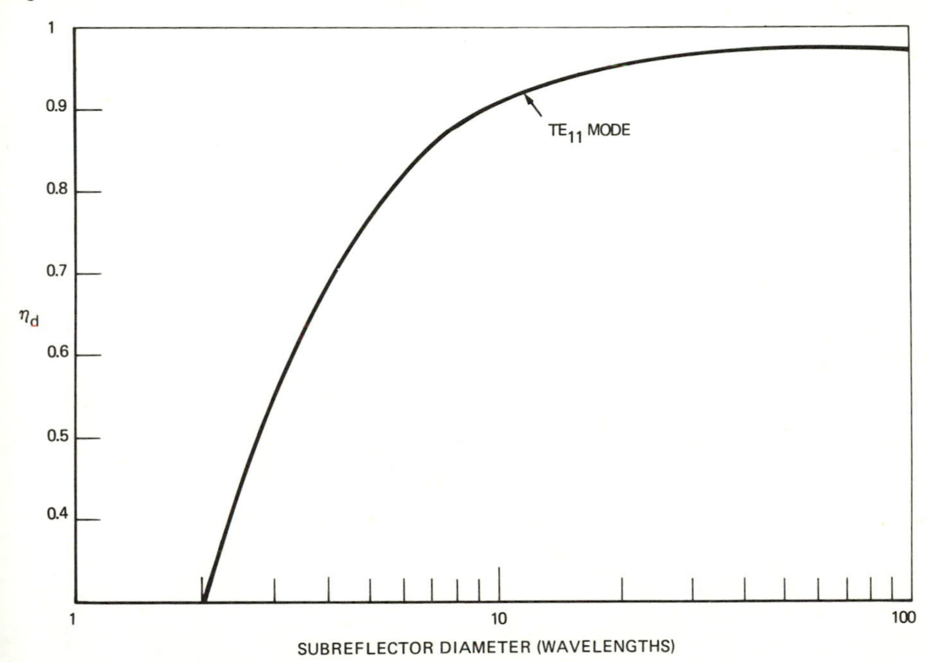

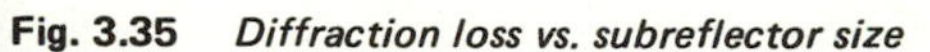

Fig. 3.35 *Diffraction loss vs. subreflector size*

(*a*) The blockage loss for $d/D < 0.1$ or 1% area blockage is less than 0.1 dB, but increases rapidly above 1% blockage as shown in Fig. 3.34.

(*b*) If the subreflector size in wavelengths, d/λ, becomes too small, an excessive spillover loss past the main reflector is encountered (Fig. 3.35). In such a case, a prime-focus feed should be considered. However, the I^2R loss of feeder line to the prime-focus, which adds noise in addition to gain loss, must also be considered.

(*c*) The minimum size antenna for which a Cassegrain design is practical for a given application requires a judgment based on overall efficiency requirements. Figs. 3.5 and 3.36 provide an aid in its determination. If it is not possible to make the subreflector electrically large, $d/\lambda > \approx 10$, a significant spillover loss past the main reflector results (Fig. 3.35). Also if the main reflector is not electrically large, i.e. has a narrow beamwidth, the blockage loss of the subreflector, even for minimum blockage, may become larger than desired (Fig. 3.36).

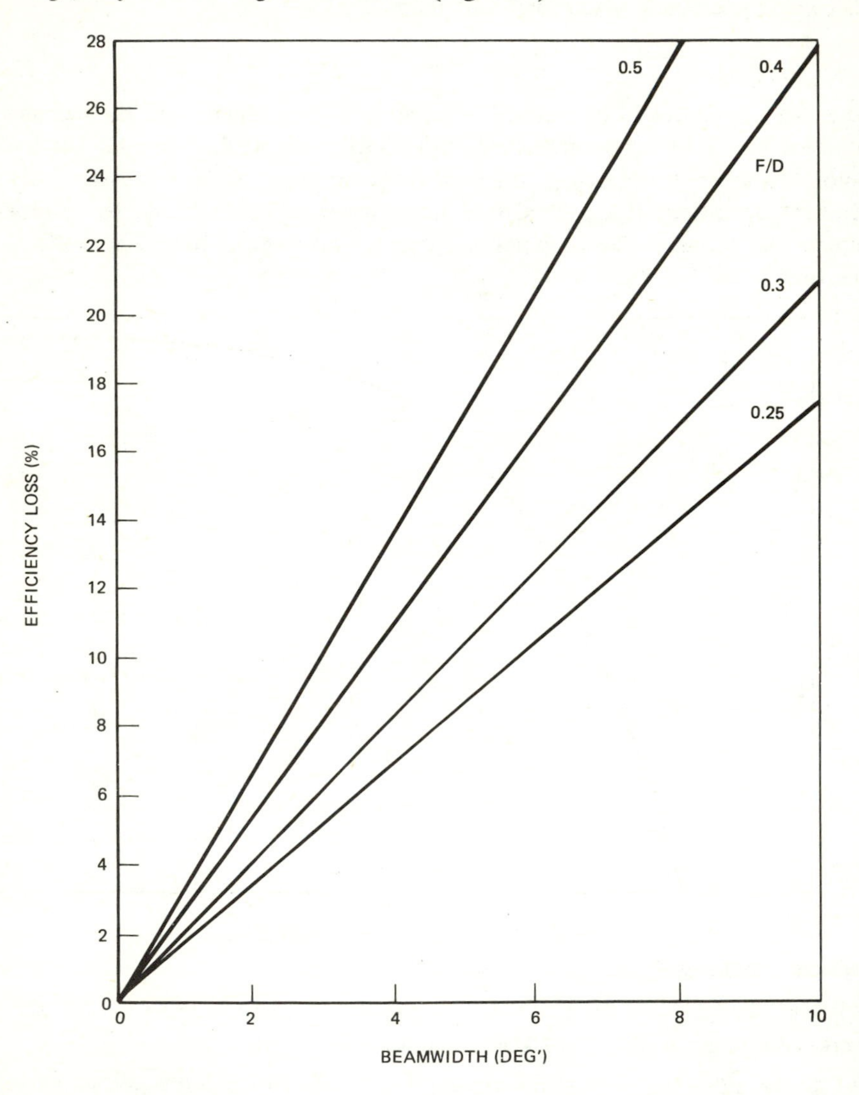

Fig. 3.36 *Minimum blockage loss vs. antenna beamwidth*

(*d*) Using Figs. 3.34 and 3.36, the best compromise for efficiency may be determined, but this may not be as important as the physical constraints on the subreflector and feed sizes. Also, the feed efficiency may be significantly affected by its aperture size. Thus, the feed design may have to be formulated before choosing a final subreflector diameter.

(*e*) For antennas large in terms of wavelength, the subreflector blockage loss and diffraction or spillover loss is usually small. The final design is usually governed by the feed design.

(*f*) The best design is achieved by using measured feed patterns to optimise the overall antenna efficiency before choosing the final subreflector dimensions.

(ii) Hyperboloid focal length and feed design

The feed location, and therefore aperture size, is related to the Cassegrain geometry by

$$f = \frac{d}{2}\,(\text{ctn}\,\alpha + \text{ctn}\phi_0). \qquad (3.58)$$

For a given F/D ratio and subreflector diameter, the focal length f and feed beamwidth 2α required for efficient illumination of the hyperboloid are inversely related. Several considerations enter into the choice of these parameters:

(*a*) Restrictions on feed location may include length or weight of a feed support cone, space available behind the paraboloid vertex, the lateral space available for the feed aperture and circuitry.

(*b*) Many feeds, especially multimode types, provide a higher aperture efficiency for relatively broad 10 dB beamwidths, e.g. greater than 25°.

(*c*) The smaller the required beamwidth, the larger the feed aperture becomes, and in turn the horn length required to maintain reasonable horn aperture phase error increases. This may result in a very difficult or expensive horn fabrication, which might be avoided if the feed design is considered when choosing the feed position.

The feed design and location usually becomes an iterative process to arrive at an optimum design. A detailed consideration of the feed design is desirable before choosing the feed location and the optimum subreflector size. Once these choices have been made, the remainder of the Cassegrain design parameters are easily determined as illustrated in Fig. 3.33.

Alignment tolerances

Once the Cassegrain geometry has been determined, it is desirable to know the alignment tolerances[33,34] required for fabrication and assembly of the antenna.

This information also provides guidance in choosing the amount of adjustment to allow for alignment. The two main items considered in alignment of a Cassegrain antenna are those factors resulting in defocus and beam pointing errors.

(i) Defocus

Two sources of defocus arise from alignment errors along the reflector axes. The most critical is positioning of the hyperboloid focus on the best-fit paraboloid focus. Of somewhat less importance is locating the feed phase centre at the second hyperboloid focus. Defocus phase error due to displacement of the subreflector is given approximately by:

$$\Delta\phi_s = \frac{2\pi}{\lambda}(2 - \cos\alpha - \cos\phi)\Delta_t \tag{3.59}$$

1. Hyperboloid Rotation

$$\theta_R = K_1 \tan^{-1} \frac{f}{F} \frac{2\beta}{M+1}$$

2. Hyperboloid Translation

$$\theta_T = -k_2 \tan^{-1} \frac{h}{F} \frac{M-1}{M}$$

3. Feed Translation

$$\theta_F = k_3 \tan^{-1} \frac{\delta}{MF}$$

4. Paraboloid Rotation

$$\theta_p = K\gamma$$

5. Paraboloid Translation (Vertex Shift)

$$\theta_s = K \tan^{-1} \delta_{v/F}$$

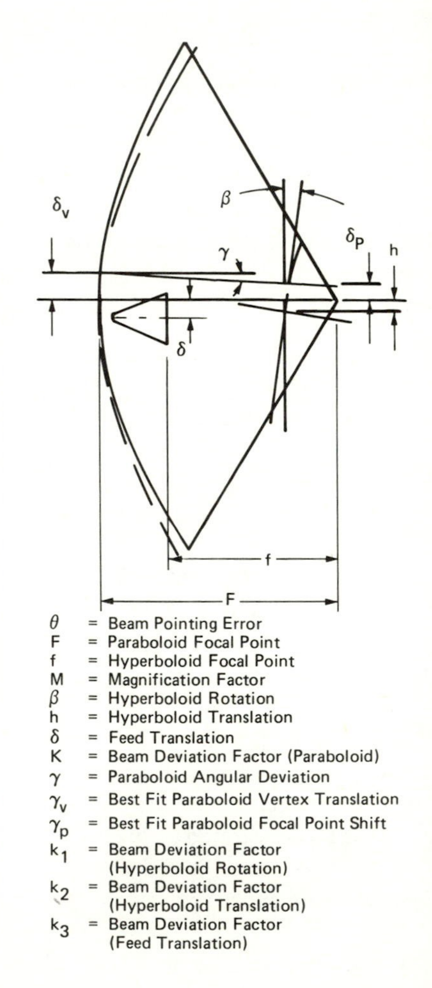

Fig. 3.37 *Cassegrain antenna beam pointing error analysis*

where $\triangle_t$ is the axial difference between the paraboloid and hyperboloid foci. For feed displacement, the defocus phase error is given approximately by:

$$\triangle\phi_f = \frac{2\pi}{\lambda}(1 - \cos\alpha)\,\delta_0 \tag{3.60}$$

where δ_0 is the axial feed displacement from the focus. The relative magnitude of the phase errors generated by equal movements of the feed and subreflector is:

$$\frac{\triangle\phi_s}{\triangle\phi_f} = 1 + \frac{1 - \cos\phi}{1 - \cos\alpha}. \tag{3.61}$$

For a typical Cassegrain antenna $\phi = 62°$, $\alpha = 15°$ and, consequently $\triangle\phi_s/\triangle\phi_f \approx 16$; thus movement of the subreflector results in 16 times more phase error across the antenna than an equal movement of the feed.

(ii) Beam pointing errors

Misalignments of the reflectors and feed transverse to the antenna axis or rotations about the foci result in phase errors causing beam pointing errors and pattern asymmetries, along with a loss in antenna gain. The principal misalignments are illustrated in Fig. 3.37 along with equations for determining the resulting beam-pointing error. Pointing errors resulting from combinations of these factors may be calculated by superposition, using care to determine the direction of beam motion for each effect.

3.2.2 Efficiency analysis

A principal consideration in the design of Cassegrain and other type antennas is the antenna gain obtainable with a given reflector and feed system. The parameters influencing the gain of an antenna can be treated as efficiency factors by which the maximum theoretical gain is multiplied to yield the actual gain. This approach permits the designer to determine which parameters are most significant in limiting the gain, and therefore which factors can most benefit by improved design.

The major factors considered in evaluating antenna efficiency are listed in Table 3.1. The organisation shown permits one to separately evaluate the effect of the feed and the Cassegrain reflector geometry on the overall efficiency.

Primary feed

(i) Spillover

To most effectively utilise the large area of a reflector antenna, the energy radiated from the feed must be distributed over the aperture with a reasonable degree of uniformity. With most primary feeds this results in a significant amount of energy

Table 3.1 *Antenna efficiency analysis*

Efficiency factor		Comments	Value
A *Primary feed*			
Spillover	η_s		
Illumination	η_i	From primary pattern integration	
Cross polarisation	η_x		
Feed insertion loss	η_1	$\eta_1 = 10^{-L/10}$	L = dB
VSWR	η_v	$\eta_v = 4\,V/(1+V)^2$	V =
Phase error loss			
1 Quadratic	η_p		$\beta 2 =$
2 Astigmatism	η_{fa}	$\eta_{fa} = 1/(1 + .25\,\eta_i \beta_a{}^2)$	$\beta_a =$
Feed efficiency	η_f		
B *Reflector*			
Blockage			
1 Feed/subreflector	η_b	$\eta_b = (1 - B^2/\eta_i)^2$	$B^2 =$ %
2 Strut	η_{sb}	$\eta_{sb} = (1 - A_s)^2$	$A_s =$ %
Surface tolerance	η_r	$\eta_r = \exp[-16\pi^2(\sigma_m{}^2 + \sigma_s{}^2)/\lambda^2]$	
		Paraboloid	$\sigma_m =$
		Subreflector	$\sigma_s =$
Astigmatism	η_{ra}	$\eta_{ra} = 1/(1 + .25\eta_i\beta_{ra}{}^2)$	$\beta_{ra} =$
Edge diffraction	η_d	Cassegrain only	
Reflector efficiency	η_R		
C *Radome loss*	η_r		
Expected efficiency	η_e	$\eta_e = \eta_F \times \eta_R \times \eta_r$	
Measured efficiency	η_m		

radiating in angular regions outside of the subtended angle of the reflector. The spillover efficiency η_s is defined as that percentage of the total energy radiated from the feed that is intercepted by the subreflector.

Spillover is one of the most significant factors in feed efficiency, and is somewhat difficult to evaluate due to edge discontinuities. It is normally calculated on a geometrical-optics basis using computer analysis on the measured or computed primary radiation patterns. This factor is expressed analytically as:

$$\eta_s = \frac{\int_0^{2\pi}\int_0^{\psi} f^2(\psi,\phi)\sin\psi\, d\psi\, d\phi}{\int_0^{2\pi}\int_0^{\pi} f^2(\psi,\phi)\sin\psi\, d\psi\, d\phi} \tag{3.62}$$

where $f(\psi, \phi)$ is the feed pattern.

(ii) Illumination

The more nearly uniform the energy distribution across an aperture, the more effectively the aperture area is utilised. Primary feeds almost always provide a tapered illumination at the dish edge. The illumination efficiency, η_I, is a measure of the uniformity of the aperture illumination, and expresses the reduction in gain below a uniformly illuminated aperture caused by the pattern taper. This efficiency as given by Silver[35] is:

$$\eta_I = \frac{\left| \int_A I(\zeta, \gamma)\, dA \right|^2}{\int_A \left| I(\xi, \gamma) \right|^2 dA} \tag{3.63}$$

where $I(\xi, \gamma)$ is the amplitude distribution over the radiating area. Again, this factor is usually determined by computer integration of the primary feed pattern.

For typical primary patterns, the illumination efficiency decreases as the spillover efficiency increases, and a compromise is required for the best overall feed efficiency. It thus becomes convenient to combine the illumination and spillover efficiencies into a common term. For a paraboloid reflector this expression is given by[35]

$$\eta_I\, \eta_s = \frac{1}{4\pi^2} \cot^2\left(\frac{\psi_0}{2}\right) \left| \int_0^{2\pi} \int_0^{\psi_0} [G_f(\psi, \varphi)]^{\frac{1}{2}} \tan \psi/2\, d\psi\, d\varphi \right|^2 \tag{3.64}$$

where $G_f(\psi, \varphi)$ is the feed system gain function. Potter[36] elaborates on a simple graphical technique for evaluating both this expression and the spillover efficiency alone. For a small aperture angles encountered in a Cassegrain system, this calculation performed using a TE_{11} mode horn shows the maximum feed efficiency is about 75% (Fig. 3.38). Table 3.2 illustrates the maximum $\eta_s\, \eta_I$ efficiency attainable for several typical feed configurations.

The above calculation neglects loss due to quadratic error in the horn caused by the horn flare angle. When feed efficiency is calculated using theoretical primary patterns this effect must be considered. This may be done by including the phase error when calculating the primary patterns, or by use of a horn quadratic phase error efficiency factor η_p. Fig. 3.39 illustrates this loss for a conventional TE_{11} mode conical horn. When measured primary patterns are used for computation of $\eta_I\, \eta_s$, this term is already included by its effect on the radiation pattern.

(iii) Cross-polarisation

Cross-polarised radiation from a Cassegrain antenna can result from both the primary feed and reflector curvature. Reflector curvature primarily becomes a problem for small F/D ratios[36]. For Cassegrain antennas with $M > 2$, the loss due to reflector-generated cross-polarisation is very small and generally ignored, even with deep primary reflectors. Fig. 3.40 shows the reflector-generated cross-polarisation loss versus subtended angle[36,37], illustrating the very low loss encountered with half angles less than 30°.

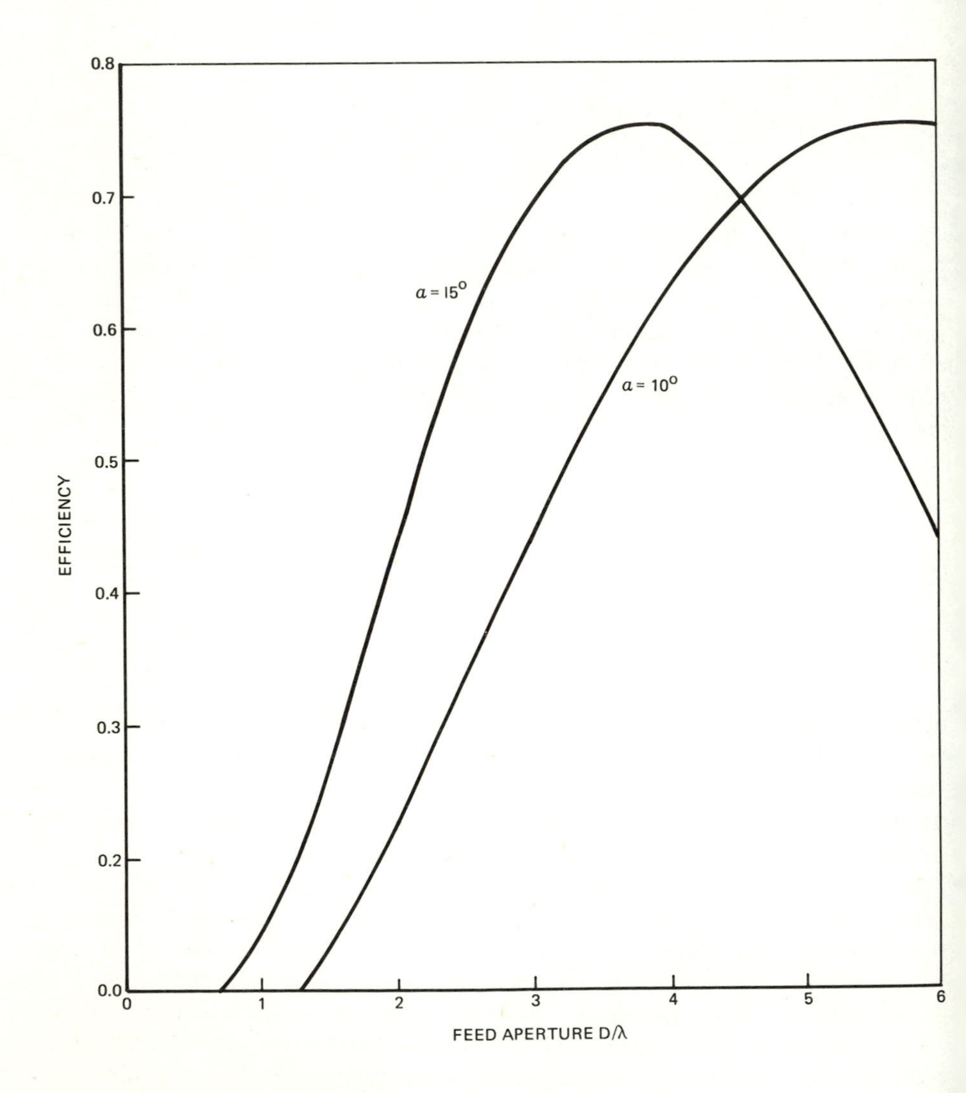

Fig. 3.38 *Feed efficiency vs. size TE_{11} mode conical horn*

Table 3.2 *Cassegrain antenna: Feed system maximum efficiencies*

	$(\eta_I \times \eta_s)$	η_x	η_{max}
Conical horn TE_{11} mode	0.751	0.96	0.720
Conical multimode TE_{11} + TM_{11} mode	0.831	0.998	0.830
Nine horn centre multimoded	0.879	0.987	0.868
Rectangular horn TE_{10} only			
b/a = 1	0.700	0.98	0.685
b/a = .79	0.744	0.98	0.729
Rectangular multimode TE_{10} + TE_{30} + TE_{12}	0.828	0.998	0.827
Four horn	0.578	0.98	0.567

η_I = illumination efficiency
η_s = spillover efficiency
η_x = crosspolarisation efficiency

Cross-polarisation generated by the feed horn generally results in efficiencies of about 96–99% depending on the feed type. Fig. 3.41 shows the amount of cross-polarised energy in a conical multimode horn propagating the TE_{11} and TM_{11} modes versus the relative amount of TM_{11} excitation.[38]

Typical computer programs normally used to calculate feed efficiency will calculate that component of cross-polarisation loss generated by non-rotational primary amplitude and phase patterns. This component of cross-polarisation does not necessarily include the total cross-polarisation loss of the feed. If measured or computed cross-polarised radiation patterns, normalised to the principle radiation pattern gain, are available for a feed, it is then possible to determine by integration the amount of cross-polarisation energy lost.

(iv) Feed phase error

There are four sources of possible loss due to the feed phase characteristics. First is the effect of horn quadratic error on its primary radiation pattern as discussed previously. This effect is principally noticed as an increased spillover loss, and is implicit in the computation if spillover efficiency calculations are performed on measured patterns.

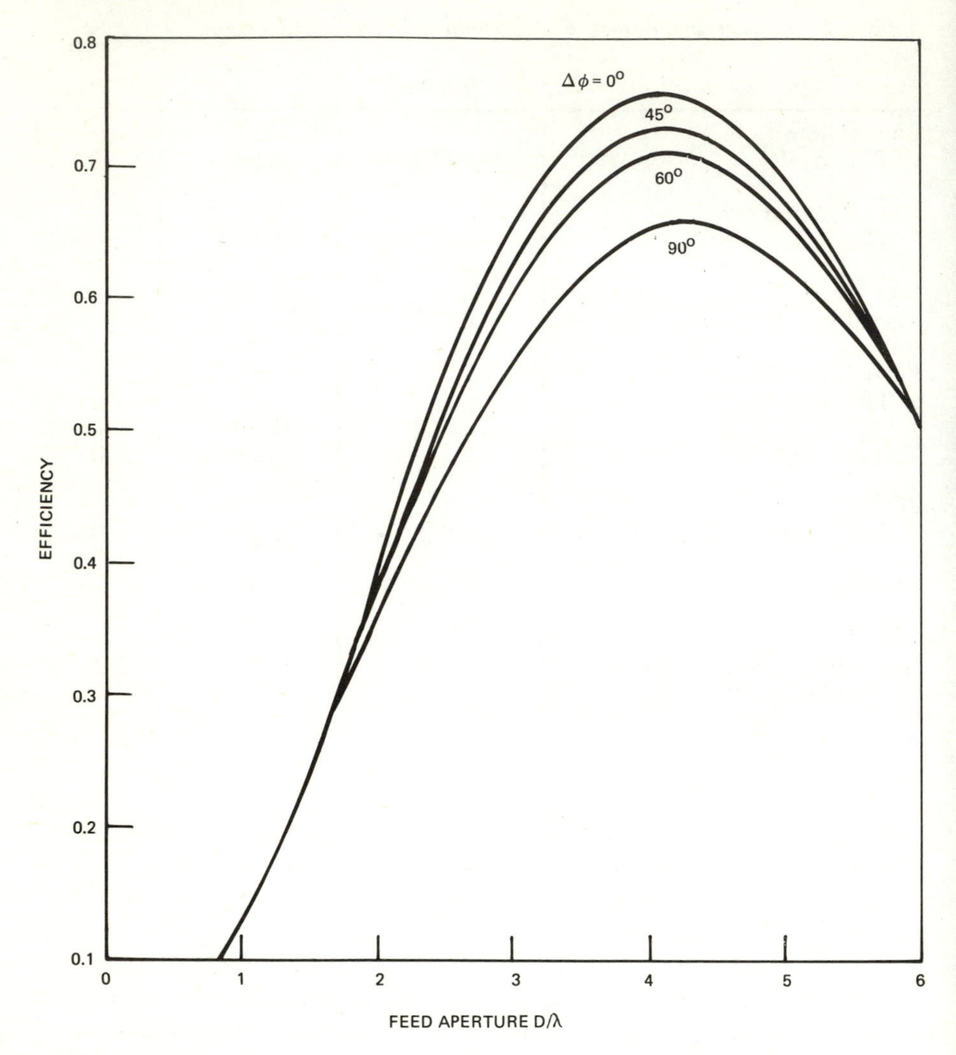

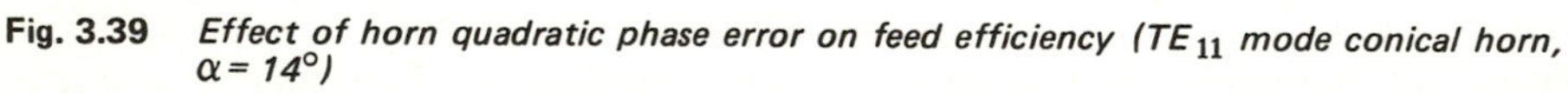

Fig. 3.39 *Effect of horn quadratic phase error on feed efficiency (TE_{11} mode conical horn, $\alpha = 14°$)*

A second source occurs if the feed phase centre is not coincident with the sub-reflector focal point. This results in a quadratic phase error across the dish whose magnitude is given approximately by

$$\Delta\phi = \frac{2\pi}{\lambda}(1 - \cos\alpha)\,\delta \qquad (3.65)$$

where α is the hyperboloid half angle and δ is the axial feed displacement from the focus. Generally, this factor is small or is removed entirely during the antenna alignment and focusing.

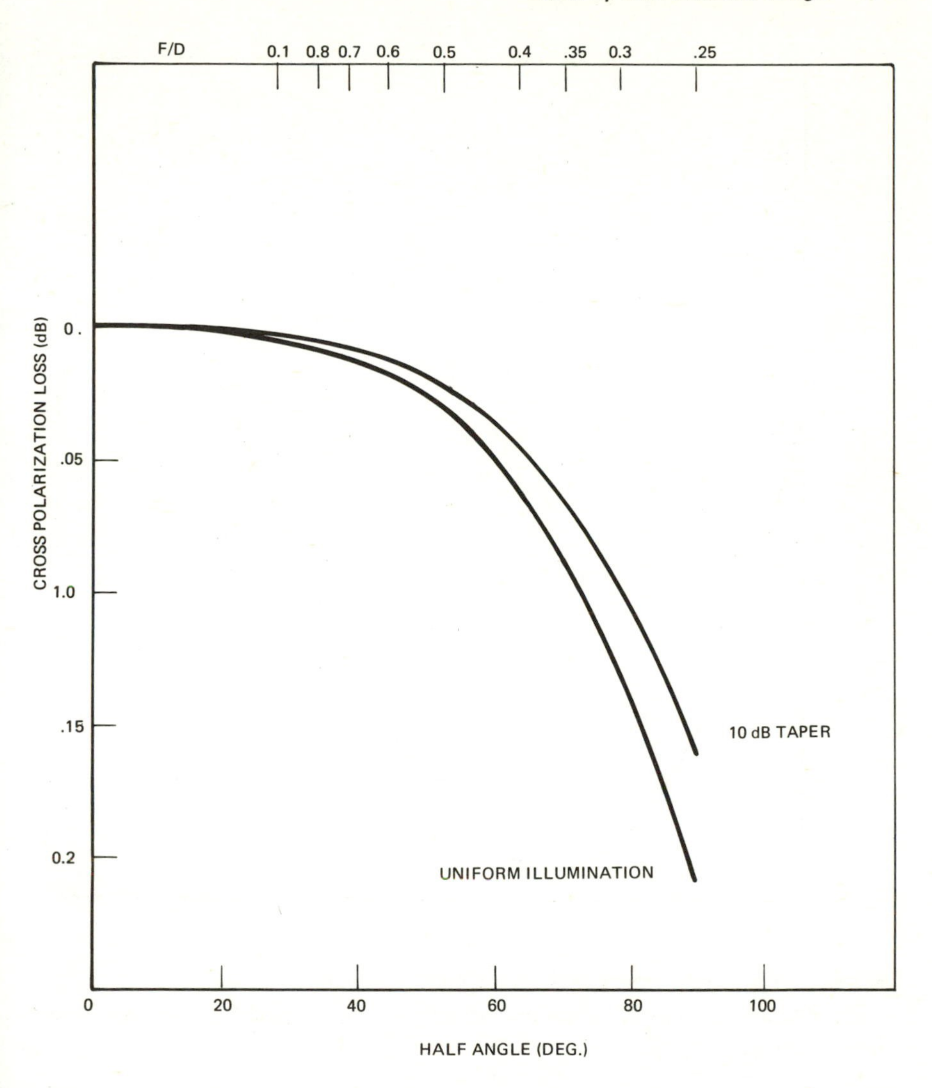

Fig. 3.40 *Cross polarization vs. subtended half angle due to reflector curvature*

The third and fourth factors are due to failure of the feed phase centre, and can be determined only by measurement of the feed phase patterns. Two cases are commonly observed in the larger feeds required for Cassegrain antennas: (*a*) lack of a defined phase centre versus angle which generally results in a quadratic phase error, and (*b*) a phase centre which varies with the plane of cut, which results in an astigmatic phase error on the dish.

The loss due to a quadratic phase error of the feed phase pattern can be partially compensated for by defocusing the feed in the dish. If this is not done, a loss is sustained as illustrated in Fig. 3.42.

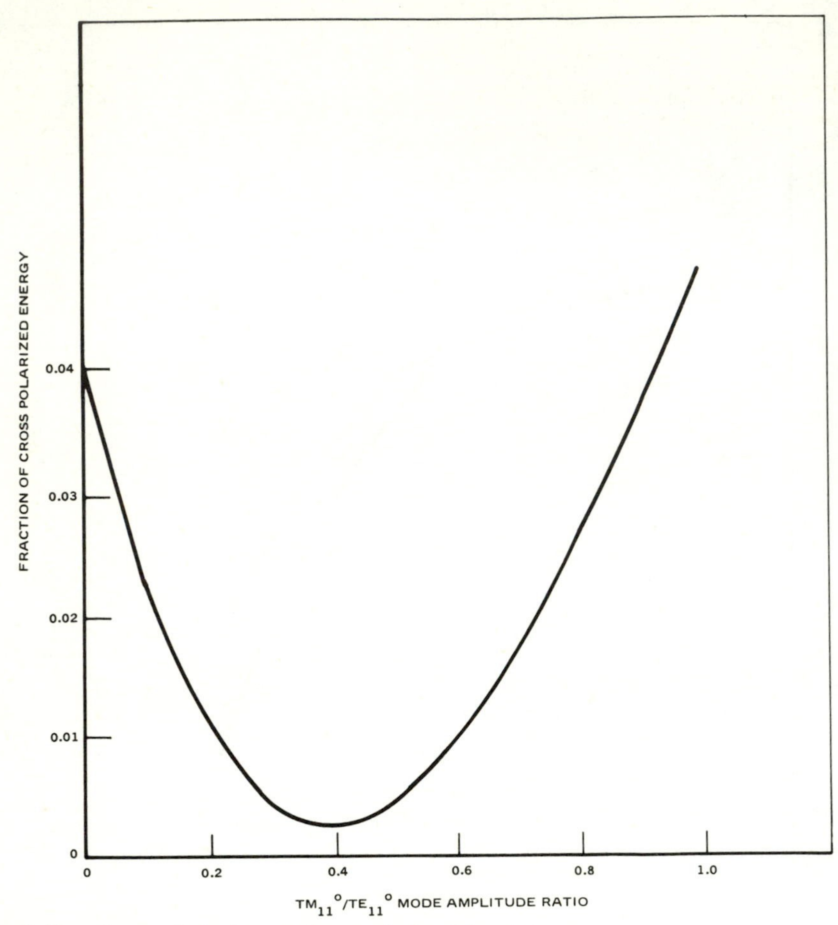

Fig. 3.41 *Cross polarised energy in a conical multimode horn*

The astigmatic phase error cannot usually be compensated for and will result in a loss in gain given by:[39]

$$\eta_{fa} = \frac{1}{1 + 0.25\,\eta_I\,\beta^2} \tag{3.66}$$

where $\pm\ \beta$ is the extreme phase error over a circular aperture caused by the astigmatism, and η_I is the illumination efficiency (Fig. 3.43).

(v) Feed insertion loss and VSWR

The insertion loss of the feed and transmission line is usually small for a Cassegrain antenna. However, it should be considered in the analysis. The efficiency for a given insertion loss is found from

$$N_l = 10^{-L/10} \tag{3.67}$$

where L = insertion loss in dB. Similarly, the loss due to mismatch is normally small. The efficiency factor for VSWR loss is given by:

$$\eta_v = 4V/(1 + V)^2 \quad (3.68)$$

where V = VSWR.

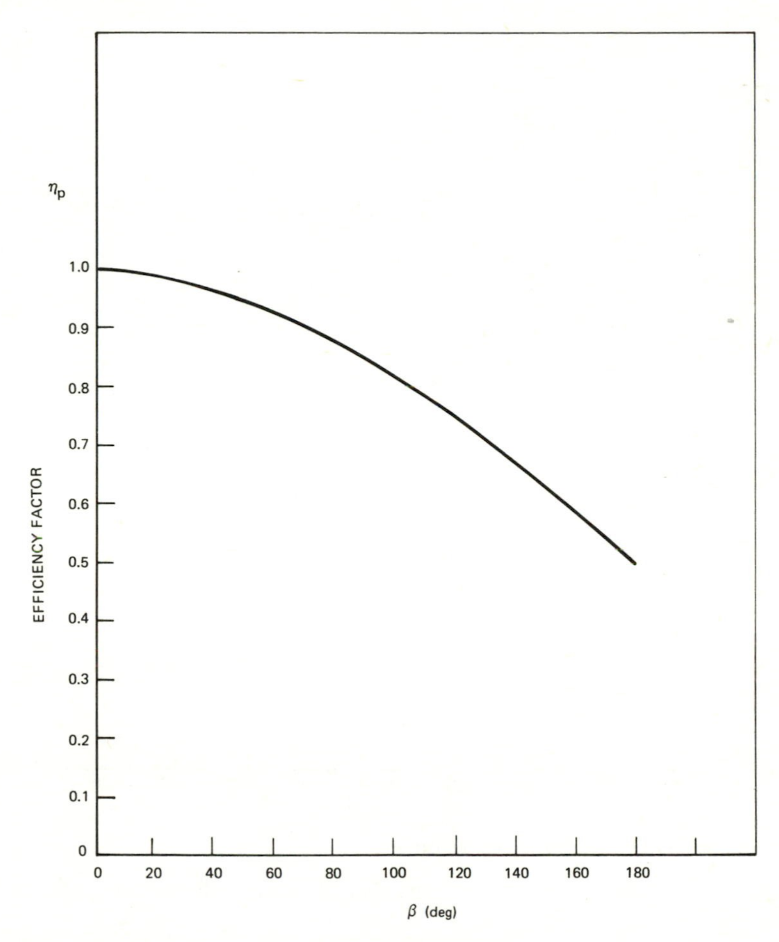

Fig. 3.42 *Efficiency factor vs. quadratic phase error*

Reflector

The energy incident on and reflected from the paraboloid reflector encounters shadowing from both the feed/subreflector central blockage, and from the subreflector support struts. This blockage has three effects:

(*a*) It reduces the total area available for capture of RF energy.

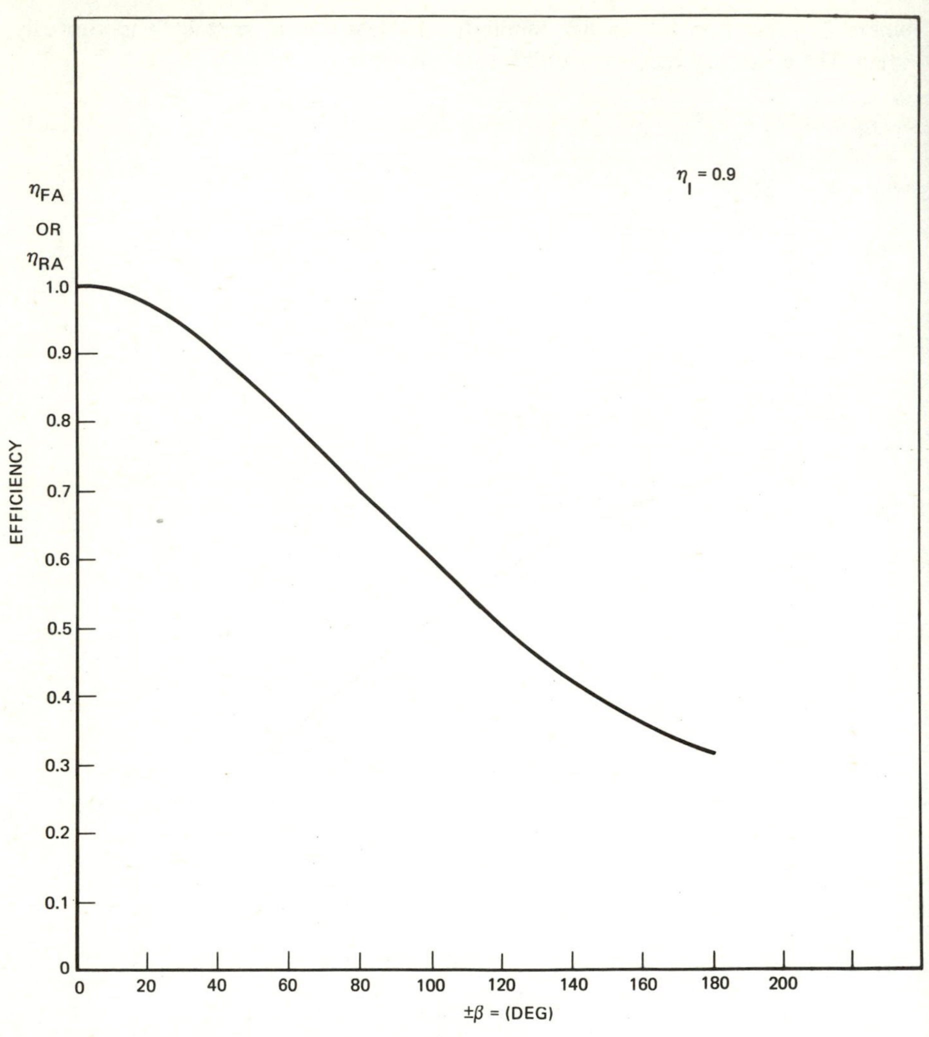

Fig. 3.43 *Efficiency factor vs. astigmatic phase error*

(*b*) It reduces the total amount of energy available for collimation in the main beam.

(*c*) It increases the sidelobes due to the discontinuous aperture distribution and scattering of energy incident on the blocked area.

The feed/subreflector blockage can be minimised by optimum feed design. When the feed and subreflector sizes are chosen such that the projected geometrical blockage of each is equal, this is the minimum blockage condition. The subreflector diameter which yields this condition is given within a few percent by eqn. 3.57.[31]

However, as discussed previously this may not represent the optimum design choice as other factors must be considered.

The resulting blockage loss expressed as an efficiency factor is given by:

$$\eta_b = \left(1 - \frac{1}{\eta_i}\,\frac{\text{area blocked}}{\text{total area}}\right) \qquad (3.69a)$$

or

$$\eta_b = \left[1 - \frac{1}{\eta_i}\left(\frac{d}{D}\right)^2\right]^2 \qquad (3.69b)$$

This expression weights the area blocked by the illumination efficiency η_I, thus relating the central area blockage to its importance in the overall gain contribution (Fig. 3.34).

The support structure blockage is a more complex area consisting of both strip blockage of the plane wave off the paraboloid, and pie-shaped regions resulting from blockage of the spherical RF wave between the subreflector and the paraboloid (Fig. 3.44). To approximate this blockage effect it is necessary to determine the total projected area blockage from the struts on the dish aperture. Also, the effective scattering cross-section[21,40] of the struts may be larger than their physical cross-section depending on the strut cross-section to wavelength ratio, and this factor should also be considered. The expression generally used for strut blockage

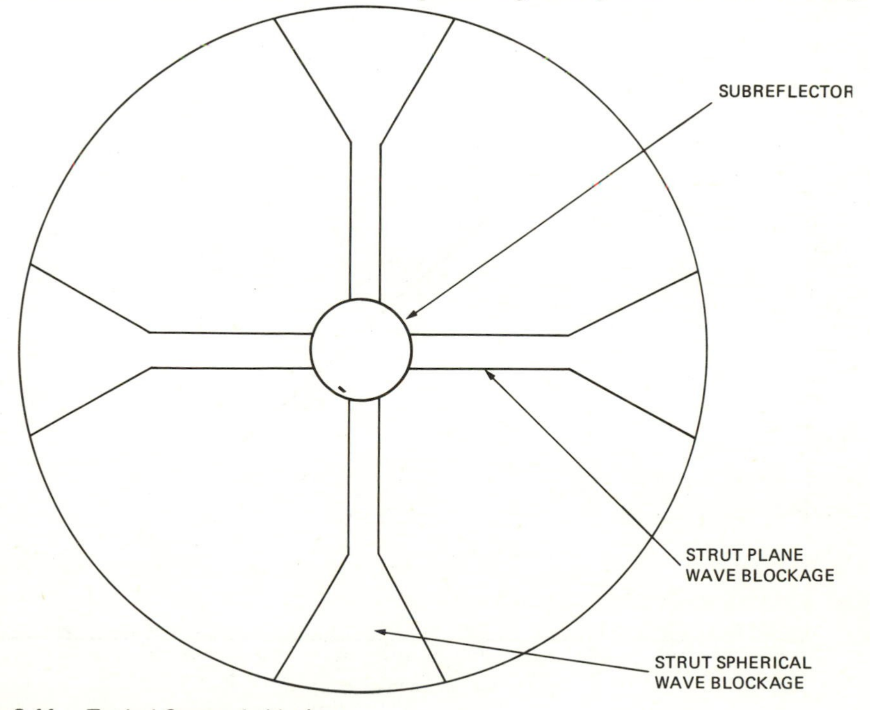

Fig. 3.44 *Typical Cassegrain blockage*

loss efficiency is:

$$\eta_{sb} = \left(1 - \frac{\text{effective area blocked}}{\text{total area}}\right)^2 \tag{3.70a}$$

or

$$\eta_{sb} = \left(1 - A_s\right)^2 \tag{3.70b}$$

and is plotted in Fig. 3.45. Weighting for illumination is normally not used because the strut blockage area is across the entire aperture. However, if the pie-shaped blockage at the dish edge becomes large, this expression will be somewhat pessimistic.

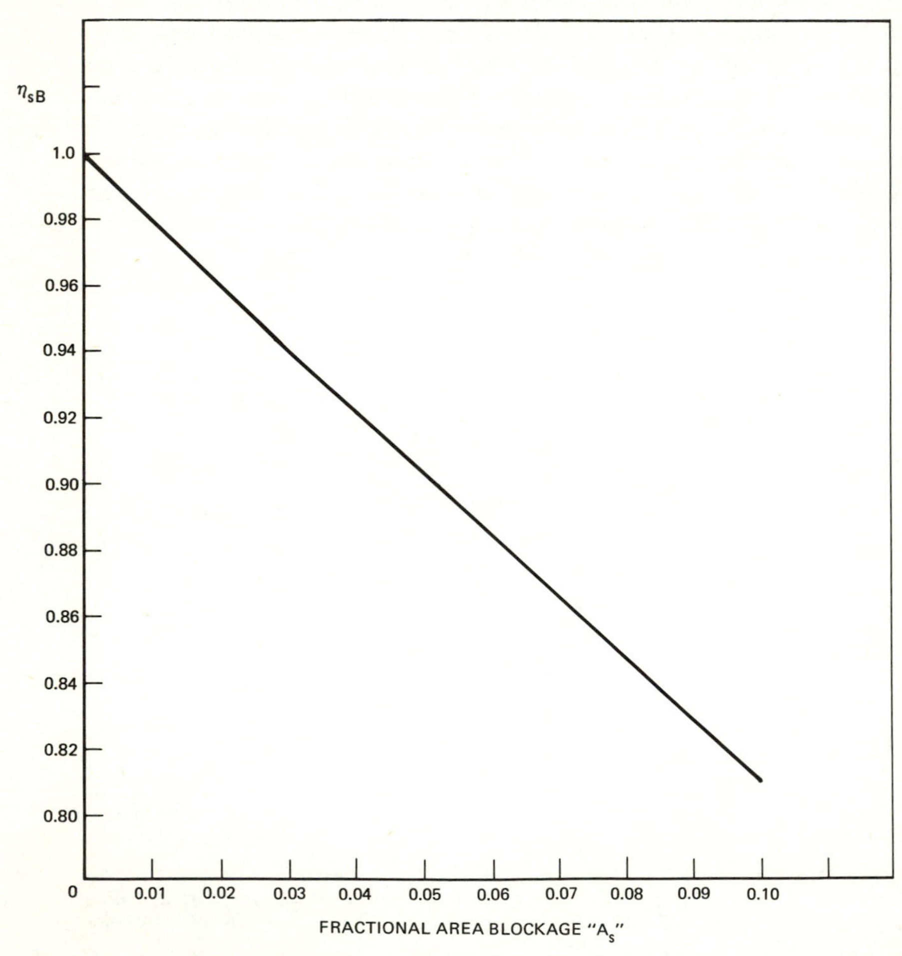

Fig. 3.45 *Strut blockage efficiency vs area blockage*

Surface tolerance

For the two reflectors of a Cassegrain system the tolerance efficiency may be approximated by (Section 3.1.2):

$$\eta_r = \exp\left[-16\pi^2 (\epsilon_m{}^2 + \epsilon_s{}^2)/\lambda^2\right] \qquad (3.71)$$

where ϵ_m is the RMS tolerance of main reflector and ϵ_s is the RMS tolerance of subreflector. A plot of surface tolerance efficiency versus frequency is shown in Fig. 3.46.

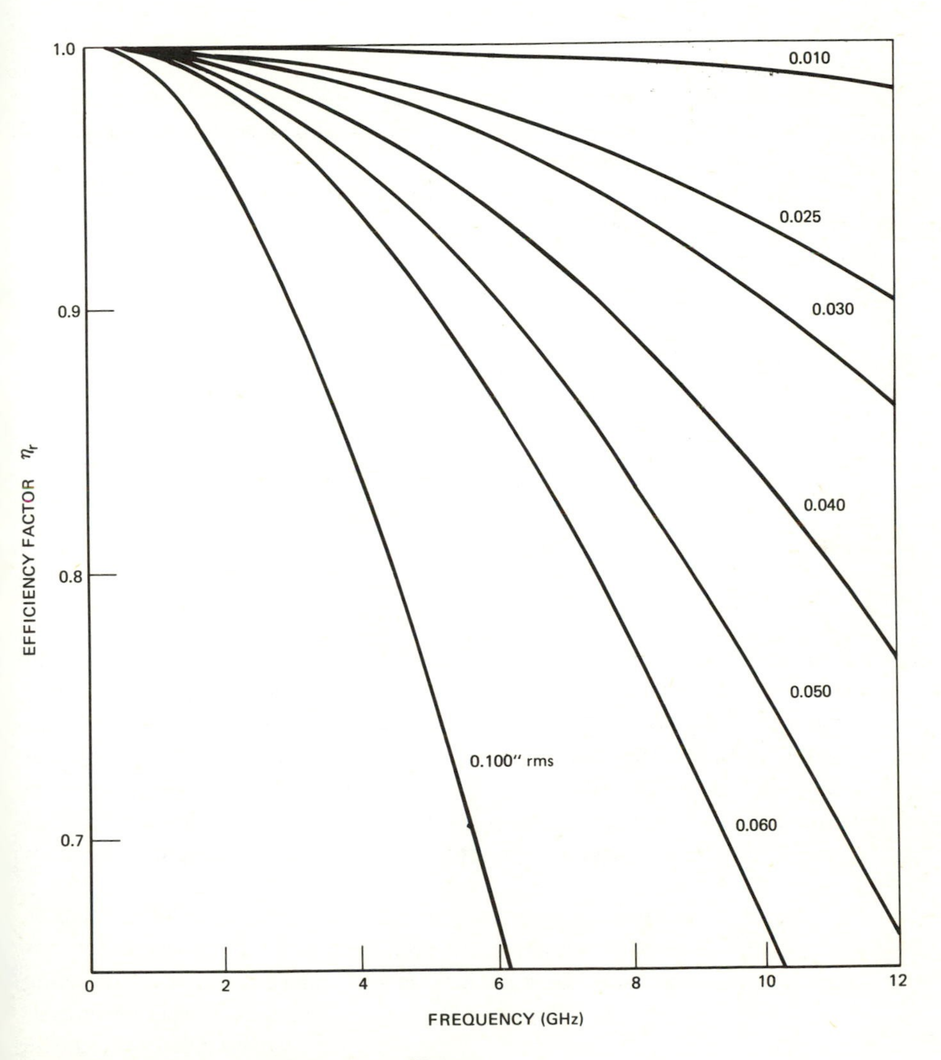

Fig. 3.46 *Reflector surface tolerance efficiency*

Astigmatism

As with the feed, it is also possible for the reflector to create an astigmatic phase error across the aperture. Reflector astigmatism results when a paraboloid deforms in such a way that the circular aperture becomes elliptical. Such forces as wind, temperature gradients, gravity loading or strut loads could cause this effect. Contours through the reflector remain parabolic, but focal lengths of the parabolas for various cuts through the dish are different lengths. If the feed is placed on axis midway between the extremes of the focal length deviations, this loss has been shown to be[39]:

$$\eta_{ra} = 1/(1 + 0.25\, \eta_I\, \beta^2) \qquad (3.72)$$

where $\pm\beta$ is the extreme phase error over a circular aperture caused by the astigmatism, and η_I is the illumination efficiency. This function was plotted in Fig. 3.43 for the feed astigmatic contribution.

If both the feed and the reflectors have astigmatic errors, it is possible for the phase errors to add, and the combined efficiency should be calculated for the total resultant phase error rather than as separate effects.

Diffraction

This factor accounts for spillover losses of the reflector system that are neglected when assuming that ideal geometrical optics govern the reflections. This loss is of two types:

(i) diffraction around the subreflector
(ii) spillover past the main reflector

The first loss is small compared to the normally calculated spillover efficiency and appears as low level sidelobes within $\pm$ 10 to $\pm$ 20° of boresight. The second loss can be significant and occurs because the subreflector is finite and cannot produce an infinite beam cutoff beyond the main reflector subtended angle. This factor approaches unity as the subreflector increases in size ($d/\lambda \gg 10$), and also as the edge illumination taper approaches zero (no edge currents). Calculation is difficult and is only approximated with the present vector diffraction computer programs (Fig. 3.35).

Other miscellaneous losses may be encountered in particular systems such as reflector leakage, polarisation alignment, etc., and should be considered as applicable. However, the above factors are those of primary importance, and using care in predicting these effects will generally provide a reasonably accurate analysis of the antenna efficiency. Additional References[41–52] are included to indicate other areas of particular relevance to Cassegrain performance.

3.2.3 Improved design techniques

Designing a Cassegrain antenna by geometric optics usually results in a system which is less efficient than can be achieved by using state-of-the-art design techniques. In 1962, Potter[53] recognised the advantage of using a modified subreflector to achieve higher efficiency. Cramer[54] realised that the geometrically designed Cassegrain geometry always yields a smaller effective subreflector than required for optimum performance. Both of these authors realised some increase in efficiency, but did not achieve the maximum possible improvement. A combination of techniques makes it possible to design a Cassegrain system which essentially makes up for the diffraction loss due to the subreflector and is particularly attractive on antennas less than 100 wavelengths in diameter (i.e. subreflectors $\simeq$ 10 wavelengths diameter), where the diffraction loss becomes a significant loss factor in the antenna design.

Other methods of improving the efficiency of dual reflector antennas include Galindo's[55] and Williams'[56] techniques of shaping both reflectors, and the various diffraction designs as discussed by Wood[57]. These techniques are considered in Section 3.4.1.

3.2.4 Equivalent parabola

The concept of an equivalent parabola is sometimes useful in the analysis of Cassegrain antennas[31]. As shown in Fig. 3.47 the subreflector and main dish are replaced by a new prime-focus feed paraboloid with the same diameter as the original paraboloid, but with a focal length of $M \times F$, which results in the same feed subtended angle. This results in a surface which would focus an incoming wave to the same feed as the original Cassegrain system. This concept simplifies certain aspects of analysis (e.g. approximating the amplitude taper across the dish and small-angle beam-scan characteristics).

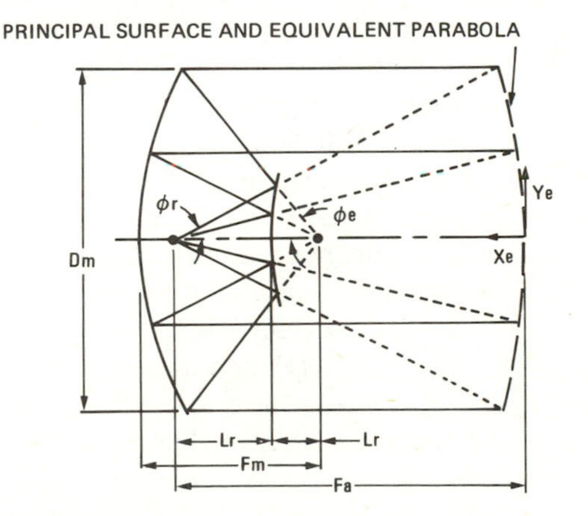

Fig. 3.47 *Equivalent-parabola concept*

3.2.5 Twist reflectors

If an antenna system need only operate for one sense of linear polarisation, it is possible to considerably reduce aperture blockage by using a polarisation twist reflector technique.[31,58,59] As shown in Fig. 3.48, the subreflector is composed of a horizontal grating which reflects an incident horizontally polarised wave from the feed back to the paraboloid. The main reflector surface 'twists' the reflected polarisation to vertical, which passes through the horizontally polarised subreflector essentially unaffected. The unavoidable feed blocking can be reduced by decreasing the feed size to illuminate a subreflector which can now be larger than the minimum blockage design because it is transparent to the wave from the paraboloid.

The idealised performance described above cannot be totally realised in a three-dimensional reflector. However, significant blockage reduction can be achieved.

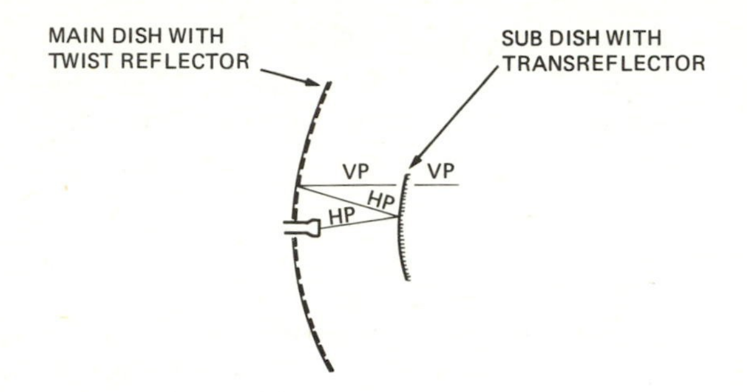

Fig. 3.48 *Polarization twist for non-blocking sub-dish*

3.2.6 Frequency-selective subreflectors

Cassegrain systems may be required to operate simultaneously at two frequencies. One solution to this requirement is a dual-frequency feed, although, if the frequencies are too widely separated, it may be virtually impossible to achieve efficient performance at both frequencies with a single antenna geometry (cf. Sections 3.2.1 and 3.2.2). An alternative solution is to construct the subreflector from dichroic material[60] which is transparent at the frequency of a prime-focus feed which is placed behind the subreflector and radiates through it. Simultaneously, the subreflector is highly reflective at the frequency of an externally placed feed which illuminates it in the conventional Cassegrain mode.

A frequency-selective subreflector (FSS) was successfully used for the 2115/8448 MHz shaped, Cassegrain antenna on the Voyager spacecraft.[61,62] The subreflector was fabricated from X band resonant aluminium crosses etched on a Mylar sheet which was bonded to a Kevlar/epoxy skin (Fig. 3.49). Both the X band reflection loss of the subreflector and its S band transmission loss were less than 0.1 dB.

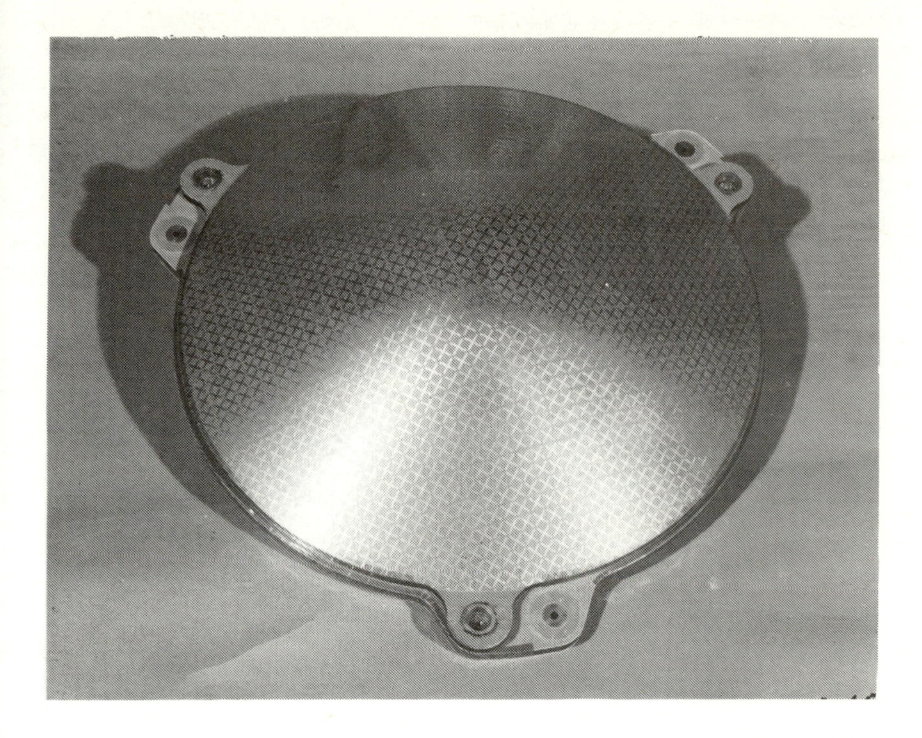

Fig. 3.49 *FSS subreflector*

3.3 Offset parabolic reflector antennas (Contribution by A.W. Rudge)

3.3.1 Introduction

Fundamental advantages and disadvantages

The offset-parabolic reflector has found applications as an antenna for many years and was certainly receiving some attention during the 1940s. However, it is only in comparatively recent times that analytical and numerical models have been developed for this device which can provide reliable predictions of its electrical properties. Although the basic analytical techniques were available at the end of World War II, the offset-reflector geometry did not readily lend itself to analysis without the aid of a digital computer. Hence, it was not until the 1960s that development in digital-computer technology provided a readily available and convenient means for accurate modelling and optimisation of the offset antenna's electrical performance.

Since the offset-parabolic reflector is a somewhat more complicated structure to deal with both structurally and analytically, it will be as well to briefly review its principal advantages and disadvantages as an antenna. First and foremost, the offset-reflector antenna reduces aperture-blocking effects. This fact, which is illustrated in Fig. 3.50, represents a very significant advantage for the offset configuration over its comparable axisymmetric counterparts. Aperture blocking by a primary feed or a subreflector, with their supporting struts, leads to scattered radiation which results in a loss of system gain on the one hand and a general degradation in the suppression of sidelobe and cross-polarised radiation on the other. These latter effects are becoming increasingly important as antenna spurious radiation specifications tighten and frequency reuse requirements demand higher levels of isolation between orthogonal hands of polarisation.

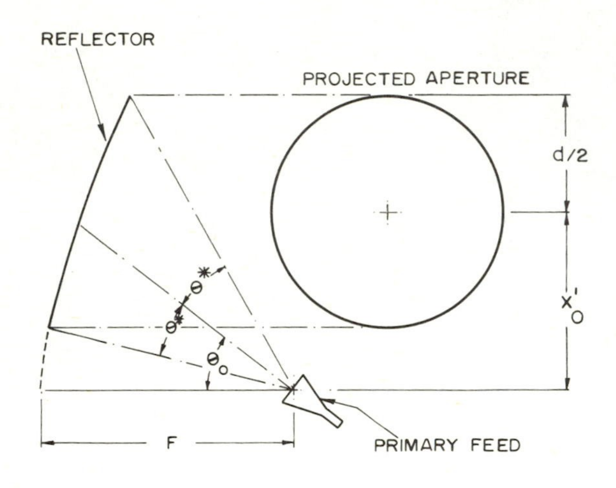

Fig. 3.50 *Basic single-offset-reflector configuration*

A second major advantage of the offset configuration is that the reaction of the reflector upon the primary-feed can be reduced to a very low order. The excellent isolation between reflector and primary-feed which can be achieved implies that the primary-feed VSWR can be made to be essentially independent of the reflector. When multiple-element or dual-polarised primary-feeds are to be employed, the mutual coupling occurring between feed elements via the reflector can be reduced to an insignificant level.

Compared to an axisymmetric paraboloid, the offset configuration leads to the use of larger focal-length to diameter ratios (F/d) while maintaining an acceptable structural rigidity. As a consequence, the offset-reflector primary-feeds employ relatively larger radiating apertures which, in the case of multiple-element primary-feeds, will result in lower direct mutual coupling between adjacent feed elements. The use of larger aperture primary-feed elements in certain cases can also provide an opportunity for improved shaping of the primary-feed radiation pattern and better suppression of the cross-polarised radiation emanating from the feed itself.

The offset-reflector configuration also has its disadvantages. When illuminated by a conventional linearly polarised primary-feed, the offset reflector will generate a cross-polarised component in the antenna radiation field. When circular polarisation is employed, the reflector does not depolarise the radiated field, but the antenna beam is squinted from the electrical boresight. For small offset reflectors this squinting effect has also been observed with linear polarisation.

Structurally the asymmetry of the offset reflector might be considered as a major drawback, although there are many applications where its structural peculiarities can be used to good advantage. In the design of spacecraft antennas, for example, an offset configuration can often be accommodated more satisfactorily than an axisymmetric design, particularly when it is necessary to deploy the reflector after launching. Nevertheless, it is clear that the offset geometry is more difficult to deal with and generally more costly to implement. For these reasons its use in the past has tended to be restricted to applications where electrical performance specifications have been severe. The rapid growth in the use of offset antenna systems in more recent times is an indication of the increasing demands which are being made upon antenna performance.

Single and dual-reflector systems

As for its axisymmetric counterpart, the offset-parabolic reflector can be utilised as a single reflector fed from the vicinity of its prime-focus, or arranged in a dual reflector system where the main offset reflector is illuminated by the combination of a primary-feed and subreflector. By this means offset Cassegrainian and offset Gregorian systems can be designed. A further dual-reflector system will be dealt with here in which the primary feed illuminates an offset section of a hyperboloid, and the combination feed an offset parabolic reflector. The geometry of this configuration can be adjusted such that no blocking of the optical path occurs either by the primary feed or the subreflector. The primary feed in this case is located below the main parabolic reflector. This arrangement contrasts with the open Cassegrainian configuration originated by Bell Laboratories in which the primary feed protrudes through the main reflector. To distinguish between these configurations, the no-blocking case will be termed a double-offset reflector antenna, while the general case will be referred to as dual-reflector offset antennas. Fig. 3.51 illustrates some of the configurations which are of particular interest.

Background

Much of the initial difficulty in dealing with the offset-parabolic reflector can be attributed to its asymmetric geometry. This geometry is the key to the analysis of the offset antenna and to ultimately understanding its electrical properties. One of

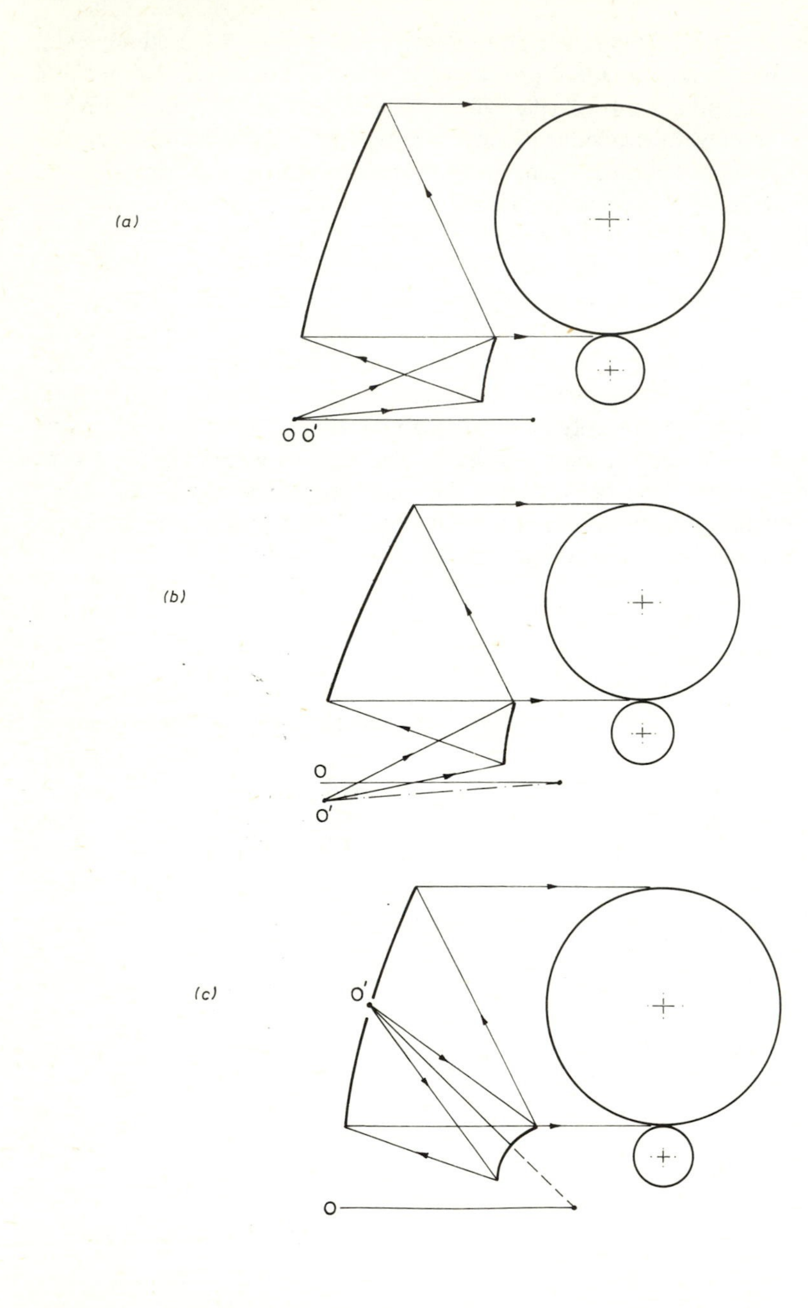

Fig. 3.51 *Dual-offset-reflector configurations:*
a Double-offset system;
b Optimised double offset;
c Open Cassegrainian system.
Paraboloid vertex at 0 and feed phase centre located at 0'

the best analyses of the offset-reflector geometry can be found in a monograph issued by the Bell Telephone System.[63] This work by Cook *et al.*, which was concerned with the analysis of a dual-reflector open Cassegrainian system, was published in 1965.[64] Much of the subsequent analysis of the offset reflector either makes use of this geometry or, if performed independently, follows a similar approach to that established by these authors.

The depolarisation properties of asymmetric antennas have deservedly received considerable attention in the literature. While the polarisation characteristics of the offset reflector were subject to the independent study of a number of authors, including the original work by Cook *et al.*,[64] Chu and Turrin[65] first published detailed graphical data and provided a clear insight into the beam-squinting properties of the circularly polarised prime-focus-fed offset reflector. The radiation properties of offset-reflector antennas with off-axis feeds were studied by Rudge *et al.*,[66-69] while Dijk *et al.*[70,71] performed an in-depth analysis of the polarisation losses of offset paraboloids. The low cross-polar radiation achievable with large F/d ratio offset reflectors was confirmed by Gans and Semplak.[72]

The optimisation of the geometry of dual-reflector offset antennas to reduce or eliminate cross-polarised radiation was first demonstrated by Graham[73] and confirmed theoretically by Adatia[74] in the UK. Working independently, Tanaka and Mizusawa[75] established a simple geometric-optics-based formula for this optimisation process.

The reduction of offset reflector cross-polarisation by use of a field-matching primary-feed technique was proposed and demonstrated by Rudge and Adatia.[76,77] Jacobsen[78] has made the point that, in principle, similar results could be achieved with an array of Huygen sources directed toward the vertex of the parent paraboloid and phased to direct the energy into the cone of angles subtended by the offset portion of the reflector. A comparison of the radiation pattern and impedance properties of offset-Cassegrainian and offset-Gregorian antennas with their symmetrical counterparts has been performed by Dragone and Hogg.[79] Their results confirm the anticipated advantages of the offset structures with regard to both sidelobe radiation levels and VSWR. The use of offset-reflector antennas in applications where very low sidelobe radiation is mandatory has also received attention[80,98]

The avoidance of aperture blockage implies that offset reflectors should offer good potential as multiple-beam antennas. This possibility has been investigated by a number of workers. Rudge *et al.*[66-69,81,82] have studied the use of small clusters of feed elements, both linearly and circularly polarised, in conjunction with single offset reflectors. Ingerson[83] has also investigated the off-axis scan characteristics of offset reflectors, and Kaufmann and Croswell[84] have considered the effect of large axial displacements of the offset-reflector primary-feed. Ohm[85] has analysed a proposed multiple-beam earth station based upon a dual-reflector offset system.

The use of offset-reflector systems to provide shaped or contoured beams for satellite communications has also received attention. Shaped beams have been achieved either by deforming the offset-reflector surface, as described by Wood *et al*,[86] or by the use of a weighted array or primary-feed elements, as favoured by Han[87] and his colleagues for the Intelsat V communications satellite.

Although their low sidelobe potential makes offset-reflector antennas attractive for many radar applications, difficulties were experienced in the past when a precision tracking capability was required. These difficulties, which arise as a result of the offset-reflector depolarisation, are now well understood, and the means of compensating for these effects by use of improved monopulse primary feeds have been recently demonstrated.[88]

3.3.2 Single offset-reflector analysis

Basic techniques

Either the surface-current (cf. eqn. 2.66) or aperture-field (cf. eqn. 2.67) version of physical optics may be used to determine the radiated field.[95, 96] In the former case, the geometrical-optics approximation of the surface-current density is integrated over the reflecting surface S_1 (Fig. 3.52); in the latter, the geometrical optics approximation of the aperture field is integrated over the projected aperture S_2. For projected aperture diameters exceeding 20 wavelengths, the two techniques provide virtually identical predictions for the copolarised fields over the main lobe and first four or

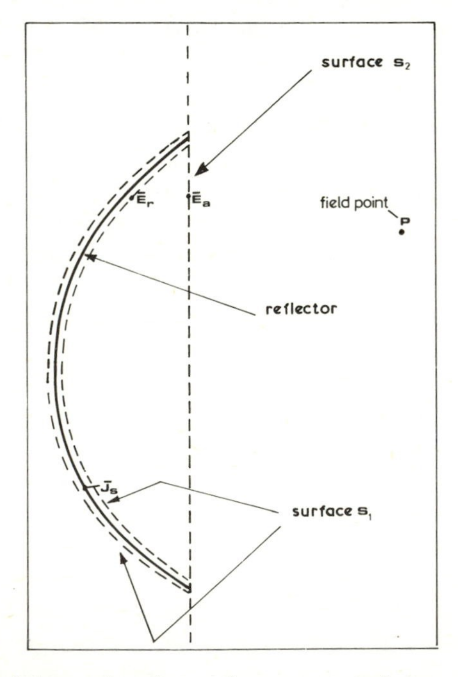

Fig. 3.52 *Surfaces of integration for surface-current technique, S_1, and aperture-field method, S_2*

five sidelobes. At wide angles from the boresight, the predictions differ and these differences tend to increase with increasing reflector curvature. For the cross-polarised radiation, the two methods differ significantly only when the peak levels of this radiation are very low (i.e. less than −50 dB relative to the peak value of the main beam copolarised field). The electric-current method is generally considered to be the more accurate of the two methods, but for most practical purposes the differences between the predictions tend to be insignificant. For small offset parabolic reflectors (i.e. projected aperture diameters of less than 20 wavelengths) the discrepancies between the predictions of the techniques become more discernable. Boswell and Ashton[89] have shown that a beam-squinting effect can occur with small linearly polarised offset reflectors. This effect is predictable using the surface-current technique but not with aperture fields. In their example a beam squint of 0.03 of a beamwidth occurred with a reflector of 6-wavelengths diameter. This result is particularly interesting in that it is a comparatively rare example of experimental confirmation of the accuracy of the surface-current technique for small parabolic reflectors.

In dealing with the radiation from large offset-parabolic-reflector antennas in a moderate cone of angles about the antenna boresight and over a dynamic range of the order of 50–60 dB, there is, for most practical purposes, little significant variation between the predictions obtained by the two methods. Under these circumstances the technique which is more convenient analytically and computationally can be employed. On this basis the aperture-field method, which involves an integration over a planar surface, results in generally more simple mathematical expressions and thus offers some saving in computational effort.

Offset-reflector geometry

The geometry of the single-offset-parabolic reflector is shown in Fig. 3.53. The basic parameters of the reflector are shown as the focal length F of the paraboloid, the offset angle θ_0, and the half angle θ^* subtended at the focus by any point on the reflector rim. With θ^* maintained constant, a rotation about the included z axis will generate a right circular cone with its apex at the reflector geometric focus. If the boundary of the parabolic surface is defined by its intersection with the cone, then the resultant reflector will have an elliptical contour, while the projection of this contour onto the $x'y$ plane will produce a true circule.

To deal with this offset geometry it is desirable to obtain a coordinate transformation from the primary spherical coordinate system of the symmetrical parent paraboloid, to an offset spherical coordinate system about the inclined z axis. The reflector parameters which are readily expressed in terms of the symmetrical primary coordinates can then be transformed into the offset coordinate system. If the symmetrical coordinates are defined conventionally as ρ', θ', ϕ' with associated Cartesian coordinates x', y', z', then their relationships to the unprimed offset coordinates ρ, θ, ϕ and x, y, z are obtained by simple geometry as:

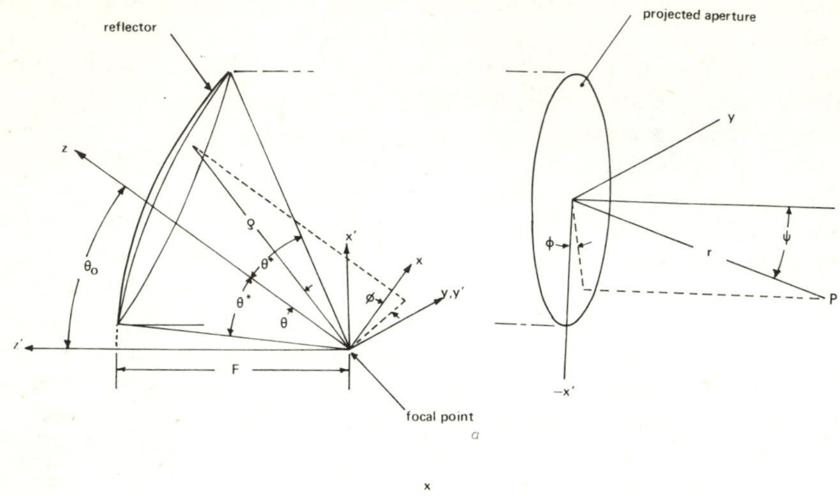

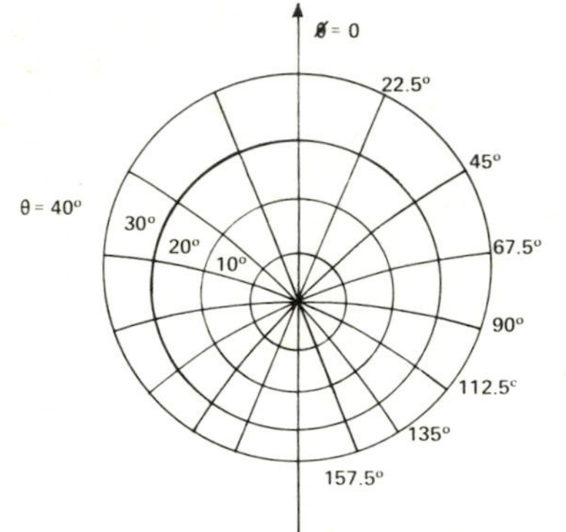

Fig. 3.53 *a* Single-offset-reflector coordinate system
b Constant θ, ϕ contours on projected-aperture plane

$$\rho' = \rho \tag{3.73a}$$

$$\cos\theta' = \cos\theta\cos\theta_0 - \sin\theta\sin\theta_0\cos\phi \tag{3.73b}$$

$$\sin\theta'\sin\phi' = \sin\theta\sin\phi \tag{3.73c}$$

$$\sin\theta'\cos\theta' = \sin\theta\cos\theta_0\cos\phi + \cos\theta\sin\theta_0 \tag{3.73d}$$

$$\tan\phi' = \sin\theta\sin\phi/(\cos\theta\sin\theta_0 + \sin\theta\cos\theta_0\cos\phi) \tag{3.73e}$$

and these equations provide the basis for the transformations. Applying the transformations, the distance from the reflector focus to a point on the parabolic surface, ρ, is given by:

$$\rho = 2F/(1 + \cos\theta')$$

$$= 2F/(1 + \cos\theta \cos\theta_0 - \sin\theta \sin\theta_0 \cos\phi). \quad (3.74)$$

The diameter of the projected aperture of the offset reflector d is:

$$d = 4F \sin\theta^*/(\cos\theta_o + \cos\theta^*) \quad (3.75)$$

Similarly the distance x_0' from the axis of the parent paraboloid to the centre of the projected apertures is

$$x_o' = 2F \sin\theta_o/(\cos\theta_o + \cos\theta^*). \quad (3.76)$$

Distances from the reflector focus in the $z' = 0$ plane can be written

$$x' = \rho\,(\cos\theta_o \sin\theta \cos\phi + \sin\theta_o \cos\theta) \quad (3.77a)$$

$$y' = \rho \sin\theta \sin\phi \quad (3.77b)$$

The Jacobian for a surface element $dx'dy'$ can be obtained from the equations given as

$$dx'dy' = \rho^2 \sin\theta'\, d\theta'\, d\phi'$$

$$= \rho^2 \sin\theta\, d\theta\, d\phi \quad (3.78)$$

The unit normal to the reflector parabolic surface ($\hat{a}_n$) is given by

$$\hat{a}_n = -\sqrt{(\rho/4F)}\ (\sin\theta \cos\phi - \sin\theta_0)\,\hat{a}_x$$

$$+\sin\theta \sin\phi\, \hat{a}_y + (\cos\theta + \cos\theta_0)\,\hat{a}_z \quad (3.79)$$

Copolar and cross-polar definitions

For offset-reflector antennas which are operated in a predominantly linearly polarised mode, the definition which is preferred by the author is the third definition presented by Ludwig[90] in the referenced paper. This definition offers excellent correlation with standard antenna-range measurement practice[91] and its advantages are discussed in Chapter 4.

In the primary spherical coordinate system ρ, θ, ϕ shown in Fig. 3.53 the primary-feed spherical-coordinate fields will be defined as E_θ, E_ϕ. If the primary-feed antenna has its principal electric vector along the y axis then the feed copolar measured-field component e_p and the cross-plar component e_q can be simply defined by:[69,90]

$$\begin{bmatrix} e_p \\ e_q \end{bmatrix} = \begin{bmatrix} \sin\phi & \cos\phi \\ \cos\phi & -\sin\phi \end{bmatrix} \begin{bmatrix} E_\theta \\ E_\phi \end{bmatrix} \tag{3.80}$$

Similarly, in terms of the secondary coordinate system, r, Ψ, Φ, the overall radiation fields from the antenna will be defined as E_Ψ, E_Φ and the copolar (E_p) and cross-polar (E_q) can be obtained from the right-hand side of eqn. (3.80) with Ψ, Φ, replacing θ, ϕ, respectively.

It is worth noting that a zero cross-polarisation primary feed by the definition of eqn. (3.80) will produce a purely linearly polarised field in the projected aperture plane of an axisymmetric paraboloidal reflector. This condition will in turn result in a low level of cross-polarised radiation in the overall antenna far-field, provided that the reflector aperture is large with respect to the wavelength and that the blockage effects are small. However, the field distribution in the mouth of the primary feed will not be purely linearly polarised but must exhibit some field-line curvature to establish this desired radiation condition.[92]

Projected-aperture fields

The tangential electric-field distribution in the aperture plane of the offset reflector ($\bar{E}_a$) can be approximated by assuming an optical reflection of the incident primary-feed fields. If the incident primary field at the reflector is taken as the radiation field of the primary-feed, then this can be expressed as

$$\bar{E}_i = \frac{1}{\rho}[A_\theta(\theta,\phi)\hat{a}_\theta + A_\phi(\theta,\phi)\hat{a}_\phi)]\exp(-jk\rho) \tag{3.81}$$

where A_θ, A_ϕ are the normalised spherical coordinate components of the primary feed radiation pattern and $k = 2\pi/\lambda$. Optical reflection then yields the offset-reflector projected-aperture electric-field distribution expressed in Cartesian components, E_{ax}, E_{ay}:

$$\begin{bmatrix} E_{ay}(\theta,\phi) \\ E_{ax}(\theta,\phi) \end{bmatrix} = K \begin{bmatrix} -S_1 & C_1 \\ C_1 & S_1 \end{bmatrix} \begin{bmatrix} A_\theta(\theta,\phi) \\ A_\phi(\theta,\phi) \end{bmatrix} \tag{3.82}$$

where

$$S_1 = (\cos\theta_0 + \cos\theta)\sin\phi \tag{3.83a}$$

$$C_1 = \sin\theta\sin\theta_0 - \cos\phi(1 + \cos\theta\cos\theta_0) \tag{3.83b}$$

$$K = \exp(-j2kF)/2F \tag{3.83c}$$

For a circularly polarised primary feed the reflector aperture fields are given by

$$\begin{bmatrix} E_{aR}(\theta,\phi) \\ E_{aL}(\theta,\phi) \end{bmatrix} = \frac{2FK}{\rho} \begin{bmatrix} \exp(j\Omega) & -j\exp(j\Omega) \\ \exp(j\Omega) & j\exp(-j\Omega) \end{bmatrix} \begin{bmatrix} A_\theta(\theta,\phi) \\ A_\phi(\theta,\phi) \end{bmatrix} \tag{3.84}$$

where E_{aR} and E_{aL} are the right- and left-handed components and

$$\Omega(\theta\ \phi) = \arctan S_1/C_1 \tag{3.85}$$

If the circularly polarised primary-feed has a normalised radiation pattern of the form

$$\bar{A}_n(\theta\ \phi) = [A_\theta(\theta)\,\hat{a}_\theta - jA_\phi(\theta)\,\hat{a}_\phi]\exp(-j\phi) \tag{3.86}$$

where the functions A_θ and A_ϕ are independent of ϕ, then the offset-reflector aperture-plane fields have the form

$$E_{aM}\ \frac{2FK}{\rho}\ [A_\theta(\theta) \pm A_\phi(\theta)] \tag{3.87}$$

where M is either R or L and L takes the upper sign. Eqn. 3.87 can be satisfied by many practical types of circularly polarised conical horns including fundamental mode, dual-mode (Potter), and corrugated types.[92, 93] It is apparent that when $A_\theta(\theta) = A_\phi(\theta)$, which is the condition for zero cross-polarised radiation from the conical feed, then the reflector aperture plane will be purely copolarised with a beam-squinting phase distribution given by $\phi + \Omega$. This can be contrasted with the linearly polarised case in which the introduction of a zero cross-polarised primary feed fails to reduce the projected-aperture field to a single linear polarisation, unless the offset angle θ_0 is put to zero.

The off-focus primary-feed

When the phase centre of the primary feed is coincident with the reflector geometric focus, the incident electric fields at the reflector are described by eqn. 3.81. This equation assumes that the reflector is in the far-field of the primary-feed. In general, movements of the primary feed can be accounted for by means of a suitable coordinate transformation between a set of coordinates employed to define the feed radiation and a set describing the offset reflector surface. However, if the primary-feed offsets are small relative to the reflector-to-feed dimension ρ, then the variation in the amplitude terms in eqn. (3.81) will be small, and the effect of such amplitude changes upon the predicted radiation fields of the overall antenna will be negligible for most practical purposes. Similarly, providing the orientation of the principal polarisation vector remains aligned with one of the principal axes of the focal plane, the depolarising effects of the feed offset will also be negligible. However, the off-focus location of the primary-feed phase centre will modify the phase terms in eqn. 3.81 and this will significantly change the overall radiation fields.

Referring to Fig. 3.54 the location of an offset primary-feed phase centre can be described in the offset coordinate system x, y, z by the vector $\bar{\Delta}$ or the cylindrical coordinates Δ_t, Δ_z and ϕ_o, where Δ_t is a small transverse offset, Δ_z a small axial offset, and ϕ_0 the polar angle to the x axis in the x, y plane. Employing this geometry, a compensating phase term can be derived which, when used as a multiplying term on eqn. 3.81, effectively corrects the phase characteristics for the off-axis feed location.

Thus, assuming $\Delta^2/\rho^2 << 1$, $\Delta_t/\rho < 1$, and $\Delta_z/\rho < 1$, the field of the feed can be assumed to be

$$\bar{E}_i \cong \frac{1}{\rho} [A_\theta (\theta, \phi)\, \hat{a}_\theta + A_\phi (\theta\ \phi)\, \hat{a}_\phi] \exp [jk(R_1 - \rho)] \tag{3.88}$$

where

$$R_1 (\theta\ \phi) = \Delta_t \sin\theta \cos(\phi - \phi_o) + \Delta_z \cos\theta \tag{3.89}$$

Similarly, the electric field in the offset-reflector projected aperture plane can be simply modified to account for the off-axis feed by multiplying eqns 3.82 and 3.84 by $\exp [jkR_1]$.

Imbriale *et al.*[94] have compared the predictions obtained from a scalar equation for the far-field radiation of a paraboloid (which included a phase-compensation term similar to eqn. 3.89) with a complete vector formulation employing coordinate

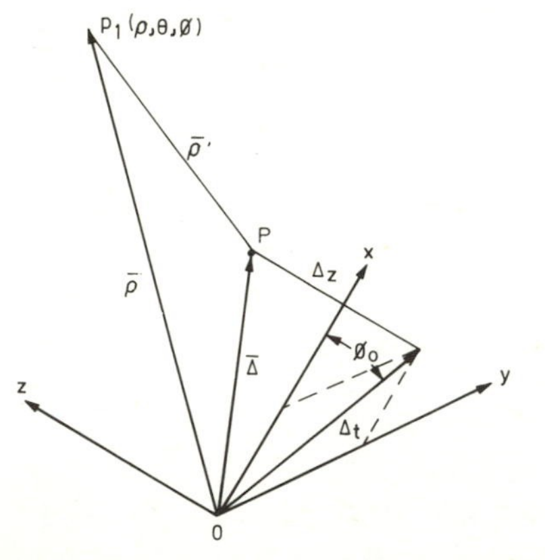

Fig. 3.54 *Coordinate geometry for offset primary-feed. P_1 is a point on the reflector surface and P a point in the vicinity of the reflector geometric focus (O)*

transformations. Their results support this author's experience that for small feed offsets, at least up to the point where the radiation-pattern degradation would be unacceptable for most practical applications, the quality of the approximate expression is very good. The use of eqns. 3.88 and 3.89 in a vector formulation of the type shown here thus offers a very reasonable compromise between accuracy and analytical and computational simplicity.

Field integrals

Making use of the physical-optics approximation and well-established vector-potential methods, mathematical models can be established for offset-reflector antennas. For example, employing the tangential aperture-field approach, the normalised copolar (E_{pn}) and cross-polar (E_{qn}) radiation patterns of a linearly polarised offset-reflector antenna can be expressed as

$$\begin{bmatrix} E_{pn}(\Psi,\Phi) \\ E_{qn}(\Psi,\Phi) \end{bmatrix} = \frac{1+\cos\Psi}{2F_p(0,0)} \begin{bmatrix} 1-t^2\cos 2\Phi & t^2\sin 2\Phi \\ t^2\sin 2\Phi & 1+t^2\cos 2\Phi \end{bmatrix} \begin{bmatrix} F_p(\Psi,\Phi) \\ F_q(\Psi,\Phi) \end{bmatrix} \tag{3.90}$$

where F_p, F_q are the spatial Fourier transforms of the copolar and cross-polar tangential electric-field distribution in the projected-aperture plane of the offset reflector, and $t = \tan \Psi/2$.

Similarly, for a circular polarised antenna the far-field radiation pattern can be expressed in terms of its normalised right- and left-hand components by:

$$\begin{bmatrix} E_{Rn}(\Psi\ \phi) \\ E_{Ln}(\Psi\ \Phi) \end{bmatrix} = \frac{1+\cos\Psi}{2F_R(0,0)} \begin{bmatrix} 1 & t^2\exp(j\,2\Phi) \\ t^2\exp(-j2\Phi) & 1 \end{bmatrix} \begin{bmatrix} F_R(\Psi\ \Phi) \\ F_L(\Psi\ \Phi) \end{bmatrix} \tag{3.91}$$

where F_R and F_L are the spatial Fourier transformations of the right- and left-hand components of the projected-aperture tangential electric field.

The general form of the transform in terms of the geometrical parameters specified in eqns. 3.75-3.79 is given by

$$F = \iint_{x',y} E_a(x',y)\exp jkR'(x',y,\Phi,\Psi)\,dx'\,dy \tag{3.92}$$

where

$$R'(x',y,\Psi,\Phi) = (x'_o - x')\sin\Psi\cos\Phi + y\sin\Psi\sin\Phi \tag{3.93}$$

and E_a is the tangential aperture field as specified by either eqn. 3.82 or 3.84.

To predict the offset antenna radiation pattern it is necessary to set values for the reflector parameters (F, θ_0, and θ^*), to specify the primary-feed radiation fields at the reflector (E_θ, E_ϕ), to compute the two-dimensional transform integrals, and finally to evaluate eqns. 3.90 and 3.91. For accurate predictions, the choice of primary-feed model is critical, while the evaluation of the two-dimensional integrals

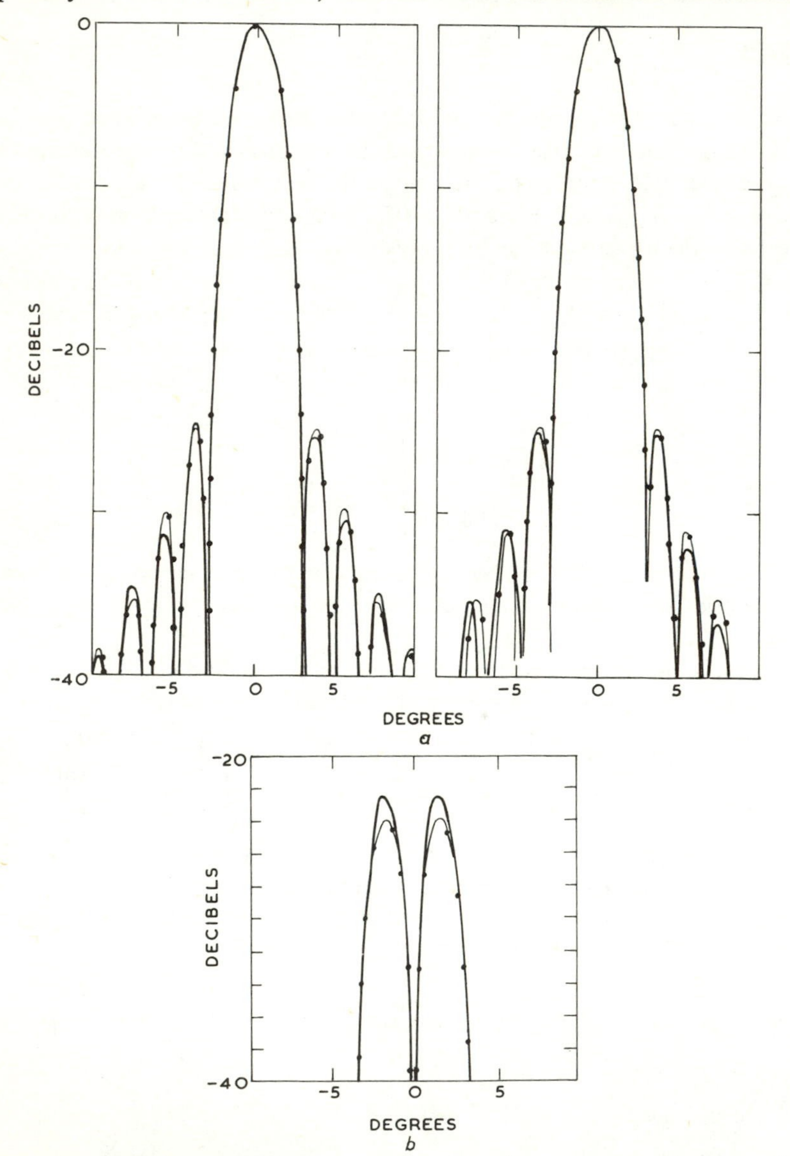

Fig. 3.55 *Radiation fields from a K-band offset-reflector antenna. Copolar radiation in (a) plane of asymmetry ($\Phi = \pi/2$) and (b) plane of symmetry ($\Phi = 0$). Cross-polar radiation in (c) plane of asymmetry. Measured . Predicted*

clearly represents the crux of the computational problem. Numerical integration is discussed in Section 2.2.4.

3.3.3 Electrical performance of the single offset-reflector antenna

Linear polarisation

The literature provides evidence of a number of analytical models, largely based upon physical-optics techniques, which have been applied to study the radiation characteristics of the offset parabolic reflector. Precise experimental data is somewhat more sparse, but sufficient material has been published to validate the main conclusions drawn from the analytical studies.

When fed by a purely linearly polarised primary-feed (as defined by eqn. 3.80) the offset-parabolic reflector exhibits a characteristic depolarising effect which results in the generation of two principal cross-polarised lobes in the plane of asymmetry (i.e. the yz' plane or $\Phi = \pi/2$ plane in the coordinate system of Fig. 3.53). Fig. 3.55 shows measured data obtained with a precisely machined offset reflector fed by a fundamental-mode rectangular horn with a 12 dB illumination taper at the reflector rim.[67,69] Predicted data obtained from a numerical model has been superimposed, and, in general, an excellent correlation is observed. The cross-polar correlation is slightly inferior; and, in fact, better correlation was observed with later unpublished results. These radiation patterns are characteristic of the linearly polarised offset reflector, and similar results have been published by Cook *et al.*,[63,64] Chu and Turrin,[65] and others. The cross-polar lobes which arise as a

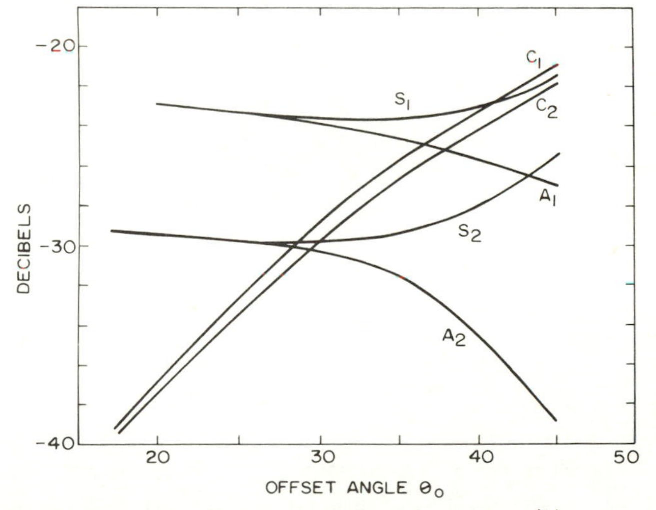

Fig. 3.56 *Peak copolar sidelobe levels in planes of symmetry (S) and asymmetry (A) for offset reflectors with $\theta_0 = \theta^* + 5^\circ$, fed by uniformly illuminated circular aperture feeds producing −10 dB (subscript 1) and −20 dB (subscript 2) illumination tapers at $\theta = \theta^*$. Peak cross-polar levels (C) occurring in plane of asymmetry are also indicated*

consequence of the reflector asymmetry have peak values in the vicinity of the 6 dB contour of the main copolarised beam, and the next subsidiary lobes are typically a further 20 dB down below the peaks. The copolar sidelobe radiation has the form of a well-defined diffraction pattern, with a comparatively rapid decay of sidelobe levels with increasing angle from boresight. Numerical models of the type discussed above reproduce these characteristics reliably and can be used profitably in performing parameter studies. For example, Fig. 3.56 illustrates some general sidelobe trends for fully offset antennas in which the reflector parameters are chosen such that the primary-feed hardware does not protrude into the projected aperture of

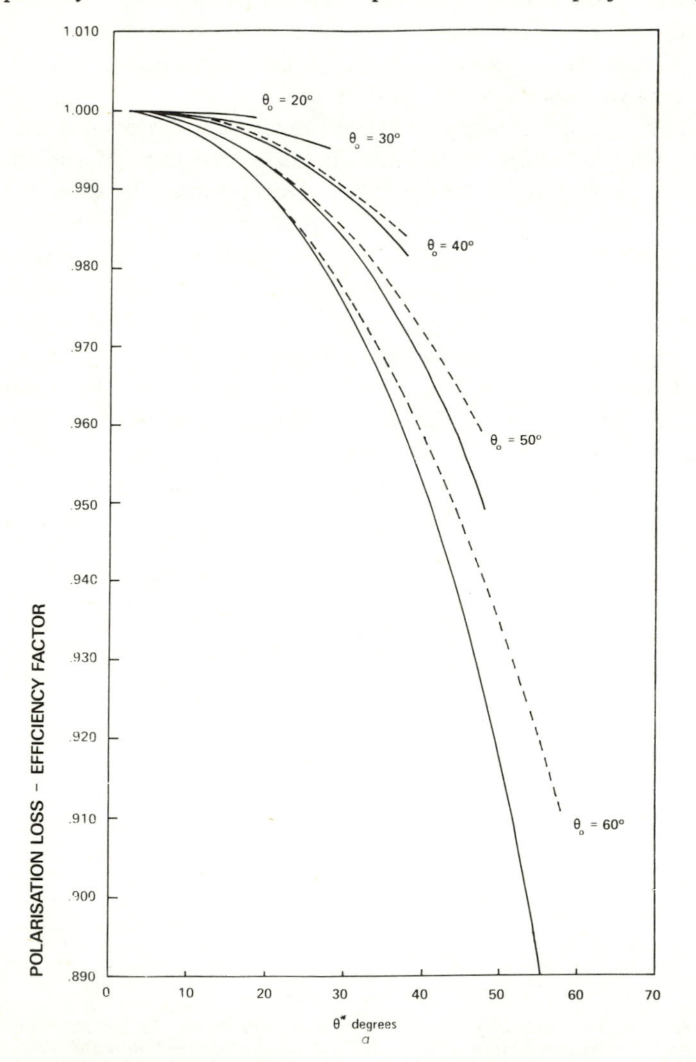

Fig. 3.57 *a* Polarisation loss efficiency factor of offset paraboloid reflector, offset angle θ_0, being a parameter, illuminated by ———— electric dipole oriented along x' axis and Huygens source and — — — — — electric dipole oriented along y' axis.

the reflector.

It is evident from Fig. 3.56 that the peak value of the reflector cross-polarisation is dependent very largely upon the parameters θ_0, θ^* and is relatively insensitive to the feed-imposed illumination taper. It is also clear that large values of θ_0 and θ^* result in higher peak levels of copolar and cross-polar radiation. However, these results should not be assumed to indicate the low-sidelobe limitations of the offset system, since at low levels this is strongly dependent upon the illumination taper introduced by the primary-feed antenna. In the sample shown, the primary-

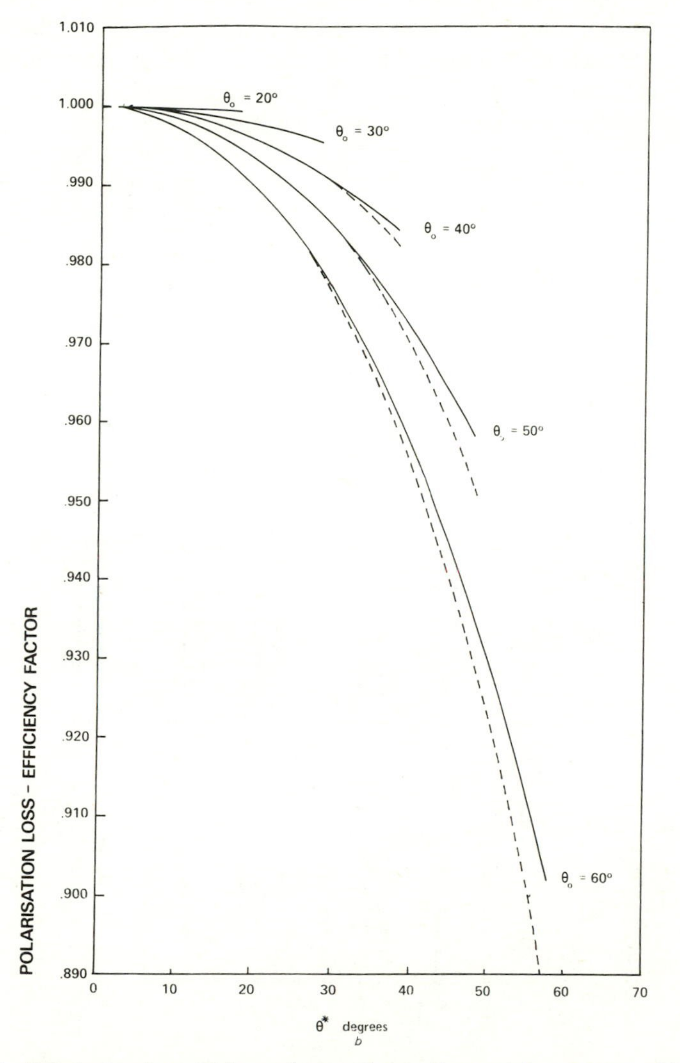

b Polarisation loss-efficiency factor of open Cassegrainian antenna, offset angle θ_0 being a parameter, illuminated by ———— electirc dipole oriented alonge x' and y' axis and — — — — —Huygens source.

feed comprises a uniformly excited circular aperture which does not produce a good illumination characteristic when low sidelobes are a major concern.

The loss of aperture efficiency, arising from offset reflectors' depolarisation, tends to be small for values of θ_0, θ^*, less than 45°, but can become significant for larger angles. Some computed data due to Dijk *et al.*[70,71] are shown in Fig. 3.57.

Fig. 3.58 shows the predicted and measured radiation patterns of a linearly polarised offset reflector fed by an off-axis primary feed.[66,69] These results provide further confidence in the quality of the analytical predictions and also serve to illustrate the comparative insensitivity of the peak cross-polar lobes to small transverse offsets in the primary-feed location. The formation of the copolarised coma lobe is the most evident source of pattern degradation. Studies have shown that the loss of gain suffered by an offset-reflector antenna with offset feed is essentially independent of the transverse plane in which the feed is offset. However, the

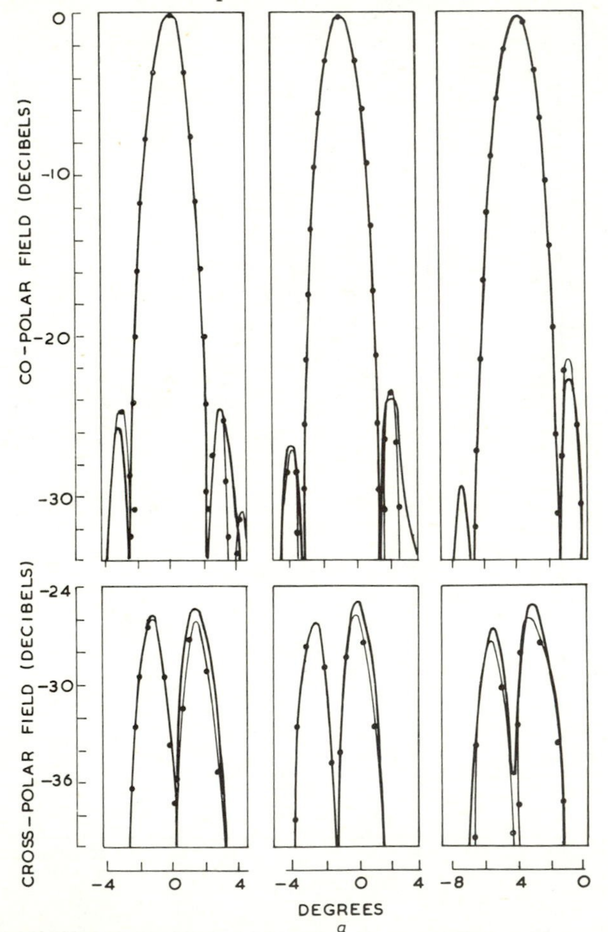

Fig. 3.58 *a* Radiation fields from offset reflector (F = 30.4λ, θ_0 = 35°, θ^* = 30°) with lineary polarised pyramidal-horn primary feed offset transversely along y axis by θ, 0,83λ, and 2.5λ, respectively.

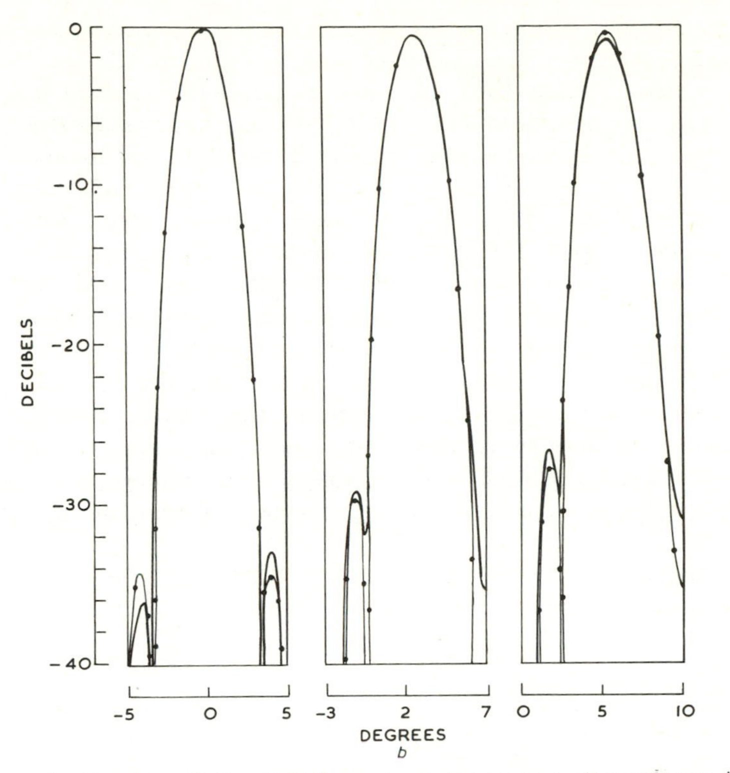

b Copolar radiation field from circularly polarised offset reflector (F = 22.7λ, θ_0 = 44°, θ^* = 30°) with scalar-horn feed offset transversely by 0, 1.4λ and 2.8λ, respectively, ———— Measured, ——●—— Predicted

general radiation pattern deterioration appears to be more pronounced when feeds are offset in the plane of symmetry rather than the plane of asymmetry.[81–83] Some relevant computed data on gain and sidelobe performance are shown in Fig. 3.59.[81] However, no distinction as to the plane of offset is made for this data. The off-axis performance of an offset-reflector antenna is dependent upon the offset angle θ_0 and is inferior to an ideal (unblocked) axisymmetric antenna with the same semi-angle θ^*. Approximate formulas for the beam deviation characteristics of offset reflectors have been derived, which illustrate the role played by the offset angle θ_0.[81,82,107]

The cross-polar characteristics of the offset-parabolic reflector are not unduly sensitive to small reflector profile errors. Small profile errors do not themselves generate a significant cross-polarised contribution to the radiated field but rather act as a phase-error distribution in the reflector aperture plane. As such, these errors will redistribute the existing copolar and cross-polar radiation in the far field of the antenna. However, like the main copolar lobe, the two main lobes of the cross-polarised field are relatively insensitive to the effects of small phase errors; and, although some increase in levels will occur in the subsidiary cross-polar lobes, the peak lobes will remain substantially unchanged.[99]

The principal cross-polarised lobes radiated from a linearly polarised offset-reflector antenna are in phase quadrature with the main copolar beam and are essentially contained within the copolar beam envelope. The polarisation over the main beam varies from linear at the beam centre to an elliptical polarisation in the region of the cross-polarised lobes. In some applications this will not constitute a serious drawback; but, if linear polarisation is required over the main beam region, then the inherent limitations of an offset parabolic antenna with a conventional primary-feed are obvious. The peak levels of the cross-polarised lobes can be reduced to low levels (i.e. below −35 dB), if the offset angle is made very small. This is evident from examination of Fig. 3.60 and Gans and Semplak[72] have described an antenna of this type with peak cross-polar levels of below −37 dB. However, if aperture blocking is to be avoided, then $\theta^* < \theta_0$, and this implies a long reflector structure which may be impractical in many cases. It is worth noting, however, that a dual linearly polarised offset antenna can radiate two orthogonal elliptically polarised signals which are resolvable into their orthogonal components by a suitably elliptically polarised receiving antenna.[76] Hence frequency reuse by

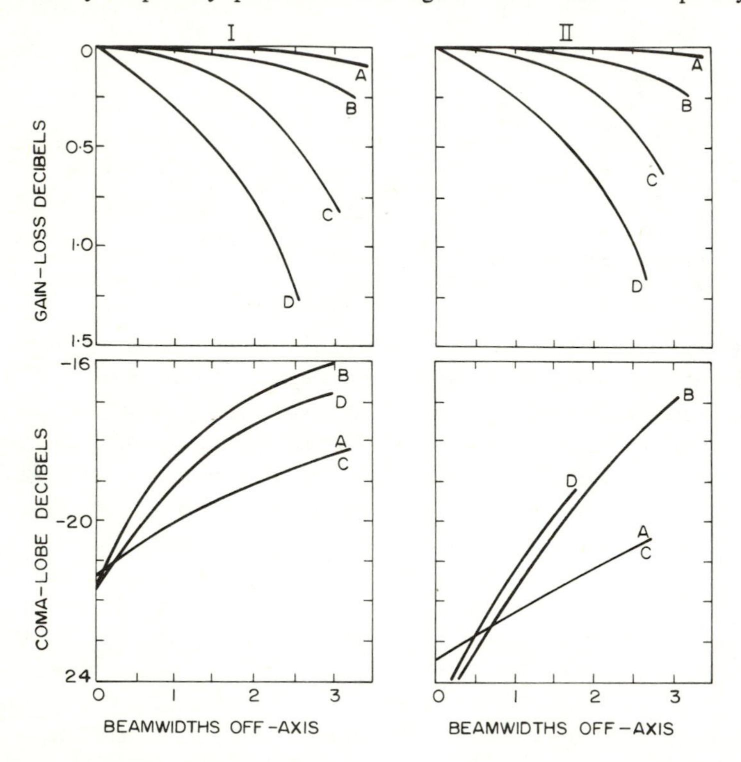

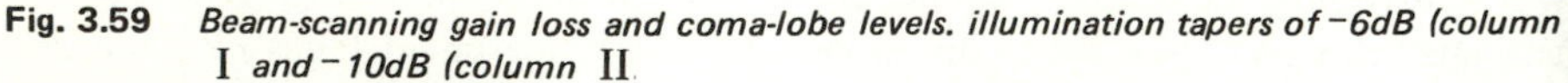

Fig. 3.59 *Beam-scanning gain loss and coma-lobe levels. illumination tapers of −6dB (column I and −10dB (column II*

a $\theta^* = 30^\circ, \theta_0 = 0$
b $\theta^* = 45^\circ, \theta_0 = 0$
c $\theta^* = 30^\circ, \theta_0 = 45^\circ$
d $\theta^* = 45^\circ, \theta_0 = 45^\circ$

polarisation diversity is feasible even with a conventionally fed offset parbolic reflector. However, the pointing accuracy of the antenna must be maintained to a high order of accuracy if the use of a fast-operating adaptive polariser is to be avoided.

Circular polarisation

With a purely circular-polarised primary-feed illumination, the offset parabolic reflector does not depolarise the signal. On reflection, each of the linearly polarised components of the incident wave effectively generates a cross-polarised component. However, when the phase-quadrature relationship is introduced between the linear components it is found that the combination of the two orthogonal copolarised vectors and the phase-asymmetric pair of cross-polarised vectors have the same direction of rotation. Hence the sum of these two signals results in a purely circularly polarised radiation, but the sum of the symmetric and asymmetric components results in a squinting of the radiation pattern from its boresight axis. The beam-squint effect acts to move the beam either toward or away from the axis

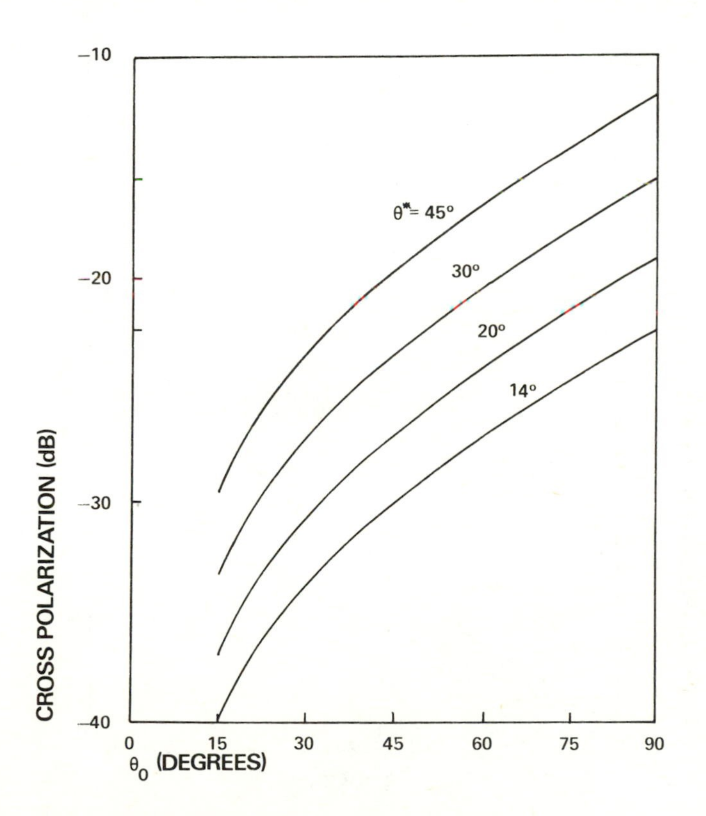

Fig. 3.60 *Peak cross-polar levels radiated in the plane of asymmetry ($\Phi = \pi/2$) as a function of the offset-reflector parameters θ_0 and θ^**

of symmetry. The direction of the movement is dependent upon the hand of polarisation.

Chu and Turrin[65] first published computed graphical data showing the effect of the reflector parameters θ_0, θ^* upon the magnitude of the beam squint, and some key results are shown ing Figs. 3.61 and 3.62. Adatia and Rudge[101] have derived an approximate formula which gives the beam-squint angle Ψ_s simply as

$$\Psi_s = \arcsin \frac{(\lambda \sin \theta_0)}{4\pi F} \tag{3.94}$$

This formula has been tested against computed and measured data and has been found, in all cases, to be accurate within 1.0% of the antenna half-power beamwidth.

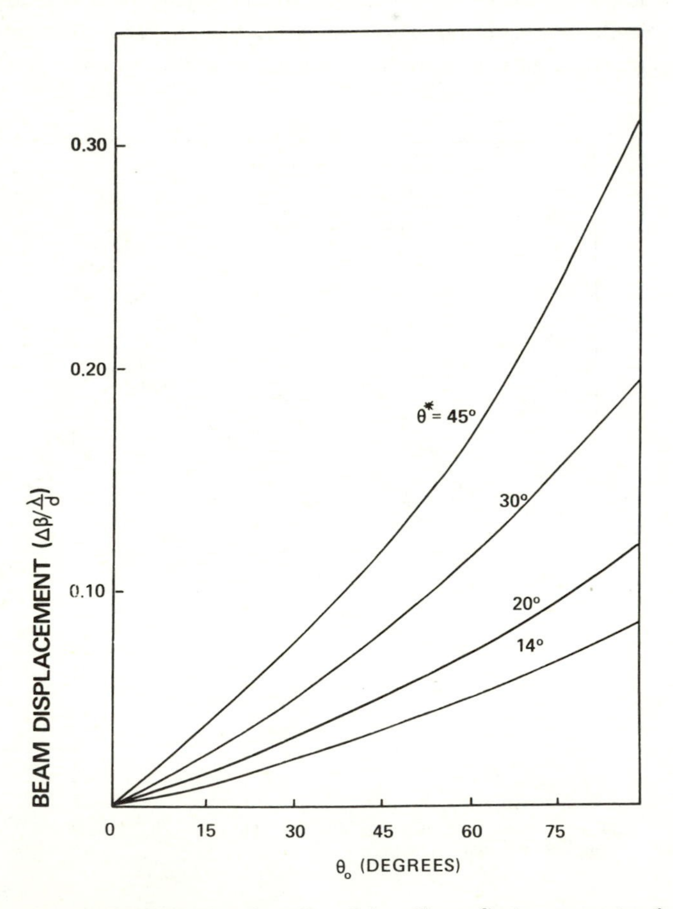

Fig. 3.61 *Beam displacement as a function of the offset-reflector parameters θ_0 and θ^*. $\Delta\beta$ is the angular squint of the beam in radians*

Practical applications

The fully offset parabolic reflector is attractive for many applications, in that it offers the possibility of improved sidelobe performance and higher aperture efficiency. Some typical applications are considered below. In these cases, unless stated otherwise, the primary feed has been assumed to be of conventional design and with good polarisation properties. Newer feed concepts are described in 4.1.

(i) *Point-to-point communications:* With circular polarisation the boresight gain reduction arising from beam-squinting effects need not be a major problem. For systems employing a single hand of polarisation the beam squint is simply compensated in the antenna alignment. For dual-polarised applications, where the direction of squint is reversed for opposite hands of polarisation, the boresight-gain loss can be readily reduced to less than 0.03 dB per beam by suitable choice of reflector parameters. This loss corresponds to a beam-squint angle of less than 10% of the antenna half-power beamwidth.

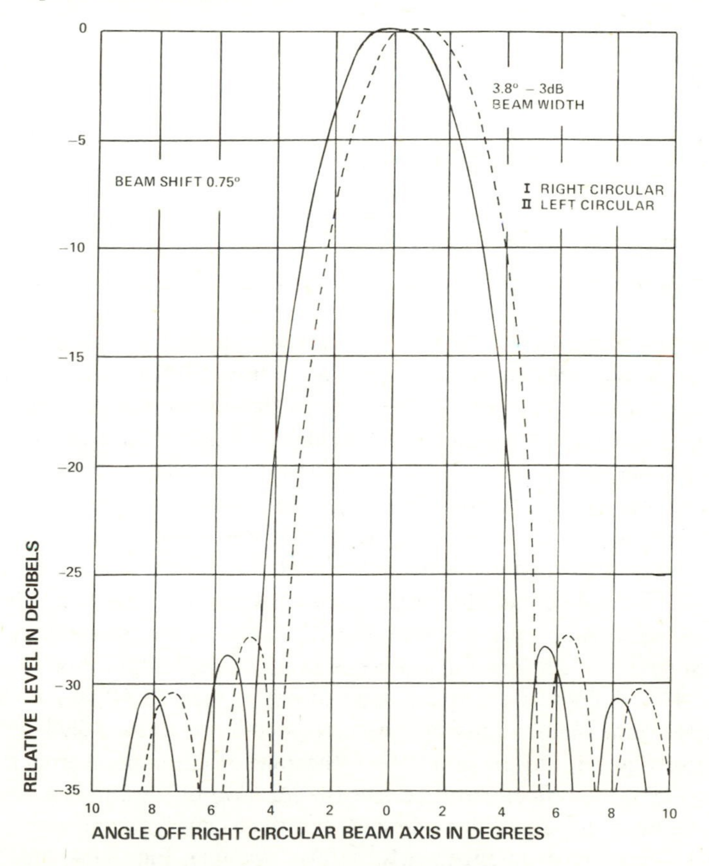

Fig. 3.62 *Measured radiation patterns of an offset-reflector antenna operated with circular polarisation at 18.5 GHz ($F = 9.4\lambda$, $\theta_0 = 45°$, $\theta^* = 45°$)*

With linear polarisation, a very exact alignment of the radiated beam is called for to remain within the boresight null of the cross-polarised radiation pattern. Practical dual-polarised operation, therefore, demands the use of a moderately large reflector *F/d* ratios.[72] Alternatively, good cross-polarisation suppression can be achieved for a single linear polarisation by use of a polarisation-selective grid, either in the aperture plane or at the surface of the reflector.[102,103] A dual-polarised system can then be formed by interleaving two orthogonal polarisation-sensitive reflector surfaces with separate foci. This is illustrated in Fig. 3.63[102,104]

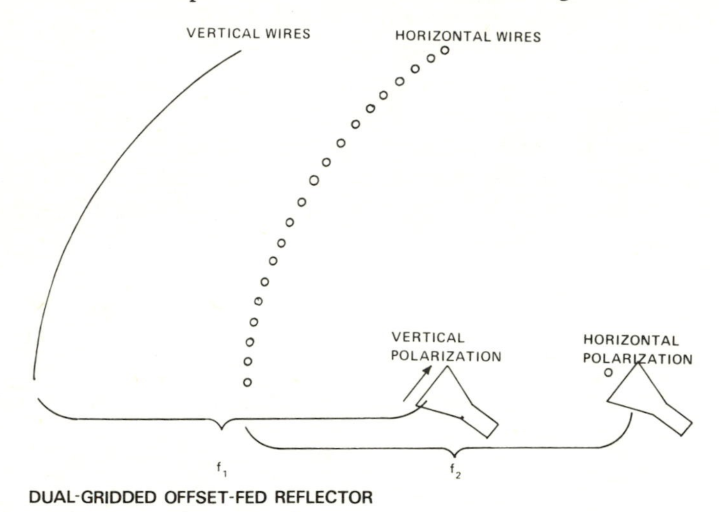

Fig. 3.63 *Illustrating a dual-polarised offset-reflector antenna employing polarised grids*

(ii) *Shaped or contoured beams:* For certain spacecraft applications, where circularly polarised area coverage is required, even a 5% beam squint can produce significant changes in the gain at the edge of the coverage zone. This problem is simply overcome for a single hand of circular polarisation, but for dual-polarised applications, the *F/d* of the reflector must be made as large as possible, consistent with the volumetric constraints. When the offset-reflector profile is deformed to provide a shaped or contoured beam, the beam squint compounds the design difficulty; numerical techniques are virtually essential to achieve a desired beam-shape. Wood *et al.*[86] have described the design and evaluation of a dual circularly polarised offset antenna intended for the European Communication Satellite System. The reflector surface was deformed in one plane to provide an elliptical coverage pattern of 8.6° x 4.9°. A copolar/crosspolar ratio of better than 33 dB was predicted over the coverage zone. Limitations in the corrugated-horn primary-feed performance reduced this figure to 30 dB over 95% of the zone on the experimental model. The tolerances upon the reflector profile and the location of the primary feed are also likely to be more critical in these designs, and the beam shape and gain can be very sensitive to small feed misalignments.[105] For example, Fig. 3.64 which is taken from Reference 105 shows the variations in the peak gain of an offset reflector with small primary-feed displacements. This reflector has an F/d ratio of 0.52 and is shaped in the plane of asymmetry to provide an elliptical beam with an aspect ratio of approximately 1.8:1.

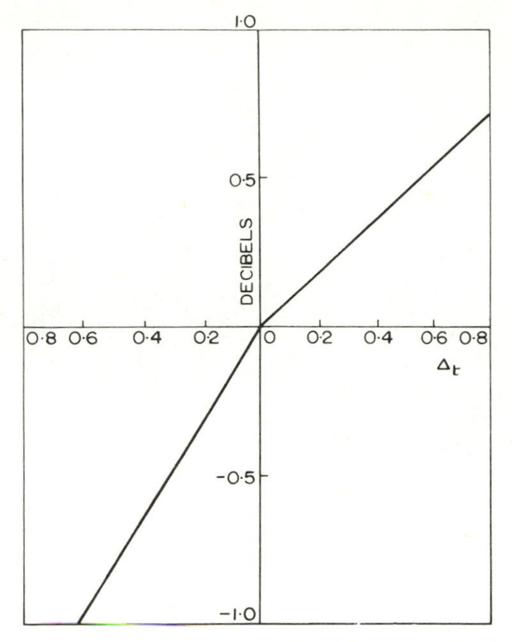

Fig. 3.64 *Variations in the peak gain of the offset antenna (F/d = 0.52) with primary-feed displacement by Δ_t wavelengths along the y axis*

For linear polarisation, the effective phase error, introduced by the departure of the shaped reflector from true parabolic, results in a spatial redistribution of the cross-polar lobes generated by the reflector. Unless the reflector offset angle is made small, the antenna will exhibit very poor polarisation purity in the coverage zone. Fasold[106] has described a linearly polarised offset-reflector system in which the perimeter of the reflector was contoured to provide an elliptical beam shape. However, the cross-polarisation from this antenna was predictably high, with peak levels close to −20dB. For a single linear polarisation, good polarisation purity can be restored by use of polarisation-sensitive grids; for dual polarisation, a larger F/d ratio is necessary.

Contoured beam shapes can also be achieved by the use of offset parabolic reflectors with multiple-element feed clusters. The feed elements are combined in amplitude and phase to generate a desired footprint over the coverage zone. Ultimately this technique offers considerably more flexibility in that the beam can be reconfigured to a variety of coverage patterns by changing the relative excitation coefficients of the feed array. The implications on the offset-reflector design are as for the previous cases, and a relatively large F/d ratio must be employed for dual-polarised applications. For applications demanding good polarisation purity there are additional complications in that the feed array must itself exhibit good polarisa-

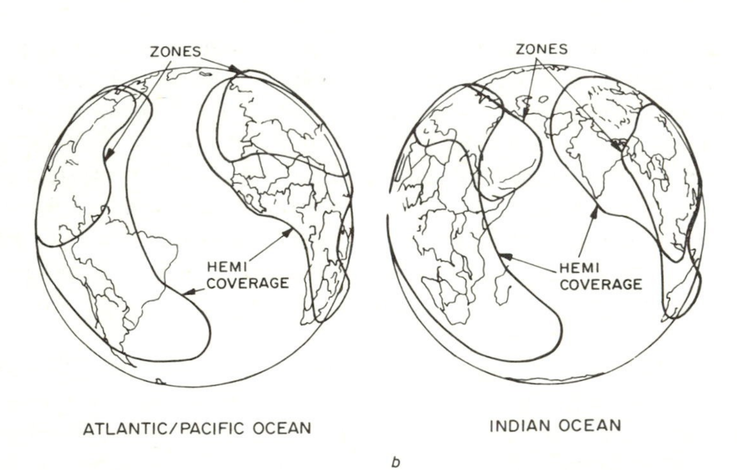

Fig. 3.65 *a* Breadboard model of a multiple-element contoured-beam antenna for satellite communications
b Required coverages for Intelsat *V* hemi/zone antenna

tion properties.[100] The combining network for the multiple elements will also be complex and is likely to be a critical feature of any practical design.

A multiple-feed offset antenna of this class has been developed for the Intelsat V spacecraft. The circularly polarised system described by Han[87] employs an offset parabolic reflector with an F/d ratio of approximate unity. The feed cluster has 78 square waveguide horns and produces four hemispheric-shaped beams, two having right-hand and two left-hand polarisation. The axial ratio of these beams within their respective coverage zones is less than 0.75 dB and, the isolation between beams is better than 27 dB. The power distribution networks for each beam employ low-loss air stripline with switches to provide some reconfiguration of the beams for different subsatellite locations. Fig. 3.65 shows a breadboard model of the antenna with a contour plot of the required coverage zones.

(iii) *Multiple spot beams:* The performance of the single offset reflector as a multiple-spot-beam antenna was studied in some detail by Rudge *et al.*[81,82] In this investigation a three-beam circularly polarised configuration was designed and

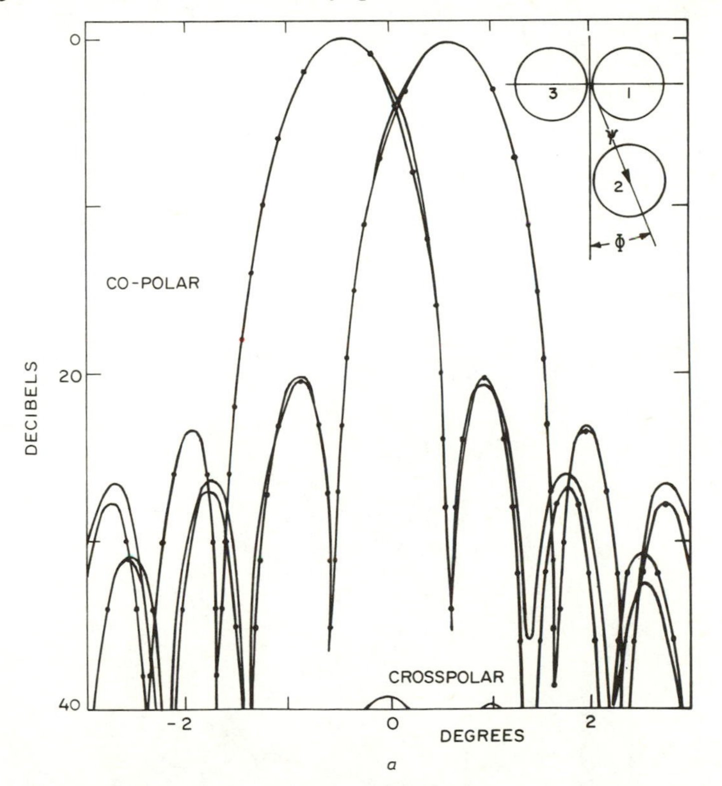

Fig. 3.66 *Copolarised and cross-polarised radiation patterns of multiple-spot beam antenna at 30 GHz. Measured ——— . Predicted —●— .*
a Cut through beams 1 and 3: $\Phi = 90°$
b Cut through beam 2: $\Phi = 23.5°$
c Photograph of breadboard spot-beam antenna

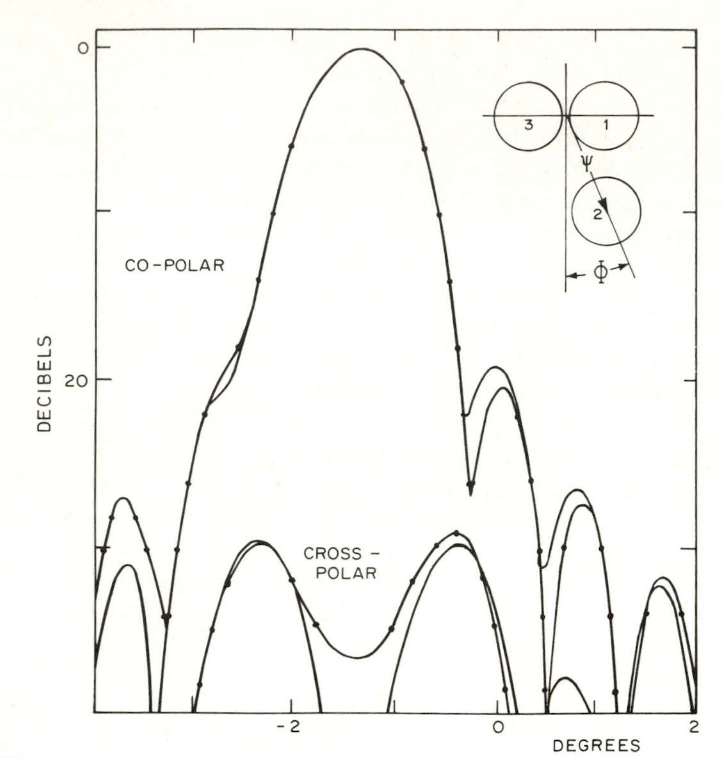

Fig. 3.66*b*

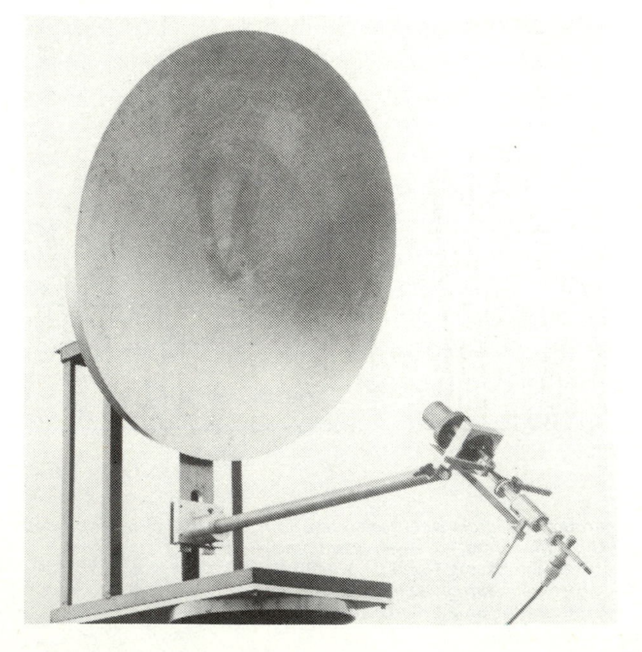

Fig. 3.66*c*

optimised using numerical techniques, and a detailed experimental evaluation was performed on a precisely made breadboard model. Since frequency reuse within each beam was not required, the antenna design made use of the beam-squint effect to reduce the beam spacing between the most closely spaced pair of beams. These beams were orthogonally polarised and had a minimum beam spacing of 1.1 HPBW with a beam efficiency of 60%. The feed elements were conical-horn elements with optimum dimensions for cross-polar suppression and exhibited an isolation between beamports of better than 40 dB. Fig. 3.66*a* shows the predicted and measured radiation patterns for the two beams. Fig. 3.66*b* shows the radiation characteristics of the third beam. A beam-to-beam isolation of better than 30 dB in the coverage zone was achieved for this antenna, which is shown in Fig. 3.66*c*. A means of achieving an outline design of antennas of this type is described in Section 3.3.4.

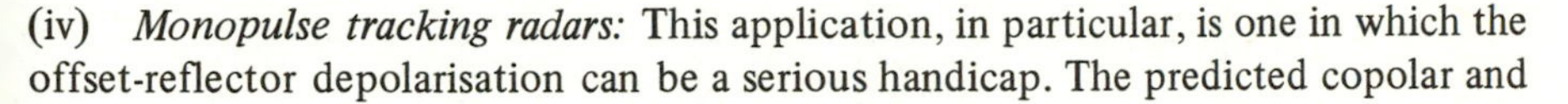

(iv) *Monopulse tracking radars:* This application, in particular, is one in which the offset-reflector depolarisation can be a serious handicap. The predicted copolar and

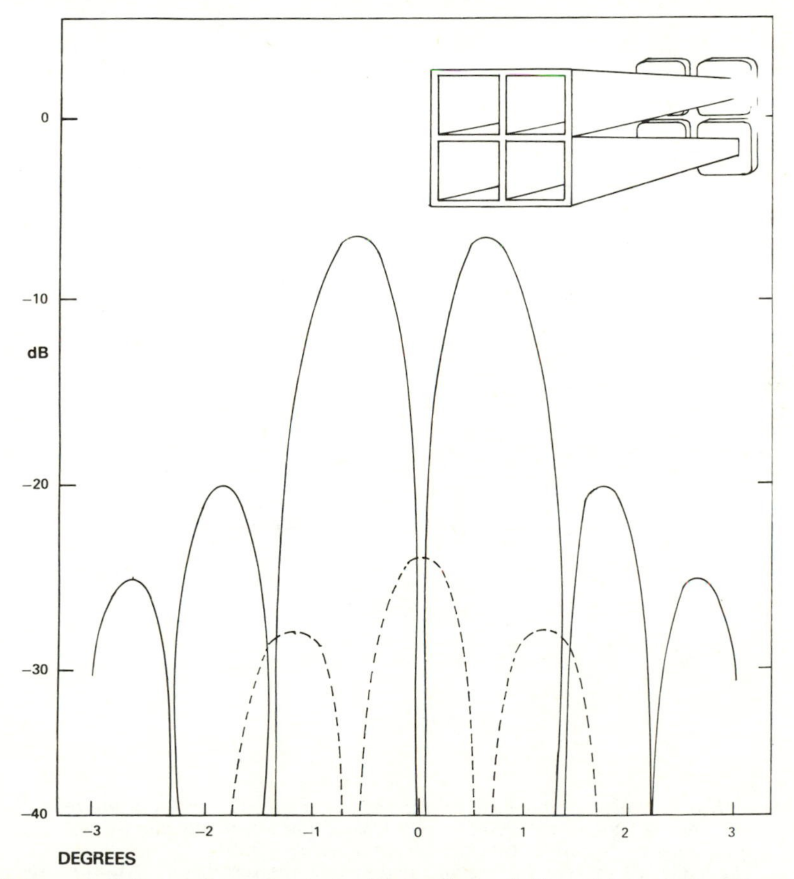

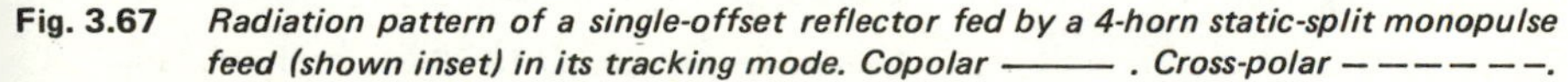

Fig. 3.67 *Radiation pattern of a single-offset reflector fed by a 4-horn static-split monopulse feed (shown inset) in its tracking mode. Copolar ——— . Cross-polar – – – – – –.*

cross-polar radiation fields of an offset reflector fed by a 4-horn static-split monopulse feed in its tracking mode are illustrated in Fig. 3.67. The cross-polar field can be seen to have a peak on the boresight axis. Randomly polarised signals from a radar target will result in an output from the tracking channel in which the boresight location appears to shift sporadically. Its precise location at any time will be dependent upon the relative magnitudes of the orthogonally polarised components in the returning signal. This boresight uncertainty, which is sometimes termed boresight jitter, can impose serious limitations upon radar tracking accuracies.

Since the cause of boresight-jitter is the reflector depolarisation, one cure (with is to employ larger F/d-ratio reflectors or improved primary feeds (Section 4.11).

(v) *Low-sidelobe antennas:* One of the major attractions of the offset reflector antenna is the possibility of lower sidelobe radiation. This feature has become

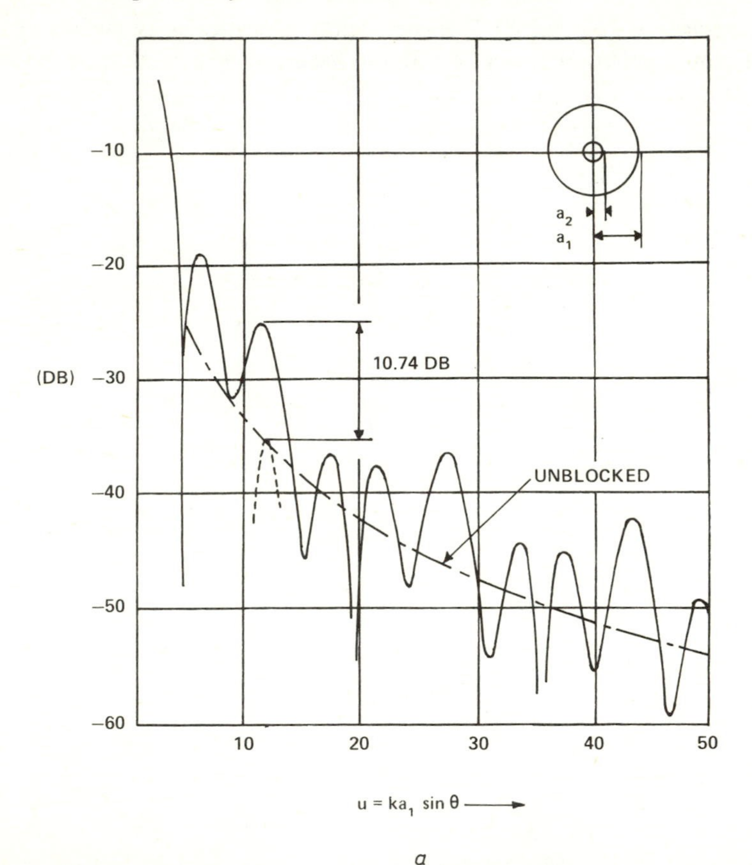

Fig. 3.68 *Effects of aperture blockage on the radiated sidelobe levels and the reflection coefficient at the primary feed for symmetrical and offset antennas*

a Radiation pattern of blocked aperture ($a_2/a_1 = 0.2$)

b Radiation pattern for marginally blocked aperture ($a_2/a_1 = 0.2$) in horizontal and vertical planes

c Reflection coefficients for illumination taper of 13 dB at subreflector edge

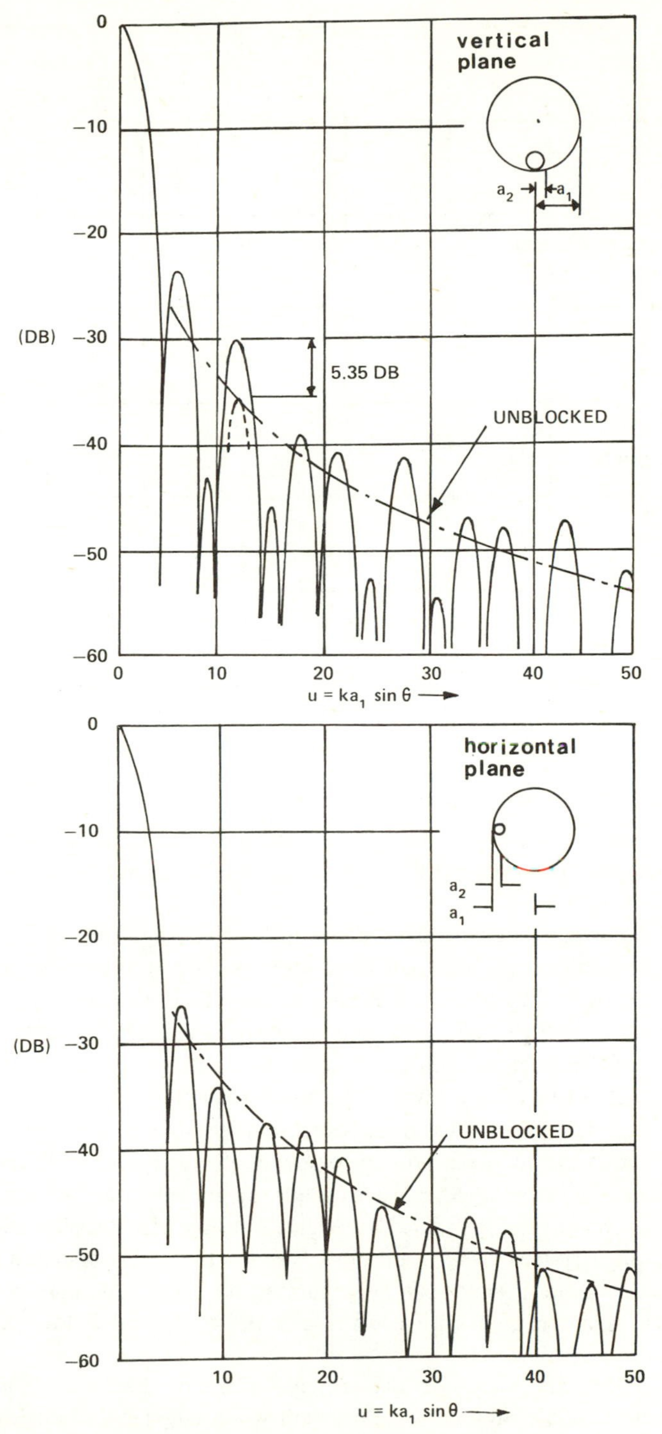
vertical plane
a_2
a_1
5.35 DB
UNBLOCKED
(DB)
0
−10
−20
−30
−40
−50
−60
0
10
20
30
40
50
u = ka₁ sin θ
horizontal plane
a_2
a_1
UNBLOCKED
(DB)
0
−10
−20
−30
−40
−50
−60
10
20
30
40
50
u = ka₁ sin θ

Fig. 3.68b

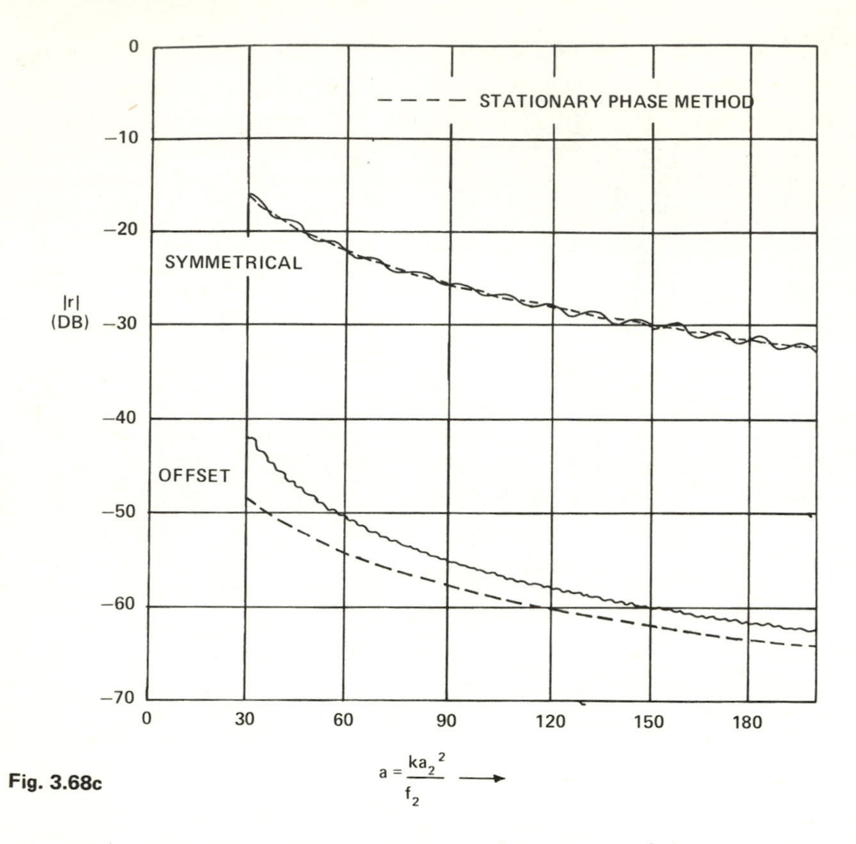

Fig. 3.68c

increasingly important in recent years as a consequence of the pressure upon the available frequency spectrum and the need to avoid interference from both friendly and unfriendly sources.

Dragone and Hogg[79] have shown theoretically that the fully offset parabolic reflectors offer significant advantages over their (blocked) axisymmetric counterparts. Improvements of up to 10 dB in near-in sidelobe levels can be inferred from their results given in Fig. 3.68. The data shown in Fig. 3.56 indicates that first sidelobe levels of the order of −30 dB can be realised with illumination tapers of the order of −20 dB. In fact more favourable tapers than those provided by the feed employed in Fig. 3.56 can reduce these levels considerably. Using corrugated horn feeds, pencil beams can be produced with first sidelobes at the −33 dB level coupled with aperture efficiencies of 70%. An example of an antenna with this order of performance is shown in Fig. 3.69. Lower sidelobes are also feasible with a more sophisticated primary-feed design, although some reduction in aperture efficiency will be implied. Elliptical beams can be generated by offset antennas in which the first sidelobes in a specified critical plane can be suppressed below the −40 dB level.[98]

The illumination asymmetry in the principal plane of the offset reflector does introduce an undesirable shoulder on the beam when larger offset angles are used.

This effect can be detected on the data shown in Fig. 3.56 and cannot be completely obviated except by the use of larger F/d ratios or more sophisticated primary-feeds.

Linearly polarised offset antennas will, of course, generate significant levels of cross-polar radiation unless the reflector has a large F/d ratio. However, for most applications it is found that the reflector depolarisation lies below the copolarised sidelobe envelope. Hence, the offset reflector depolarisation does not preclude the

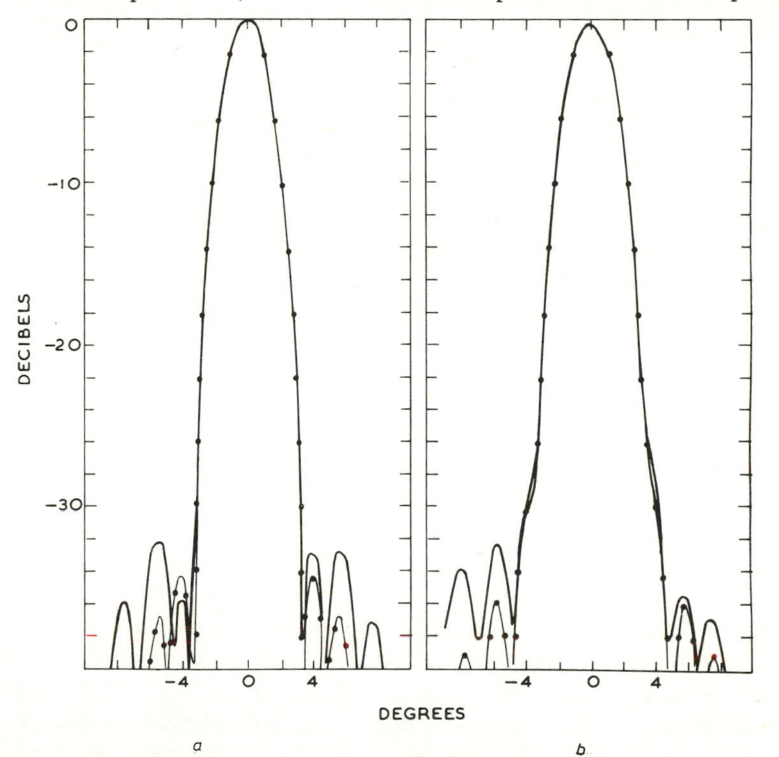

Fig. 3.69 *Radiation patterns in principal planes for an offset reflector ($F = 22.7\lambda$, $\theta_0 = 44°$, $\theta^* = 30°$) fed by a circularly polarised corrugated horn with a −18dB illumination taper at $\theta = \theta^*$. Frequency = 30 GHz. Antenna efficiency 70% neglecting ohmic losses*
a $\Phi = \pi/2$
b $\Phi = 0$
Measured ——— . Predicted —●— .

use of this antenna in a low sidelobe role. For smaller offset reflectors, primary-feed spillover constitutes the main limitation on the overall sidelobe performance. These effects can be alleviated by good primary-feed design and some use of shields or blinders about the antenna aperture.

3.3.4 Design procedure for multiple spot-beam antennas

Introduction

In the electrical design of a multiple spot-beam offset-reflector antenna a large number of parameters must be taken into account and a considerable degree of interaction must be anticipated between these parameters. To achieve an optimised design it is obviously desirable to make use of numerical techniques to model the antenna configuration. However, even when sophisticated numerical models are available, a case can still be made for a more prosaic design procedure to produce an outline design. By this means the ranges of the parametric variables in the computations can be reduced and a better understanding obtained of the trade-offs and compromises involved.

To illustrate this process a step-by-step procedure will be described which can be employed to establish the outline design of a circularly-polarised offset reflector with multiple spot beams. A single spot beam can be designed using this process but the graphical data will not be accurate for very low sidelobe designs. The design of a linearly-polarised antenna follows a similar procedure but in this case the asymmetric reflector will not generate a beam-squint and the usual cross-polar lobes will appear in the plane of asymmetry. For linearly-polarised antennas the peak value of such lobes can be obtained from Fig. 3.60.

The design parameters

The step-by-step procedure takes as its starting point a required beam-coverage specification for a set of n circularly polarised beams. The beam-coverage specification is expressed as the half-power beamwidth Ψ_{on} of each beam with its associated beam-pointing angles Ψ_n, Φ_n. These angles refer to fixed values in the secondary co-ordinate system of Fig. 3.53. Since the maximum dimensional requirements of the antenna system are often an important parameter in a practical design, a volumetric constraint in the form of a bounding cylinder has been assumed. The bounding cylinder, which is described by its diameter D and height h_t, is assumed to contain both the reflector and its primary-feed system. The general configuration is illustrated in Fig. 3.70. Other geometric constraints are obviously possible and could be treated similarly by a suitable geometric analysis.

Referring to Fig. 3.70, the defining parameters of the offset reflector antenna to be specified by the design procedure are the focal length of the parent paraboloid F, the offset angle θ_0 and the maximum semi-angle subtended by the reflector at the focus. Two secondary parameters, which are of particular interest here, are the diameter of the offset reflector projected aperture d and the clearance distance d_c between the parent paraboloid axis and the lower edge of the projected aperture.

The primary-feed is defined by type, diameter $2b$ and length l. Three feed types are considered here; the conical corrugated horn, the dual-mode (Potter) horn and the fundamental-mode conical horn.

The locations of the primary-feed phase centres, to satisfy the desired beam-pointing criteria, are given in polar co-ordinates (Δ_{tn}, ϕ_n) in the xy plane (see Fig 3.54), with origin at the reflector focus. The polar angle ϕ_n is measured from the x axis.

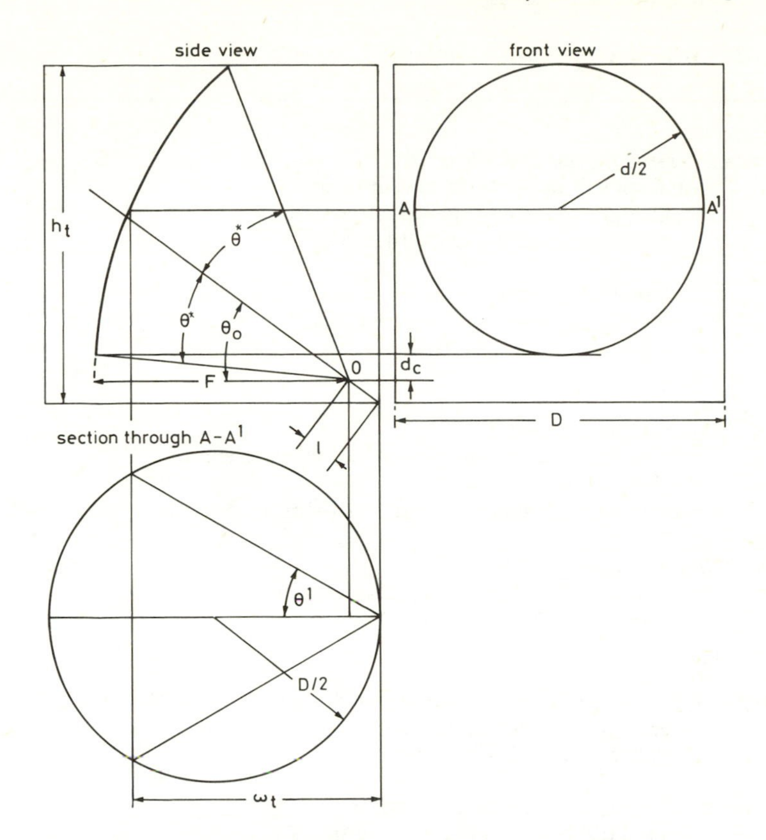

Fig. 3.70 *Geometry of the bounding cylinder*

Having established values for each of the parameters given above, data can be extracted from the graphs provided in this Section to obtain an outline performance estimate for each beam. In addition to the achievable beam-pointing specification, which may differ from that desired initially, estimates can be made of the peak coma-lobe level, the gain loss due to off-axis feed locations, the spillover loss and the peak levels of the cross-polarised lobes. In Table 3.3 the input, output and performance parameters are summarised.

The design procedure

The design procedure commences from a given beam-coverage specification and the dimensions of the bounding cylinder about the antenna. The procedure is outlined in Table 3.4. For circularly polarised offset reflector antennas, the primary-feed location must account for the beam-squinting effects introduced by the asymmetric geometry. This has been dealt with here by use of a modified 'beam deviation factor', which is derived in the text.

Table 3.3 *Step-by-step design parameters*

Input data	
(i) *Desired beam-coverage specification*	
$-\psi_{on}$	= half-power beamwidth of nth beam
ψ_n	= boresight pointing-angle of nth beam
Φ_n	= azumuthal pointing-angle of nth beam
(ii)	*Volumetric constraints*
D	= diameter of bounding cylinder
h_t	= height of bounding cylinder
Output data	
(i) *Basic reflector parameters*	
F	= focal length of parent paraboloid in wavelengths
θ_0	= offset angle
θ^*	= semiangle subtended by reflector rim from focus
(ii) *Secondary reflector parameters*	
d	= diameter of reflector projected aperture
d_c	= clearance distance reflector-to-feed
(iii) *Feed parameters*	
Feed type	= hybrid-mode, dual-mode or fundamental-mode conical horn
$2b$	= feed-horn aperture diameter
l	= overall length of feed system
(iv) *Feed phase-centre location in x y plane*	
Δ_{tn}	= radial distance from geometric focus
ϕ_n	= polar angle from x axis
(v) *Performance parameters*	
Achievable beam-coverage specification	
Peak coma-lobe levels	
Off-axis gain loss	
Spillover loss	
Peak crosspolarised lobe level	

Step 1: Reflector projected aperture

The minimum half-power beamwidth (ψ_{min}) required to satisfy a given antenna specification will define the minimum projected-aperture area which must be made available. For a moderate degree of sidelobe suppression (i.e. maximum sidelobe levels below -20 dB) a minimum projected-aperture diameter of the order of $1.1\lambda/\psi_{min}$ will be necessary. To achieve a sidelobe specification of the order of -30 dB a minimum projected-aperture diameter of at least $1.2\lambda/\psi_{min}$ will be required. The minimum beam-spacing which can be achieved with a simple primary-feed configuration is of the order of one half-power beamwidth, and to achieve this limit with a single reflector implies a compromise with regard to both the maximum sidelobe levels and the spillover loss, which can be tolerated. To provide some guide

Table 3.4 *Design procedure*

Step	Operation	Output
1	Specify projected aperture diameter	d
2	Specify reflector parameters	F, θ_0, θ^*
3	Specify primary-feed parameters	b_n, l
4	Determine primary-feed locations	Δ_{tn}, ϕ_n
5	Test for feed spacing	
6	Outline performance estimate	half-power beamwidth beam-pointing angles coma-lobe level off-axis gain loss spillover loss crosspolar lobe level

to this trade-off, note that to achieve a beam-spacing approaching one beamwidth implies sidelobe levels of the order of −20 dB for a beam close to the antenna boresight, and a corresponding spillover loss of the order of 1.5 to 2.0 dB. From the data and trends indicated, an initial value for the projected aperture d can be selected.

Step 2: Reflector parameters

In many practical applications the volumetric constraints upon the overall antenna structure will impose significant constraints upon the range of the reflector parameters, F, θ_0, and θ^*. In the procedure established here the bounding surface of the antenna structure is taken as a cylinder of height h_t and diameter D. The geometry is illustrated in Fig. 3.70. The diameter of the projected-aperture of the offset reflector is related to the other parameters by eqn. 3.75, and from the geometry we find

$$h_t = 2F \tan \frac{\theta_o + \theta^*}{2} + l \sin \theta_0 \tag{3.95}$$

$$D = \frac{d^2/4 + W_t^2}{W_t} \tag{3.96}$$

where

$$W_t = F\left[1 - \left(\frac{\sin \theta_o}{\cos \theta_o + \cos \theta^*}\right)^2\right] + l \cos \theta_0 \tag{3.97}$$

The feed clearance distance d_c is given by

$$d_c = 2F \tan \frac{\theta_o - \theta^*}{2} \tag{3.98}$$

To avoid aperture-blocking effects, and to minimise the reflector reaction upon the primary-feed, it is desirable that d_c be large enough to provide clearance for the primary-feed hardware.

When illuminated with circularly-polarised radiation the offset reflector will produce a beam-squinting effect, the magnitude of which is a function of the reflector parameters. With respect to the offset reflector axis of symmetry, the direction of the squint is toward the left for right-handed circular polarisation and toward the right for left-handed polarisation.* In certain cases it may be desirable to maximise this effect, with the objective of squinting two orthogonally-polarised beams toward each other. In other cases it may be necessary to minimise the effect to prevent movement of a beam which is to be used simultaneously for two hands of polarisation. From eqn. 3.94 the angle squinted through (ψ_s) can be obtained to an accuracy of better than $0.01\lambda/d$.

Having established an initial value for d from step 1, eqns. 3.94-3.98 can be employed to tabulate the values of h_t, D, d_c and ψ_s as a function of only θ_0 and θ^*. An initial value of between 5 and 10 wavelengths should be allowed for the feed length l. To provide an example, Fig. 3.71 shows these values for a constant projected-aperture d of 70λ and $l = 10\lambda$. Given the volumetric constraints in terms of h_t and D a range of acceptable parameter combinations may be obtained from these data. The range can be expressed, for example, as the combination which maximises θ_0 and that which minimises θ_o while satisfying the volumetric constraints and providing an acceptable feed-clearance distance d_c. Initial values for F, θ_o and θ^* can thus be established as one, or both, of the limiting values of this range, or alternatively, a combination which corresponds to a suitable mid-point in the θ_o spread. In making this choice the following comments may be found useful:

If two polarisations are to be employed on a single-beam ψ_s should be made as small as possible. If dual polarisation on a single beam is not contemplated and two orthogonally polarised beams are to be closely spaced, then the beam squint effect may be utilised usefully to reduce the spacing of the beams.

Large values of $\theta_o + \theta^*$ tend to produce increased distortion of off-axis beams and some disparity in the sidelobe performance in the two principal planes. Fig. 3.56 presents some computed data showing the sidelobe trend for on-axis beams.

*A wave receding from an observer having clockwise rotation of the electric field is taken as right-handed circular polarisation. (see Section 1.7)

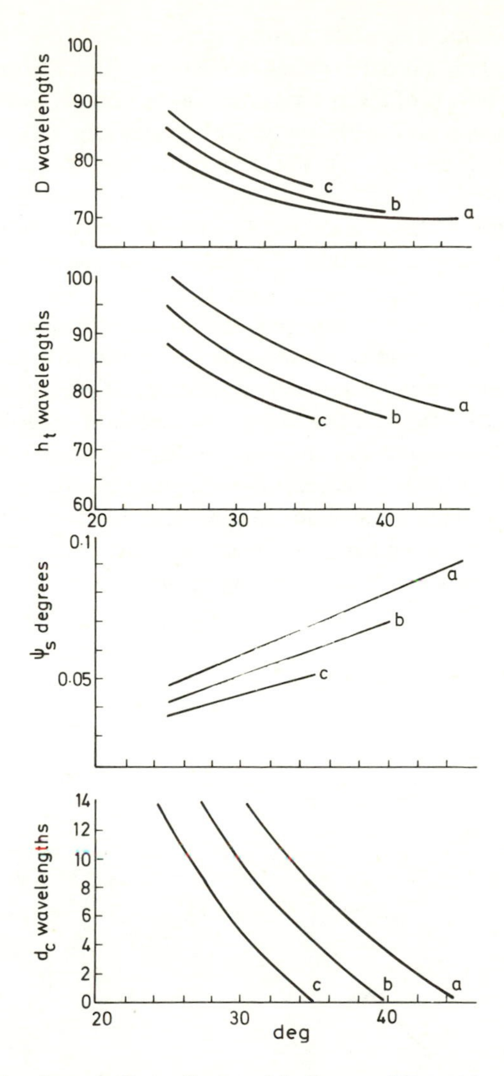

Fig. 3.71 *Parameters for an offset reflector with diameter 70λ and feed length 10λ, contained within a cylinder a $\theta_0 = 45°$ b $\theta_0 = 40°$ c $\theta_0 = 35°$*

Step 3: Primary-feed parameters

Three practical primary-feed types are considered here as suitable feed elements for an offset reflector antenna:

(i) The corrugated or hybrid-mode conical horn
(ii) The 'Potter' or dual-mode conical horn
(iii) The fundamental mode (TE_{11}) conical horn.

Both the hybrid-mode and dual-mode horns can be considered to be high-performance feeds for the circularly-polarised offset-reflector antenna. The cross-polarisation generated by these devices is of a very low level (i.e. < -30dB) and the co-polarised beam is rotationally symmetric with low sidelobe radiation; VSWRs in both cases can be made very low (i.e. $< 1.1:1$). The dual mode horn is a somewhat narrow-band device with a bandwidth of the order of 5%. The hybrid-mode feed can be operated over significantly larger bandwidths and its drawbacks in the multiple-beam application are largely a function of its higher mass and that the corrugations make its outer dimensions significantly larger than its radiating aperture area. In addition, both the corrugated horns and the dual-mode horn have a tapered aperture-illumination which implies a lower gain for a given aperture diameter. This feature can be very undesirable when closely-spaced beams are required.

The small flare-angle fundamental-mode conical-horn will operate over a reasonably wide bandwidth and has an aperture diameter which is approximately 11% smaller than that of either the hybrid-mode or dual-mode horn for approximately the same shape main-beam illumination characteristics. The flanged TE_{10} mode horn generates a cross-polar component, however, which has a minimum value when the horn aperture is 1.14 wavelengths in diameter.[92] The effect of the horn cross-polar radiation is illustrated in Fig. 3.72 which shows typical values of

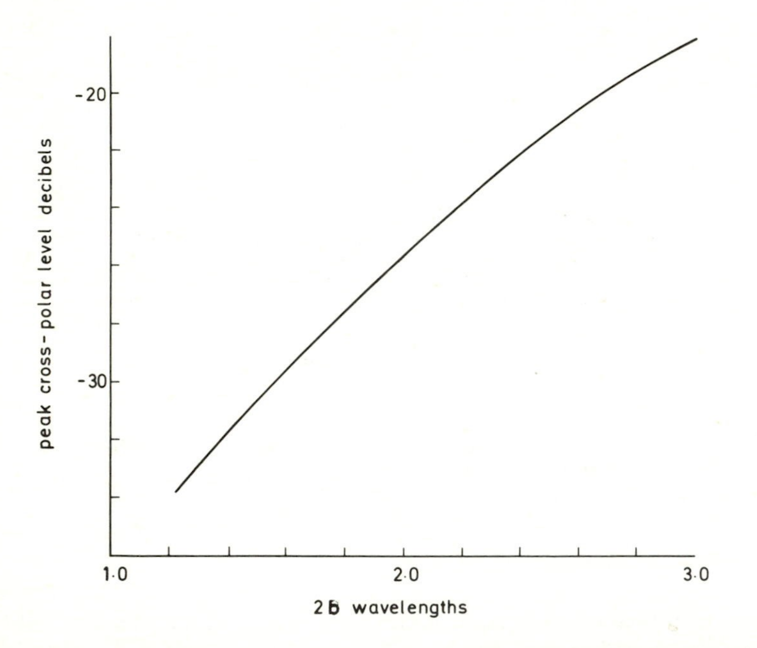

Fig. 3.72 *Computed radiated cross-polar peaks from offset reflector antennas with θ_0 values in range 40-45° and θ^* values in the range 30-35° when fed by circularly-polarised fundamental-mode conical horns of diameter 2B*

radiated cross-polarisation peaks from offset reflector antennas fed by TE_{11} mode horns of diameter $2b$. The peak values occur in a crest around the main co-polar beam, at points corresponding to the −10 to −20 dB level of the main co-polar beam, and are theoretically similar in any plane.

Either a hybrid-mode horn or a dual-mode horn is generally preferable from the point of view of illumination characteristics. However, for closely-spaced beams it may be necessary to employ the physically smaller, higher gain, fundamental-mode horn. Small diameter unflanged horns of this type can also produce good cross-polarisation characteristics.[97]

Having established values for θ_0 and θ^* in step 2 the aperture radius b of either a hybrid-mode, dual-mode or fundamental-mode horn can be determined to provide a specified sidelobe performance for an on-axis beam. From Fig. 3.73 the necessary aperture-illumination taper T to produce a given sidelobe level is illustrated for several values of θ^*. The illumination taper on an asymmetrical reflector differs in

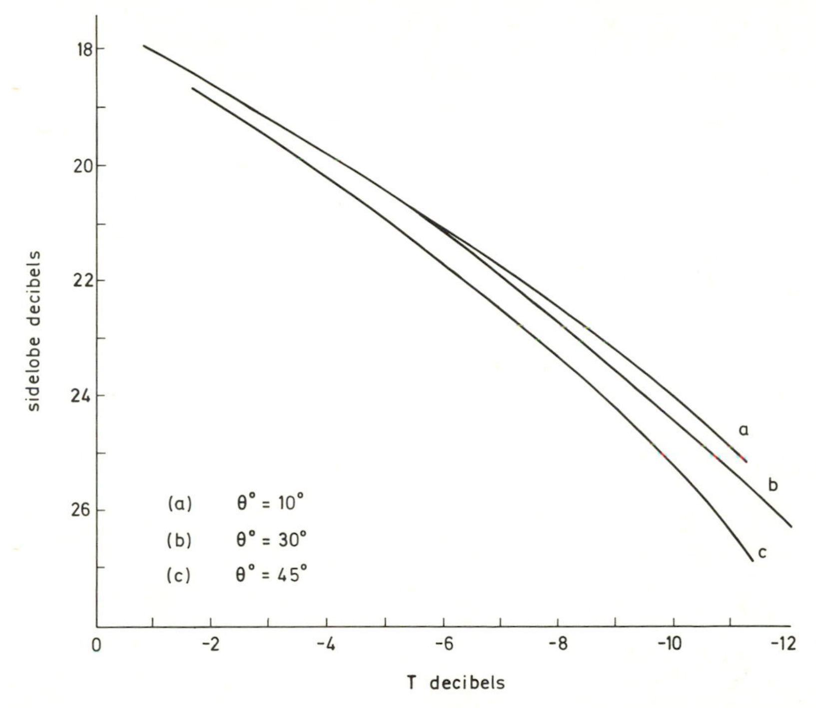

Fig. 3.73 *First radiated sidelobe levels from reflectors, as a function of reflector illumination taper T*

its principal planes and this is accounted for in Figs. 3.74 and 3.75. The illumination taper T is composed of a symmetrical component G, which is a function of the feed aperture diameter $2b$, plus a space-factor S which is dependent upon θ_0 in the plane of symmetry (S_s) but essentially independent of θ_0 in the plane of asymmetry (S_a). The decibel sum of either $G + S_s$ or $G + S_a$ provides an effective illumination taper T for each principal plane which is related to the first sidelobe levels in these planes in Fig. 3.73. The illumination taper $G + S_a$ is in fact the mean value of the edge illumination, which varies around the reflector periphery. The taper $G + S_s$

corresponds to the minimum taper value and the correspondence between this factor and the resultant sidelobe prediction has been tested against computed data and found to be accurate within 1 dB for illumination tapers of not more than 10 dB.

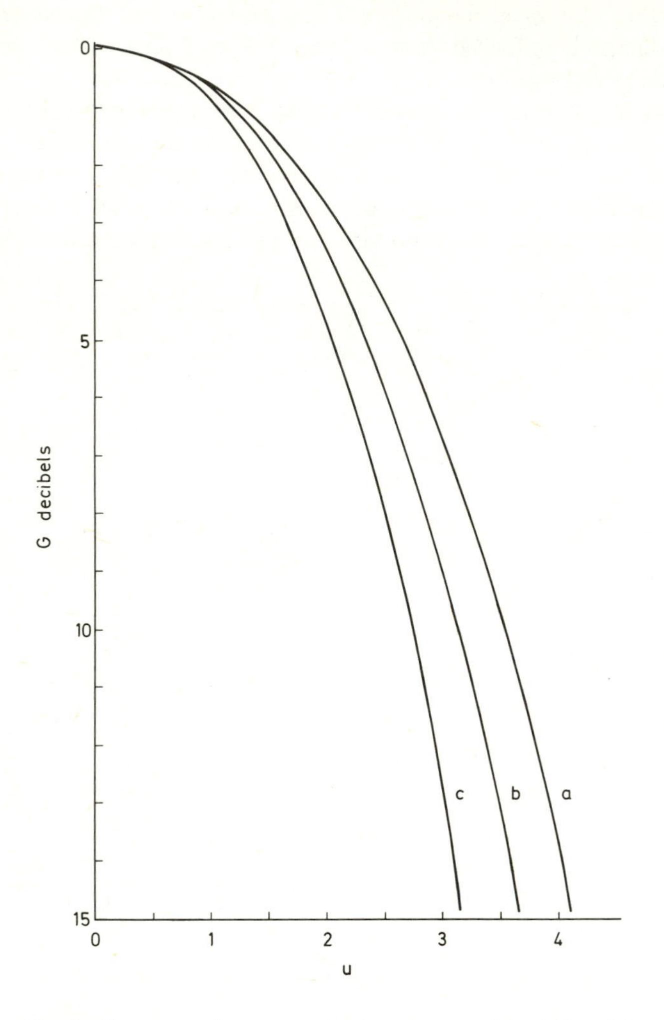

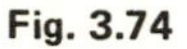
Fig. 3.74 *Illumination taper due to pre-dominant terms of feed directivitv as a function of $u = 2\pi b \sin\theta$*, where b is the feed aperture radius in wavelengths.*
a Hybrid-mode or dual-mode feeds
b Circularly-polarised fundamental-mode feed
c Uniformly illuminated circular-aperture feed for comparative purposes

While employing Figs. 3.73-3.75 to select a suitable feed diameter it may also be desirable to be aware of the beam-broadening effect of the edge illumination taper upon the radiated beam. In Fig. 3.76 a beam-width factor N is shown as a function of the illumination taper T. The half-power beamwidth of the radiated beam from the offset reflector can be obtained approximately from

$$\psi_{on} = \frac{N\lambda}{d} \tag{3.99}$$

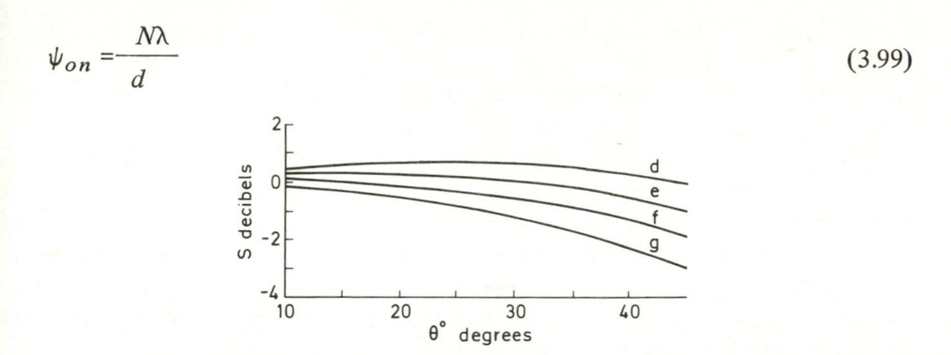

Fig. 3.75 *Space attenuation factor for offset reflectors, including additional ½(1 + cos θ*) term from feed expression. For plane of symmetry use (d) $\theta_0 = 45°$, (e) $\theta_0 = 40°$, (f) $\theta_0 = 35°$. For plane of asymmetry use curve (g) for any offset angle*

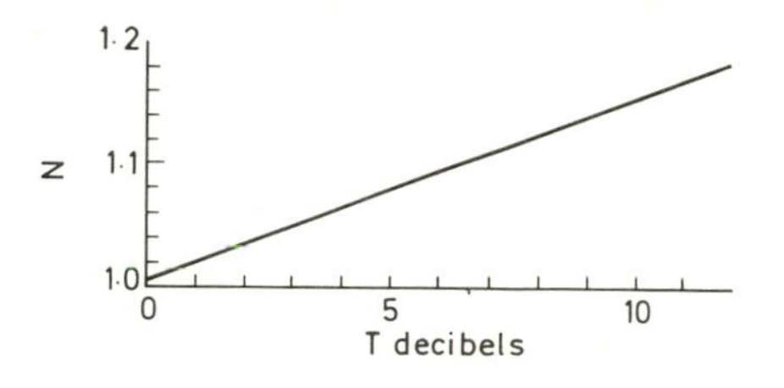

Fig. 3.76 *Beamwidth factor N as a function of reflector illumination taper*

The length of the primary-feed elements can then be estimated by assuming a typical flare for the feed of the order of 10° semi-angle. Having determined the aperture radius b, the feed length can be simply derived, but on specifying l an additional allowance must be made for the location of such components as polarisers, orthomode transducers or transitions if they are to be employed.

Step 4: Primary-feed location

Since the offset reflector antenna is an asymmetric structure, some decision must be taken with regard to the relative alignment of the reflector axis of symmetry and the beam cluster to be generated. In certain cases this decision may be dictated by other considerations, but, where a freedom of choice exists, the following general points are worthy of consideration:

Gain loss due to off-axis beam locations is virtually independent of the plane of offset.

The beam-squint effect can best be used to squint beams together, or apart, by locating orthogonally polarised beams on opposite sides of the reflector axis of asymmetry (i.e. about the plane $\Phi = 0°$).

The quality of the radiation pattern appears to be best maintained by locating off-axis beams in the plane of asymmetry ($\Phi = 90°$). Off-axis beams in the plane of symmetry ($\Phi = 0$) tend to exhibit a 'shoulder' on the main beam for relatively small off-axis shifts.

Having aligned the reflector axis cf symmetry with the required beam cluster, beam locations can be specified in terms of their pointing angles ψ_n, Φ_n. From Steps 1-3 initial parameter values have been assigned for both the reflector and the primary-feed. The primary-feed aperture radius for any specified beam b_n has been selected initially purely on the desired illumination characteristics for an on-axis beam. When several beams are to be generated from a single reflector this initial value may well be modified.

From a knowledge of the initial estimate of the feed radius b_n Figs. 3.74 and 3.75 can be employed to determine the mean illumination taper $(G + S_a)$ dB. This illumination taper can then be applied in Fig. 3.77 to obtain the relevant beam deviation factor (B_{df}) for this reflector and feed configuration.

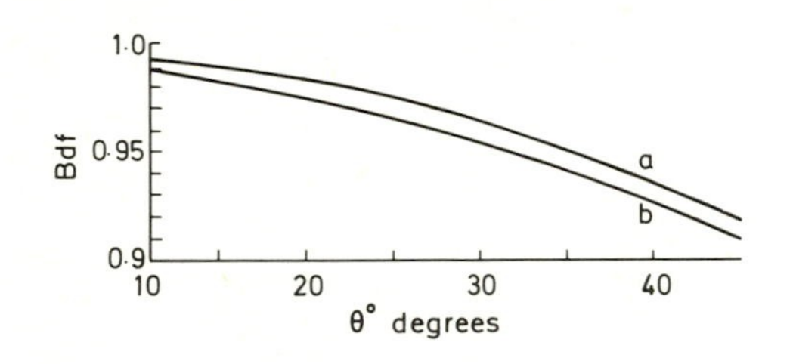

Fig. 3.77 *Beam deviation factor with mean reflector illuminations of a −10dB and b −6dB*

Given the beam-pointing coordinates (ψ_n, Φ_n) the location of the primary-feed phase-centre is to be specified in terms of the focal region coordinates (Δ_{tn}, ϕ_n). The derivation of Δ_{tn}, ϕ_n, is complicated by the beam-squinting effect but this effect has been accounted for in the following equations. The direction of the squint is a function of the hand of polarisation and in eqns. 3.100 and 3.101 the choice of the upper sign implies a beam-squint toward the axis of symmetry of the antenna ($\Phi = 0$) while the lower sign should be taken for a squint away from the symmetry axis. The equations provided have been tested against computed data for high-gain offset-reflector antennas (d greater than 20 wavelenths) and found to produce good results for off-axis beams within 10° of the antenna boresight, providing that the comatic distortion of the radiation pattern is not excessive.

To determine Δ_{tn}, Φ_n, the following procedure is followed:

The beam-squint angle ψ_s can be obtained from eqn. 3.94. The azimuthal feed-location angle (ϕ_n) is then given by

$$\phi_n = \arctan\left[\frac{\sin\psi_n \sin\Phi_n \pm \sin\psi_s}{\sin\psi_n \cos\Phi_n}\right] \tag{3.100}$$

The effective focal-length for the offset reflector is given by:

$$F_e = F \frac{1 + \cos\theta^*}{\cos\theta_0 + \cos\theta^*} \quad (3.101$$

and the radial feed location distance Δ_{tn} is then given by

$$\Delta_{tn} = \frac{F_e}{B_{df}} (\sin^2\psi_n + \sin^2\psi_s \pm 2\sin\psi_n \sin\psi_s \sin\Phi_n) \tag{3.102}$$

Step 5: Test for feed-spacing

Having followed steps 3-4 for the beam-pointing angles of interest, a test must be carried out to ensure that the feed aperture-dimensions and the feed locations are physically compatible. In the event that the configuration fails this test, the dimensions of one or more of the feeds must be reduced and steps 3-5 reiterated until an acceptable compromise is reached.

Step 6: Outline electrical performance

The outline electrical performance of the multiple-beam antenna configuration derived from steps 1-5 can be estimated by reference to the data provided here. Commencing with the primary-feed radius for the nth beam b_n, Figs. 3.74 and 3.75 can be employed to determine the feed mean illumination taper T, given by $(G + S_a)$ dB. From Fig. 3.76 the half-power beamwidth for the beam can then be obtained. Although this data was computed for on-axis beams, it has been found that the half-power beam-broadening effect due to a small shift from the focus is small for offset-reflectors with θ^* not greater than 45°. An estimate of the spillover loss for this combination of feed and reflector can be obtained from Fig. 3.78. For a fundamental-mode conical horn a guide to the peak cross-polarisation level in the antenna far-field can be obtained from Fig. 3.72 which was compiled from computed data for offset reflectors with offset angles in the range 35-45° and θ^* in the range 30-35°. The E-field model for the radiation from an open-ended circular waveguide was employed in the computations and it was found that the predicted levels were, to a first order accuracy, independent of the reflector parameters over the range indicated. It has also been noted that the peak levels of the cross-polarised radiation are relatively insensitive to small beam offsets.

Finally, the curves provided in Fig. 3.59 give some indication of the reduction in gain which can be expected for an off-axis beam and the level of the peak coma-lobe. Interpolation between these computed characteristics will provide an indication of the antenna performance, but extrapolation should be applied with caution, since radiation patterns which involve significant gain losses due to off-axis beams may also exhibit other distortions than the coma-lobe characteristic indicated. Any significant discrepancies between the desired antenna performance and that realised will demand a reiteration of the process from the relevant step in the procedure.

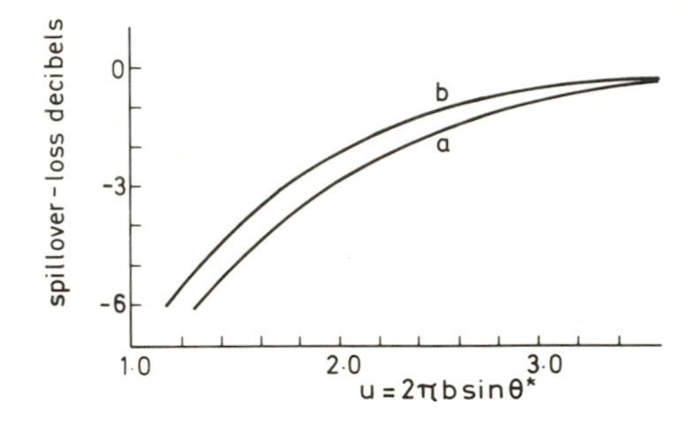

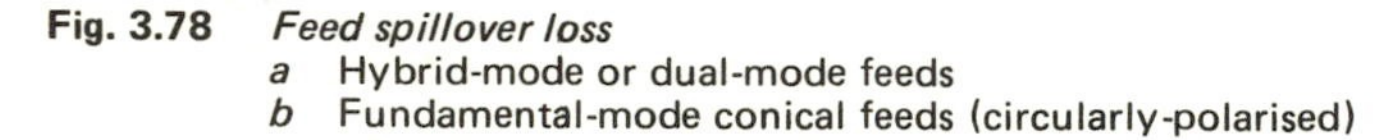

Fig. 3.78 *Feed spillover loss*
a Hybrid-mode or dual-mode feeds
b Fundamental-mode conical feeds (circularly-polarised)

Beam deviation factor for offset reflector antennas

For an axisymmetric parabolic reflector, given a beam-pointing specification (ψ', Φ') the corresponding primary-feed location can be expressed in terms of focal plane polar coordinates Δ_{tn}, ϕ_n,

$$\Delta_{tn} = \frac{F' \sin \psi'}{B_{df}} \tag{3.103a}$$

$$\phi_n = \Phi' \tag{3.103b}$$

where B_{df} is termed the *beam deviation factor* and is a function of the reflector curvature and the feed illumination taper, and F' is the focal length of the axisymmetric paraboloid.

For offset reflector antennas with offset angles not exceeding 45° it has been found that, in the absence of beam-squinting effects, the relationships expressed by the above equations can provide useful results by applying a concept of an 'equivalent' symmetrical reflector. Equivalence is taken here to imply a symmetrical reflector which subtends the same semiangle θ^* and has the same value of projected-aperture diameter as the offset reflector of interest. Equating projected-apertures for offset and symmetrical reflectors of the same semiangle θ^*, we find that the focal-length of the equivalent reflector is given by

$$F_e = \frac{1 + \cos\theta^*}{\cos\theta_o + \cos\theta^*} F \qquad (3.104)$$

where F is the focal length of the offset reflector as defined in Fig. 3.50. The dimension F_e, which will be a constant for any given offset reflector, could also be interpreted as the effective focal length of the offset reflector.

Substitution of the effective focal length of F' in eqn. 3.103*a* will provide a relationship between the primary-feed location and the resultant beam-pointing direction, which is valid for offset reflectors in the absence of beam-pointing effects. From an examination of the beam-squint phenomenon in high-gain offset reflector antennas, it has been observed that, within the small cone of angles of interest here, the beam-squint acts similarly on every radiated beam to either shift this beam

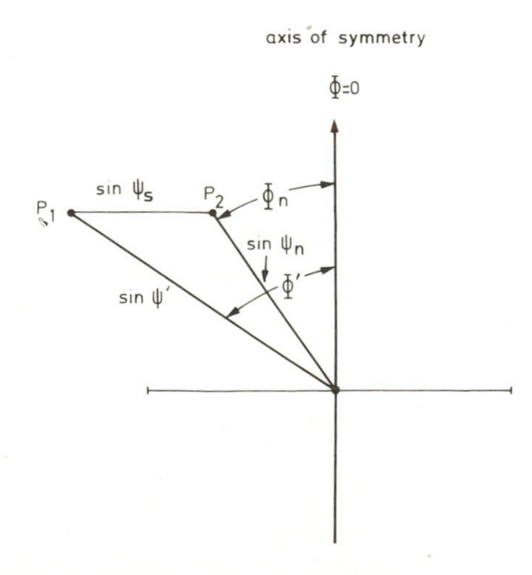

Fig. 3.79 *Beam-squint diagram, where P_2 and P_1 refer to the location of the beam centre with and without the beam-squint effect*

toward or away from the axis of symmetry of the reflector. On a contour plot of the antenna beam coverage (in sin ψ space) the direction of this shift in thus perpendicular to the $\Phi = 0$ axis, with a magnitude given by $\sin \psi_s$. This is illustrated in Fig. 3.79 where the primed coordinates ψ', Φ' indicate the position of the beam prior to the action of the beam-squint effect and ψ_n, Φ_n denotes the final location of the beam. In the case illustrated the hand of polarisation has been chosen to squint the beam toward the antenna axis of symmetry. Use of the opposite hand of polarisation would result in a squint of the same magnitude, still acting perpendicular to the axis of symmetry, but acting in the opposite sense to that shown and thus leading to $\psi_n > \psi'$ and $\Phi_n > \Phi'$.

Making use of eqn. 3.94 and the geometry of Fig. 3.79, the relationship between the unsquinted and the squinted locations of the radiated beams can be resolved into radial and azimuthal components as follows:

$$\sin \psi' = \sqrt{(\sin^2 \psi_n + \sin^2 \psi_s \pm 2 \sin \psi_n \sin \psi_s \sin \Phi_n)} \tag{3.105a}$$

$$\tan \Phi' = \frac{\sin \psi_n \sin \Phi_n \pm \sin \psi_s}{\sin \psi_n \cos \Phi_n} \tag{3.105b}$$

where a choice of the upper implies a shift toward the axis of symmetry and the lower sign indicates a shift away from the axis. Substitution of eqns. 3.105 into eqns. 3.103 provides an approximate offset reflector beam-deviation factor, inclusive of beam-squint effects.

3.3.5 Dual-reflector offset antennas

The open Cassegrainian antenna

For applications involving complex primary-feed structures, the use of a Cassegrainian feed system has some obvious advantages. In particular, the Cassegrainian configuration allows the feed elements and the associated circuitry to be located close to the main reflector surface, possibly avoiding long RF transmission paths and the need for extended feed support structure, while the forward pointing feed format can be a desirable attribute for applications requiring low noise performance.

Of the variety of offset Cassegrainian systems proposed in the literature, perhaps the best known is the open Cassegrainian antenna introduced in 1965 by the Bell Telephone Laboratories.[63,64] The antenna, which is illustrated in Fig. 3.51, comprises an offset section of a paraboloid and an offset hyperboloid subreflector, fed by a primary feed which protrudes from an aperture in the main reflector surface. With this configuration it is possible to design the antenna such that the subreflector does not block the aperture of the main reflector. However, as a direct consequence of the positions of the primary feed, some aperture blockage due to the feed system is unavoidable.

The analysis of the complete antenna can be performed by means of the physical-optics-based current-distribution technique for the subreflector and the aperture-field integration method for the main reflector. Expressed in spherical coordinate components (E_θ, E_ϕ), the subreflector fields thus obtained can be inserted into eqn. 3.82 or 3.84 to determine the tangential aperture fields of the main reflector, and, hence, via eqn. 3.90 or 3.91 to determine the far fields of the overall antenna. Thus the analysis essentially involves the evaluation of four two-dimensional diffraction integrals at each field point. Under certain circumstances, use can be made of the axes of symmetry afforded by the subreflector geometry to eliminate the azimuthal dependent integrals, thereby alleviating the computational problem. Ierley and Zucker[111] have also described a technique for reducing the double integrals associated with the main reflector into a more convenient one-dimensional

ig. 3.80 *6-m open Cassegrainian antenna located at the University of Birmingham, England*

form. The technique, which is based upon an application of the stationary-phase approximation in the azimuthal part of the integral, allows more economical predictions of both the near-in and the far-out sidelobe performance of the open Cassegrainian antenna.

In general, the basic radiation characteristics of the open Cassegrainian antenna do not differ significantly from those of an equivalent single-offset-reflector antenna. To avoid aperture blockage, the open Cassegrainian antenna must employ large offset angles and, when fed by conventional primary feeds, exhibits beam squinting and depolarising characteristics which are similar to the single offset reflector. However, for applications where these particular performance parameters are not of major concern, the open Cassegrainian configuration offers excellent potential for realising high overall efficiency and low wide-angle sidelobe radiation. Fig. 3.80 shows an open Cassegrainian antenna of the Bell Laboratories design located at the University of Birmingham in England. Fig. 3.81 shows measured and predicted data made on a precision model of this design at the Bell Laboratories. This antenna had a computed efficiency of better than 65% (including spillover and scattering losses, but not ohmic loss).

Double-offset-reflector antenna

An alternative dual-offset-reflector configuration, whch offers a number of attractive features, is the so-called double-offset antenna shown in Fig. 3.51*a*. This antenna,

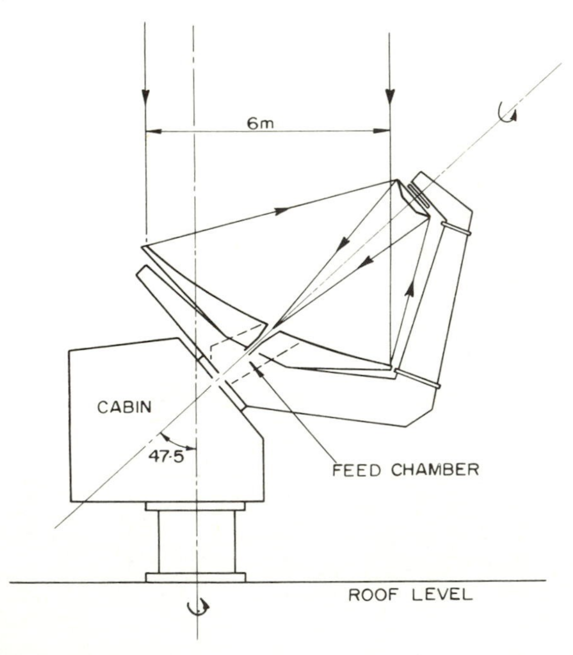

Fig. 3.80 (cont)

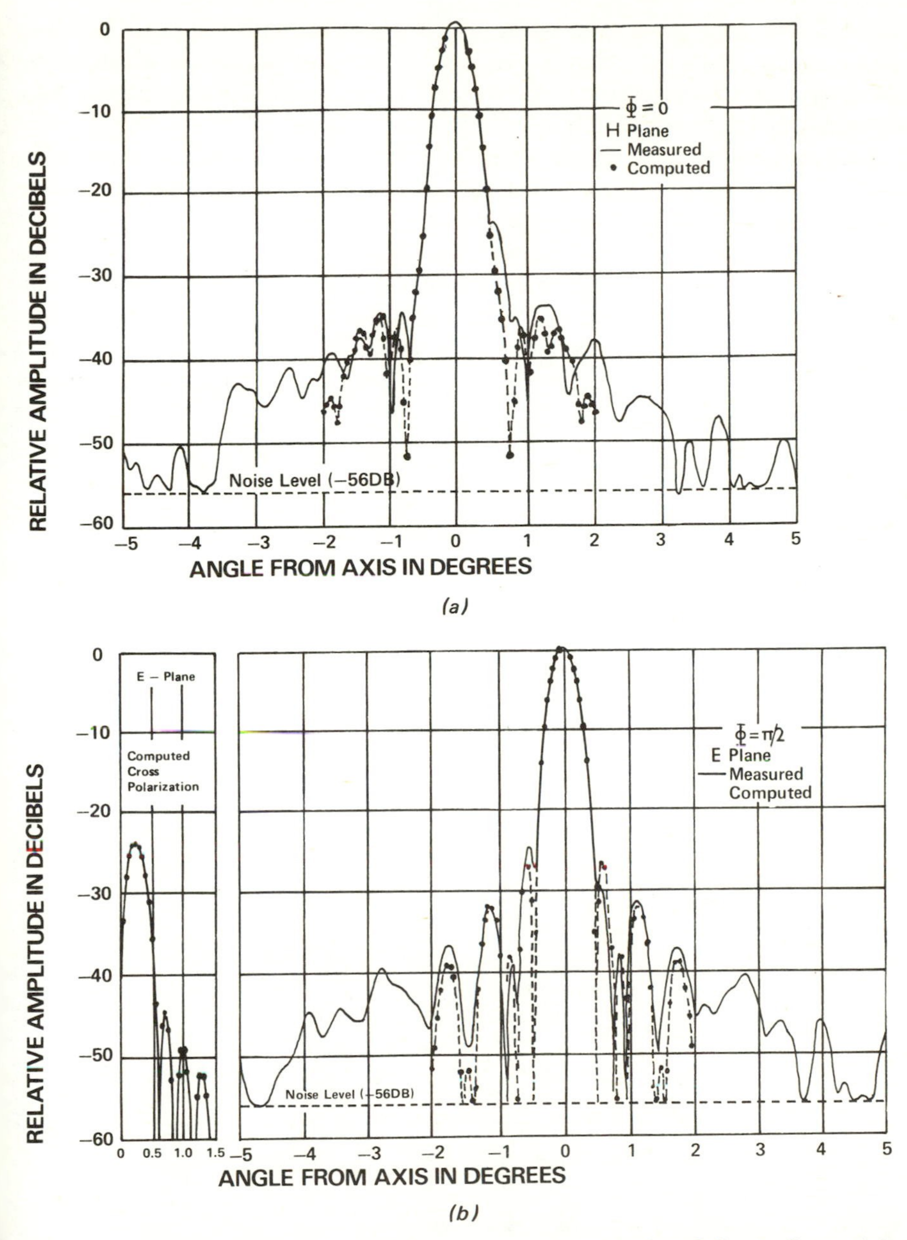

Fig. 3.81 *Radiation patterns of a 60 GHz experimental model of the Bell open Cassegrainian antenna with parameters $F = 152\lambda$ $\theta_0 = 47.5°$, $\theta^* = 30.5°$, $\theta_1 = 7.5°$, and a subreflector diameter of 40λ (a) H plane (b) E plane*

which was first implemented by Graham[73] provides a convenient location for the primary feed hardware by use of an offset section of a hyperboloidal subreflector in a Cassegrainian arrangement. Two variations of the double offset are illustrated in the Figure. A Gregorian version, in which the subreflector comprises an offset portion of an ellipsoidal reflector, is also feasible and has been considered by Mizugutch *et al.*[112] For either of the versions shown the overall antenna geometry can be designed to be completely free of aperture blockage.

Analyses performed by several workers,[74,75,112,113] has shown that the double-offset antenna can be designed such that, when fed by a conventional linearly polarised primary-feed, the depolarisation arising from the two offset reflectors can be made to cancel, thus providing an overall low cross-polar characteristic. This performance is achieved by matching the scattered radiation fields from the sub-reflector to the main reflector. The principle is essentially similar to the focal-plane matched-feed approach described for single offset reflectors in Section 4.11 (see refs 76, 77 and 107-110), and, in theory, the technique offers a greater potential for broad-band performance.

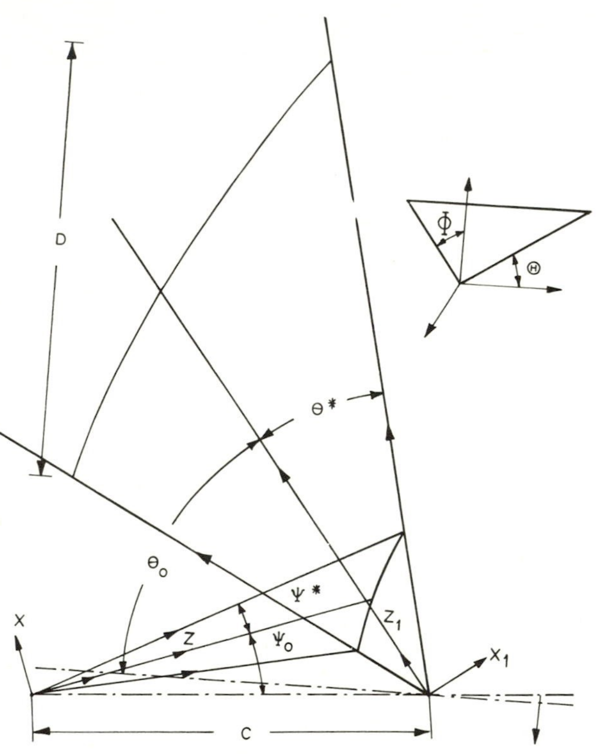

Fig. 3.82 *Double-offset antenna geometry and parameters*

Approximate techniques based upon the use of geometric optics[75,112,113] indicate that a perfect match can be achieved (i.e. giving zero cross-polar fields in the main reflector aperture) when the axis of the parent subreflector surface is depressed by an angle α from the axis of the parent paraboloid. This condition is illustrated in Fig. 3.51*b*. A mathematical expression relating the depression angle

α to the parameters of the subreflector has been derived by Mizugutch *et al.*[112] In its simplified form this can be expressed as[114]

$$\tan\frac{\alpha}{2} = \frac{1}{M}\tan\frac{\psi_0}{2} \tag{3.106}$$

where $M = (1 + e)/(1 - e)$ and e is the eccentricity of the subreflector, ψ_0 is the feed offset angle, and the geometry is illustrated in Fig. 3.82.

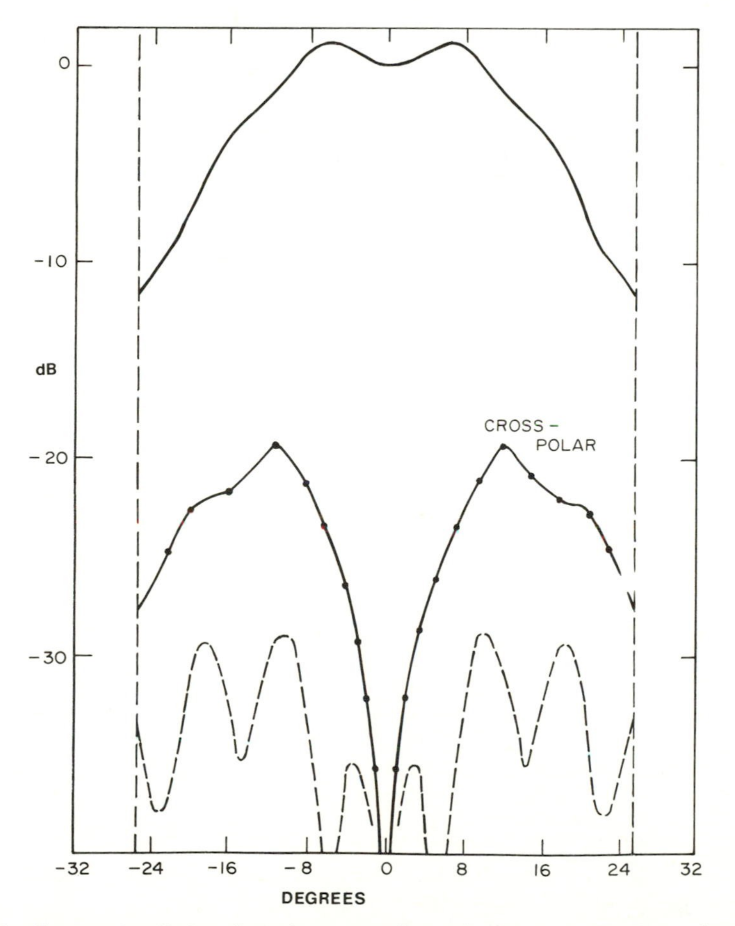

Fig. 3.83 *Scattered radiation fields from the offset subreflector in the plane of asymmetry. The broken curve shows the cross-polar distribution in the aperture of the main reflector*

Structurally, the optimised geometry is not very compact, and this can be a drawback for applications where the volumetric constraints are severe. Earlier experimental results described by Graham[73] and the rigorous diffraction analysis performed by Adatia[74] confirmed the general validity of the optimisation formula, although the level of cross-polar suppression is not independent of the subreflector dimensions and curvature. Later results obtained by Adatia[114] indicated that the major limiting factor in the realisation of polarisation purity with double-offset antennas is the diffraction effects introduced by the finite-sized subreflector. Diffraction analysis shows that the magnitude of the diffraction-generated cross-polar field component is primarily a function of the transverse dimensions of the subreflector in wavelengths, the eccentricity of the subreflector surface, and the primary-feed illumination taper. For good cross-polar suppression, subreflectors with large transverse dimensions and high values of eccentricity are favoured.

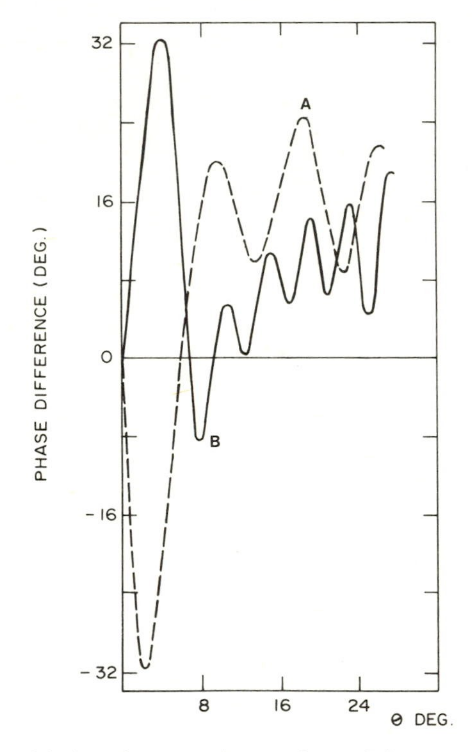

Fig. 3.84 *Differential phase between the copolar and the cross-polar components of fields scattered from the offset subreflector*
a subreflector diameter 10λ
b subreflector diameter 20λ

In Fig. 3.83 the copolar and cross-polar components of the fields scattered from an offset subreflector with a diameter of 10λ are shown for the plane of asymmetry. The associated differential phase characteristics are shown in Fig. 3.84. The deviation from the ideal in-phase characteristics is a consequence of the nonoptical scattering by the finite subreflector. This phase deviation essentially limits the polarisation purity of the overall antenna, since the cancellation of the depolarised field generated at the main reflector is impaired. This is apparent in Fig. 3.83 which also shows the residual cross-polar field distributions in the main offset-reflector projected aperture. The multilobe structure of this distribution is directly attributable to the nonuniform phase distributions of the subreflector fields. Fig. 3.85 shows the predicted secondary field characteristics in the principal planes for a double-offset antenna with the parameters given in Table 3.5 and a subreflector diameter of 10λ. The cross-polar fields are well above the levels indicated by geometric optics and can be reduced to below −40 dB only by use of a subreflector diameter of greater than 25λ or a compensating primary-feed[76,77]

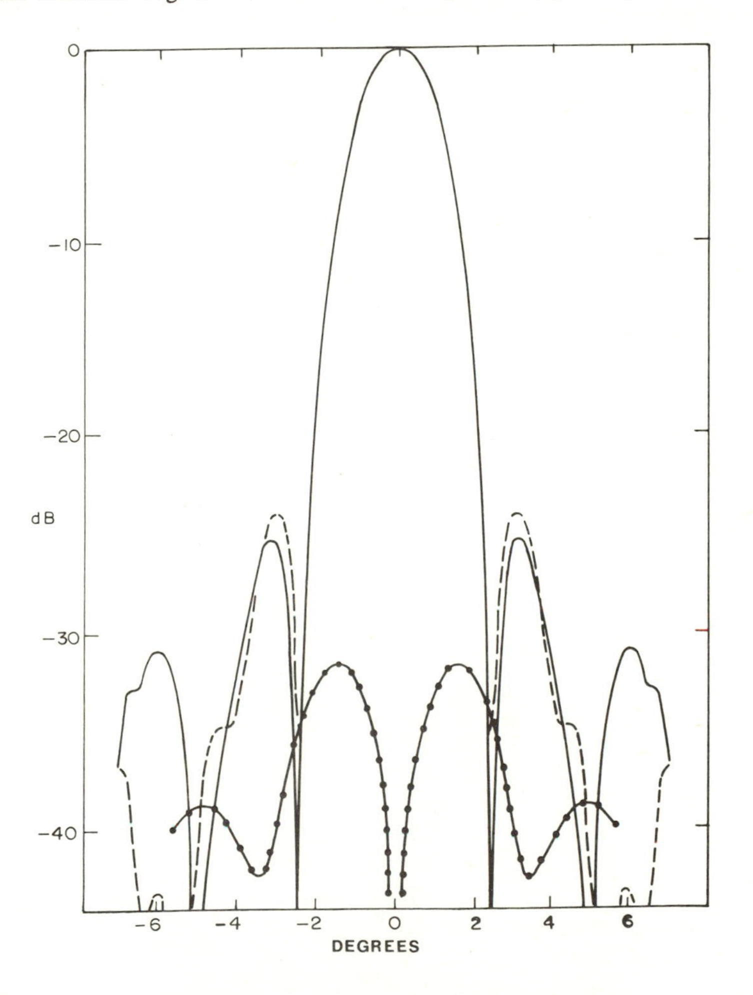

Fig. 3.85 *The radiation fields from the double-offset reflector antenna*

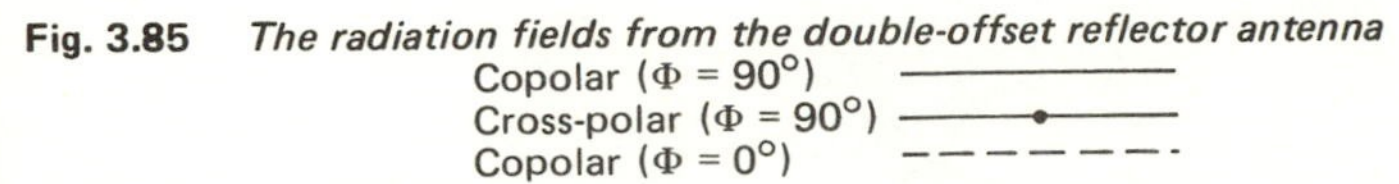

An alternative approach to the elimination of cross polarisation from dual-offset reflector systems has been followed by Albertsen.[115,116] In this approach the two reflecting surfaces are shaped to provide the desired aperture-field conditions in the main reflector. The mathematical approach involves the solution of simultaneous partial differential equations with certain specified initial conditions. The solutions provide the profiles of the two reflector surfaces under constraints which provide for an in-phase cross-polar free-field distribution in the projected aperture of the main reflector. One of the principal attractions of this general approach is that the problem may be formulated for any prescribed position of the main and sub-

Table 3.5 *Parameters of a dual offset reflector antenna with geometry optimised for optimum crosspolar cancellation*

Main offset reflector	Sub reflector
$\theta_0 = 52.3°$	$\bar{\Psi}_0 = 16°$
$\theta^* = 24.8°$	$\Psi^* = 8°$
$d = 36\lambda$	$\alpha = 4°$

Magnification $M = 4.03$

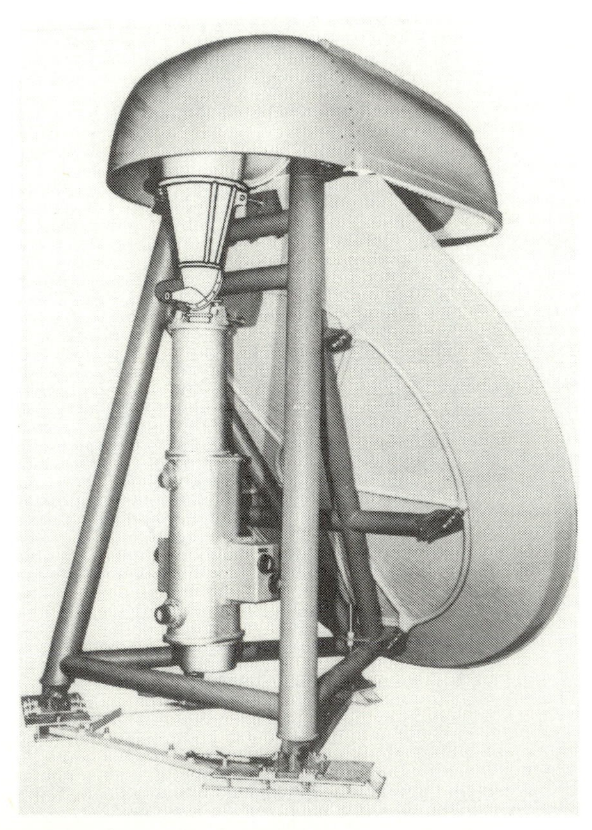

Fig. 3.86 *a* Photgraph of a four-band offset Cassegrainian antenna (Photograph reproduced with the permission of Thomson CSF).

reflector surfaces. Hence, at least in principle, highly compact dual-offset configurations can be realised, although the subreflector diffraction effects will still limit the cross-polar performance when small subreflectors are employed. An additional advantage is that, by appropriate choice of initial conditions, the method can provide solutions for elliptically contoured beams of any aspect ratio.[115,116] Hence, state-of-the-art axisymmetric low cross-polar primary-feeds, such as the cylindrical corrugated horn,[93] could be employed to efficiently illuminate a dual-reflector system generating an elliptical main beam with good polarisation purity. However, the full implications of this approach have yet to be investigated and confirmed experimentally.

A further variation of the double-offset-reflector configuration was considered by Cha *et al.*[117] In this work the offset subreflector was designed as a frequency selective filter. The double-offset antenna was thus fed from both the prime focus of the main reflector and via the subreflector, which comprised an array of conformal printed-circuit crossed dipoles. The Cassegrainian geometry was utilised in the resonant band of the subreflector surface. An overall efficiency of 50% was estimated for the Cassegrainian system, but the sidelobe performance was poor.

Applications

Dual-offset-reflector antennas can be designed to avoid (or minimise) aperture-blockage effects and thereby offer a good range of compromise between high

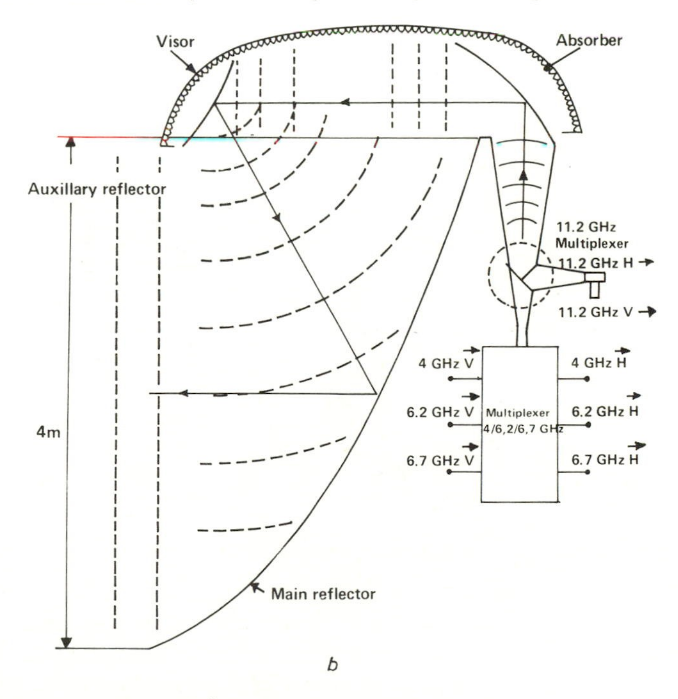

b Schematic diagram of a four-band offset Cassegrainian antenna

efficiency and low sidelobes. The use of a subreflector provides a mechanical advantage for some applications in that the antenna system can be folded to locate the primary feed either within or below the main reflector. Electrically the dual-reflector system offers the designer some additional degrees of freedom, which can be employed to compensate for the depolarisation effects introduced by the asymmetric reflectors. However, in this optimised configuration the dual-offset system is not especially compact. In addition, the very low cross-polar performance (i.e. below −40 dB), which is predicted by geometric optics, can only be achieved if the subreflector dimensions are greater than about 20 wavelengths. Hence, when good polarisation purity is desired, dual offset reflector systems are best suited for applications where their dimensions can be made large with respect to the operating frequency. Earth stations, radio telescopes, terrestrial communications, and ground radar systems are likely examples.

Fig. 3.87 *Dual-offset radar reflector antenna.*
(Photograph reproduced by permission of Marconi Elliott Avionic Systems Ltd.)

The Bell System open Cassegrainian design has been implemented with a 6 m main reflector and has been used in satellite communication experiments with the UK Post Office and the University of Birmingham (UK). The design has operated successfully although it exhibits the predictably poor crosspolar performance.

Magne and Bui-Hai[118] have described a 4-band dual-polarised double-offset antenna for terrestrial microwave links. Their design in illustrated in Fig. 3.86. One example of the use of dual offset reflectors in a radar antenna is that manufactured by Marconi Avionics in the UK, which is illustrated in Fig. 3.87. Mizusawa *et al.*[119] have described the geometric-optics design and experimental modelling at 24 GHz of a dual-shaped reflector configuration for a circular polarised shaped-beam antenna. The design was intended for a search radar and provided a very narrow beam in the azimuth plane with a cosecant-squared beam in the elevation plane. This fully offset antenna incorporated an axisymmetric primary-feed illumination which was converted by the shaped sub-reflector to feed the main elliptically contoured shaped main reflector. The measured azimuthal-plane sidelobes of this antenna were only − 18 dB down, and the axial ratio of the circular polarisation was also poor. Both performance parameters were attributed to a nonoptimum primary-feed horn design.

Semplak[120] has described measurements (at 100 Ghz) on a multiple- beam offset antenna which employs an interesting three-reflector configuration. Although the author indicates that the design is suitable for both satellites and Earth stations, the structure has a projected-aperture diameter of more than 200 wavelengths with an F/d ratio of 1.9. Without deployment this would be difficult to accommodate on a satellite for frequencies lower than the 30 GHz possibility mentioned in the text. The three-reflector design exhibits good wide-angle beam performance.

Much of the analysis of the dual-offset reflector antenna has been performed in connection with either future earth-station antennas[75,112,113] or spacecraft applications.[114–116] The spacecraft application tends to be limited by the volumetric constraints imposed by the launcher. Without deployment, the cross-polar performance of the antenna tends to be limited by diffraction effects arising from the relatively small subreflector. As an earth-station antenna, however, the electrical performance of the optimised double offset reflector antenna is very attractive. The increasing demands made upon the radiation performance of earth-station antennas, and particularly the sidelobe and cross-polar specification, suggest that optimised double-offset reflector antennas may well be necessary to satisfy the electrical requirements of the next generation of large earth stations.

The more general application of the dual offset reflector antenna may be hampered by the considerable computationsl effort which is involved in performing design and optimisation with physical-optics techniques. Approximate techniques, based upon geometric optics and the GTD, are adequate for large reflectors, but subreflector diffraction must be accurately modelled for other cases. Analytical techniques of the type introduced by Galindo-Israel and Mittra,[121,122] the GTD methods of Pontoppidan[123] and the analytical solutions of Westcott *et. al.*[124] may well prove to be useful in optimising dual offset reflector designs. It is worth noting that much of the antenna optimisation can be performed by examining the vector fields in the aperture plane of the main offset reflector, and thus minimising the need for the second (and costly) two-dimensional integration over the reflector surface.

3.4 Other reflector types (Contribution by W.V.T.Rusch)

3.4.1 Shaped-beam optical reflectors

The paraboloid and its variations considered in the previous sections radiate focused, diffraction-limited pencil beams of elliptical or circular cross-section. Other applications may demand reflector radiation patterns of a substantially different cross-section, e.g. the subreflector in a Cassegrain antenna. Such reflectors are generally contoured in accordance with ray-optical principles. Both shaped single reflectors and shaped dual reflectors are considered below.

Shaped single reflectors

The earliest optically shaped reflectors were probably surveillence-search azimuth radar antennas with a narrow azimuth pattern and a broad, shaped elevation pattern.[125–129] Designed to have a constant echo strength from a target flying at constant altitude h, the voltage pattern of the antenna is proportional to the secant of the evaluation angle e, to compensate for the $1/r$ falloff with $r = h$ times the cosecant of e. Such designs have been achieved by effectively decouping the design of the reflector's central-section curve from the design of its transverse sections. Application of Snell's law to the geometry of the central-section in Fig. 3.88a yields at point P

$$\frac{1}{\rho_c}\,\frac{d\rho}{d\psi} = \tan\frac{\beta(\psi)}{2} \tag{3.107}$$

when $\beta(\psi) = \psi + e(\psi)$. This may be integrated to yield

$$ln\,(\rho_c/\rho_{c1}) = \int_{\psi_1}^{\psi} \tan\,[\beta(\psi)/2]\; d\psi. \tag{3.108}$$

The dependence $e(\psi)$ is then obtained from a two-dimensional energy conservation requirement

$$G(e)\,de = [I(\psi)/\rho_c(\psi)]\; d\psi \tag{3.109}$$

where $I(\psi)$ is the power pattern of the feed in the central plane, and $G(e)$ is the desired elevation pattern (power), e.g. the cosecant-squared pattern discussed previously.

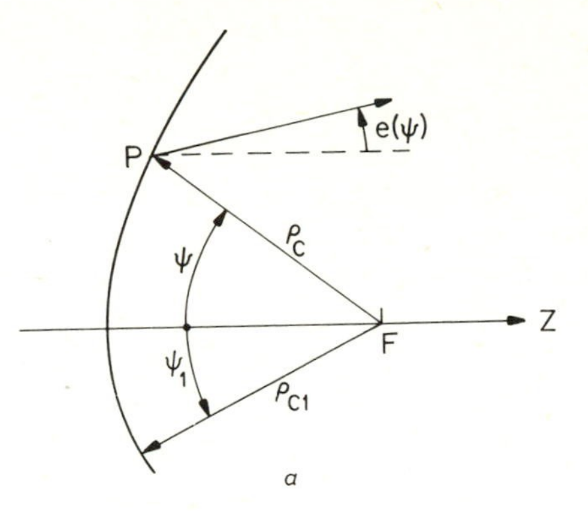

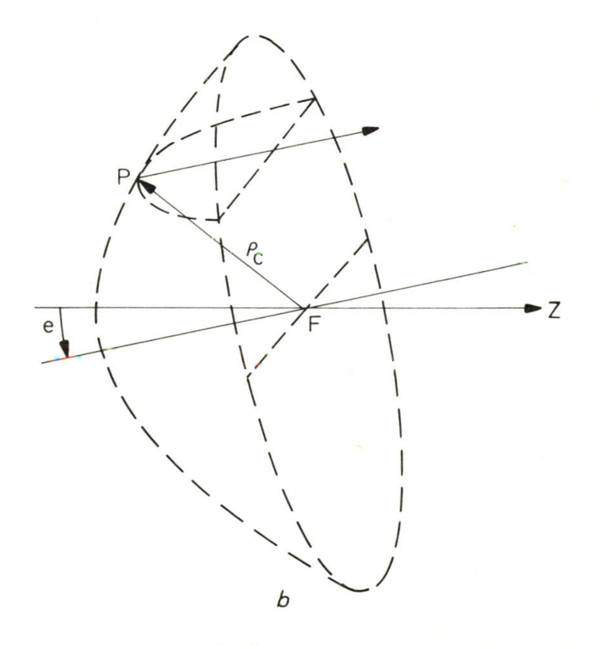

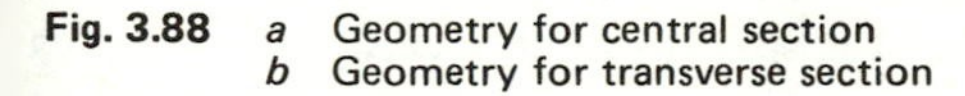
Fig. 3.88 *a* Geometry for central section
b Geometry for transverse section

The reflector contour in transverse planes through the central section at P must be such that all rays are reflected parallel to the central ray. Thus, as shown in Fig. 3.88*b*, each transverse section effectively lies on a paraboloid which has been tilted by an angle $e(\psi)$ and with a focal length

$$F(\psi) = \frac{1 + \cos \beta(\psi)}{2} \rho_c(\psi). \tag{3.110}$$

Having completely specified the reflector contour, the radiation patterns can be determined using physical-optics or the GTD.

The design procedure described above is essentially a two-step, one-dimensional procedure whereby the radiation pattern in one plane is diffraction limited. This method has also been applied to the design of elliptic-beam satellite antennas.[130] A general two-dimensional procedure was developed in the process of designing a

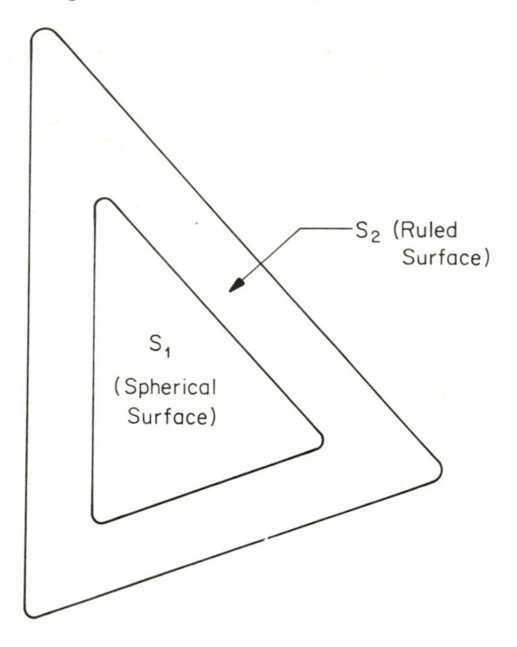

Fig. 3.89 *Idealized Contoured Beam*

satellite antenna with a beam shape conforming to the islands of Japan.[131,132] As a first step, the shape of a portion of a spherical wavefront near the antenna aperture is selected to correspond to the desired beam shape (S_1 in Fig. 3.89). The energy from this portion of the wavefront is spread over the desired service area Secondly, the contoured wavefront is surrounded by a ruled-surface 'flange' which maintains a constant width over the propagation path (S_2 in Fig. 3.89). The directrix of this surface is the contour of the inner part. Thus the energy crossing the outer surface is concentrated on the edge of the target area. Finally, the reflector contour itself is determined by requiring a constant path length along all rays from the phase centre of the feed to the reflector and thence to the wave front.

Shaped multiple reflectors

Geometrical-optics techniques may be used[133–135] to design axially symmetric dual-reflector systems which satisfy the boundary conditions imposed by (i) on assumed primary-feed field distribution, and (ii) an arbitrary amplitude and phase distribution in the output aperture of the system (Fig. 3.90). Thus:

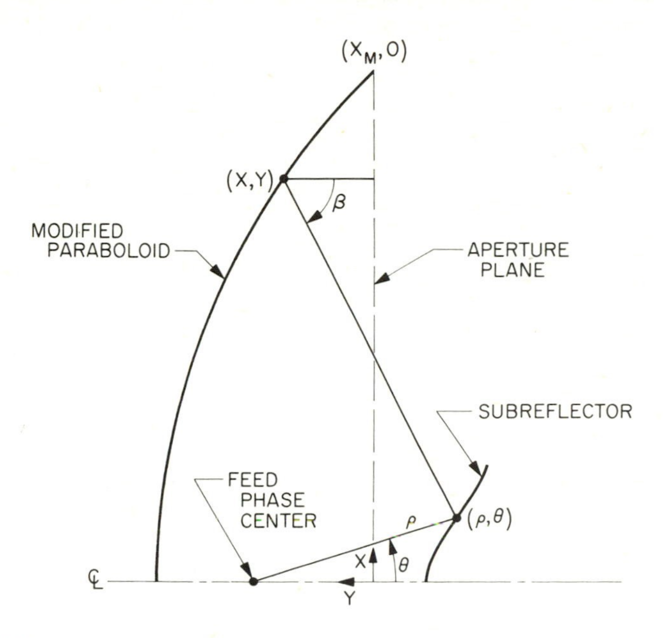

Fig. 3.90 *Shaped dual-reflectoe systems*

(*a*) At each reflector the incident and reflected rays and the surface-normal are coplanar, and the angle of incidence equals the angle of reflection (Snell's law). Hence,

$$\left(\frac{1}{\rho}\right)\frac{d\rho}{d\theta} = \tan\,[(\theta + \beta)/2] \tag{3.111}$$

$$\frac{dy}{dx} = -\tan\,(\beta/2). \tag{3.112}$$

(*b*) Energy flow along each differential tube of rays remains constant even when the tube undergoes reflection (conservation of energy). Hence

$$I(x)\,x dx = F(\theta)\sin\theta\;d\theta \tag{3.113}$$

where $F(\theta)$ is the known axially symmetric angular distribution of the power fron the primary feed and $I(x)$ is the desired power distribution in the aperture.

(*c*) Ray directions are normal to the constant-phase surfaces, and this conditior is maintained after reflections (theorem of Malus).

Rearranging these three simultaneous differential equations in the unknowns ρ, θ and y yields a form suitable for numerical integration:

$$\frac{d\rho}{dx} = \rho \frac{d\theta}{dx} \tan \frac{\theta + \beta}{2} \tag{3.114}$$

$$\frac{d\theta}{dx} = \frac{xI(x)}{\sin\theta \, F(\theta)} \tag{3.115}$$

$$\frac{dy}{dx} = -\tan\frac{\beta}{2} \tag{3.116}$$

Thus, both the subreflector and main reflectors, $\rho(\theta)$ and $y(x)$, may be determined Because the procedure is based upon geometrical optics, valid solutions are limite to surfaces with large radii of curvature. Furthermore, diffraction effects are nc included.

A high-efficiency dual-reflector system generally requires that (i) a high pe centage of the feed energy be intercepted by the reflectors (i.e. reduction of spill over), and (ii) the field in the aperture of the main reflector be distributed a uniformly as possible. Ordinarily these two effects work against each other; i.e reduction of spillover requires tapering the field distribution and a uniform apertur distribution generally involves substantial spillover. Consequently, optimum pe formance generally involves a compromise which has limited efficiencies c conventional systems to above 55-60%.

The shaped dual-reflector concept permits the apparent contradiction betwee the two requirements for high efficiency to be overcome with the followin rationale: a feed is selected with a high taper at the edge of the subdish to minimis forward spillover. The subdish shape is designed to distribute the highly tapere feedhorn energy uniformly over the aperture of the main reflector. The classic hyperboloidal subreflector is transformed into an empirical contour which ha a smaller radius of curvature than a hyperboloid in the central section to deflec more of the rays to the outer part of the main reflector. In this manner, there little spillover and, at the same time, nearly a uniform aperture distribution. Th main reflector must then be slightly reshaped from its original paraboloidal contou to produce a constant-phase aperture distribution. Because of complete axi symmetry, these systems can be used with orthogonal linear or circular polarisatio and can be made compatible with a monopulse requirement.

The shaped subreflector (modified hyperboloid) resulting from a typical calcu-lation[136] is illustrated in Fig. 3.91. The central region of the modified contour has a considerably smaller radius of curvature than the hyperboloid.

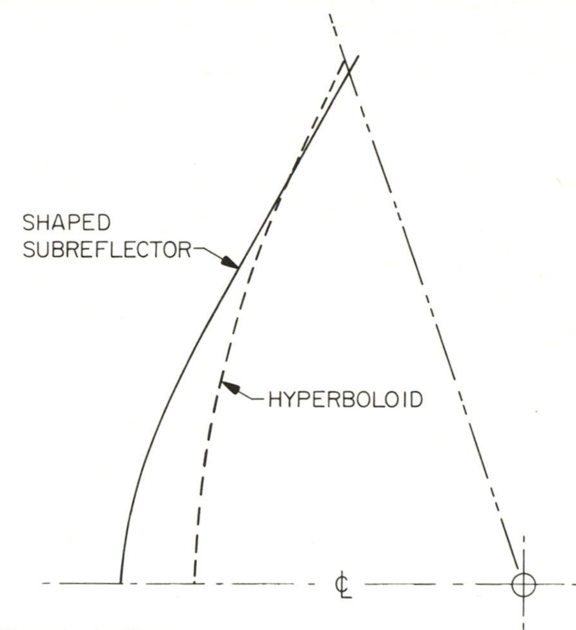

Fig. 3.91 *Shaped subreflector*

Maximum aperture efficiency is generally achieved by setting $I(x)$ equal to a constant in eqn. 3.115. The resulting significant increase in efficiency is evidenced by the performance of large operational ground systems.[137] The 1977 Voyager spacecraft used a 3.7 m shaped-dual-reflector system for high data-rate telemetry transmissions.[138] The system was optically shaped for high efficiency at X-band. The dichroic subreflector was transparent to radiation from an S-band prime-focus horn nestled behind it (Fig. 3.92).

Other distribution functions $I(x)$ may also be synthesised using this procedure, e.g. highly tapered exponential functions for ultra-low sidelobes[139] and patterns with a central 'hole' to reduce aperture blocking caused by the subreflector.[140,141]

As described above, any arbitrary amplitude and phase distribution may be synthesised in the aperture of an axially symmetric dual reflector system by the integration of simultaneous, nonlinear, ordinary differential equations 3.114-3.116. For the offset, noncoaxial geometry, however, the equations found by this method are partial differential equations which, in general, do not form a total differential. Consequently, an exact solution to this problem does not generally exist.[142] However, the partial differential equations form a nearly total differential for many important problems. In such cases, the specified aperture phase function can be synthesised exactly, and the specified aperture amplitude function can be approximated with high accuracy. Excellent results may be obtained for a wide variety of high-gain, low-sidelobe, near-field Cassegrain and beam waveguide systems.[142, 143]

Reflector shaping may also include diffraction effects. Elliot and Poulton[144] have synthesised a desired scattered diffraction pattern on a least-squares fit by

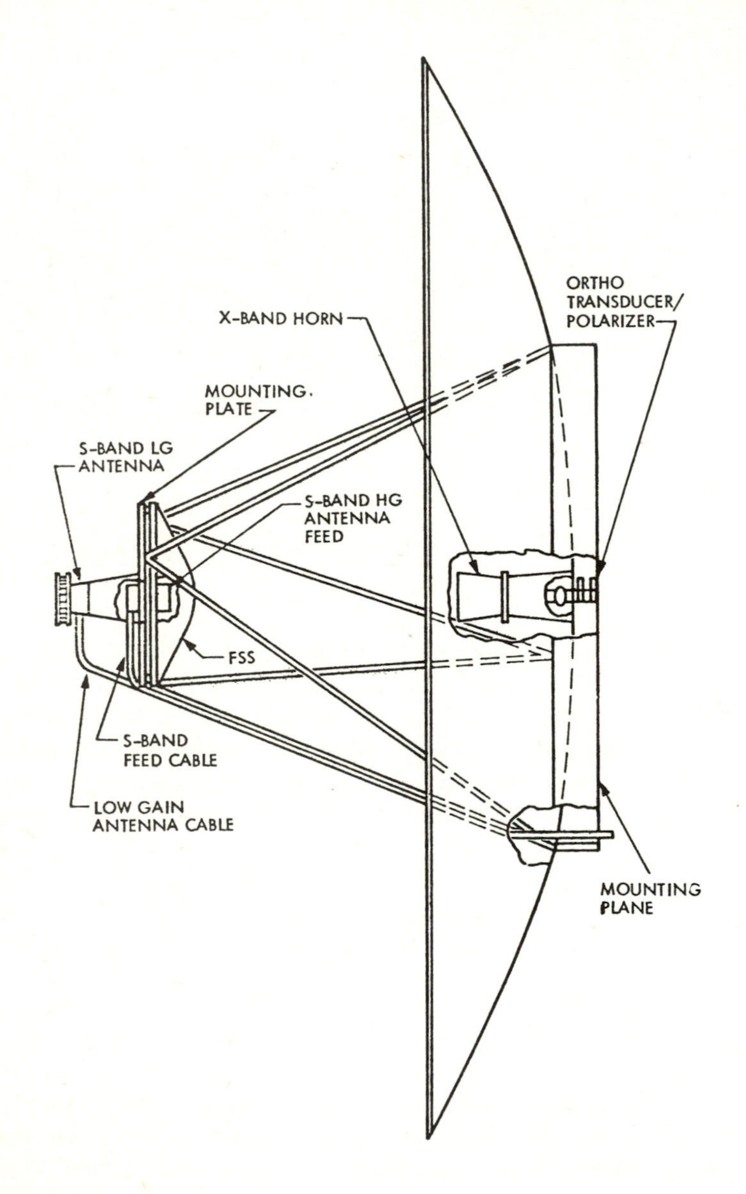

Fig. 3.92 *Voyager high-gain antenna overall layout*

iteratively perturbing the reflector profiles. Wood[145–147] has diffraction synthesised reflector profiles by seeking a conjugate match of the antenna receive and transmit fields at the reflector. This technique has been extended to the design of multiple-reflector beam waveguide systems.[148–149]

3.4.2 Spherical reflectors

The complete spherical symmetry of a spherical reflector is attractive from the viewpoint of beam scanning applications. All parallel rays incident on the sphere generate reflected rays which cross a line through the origin and parallel to the incident rays (Fig. 3.93). A typical incident ray $\overline{AB}$ is reflected through points C_1 and C_2, where C_1 and C_2 are the caustics of the reflected ray (Section 2.2.3).

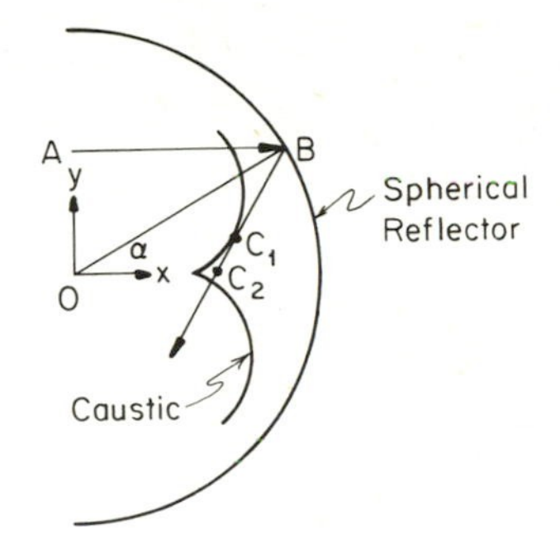

Fig. 3.93 *Ray Behaviour of Spherical Reflector*

The caustic surface containing C_2 is that portion of the x-axis for $0.5\,R_0 \leqslant x \leqslant R_0$ as $0 \leqslant \alpha \leqslant 120$ degrees. The caustic surface containing C_1 intersects any plane through the x-axis in the form of an epicycloid called the nephroid because of its resemblance to a kidney.[150–153] The coordinates of the two caustics are given by

$$\overline{OC_1} = R_0\left[\frac{1}{2}\cos\alpha\,(1 + 2\sin^2\alpha)\,\hat{a}_s + \sin^3\alpha\,\hat{a}_y\right] \tag{3.117a}$$

$$\overline{OC_2} = \frac{1}{2}R_0/\cos\alpha\,. \tag{3.117b}$$

A suitable primary feed can thus lie along a portion of the axis (phased line feed) or near the paraxial focus at $x = ½R_0$, $y = 0$ (transverse feed) where the caustic cusp and focal line and near which most of the rays pass.

Massive interest in spherical reflectors was generated by the construction of the Arecibo 1000 ft diameter fixed spherical reflector (Fig. 3.94). Both physical optics[153] and extended geometrical optics (stationary phase)[154] have been used to

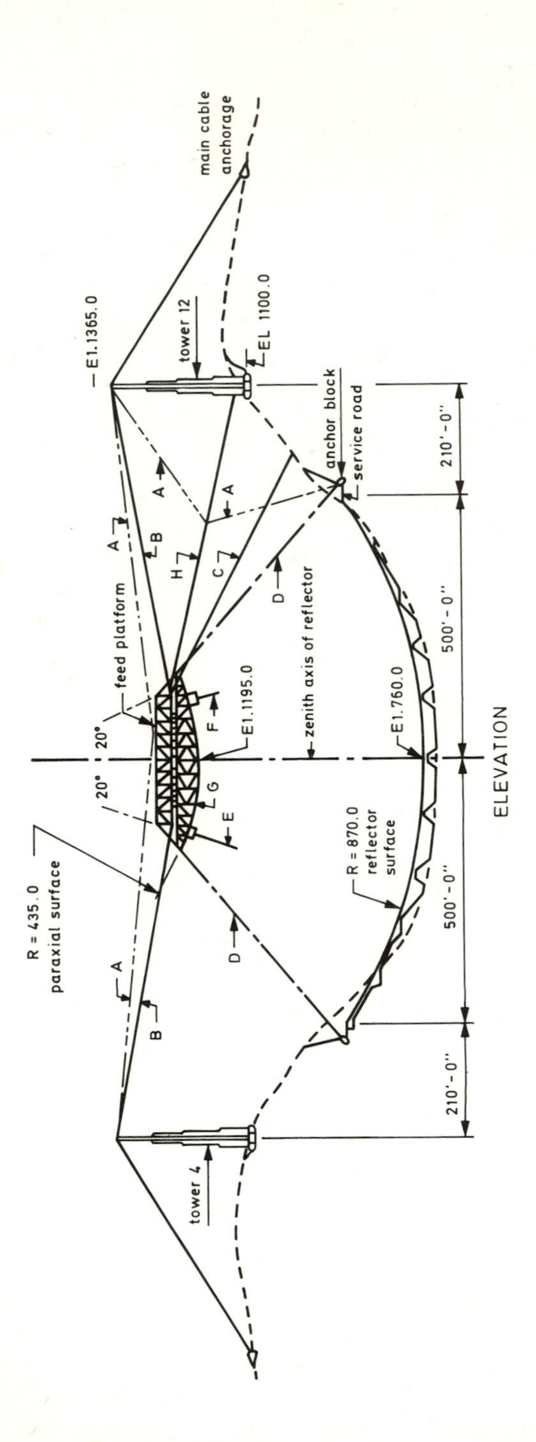

Fig. 3.94 *Arecibo Spherical Reflector*

study focal region fields of spherical reflectors, resulting in designs for both radial line source feeds[155,156] and transverse aperture feeds.[157,158] Phase-correcting Gregorian[159] and Cassegrain[160] subreflectors may also be incorporated to eliminate spherical aberration, although at the expense of introducing significant aperture blocking.

3.4.3 Non-spherical scanning reflectors

Ruze[10] demonstrated that feed movement transverse to the symmetry axis of a paraboloid causes beam scan with only moderate aberration when the beam is scanned less than 10-12 beamwidths. Nevertheless, the intrinsic geometry of a

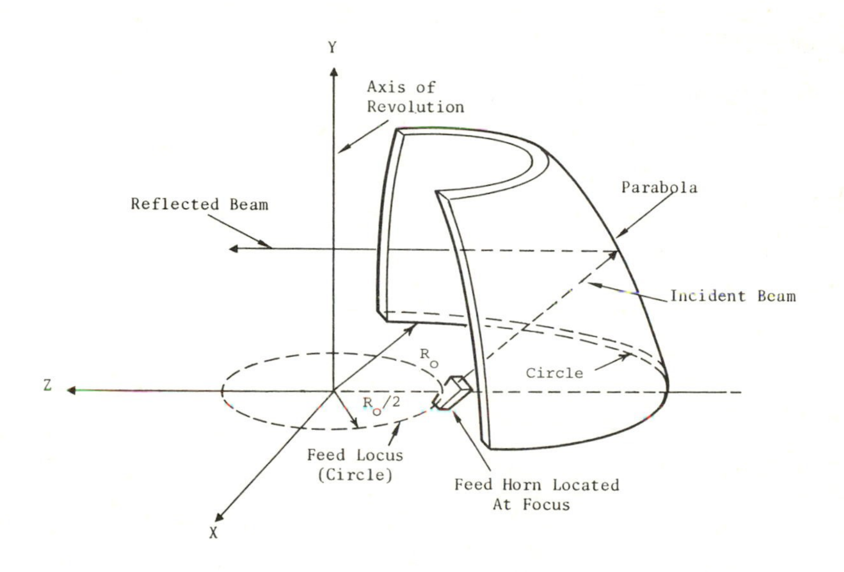

Fig. 3.95 *Parabolic torus antenna*

paraboloid is not naturally amenable to azimuth beam scanning without the penalty of scan loss, coma lobes, etc. Reflectors which are symmetric about the scan axis, such as the sphere considered in the previous Section, are more suited for wide-angle beam scan applications. Two such reflector geometries are treated in this section.

Toroidal reflectors[161–164]

The toroidal geometry shown in Fig. 3.95 is symmetric about the y-axis. In the xz-plane, the reflector is a circle of radius R_0. Any plane containing the y-axis intersects the reflector in its generating function

$$\sqrt{x^2 + z^2} = f(y, R_0) \tag{3.118}$$

Thus, to the extent that truncation of the torus can be neglected, a feed rotated about the y-axis will always encounter the same reflector environment and thus generate a beam which is independent of angle.

The paraboloidal torus represents a compromise between the focusing properties of the paraboloid and the scanning properties of a torus. In this case the generating function is

$$f(h, R_0) = y^2/2R_0 - R_0 \tag{3.119}$$

which, in an attempt to reduce aperture phase-errors, has placed the focus of the parabola at the paraxial focus, $R_0/2$, of the circle. Movement of the feed in the xz-plane on a circle of radius approximately equal to $R_0/2$ thus yields essentially equivalent patterns over a wide range of azimuth. Nevertheless, the system suffers from less-than-desirable aperture efficiency and sidelobe levels because of residual aperture phase errors. It can be shown that phase errors are reduced if the generating curve is an ellipse, although sidelobes in the diagonal planes still remain relatively high.

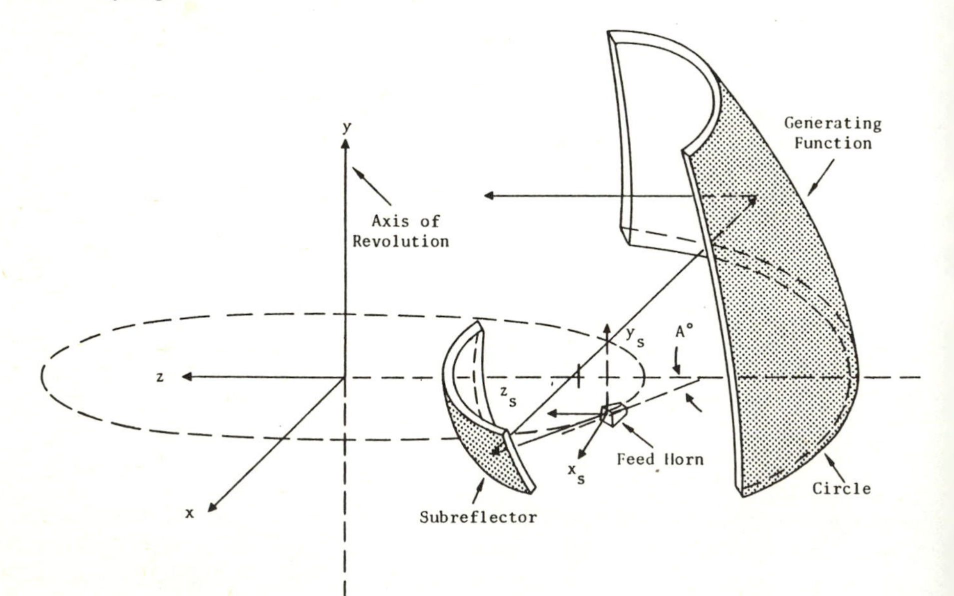

Fig. 3.96 *Phase-correcting subreflector for toroidal antenna*

Dual-reflector toroidal systems

Phase errors in the aperture of a toroidal antenna can be eliminated by a phase-correcting subreflector. Thus, a bundle of parallel rays incident on the torus can be concentrated to a perfect point focus (Fig. 3.96). The subreflector contour can be determined by geometrical optics. As shown in the geometry of Fig. 3.97 (yz-plane), a point source is located at $x = 0, y = -H, z = -F$, which is also the origin of the origin of the $x_s - y_s - z_s$ system. A concave Gregorian subreflector is located below the aperture plane, with its tip at $x = 0, y = 0, z = -S$. A ray $\hat{a}_z$ incident at B on the main reflector is reflected in the direction of $\hat{m}(x, y)$ where, by Snell's law

$$\hat{m} = -\hat{a}_z + 2n_z\hat{n} \tag{3.120}$$

where $\hat{n}(x, y) = n_x(x, y)\hat{a}_x + n_y(x, y)\hat{a}_y + n_z(x, y)\hat{a}_z$ is the outward unit normal at point x, y on the torus, determined from eqns. 3.118 and 3.119.

If $C(x, y)$ is the length of the ray from (x, y) to the subreflector, $|\bar{R}_s|$ is the length from the subreflector to the feed (focus) then the requirement that all rays from the aperture plane to the focus are equal in length yeilds

$$|\bar{R}_s| + C - Z = K \equiv \sqrt{H^2 + (F - S)^2} + 2R_0 - S. \tag{3.121}$$

But

$$\bar{R}_s = \bar{R}F + C(x, y)\,\hat{m} \tag{3.122}$$

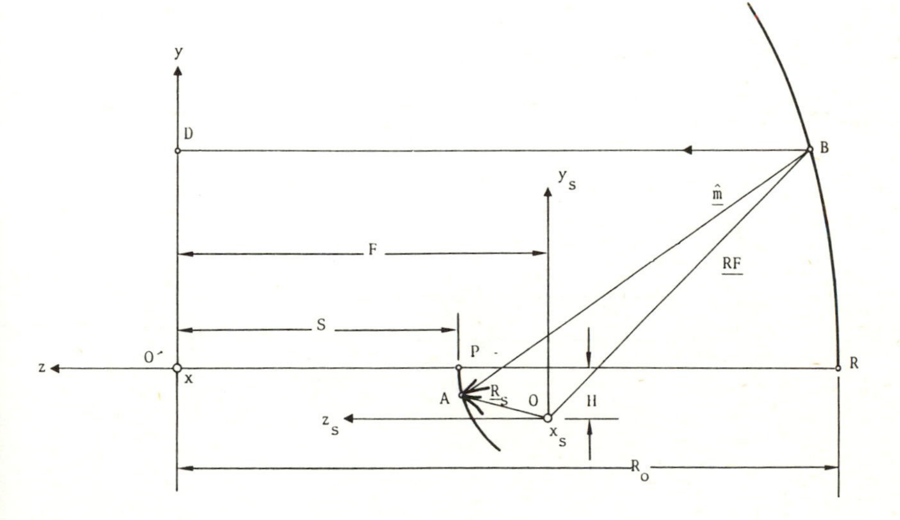

Fig. 3.97 *Gregorian-corrected toroidal antenna system geometry*

where $\bar{R}_s$ is the vector from the feed to the subreflector point and $\overline{RF}$ is the vector from the feed to the torus point on the same ray. From eqns. 3.121 and 3.122

$$C(x, y) = \frac{1}{2} \frac{(k+z)^2 - x^2 - (y+H)^2 - (z+F)^2}{(k+z) + xm_x + (y+H)\, m_y + (z+F)\, m_z} \tag{3.123}$$

Thus, in the coordinate system of the feed, the coordinates of the subreflector are:

$$x_s(x, y) = x + C(x, y)\, m_x \tag{3.124a}$$

$$y_s(x, y) = (y + H) + C(x, y)\, m_y \tag{3.124b}$$

$$z_s(x, y) = (z + F) + C(x, y)\, m_z. \tag{3.124c}$$

Similar results are obtainable for convex subreflectors.

Conical-scan dual reflectors

A modification of the phase-corrected toroidal antenna described in the previous Section is shown in Fig. 3.98.[165] The primary reflector is a surface of revolution placed symmetrically in the x'-y'-z' system, coaxial with the z'-axis. The point-source feed is located outside the column of parallel rays reflected from the primary reflector in the z-direction. With the exception of an additional transforma-

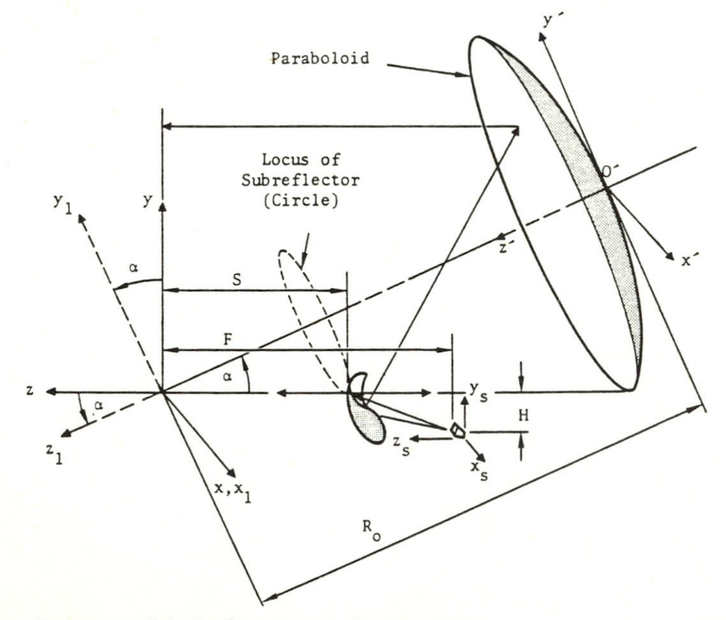

Fig. 3.98 *Geometry of dual-reflector conical scan system*

tion between the x–y–z and x'–y'–z' systems, eqns. 3.120 through 3.124 can be used to generate a subreflector contour to reflect all rays from the feed toward the main reflector in such a way that they are then all reflected parallel to the z-axis, thus creating a focused pencil beam. Because both feed and subreflector lie outside the column of reflected rays, the radiating aperture is unblocked.

This offset geometry avoids the disadvantage of most single and multiple offset reflector systems: a non-symmetric primary reflector. Furthermore, if the subreflector and feed are mechanically rotated about the z'-axis, the beam will be scanned azimuthally about the z-axis, thus creating a conical-scan capability. Because of the axial symmetry, there will be no degradation of the beam as it is scanned.

3.4.4 Other types

Horn-reflector antenna

The horn-reflector is a combination of electromagnetic horn and a reflecting section taken from a paraboloid[166] (Fig. 3.99). The apex of the horn coincides with the focus of the paraboloid. The geometry is based on equal path lengths from the vertex to the aperture. Consequently, it is broadband and is amenable to orthogonal linear and circular polarisation. Being similar to an offset paraboloid, very little energy from the apex is reflected back to the transmitter to produce impedance mismatch. The horn walls produce sufficient shielding to yield very low far sidelobes and backlobes, except for a high sidelobe near 90°, which can be reduced by blinders.[167]

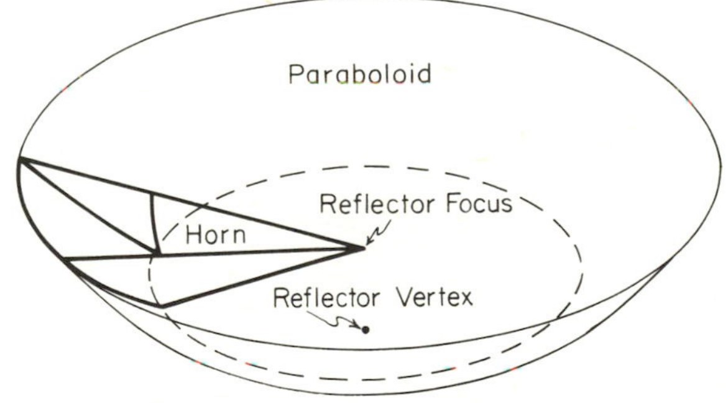

Fig. 3.99 *Horn-reflector Antenna*

The horn reflector is used extensively in microwave relay systems, because of its high front-to-back ratio.[168] It has been adapted to an azimuth/elevation mount for low-noise space communication and radio-astronomy ground antenna.[169] The contoured-beam antenna described previously was also a horn-reflector design.[131,132]

Corner-reflector antenna

The corner-reflector antenna consists of a driven element, normally a linear antenna, located in front of a reflector constructed from two planar conductors which interesect at an angle β to form a corner (Fig. 3.100*a*).[170,171] A three-dimensional version uses a monopole and ground plane as the driven element (Fig.

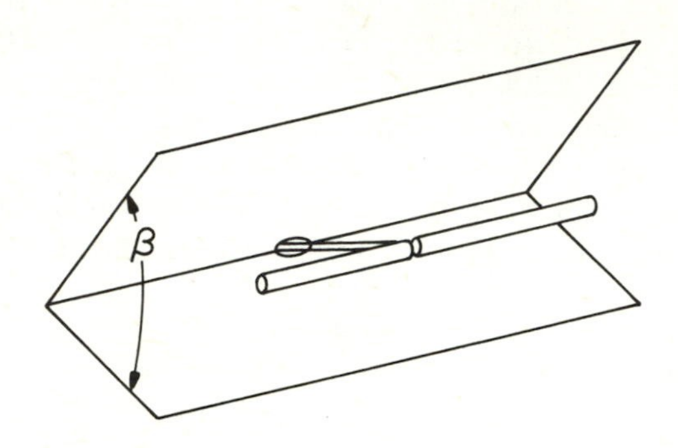

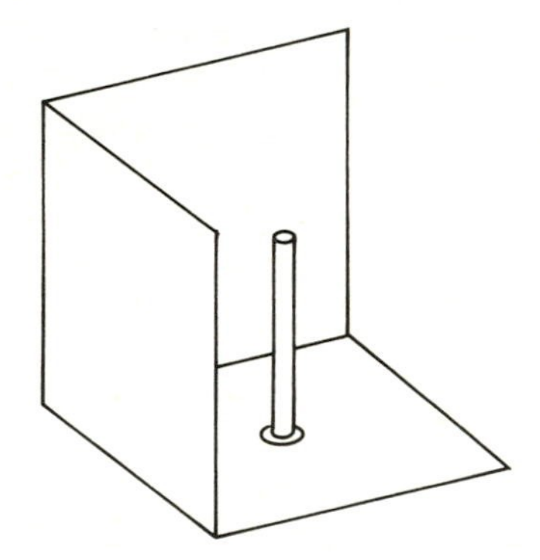

Fig. 3.100 *a* Conventional corner-reflector antenna
b Three-dimensional corner-reflector antenna

3.100*b*).[172,173] This easily fabricated reflector configuration yields a moderate-gain antenna which has been widely used. If the corner dimensions are above a minimum value, the antenna properties can be computed with sufficient accuracy by assuming the conducting sheets to be infinite in extent.[172] If the corner angle β is π/n, where n is an integer, the effects of the reflector can be replaced by $(2n-1)$ equally spaced images. The finite extent of the conductors can be accounted for by using the GTD[174] or integral-equation techniques.[175]

Conical-reflector antenna

The singly curved conical reflector (Fig. 3.101) has obvious structural and fabricational advantages. The basic structure focuses incoming rays which are parallel to the axis to a coaxial line feed (Fig. 3.101*a*).[176–181] Alternatively, a subreflector can be used to focus the incoming rays to a conventional point-source feed.[179–181] Mechanically, the conical Gregorian appears to be the most attractive design (Fig. 3.101*b*).[180–181] The subreflector consists of a rotated section of a parabola

$$\rho(\theta) = \frac{2f}{1 + \cos(2\beta - \theta)}. \tag{3.125}$$

Measured efficiencies of nearly 60% have been reported, although unavoidable aperture blocking causes inherently high near sidelobes.[182]

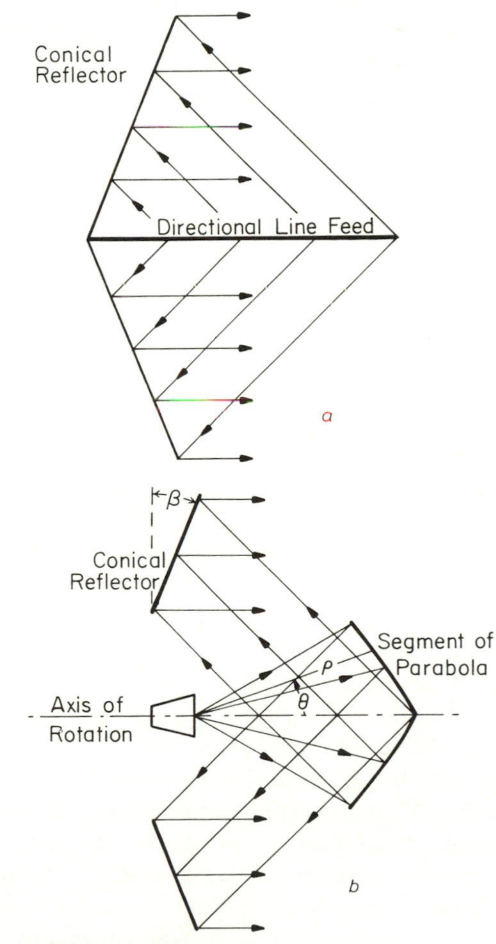

Fig. 3.101 *a* Conical reflector with line feed
b Conical-Gregorian geometry

3.5 Reflectors with contoured-beams (Contribution by W.V.T. Rusch)

An antenna which is designed to have a prescribed pattern shape, generally in the form of a geographical target, but which differs substantially from a pencil beam, is said to be a contour-beam antenna. Most communication-satellite antennas require contoured beams in order to: (i) increase the minimum flux density over the coverage area; (ii) reduce radiation interference into contiguous areas; and (ii) conserve available RF spectrum by spatial isolation or polarisation isolation. The material in this Section will deal with contour-beam antennas consisting of a reflector and an array feed.

3.5.1 Design of contoured-beam reflector antennas

The principal RF parameters in the design of a contour-beam antenna are gain coverage, gain ripple over the coverage area, fall-off at the edge of the coverage area, sidelobes, polarisation purity, and bandwidth. The following general observations concern the feed element, array pattern, and reflector:

Feed element: The feed element is generally selected prior to the contour design on the basis of pattern, polarisation characteristics, mutual-coupling properties, etc. Probably a particular company's experience and degree of success with a certain type of feed is an important factor. Dielectric loading is sometimes considered to reduce aperture size[183] and mutual coupling effects.[184]

Array configuration: Successful systems have employed contiguous elements. The crossover levels of neighbouring secondary beams may be from 4 to 6 dB. When circular elements are used, design methodology may be facilitated by means of an equilateral-triangle lattice considered to be made up of overlapping hexagonal subarrays.[185,186]

Reflector: All designs to date have used unblocked apertures, either offset-fed paraboloids or horn-reflectors. Neither synthetic nor analytic techniques to treat aperture blocking have advanced to the degree needed by contoured-beam systems, particularly with regard to cross-polarisation and sidelobes. Aperture size is dictated by falloff and sidelobe requirements. As a compromise between falloff and gain ripple, it may be shown that the reflector edge should lie approximately halfway between the principal peak and the first grating lobe of the array pattern.[187]

Fig. 3.102 shows the block diagram of a general design procedure, features of which are found in most contour-beam antenna designs. The feed element, reflector configuration, *and array configuration* generally are determined *prior* to the design, on the basis of some of the factors considered earlier. The starting point for the design loop or iterative procedure generally consists of the patterns of each elementary beam, which may come from simple mathematical models and rules of thumb such as 'beam-deviation-factors' etc. or which may be derived from physical optics. Mutual impedance and defocusing aberrations may or may not be included in these determinations. An initial set of excitation coefficients will then serve to superimpose these elementary beams into a form which may then be compared with the desired coverage, isolation, etc. An optimisation algorithm may then be

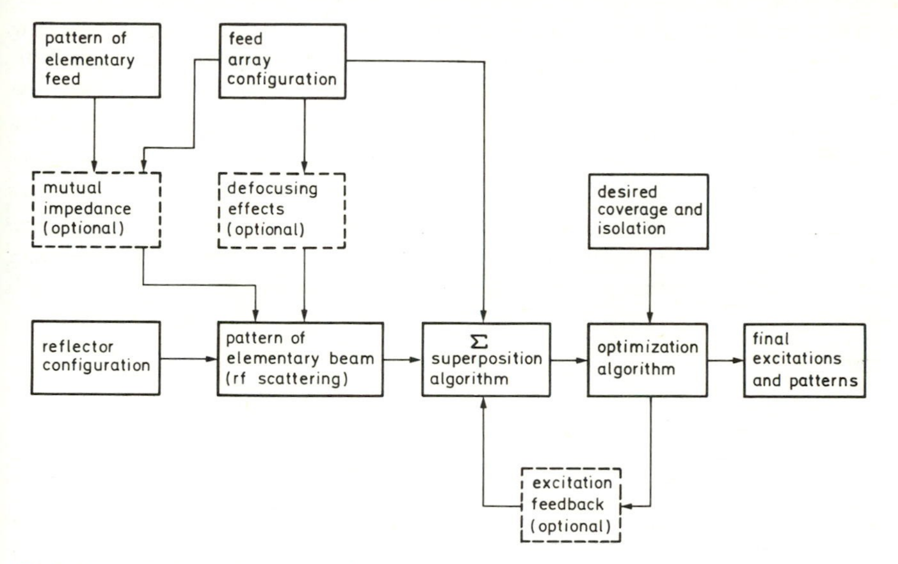

Fig. 3.102 *Basic synthesis procedure*

used to refine the excitation coefficients into successively better and better approximations to the desired results. Engineering intuition and experience generally govern geometrical layout and physical parameters, whereas automatic mathematical optimisation has, to date, been used only in improving the element excitation voltages. A completely automated design, e.g. number and position of array elements, reflector configurations, etc., as well as the excitation matrix, has, to date, been beyond the capability of modern optimisation algorithms.

Inclusion of mutual-coupling effects to determine the secondary beams of tightly packed feed-array elements has not been possible in general. Measurement of feed-element primary patterns, both alone and also embedded in the array with the other elements terminated with matched loads, indicates some differences in pattern characteristics,[184] although whether these differences are significant or not depends on the figure-of-merit and various weighting factors used. In any case, a relatively rigorous analysis of mutual coupling between a large but finite number of array elements is possible only if the elements are simple wire structures.[188] As many as 100 wire subsections have been evaluated for wire-feed structures illuminating reflectors. For aperture-type elements (i.e. horns), mutual coupling has been included in the analysis either by assuming each element to be embedded in an infinite periodic array, or by empirical determination of each element pattern (magnitude and phase of each vector component) with the remaining elements terminated.[189] The planar near-field probing technique[186] has provided a means whereby the effects of mutual coupling can be significantly reduced by the insertion of additional tuning into the feed lines of appropriate elements.

The stringent requirements of a high-performance contoured-beam antenna system require the development of considerable attendant technology of a non-RF nature. For example, the polarisation technology of the etched, overlapping reflectors on the RCA Satcom satellite[190] and the Kapton polarising screens on the Comstar satellite were instrumental to the frequency-reuse capability of those systems. No contoured-beam system would be possible without interactive timeshare computational facilities capable of large-core, high-speed number-crunching plus an extensive, highly specialised software library for RF scattering, optimisation, contour plots, statistical and matrix manipulations, orthographic projections, etc. Virtually new technical disciplines have arisen to handle the thermal, mechanical and vibrational problems of the launch and space environments.

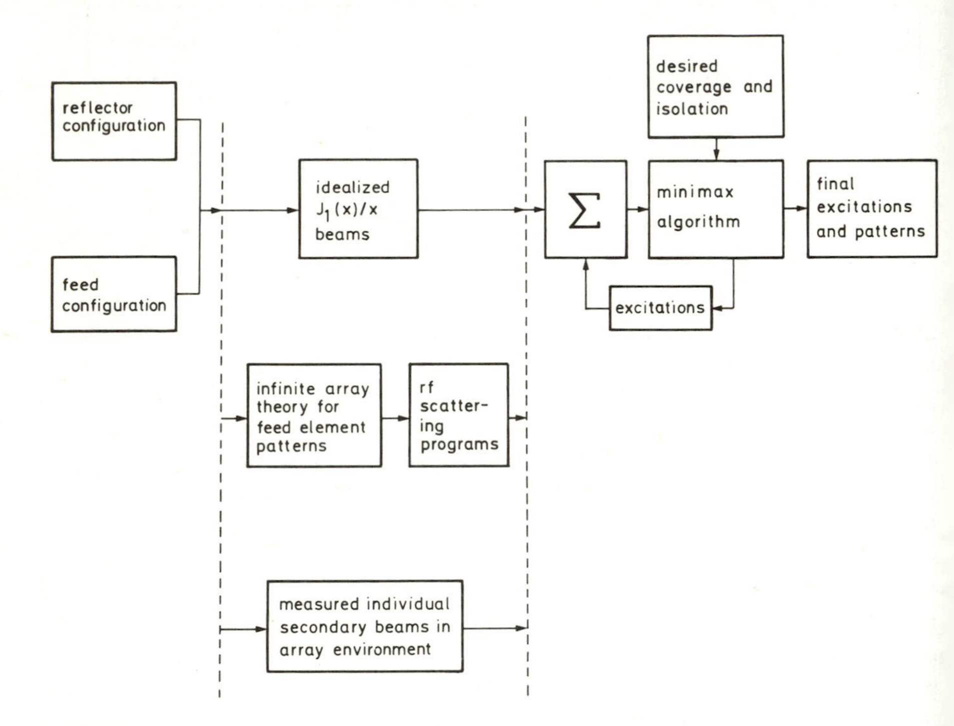

Fig. 3.103 *Ford-aerospace Intelstat V synthesis procedure*

Several block diagrams describing actual design procedures used in practice will now be considered. Fig. 3.103 represents an abbreviated version of the design approach used by Ford Aerospace in the design of the Intelsat V spacecraft.[189] The reflector parameters were determined on the basis of gain, beamwidth and sidelobe requirements. The feed element type and spacing were selected from the desired coverage, polarisation and isolation. The corresponding secondary beams were then modelled in three different ways: (i) using a highly idealised $J_1(x)/x$ model (thus excluding scanning aberrations and mutual coupling); (ii) using infinite array theory for the primary field of each feed element, and thus, by means of an RF scattering program, to the corresponding secondary beams (including scanning aberrations and

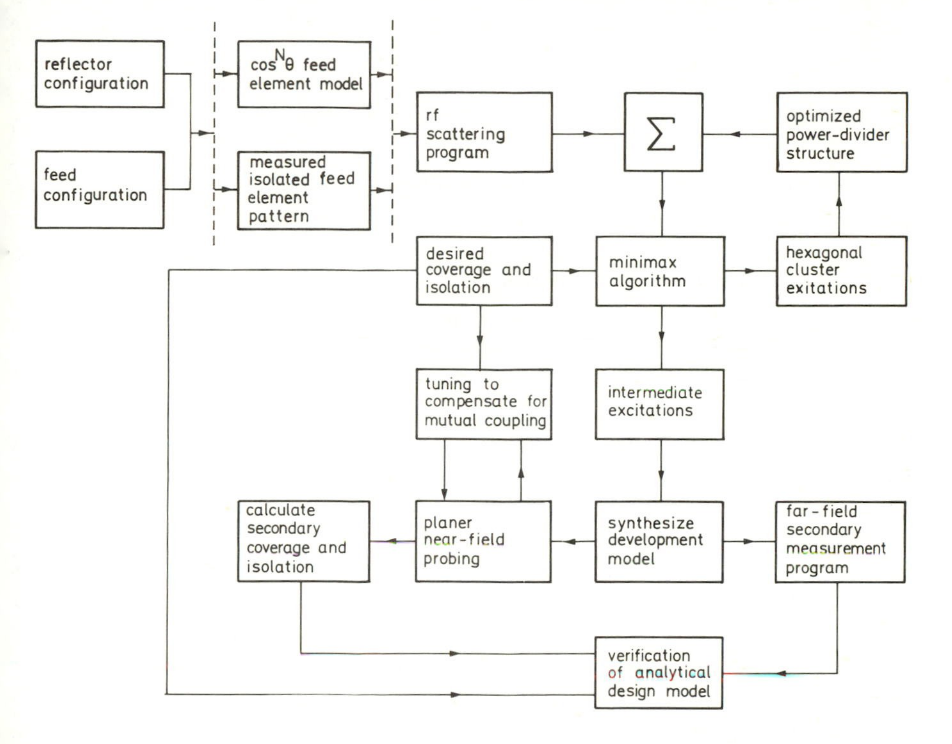

Fig. 3.104 *TRW Intelsat V synthesis procedure*

approximating mutual coupling analytically); and, finally, in hardware development, (iii) measuring the secondary beam of each array element (thus including both aberrations and mutual coupling). After each of these stages, an interpolation/superposition algorithm was used in conjunction with a minimax optimisation scheme to refine the excitation coefficients of the feed elements.

Fig. 3.104 represents an abbreviated version of a similar TRW design approach. Two successive techniques were used to determine the secondary beams; (i) a simplified $\cos^n\theta$ element pattern, and (ii) the measured primary fields of an isolated feed element. The resulting calculated secondary beams then entered a superposition/minimax loop to determine the element excitations. The feedback loop operated on the hexagonal clusters as the basic unit, and also included a program to optimise the actual power-divider structure in a realistic manner. Having determined intermediate element excitations in this manner, a development model was synthesised for measurements on a far-field range as well as on a planar-near-field range. The near-field facility, which provided an opportunity to compensate for mutual coupling, then yielded near-field results which were used to calculate the secondary performance of the antenna. Comparisons of these calculated results with the previously measured performance then verified the reliability of the analytical design model.

3.5.2 Examples of contoured-beam reflector technology

The following examples of contoured-beam flight hardware illustrate many of the principles discussed in the previous section.

RCA Satcom communications satellite

The RCA Satcom Communications Satellite was built for RCA American Communications Inc., for use in their commercial system for US domestic service. The two flight spacecraft were successfully launched on 12 December 1975 and 26 March 1976. The mission of Satcom involves coverage of all 50 states, with Hawaiian service provided by an offset spotbeam and the primary beam covering the contiguous 48 states and Alaska (Fig. 3.105).

Alignment stability precluded deployment of any feed or reflector. Polarisation isolation requirements led to separate feed/reflector pairs for each polarisation and for each of the two transponder output ports of each polarisation. The reflectors consisted of grids of parallel wires embedded in a low-loss dielectric substrate. The grids of the two cross-polarised antennas were orthogonal. Gain and coverage requirements, within the volume constraints of the launch vehicle fairing, resulted in an overlapping configuration (Fig. 3.106). The cross-gridded reflectors were virtually transparent to an orthogonally polarised wave. Furthermore, the feed of each antenna was displaced from the focus of the orthogonally polarised reflector. Consequently, the polarisation isolation over the full beam area was better than 33 dB for all frequencies.[190,191]

All four feed/reflector pairs generated primary beams covering the 49 continental states. Offset feeds on the two west antennas generated the Hawaii beams. Utilisation of a high-strength, lightweight dielectric for the reflectors, plus graphite-fibre epoxy composite for the feed tower, waveguides, and feed horns resulted in a total weight of 50 lb for the four-reflector, six-feed-horn antenna structure.

The beam was an 8.4° x 3.2° elliptical beam rotated 20.5° from the NS axis. A total of four feed horns (two 'EW' polarised and two 'NS' polarised), each carrying six transponder channels, provided frequency-reuse coverage of the contenential USA from 3.7 to 4.2 GHz on the downlink and from 5.925 to 6.425 GHz on the uplink. No evidence was found of resonances between the overlapping and overlapped surfaces.

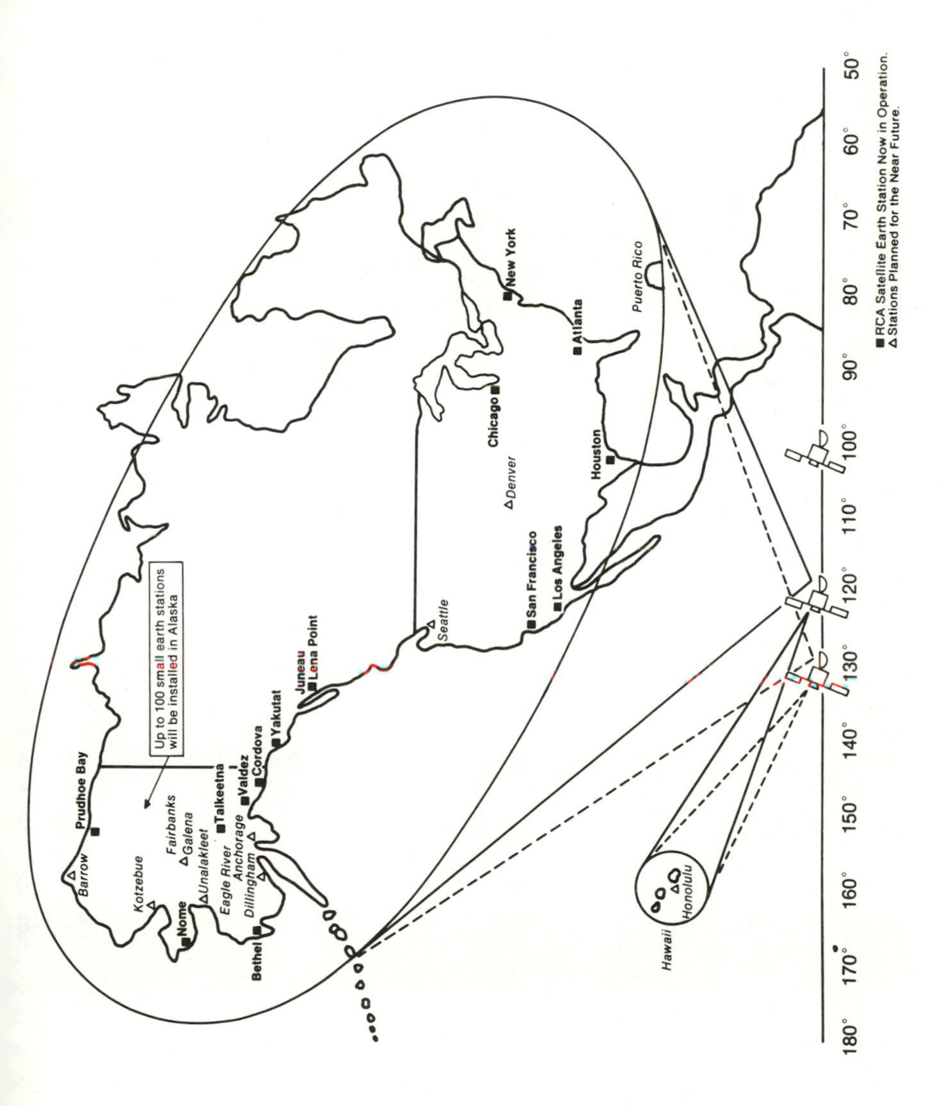

Fig. 3.105 *RCA Satcom Earth coverage*

Westar/Telesat (Anik) communications satellite

The world's first domestic communications satellite in synchronous orbit was the Anik satellite launched for Telesat of Canada. The communications capacity was 5760 voice circuits, and the launch weight was 1242 lb. Its antenna system was the first shaped/contoured beam from synchronous orbit. Westar, owned and operated by Western Union, was a modified version of the Telstar satellite, with beam coverage of the contiguous USA, Alaska and Hawaii. Telesat was launched in 1972, and Westar was launched in 1975.

Fig. 3.106 *RCA Satcom satellite*

The reflector consisted of two parts: a skeleton structure onto which an RF reflective metallic mesh material was stretched, and a three-rib support structure for the skeleton (Fig. 3.107). The mesh material was partially transparent to the solar wind. Both the skeleton and support ribs were fashioned from an aluminium honeycomb, graphite fibre composite.[187]

An early version of the Westar transmit/receive network is shown in Fig. 3.108. The feed was optimised to provide maximum gain at 4 and 6 GHz throughout the three coverage zones of contiguous states, Alaska, and Hawaii. The vertically polarised received signals at 6 GHz and the horizontally polarised transmit signals at 4 GHz utilised a common hornfeed array. The offset horn was for the Hawaii spot-beam.

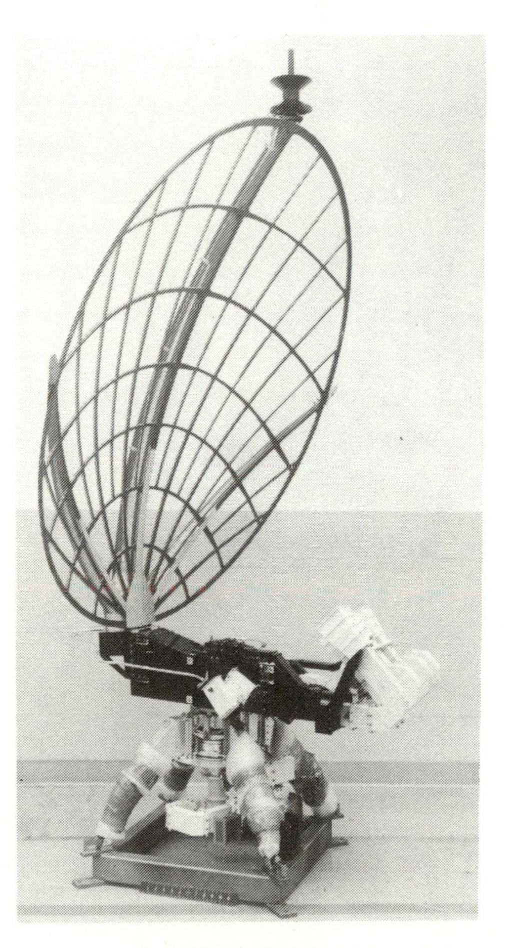

Fig. 3.107 *Westar communications antenna*

Intelsat IV-A

The Intelsat IV-A antenna system (Fig. 3.109) consisted of three offset-fed paraboloids, each fed by a multiple-horn array. Two were used for transmit and a third for receive. Frequency reuse was achieved through spatial beam isolation, where the sidelobes of each beam were controlled to minimise interference with adjacent beams operating at the same frequencies. To achieve nearly a doubling of the Intelsat IV capacity, the Intelsat IV-A satellite reused the available 500 MHz spectrum at the 5-4 GHz band. The first of the Intelsat IV-A series was launched in September 1975.

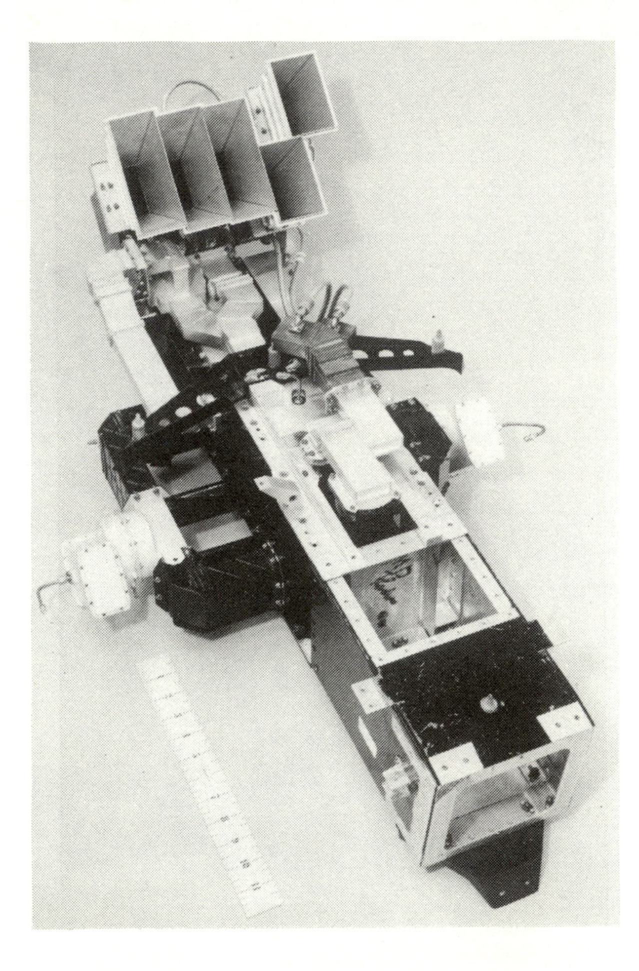

Fig. 3.108 *Westar transmit/receive feed network*

The contoured beam receive network (Fig. 3.110) was designed to operate from 5.9 to 6.4 GHz. The antenna received LHCP with an ellipticity of less than 3 dB over the coverage area. The 34 horns had integrated polarisers energised with a TEM-mode 'squarax' transmission-line power divider network. 17 horns provided the east contour coverage and 17 provided the west contour coverage.[192]

Fig. 3.109 *Intelsat IV-A communications antenna*

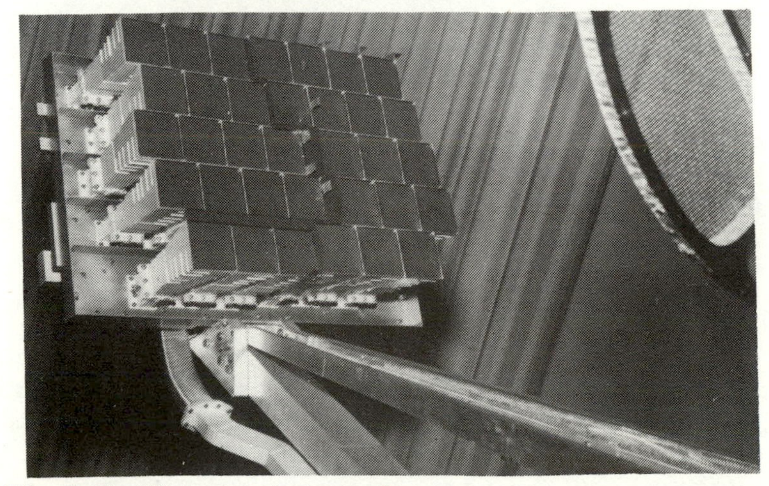

Fig. 3.110 *Intelsat IV-A receive feed network*

Comstar I

The Comstar I communication satellite (launched in 1976) had a pair of high-gain, narrow-beam antennas which provided communication service within the contiguous USA, Alaska, Hawaii and Puerto Rico (Fig. 3.111). Polarisation diversity was employed to double the available bandwidth through frequency reuse. Polarisation screens were placed in the aperture plane in front of each of two offset parabolas. The screens consisted of parallel gratings of conducting strips. The measured isolation between the two antennas was better than 33 dB over both frequency bands.[193]

Fig. 3.111 *Comstar I communication antenna*

The horizontal-polarisation feed assembly (Fig. 3.112) consisted of six feed-horns, and separate feed networks for both transmit and receive. These horns, which both transmit and receive, formed beams for the contiguous USA, Hawaii, and Puerto Rico.

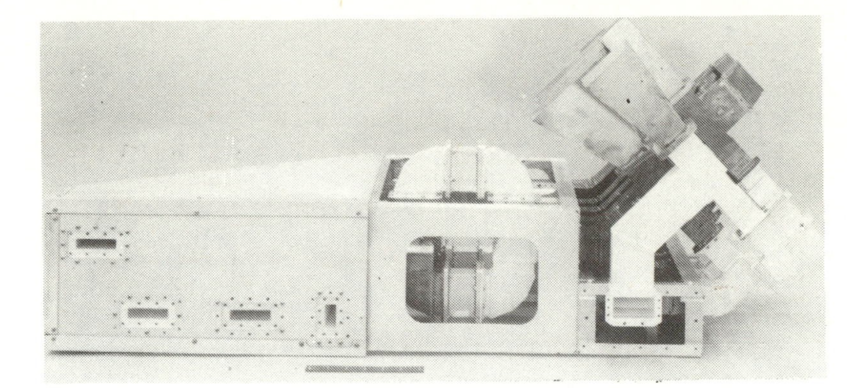

Fig. 3.112 *Comstar I transmit/receive feed network (horizontal polarisation)*

Fig. 3.113 *Prototype antenna model for the Japanese communications satellite*

Japanese CS Satellite

The Japanese CS Satellite (launched 14 December 1977) was designed to provide six channels in the 20-30 GHz band and two channels in the 4-6 GHz band, with each channel 200 MHz wide. It was built by Ford Aerospace for MELCO under

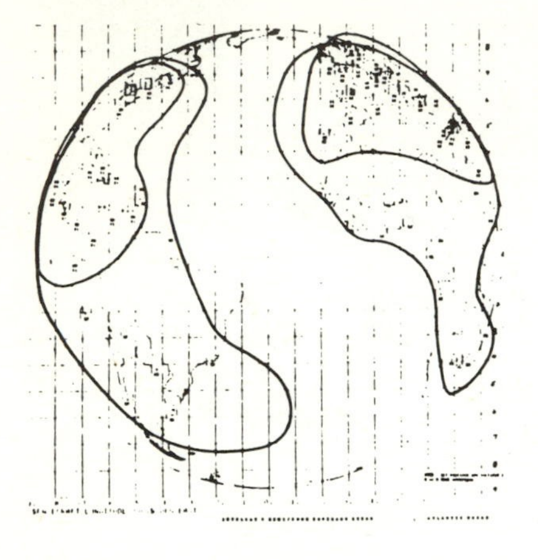

A. Atlantic Ocean

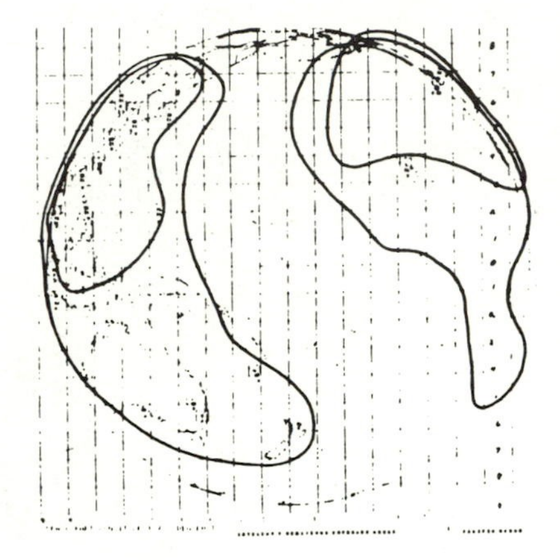

B. Pacific Ocean

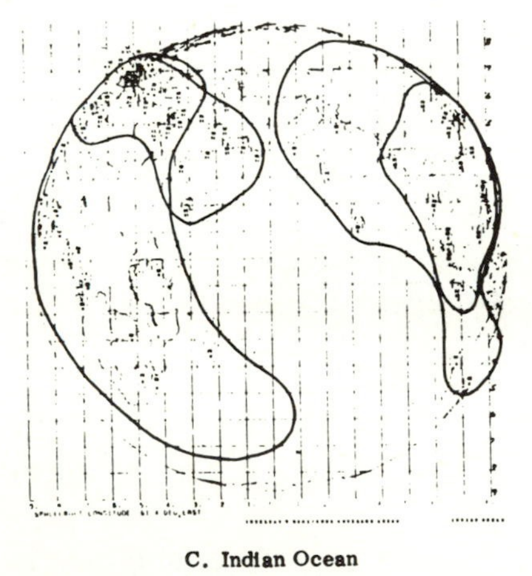

C. Indian Ocean

Fig. 3.114 *Intelsat V hemi-zone antenna coverage contours*

contact with NASDA. It was the experimental version of a commercial communications satellite being designed in Japan.[194]

A mechanically despun, wide-band, contoured-beam horn-reflector antenna provided a high-gain communication link with the Japanese Islands in the 4-6 and 20-30 GHz bands (Fig. 3.113). The antenna design extended existing horn-reflector technology by (i) the development of a single-feed system capable of operating over four separate frequency bands, and by (ii) synthesising a doubly curved, shaped-reflector providing radiation pattern characteristics that closely resemble the outline of the Japanese Islands at the 20-30 GHz band. The initial electrical design concept for this antenna was developed by Nippon Telegraph & Telephone Public Corporation and Mitsubishi Corporation. The final antenna design and its manufacture were undertaken by Ford Aerospace WDL.[195] The principles of this design are outlined in Section 3.4.1.

Intelsat V

The Ford Aerospace Intelsat V spacecraft is the largest commercial communications satellite ever put in orbit (1981).[196,197] It covers four frequency bands from 4-6 GHz and from 11-14 GHz. It replaced an Intelsat IV-A as a primary satellite,

Fig. 3.115 *Range testing Ford Aerospace Intelsat V 4 GHz reflector and feed*

providing an 11-12 thousand channel capacity compared to the previous 6000 of Intelsat IV. The required antenna coverage contours are shown in Fig. 3.114.

Fig. 3.115 shows the breadboard model of the 4 GHz reflector during range testing. The feed array consists of a cluster of square waveguide elements arranged in a closely packed planar configuration located on the optimum scan plane for the offset reflector. Each feed element consists of a coax-to-waveguide transition, a septum polariser, and a step transformer with dual input ports that provide for both senses of circular polarisation.

The TRW concept for Intelsat V introduced some significant new contoured-beam technology (Fig. 3.116). The offset reflector was illuminated by a planar array of dual-circularly-polarised cup turnstile elements. The entire array feed, consisting of 45 elements, is shown on a planar near-field test facility. The 20-element subset for the Western-hemisphere beam comprised the top of the feed. The remaining 25 elements, which simulated an Eastern-hemisphere feed, were terminated in matched loads for the measurements in order to evaluate mutual coupling effects. By probing the near-field on a plane in front of the feed, mutual impedance effects could be detected and significantly reduced by means of

Fig. 3.116 *TRW Intelsat V feed array on near-field test range*

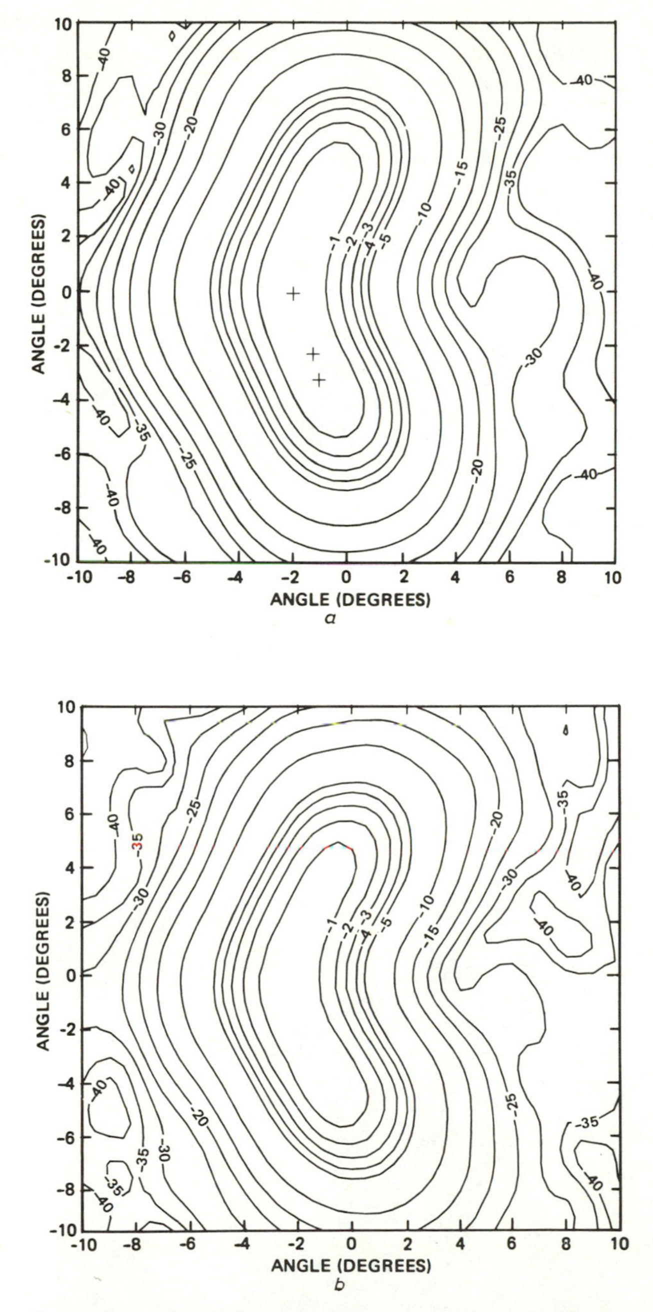

Fig. 3.117 *Comparison of calculated (based on near-field probing) and measured far-field patterns for TRW Intelsat V antenna*

a Reflector radiation pattern calculated from near-field feed measurements at 3.83 GHz

b Measured far-field pattern of reflector at 3.83 GHz

additional feed-element tuning.[186] Fig. 3.117 shows a comparison of the far-field contours calculated from near-field feed measurements at 3.83 GHz with the corresponding contours measured on a far-field range.

3.6 Reflector structures and materials (Contribution by W.V.T. Rusch)

The first reflector-antenna structure was the 2m-high, 1.2m-wide, wood-frame supported, sheet-zinc parabolic cylinder used by Hertz in his original 66 cm experiments in 1888.[198–200] Marconi also used a deep parabolic cylinder for this 25 cm experiments.[201] The first large-aperture, doubly-curved paraboloidal reflector was constructed by Reber in 1937.[202] This 31.5ft-diameter, 0.6 F/D paraboloid consisted of thin aluminium plates on a framework of 72 radial wooden rafters. Since that time, reflector structures and materials have undergone many significant developments, the most recent of which are described in this section.

3.6.1 Ground-based antenna systems

The requirements for a cost-effective, high G/T earth-based satellite tracking antenna have led to three basic structural types: (i) the kingpost structure (Fig.

Fig. 3.118 *Kingpost Configuration*

3.118) which employs a kingpost arrangement for the primary azimuth rotating mechanism; (ii) the wheel-and-track structure (Fig. 3.119) which facilitates an elevated RF equipment room; (iii) the beam waveguide structure (Fig. 3.120) which utilises four mirrors to guide the RF signals between the reflector and a stationary equipment room on the ground. Thus cryogenic receivers can be housed on the ground without incurring excessive losses due to rotary joints or long transmission on-live runs.[203]

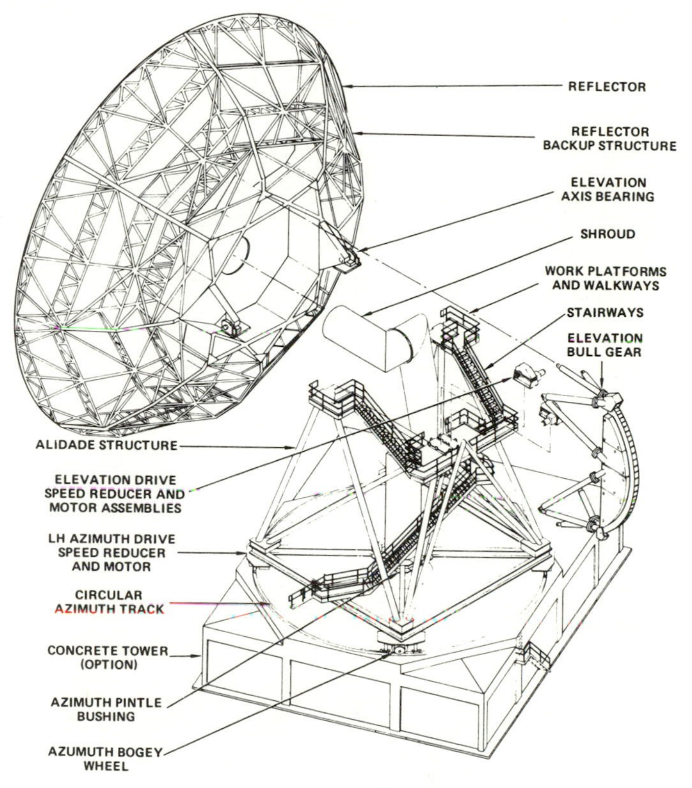

Fig. 3.119 *Wheel-and-track Configuration*

State-of-the-art, ultra-large D/λ ground antennas have been built on the maximum stiffness/minimum deflection principle. Two examples are:

(*a*) The Five Colleges Radio Astronomy Observatory (FCRAO) 45 ft (13.7 m) diameter symmetric Cassegrain reflector system, which is enclosed in a 68ft-diameter radome (Fig. 3.121).[204,205] Predecessors of this unique mm-wave reflector design are a 66 ft antenna in Sweden,[206] and a 45 ft antenna in Brazil,[207] both radome-enclosed. The FCRAO paraboloidal mirror consists of 40 machined aluminium subpanels. The combined manufacturing, setting and backstructure tolerance of the antenna permits operation at frequencies as high as 300 GHz, at which the reflector D/λ is 13700.

(*b*) The Bell Telephone Laboratories (BTL) 7m-diameter offset Cassegrain antenna is another state-of-the-art ultra-large D/λ mm-wave reflector constructed on the maximum stiffness/minimum deflection principle.[208] Stringent satellite-tracking requirements for low near-in sidelobes and low cross-polarisation demand the elimination of virtually all aperture blocking, thus leading to the offset

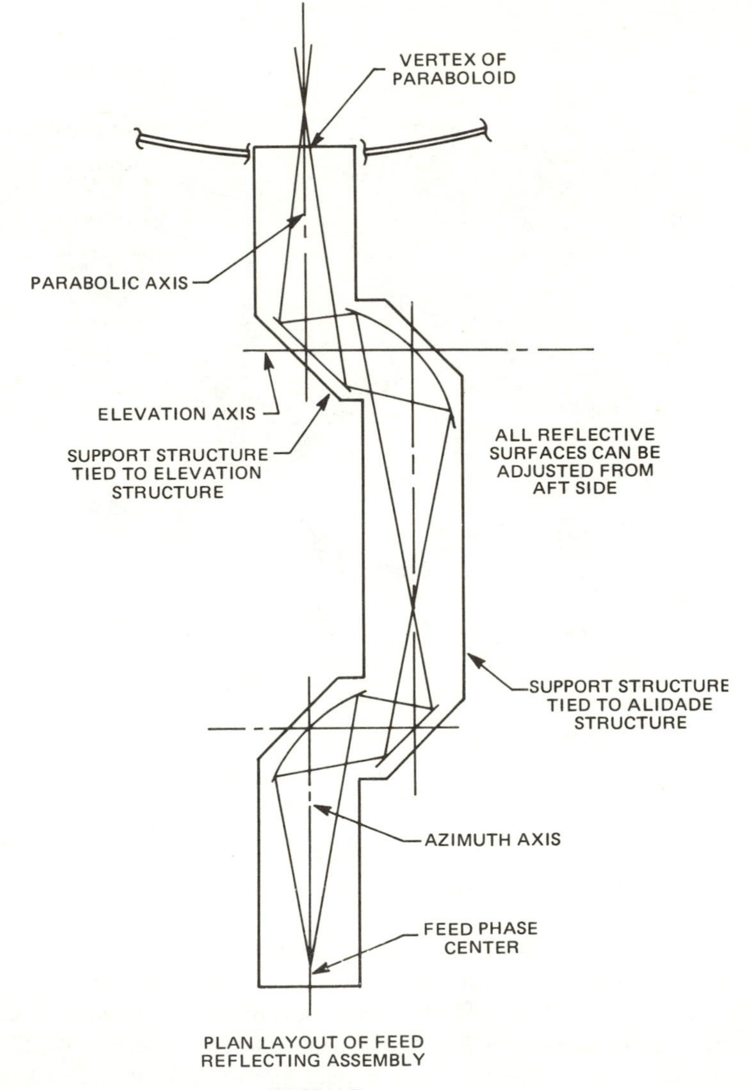

Fig. 3.120 *Beam waveguide optics*

Cassegrain configuration. A quasi-optical frequency/polarisation diplexer uses offset ellipsoids with corrugated horns. The primary mirror consists of 27 aluminium-cast surface panels with 17-cm-deep ribs on the back sides near the edges. Alignment of the surface panels was accomplished using a sweep template. The overall surface tolerance also permits operation up to 300 GHz.

Fig. 3.121 *A240 synergised system. Backstructure, pickup arms and pedestal*

On the other hand, the Max Planck Institut fuer Radioastronomie (MPIFR) 100m-diameter paraboloid (Fig. 3.122) is designed on the homologous deformation principle, by which controlled elastic deformations of the steel back-structure is permitted in such a way that the surface converts into a continuous family of paraboloidal shapes as the antenna tilts about the elevation axis.[209] The resulting displacement of the focal point is tracked by computer-controlled movement of the feed. The inner 80m of the surface consist of solid panels with a tolerance better than one mm RMS; the next 2.5m ring is perforated with 6-mm-diameter holes; the next 5-m ring is 6-mm square mesh, and the outer 2.5-m ring is 8-mm mesh. The fully steerable reflector, with both prime focus and gregorian feeds (Fig. 3.123), has been operational since 1972 at frequencies as high as 25 GHz.

steerable reflector, with both prime focus and gregorian feeds (Fig. 3.123), has been operational since 1972 at frequencies as high as 25 GHz.

Fig. 3.122 *Max Planck Institute 100-m Radiotelescope*

3.6.2 Satellite-borne reflector systems

Spacecraft reflectors are inherently broadband, relatively easy to fabricate and test, and light in weight (1-5 kg/m^2).[210] They consequently constitute the largest class of spacecraft-borne high-gain antennas.

Configurations

The classical prime-focus reflector, which creates a large D/λ, co-phasal aperture distribution, has been successfully employed on numerous missions, among which are the Viking Orbiter 1975 high-gain, dual-frequency, 1.5 m paraboloid (Fig. 3.124) and the ATS F&G 30 ft gored-parabolic reflector (Fig. 3.125). The Viking antenna was of sandwich construction with a 1.27cm-thick honeycomb core and 0.25 mm uncoated graphite epoxy face skins.[211] The ATS antenna was a self-deploying, 30 ft reflector, stored in a 58in-diameter torus, that deployed by means of the stored strain energy in 48 flexible cambered ribs.[212] While mechanically relatively simple,

Fig. 3.123 *100-m Radiotelescope gregorian optics*

the prime-focus configuration suffers from gain, sidelobe, scanning, and cross-polarisation degradation due to the RF blocking by the feed and support structures in front of the radiating aperture.

Offset-fed single reflectors have been employed to eliminate the blocking aberrations mentioned above. Examples are the Telesat (Anik)/Westar and Intelsat IV-A rib-supported mesh reflectors shown in Figs.3.107 and 3.109 and the solid Intelsat V reflector shown in Fig. 3.115. The increased effective F/D values of offset-fed reflectors enhances scanning performance. However, significant cross-polarisation may be introduced by the reflector asymmetry if the system is linearly polarised, and circularly polarised beams are slightly squinted off the boresight direction, causing a slight pointing loss if simultaneous dual polarisation is used.

The classical symmetric dual-reflector system employs confocal paraboloid/hyperboloid/(Cassegrain) or paraboloid/ellipsoid (Gregorian) combinations. Because of an increased effective focal length, the scan performance of these configurations is superior to the prime-focus arrangement. A rigid, 12 ft symmetric Cassegrain high-gain antenna was used on the Voyager spacecraft (Fig. 3.126).[211] The two 16 ft gore/mesh deployable antennas on the TDRSS spacecraft were also Cassegrain. Both the Voyager and TDRSS antennas employ geometrical shaping to enhance aperture efficency. Other proposed symmetric dual-reflector systems are micro-

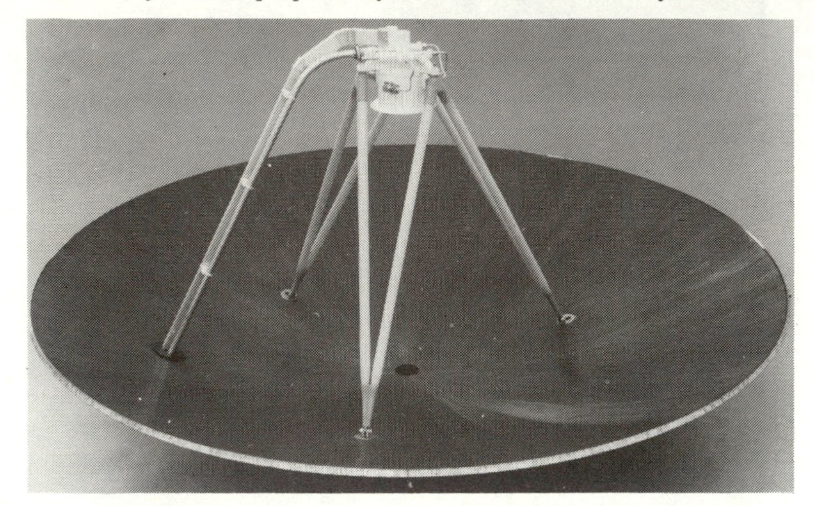

Fig. 3.124 *Viking Orbiter Antenna*

Fig. 3.125 *ATS F/G Antenna*

wave analogs of coma-corrected, dual-mirror optical systems[213] and other types of numerically generated non-classical reflector combinations.[214]

Offset-fed versions of dual-reflector systems are intended to reduce or eliminate

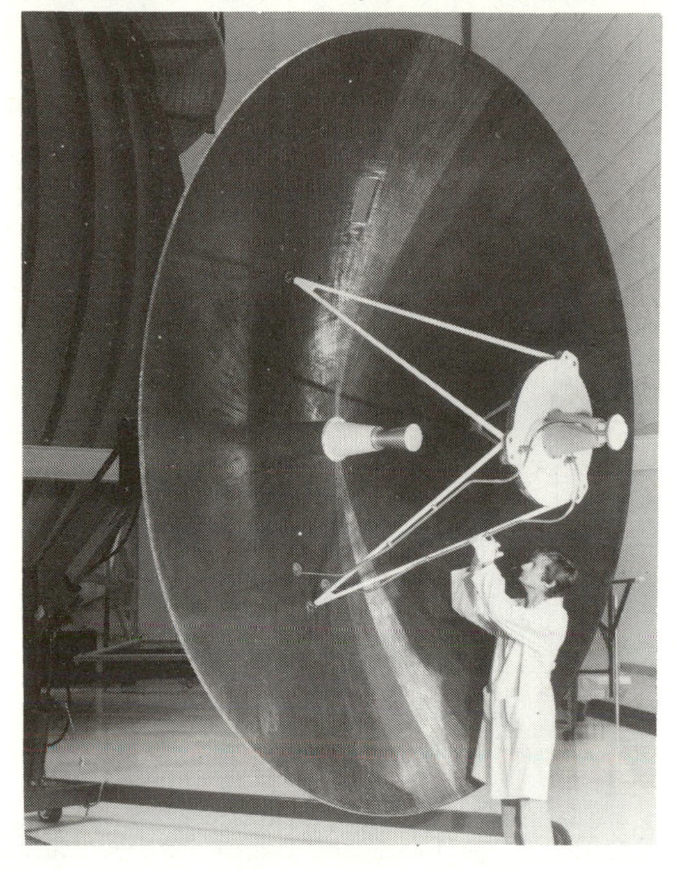

Fig. 3.126 *Voyager High-gain Antenna*

aperture blocking, e.g., the offset Cassegrain (Fig. 3.127*a*) and the open Cassegrain (Fig. 3.127*b*). The latter configuration, which eliminates blockage by the sub-reflector and supports but not by the feed, permits conical scanning. Another potentially useful offset configuration consists of two confocal paraboloids with a plane-wave feed (Fig. 3.128).[215]

Structural considerations for solid-surface reflectors

The maximum diameter of a spacecraft-borne solid-surface reflector currently is limited by the size of the launch vehicle or shuttle bay to about 4.4 m. Larger apertures can only be achieved by deploying, erecting, or manufacturing the

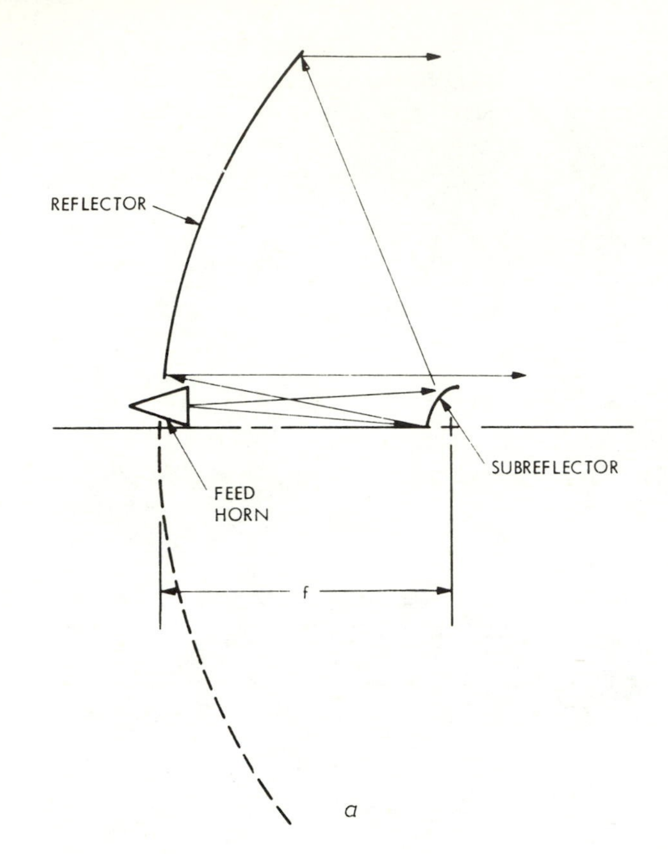

Fig. 3.127 *a* Offset Cassegrain Dual Reflector antenna

antenna in space, and are excluded from the present section.

The harsh demands of the launch and in-orbit space environments have established unusual material and structural design requirements:[216]

(*a*) high strength to survive the loads during launch
(*b*) high stiffness to maintain high structural frequencies and low deflections
(*c*) low mass to maintain spacecraft performance and permit larger payloads
(*d*) low thermal expansion coefficient to maintain reflector contour in the presence of thermal excursions and gradients up to 200°C
(*e*) high manufactured contour accuracy for optimum RF performance
(*f*) long-term (5-10 years) materials stability in the space environment
(*g*) structural compatibility with post-fabrication adjustments.

In addition, low material and fabrication costs are a desirable goal, although the current level of such designs are sufficiently advanced that further improvements will be achieved at significantly increased costs.[216]

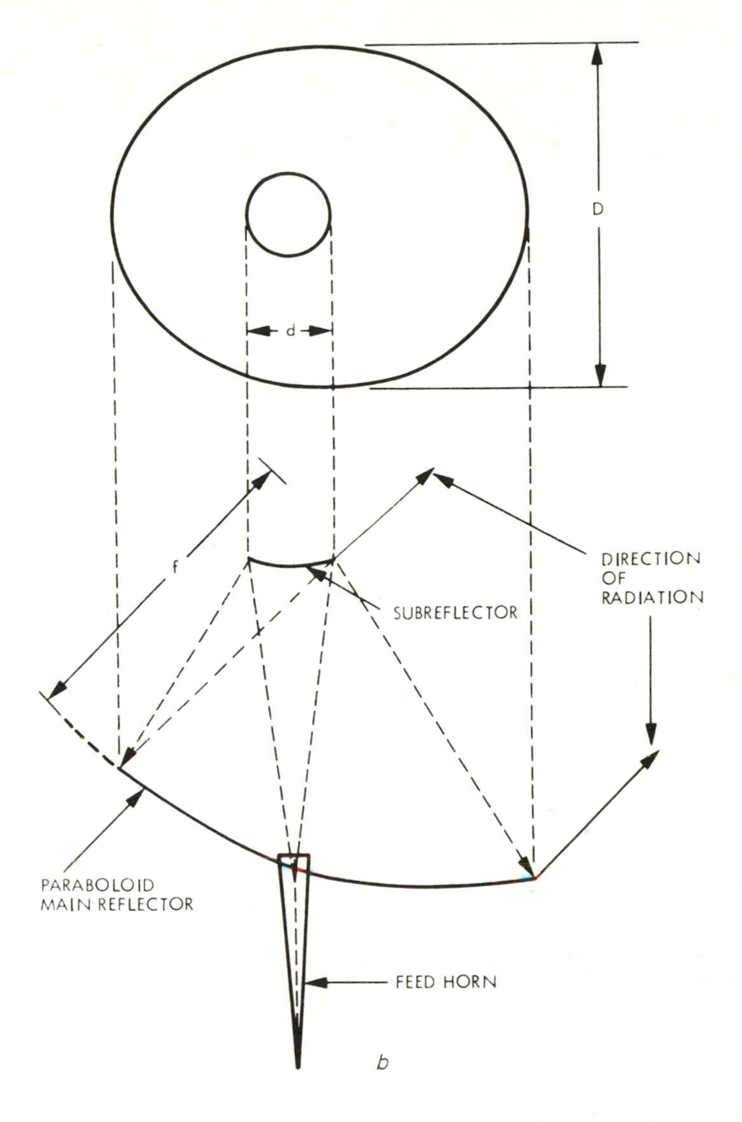

b 'Open Cassegrain' geometry

Conventional materials for aircraft and spacecraft structures have been aluminium, titanium, magnesium, beryllium, stainless steel, and fibreglass. Recently advanced composite materials have received wide acceptance.[217,218] Similar to fibreglass, for which glass fibres are embedded in a low-strength adhesive matrix, the advanced composite materials employ stronger and stiffer fibres of graphite, Kevlar, or boron embedded in an epoxy adhesive matrix. The fibres may be unidirectional, woven, or random. A wide range of mechanical properties can be achieved by different combinations of fibre and ply.

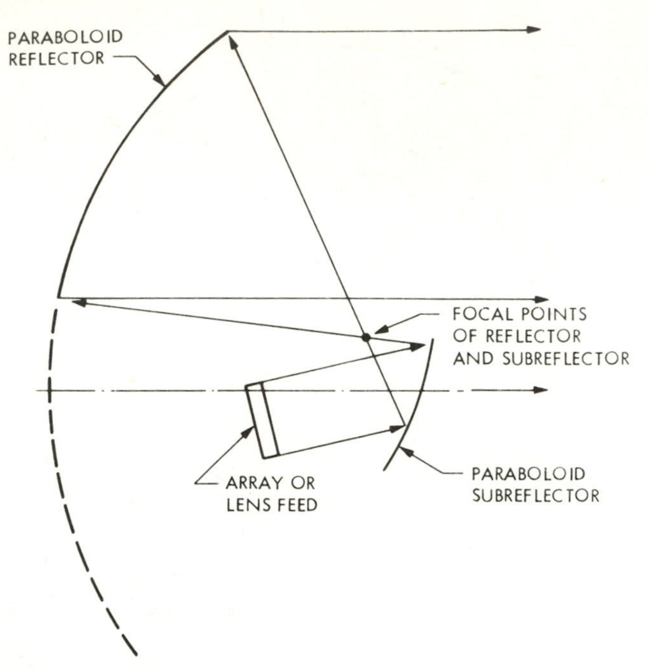

Fig. 3.128 *Offset near-field Gregorian antenna*

Table 3.6 *Material properties*

	Specific strength Nm/kg	Specific stiffness Nm/kg	Density kg/m^3	Thermal expansion m/m deg C	Thermal conduction $Wm/m^2 deg C$
	$\times 10^5$	$\times 10^7$	$\times 10^3$	$\times 10^{-6}$	
Graphite-epoxy					
HTS, unidirectional	5.0	8.7	1.6	−0.3	–
HTS, isotropic	1.7	3.0	1.6	+1.0	–
GY70, unidirectional	2.5	16.6	1.6	−1.1	12.1
GY70, isotropic	0.9	6.2	1.6	−0.1	26
Chopped, random mat	0.8	5.4	1.6	1.4	–
Kelvar					
Unidirectional	1.4	5.2	1.4	−3.6	1.7
Isotropic	0.7	1.9	1.4	−1.1	0.5
Fibreglass					
Unidirectional	4.0	2.7	2.0	7.2	0.2
Isotropic	1.3	1.0	2.0	7.2	0.1
Aluminium	1.2	2.5	2.8	23	132
Beryllium	1.5	15.8	1.8	11	180
Invar	0.3	1.8	8.0	1.3	11
Magnesium	1.2	2.5	1.8	25	87
Titanium	2.0	2.5	4.4	9.5	7
Stainless steel	1.2	2.5	8.1	11	17

Mechanical properties of the composite materials are compared in Table 3.6 with the metals.[217] Clearly, the properties depend upon fibre arrangement. The unidirectional configuration yields outstanding strength and stiffness in the direction of the fibres, but low values transverse to the fibres. Unidirectional fibres are practical in members where internal forces act only in one direction, as in truss members. When multi-axial stress conditions dominate, strength and stiffness are achieved by multiple orientation of unidirectional fibre plys.

A figure of merit in material selection is the relationship of stiffness divided by density and coefficient of thermal expansion. Table 3.7 shows the dominance of the advanced composites using this criterion. RF reflectance of uncoated graphite is similar to copper at frequencies up to K-band. At higher frequencies, a vapour-deposited high-conductance metallic film is needed to avoid gain degradation.[219]

Table 3.7 *Figure of Merit*

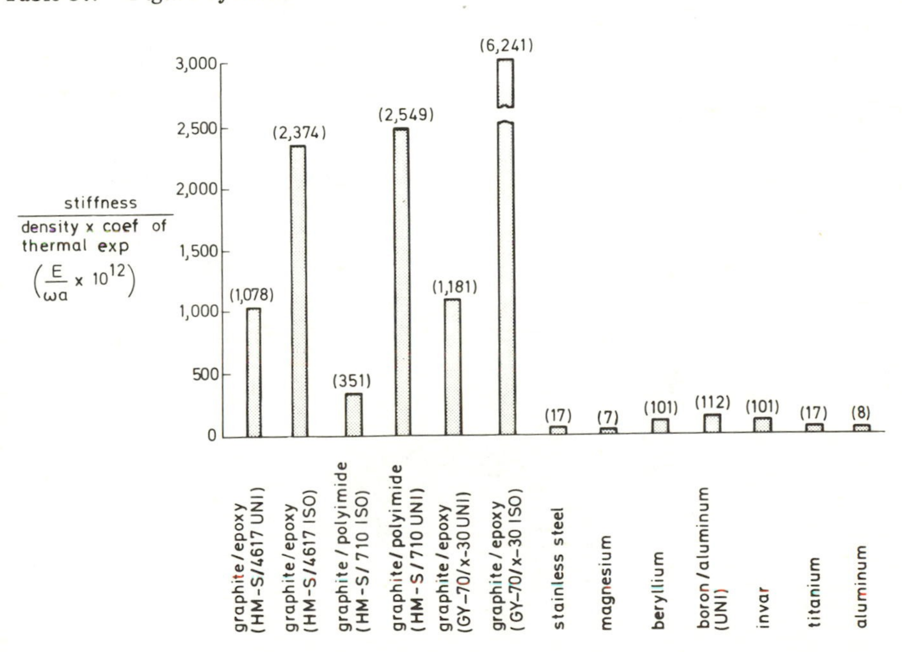

Structural configurations

Structural configurations which have proven most compatible with the advanced composite materials are (Fig. 3.129):

(*a*) Thick sandwich unstiffened shell, such as the 1.5 m Viking Orbiter Antenna in Fig. 3.124. This sandwich reflector has 0.5 mm graphite epoxy face skins and 1.27 cm aluminium honeycomb core;[217]

(*b*) Truss or rib-stiffened thin membrane shell, as exemplified by a General Dynamics 8 ft, 200 GHz, graphite, trussed-rib supported reflector with a face-skin thickness of 0.8 mm.[219] This configuration was felt to provide less thermal distortion than the sandwich configuration, and was more amenable to surface adjustment during fabrication. A similar 88 cm reflector was developed at the European Space Technology Centre.[220]

(*c*) Rib stiffened thin sandwich shell. The Voyager antenna (Fig. 3.126) was also of sandwich construction, with thin graphite-epoxy face skins and aluminium honeycomb design. The reflector was supported by a 1.7m-diameter graphite epoxy sandwich ring,[217] although other similar antennas use rib or truss supports.

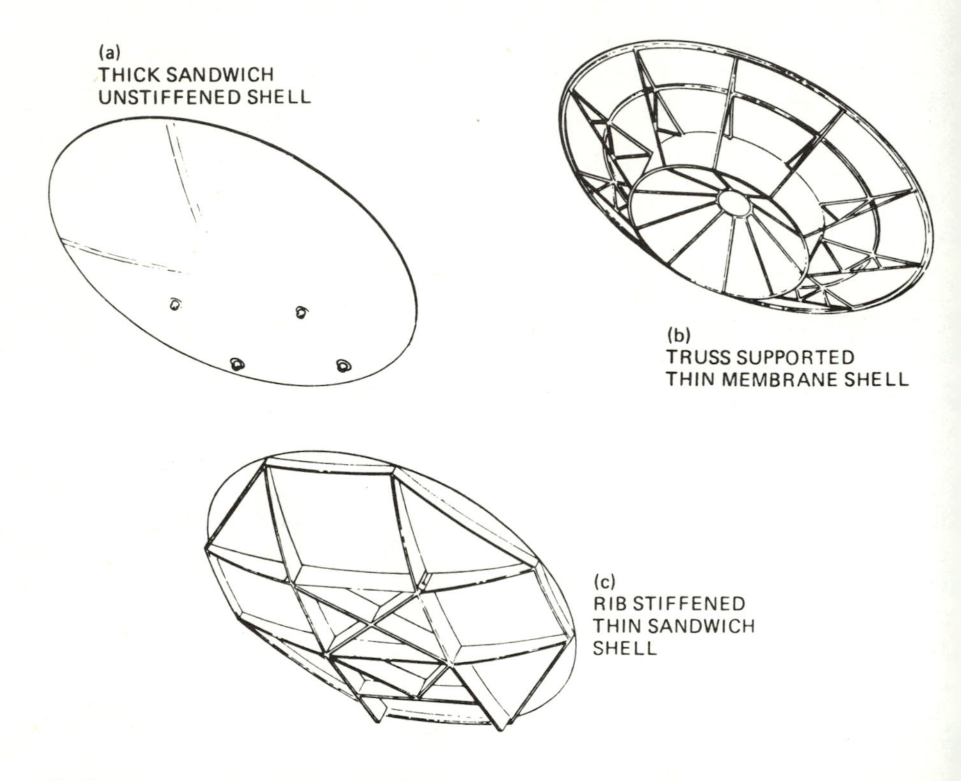

Fig. 3.129 *Classical reflector configurations*

Testing and performance

Verification of a solid-surface reflector under thermal and vibrational conditions to simulate the launch and space environments constitutes an important part of the qualification process. The reflector profile can be verified using template/micrometer techniques or with photogrammetric or holographic techniques under various conditions of thermal cycling, solar simulation, sinusoidal vibration, or acoustic vibration.[220] RF gain measurements are also mandatory. At 94 GHz, the General Dynamics 8 ft reflector is the largest D/λ (750) flight-qualified, solid-surface reflector known today. The manufacturing tolerances of the GFRP surface are 0.07 mm RMS, which degrades to 0.13 RMS under worst-case thermal distortion, with approximately 1 dB loss in gain.[221]

3.6.3 Deployable spacecraft reflectors

Three types of large deployable reflectors have been sufficiently developed to be considered flight qualified; these configurations are considered below. Potential candidates for new development are treated in Section 3.6.4.

Existing large deployable reflectors

(i) *Rigid radial-rib reflector*

The prime example of this configuration was the 16 ft reflector for the single-access antenna on the TDRSS spacecraft (Fig. 3.130). 1.2 mil-diameter gold-plated molybdenum mesh was supported by adjustable standoffs on 18 GFRP ribs. The rigid ribs were tapered in cross-section and circular in diameter. The number of ribs and size of the mesh represent a compromise between weight and RF performance. The reflecting surface profile was set with Invar ties to a secondary drawing surface. A series of drive-screw driven push-rods mechanically deployed the ribs, which were hinged at the central hub. Thermal control of the ribs was provided by a multilayered blanket of aluminised kapton. The worst-cast reflector distortion, both manufacturing and thermal, was 0.02 in RMS surface tolerance and 0.11 in feed defocusing.[222]

(ii) *Flexible radial wrap-rib reflector*

This reflector configuration consists of a toroidal hub to which are attached a series of flat radial ribs, between which is stretched a lightweight reflecting mesh. The hub provides stowage area for the furled ribs and mesh, as well as support for the in-space deployment (and possible refurl) mechanism, and an in-space surface-contour evaluation and adjustment system. To furl the reflector, the ribs are wrapped around the hub with the mesh folded between them, thus achieving a furled diameter. Deployment is accomplished by releasing the stored elastic strain energy of the wrapped ribs. The gore panels stretched between the deployed ribs provide an approximation to a smooth paraboloid.[223]

The 30 ft wrap-rib ATS F&G reflector (Fig. 3.125) had 48 ribs, weighed 180 lbs, and stowed into a 58 in ID by 78 in OD by 8 in thick torus.

(iii) *Paraboloid expandable truss antenna (PETA)*

The basic truss element of the PETA configuration is an expandable, hinged pentahedron, the struts of which are rigid CFRP tubes attached by hinges at each end to six-legged spider members. A series of these elements forms the truss structure, which collapses radially inward for stowage. The forward face of the deployed truss is a faceted approximation to the desired paraboloid, which supports a series of flat, reflecting mesh panels. Thermal control of the truss members is accomplished by means of deposited coatings.[222, 224]

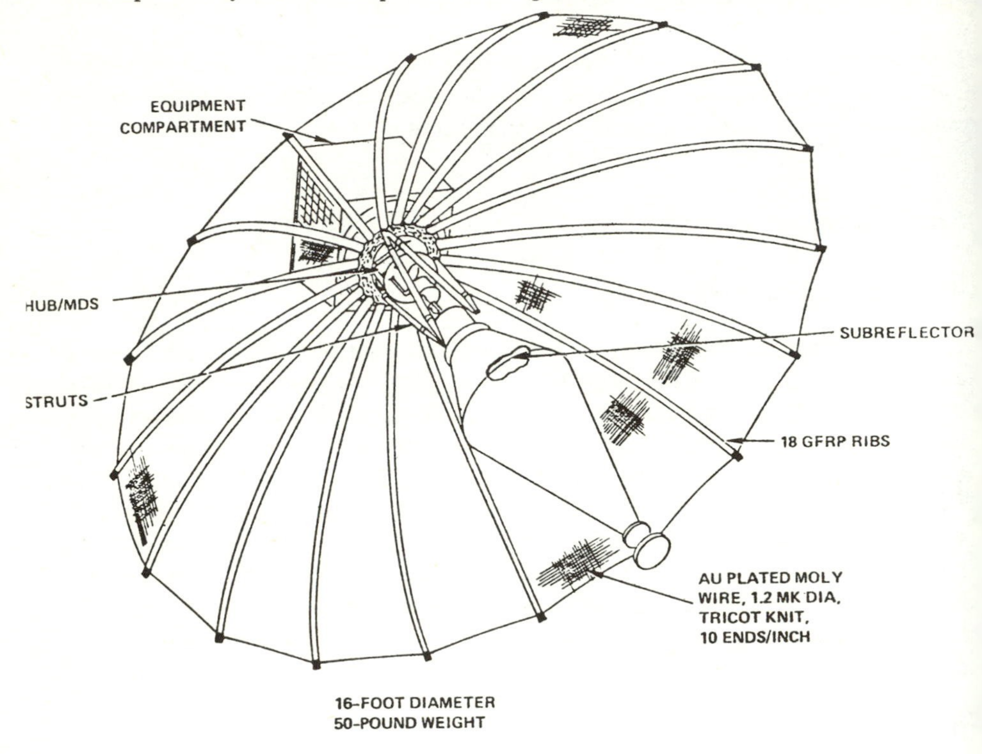

Fig. 3.130 *TDRS single access antenna*

Unique design features of deployable reflectors

Large deployable mesh reflectors exhibit several unusual design features unique to this type of antenna:[222]

(*a*) The reflector mesh has a tendency to 'pillow' toward the feed. This pillowing effect can be reduced by reducing the stiffness of the radial elements with strain-relief features or by increasing the circumferential pretension, which, however, increase the surface sensitivity to thermal forces, or by means of a secondary drawing surface;

(*b*) The surface contour should be 'structure-dominated' rather than 'mesh-dominated' in order to reduce thermal sensitivity.

(*c*) Materials with low coefficients of thermal expansion are mandatory, as well as, where possible, thermal control by means of coatings or thermal blankets.

(*d*) The RF and mechanical characteristics of the mesh must be accurately described.

(*e*) Post-assembly contour adjustment is highly desirable from a cost-effective standpoint.

(*f*) The structure contour and deployment should be verifiable in the presence of gravity.

(*g*) The thermal and structural aspects of the design must be accurately predictable by numerical analysis and computation.

3.6.4 Future deployable antennas

Availability of large launch vehicles such as the Space Shuttle will create some new technological considerations among which are (*a*) potentially greater acoustic and structural loadings during launch;[222] (*b*) a deployment check before release and/or retrieval capability; (*c*) in-space surface contour evaluation and adjustment;[225] and (*d*) increased reflector diameters, facilitating several new types of structures.

Two promising candidates for the super-large-diameter deployable satellite-borne reflectors are the polyconic and maypole configurations. The surface of the polyconic reflector consists of a series of circular conical mesh segments. The conical segments are positioned by mesh ribs and a series of radial booms mounted to a central hub. The booms and mesh surface are folded like an umbrella. For sufficiently large diameters, the booms must be folded in order to fit into the launch vehicle. Fine-surface control is made by catenary-member control on the mesh ribs.[225]

The 'maypole' or 'hoop/column' antenna consists of a long central hub or mast. and a rigid outer rim or hoop, and a series of interconnecting cables or stringers (Fig. 3.131).[222, 225] The reflecting mesh is hung between the hoop and the mast and supported by mesh tensioning stringers. The deployed hoop is sufficiently rigid to withstand the tension in the control stringers. The telescoping central mast can contain the electronic payload and control mechanisms. The deployment sequence is shown in Fig. 3.132. After extension of the mast and lower stringers, the hoop hinges outward to form a multi-sided polygon. The future usefulness of this structure depends upon the development of materials with extremely low thermal coefficients of expansion for the mesh, hub, hoop and tie stringers. It is projected that shuttle-launched, S-band maypole reflectors will be feasible up to 11000 ft in diameter.[225]

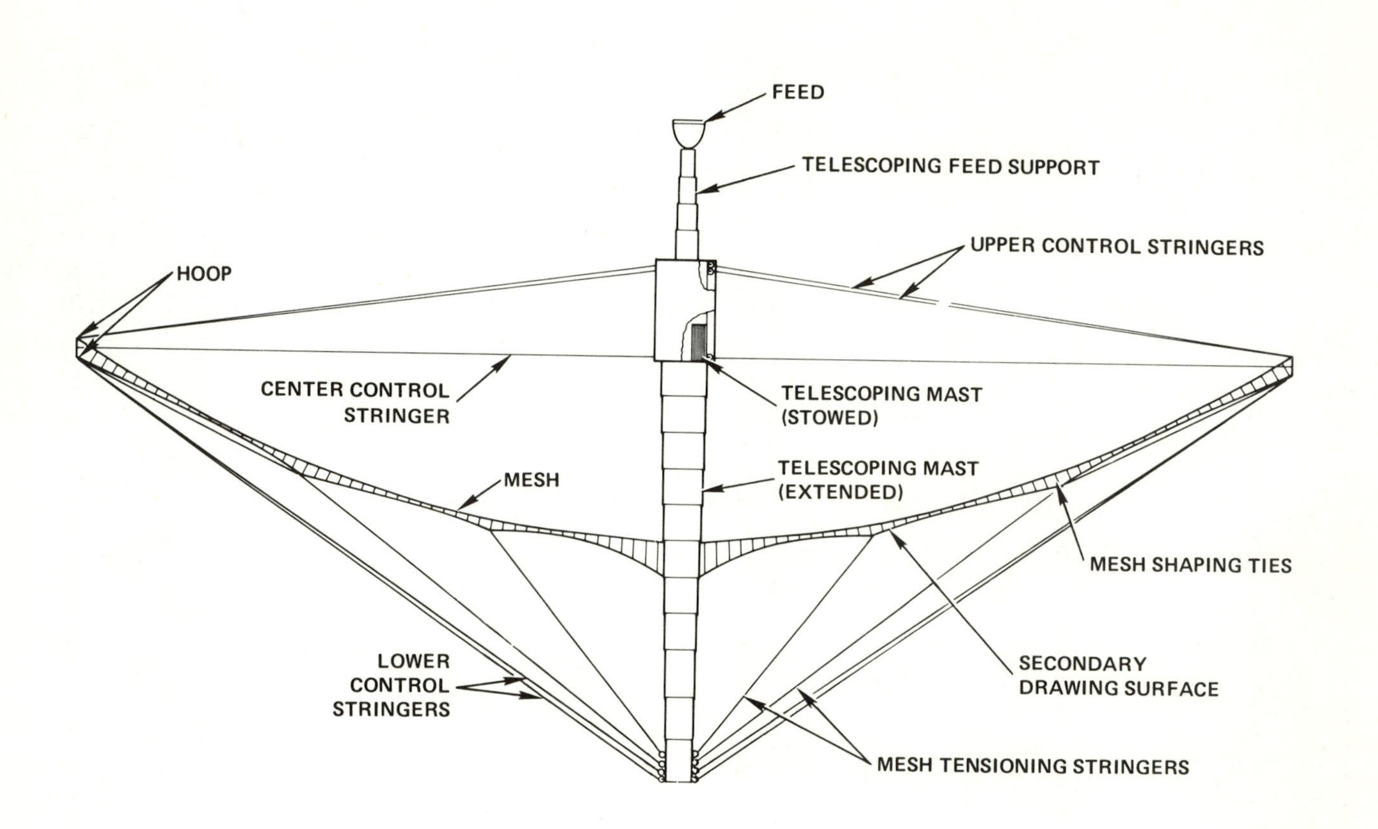

Fig. 3.131 *Hoop/column concept*

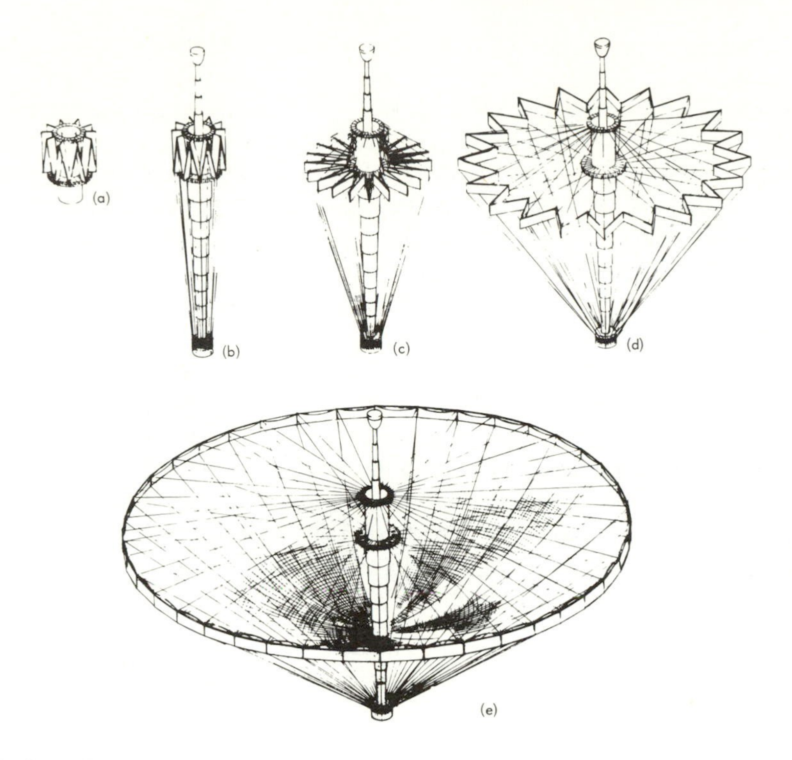

Fig. 3.132 *Deployment sequence*

3.7 Lens antennas (Contribution by A.R. Dion)

Up to some years ago, the principal use of microwave lenses was in antennas where a highly directive pattern with low sidelobes was required and where size and weight of the antennas were not important factors. In recent years, however, the emergence of multiple-beam antennas has promoted new applications for lens antennas. Multiple-beam antennas (discussed in Chapter 6) require focusing devices that present as little aperture blockage as possible and that exhibit the least amount of aberrations for off-axis operations. Both these requirements are best satisfied using a lens to collimate the beam. An important application of multiple-beam antennas is in satellite-communications systems where the satellite antennas must have the capability of producing a variety of coverages on earth. Several new, low-weight lenses, all of the constrained type, were conceived for this application; namely, the minimum-thickness zoned waveguide lens, the minimum-phase error zoned waveguide lens, the halfwave plate lens, the helix lens, the compound waveguide lens, and the printed-circuit bootlace lens. This Section is devoted to the description of these lenses and to the formulation of their properties. Other lenses have received comprehensive treatments by Jasik,[226] Collins,[227] and others[228,229] and will not be discussed here.

3.7.1 Constrained lenses for multiple-beam antenna applications

The lenses being considered are of axial symmetry and are constituted of elements capable of propagating a circularly polarised wave and having axis parallel to the lens axis. The lens design consists in the determination of the shape of the lens surfaces, of the lens thickness, and of the characteristics of propagation within the lens elements. The inner surface of each of the lenses considered is a sphere centred at the focal point. This shape is chosen to satisfy the Abbe sine condition[230] which ensures that for small displacements of the feed perpendicular to the lens axis the system will be free of aberrations. For an optical system focused at infinity, as is the case here, the sine condition is satisfied if the intersection of the incident rays and of the transmitted rays (parallel to the axis) all lie on a circle, called the Abbe circle, centred at the focal point. For the constrained lenses under consideration, propagation within the lens is parallel to the axis and, therefore, the sine condition obviously requires the inner surface of the lens to be a sphere centred at the focus. There remains to determine the length of each element of the lens given the characteristics of propagation within it. The lens outer surface necessarily follows from the determination of this length.

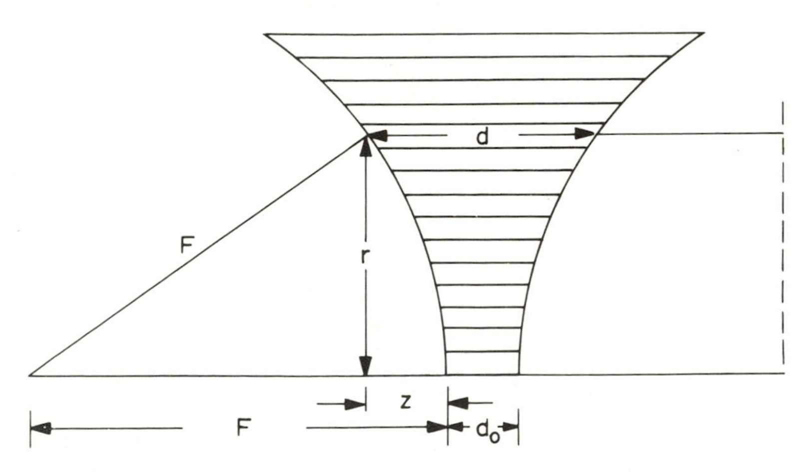

Fig. 3.133 *Unzoned waveguide lens*

Waveguide lenses

The elements of waveguide lenses are sections of square or circular waveguides. The length of an element is determined by the condition that the path length of a general ray from a source at the focus to a plane perpendicular to the lens axis is equal to the path length of the central ray, at the design frequency. This condition depicted in Fig. 3.133 is satisfied with the length d of a general element given by

$$d = d_0 + z/(1 - \eta) \qquad (3.126)$$

where d_0 is the lens thickness at the centre, $\eta = [1-(\lambda/\lambda_c)^2]^{1/2}$ is the refractive index at the free-space wavelength λ with λ_c the waveguide cutoff wavelength

$$z = F - (F^2 - r^2)^{1/2} \simeq r^2/2F \tag{3.127}$$

where r is the distance of the element axis to the lens axis and F is the focal length.

Since the refractive index is smaller than 1, the lens thickness increases with radial distance. The bandwidth of the waveguide lens is limited principally by dispersion in the lens elements. At a frequency spaced Δf from the design frequency f, dispersion causes an aperture phase error given by

$$\phi_e = \frac{2\pi z}{\lambda} \frac{(1+\eta)}{\eta} \frac{\Delta f}{f}. \tag{3.128}$$

The aperture phase error is a linear function of the fractional bandwidth and a quadratic function of the radial distance (since $z \simeq r^2/2F$). The frequency band over which the phase error is smaller than $\pi/4$ (corresponding to deviation of $\pm\lambda/16$ from a plane wavefront) is, for $F/D > 1$,

$$\text{Bandwidth}, \simeq \frac{200\,\eta(F/D)}{(1+\eta)(D/\lambda)} \text{ percent.} \tag{3.129}$$

For practical values of F/D and D/λ, the bandwidth of the conventional waveguide lens is only a few percent.

Minimum-thickness waveguide lenses

The thickness of the waveguide lens is reduced, and its bandwidth is increased by zoning.[231] The lens is zoned by reducing the length, d, (eqn. 3.126) of the waveguide elements by

$$md_\lambda = m\lambda/(1-\eta), \; m = 1,2,3,\ldots \tag{3.130}$$

wherever d exceeds $m\,d_\lambda$. Since a signal transmitted through an element of length $d_\lambda = \lambda/1-\eta$ is delayed 2π radians less than is a signal transmitted through an equal length in free space, shortening the waveguide elements by a multiple of d_λ does not affect phase in the lens aperture. The location of the zones is obtained by making d in eqn. 3.126 equal to $d_0 + md_\lambda$, yielding

$$z_m = m\lambda \tag{3.131}$$

and providing for the simple geometric construction illustrated in Fig. 3.134. The maximum thickness of the lens is independent of the lens diameter and is equal to

d_λ plus a minimum thickness d_0 required for structural rigidity. The resulting zoned waveguide lens is of minimum thickness and is often called minimum-weight waveguide lens.

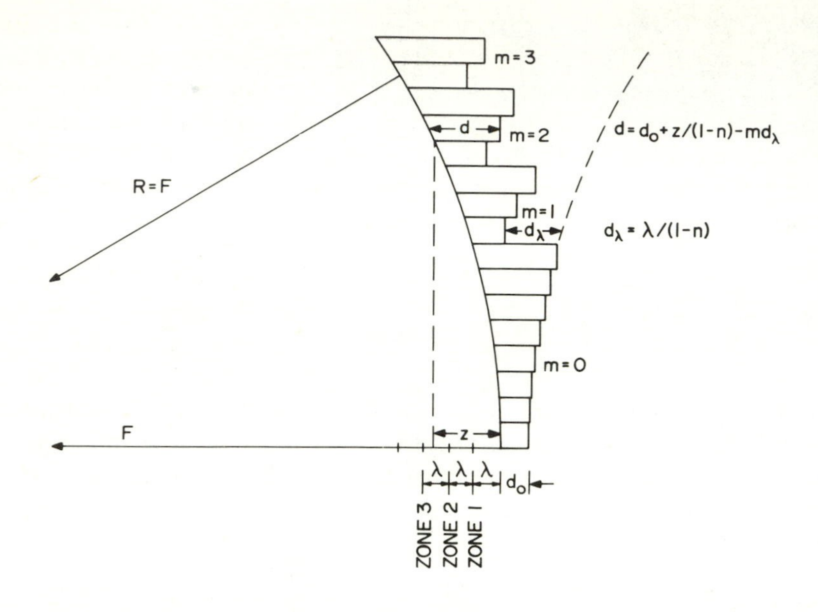

Fig. 3.134 *Waveguide lens zoned for minimum thickness*

The aperture phase error of the minimum-thickness zoned lens is

$$\phi_e = \frac{2\pi z}{\lambda} \frac{(1+\eta)}{\eta} \left[1 - \frac{m\lambda}{(1+\eta)z} \right] \frac{\Delta f}{f} \tag{3.132}$$

where m = INT(z/λ) (INT meaning the integral part of) denotes the zone number with $m = 0$ for the central zone. At the start of each zone $z = m\lambda$ and, therefore, $\phi_e = 2\pi m(\Delta f/f)$. The normalised phase error $\phi_e/(2\pi\Delta f/f)$ is plotted in Fig. 3.135 where it is compared to that of the unzoned lens for a practical value of η. The bandwidth of the minimum-thickness zoned lens is

$$\text{Bandwidth} \cong \frac{200\,\eta(F/D)}{(1+\eta)(D/\lambda)} \quad \frac{1}{1 - \dfrac{8\,M(F/D)}{(1+\eta)(D/\lambda)}} \text{ per cent} \tag{3.133}$$

where $M = \text{INT}[D/\lambda)/8(F/D)]$ is the number of zones. For M large, the bandwidth approximates to

$$\text{Bandwidth} \simeq 200(F/D)(D/\lambda) \text{ percent}. \tag{3.134}$$

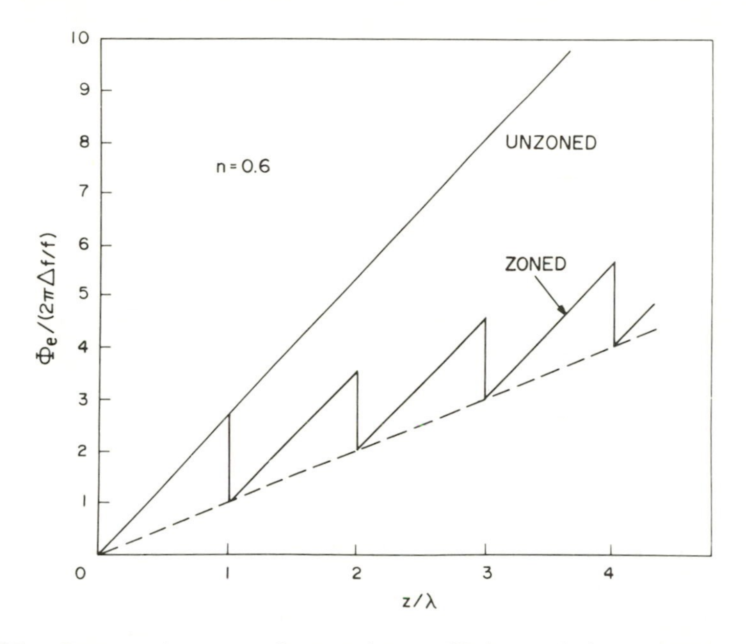

Fig. 3.135 *Aperture phase error of unzoned waveguide lens and of lens zoned for minimum thickness*

Fig. 3.136 *Multiple beam antenna with a waveguide lens zoned for minimum thickness*

i.e. $(1 + \eta)/\eta$ greater than that of the unzoned lens. A photograph of a minimum-thickness zoned lens is shown in Fig. 3.136. The lens designed for operation at a centre frequency of 7.68 GHz has a diameter of 30 in and a focal length of 30 in. It comprises approximately 700 titanium waveguide elements spot-welded together. Each element has a 1 x 1 inch cross-section and a 0.005 in wall thickness. There are three zones in the outer surface of the lens.

Minimum phase error waveguide lenses

The phase-error curve of the minimum-thickness zoned lens (Fig. 3.135) is a sawtooth curve with the amplitude of the flyback at a step smaller than the amplitude of the phase error immediately preceding the step, thus causing the peak error to progressively increase with radial distance (or with axial coordinate z). By appropriately changing the location of the steps, Colbourn[232] caused the amplitude of the phase error to become independent of the lens diameter. The step location required to achieve this result is obtained by making the two terms inside the bracket of eqn. 3.132 equal to each other, or

$$z_m = m\lambda/(1 + \eta),\ m = 1,2,\ldots \qquad (3.135)$$

The number of zones in the resulting lens is then

$$M = \mathrm{INT}[z_{max}(1 + \eta)/\lambda] = \mathrm{INT}[(D/\lambda)(1 + \eta)/8(F/D)] \qquad (3.136)$$

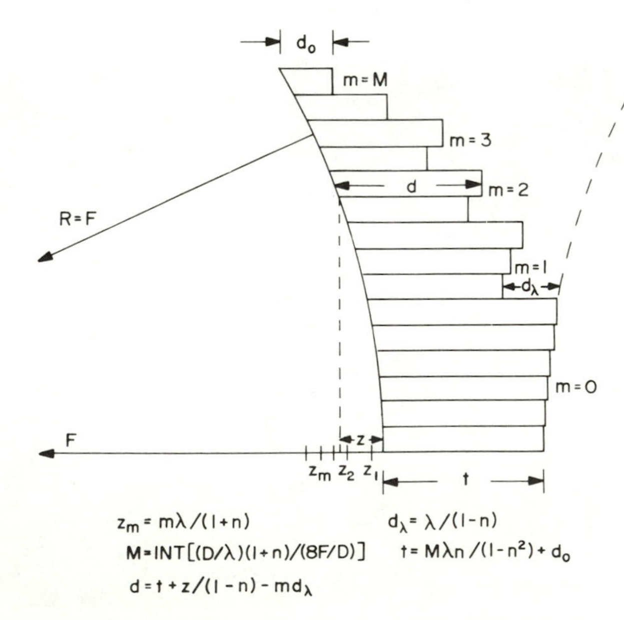

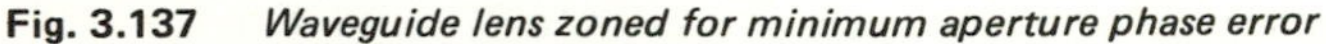

Fig. 3.137 *Waveguide lens zoned for minimum aperture phase error*

or $(1 + \eta)$ times the number of zones in the minimum-thickness zoned lens. The thickness at the centre of the minimum-phase-error zoned lens is

$$\text{Thickness} = M\lambda\eta/(1 - \eta^2) + d_0 \qquad (3.137)$$

where d_0, the minimum thickness in the last zone, is determined by structural requirements.

The cross-section of the minimum-phase-error lens is shown in Fig. 3.137. The formula for the aperture phase error of this lens is identical to that of the minimum-thickness waveguide lens, given eqn. 3.132, but with $m = \text{INT}[z(1 + \eta)/\lambda]$ and is plotted in Fig. 3.138. The peak phase error is

$$\phi_{max} = \frac{2\pi\Delta f}{\eta f} \qquad (3.138)$$

and the bandwidth corresponding to $|\phi_{max}| < \pi/4$ is Bandwidth = 25η percent.

Since the bandwidth is independent of the lens diameter, this design is particularly useful for large D/λ.

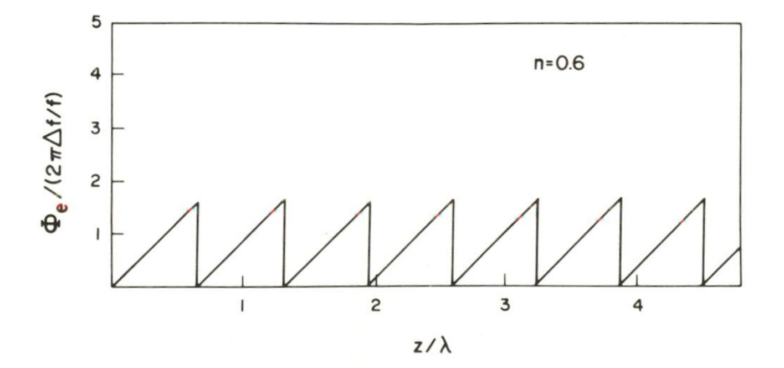

Fig. 3.138 *Aperture phase error of waveguide lens zoned for minimum phase error*

Zoning of waveguide lenses increases bandwidth, but, unfortunately, it also causes a reduction of aperture efficiency. The radiation pattern and gain of unzoned waveguide lenses are readily calculated following methods described by Silver.[228] However, for zoned lenses, it is necessary to make use of computer simulation to obtain acceptable accuracy. A typical simulation is presented in Section 3.7.2 where the adverse effects of zoning are described and evaluated. A photograph of a minimum-phase-error lens is shown in Fig. 3.139. The lens designed for a centre frequency of 8.15 GHz has a diameter of about 45 in and an $F/D = 1$. It is fabricated of about 1600 waveguide elements and its outer surface is formed of 7 zones. The square waveguide elements are 1 x 1 inches ID with a wall thickness of 0.016 in. The waveguide elements were formed by extruding circular aluminium pipe, and the wall thickness was reduced to 0.016 in by chemical milling once the lens was assembled and bonded by epoxy. The weight of the lens including a supporting ring is ~20 lbs.

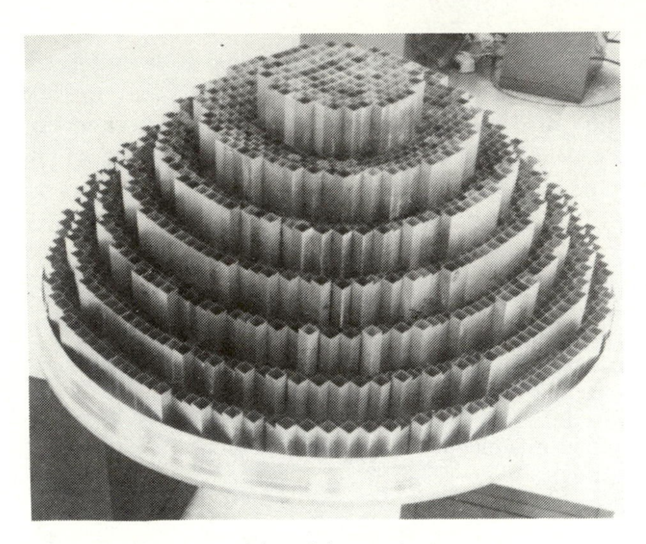

Fig. 3.139 *Photograph of lens zoned for minimum phase error*
(Courtesy of General Electric Co., Space Division, Valley Forge, PA)

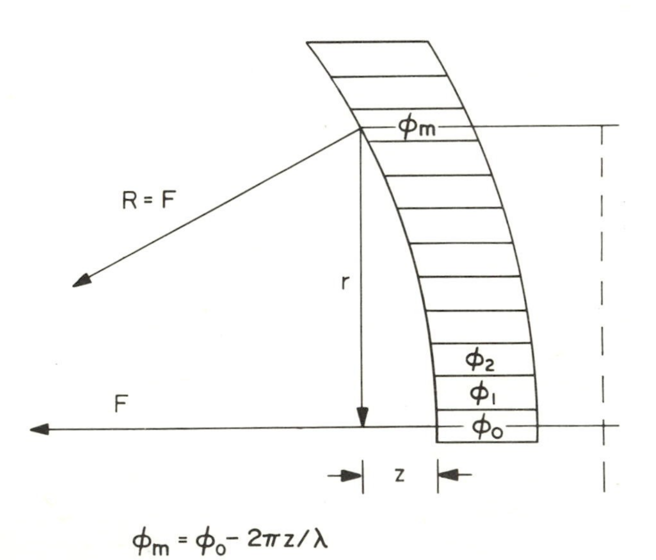

Fig. 3.140 *Constant-thickness, variable-phase-shift lens*

Waveguide lenses with phase shifters in the elements

The previously described waveguide lenses transform a spherical wave to a plane wave by means of waveguide elements of appropriate lengths. In the following waveguide lenses, this transformation is produced totally or in part by means of phase shifters in the waveguide elements.

(i) *Halfwave-plate lens*

Consider the lens of Fig. 3.140 with inner surface of radius F centred at the focus, and with waveguide elements of constant length and parallel to the lens axis. The path length of a general ray is longer than the path length of the central ray by an amount equal to z. To accomplish focalisation, a phase shifter is placed in each waveguide element, producing a differential phase shift, referred to the centre element, given by

$$\phi = -2\pi z/\lambda + 2m\pi,\ m = 0, \pm 2, \ldots \tag{3.139}$$

where λ is the design free-space wavelength and, as before, $z \simeq r^2/2F$ for $F/D > 1$. At a frequency Δf from the design frequency, an aperture phase error given by

$$\phi_e = \frac{2\pi z}{\lambda}\left[\frac{\Delta f}{f}\right] \tag{3.140}$$

is produced, assuming frequency-independent phase shifters in the elements. The aperture phase error of this type of lens is identical to the lower bound of the phase error for the minimum-thickness zoned lens and is shown as the dashed curve in Fig. 3.135. The bandwidth of the halfwave-plane lens is

$$\text{Bandwidth} \simeq 200(F/D)/(D/\lambda) \text{ percent.} \tag{3.141}$$

In applications with circularly polarised waves, a practical phase shifter for the lens elements makes use of the halfwave plate as first described by Fox.[233] A halfwave plate is a section of transmission line exhibiting different propagation constants for waves polarised parallel and perpendicular to the plate, and of length such as to introduce a phase difference of 180° between these two waves. Halfwave plates are identical in principle and in construction to the quarterwave plates widely used in circular polarisers.[235] The configurations most frequently used are flat dielectric slabs or a series of in-line capacitive pins or other metallic objects centred in a circular waveguide. A circularly polarised wave propagating in the halfwave-plate section of a waveguide element undergoes a phase shift of magnitude determined by the orientation of the plate. The differential phase shift is equal to twice the differential rotation of the plate and is independent of frequency. This is true even if the differential phase shift of the halfwave plate is not equal to 180°.[237] The effect of deviations from 180° is to give rise to a cross-polarised component, of power level equal to $\cos^2(\phi/2)$, where ϕ is the differential phase shift of the half-wave plate.

Fig. 3.141 *Halfwave-plate lens*
Courtesy of Hughes Aircraft Co., Ground Systems Group, Fullerton, CA)

A picture of a halfwave-plate lens is presented in Fig. 3.141. The lens[239,240] is made of circular aluminium waveguide sections of equal lengths, spot-welded together. The halfwave-plate phase shifter in each waveguide section is an array of metallic elements etched on Kapton film clad with copper, and supported by a polyurethane foam frame. The halfwave plates are positioned within the waveguide sections with an angle, relative to the centre element halfwave plate, equal to 1/2 the phase shift given by eqn. 3.139.

(ii) *Compound lens*

The variable-phase-shift constant-thickness lens exhibits a relatively narrow band over which the aperture phase error remains small. However, by combining this lens with a conventional waveguide lens, the bandwidth can be considerably increased.[236,237] This technique is well known in optics, accounting for the development of the achromatic doublet. A practical design of a waveguide compound lens is shown in the profile of Fig. 3.142. The lens consists of a conventional waveguide lens, with elements of length d, in contact with another lens of constant thickness and whose elements contain variable shifters. The phase shift produced by the phase shifters is assumed independent of frequency and may be realised with half-wave-plate phase shifters such as are described above. The waveguide elements of

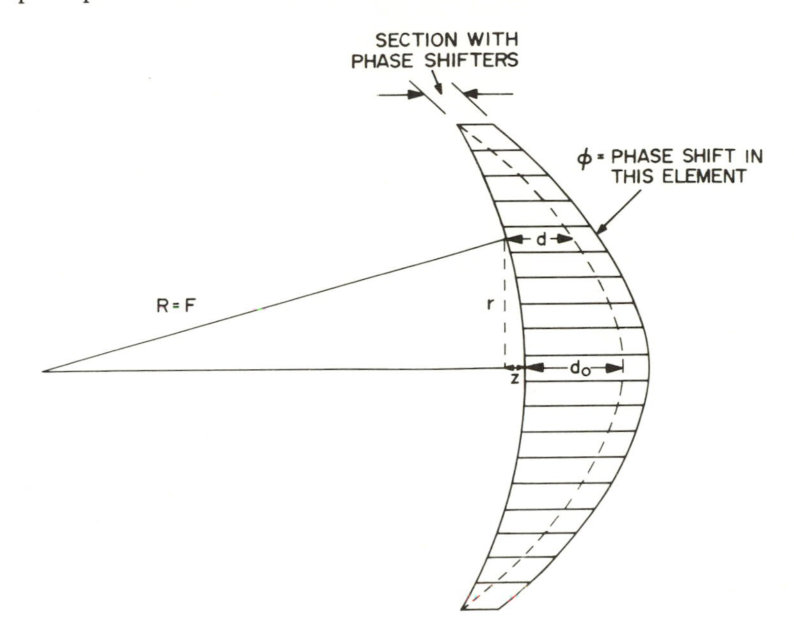

Fig. 3.142 *Broadband, compound waveguide lens*

the conventional lens and those of the halfwave plate lens have the same cross-section. The length d of the elements of the conventional waveguide lens and the phase shift ϕ in the elements of the halfwave-plate lens are functions of the radial distance r from the element axis to the lens axis. It should be noted that the compound lens is considered being made of two different lenses to simplify analysis but that since the axial location of the phase shifters in the waveguide element is immaterial, no two unique separate lenses can be defined. To satisfy the Abbe sine condition, the inner surface of the lens is spherical and centred at the focus. The condition for focalisation is again obtained by equating the path length of a general ray, from a focal source to a plane wavefront perpendicular to the axis, to that of the centre ray, at the design frequency. The condition for focalisation is

$$(1 - \eta)(d - d_o) - \phi\lambda/2\pi - z = 0 \qquad (3.142)$$

where $\eta = [1 - (\lambda/\lambda_c)^2]^{1/2}$ is the lens refractive index and $z = F - (F^2 - r^2)^{1/2} \simeq r^2/2F$ is the axial coordinate of an element on the lens inner surface.

Since two variables have to be determined, d and ϕ, eqn. 3.142 may be satisfied at two frequencies (two-frequency design), or a second condition may be imposed by making eqn. 3.142 invariant with frequency, and this second condition, together with the focalisation condition, may be satisfied at one frequency (single-frequency design).

(a) *Single-frequency design*

Making eqn. 3.142 invariant with frequency yields the second condition

$$(d - d_o)(\eta^2 - 1)/\eta\lambda + \phi/2\pi = 0 \tag{3.143}$$

which together with eqn. 3.142 yield

$$d = d_o - \eta z/(1 - \eta) \tag{3.144}$$

and

$$\phi = -2\pi(1 + \eta)z/\lambda \pm 2m\pi,\ m = 0, 1, 2, \ldots \tag{3.145}$$

The thickness at the centre of the lens, exclusive of the thickness of the phase-shifter section, is obtained by making $d = 0$ for $z = z_{max} \simeq D^2/8F$, and is

$$d_o = \frac{\eta}{1 - \eta}\,\frac{D^2}{8F}\,. \tag{3.146}$$

At a frequency separated Δf from the design frequency the aperture phase error is

$$\phi_e = \frac{2\pi z}{\lambda}\,\frac{(1 + \eta)}{2\eta^2}\left(\frac{\Delta f}{f}\right)^2 \tag{3.147}$$

yielding for the bandwidth of the single-frequency-design compound lens

$$\text{Bandwidth} \cong 200\left[\frac{2\eta^2\ (F/D)}{(1 + \eta)(D/\lambda)}\right]^{1/2} \text{percent.} \tag{3.148}$$

(b) *Two-frequency design*

The focusing condition (3.142) is satisfied at two frequencies with

$$d = d_o + \frac{z(\lambda_1 - \lambda_2)}{(1 - \eta_2)\lambda_1 - (1 - \eta_1)\lambda_2} \tag{3.149}$$

and

$$\phi = \frac{2\pi(\eta_2 - \eta_1) z}{(1 - \eta_2)\lambda_1 - (1 - \eta_1)\lambda_2} \pm 2m\pi \tag{3.150}$$

where λ_1, λ_2 are the free-space wavelengths and η_1, η_2 are the refractive indices corresponding to the two design frequencies. The aperture phase error at a different frequency is

$$\phi_e = \frac{2\pi z}{\lambda} \frac{\lambda_2(\eta_1 - \eta) - \lambda_1(\eta_2 - \eta) - \lambda(\eta_1 - \eta_2)}{(1 - \eta_2)\lambda_1 - (1 - \eta_1)\lambda_2} \tag{3.151}$$

where λ and η are the free-space wavelength and the refractive index at this frequency.

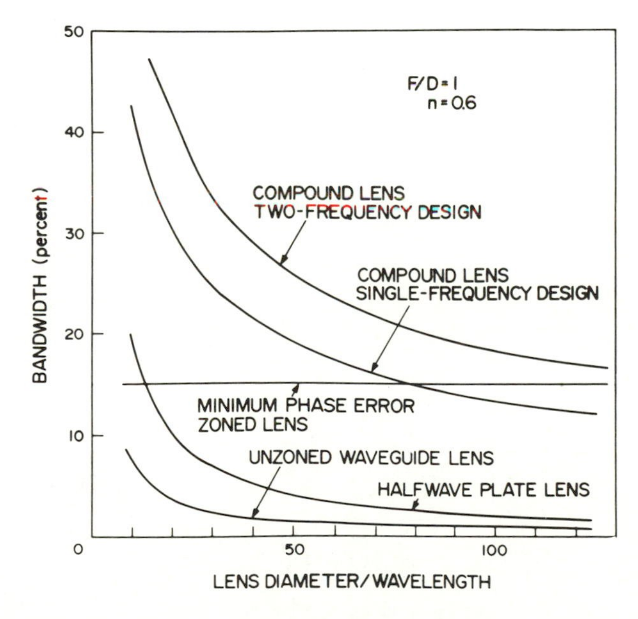

Fig. 3.143 *Bandwidth comparison of various constrained lenses*

The bandwidth of the compound lens (plotted in Fig. 3.143) is about an order of magnitude greater than the bandwidth of either of its component lenses, and the two-frequency design provides about 40% more bandwidth than the single-frequency design. The two-frequency design, it should be pointed out, is equivalent to a defocused single-frequency design. This is the case because the aperture phase error of the compound lens (eqn. 3.147) is a quadratic function of both frequency and radial distance ($z = r^2/2F$). Since defocusing also causes a quadratic phase error function of radial distance, the two errors may be made to cancel each other at the two frequencies $f + \Delta f$ and $f - \Delta f$. The amount of defocusing required to achieve this result is

$$\delta = \frac{F(1 + \eta)}{2\eta^2} \left(\frac{\Delta f}{f}\right)^2. \tag{3.152}$$

A compound lens fabricated of circular waveguide, and utilising for phase shifters the halfwave-plate phase shifters described previously, was reported by Ajioka.[237]

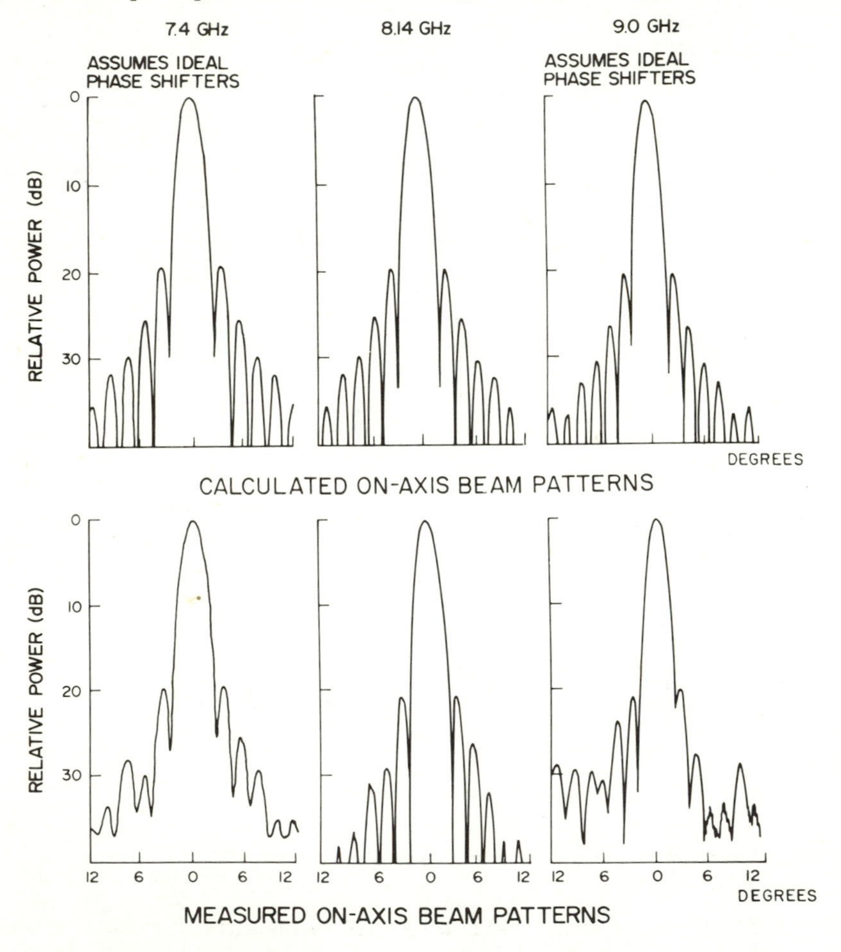

Fig. 3.144 *Calculated and measured radiation patterns of broadband compound lens* (After Ajioka[237])

The surfaces of the lens were matched to free space by an appropriate choice of the waveguide diameter, following a technique used for aperture matching of phased arrays.[238] The radiation patterns of this lens antenna, shown in Fig. 3.144, are observed to be devoid of the effects of quadratic aperture phase error over a 7·4 to 9 GHz frequency range, and compare well with expected performance. The

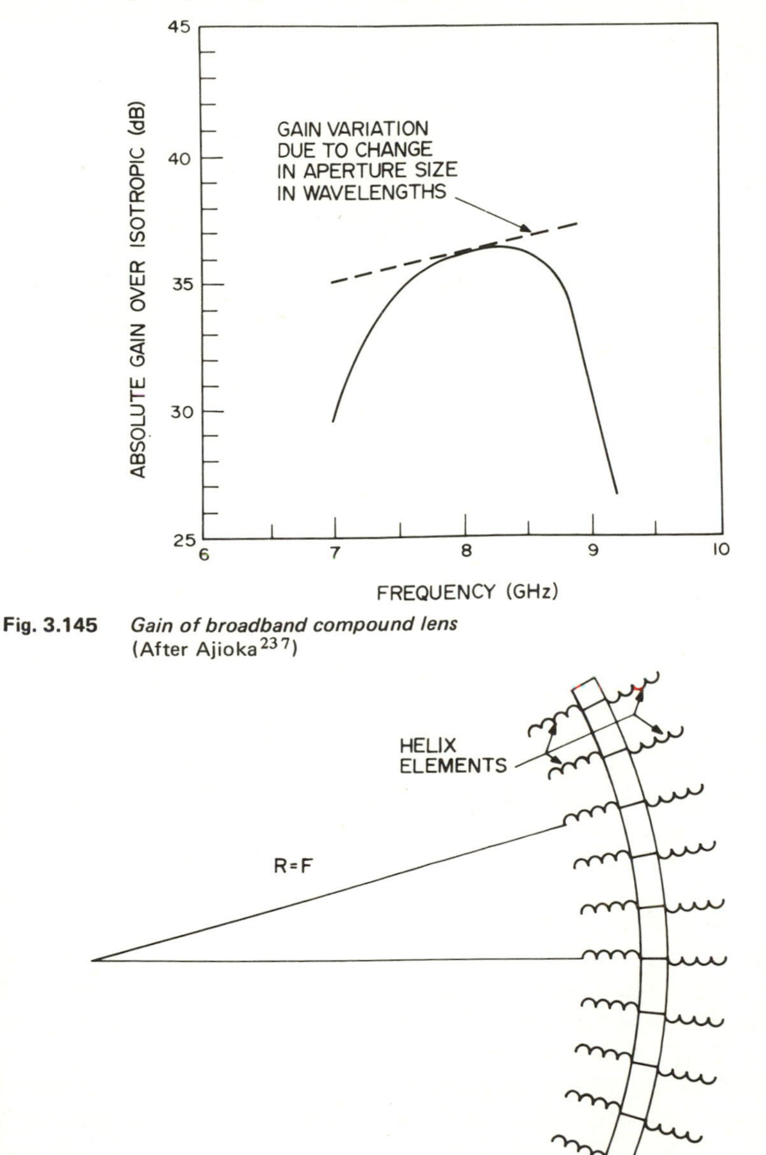

Fig. 3.145 *Gain of broadband compound lens*
(After Ajioka[237])

Fig. 3.146 *Helix lens*

parameters of the lens are $D = 46$ in, $F = 72$ in. The lens is comprised of 1573 circular aluminium waveguide sections spot-welded together. The waveguide inside diameter is 1.061 in, and the wall thickness is 0.006 in. The lens thickness at the centre is 7.08 in. The radiation patterns of Fig. 3.144 were obtained with a feed producing an illumination taper of $\cong -5$ dB at the edge of the lens. The gain versus frequency of this lens is given in Fig. 3.145 where it is compared to the expected gain variation due to change in aperture size (area/λ^2) with frequency. The loss of gain at the lower frequencies is mostly due to increasing surface reflections as the waveguide cutoff frequency is approached, while the loss of gain at the higher frequencies is mostly due to the increasing level of cross-polarised energy generated as the bandwidth of the halfwave plate phase shifter is exceeded.

Helix lens

The helix lens shown schematically in Fig. 3.146 is an array of back-to-back helices mounted on a shell. The shell is spherical with centre at the focus, again to satisfy the sine condition. The elements form a triangular lattice having a spacing somewhat smaller than one wavelength to avoid grating lobes and to provide a broad element pattern. The circularly polarised helices on one side of the shell absorb energy from an impinging wave. This energy is transmitted by coaxial lines to the helices on the other side of the shell where it is radiated. The lens is focused by rotating each element with respect to the centre element, to compensate for the path-length differential between the centre ray and the ray passing through each element. A similar phasing technique has been described by Brown and Dodson[234] for an array of spiral elements. Rotation of a helix element causes the circularly polarised wave transmitted through this element to be phase shifted; the differential phase shift is equal to the differential rotation. Rotation by an equal amount of both the receiving and transmitting helices thus causes a differential phase shift equal to twice the differential rotation. With back-to-back helices physically bound together and, therefore, subjected to rotation in the same direction, the helices must be wound in opposite directions. The phase shift required at each element and the aperture phase error of the helix lens are almost identical to those of the halfwave plate lens as given by eqns. 3.139 and 3.140, respectively. For optimum efficiency, the directivity G_e of each helix element must be equal to the area gain of the unit cell, i.e. $G_e = 4\pi A/\lambda^2$ where A, the area of the unit cell, is equal to the area of the lens aperture divided by the number of elements. In addition, the radiation impedance of each helix, within the lens environment, must be matched to the characteristic impedance of the coaxial line coupling each helix pair. The photograph of Fig. 3.147 shows the outer face of a helix lens used in a multiple-beam antenna.[239,240] The lens is comprised of 1213 back-to-back helix elements mounted on a spherical honeycomb shell. The element locations form a triangular lattice having a 0.86λ spacing. The 50 in-diameter lens has a focal length of 76 in and a centre frequency of operation of 8.15 GHz. The components of the helix elements are shown in Fig. 3.148. The back-to-back helices are coupled through a coaxial line and are wound in opposite sense.

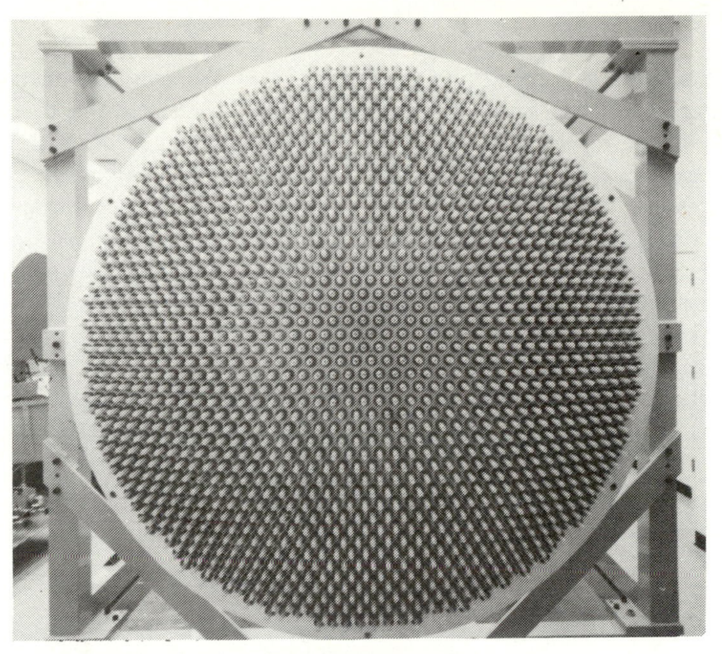

Fig. 3.147 *Photograph showing outer face of helix lens*
(Courtesy of Hughes Aircraft Co., Space and Communications Group, El Segundo CA)

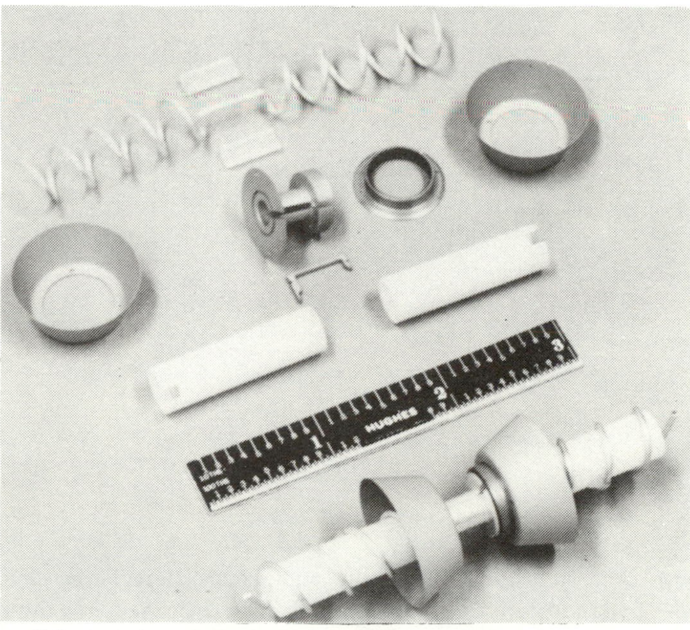

Fig. 3.148 *Helix lens element and its components*
Courtesy of Hughes Aircraft Co., Space and Communications Group, El Segundo, CA)

Bootlace lens

The bootlace lens is a circuit analogue of the dielectric lens. On one surface of the bootlace lens is an array of receiving elements connected one-to-one to the elements of an array of similar transmitting elements on the other surface, by a length of non-dispersive transmission line (Fig. 3.149). The length of the connecting line is chosen so that the path length from a focal source to a radiating element is the same for all elements. The bootlace lens is fundamentally non-dispersive. Its bandwidth is limited by the bandwidth of its elements and by the impedance mismatch between radiators and the interconnecting transmission lines. The spacing between elements is generally chosen slightly smaller than one wavelength to avoid grating lobes and to permit beam offsetting with little loss of gain. In recent years lightweight bootlace lenses have been fabricated using printed circuit transmission lines and radiators as elements.[241]

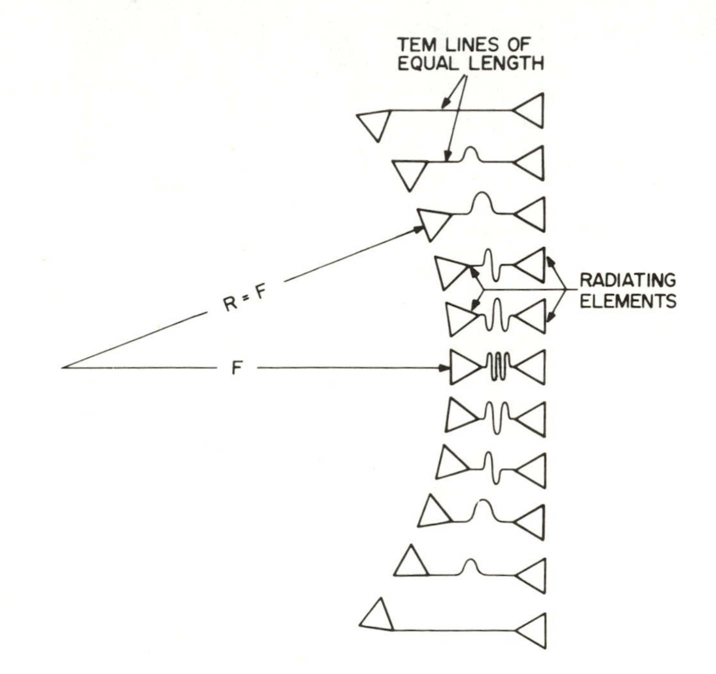

Fig. 3.149 *Bootlace lens (also called TEM lens)*

3.7.2 Computer modelling of constrained lenses

A constrained lens is an array, generally non-planar, of radiating elements, and therefore the determination of its radiation characteristics requires the knowledge of the complex amplitude of excitation and of the radiation characteristics of each element. The latter characteristics may be determined from measurements or, as will be described for the waveguide constrained lens, they may be assumed to exhibit some approximate analytical form. A computer model[242] successfully used to predict the performance of waveguide lenses is based on the geometry given in Fig. 3.150. The lens and feed geometry are specified by the rectangular coordinates X, Y, Z centred at 0_1 while the radiation pattern is specified by the spherical coordinates θ, ϕ centred at 0_2. The waveguide elements have a square cross-section of side dimension a, and are positioned by the subscripts indicating the mth row and the nth column. The coordinates of the centre of each waveguide aperture on the inner surface, $Z_1\,(m,n)$, and on the outer surface $Z_2(m,n)$ are determined from the formulas describing a particular lens as derived in the preceding Subsection. Considering the lens antenna in the transmission mode, the waveguide aperture centred at $Z_1(m,n)$ receives energy which propagates through element m,n and is radiated by the aperture centred at $Z_2(m,n)$. Reflections at the lens surface are neglected and propagation within the waveguide elements is assumed lossless. Under these conditions, the excitation of each waveguide element is determined by the power absorbed by each waveguide. The receiving cross-section of the waveguide elements is $a^2/\cos\alpha$ and its receiving radiation pattern is assumed to be a cosine pattern with axis along the normal to the receive aperture such as depicted in Fig. 3.150. Thus with an offset feed such as shown in the Figure, the power absorbed by element m,n is proportional to $a^2\cos\beta/\cos\alpha$.

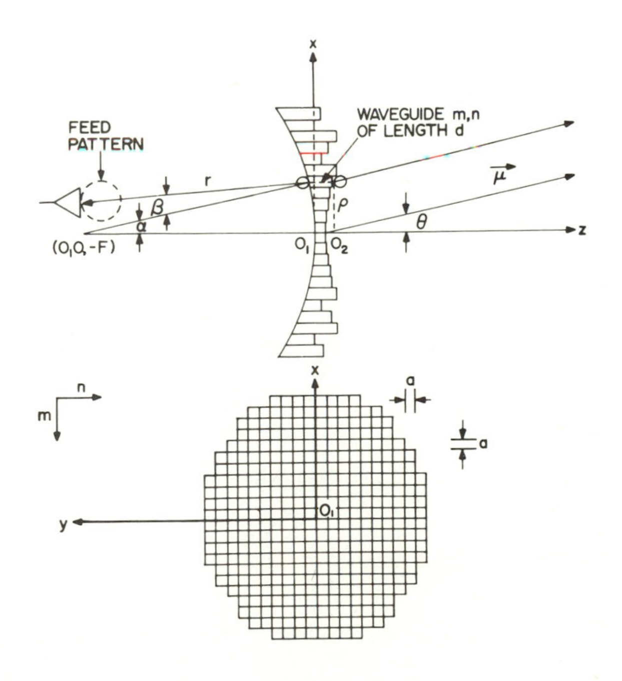

Fig. 3.150 *Waveguide lens and feed geometry for computer model*

There remains to determine the radiation characteristics of the transmit apertures. The radiation pattern of these apertures is assumed to be cos θ, and their directivity is assumed to be equal to the area gain of each aperture, i.e. $4\pi a^2/\lambda$. However, for zoned waveguide lenses, these assumptions have to be modified for waveguide elements adjacent to a step. Radiation from waveguide elements adjacent to a step is partly reflected, partly diffracted by the step, causing the radiation

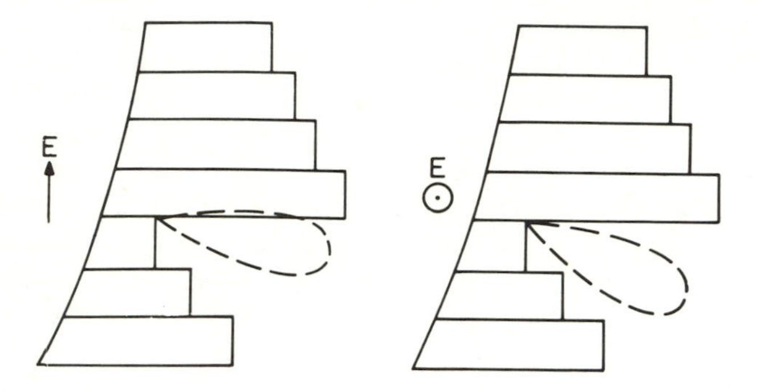

Fig. 3.151 *Effect of zoning on radiation pattern of elements adjacent to a step*

pattern to take the shapes illustrated in Fig. 3.151 for parallel and perpendicular polarisations. In the computer model the radiation pattern of these elements is taken to be the pattern obtained assuming that the elements are perfectly mirrored in the plane formed by the step. More specifically, the radiation pattern $h(\theta,\phi)$ at the exit apertures is taken to be

$$h(\theta,\phi) = \cos\theta \tag{3.153}$$

for all elements except those elements adjacent to a step where it is

$$h(\theta, \phi) = 2\sin^2 U + 2j\sin U\cos U \tag{3.154}$$

with $U = (\pi a/\lambda)\sin\theta\sin\phi$ for the case where the electric field is parallel to the surface of a step or

$$h(\theta,\phi) = 2\cos^2 V - 2j\cos V\sin V \tag{3.155}$$

with $V = (\pi a/\lambda)\sin\theta\cos\phi$ for the case where the electric field is perpendicular to the surface of a step and

$$h(\theta,\phi) = 0 \tag{3.156}$$

for directions behind a step.

Under these assumptions the directive gain of the lens antenna is

$$G(\theta,\phi) = \frac{(4\pi A/\lambda^2)G_0}{16(F/D)^2} |F(\theta,\phi)|^2 \qquad (3.157)$$

where A is the area of the lens aperture, D is the diameter, F is the lens focal length, G_0 is the directivity of the feed and

$$F(\theta,\phi) = \frac{1}{N} \sum_n \sum_m \frac{h(g \cos\beta \cos\alpha)^{1/2}}{(r/F)} e^{-jk(r + \eta d - \vec{\rho} \cdot \vec{\mu})} \qquad (3.158)$$

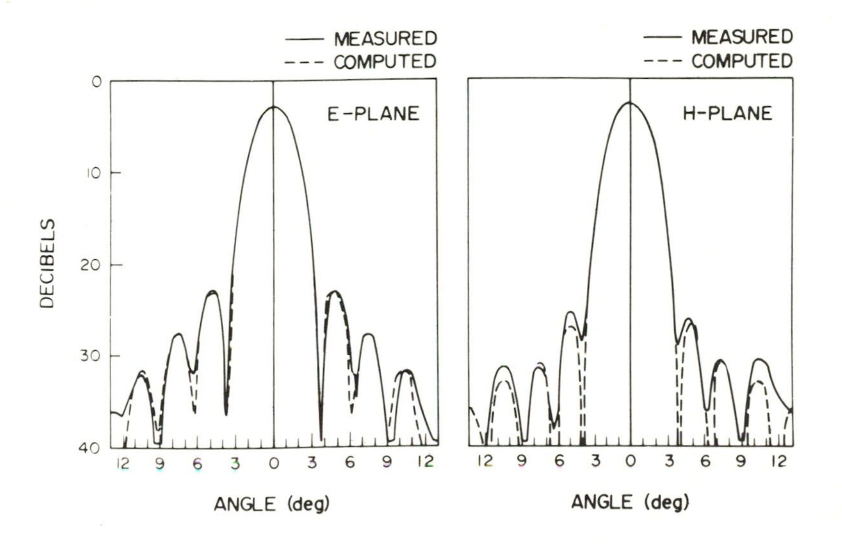

Fig. 3.152 *Measured and computed radiation patterns of waveguide lens zoned for minimum thickness*

where g is the feed directive gain (normalised to G_0) in the direction of waveguide m,n; $\vec{\rho}$ is the vector position of exit aperture m,n referred to origin 0_2; $\vec{\mu} = \sin\theta \cos\phi\, \vec{\mu}_x + \sin\theta \sin\phi\, \vec{\mu}_y + \cos\theta\, \vec{\mu}_z$ is the unit direction vector, N is the total number of elements, η is the lens refractive index and other parameters are defined in Fig. 3.150. Radiation patterns of a minimum-thickness waveguide lens computed with this model are compared to measured patterns in Fig. 3.152. The parameters of the lens designed for operation at a centre frequency of 7·8 GHz were: D = 30 in, F/D = 1·0 and a = 1 in. A 2 in-diameter circular horn excited with a TE_{11} mode illuminated the lens with a taper of $\simeq$ 6 dB in the E-plane and a taper of $\simeq$ 5 dB in the H-plane. The measured gain and pattern of the feed were used as input to the lens program. The good agreement observed in Fig. 3.152 lends support to the assumptions associated with the model. The measured and calculated directivities agreed within 0·5 dB.

Effect of zoning on directivity

Zoning of a lens has adverse effects on radiation characteristics, the principal one being a reduction of gain. This reduction was calculated for the minimum-thickness lens and also for the minimum-phase-error lens, using the computer model described above. By comparing the directivity of a zoned lens to that of an unzoned lens, the magnitude of the gain loss was obtained. The calculation was made for an illumination taper over the lens of about 6 dB, which is approximately the taper of optimally designed, multiple-beam antennas. The results are given in Fig. 3.153 as a function of F/D. The reduction of directivity for a given type of zoning is a function of F/D only and is not a function of the lens diameter even though a larger lens has more zones. This is the case because the reduction is proportional to the ratio of the number of waveguide elements adjacent to a step divided by the total number of waveguide elements in the lens, and this ratio is independent of the lens diameter but is inversely proportional to the F/D ratio. The loss of directivity due to zoning is given approximately by $L \simeq 1/(F/D)$ dB for the minimum-thickness lens and $L \simeq 2/(F/D)$ dB for the minimum-phase-error zoned lens.

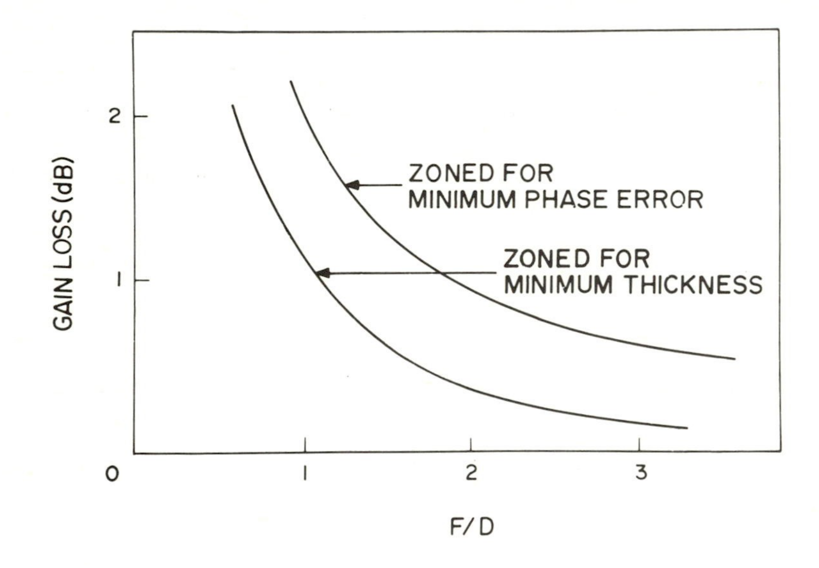

Fig. 3.153 *Loss of gain due to zoning*

Effect of size of waveguide elements

The lower limit of the waveguide size is determined by the reflection at the lens surfaces, which becomes prohibitive as the waveguide cutoff frequency is approached. For lenses made of square waveguide elements, the reflection coefficient is approximately $(1 - \eta)/(1 + \eta)$ where η is the refractive index and plotted in Fig. 3.154 together with the loss due to surface reflection, as a function of $\lambda/2a$, where a is the side dimension of the waveguide. A practical lower bound for η is 0·6 to which corresponds a transmission loss of about 0·5 dB. The higher limit of the waveguide size is determined by allowing only the fundamental mode to propagate yielding $a < 0{\cdot}707\ \lambda$ which corresponds to refractive indices $< 0{\cdot}7$. Thus the waveguide size of lenses fabricated of square waveguide elements is pretty well limited to values between $0{\cdot}6\lambda$ and $0{\cdot}7\lambda$. Additional factors bearing on the choice of waveguide size are the lens thickness which varies as $1/(1-\eta)$ and the number of waveguide elements which varies as $1 - \eta^2$.

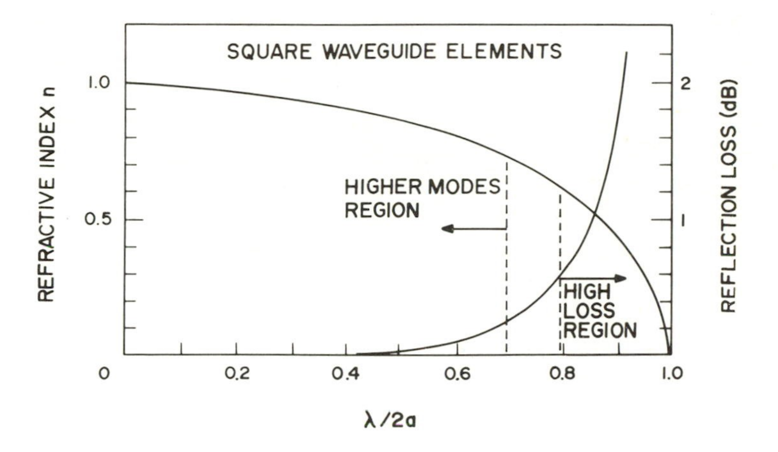

Fig. 3.154 *Refractive index and surface reflection loss of waveguide lens*

Effect of F/D ratio

As shown previously, the aperture phase error due to dispersion is smaller the larger the F/D ratio. This is also true of aperture phase error resulting from off-axis operation. In addition, the loss of gain due to zoning also decreases with increasing F/D ratios. Thus lenses with large F/D have better performance. However, to limit focal lengths to practical values lenses are generally built with F/D ratios of between 1 and 1·5.

3.8 **Low noise antennas** (Contribution by T. S. Chu)

3.8.1 Introduction

The system noise temperature of a receiving system is the sum of the receiver noise temperature and the antenna temperature. The antenna temperature is contributed[243] by the cosmic radio noise, atmospheric absorption, thermal radiation from the ground and transmission line loss. The cosmic and atmospheric noise, which come through the main beam as well as the sidelobes, are basic limitations. The downward-looking antenna on an orbiting space vehicle will also receive the ground radiation via the main beam and near-in sidelobes, whereas the pick up of the ground radiation for a radio telescope or a space communication earth station can be greatly reduced by suppressing the back lobes. The transmission-line loss can be minimised by cooled waveguide or Gaussian-beam wave transmission. Section 3.8.2 will describe each thermal noise component and the G/T parameter of a receiving system. Section 3.8.3 will discuss two types of earth station antennas: the horn reflector, which has the lowest antenna noise, and the symmetrical Cassegrain, which has been the most popular type of large reflector antenna.

A communication system was usually limited by the thermal noise until the recent past. However, interference noise has increasingly become the dominating limit for both satellite and terrestrial communication systems. Here priority should be given to the minimisation of sidelobes in the directions of interference. Section 3.8.4 will interpret the CCIR minimum pattern requirement and address the interference problem by explaining three methods: offset configuration, spatial filtering and aperture shaping.

Substantial reduction of thermal noise can be achieved by beam-switching techniques in radio astronomy and radiometric measurements. Adaptive cancellation of interference by an auxiliary antenna can be useful in some interference problems. These cancellation techniques will be the topic of the final subsection.

3.8.2 Thermal noise components

Receiver noise

The advance in receiver technology has led to decreasing noise contributions from the receiver. The current state-of-the-art performance of receiver noise is given below to provide a background for the discussion of noise sources associated with antennas. At S band[244], a maser equivalent input temperature of 2·1K was achieved to give the 8·4K total zenith noise of a horn receiving system. At K band[245] the noise temperatures 13K and 30K were obtained, respectively, for a maser and a complete horn receiving system. When the receiver is mounted on a symmetrical Cassegrain antenna (see Section 3.8.3), the system noise[246] is typically a factor of 1·5 to 2 times that of a horn receiving system.

At higher frequencies, a cryogenic (18K) millimetre-wave receiver[247] operating on an offset Cassegrainian antenna has shown measured total single-sideband system temperature of 310-480K in the 60-90 GHz band. The minimum of 310K occurs at 81 GHz where the mixer temperature is 197K. A noteworthy advance in millimetre wave mixers is the subharmonic pump.[248]

Transmission-line loss

It is often necessary to feed the antenna with a substantial length of transmission line. The effect of transmission-line loss will be considered here in some detail because this also serves as a simplified model for the atmospheric noise. Let us take a semi-infinite single-mode transmission line with a power absorption coefficient α, at ambient temperature T_0 = 290K. Since the noise appearing from the direction of infinity at both end points of a section ΔX is expected to be kT_0B, the attenuation by the transmission line section ΔX must have been exactly compensated by the contribution ΔT from ΔX:

$$kT_0Be^{-a\Delta X} + k\Delta TB = kT_0B \tag{3.159}$$

$$\Delta T \approx \alpha T_0 \Delta X. \tag{3.160}$$

Integrating over a line of length d yields the effective noise temperature at $X = 0$:

$$T_l = \int_0^d \alpha T_0 \, e^{-\int_0^x a dx} dx \tag{3.161}$$

If the transmission line loss is only 0·1 dB (αd = 0·025), $T_l \approx T_0\alpha d$ = 7·3°K.

The physical temperature T_0 of a waveguide can be lowered by refrigeration to reduce the effect of transmission-line loss. Gaussian-beam wave transmission[249] by a sequence of offset reflectors[250] has been found to be a very low-loss method of connecting widely separated antenna feed and receiver.

Any impedance mismatch between antenna and receiver will add to the system noise temperature by reflecting part of the receiver noise. Furthermore, multiple reflection between the antenna and the receiver via a long transmission path can cause base-line ripple in radio-astronomy measurements and introduce intermodulation noise in radio communication systems.

Cosmic noise

To minimise the extraterrestrial noise, the antenna beam should avoid pointing near discrete radio sources such as the sun, moon and centre of the galaxy. The galactic noise diminishes as the frequency increases toward the centimetre-wavelength region. However, a more basic limitation is an isotropic background of about 3·5K.[251] This noise has been identified with the origin of the universe and found to be independent of beam direction. Thus, it is a constant term in calculating the sky temperature of all microwave antennas.

Atmospheric noise

Since oxygen and water vapour in the atmosphere exhibit absorption loss at microwave frequencies, the atmospheric noise seen by the antenna beam at an elevation angle θ can be obtained in a similar manner as a lossy transmission line.

$$T_a(\lambda, \theta) = \int_0^\infty \alpha(\lambda\, r)T(r) \exp \left| - \int_0^r \alpha(\lambda\, r)dr \right| dr \qquad (3.162)$$

where T is the temperature of the air and α is the absorption coefficient. Using an atmosphere model for the typical clear conditions at midlatitude, the noise temperature contributed by oxygen and water vapour can be calculated[252] for various elevation angles as shown in Fig. 3.155. The measured points in the zenith direction (about 2·5K at 4 GHz) have shown good agreement with the calculated curves.

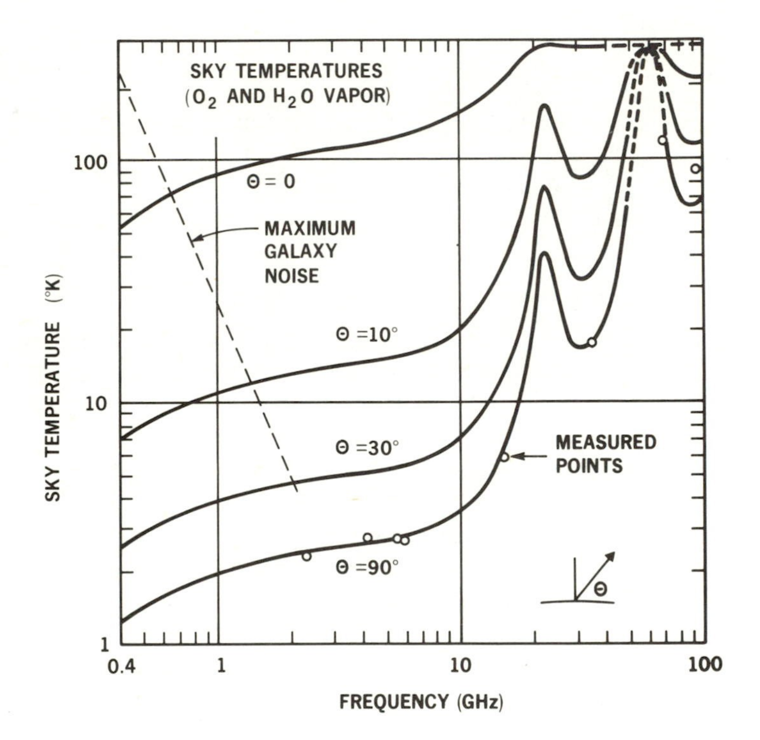

Fig. 3.155 *Maximum galaxy noise and calculated atmospheric sky noise for various elevation angles as a function of signal frequency. Measured points agree with calculated values*

Very little frequency dependence is found at lower microwave frequencies until about 8 GHz, beyond which the temperature increases appreciably with the frequency. Since absorption in the clear atmosphere is small enough up to about 16 GHz, this sky temperature component for directions other than those indicated in Fig. 3.155 can be obtained from zenith values by $T(\theta) = T_z \csc \theta$, valid down to $\theta=5°$. The measured sky noise will vary with water-vapour content and will increase sharply in the presence of precipitation. Microwave attenuation by rain and clouds plays a major role in the design of satellite communications, especially at frequencies above 10 GHz.[253] If a satellite communication system has a typical rain fading margin of 10 dB, the noise temperature due to rain will be about 260K.

Noise via the sidelobes

The total antenna noise can be obtained by integrating the complete radiation pattern over the brightness temperature distribution of the environment

$$T_A = \int G(\psi,\phi)T(\psi,\phi)d\,\Omega\,. \tag{3.163}$$

The cosmic and atmospheric noise contributions received by the main beam have been discussed earlier. A substantial part of the noise will be contributed by the sidelobes, especially the back lobes. For a zenith-pointed antenna, the sidelobes in the upper hemisphere will see the atmospheric noise in the relatively cold sky. These wide-angle sidelobes will give rise to a total (which is insensitive to frequency below 8 GHz) contribution of about 1K. The back lobes may receive radiation from the absorbing ground. If the average level of the back lobes is 10 dB below the isotropic level (G_b = 0·1) the warm earth (T_g = 300K) below the antenna may contribute 15K. This noise from the back lobes will be reduced to 8K and 1K, respectively, if the site surroundings are water and perfect reflector.[243]

The sidelobes of a reflector antenna can be classified into four contributors: (*a*) edge diffraction, (*b*) feed spillover, (*c*) scattering by aperture blocking, (*d*) surface tolerance. Both edge diffraction and feed spillover can be reduced by stronger illumination taper at the expense of moderate gain reduction. Edge diffraction sidelobes, which decay rapidly away from the main beam, contribute relatively little to the antenna temperature provided that the antenna pointing is well above horizon. Although shallow reflectors can be more conveniently illuminated by a prime-focus feed, are more easily constructed, and produce less depolarisation, a deep reflector more effectively shields the feed from the environment, thereby reducing the noise temperature due to spillover. The effect of spillover and aperture blocking in axisymmetrical Cassegrainian configuration is discussed in Section 3.2.2. In practice, surface roughness of small correlation length is often an important contributor to wide-angle sidelobes. An RMS surface tolerance of better than 0·01λ is needed to approximately realise the calculated pattern assuming perfect surface.[254]

G/T parameter

A convenient figure of merit for the sensitivity of a thermal noise limited receiving earth station is the value of *G/T*, which is simply the ratio of the antenna gain to the system noise temperature expressed in degrees Kelvin. In a typical 4/6 GHz satellite communication earth station with a Cassegrainian antenna, the system noise temperature is about 80K, where the low noise amplifier accounts for 20K, the feed line loss gives 30K, and the external antenna noise (at 20° elevation angle) is 30K. If the antenna gain is 61 dB, then the G/T value is about 42 dB.

The measured G/T is obtained [255] from the ratio of output noise powers when the antenna is pointed at a radio star such as Cassiopeia A, and then pointed to the nearby cold sky. The accuracy of the G/T measurement depends strongly on the flux calibration of radio stars.[256,257] An estimate of ±5% is given as the error limit for the flux of CAS A which is also known to be decreasing with time at a rate of about 1·1% per year.

3.8.3 Earth station antennas for space communication

Horn reflector antenna

The horn-reflector antenna generates the lowest antenna noise due to sidelobes among existing space communication earth station antennas. This antenna is the combination of an offset paraboloid and a horn which extends all the way from the reflector focus to the aperture. The horn proper can be either pyramidal or conical. The pyramidal horn-reflector antenna of 6-m aperture[258] shown in Fig. 3.156 was used in the Echo and Telstar™ Satellite project. Later, its extremely low sidelobe noise (< 0·2K at zenith) led to the discovery of 3·5K isotropic background noise.[251]

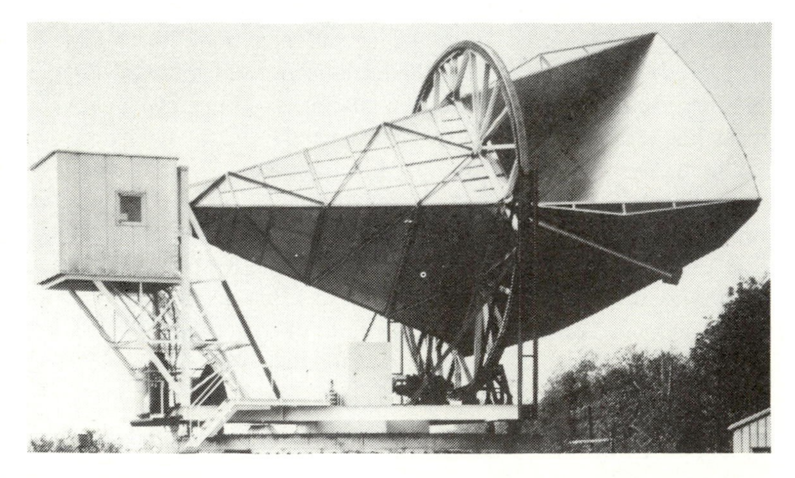

Fig. 3.156 *A pyramidal horn reflector antenna of 6 meter aperture. The reflector is constructed from a segment of generating paraboloid*

The offset configuration is completely free from aperture blocking, and the shielding horn reduces the feed spillover to negligible values. Therefore, the horn reflector has extremely low back-lobes of less than −30 dB with respect to the isotropic level. The absence of aperture blocking also brings an excellent impedance matching of −40 dB return loss. The horn reflector is also capable of very broad band feeding and very low ohmic loss. Antenna noise of a horn reflector due to sidelobe and dissipative loss can be easily kept below 1K.

Wind loading and design economy sometimes lead to a radome surrounding a large reflector antenna. No radome is employed in Fig. 3.156 where the design and the modest size permit accurate pointing in high winds. However, a radome noise of about 10 K is encountered at 4 GHz in the case of the AT&T/Comsat 20m conical horn reflector antenna at Andover, Maine.[243]

Symmetrical Cassegrain

A symmetrical Cassegrain antenna has been, so far, the most popular space-communication earth-station antenna. Excessive length of transmission line is often a major drawback for the prime-focus fed reflector. The Cassegrainian (or Gregorian) configuration offers the convenience of locating equipment near the vertex and eliminates the long feeding transmission line. For a given aperture size the symmetrical Cassegrain configuration is more economical and has less wind loading than a horn reflector. A radome has been avoided for the NASA/JPL 64m Cassegrainian antenna as shown in Fig. 3.157 at Goldstone, California.[246] The improved

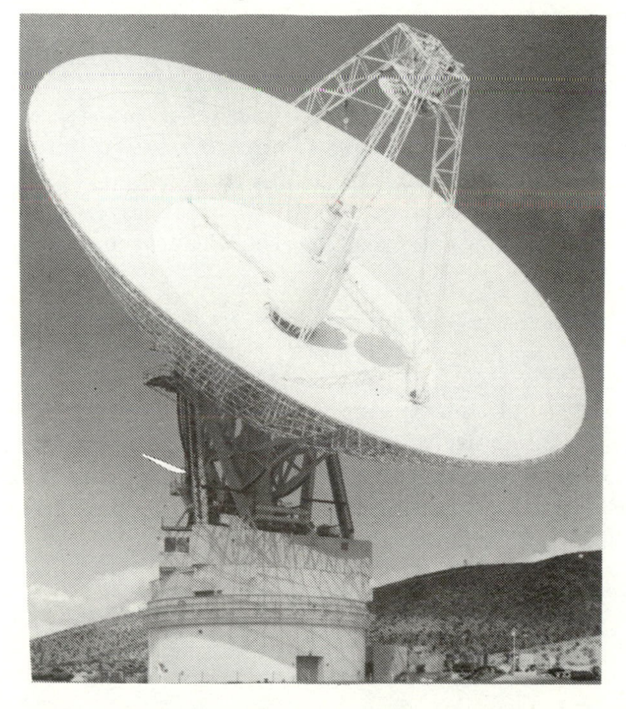

Fig. 3.157 *The 64m Cassegrainian antenna at Goldstone, California. A slightly asymmetrical hyperbolic subreflector is capable of rotation around the axis of the main reflector to be aligned with several feeds*

feeds using dual mode horn or corrugated horn have circularly symmetrical patterns with very low sidelobes, and thus greatly reduce the feed spill-over problem. The feed illumination at the edge of the subreflector is usually in the range of 10 to 15 dB down from the feed pattern maximum; however, the spill-over in the vicinity of the subreflector looks toward the cold sky for an antenna pointing at high elevation angle and contributes little to the noise temperature. The feed illumination reflected back from the subreflector toward the main reflector will have a weaker edge illumination at the main reflector than geometrical optics ray tracing, thereby reducing the spillover toward the ground. The wide-angle feed spillover is reduced in a near field Cassegrain configuration[259] which employs a large feed aperture to keep the feed illumination collimated in reaching the subreflector.

The only major shortcoming of a symmetrical Cassegrain antenna is the aperture blocking due to the subreflector and its supporting struts. The latter scatters not only the plane wave from the main reflector, but also the spherical wave from the subreflector. To minimise the blocking of the high-density spherical wave, the bases of the struts should be located well away from the centre of the main reflector. The pick-up of ground radiation by an axisymmetric Cassegrain configuration can be estimated as follows.[260] The scattering from the subreflector and its support struts gives 3–10K (depending on strut structure); the diffraction from the edge of the subreflector going beyond the main reflector, picks up about 1K (depending on size and edge taper). At low elevation angles, the feed spill-over beyond the subreflector may go into the ground, contributing up to 4K (depending on feed taper). The total ground radiation may contribute 4 to 15K.

The aperture blocking can be completely eliminated in an offset configuration. However, it is not necessarily always cost effective to reduce thermal noise by the offset geometry. The main advantages of an offset dual reflector antenna are lower sidelobes for interference reduction, which will be discussed in the following Section, as well as perfect impedance matching and wide scanning capability.

3.8.4 Low sidelobes for interference reduction

CCIR reference pattern

In response to the need of a mathematical model for interference calculations, the international radio consultative committee (CCIR) of the International Telecommunication Union made a survey of measured antenna patterns of existing communication satellite earth stations in the late sixties. A proposal based upon this survey is the CCIR pattern with respect to the isotropic level[261]

$$G = \begin{cases} (32 - 25 \log \theta) \text{ dB} & \text{for } \theta < 48^\circ \\ -10 \text{ dB} & \text{for } \theta > 48^\circ \end{cases} \tag{3.164}$$

This pattern has become the minimum specification for an earth-station antenna. One notes that the asymptotic decay for the sidelobe envelope of the radiation pattern of a circular aperture should be θ^{-3}. The slower decay rate of $\theta^{-2.5}$ in the CCIR pattern reflects the degrading effect of feed spill-over, surface tolerance and aperture blocking in a practical reflector antenna. The improvements of feed design during the last decade led to a lower sidelobe envelope than the CCIR requirement for earth-station antenna. The trend toward smaller earth-station antennas in forthcoming satellite communication systems will inevitably impose more stringent requirements on the sidelobe envelope in order to keep interference noise under control.

Fig. 3.158 *The Crawford Hill 7m offset Cassegrainian millimetre wave antenna. The subreflector is mounted below antenna aperture, and key structural parts are covered with insulation*

Methods of increasing immunity to interference

Owing to the rapid growth of communication traffic, receiving systems have now become increasingly limited by interference noise instead of thermal noise. Proliferation of synchronous communication satellites will narrow down the orbital spacing toward perhaps one degree. (The current spacings are 4° for 4/6 GHz satellite and 3° for 12/14 GHz satellite.) This prospect implies much more stringent sidelobe requirements than the CCIR reference pattern, and thus motivates the offset dual reflector antenna. The elimination of aperture blocking leaves edge diffraction and surface tolerance as the only contributors to the sidelobes within a few degrees of the main beam. The cross-polarisation of the offset reflector can be overcome by the cancellation between those of the main reflector and the sub-reflector. The feasibility of a large offset Cassegrainian antenna has been demonstrated by the Crawford Hill 7m millimetre-wave antenna[262] as shown in Fig. 3.158. Measured low sidelobe level ($\lesssim -40$ dB) at one degree off the main beam and measured low cross-polarisation ($\lesssim -40$ dB) throughout the main beam are achieved using a quasi-optical 19/28·5 GHz feed system that also demonstrates very low multiplexing loss (~ 0·1 dB).

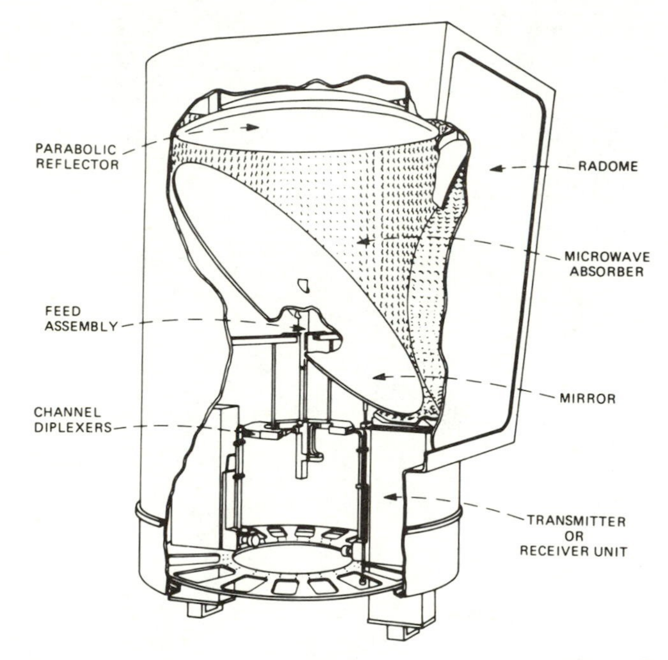

Fig. 3.159 *Schematic depiction of a shielded inverted-periscope antenna. The parabolic reflector is imaged by the 45° plane reflector into an effective location behind the plane reflector*

A fence around an earth station may reduce terrestrial interference by shielding the antenna from direct interference rays. A short cylindrical shield is often built around the edge of the main reflector to suppress the back-lobe radiation. Another remarkable demonstration of this spatial filtering concept is the very long equivalent shield provided by the inverted periscope configuration[263,264] in Fig. 3.159, where the parabolic reflector is imaged by the 45° plane reflector into an effective position located behind the aperture by a length of more than the aperture diameter. Since the feed becomes invisible for geometric optical rays beyond a certain angle (~ 40° in Fig. 3.159 from the boresight, a sharp drop of sidelobe level was observed at this angular boundary. This property is very desirable for microwave repeater antennas at a metropolitan junction where good discrimination against interfering stations is a necessity.

Sources of interference noise for a microwave repeater receiver or earth-station receiver are generally confined to the horizontal plane. Antenna design can take advantage of this fact to minimise the transmitting or receiving sidelobes in the azimuthal plane at the expense of those in other directions. For example, when a pyramidal horn reflector with vertical horn axis is used as a microwave repeater antenna, the corner diffraction of its forward-tilt aperture of trapezoidal shape gives a general asymptotic power decay of θ^{-4} for the sidelobe level in the azimuthal plane pattern. However, an extra remedy such as a multiple-edge blinder[265] is needed for certain horizontal directions which are perpendicular to the aperture edges and have an asymptotic power decay of only θ^{-2}. A diamond-shaped aperture[266] was also proposed for the horn reflector to obtain lower sidelobes in the horizontal plane.

3.8.5 Noise cancellation techniques

In radio-astronomy and radiometry measurements, effective cancellation of atmospheric noise can be achieved by a beam switching technique. Two neighbouring feeds of a large reflector antenna radiate two distinct far-zone beams which intercept essentially the same region of atmosphere in the Fresnel zone. An on-axis beam receives both the desired signal and the undesired atmospheric noise originating in the Fresnel zone, whereas the off-axis beam points away from the source,

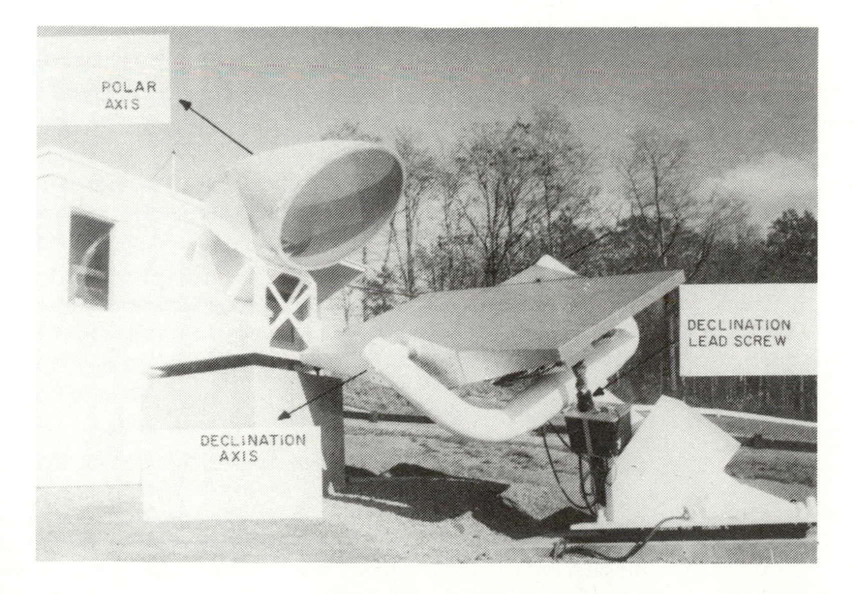

Fig. 3.160 *A microwave suntracker using a polar-mounted flat reflector that nods about the declination axis for beam switching*

but receives essentially the same atmospheric noise. A differential comparison of power in the two beams on a time-average basis has been found effective in suppressing the atmospheric noise.[267]

As an alternative to two independent feeds, two modes TE_{11} amd TM_{01}, in a conical horn reflector with a mode selective coupler have been suggested[268] to provide the two beams. To preserve the broad bandwidth of the horn reflector in the beam switching scheme, a nodding flat reflector can be used as shown in the microwave suntracker[269] of Fig. 3.160. This suntracker was used to measure the rain attenuation simultaneously at 16 and 30 GHz. The flat reflector moves about the polar axis with time of day, reflecting the solar radiation into the fixed antenna; but it is also mounted on bearings on the orthogonal axis about which it oscillates with an angular amplitude of 2.5° at a rate of 1 Hz. Nodding the subreflector of a Cassegrainian antenna is another method of mechanically switching the beam.[270] An alternate technique is used in a quasi-optical millimeter-wave feed system[271] for an offset Cassegrainian configuration, where two feed beams are alternately transmitted and reflected by a mechanical chopper.

The interference problem of earth stations is often caused by a particular neighbouring microwave repeater. Sometimes this interfering signal can be cancelled by the output from an auxiliary antenna.[272] It is difficult for the auxiliary antenna output to match the frequency characteristic of an interfering signal which comes from a direction many beamwidths away from the main beam. If the interference comes from a direction close to the main beam, the interfering signal will not be very frequency sensitive, and thus the cancellation scheme can be easily implemented for a useful bandwidth.

Deep nulls at arbitrary angular locations in the radiation pattern of a phased array can be formed by adaptive control of phase and/or amplitude. This technique can overcome interference from a jamming source and will be treated in Chapter 13, Volume II.

3.9 References

1 SILVER, S. (ed): Microwave antenna theory and design'. McGraw-Hill, 1949, Chap. 12

2 HANSEN, R.C. (ed): 'Microwave scanning antennas', Academic Press, 1964, pp. 64-79

3 RUSCH, W.V.T., and POTTER: 'Analysis of reflector antennas'. Academic Press, 1970

4 SKOLNIK, M.I.: 'Radar handbook', McGraw Hill, 1970, Chap. 9

5 RUSCH, W.V.T.: 'Antenna notes', Vol. II, NB846, Technical University of Denmark, Aug. 1974, pp. 1-83

6 RUZE, J.: 'The effect of aperture errors on the antenna radiation pattern, Suppl. of *Nuovo Cimento*, **9**, No. 3, 1952, pp. 364-380

7 RUZE, J.: 'Antenna tolerance theory – A review', *Proc. IEEE,* **54**, No.4, April 1966, pp. 633-640

8 SCHANDA, E.: 'The effect of random amplitude and phase errors of continuous apertures', *IEEE Trans.* **AP-15**, May 1976, pp. 471-473

9 INGERSON, P.G. and RUSCH, W.V.T.: 'Radiation from a paraboloid with an axially defocused feed', *IEEE Trans.* **AP-21**, Jan. 1973, pp. 104-106

10 RUZE, J.: 'Lateral-feed displacement in a paraboloid', *IEEE Trans.,* **AP-13,** Sept. 1965, pp. 660-665

11 RUSCH, W.V.T.: 'Analysis of paraboloidal-reflector systems', Chapter 1.3 in Methods of experimental physics', Vol.12, Astrophysics, Part B, 1976, Academic Press

12 RUSCH, W.V.T. and LUDWIG, A.C.: 'Determination of the maximum scan-gain contours of a beam-scanning paraboloid and their relation to the Petzval surface', *IEEE Trans.,* **AP-21,** pp. 141-148, March 1973

13 RUSCH, W.V.T., and POTTER, P.D.: *Op. Cit.,* p.96
radioteleskops effelsberg', *Archiv für Elektronik und Ubertragungstechnik,* 30, 1976, pp.

14 GRAY, C.L.: 'Estimating the effects of feed support member blocking on antenna gain and sidelobe level', *Microwave J.*, March 1964, pp. 88-91

15 WESTED, J.H.: 'Shadow and diffraction effects of spars in a Cassegranian system', P2118, Microwave Laboratory, Danish Academy of Technical Sciences, Copenhagen, March 1966

16 RUDGE, A.W. and SHIRAZI, M.: 'Investigation of reflector-antenna radiation', Interim Report, ESRO/ESTEC Contract No. SC/11/73/HQ, 10 October 1973

17 DIJK, J., JEUKEN, M. and MAANDERS, E.J. : 'Blocking and diffracting in Cassegrain antenna systems, *De Ingenieur,* No.27, Technisch Wetenschappelijk Onterzoek 7, 5 July 1968, pp. 79-91

18 RUSCH, W.V.T. and POTTER, P.D.: 'Academic Press, 1971, Chap. 3

19 KOEHLER, J.: 'Das fernfeld einer verblockten kreisapertur am beispiel des 100-m-radioteleskops effelsberg', *Archiv für Elektronik und Ubertragungstechnik,* 30, 1975, pp. 409-412

20 KAY, A.F.: 'Electrical design of metal space frame radomes', *IEEE Trans.,* **AP-13,** No. 2, March 1965, pp. 188-202

21 RUZE, J.: 'Feed support blockage loss in paraboloic antennas', *Microwave J.,* **11,** No. 12, 1968, pp.76-80

22 RUSCH, W.V.T.: 'Application of two-dimensional integral equation theory to reflector antenna analysis', TM 33-478, Jet Propulsion Laboratory, Pasadena, California, 15 May 1971

23 RUSCH, W.V.T.: 'Analysis of blockage effects on gain and sidelobes by struts of arbitrary cross-section in a dual-reflector antenna', 1971 G-AP International Symposium Digest, UCLA, Los Angeles, Sept. 22-24, 1971, p. 211

24 RUSCH W.V.T. and SØRENSEN, O.: 'Aperture blocking of a focused paraboloid', R126, Technical University of Denmark, Lyngby, Denmark, 31 July 1974

25 RUSCH W.V.T. and SØRENSEN, O.: 'Strut blocking of a focused paraboloid', 1975 G-AP International Symposium Digest, Urban, Illinois, 2-4 June 1975, pp. 109-112

26 RUSCH, W.V.T., APPEL-HANSEN, J., KLEIN, C.A. and MITTRA, R.: 'Forward scattering from square cylinders in the resonance region with application to aperture blockage, *IEEE Trans.* **AP-24,** No. 2, March 1976, pp. 182-189

27 LEE, S.H., RUDDUCK, R.C., KLEIN, C.A. and KOUYOUMJIAN, R.G.: 'A GTD analysis of the circular reflector antenna including feed and strut scatter. T.R. 4381-1, Electrical Engineering Department, The Ohio State University, Columbus, Ohio 43210, 25 May 1977

28 LEE, S.H., RUDDUCK, R.C., BURNSIDE, W.D. and WONG, N.: 1979 G-AP International Symposium Digest, Seattle, Washington, 18-22 June 1979, pp. 63-66

29 HARTSUIJKER, A.P., BAARS, J.W.M., DRENTH, S. and GELATO-VOLDERS, L.: 'Interferometric measurement at 1415 MHz of radiation pattern of paraboloidal antenna at Dwingeloo Radio Observatory', *IEE Trans.* **AP-20,** No. 2, March 1972, pp.166-176

30 SHEFTMAN, F.I.: 'Experimental study of subreflector support structures in a Cassegrainian antenna'. MIT, Lincoln, Massachusetss, Rpt. 416, 23 Sept. 1966

31 HANNAN, P.W.: 'Microwave antennas derived from the Cassegrain telescope', *IRE Trans.* **PGAP-9,** March 1961, pp. 140-153

32 JENSEN, P.A.: 'Designing Cassegrain antennas', *Microwaves*, Dec. 1962, pp. 12-19
33 BICKEL, S.H.: 'Cassegrainian error study'. Hughes Aircraft Co., Internal Correspondence, 12 Sept. 1960
34 ISBER, A.W.: 'Obtaining beam pointing accuracy with Cassegrain antennas', *Microwaves,* Aug. 1967, pp. 40-44
35 SILVER, S.: 'Microwave antenna theory and design', MIT, Rad. Lab Series, Volume 17, McGraw-Hill, New York, 1949
36 POTTER, P.D.: 'The aperture efficiency of large paraboloid antennas as a function of their feed system radiation characteristics', JPL Technical Report No. 32-49, 25 Sept. 1961
37 DELL-IMAGINE, R.A.: 'Calculation of cross-polarisation loss of paraboloidal reflector', Hughes Aircraft Co., Internal Correspondence IDC 72/1422.00-9, 21 Feb. 1972
38 DELL-Imagine, R.A.: 'Derivation of cross-polarisation loss for a multimode conical horn', Hughes Aircraft Co., Internal Correspondence IDC 70/1422.00-80, 12 March 1970
39 BRACEWELL, R.N.: 'Tolerance theory of large antennas', *IRE Trans.,* **PGAP-9,** No. 1. 1961, pp. 1588-1594
40 MEI, K.K. and VAN BLADEL, J.G.: 'Scattering by perfectly conducting rectangular cylinders', *IEEE Trans.,* **AP-11,** March 1963, pp. 185-192
41 SCIAMBI, A.F.: 'Effect of aperture illumination on circular aperture antenna pattern characteristics', *Microwave J,* **8,** No. 8, August 1965
42 DELL-IMAGINE, R.A.: 'Effect of horn phase error on patterns and efficiency', Hughes Aircraft Co., Internal Correspondence IDC 71/1422.00-16, 30 March 1971
43 CROMPTON, J.W.: 'On the optimum illumination taper for the objective of a Microwave aerial', *Proc. IEE,* **101**, Part III, 1954, pp. 371-382
44 HANNAN, P.W.: 'Optimum feeds for three modes of a monopulse antenna', *IRE Trans.,* **PGAP-9,** Sept. 1961, p. 458
45 DELL-IMAGINE, R.A.: 'Efficiency calculations for conical horns', Hughes Aircraft Co., Internal Correspondence, 18 Sept. 1972
46 DOIDGE, F.G.: 'Antenna gain as it applies to satellite communication earth stations', United States Seminar on Communications Satellite Earth Station Technology, Washington, D.C., 16-27 May 1966, p. 55
47 WONG, W.C.: 'On the equivalent parabola technique to predict the performance characteristics of a Cassegrainian system with an offset feed', *IEEE Trans.,* **AP-21,** No. 3 May 1973, pp. 335-339
48 BYARS, M.: 'An approximate method of evaluating the effects of beam steering by subreflector tilting in a Cassegrain system, earth station technology', IEEE Conference Publ. 72, 1970, pp. 144-149
49 SORENSEN, O. and RUSCH, W.V.T.: 'Application of the geometrical theory of diffraction to Cassegrainian subreflectors with laterally defocused feeds', *IEEE Trans.* **AP-23,** Sept. 1975, pp. 698-702
50 MENTZER, C.A. and PETERS, L., Jr.: 'A GTD analysis of the far-out sidelobes of Cassegrain antennas', *IEEE Trans.,* **AP-23,** Sept. 1975, pp. 702-709
51 RUSCH, W.V.T.: 'Phase error and associated cross-polarization effects in Cassegrainian-fed microwave antennas', Tech. Report 32-610 JPL, CIT, Pasadena, California, 30 May, 1964
52 RUSCH, W.V.T.: 'Scattering from a hyperbola reflector in a Cassegrain feed system', *IEEE Trans.,* **AP-11,** No. 4, pp. 414-421
53 POTTER, P.D.: 'A simple beamshaping device for Cassegrainian antennas', JPL Technical Report 32-214, Jet Propulsion Laboratory, Pasadena, California, 31 Jan. 1967
54 CRAMER, P.W.: 'Large spacecraft antennas: Performance comparison of focal point and Cassegrainian antennas for spacecraft applications', JPL Space Programs Summary 37-62, Volume III, pp. 80-87
55 GALINDO, V.: 'Design of dual-reflector antennas with arbitrary phase and amplitude distribution', *IEEE Trans.,* **AP-12,** July 1964, pp. 403-408

56 WILLIAMS, W.F.: 'High efficiency antenna reflector', *Microwave J.*, July 1965, pp. 79-82
57 WOOD, P.J.: 'Reflector profiles for the pencil-beam Cassegrain antenna', *Marconi Rev.*, 2nd Quarter, 1972, pp. 121-138
58 HANNAN, P.W.: 'Design for a twist reflector having wideband and wide-angle performance'. Armed Services Technical Information Agency, Dayton. Ohio, Report AD 306000, 1955
59 JOSEFSSON, L.G.: 'A broad-band twist reflector', *IEEE Trans.*, **AP-19**, July 1971, pp. 552-554
60 SCHENNUM, G.H.:'Design of frequency selective surfaces using a waveguide simulator technique', IEEE International Antenna and Propagation Symposium, University of Illinois, Urbana, June 1975
61 STONIER, R.A.: 'Development of a low expansion, composite antenna subreflector with a frequency selective surface', 9th National SAMPE Technical Conference, Atlanta, Georgia, Volume 9, October 1977
62 BREJCHA, A.G. and SMITH, C.A.: 'Telemetry antennas for deep space probes', International Telemetry Conference, Los Angeles, October 1977
63 COOK, J.S., ELAM, E.M., *et al.:* 'The open Cassegrain antenna', *Bell Telephone Syst. Tech.Publ.,* Monograph 5051, 1965
64 COOK, J.S., ELAM, E.M. and ZUCKER, H.: 'The open Cassegrain antenna – Part 1: Electromagnetic design and analysis', *Bell Syst. Tech. J.*, **44**, 1965, pp. 1255-1300
65 CHU, T.S. and TURRIN, R.H., 'Depolarisation properties of offset reflector antennas', *IEEE Trans.*, **AP-21**, May 1973, pp.339-345
66 RUDGE, A.W. and SHIRAZI, M.: 'Multiple beam antennas: Offset reflectors with offset feeds', Univ. Birmingham, England, July 1973, Final Report ESA Contract 1725/ 72PP
67 RUDGE, A.W., PRATT, T. and SHIRAZI, M.: 'Radiation-fields from offset reflector antennas', in *Proc. European Microwave Conf.* (Brussels, Belgium), 1973, vol. C3.4, pp. 1-4
68 RUDGE, A.W.: 'Multiple-beam offset reflector antennas for spacecraft', in *Proc. Inst. Elec. Eng. Int. Conf. on Antennas for Aircraft and Spacecraft* (London, England), 1975, Conf. Publ. 128, pp. 136-141
69 –, 'Multiple-beam antennas: Offset reflectors with offset feeds', *IEEE Trans.,* **AP-23**, May 1975, pp. 317-322
70 DIJK, J., VAN DIEPENBEEK, C.T.W., *et al.,* 'The polarisation losses of offset antennas', Eindhoven Univ. Tech., The Netherlands, June 1973, TH Rep 73-E-39
71 –, 'The polarisation losses of offset paraboloid antennas', IEEE Trans. **AP-22**, July 1974, pp. 513-520
72 GANS, M.J. and SEMPLAK, R.A., 'Some far-field studies of an off-set launcher', *Bell Syst. Tech. J.,* **54**, 1975, pp. 1319-1340
73 GRAHAM, R.: 'The polarisation characteristics of offset Cassegrain aerials', in *Proc. Inst. Elec. Eng. Int. Conf. on Radar Present and Future,* 1973, Conf. Publ. 105, pp.23-25
74 ADATIA, N.A.: 'Cross-polarisation of reflector antennas', Ph.D. dissertation, Univ. Surrey, England, Dec. 1974
75 TANAKA, H. and MIZUSAWA, M. 'Elimination of cross-polarisation in offset dual reflector antennas', *Elec. Commun.* (Japan), **58**, pp. 71-78, 1975
76 RUDGE, A.W. and ADATIA, N.A.: 'New class of primary-feed antennas for use with offset parabolic-reflector antennas', *Electron. Lett.,* **11**, 1975, pp. 597-599
77 –, 'Matched-feed for offset parabolic reflector antennas', in *Proc. 6th European Microwave Conf.* (Rome, Italy), Sept. 1976, pp. 143-147
78 JACOBSEN, J.: 'On the cross polarisation of asymmetric reflector antennas for satellite applications', *IEEE Trans.* **AP-25**, Mar. 1977, pp. 276-283,
79 DRAGONE, C. and HOGG, D.C.: 'The radiation pattern and impedance for offset and

symmetrical near-field Cassegrainian and Gregorian antennas', *IEEE Trans.*, **AP-22**, May 1974, pp.472-475

80 COLEMAN, H.P. *et al.*: 'Paraboloidal reflector offset feed with a corrugated conical horn', *IEEE Trans.*, **AP-23**, Nov. 1975, pp. 817-819

81 RUDGE, A.W., FOSTER, P.R. *et al.*: 'Study of the performance and limitations of multiple-beam antennas', ERA (RF Technology Centre, England), Sept. 1975, Rep. ESA Con. 2277/74HP

82 RUDGE, A.W. and WILLIAMS, N.: 'Offset reflector spacecraft antennas: Design and evaluation at 30 GHz,' in *Proc. Symp. on Advanced Satellite Communications Systems* (Genoa, Italy), Dec. 1977, ESA Publ. SP-138, pp.105-113

83 INGERSON, P.G.: 'Off-axis scan characteristics of offset fed parabolic reflectors', in *IEEE Int. Symp. Digest AP-S* (Urbana, IL), June 1975, pp.382-383

84 KAUFMANN, J.F. and CROSWELL, W.F.: 'Off-focus characteristics of the offset fed parabola', in *IEEE Int. Symp. Digest AP-S* (Urbana, IL), June 1975, pp.358-361

85 OHM, E.A.: 'A proposed multiple-beam microwave antenna for earth stations and satellites', *Bell Syst. Tech. J.*, **53**, Oct. 1974, pp. 1657-1665

86 WOOD, P. and BOSWELL A.G.P. *et al.*: 'Elliptical beam antenna for satellite applications', in *Proc. Inst. Elec. Eng. Int. Conf. on Antennas for Aircraft and Spacecraft* (London, England), June 1975, Conf. Publ. 128, pp. 83-94

87 HAN, C.C.: 'A multifeed offset reflector antenna for the Intelsat V Communications Satellite', in *Proc. 7th European Microwave Conf.* (Copenhagen, Denmark), Sept. 1977, pp. 343-347

88 RUDGE, A.W. and ADATIA, N.A.: 'Primary-feeds for boresight-jitter compensation of offset-reflector radar antennas', in *Proc. Inst. Elec. Eng. Int. Conf. on Radar 77* (London, England), Oct. 1977, Conf. Publ. 155, pp. 409-413

89 BOSWELL, A.G.P., and ASHTON, R.W.: 'Beam squint in a linearly polarised offset reflector antenna', *Electron. Lett.*, **12**, Oct. 1976, pp. 596-597

90 LUDWIG, A.C.: 'The definition of cross polarisation', *IEEE Trans.*, **AP-21**, Jan. 1973, pp. 116-119

91 SILVER, S.: 'Microwave Antenna Theory and Design', New York: McGraw-Hill 1949

92 ADATIA, N.A., RUDGE, A.W. and PARINI, C.: 'Mathematical modelling of the radiation fields from microwave primary-feed antennas', in *Proc. 7th European Microwave Conf.* (Copenhagen, Denmark), Sept. 1977, pp. 329-333

93 CLARRICOATS, P.J.B. and POULTON, G.T.: 'High efficiency microwave reflector antennas – A review', *Proc. IEEE*, **65**, Oct. 1977, pp. 1470-1504

94 IMBRIALE, W.A. *et al.*: 'Large lateral feed displacements in a parabolic reflector', *IEEE Trans.*, **AP-22**, 1974, pp. 742-745

95 LUDWIG, A.C.: 'Calculation of scattered patterns from asymmetrical reflectors', Jet Propulsion Lab., (Passadena, CA), 1970, Tech. Rep. 32-1430

96 RUDGE, A.W., PRATT, T. and FER, A.: 'Cross-polarised radiation from satellite reflector antennas', in *Proc. AGARD Conf. on Antennas for Avionics* (Munich, Germany), Nov. 1973, **16**, pp. 1-6

97 HANSEN, J.E. and SHAFAI, L.: 'Cross-polarised radiation from waveguides and narrow-angle horns', *Electron. Lett.*, May 1977, pp. 313-315

98 FOSTER, P.R. and RUDGE, A.W.: 'Low sidelobe antenna study: Part I: Literature survey and review', ERA RF Technology Centre, England, Oct. 1975, Rep. 190476/1

99 ADATIA, N.A., FOSTER, P.R. and RUDGE, A.W.: 'A study of the limitations in RF sensing signals due to distortions of large spacecraft antennas', ERA (RF Technology Centre, England), Sept. 1975, Rep. ESA Con. 2330/74 AK

100 DIFONZO, D.F., ENGLISH, W.J. and JANKEN, J.L.: 'Polarisation characteristics of offset reflectors with multiple-element feeds', in *IEEE Int. Symp. Digest* PGAP (Boulder, Colorado), 1973, pp. 302-305

101 ADATIA, N.A. and RUDGE, A.W.: 'Beam-squint in circularly-polarised offset reflector

antennas', *Electron. Lett.,* Oct. 1975, pp. 513-515

102 GRUNER, R.W. and ENGLISH, W.J.: 'Antenna design studies for a U.S. domestic satellite', *Comsat Tech. Rev.,* **4**, Fall 1974, pp. 413-447

103 RAAB, A.R.: 'Cross-polarisation performance of the RCA Satcom frequency re-use antenna', *IEEE Int. Symp. Digest AP-S* (Amherst, MA) Oct. 1976, pp. 100-104

104 LANG, K.C., EICK, M.K. and NAKATANI, D.T.: 'A 6/4 and 30/20 dual foci offset paraboloidal reflector antenna', in *IEEE Int. Symp. Digest AP-S* (Urbana, IL), June 1975, pp. 391-395

105 ADATIA, N.A. and RUDGE, A.W.: 'High performance offset-reflector spacecraft antenna development study, ERA RF Technology Centre, England, June 1976, Rep. ESA Con. 2654/76/NLSW

106 FASOLD, D,:'Rechnergestützte Optimierung und Realisierun einer Offset-Reflektorantenne für Satelliten', *NTG-Fachber,* **57**, 1977, pp. 124-133

107 BEM, D.J.: 'Electric field distribution in the focal region of an offset paraboloid, *Proc. IEE*, **116**, 1974, pp. 579-684

108 VALENTINO A.R. and TOULIOS, P.P.: 'Fields in the focal region of offset parabolic reflector antennas', *IEEE Trans.,* **AP-24**, Nov. 1976, pp. 859-865

109 INGERSON, P.G. and WONG, W.C.: 'Focal region characteristics of offset fed reflectors' in *IEEE Int. Symp. AP-S* (Georgia), 1974, pp. 121-123

110 WATSON, B.K., RUDGE, A.W. and ADATIA, N.: 'Dual-polarised mode generator for cross-polar compensation in offset parabolic reflector antennas', presented at 8th European Microwave Conf., Paris, France, Sept. 1978

111 IERLEY, W.H. and ZUCKER, H.: 'A stationary phase method for computation of the far-field of open Cassegrain antennas', *Bell Syst. Tech. J.,* **49**, Mar. 1970, pp. 431-454

112 MIZUGUTCH, Y., AKAGAWA, M. and YOKOI, H.: 'Offset dual reflector antenna), in *IEEE Int. Symp. AP-S* (Amherts, MA), Oct. 1976, pp. 2-5

113 MIZUSAWA, M. and KATAGI, T.: 'The equivalent parabola of a multi-reflector antenna and its application', *Mitsubishi Elec. Eng.,* **No. 49**, Sept. 1976, pp. 25-29

114 ADATIA, N.A.: 'Diffraction effects in dual offset Cassegrain antenna', IEEE Int. Symp. **AP-S**, Washington, May 1978

115 ALBERTSEN, N.C.: 'Shaped-beam antenna with low cross-polarisation' in *Proc. 7th European Microwave Conf* (Copenhagen, Denmark), Sept. 1977, pp. 339-342

116 –, 'Dual offset reflector antenna shaped for low cross polarisation', TICRA (Copenhagen, Denmark), Mar. 1977, Rep. S-53-01

117 CHA, A.G., CHEN, C.C. and NAKATANI, D.T.: 'An offset Cassegrain reflector antenna system with frequency selective sub-reflector', in *IEEE Int. Symp. Digest AP-S* (Urbana, IL), 1975

118 MAGNE, P. and BUI-HAI, N.:'A modular offset Cassegrain antenna operating simultaneously in four frequency bands', in *IEEE Int. Symp. Digest AP-S* (Amherst Mass.), Oct. 1976, pp.10-19

119 MIZUVAWA, M., HETSUDAN, S., *et al.*: 'The doubly-curved reflectors for circularly polarised shaped-beam antennas', *IEEE Int. Symp. Digest*: **AP-S** (Georgia), 1974, pp. 249-252

120 SEMPLAK, R.A.: '100-GHz measurement on a multiple-beam offset antenna', *Bell Syst. Tech. J.,* **56**, 1977, pp. 385-398

121 GALINDO-ISRAEL, V. and MITTRA, R.: 'A new series representation for the radiation integral with application to reflector antennas', *IEEE Trans.,* **AP-25**, Sept. 1977, pp. 631-641

122 GALINDO-ISRAEL, V., MITTRA, R. and CHA, A.L.: 'Aperture amplitude and phase control of offset dual reflectors', *IEEE Trans.,* **AP-27**, pp. 154-164

123 PONTOPPIDAN, K.: 'General analysis of dual-offset reflector antennas', TICRA **AP-S**, (Denmark), Oct. 1977, Final Rep. S-66-02

124 WESTCOTT, B.S. and BRICKELL, F.: 'Exact synthesis of offset dual reflectors' *Electron. Lett.,* **16** Feb. 1980, pp. 168-169

125 DUNBAR, A.S.: 'Calculation of doubly curved reflectors for shaped beams', *Proc. IRE*, **36**, Oct. 1948, pp. 1289-1296

126 VAN ATTA, L.C. and KEARY, T.J.: 'Shaped-beam antennas', in 'Microwave antenna theory and design', S. Silver, (ed.), McGraw Hill Book Co., New York, 1949, pp. 465-509

127 CARBERRY, T.F.: 'Analysis theory for the shaped-beam doubly curved reflector antenna', *IEE Trans.*, **AP-17**, March 1969, pp.131-138

128 BRUNNER, A,: 'Possibilities of dimensioning doubly curved reflectors for azimuth-search radar antennas', *IEEE Trans.*, **AP-19**, Jan. 1971, pp. 52-57

129 WINTER, C.F.: 'Dual vertical beam properties of doubly curved reflectors', *IEEE Trans.*, **AP-19**, March 1971, pp. 174-180

130 DORO, G. and SAITTO, S.: 'Dual polarization antennas for OTS', Int. Conf. on Antennas for Aircraft and Spacecraft, 3-5 June 1975, London, pp. 76-82

131 KATAGI, F. and TAKEICHI, Y.: 'Shaped horn-reflector antennas', *IEEE Trans.*, **AP-23**, 1975, pp. 757-763

132 WICKERT, A.N., HAN, C.C., FORD, D., KUDO, M. and NOMATO, Y.: 'The design of the communication antenna for CS', AIAA 6th Communication Satellite Systems Conference, Montreal, 1976

133 KINBER, B.Y.: 'On two reflector antenna', *Radio Eng. Electron. Phys.*, **6**, June 1962

134 GALINDO, V.: 'Design of dual reflector antennas with arbitrary phase and amplitude distribution', *IEEE Trans.*, **AP-12**, 1964, pp. 403-408

135 WILLIAMS, W.F.: 'High efficiency antenna reflector', *Microwave J.*, 8, 1965, p. 79

136 LUDWIG, A.C.: 'Shaped reflector Cassegrainian antennas', SPS No. 37-35, Vol. IV, pp. 266-268, Jet Propulsion Laboratory, Pasadena

137 LINDSEY, R.: 'Italy sets up first independent Comsat link', *Aerospace Technol.*, 6 Nov. 1965, pp. 28-30

138 BREJCHA, A,G. and SMITH, C.A.: 'Telemetry antennas for deep space probes', International Telemetry Conference, Los Angeles, California, October 18020, 1977

139 KITSUREGAWA, T., and MIZUSAWA, M.: 'Design of the shaped-reflector Cassegrainian antenna in consideration of the scattering pattern of the subreflector', IEE Group on Antennas and Prop., International Symp. Digest, Boston, Mass., Sept. 9-11 1968, pp. 891-896

140 DIJK, J. and MAANDERS, E.J.: 'Optimizing the blocking efficiency in shaped-grain systems', *Electron. Lett.*, **4**, Sept. 1968, pp.372-373

141 WILLIAMS, W.F.: 'DSN 100-meter X- and S-Band microwave antenna design and performance', TR 78-65, Jet Propulsion Laboratory, Pasadena, California, Aug. 1 1978

142 GALINDO-ISRAEL, V., MITTRA, R., and CHA, A.G.: 'Aperture amplitude and phase control of offset dual reflectors', *IEEE Trans.*, **AP-27**, March 1979, pp. 154-164

143 MIZUSAWA, M., BETSUDAN, S.I., URASAKI, S. and IIMORI, M.: 'A dual doubly curved reflector antenna having good circular polarization characteristics', *IEEE Trans.*, **AP-26**, May 1978, pp. 455-458

144 ELLIOT, R.D. and POULTON, G.T.: 'Diffraction optimized shaped beam reflector antennas, *Electron. Lett.*, 1977, pp. 325-326

145 WOOD, P.J.: 'Field correlation theorem with application to reflector aerial diffraction problems', *Electron. Lett.*, 1970, **6**, pp. 326-327

146 WOOD, P.J.: 'Reflector profiles for the pencil beam Cassegrain antenna', *Marconi Rev.*, 1972, **35**, pp. 121-138

147 WOOD, P.J.: 'Near field defocusing in the pencil beam Cassegrain antenna', *Marconi Rev.*, 1973, **36**, pp. 201-223

148 BRAIN, J.R., *et al.*: 'An 11/14 GHz 19 m satellite earth station antenna', IEE International Conference on Antennas and Propagation, Nov. 1978, pp. 384-388

149 WOOD, P.J., *et al.*, 'A beam waveguide feed for an 11/14 GHz antenna,' *ibid.*, pp. 223-229

150 SPENCER, R.C. and HYDE, G.: 'Studies of the focal region of a spherical reflector: Geometrical optics', *IEEE Trans.*, **AP-16**, May 1968, pp. 317-324

151 HYDE, G. and SPENCER, R.C.: 'Studies of the focal region of a spherical reflector: Polarization effects, *IEEE Trans.* **AP-16**, July 1968, pp. 399-404

152 HYDE, G.: 'Studies of the focal region of a spherical reflector: Stationary phase evaluation', *IEEE Trans.* **AP-16**, Nov. 1968, pp. 646-656

153 SLETTEN, G.J.: 'Reflector antennas', Chap. 17 in 'Antenna Theory', McGraw Hill, New York, COLLIN, R.E. and ZUCKER, F.J., (eds.)

154 KAY, A.F.: 'A line source for a spherical reflector', USAF Cambridge Research Labs, Bedford, Mass., T.R. AFCRL-529, May 1961

155 RAMSEY, V.: 'On the design and performance of feeds for correcting spherical aberration, *IEEE Trans.* **AP-18,** 1970, pp. 343-351

156 LOVE, A.W.: 'Scale model development of a high efficiency dual-polarized line feed for the Arecibo spherical reflector', *IEEE Trans.* **AP-21,** Sept. 1973, pp.628-639

157 THOMAS, B. MacA., MINNETT, H.C., and VU, T.B.: 'Fields in the focal region of a spherical reflector', *IEEE Trans.* **AP-17**, March 1969, pp. 229-232

158 VU, T.B., VU, Q.H. and DOAN, D.L.: 'High-efficiency spherical reflector antenna – A case study, *IEEE Trans.*, **AP-25**, May 1977, pp. 351-355

159 DOAN, D.L., and VU, T.B.: 'Study of efficiency of spherical Gregorian reflector', *IEEE Trans.* **AP-23**, Nov. 1975, pp. 819-824

160 ISHIMARU, A., SREENIVAPIAH, I., and WONG, V.K.: 'Double spherical Cassegrain reflector antennas', *IEEE Trans.* **AP-21**, Nov. 1973, pp. 774-780

161 PEELER, G.D.M. and ARCHER, D.H.: 'A toroidal microwave reflector', IRE National Convention Record, Part I, 1956

162 DOLAN, L.J.: 'Recent studies of torus reflectors', Radiation Eng. Lab., Maynard, Mass., Final Rept. TR-59-231, Contract AF 30 (602)-1527, RADC, 1 December 1959

163 GUSTINCIC, J.J.: 'Spherical reflector feasibility study', Final Report on JPL Contract No. 560080, December 1974

164 YOUNG, F.A.: 'Gregorian-corrected toroidal scanning antenna', Ph.D. Dissertation, University of Southern California, June 1978

165 CHANG, D.C., RUSCH, W.V.T. and YOUNG, F.A.: 'Gregorian corrected offset-fed reflector with conical-scan capability. IEEE APS International Symposium Digest, 18-22 June 1979, Seattle, Washington, pp. 258-261

166 FRIIS, H.T. and BECK, A.C.: U.S. Patent 2, 236,393

167 THOMAS, D.T.: 'Design of multiple-edge blinders for large horn reflector antennas, *IEEE Trans*. **AP-21**, March 1973, pp. 153-158

168 FRIIS, R.W. and MAY, A.S.: 'A new broad-band microwave antenna system', *AIEE Trans.*, Part I, **77**, 1958, p.97

169 CRAWFORD, A.B., HOGG, D.C. and HUNT, L.E.: 'A horn reflector antenna for space communication', *Bell System Tech. J.*, **40**, 1961, pp. 1095-1116

170 KRANS, J.D.: 'The corner-reflector antenna', *Proc. IRE*, **28**, Nov. 1940, p.513

171 KLOPFENSTEIN, R.W.: 'Corner reflector antennas with arbitrary dipole orientation and apex angle', *IRE Trans.*, **AP-5**, July 1957, pp. 297-305

172 SCHELL, A.C.: 'The corner array'. AFCRL, Bedford, Mass., TR-59-105, June 1959

173 INAGAKI, N.: 'Three-dimensional corner reflector antenna', *IEEE Trans. Ant. Prop.*, July 1974, pp.580-582

174 OHBA, Y.: 'On the radiation of a corner reflector finite in width', *IEEE Trans.* **AP-11**, March 1963, pp. 127-132

175 HARRINGTON, R.F.: 'Field computation by moment methods', New York, Macmillan, New York, 1968

176 Compagnie Generale de Telegraphie: 'Improvements in or relating to systems for radiating ultra high frequency waves', UK Patent 801 886, Sept. 1958

177 Members of the Technical Staff, Hughes Aircraft Co.: 'Manned spacecraft deep space

antenna study', Final Eng. Rept. P64-51, NASA Contract NAS 9-2099, Culver City, California, April 1964, pp. 65-90

178 TORADA, T., NAGAI, K. and UCHIDA, H.: 'Core reflector antenna'. *Denki Tsushin Gakki Zasshi,* **50**, Feb. 1967, pp. 9-10, (in Japanese).

179 DICKINSON, R.M. and CRAMER, P.W.: 'A survey of unfurlable spacecraft antennas', 1966 NEREM Record, pp. 222-223

180 LUDWIG, A.C.: 'Spacecraft antenna research' Large spacecraft antennas (non-paraboloidal reflector)', Jet Propulsion Laboratory Space Program Summary 87-59, **3**, 31 Oct. 1969, pp. 55-57

181 LUDWIG, A.C.: 'A new geometry for unfurlable antennas', *Microwaves*, Nov. 1970, pp. 41-42

182 LUDWIG, A.C.: 'Conical-reflector antennas', *IEEE Trans.,* **AP-20**, March 1972, pp. 146-152

183 MARTIN, A.G.: 'Radiation from dielectric sphere loaded horns', *Electron. Lett.,* **14**, 5 Jan. 1978, pp. 17-18

184 ERICKSON, E.F., WARD, H.T. and BAILEY, M.C.: 'Development of a multiple shaped beam antenna – Final Report', Fairchild Industries, Inc., Fairchild Space and Electronics Division, Sept. 1972

185 MITTRA, R: 'Efficient computation of radiation patterns and contour beam synthesis using reflector antennas'. Short Course Notes, Section 4.2

186 DUNCAN, J.W., HAMADA, S.J. and INGERSON, P.G.: 'Dual-polarization multiple-beam antenna'. Progress in Astronautics and Aeronautics, **55**, 1977, pp. 223-243

187 NAKATANI, D.T., TAORMINA, F.A., KUHN, G.G. and McCARTY, D.K.: 'Design aspects of commercial satellite antennas'. Communication Satellite Antenna Technology Short Course, UCLA, March 1976, p. 13

188 IMBRIALE, W.A.: 'Applications of the method of moments to thin-wire elements and arrays', Chapter 7 in Topics in applied physics, **3**, Numerical and Asymptotic Techniques in Electromagnetics, 1975

189 HAN, C.C., SMOLL, A.E., BILENKO, H.W., CHUANG, C.A. and KLEIN, C.A.: 'A general beam shaping technique – multiple-feed offset reflector antenna system', AIAA CASI 6th Communications Satellite Systems Conference, 5-8 April, 1976, Montreal, Canada

190 RAAB, A.R.: 'Cross-polarization performance of the RCA Satcom Frequency reuse antenna', APS Symposium, Oct. 1976, pp. 100-104

191 KEIGLER, J.E. and HUME, C.R.: 'RCA Satcom – maximum communication capacity per unit cost', RCA Government and Commercial Systems Astro-Electronics Division, PO Box 800, Princeton, New Jersey 08540

192 TAORMINA, F., McCARTY, D.K., CRAIL, T. and NAKATANI, D.: 'Intelsat IV – A communications antennas – Frequency reuse through spatial isolation', 1976 International Conference on Communications

193 NAKATANI, D.T. and KUHN, G.C.: 'Comstar I antenna system', AP-S International Symposium, Palo Alto, 1977, pp. 337-340

194 CUCCIA, C.L.: 'Japan bids for leadership in space communications', *MSN*, March 1978, pp. 57-74

195 KUDO, M., NEMOTO, Y, WICKERT, A.N., HAN, C.C. and FORD, D.: 'Communications antenna design for the Japanese communications satellite'. Progress in Astronautics and Aeronautics, **55**, 1977, pp. 271-282

196 *Microwave System News*, Oct./Nov. 1976, p. 15

197 HEEBER, C.W.: 'Intelsat V system design'. Wescon '77 Record, pp. 2-11

198 LOVE, A.W.: 'Some highlights in reflector antenna development', *Radio Sci.,* **11**, No. 8-9, Aug-Sept. 1976, pp. 671-684

199 RAMSAY, J.F.: 'Microwave antenna and waveguide techniques before 1900', *Proc. IRE*, **46**, Feb. 1958, pp. 405-415

200 HERTZ, H.: 'Electric waves', Macmillan and Co., Ltd., London, 1893
201 MARCONI, G.: British Patent Specification No. 12039, HM Stationery Office, 2 June 1896
202 National Radio Astronomy Photo GB-62-802
203 MIZUSAWA, M. and KITSUREGAWA, T.: 'A beam waveguide feed having a symmetric beam for Cassegrain antennas', 1972 IEEE-AP Symposium Digest, pp. 319-322
204 O'BRYANT, M.: 'Radio astronomy MM Antenna dedicated', *Microwave J.* **19**, Nov. 1976, p. 30
205 HENSEL, S.L. and RHOADES, L.E.: 'The development and manufacture of ESSCO precision MM wave radio telescopes', 2 May 1977, MTP-77-0008, Electronic Space Systems Corporation, Concord, Mass. 01742
206 MENZEL, D.H.: 'A new radio telescope for Sweden', *Sky and Telescope,* **52**, 4, Oct. 1976
207 KAUFMANN, P. and D'AMATO, R.: 'A Brazilian radio telescope for millimeter wavelengths', *Sky and Telescope*, **45**, March 1973
208 CHU, T.S.; WILSON, R.W., ENGLAND, R.W., GRAY, D.A. and LEGG, W.E.: 'The Crawford Hill 7-meter millimeter wave antenna', *BSTJ,* **57**, May-June 1978, pp. 1257-1288
209 HACHENBERG, O., GRAHL, B.H. and WIELEBINSKI, R.: 'The 100-meter radio telescope at Effelsberg', *Proc. IEEE,* **61**, Sept. 1973, pp. 1288-1295
210 SMITH, C.A.: 'A review of the state of the art in large spaceborne antenna technology', Publication 78-88, JPL, Pasadena, California, 15 Nov. 1978
211 BREJCHA, A.G. and SMITH, C.A.: 'Telemetry antennas for deep space probes'. International Telemetry Conference, Los Angeles, California, 18-20, Oct. 1977
212 GERWIN, H.L. and CAMPBELL, G.K.: 'ATS 30-foot flex-rib antenna, *Telecomm. J.,* **38**, 1971, pp. 503-506, 529
213 GASCOIGNE, S.C.B.: 'Recent advances in astronomical optics', *Applied Optics,* **12**, July 1973, pp. 1419-1429
214 TONG, G., CLARRICOATS, P.J.B. and JAMES, G.L.: 'Evaluation of beam-scanning dual-reflector antennas', *Proc. IEE,* **124**, Dec. 1977, pp. 1111-1113
215 FITZGERALD, W.D.: 'Limited electronic scanning with an offset-feed near-field Gregorian system'. MIT Lincoln Lab T.R. 486, 24 Sept, 1971
216 ARCHER, J.S.: 'High performance parabolic antenna reflectors', 7th AIAA Communications Satellite Systems Conference, San Diego, California, April 1978
217 YOUNG J.W. and DOUGHERTY, T.A.: 'Application of composite materials for fabrication of spacecraft communication antennas', XXVII Congress, IAF, Anaheim, California, 10-16 Oct. 1976
218 ARCHER, J.S. and BERMAN, L.D.: 'Advanced composites in spacecraft design', AIAA/CASI 6th Communications Satellite Systems Conference, Montreal, Canada, 5-8 April 1976
219 FAGER, J.A.: 'Application of graphite composites to future spacecraft antennas'. AIAA/CASI 6th Communications Satellite Systems Conference, Montreal, Canada, 5-8 April 1976
220 KEEN, K., MOLETTE, P., PIEPER, B., HERKERT, C.M. and SCHAEFER, W.: 'Development and testing of a new CFRB antenna reflector for communications satellites', *Raumfahrtforschung,* **4**, 1976, pp. 173-181
221 'Technological priorities for future satellite communications', Office of the Director, NASA Goddard Space Flight Center, July 1978, p. 36
222 TAKERSLEY, B.C.: 'Large deployable antenna technology'. Short Course Notes, Communications Satellite Technology, Boston, Mass., 1978
223 'The wrap rib parabolic reflector antenna'. LMSC Report A969503
224 FAGER, J.A. and GARRIOTT, R.: 'Large-aperture expandable truss microwave antenna', *IEEE Trans.,* **AP-17**, July 1969, pp. 452-458

225 CAMPBELL, G.K.C.: 'Large deployable antenna forecast', LMSC Report D384788, 5 Dec. 1974
226 JASIK, H.: 'Antenna engineering handbook', Chap. 14, McGraw-Hill Book Company, Inc., New York 1961
227 COLLINS, R.E. and ZUCKER, F.J.: 'Antenna theory', Part 2, Chap. 18, McGraw-Hill Book Company, Inc., New York, 1969
228 SILVER, S.: 'Microwave antenna theory and design', MIT Radiation Laboratory Series 12, Chap. 11, McGraw-Hill Book Company, Inc., New York 1949
229 BROWN, J.: 'Microwave lenses'. Methuen & Company, Ltd., London, 1953
230 CORNBLEET, S.: 'Microwave optics'. Academic Press, London, 1976, pp. 374-376
231 SILVER, S.: 'Microwave antenna theory and design'. MIT Radiation Laboratory Series 12, Chap. 11, McGraw-Hill Book Company, Inc., New York, 1949, p. 395
232 COLBOURN, C.B., Jr.: 'Increased bandwidth waveguide lens antenna'. Report No. TOR-0076(6403-01), Aerospace Corporation, Los Angeles, CA, 8 Dec. 1975
233 FOX, A.G.: 'An adjustable waveguide phase changer', *Proc. IRE,* **35**, Dec. 1947, pp. 1489-1498
234 BROWN, R.M., Jr., and DODSON, R.C.: 'Parasitic spiral arrays', NRL Report 5497, AD-243-077, 15 Aug. 1960
235 SIMMONS, A.J.: 'A compact broadband microwave quarter-wave plate', *Proc. IRE,* **40**, 1952, pp. 1089-1090
236 DION, A.R.: 'A broadband compound waveguide lens', *IEEE Trans.,* **AP-26**, Sept. 1978, pp. 751-755
237 AJIOKA, T.S. and RAMSEY, V.W.: 'An equal group delay lens', *IEEE Trans.,* **AP-26**, July 1978, pp. 519-527
238 CHEN, C.: 'Wideband wide-angle impedance matching and polarization characteristics of circular waveguide phased arrays', *IEEE Trans.,* **AP-22**, May 1974, pp. 414-418
239 NAKATAMI, D.T. and FLATEAU, S.L.: 'Multiple beam antenna design for satellite communication systems', ICC'77 Conference Record, *2*, p. 317
240 NAKATANI, D.T. and AJIOKA, T.S.: 'Lens designs using rotatable phasing elements'. IEEE **AP-5** Int. Symp. Digest, June 1977, pp. 357-360
241 MATTHEWS, E.W., SCOTT, W.G. and HAN, C.C.: 'Advances in multibeam satellite antenna technology'. IEEE EASCON'76 Record, Sept. 1965, pp. 132A-1320
242 DION, A.R. and RICARDI, L.J.: 'A variable-coverage satellite antenna system', *Proc. IRE,* **59**, Feb. 1971, pp. 252-262
243 HOGG, D.C.: 'Ground station antennas for space communication' in 'Advances in microwaves' Vol. 3. Academic Press Inc., New York, 1968, pp. 1-66
244 CLAUS, R. and WIEBE, E.: 'Low-noise receivers: Microwave maser development'. JPL Technical Report 32-1526, Vol. XIX, pp. 93-99
245 MOORE, C.R. and CLAUS, R.C.: 'A reflected-wave ruby maser with K-band tuning range and large instantaneous bandwidth. *IEEE Trans.,* **MTT-27,** March 1979, pp. 249-256
246 REID, M.S., CLAUS, R.C., BATHKER, D.A. and STELZRIED, C.T.: 'Low-noise microwave receiving systems in a worldwide network of large antennas', *Proc. IEEE,* **61**, Sept. 1973, pp. 1330-1335
247 LINKE, R.A., CHO, A.Y. and SCHNEIDER, M.V.: 'Cryogenic millimeter-wave receiver using molecular beam expitaxy diodes', *IEEE Trans.,* **MTT-26,** Dec. 1978, pp. 935-938
248 CARLSON, E.R., SCHNEIDER, M.V. and McMASTER, T.F.: 'Subharmonically pumped millimeter wave mixers', *IEEE Trans.,* **MTT-26,** Oct. 1978, pp. 706-719
249 KOGELNIK, H. and LI, T.: 'Laser beams and resonators', *Proc. IEEE*, **54**, Oct. 1966, pp. 1312-1329
250 MIZUSAWA, M., and KITSUREGAWA, T.: 'A beam-waveguide feed having a symmetric beam for Cassegrain antennas', *IEEE Trans.,* **AP-21**, Nov. 1973, pp. 884-886
251 PENZIAS, A.A. and WILSON, R.W.: 'A measurement of excess antenna temperature at 4080 MCS,', *Astrophysics J.* **142**, 1965, pp. 419-421

252 HOGG, D.C.: 'Effective antenna temperatures due to oxygen and water vapor in the Atmosphere', *J. Appl. Phys.*, **30**, 1959, pp. 1417-1419

253 HOGG, D.C. and CHU, T.S.: 'The role of rain in satellite communications', *Proc. IEEE*, **63**, Sept. 1975, pp. 1308-1331

254 DRAGONE, C. and HOGG, D.C.: 'Wide-angle radiation due to rough phase fronts', *BSTJ*, **42**, Sept. 1963, pp. 2285-2296

255 WAIT, D.F.: 'Satellite earth terminal G/T measurements', *Microwave J.*, **20**, April 1977, pp. 49-58

256 PENZIAS, A.A. and WILSON, R.W.: 'Measurement of the flux density of CAS A at 4080 Mc', *Astrophys. J.*, **42**, 1965, pp. 1149-1155

257 BAARS, J.W.M.: 'The measurement of large antennas with cosmic radio sources', *IEEE Trans.*, **AP-21**, July 1973, pp. 461-474

258 CRAWFORD, A.B., HOGG, D.C. and HUNT, L.E.: 'A horn-reflector antenna for space communication', *BSTJ*, **40**, July 1961, pp. 1095-1116

259 HOGG, D.C. and SEMPLAK, R.A.: 'An experimental study of near-field Cassegrainian antennas', *BSTJ*, **43**, Nov. 1974, pp. 2677-2704

260 VON HOERNER, S.: 'Minimum-noise maximum-gain telescopes and relaxation method for shaped asymmetric surfaces', *IEEE Trans.*, **AP-26**, May 1978, pp. 464-471

261 RICE, P.L., THOMPSON, W.I., III and NOBLE, J.L.: 'Idealized pencil-beam antenna patterns for use in interference studies', *IEEE Trans.*, **Com.**, **18**, Feb. 1970, pp. 27-32

262 CHU, T.S., WILSON, R.W., ENGLAND, R.W., GRAY, D.A. and LEGG, W.E.: 'The Crawford Hill 7-meter millimeter wave antenna', *BSTJ*, **57**, May-June 1978, pp. 1257-1288

263 CRAWFORD, A.B., and TURRIN, R.H.: 'A packaged antenna for short-hop microwave radio systems', *BSTJ*, **48**, July-Aug. 1969, pp. 1605-1622

264 SILLER, C.A., Jr., BUTZIEN, P.E. and RICHARDS, J.E.: 'Design and experimental optimization of a canister antenna for 18 GHz operation', *BSTJ*, **57**, March 1978, pp. 603-633

265 THOMAS, D.T.: 'Design of multiple-edge blinders for large horn reflector antennas', *IEEE Trans.*, **AP-21**, March 1973, pp. 153-158

266 TAKEICHI, Y., MIZUSAWA, M. and KATAGI, T.: 'The diagonal horn-reflector antenna', IEEE Symposium Digest, pp. 279-285, Austin, Texas, Dec. 1969, pp. 279-285

267 CONWAY, R.G.: 'Measurement of radio sources at centimeter wavelengths', *Nature*, **199**, 21 Sept. 1963, p. 1177

268 LI, T.: 'Reducing-noise with dual-mode antenna', US Patent No. 3,461,453, 12 August, 1969

269 WILSON, R.W.: 'Suntracker measurements of attenuation by rain at 16 and 30 GHz', *BSTJ*, **48**, May-June 1969, pp. 1383-1404

270 SLOBIN, S.D. and RUSCH, W.V.T.: 'Beam switching Cassegrainian feed system and its applications to microwave and millimeter-wave radioastronomical observations', *Rev. Sci. Instrum.*, **41**, March 1969, pp. 439-443

271 GOLDSMITH, P.F.: 'A quasi-optical feed system for radio-astronomical observations at millimeter wavelengths', *BSTJ*, **56**, Oct. 1977, pp. 1483-1501

272 LUBELL, P.D. and REBHUN, F.D.: 'Suppression of co-channel interference with adaptive cancellation devices at communications satellite earth stations'. ICC Convention Record, 1977, p. 49.3-284

Chapter 4

Primary feed antennas

A.W. Love, A.W. Rudge and A.D. Olver

4.1 General considerations

In a quasi-optical antenna system it makes little difference, so far as the illumination characteristics of the feed are concerned, whether the collimating device is reflective (i.e. single or multiple mirrors) or refractive (i.e. a lens). In either case it is the function of the feed to provide a radiation pattern which is largely confined to the cone defined by the solid angle subtended at the focus by the optical aperture. For convenience, therefore, the following discussion of feeds will refer initially to single-mirror reflective systems in which the feed is placed at the prime focus. It will further be assumed that the system possesses symmetry about the optical axis so that the aperture is circular, as indicated in Fig. 4.5. Feed design is influenced not by the absolute size of the reflector D or its focal length F, but by the ratio F/D, since this ratio determines θ^* the maximum semi-angle subtended by the reflector surface from the focus, through the geometrical relationship

$$\tan\frac{\theta^*}{2} = \frac{D}{4F} \qquad (4.1)$$

Even if the feed were a fictitious point source, radiating isotropically in all directions, the illumination over the aperture would not be uniform, due to the $1/\rho^2$ space attenuation suffered by the spherical wave diverging from the focus. Thus, the power distribution illuminating the aperture (P_i) will be tapered in accordance with

$$P_i(\theta) = \left(\frac{F}{r}\right)^2 = \cos^4\frac{\theta}{2} \qquad (4.2)$$

The edge illumination, relative to that at the centre of the aperture, is easily seen to be given by

$$P_i(\theta^*) = \left[1 + \left(\frac{D}{4F}\right)^2\right]^{-2} \tag{4.3}$$

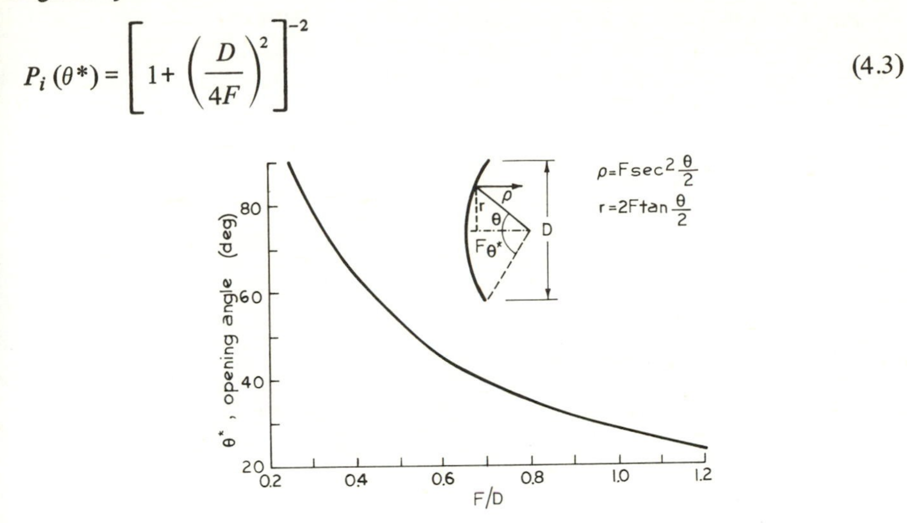

Fig. 4.1 *Geometry of the axially symmetric paraboloid*

These space attenuation factors, or path tapers, $P_i(\theta)$ and $P_i(\theta^*)$ are shown graphically in Fig. 4.2.

The space attenuation factor can be also applied to symmetrical Cassegrain system if F is interpreted as the focal length of the equivalent parabola, as defined by Hannan.[1] Since the equivalent focal length is longer than the actual focal length by the ratio of the magnification factor, $(e+1)/(e-1)$, where e is the eccentricity of the sub-reflector, Cassegrain systems have relatively large equivalent F/D ratios (unity or higher). The space attenuation factor is small for these systems and the more uniform aperture illumination which results can be advantageous if maximising the gain is of prime concern.

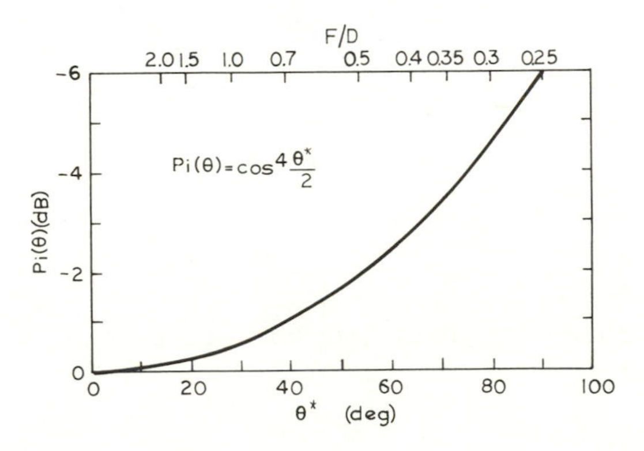

Fig. 4.2 *Space attenuation of path taper factor*

Offset fed, i.e. unsymmetrical, systems are frequently used in order to reduce or eliminate feed blockage. In this case, although the aperture may still be circular, its centre will not correspond to $\theta = 0$ in Fig. 4.1. Therefore, the path taper will not be symmetrical about the aperture centre. In the plane of offset the taper increases uniformly from edge to edge ($\theta = 0$ to $\theta = \theta^*$) across the aperture.

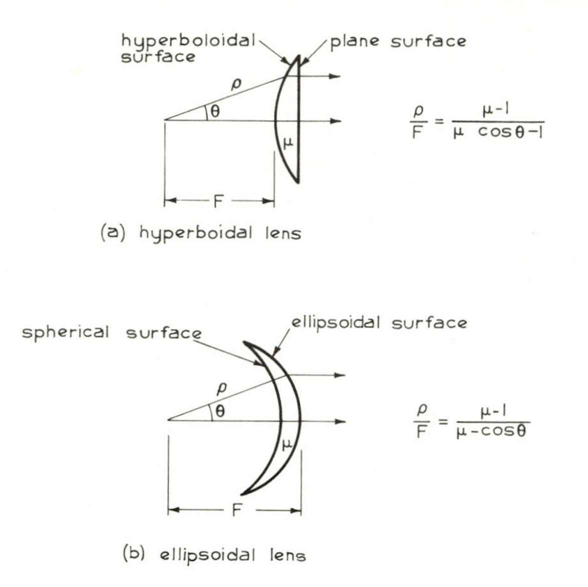

Fig. 4.3 *Two simple one-surface lenses*
a hyperboidal lens
b ellipsoidal lens

Lens antennas are also subject to path taper but in a more complex way due to the introduction of a new parameter, μ, the index of refraction, and to the fact that a lens has two surfaces, not one. Two simple forms of solid dielectric lenses are the plane- and spherical-convex types shown in Fig. 4.3. The convex surface is a hyperboloid of revolution for the former and the illumination ($r(\theta)$) is conventionally tapered in accordance with

$$r_1(\theta) = \frac{(\mu \cos \theta - 1)^3}{(\mu - 1)^2 \, (\mu - \cos \theta)} \tag{4.4}$$

For the latter, the convex surface is an ellipsoid and an inverse illumination taper results, given by

$$r_2(\theta) = \frac{(\mu - \cos \theta)^3}{(\mu - 1)^2 \, (\mu \cos \theta - 1)} \tag{4.5}$$

If the angular aperture is large enough to satisfy $\cos\theta^* = 1/\mu$ then the edge illumination relative to the centre is zero for the hyperboloid lens, but becomes infinite for the ellipsoid. This is apparent in Fig. 4.4 which shows the path tapers for the two cases. General two-surface lenses and constrained lenses in which Snell's law is not obeyed require special consideration; the reader should consult Silver[2] and Brown[3]. It should be noted that both eqns. 4.4 and 4.5 reduce to eqn. 4.2 for the paraboloidal reflector on setting $\mu = -1$.

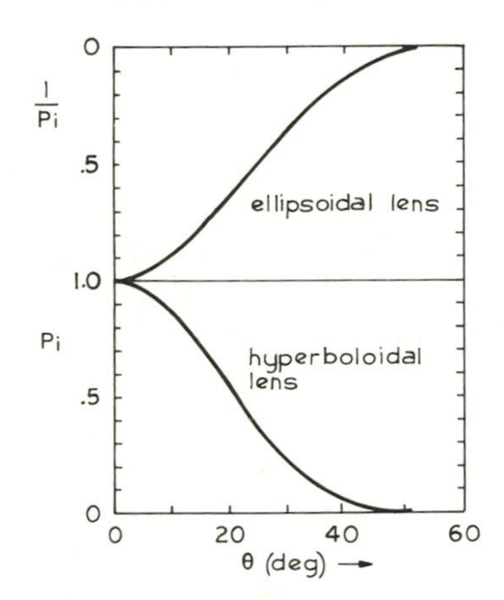

Fig. 4.4 *Space attenuation factors for Polystyrene lenses $\mu = 1{\cdot}6$, $\theta^* = 51{\cdot}3°$*

The radiation from a practical primary-feed antenna is, of course, non-isotropic, and need not exhibit rotational symmetry. The power pattern $P(\theta,\phi)$ of such a device can be combined with the isotropic space attenuation factor $P_i(\theta)$ to yield the effective illumination function (P_a)

$$P_a(\theta,\phi) = P_i(\theta)\,P(\theta,\phi) \tag{4.6}$$

If the amplitude distribution in the aperture is described in polar coordinates by $f(r,\phi)$ in the aperture plane, and the field strength pattern of the feed is given by $E(\theta,\phi)$ then

$$f(r,\phi) = \sqrt{P_i(\theta)}\;E(\theta,\phi) \tag{4.7}$$

where the radial aperture coordinate r is related to θ, in Fig. 4.1, by

$$r = 2F\tan\frac{\theta}{2} \tag{4.8}$$

A uniform phase distribution over the aperture of a paraboloidal reflector will be assured if the feed possesses a true centre of phase, thereby giving rise to a spherical wave diverging from the focus. This implies that the far field pattern of the feed $E(\theta, \phi)$ be a real function, - where the phase of the radiated field must depend only on ρ, but not on θ or ϕ.

4.2 Polarisation definitions

A clear understanding of polarisation properties is now a key requirement in modern reflector antenna design and some basic definitions will be summarised here, and, in view of their impact on primary-feed design, certain aspects will be further amplified in the Sections which follow (see also Chapter 1).

The radiation fields from an antenna can be completely specified in terms of two vector components. The definition of the two components at a point in space and their identification in terms of co-polarised and cross-polarised components can be somewhat arbitrary. Ludwig[4] clarified and discussed some of the popular choices in his 1973 paper. For reflector antennas and their primary-feeds there is much to be said in favour of the third definition given by Ludwig. With this choice the principally polarised or co-polar field of a linearly-polarised antenna is tha which would be measured by a conventional antenna-range technique; where, fo any cut through the radiation-pattern, the polarisation of the distant source i initially co-aligned with the test antenna on boresight, and remains fixed while th test antenna is rotated in the normal way to produce the radiation-pattern (Chapte 8). The polarisation of the distant antenna is then rotated through 90° and th measurement procedure repeated to obtain the cross-polarised field.

With this definition a primary-feed having cross-polarisation will generate purel linearly-polarised currents when used to feed a large paraboloidal reflector. In addi tion a 90° rotation of the polarisation vector at the primary-feed results in a direc interchange between the co-polar and cross-polar components in the primary-fee radiation field, and in the overall reflector-antenna fields, at any point in space This symmetry does not exist with a definition of cross-polarisation which is base on the orthogonal spherical co-ordinate fields (E_θ, E_ϕ), except in the principa planes of the antenna. If the principal polarisation vector is aligned with the y-axi the relationship between the measured co-polar and cross-polar field component and the conventional spherical field components is given simply by

$$\begin{bmatrix} e_p(\theta,\phi) \\ e_q(\theta,\phi) \end{bmatrix} = \begin{bmatrix} \sin\phi & \cos\phi \\ \cos\phi & -\sin\phi \end{bmatrix} \begin{bmatrix} E_\theta \\ E_\phi \end{bmatrix} \tag{4.9}$$

Where an alignment of the antenna principal plane of polarisation with the y axis of Fig. 4.5 implies that e_p will be the co-polar, and e_q the cross-polar, component of the radiated field, while alignment with the x-axis merely interchanges the co-polar/cross-polar designation of e_p and e_q.

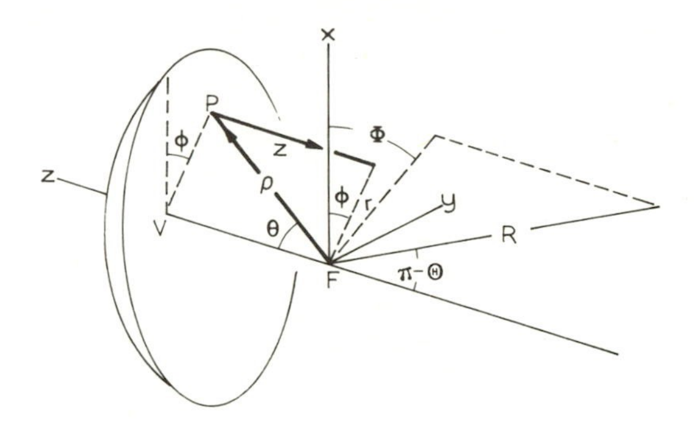

frames of reference:

rect. coordinates	x, y, z
primary coordinates	ρ, θ, φ
far-field coordinates	R, Θ, Φ

Fig. 4.5 *Coordinate systems for feed and paraboloidal reflector*

4.3 Dipole illumination and the aperture field

Although now seldom used as a feed, the simple dipole is of importance, not only historically but because it serves as a reference against which to compare other feed systems. An understanding of its shortcomings yields useful insight into the nature of the problem of feeding any quasi-optical microwave antenna. For this reason, the dipole feed will be discussed in some detail. It will be assumed that the dipole is short, possessing a simple cosine variation in field strength in the E-plane (instead of the slightly more complex form of the half-wave dipole) and a uniform pattern in the H-plane. The dipole is located at the focus of the paraboloid shown in Fig. 4.5 and is aligned parallel with the x axis.

To aid in understanding the nature of the field radiated by this element first consider the dipole in free space, as in Fig. 4.6. At a point P in the far field its radiation pattern is very simply described in terms of a spherical coordinate system in which the polar axis coincides with that of the dipole. Thus, if ψ is the polar angle of P with respect to the x axis, then the field at P has only a single component of electric field;

$$E_\psi \propto - \sin \psi \tag{4.10}$$

With this arrangement the radial and azimuthal components both vanish and if Ludwig's 2nd definition[4] were adopted, the dipole would be said to have no cross-polarisation. However, this coordinate alignment is not convenient when dealing with the dipole as a feed for the paraboloid of Fig. 4.5. In this case it is natural to use the coordinate system (ρ, θ, ϕ) whose polar axis z does not coincide with the

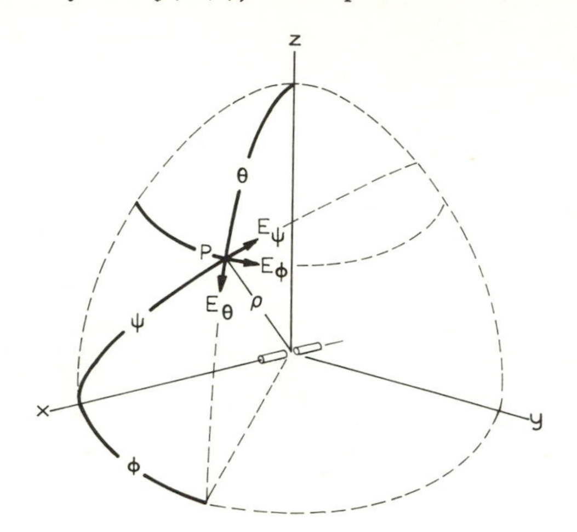

Fig. 4.6 *Dipole in free space*

axis of the dipole, and in this frame two components are needed to describe the radiated field of the dipole. Straightforward use of spherical trigonometry in Fig. 4.6, along with eqn. 4.10 shows that

$$\left.\begin{aligned} E_\theta &\propto \cos\theta\cos\phi \\ E_\phi &\propto -\sin\phi \end{aligned}\right\} \tag{4.11}$$

Two major drawbacks of the short dipole as a primary feed are immediately evident from these equations. It can be seen that E_θ varies as $\cos\theta$, but E_ϕ does not, while the dipole also radiates directly into both hemispheres. This latter difficulty can be largely overcome by placing a reflecting plate behind the dipole in order to direct its radiation towards the paraboloid. This can be idealised to a dipole situated a quarter wavelength in front of an infinite conducting plane and thence, by image theory, to a pair of anti-phased dipoles spaced $\lambda/2$ apart radiating into the half space

$$\theta \lesseqgtr \frac{\pi}{2}.$$

The field radiated toward the reflector (assumed to be in the far field of the feed system) now has the components

$$\begin{aligned} E_\theta(\theta,\phi) &\propto K(\theta)\cos\theta\cos\phi \\ E_\phi(\theta,\phi) &\propto -K(\theta)\sin\phi \end{aligned} \tag{4.12}$$

$$\text{where } K(\theta) = \sin\left(\frac{\pi}{2}\cos\theta\right), \ \theta \lesseqgtr \frac{\pi}{2}$$

$$= 0, \ \theta > \frac{\pi}{2}$$

The factor $K(\theta)$ accounts for the undirectionality introduced by the reflecting plate.

The E- and H- plane patterns of this feed, respectively, given by $\phi = 0$ and $\phi = \pi/2$, are shown in Fig. 4.7. It is clear that if the reflector opening angle is chosen to yield a reasonable taper in the E-plane of the aperture, then there will be insufficient taper in the H-plane, giving rise to undesirable spillover loss. But if the H-plane taper is chosen to be optimum there will be a serious under illumination of the reflector in the E-plane. In either of these cases the system gain will suffer, in the first example due to spillover loss and in the second due to decreased aperture illumination efficiency. This phenomenon is characteristic of all simple dipole feed systems used with circularly symmetric reflectors. If maximising gain is of prime importance, then an improvement may be realised by using a main reflector with an elliptically contoured aperture, but the secondary pattern will obviously not be axially symmetric.

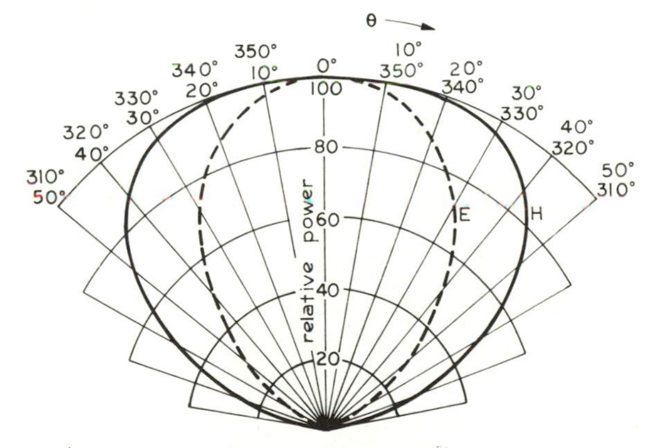

Fig. 4.7 *Patterns of short dipole λ/4 from reflecting plate*

A more serious problem with the dipole feed comes to light when the polarisation properties of the antenna are of concern. Substituting in eqn. 4.9 (with e_p and e_q interchanged) from 4.12 the dipole radiation fields have the form

$$e_p(\theta,\phi) \propto \frac{K(\theta)}{2}\left\{(1+\cos\theta)-(1-\cos\theta)\cos 2\phi\right\}$$

$$e_q(\theta,\phi) \propto \frac{-K(\theta)}{2}(1-\cos\theta)\sin 2\phi \qquad (4.13)$$

The expression for the cross-polarised field (e_q) reveals the presence of four cross-polar lobes having zero values on axis and rising to peak values in the intercardinal planes, only 15dB below the maximum of the co-polar main beam.

For our present purposes it will be instructive to examine the form of the reflected fields in the aperture of the paraboloid when it is illuminated by the dipole field. Using the conventional physical optics approximation to determine the fields reflected at the paraboloidal surface ($\boldsymbol{E}_r$) we have

$$\boldsymbol{E}_r = 2\,(\boldsymbol{a}_n \, . \, \boldsymbol{E}_i)\,\boldsymbol{a}_n - \boldsymbol{E}_i \qquad (4.14)$$

where $\boldsymbol{E}_i$ is the incident field at the reflector surface and $\boldsymbol{a}_n$ is the unit surface normal, which for a paraboloid is given by

$$\boldsymbol{a}_n = \sin\theta/2\cos\phi\boldsymbol{a}_x + \sin\theta/2\sin\phi\,\boldsymbol{a}_y + \cos\theta/2\boldsymbol{a}_z$$

After some minor algebraic manipulations we can write the result in the form

$$\boldsymbol{E}_a = \cos^2\theta/2\,(e_p\,(\theta,\phi)\,\boldsymbol{a}_x + e_q\,(\theta,\phi)\,\boldsymbol{a}_y) \qquad (4.15)$$

where $\boldsymbol{E}_a$ is the projection of the reflected field into the reflector aperture plane[2]. The form taken by the electric field lines in the reflector aperture plane is illustrated in Fig. 4.8.

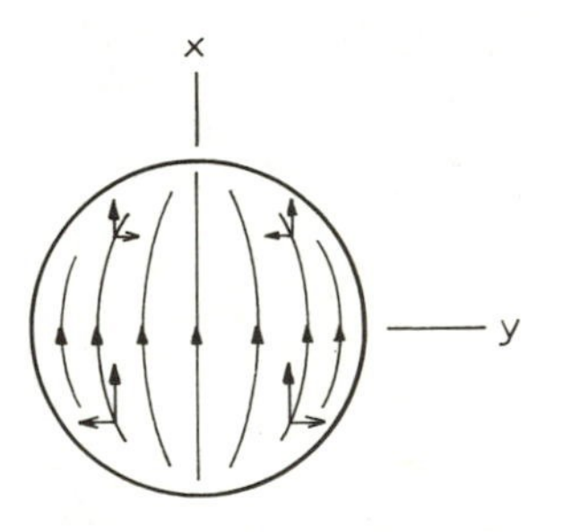

Fig. 4.8 *E-field in aperture of paraboloid fed by electric dipole along x-axis*

From eqn. 4.15 we note that the co-polar and cross-polar components of the primary-feed radiation (e_p, e_q) transform directly into the cartesian co-polar and cross-polar components in the aperture plane of the paraboloid. The additional $\cos^2\theta/2$ multiplier accounts for the space attenuation effect. This general transformation is not restricted to the case of the dipole feed and would apply similarly for any primary-feed antenna.

The dual of the electric dipole is the magnetic dipole, which can be approximated to in practice by a small loop or a narrow slot radiator. If such a dipole is aligned with the y axis in Fig. 4.6 then its distant electric field possesses only a single component in the spherical coordinate system for which y is the polar axis

When the transformation is made to the ρ, θ, ϕ frame the field is found to have components

$$E_\theta \propto \cos\phi$$
$$E_\phi \propto -\cos\theta \sin\phi \tag{4.16}$$

Substituting into eqn. 4.9 (with e_p and e_q interchanged) the measured-field components for the magnetic dipole become

$$e_p(\theta,\phi) \propto \frac{1}{2}\left[(1+\cos\theta) + (1-\cos\theta)\cos 2\phi\right]$$
$$e_q(\theta,\phi) \propto \frac{1}{2}(1-\cos\theta)\sin 2\phi \tag{4.17}$$

The resultant field pattern in the reflector aperture plane is sketched in Fig. 4.9 and a comparison with the electric dipole case in Fig. 4.8 immediately suggests that a combination of these characteristics could linearise the aperture field. This is confirmed by a comparison of eqn. 4.13 (with $K(\theta)$, the function arising from the presence of a reflecting plate, suppressed) and eqn. 4.17 which then reveals co-polar characteristics of identical distribution and of opposite sign.

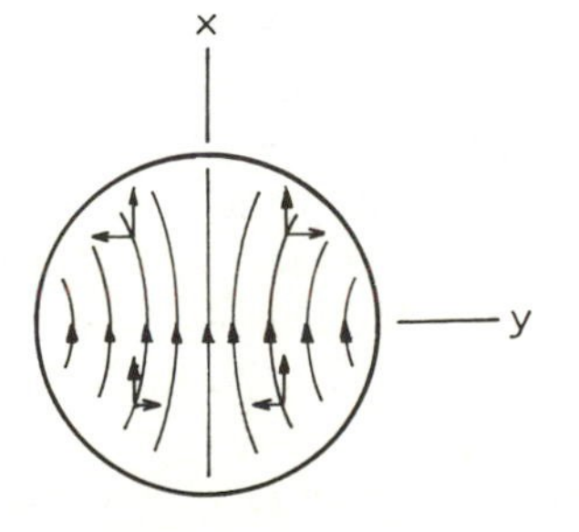

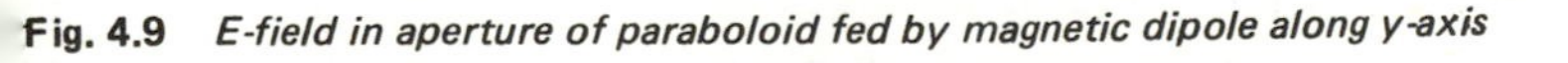

Fig. 4.9 *E-field in aperture of paraboloid fed by magnetic dipole along y-axis*

The linearisation of the reflector currents or aperture fields was first investigated by Jones[5], who deduced that the required condition is that the ratio of magnetic current (i.e. voltage) in the magnetic dipole, to electric current in the electric dipole, be made equal to the free-space impedance or 377 Ω. Jones called this combination a 'plane wave source' by analogy with the field of a plane wave in free space. He observed that the radiation from an open-ended fundamental-mode (TE_{10}) rectangular waveguide approaches such a source with its dimensions are large with respect to a wavelength, but that this condition is not satisfied by large cylindrical waveguides.

When this condition is satisfied by the dipole combination, their fields can be added to give a resultant field in the reflector aperture plane which is completely linearly polarised. Under these conditions the aperture-field cross-polar component is zero and the co-polar component is independent of the angle ϕ and thus exhibits circular symmetry.

Koffman[6] has termed this orthogonal pair of electric and magnetic dipoles a 'Huygens source' and he has generalised the concept to include reflecting surfaces formed by other conic sections. The field radiated by the Huygens source with its electric dipole aligned with the x axis is given by the linear combination of the electric and magnetic dipoles, i.e.

$$E_\theta \propto (1 + \cos\theta)\cos\phi$$
$$E_\phi \propto -(1 + \cos\theta)\sin\phi \tag{4.18}$$

and, from eqns. 4.9 and 4.15 this results in a paraboloidal aperture plane field, including the space attenuation effect, of the form

$$E_a \propto 2\cos^4\theta/2\,\boldsymbol{a}_x \tag{4.19}$$

This would represent an ideal source for the paraboloid from a polarisation purity viewpoint. As a primary-feed for a paraboloid with $F/D = 0{\cdot}25$ the Huygens source, combined with the space attenuation effect, would produce a −12db illumination taper in the reflector aperture plane.

Although the radiation from such a primary feed will generate no cross-polarisation in an axisymetric paraboloid, the curvature of the reflector surface will in itself generate a low level of cross-polarised radiation as a consequence of axial current flow (i.e. z-directed currents on the reflector). These currents clearly cannot contribute to the on-axis field in the $-z$ direction (i.e. at $\theta = 0$) but give rise to cross-polar lobes in the intercardinal planes. For a focal-plane reflector with F/D of 0·25 and a diameter exceeding 25λ, fed by a Huygens source, the level of these cross-polar lobes will be at least 44dB below the peak of the antenna main beam[6,6]. This level reduces rapidly if either the F/D ratio or the diameter of the reflector is increased[7]

Although the Huygens source is an ideal feed for an axisymmetric paraboloidal reflector where polarisation purity of the aperture field is concerned, it should be noted that this is not true for non-axisymmetric reflectors or axisymmetric lens antennas. For example, Jones has shown[5] that, for the plano-hyperboloidal lens of Fig. 4.3*a*, the simple electric dipole will be ideal in the same sense. When the dipole field given in eqn. 4.11 is incident on the convex surface of the lens, the $\cos\theta$ factor in E_θ is exactly cancelled by a factor $\sec\theta$ which appears in the Fresnel transmission coefficient for parallel polarization. This does not occur for E_ϕ because the Fresnel coefficient for perpendicular polarization involves no such $\sec\theta$ term. The lens effectively compensates for the different behaviour of E_θ and E_ϕ as functions of θ with the result that there is no cross-polarisation in the aperture field. It is easy to show that the same is true for the spherical/ellipsoidal lens of Fig. 4.3*b*. Both the magnetic dipole and the Huygens source feeds will give rise to cross-polarisation and lack of symmetry in the aperture fields of these lenses. The effect is most pronounced for the magnetic dipole.

4.4 Practical dipole type feeds

If the short electric dipole discussed in the previous Section is replaced by a half-wave dipole and provided with either a reflecting plate (preferably 3λ/4 or more in diameter) or a reflecting dipole element, then a practical feed results. Such a feed will suffer from the shortcomings already mentioned but nevertheless makes a useful feed when only modest levels of performance are required. It may be used with reflectors having F/D in the range 0·33 to 0·45. Aperture blocking is usually at a minimum with this kind of feed in which the dipole assembly is supported at the end of a coaxial-feed line which passes through the vertex of the reflector. Several examples of this simple dipole feed are described in Silver (Ref. 2, chap. 8). A circularly polarised variant, called the paradisc feed, has been described by Silberberg.[8]

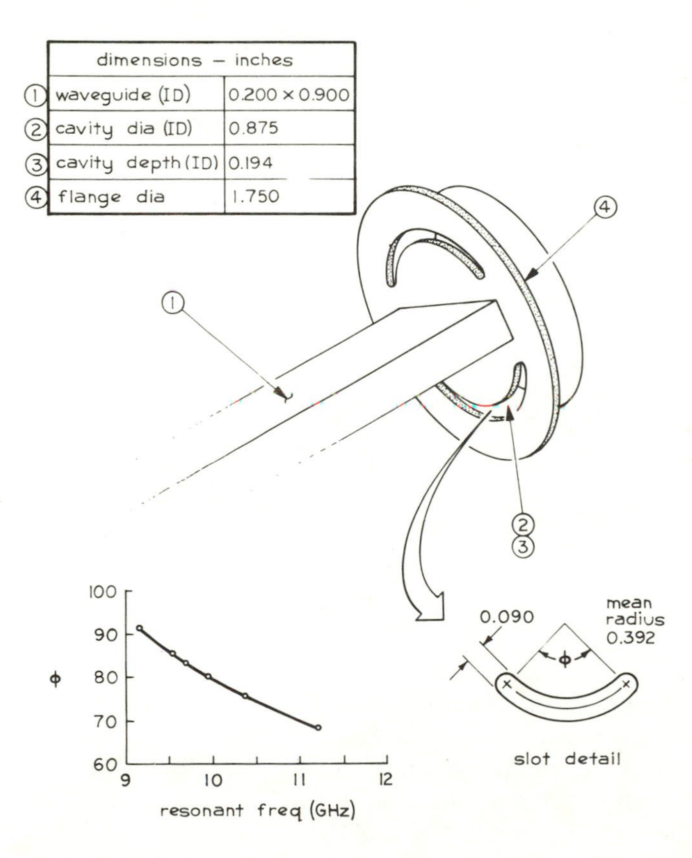

Fig. 4.10 *Rear feed: Cutler design*

The Cutler feed[9] is another example in which the feeder line enters from the reflector through the vertex. In the original version two narrow slots backed by a shallow cylindrical cavity formed the radiating elements. The slots, which act like magnetic dipoles, should be less than $\lambda/2$ apart. Since the slots lie on either side of the rectangular waveguide feeder it is necessary to reduce the narrow dimensions of the guide to about half-height in order to meet this condition. An example, which works well in a reflector with F/D = 0·38, is shown in Fig. 4.10 for use at X-band. The given dimensions may easily be scaled for operation at other frequencies. This feed differs from Cutler's original version in the use of curved slots which are part of a circular arc. The arc length affects the 'resonant' frequency, as shown in the graph accompanying Fig. 4.10. The mismatch is minimised at this frequency and can be tuned out almost completely by means of a threaded post in the cavity. The bandwidth is about 3% for a VSWR of less than 1·3.

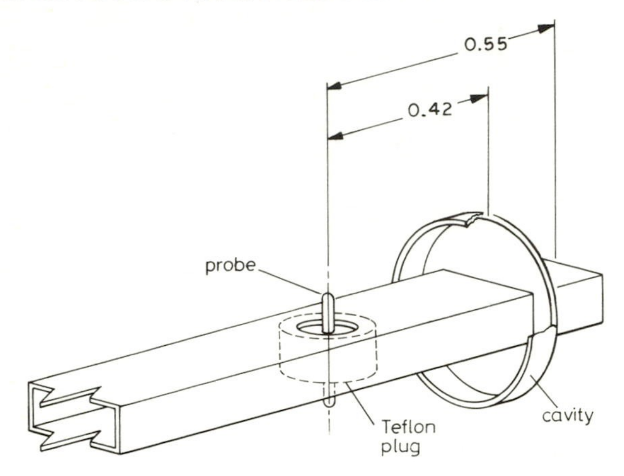

dimensions – inches	
waveguide (ID)	0.200 × 0.900
waveguide hole dia	0.350
cavity dia (ID)	1.450
cavity depth (ID)	0.350
teflon plug dia (OD)	0.500
probe length	0.900

Fig. 4.11 *Rear feed: Clavin design*

At 9·15 GHz the feed shown in Fig. 4.10 (with ϕ = 91°) gave nearly identical *E*- and *H*-plane patterns with a width of 126° at the −10dB level. Retuned to 9·25 GHz and used in a 6 ft paraboloidal reflector with 27·5 in focal length, this feed performed well. The secondary-pattern half-power beamwidth was about 1·3° (slightly more in the *E* plane and slightly less in the *H* plane) with no side lobes above −26dB. The measured gain was 42·5dB, corresponding to an aperture efficiency of 57%.

Another form of rear feed, representing one of the earliest attempts to create the ideal Huygens source, is due to Clavin[10] and is shown in Fig. 4.11. As in the Cutler feed a half-height rectangular waveguide feeder line is used. Energy is coupled out of the line by a dipole protruding symmetrically from each broad face and supported by a Teflon plug inside the guide. The latter is short-circuited at a point somewhat less than $\lambda/4$ from the dipole, giving a VSWR under 1·5 over a 12% band from 8·5 to 9·6 GHz. The dipole lies almost in the open mouth of a shallow cylindrical cavity that is supported by the short-circuited end of the waveguide. At first sight this arrangement appears to be a simple dipole-reflector combination. The cavity, however, is far more than a mere reflector. Along with the central supporting waveguide the cavity forms a short section of irregular coaxial line shorted at one end and open at the other. This line is excited parasitically in a TE_{11} mode by the nearby dipole. Radiation accordingly takes place from the open end of the line as though emitted by a pair of magnetic dipoles, one adjacent to each broad face of the waveguide. The radiating mechanism here is almost exactly the same as that of the curved slots in the modified Cutler feed described above.

In the Chlavin feed, however, the electric dipole also radiates in addition to the magnetic dipole and the combination of these complementary elements is a very good approximation to a Huygens source when the cavity length is properly chosen. To see this, suppose that the cavity is very long so that it completely engulfs the electric dipole. Then the radiation is entirely from the TE_{11} mode at the open end and has the characteristics of a magnetic dipole. On the other hand, if the cavity is shortened to zero length then it becomes nothing more than a reflecting plate and the radiation is entirely due to the electric dipole. Plausibly, then, there exists an intermediate cavity length in which the electric and magnetic dipole strengths are exactly right to yield a Huygens source.

For the dimensions shown in Fig. 4.11 the E- and H-plane patterns at 9·25 GHz are essentially identical, with a full width of 130° at the −10dB level, and back lobe radiation suppressed to −30dB or more. Used with a reflector having F/D = 0·33, this feed yielded nearly equal E and H plane patterns over the frequency range 8·4 to 9·6 GHz with sidelobe levels less than −26dB. At 9·0 GHz the measured aperture efficiency was nearly 65%. No experimental information is available on cross-polarisation in the secondary pattern but it appears likely that at least a modest degree of cross-polarisation suppression in the intercardinal planes (i.e. ± 45°) can be achieved. Circularly polarised feeds of this kind have been built using round waveguide and orthogonal dipoles. The round guide is dielectrically loaded in order to reduce its diameter to less than $\lambda/2$.

4.5 Simple prime focus horn and waveguide feeds

Multimode and hybrid mode horns radiate axially symmetric patterns with extremely low cross-polarisation. Thus, they closely approximate the ideal Huygens source, and, for this reason, they are now widely used as feeds. Nevertheless, there are circumstances in which such sophistication is unnecessary, or perhaps, even undesirable. As an example of the latter, the much larger aperture of a scalar hybrid mode horn can cause intolerable shadowing in a reflector which is only a few wavelengths in diameter. In these cases an open-ended waveguide or a small horn may make a perfectly acceptable feed.

A feed pattern with good axial symmetry in the main beam can be had from an open waveguide or small pyramidal horn with TE_{01} mode excitation if the mouth dimensions are suitably chosen to yield equal E and H plane beamwidths. There will be a substantial cross-polarised component in the ± 45° planes, however, and the aperture is generally rectangular, limiting the feed to single linear polarisation usage. Approximate dimensions for such a simple feed may be estimated from the empirical formula given by Silver [Ref. 2 p. 365] for the full −10dB beamwidths, in degrees, for horns with small flare angle (hence small phase error in the aperture).

$$\begin{aligned} &\text{E plane width } (-10\text{dB}) \simeq 88\frac{\lambda}{b}; \quad \frac{b}{\lambda} < 2{\cdot}5 \\ &\text{H plane width } (-10\text{dB}) \simeq 31 + 79\frac{\lambda}{a}; \quad 0{\cdot}5 < \frac{a}{\lambda} < 3 \end{aligned} \tag{4.20}$$

where a and b are the aperture dimensions in the magnetic and electric planes.

As an example, the opening angle for a reflector with $F/D = 0{\cdot}4$ is $\theta^* = 64°$ and its space attenuation factor is nearly −3dB as shown in Fig. 4.2. If the −10dB width of the feed pattern, assumed to be axisymmetric, is 128°, then the reflector illumination will taper to nearly −13dB at the edge. Eqns. 4.20 show that the horn aperture dimensions should be $b \simeq 0{\cdot}69\,\lambda$ (E-plane) and $a \simeq 0{\cdot}81\,\lambda$ (H-plane). For standard rectangular waveguide the b dimension lies in the range $0{\cdot}3\,\lambda$ to $0{\cdot}45\,\lambda$ while the a dimension is between $0{\cdot}6\,\lambda$ and $0{\cdot}9\,\lambda$. Thus, while no flare may be needed in the H plane the waveguide must be flared in the E plane to the desired dimension.

When maximum gain is desired the field-matching procedure described by Rudge and Withers[11] may be used for designing small flare-angle (< 10°) rectangular horns for parabolic reflectors of any F/D ratio and for rectangular, as well as circular, reflector contours. The method is based upon achieving an optimum match between the electric field distribution in the focal plane of the reflector when it is illuminated by a far field source, and the electric field in the mouth of the primary feed horn when the horn transmits. The optimum horn dimensions are thus related to the dimensions of the principal lobe in the reflector focal-plane field, which will be constant for any given F/D ratio. The method has been used by Truman and Balanis[12] for circularly contoured parabolic reflectors, and extended to include the spherical phase effects introduced by large pyramidal horns. The optimum horn dimensions versus reflector F/D ratio are shown in Fig. 4.12 for various slant lengths of the horn.

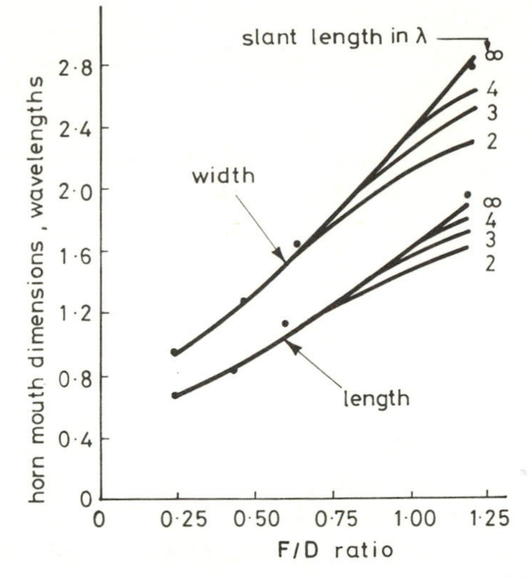

Fig. 4.12 *Optimum rectangular feed horn dimensions versus F&D for a circular paraboloid IEEE Trans.,* July 1974, T-AP22

A cylindrical waveguide or small conical horn with fundamental mode (TE_{11}) excitation has obvious advantages if there is a requirement for dual or circular polarisation. The radiation pattern of these devices is very diameter dependent. For horns which are large with respect to the wavelength these radiators generally lack pattern symmetry and have high cross-polarised lobes in the intercardinal planes. For conical horns with diameters of one wavelength or less the radiation pattern symmetry, and hence the cross-polar performance is very dependent upon the flange surrounding the horn aperture. For example when the guide wall is thin and the mouth diameter d is 0·86 λ the E and H plane patterns are closely similar over the main beam and good axial symmetry is achieved. The full −12dB width of the pattern is approximately 145°, which is satisfactory for a reflector with F/D between about 0·32 and 0·40. No flare is necessary for, with d = 0·86 λ, the dominant TE_{11} mode can propagate but all higher modes are cut off. The maximum cross-polar lobe level appears to be about −26dB[13].

For a fundamental-mode conical horn, with a large planar flange it has also been shown[14] that, for an aperture diameter d = 1·15λ, the intercardinal cross-polarised lobes theoretically vanish. In practice, levels of better than −30dB have been measured over an 8% bandwidth for horns with quite small flanges. This device is a good candidate for a dual-polarised reflector with an F/D ratio of around 0·5.

Measurements on small aperture diagonal horns with TE_{01}/TE_{10} excitation show that here, too, there is a special aperture size (d = 1·23 λ), that yields equal E and H plane pattern widths at all levels. Since the −20dB width is 130° such a horn will produce a high illumination taper when F/D = 0·4 and a more reasonable choice is F/D = 0·5.

A simple ridged waveguide horn feed due to Shimizu[15] is shown in Fig. 4.13

along with its radiation patterns at the end points of a full octave frequency band. It should be noted that the polar plots are amplitude patterns, not power patterns. At 15·1 GHz, where aperture diameter is $d = 0{\cdot}75\lambda$, the E and H plane patterns are nearly identical with the −10dB level occurring at 61·5°. One octave higher, at 30·2 GHz, where $d = 1{\cdot}5\lambda$, the E-plane pattern is somewhat sharper than that in the H-plane and both are considerably narrower than at 15·1 GHz. Sidelobes within the angular range ± 61·5° are at least 20dB down. When used to illuminate a symmetrical paraboloid with $F/D = 0{\cdot}42$ this feed gave very nearly identical patterns in both the E- and H-planes and at the two frequencies 15·1 and 30·2 GHz. In all cases the half power beamwidth was 1·2° and sidelobes were all below −23dB. The cross-polar performance of this device is likely to be poor over the full bandwidth. In the published literature the dimension d was given as 0·587 in. while the ridge width was defined as $t = a/2$. From data given by Shimizu relative to the TE_{11} and TM_{01} mode cut-offs in ridged guide, it can be inferred that the ratio l/a must have exceeded 0·4 in order to obtain single mode operation over a full octave. The authors claim that a constant beamwidth secondary pattern could be obtained over the whole octave and that operation over two octaves would be possible for $l/a > 0{\cdot}6$.

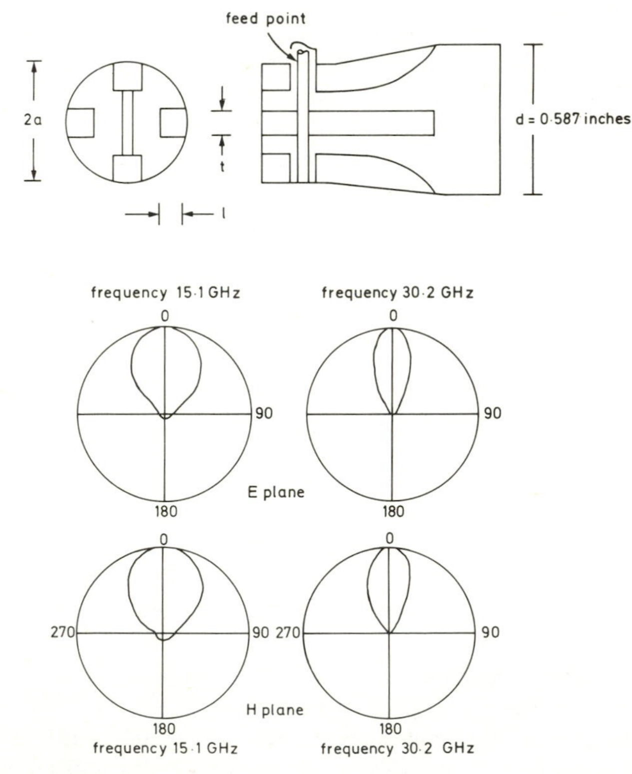

Fig. 4.13 *Octave band ridged horn with amplitude patterns at 15·1 and 30·2 GHz IEEE Trans.* **Mar. 1961 T-AP9**

4.6 Multimode horn feeds

The lack of axial pattern symmetry in large conical horns may be overcome by introducing the TM_{11} mode along with the dominant TE_{11}, as shown by Potter[16]. The presence of this mode has essentially no effect on the H-plane aperture distribution of the horn, nor on its H-plane radiation pattern. When properly phased and combined with the TE_{11} mode it can exert a profound effect on the horn E-plane aperture distribution and the corresponding radiation pattern. The electric field lines in the aperture of the horn can be linearised by the addition of the two modes in the correct phase relationship, Fig. 4.14. For large conical horns this will result in axisymmetric co-polar radiation patterns with a true phase centre and low cross-polarisation. In addition the normally high E-plane sidelobes of the conical horn are reduced at least to the level of those in the H-plane. To ensure that the higher-order mode operates well away from its cut-off frequency in the horn the feed aperture must exceed about 1·3 λ. For this reason, the dual mode horn is best suited to Cassegrain systems or as a front feed in a long focus reflector. Turrin[17] has described such a horn and presents patterns for two different aperture diameters, 1·31 λ and 1·86 λ. These patterns, shown in Fig. 4.15, indicate suitability as front feeds for reflectors having F/D radios of 0·6 and 0·8, respectively. In these horns partial conversion of TE_{11} mode energy to TM_{11} mode takes place in the flared section of the horn while the straight section of length l ensures that the two modes have the proper phase relationship at the aperture. This is maintained over a bandwidth of somewhat less than 10%.

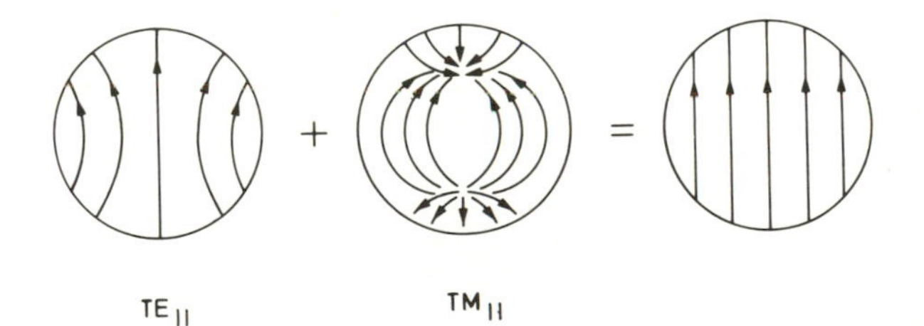

Fig. 4.14 *Illustration of the effect on the horn aperture fields of combining the TE_{11} and TM_{11} modes*

In the original dual-mode horn of Potter[16], shown in Fig. 4.16, the TM_{11} mode is generated at the step where the horn radius changes abruptly from a_0 to a_1 and is correctly phased at the aperture by proper choice of the length l in the straight section that follows the step. This horn is evidently a very satisfactory feed in a Cassegrain system having F/D equal to about 2, but its bandwidth is limited by the need to maintain the correct phase relationship between the TE_{11} and TM_{11} modes in the horn aperture. Since the modes propagate at different velocities in the horn, the phase path l will be correct at one frequency only.

The characteristics of step-discontinuity TM_{11} mode converters in round guide have been described by Agarwal and Nagelberg[22] and by English[23]. Tomiyasu[24] has analysed conversion by means of flare-angle change at the junction of a conical horn with a round waveguide.

By using coaxial circular apertures, in which TE_{11} and TM_{11} modes exist in the inner, while TE_{11} plus TM_{12} modes are excited in the annulus, Koch[25] synthesised an approximation to a sector beam and obtained increased aperture efficiency in front fed paraboloids having F/D in the range 0·3 to 0·5. A similar arrangement has been described by Kumar[26].

Jensen[18] showed that similar results could be realised in a square aperture pyramidal horn by means of a step in the throat region to cause conversion of dominant

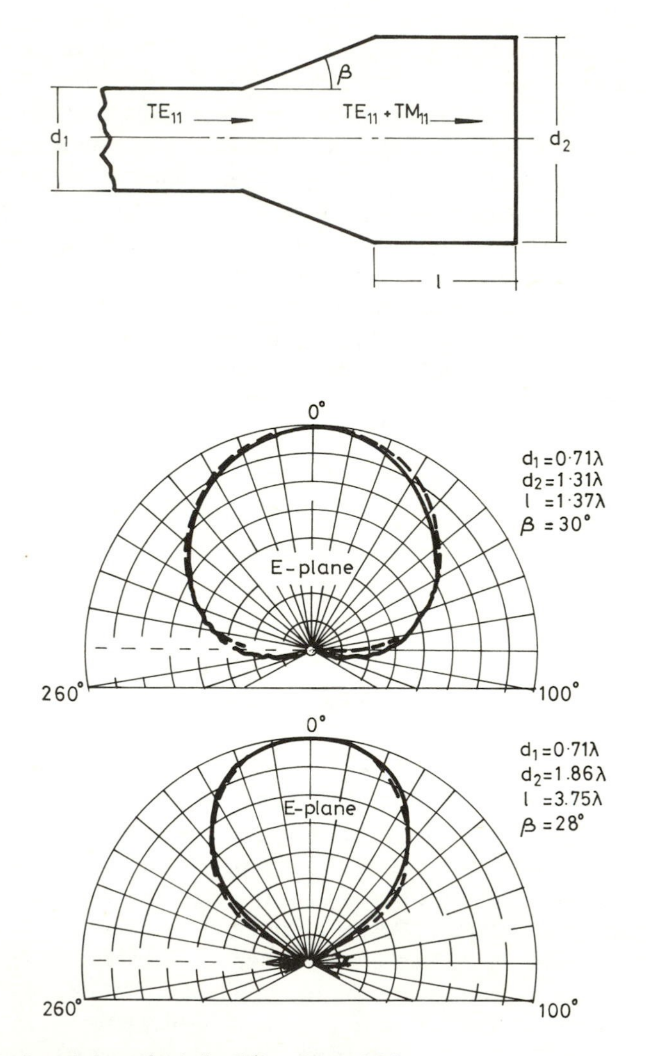

Fig. 4.15 *Dual mode horn feeds for F/D = 0·6 and 0·8, IEEE Trans.,* March 1976, T-AP15

TE_{10} mode energy to a hybrid mixture of TE_{12} and TM_{12} modes. A version of this horn, which was designed for use with a microwave radiometer[19], is shown in Fig. 4.17 along with its principal plane patterns at 2·65 GHz. Good pattern symmetry

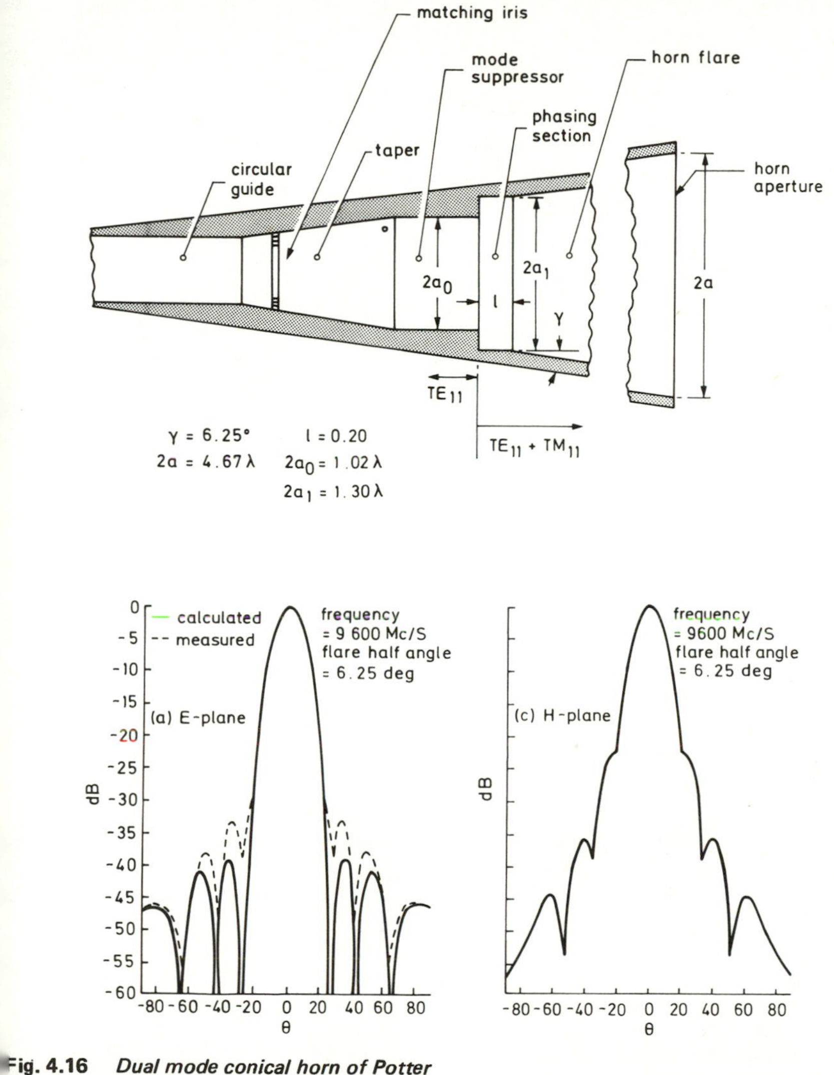

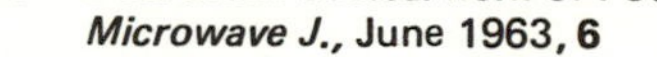

Fig. 4.16 *Dual mode conical horn of Potter*
Microwave J., June 1963, **6**

is achieved over a 100 MHz band and the horn is suitable as a Cassegrain feed for F/D between 1·3 and 1·4. Increased bandwidth may be obtained by eliminating the mode generating step in the throat region and using changes in the horn flare angle to create the desired TE_{12}/TM_{12} mixture. A fundamental mode (TE_{10}) square pyramidal horn has an H/E plane half-power beamwidth ratio 1·35:1 with principal-plane sidelobes of −23dB and −13dB respectively. By adding components of the

TE_{12} and TM_{12} modes, in the current amplitudes and phases, Cohn[20] has shown that the beamwidths of the horn in the principal planes can be equalised and that the sidelobes in these planes can be reduced below the −20dB level. Another technique that yields increased bandwidth in a dual mode horn is due to Ajioka and Harry[21].

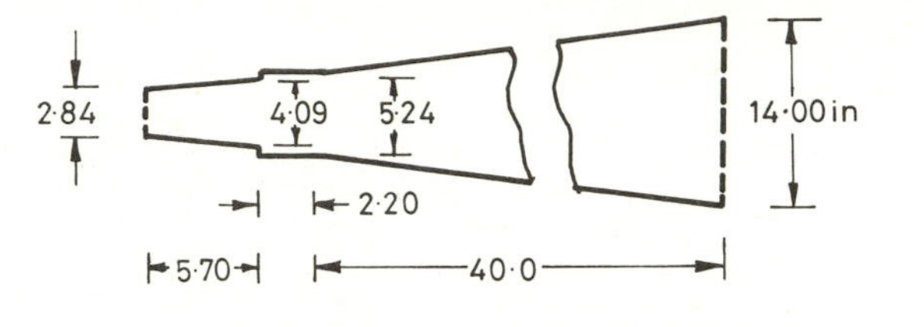

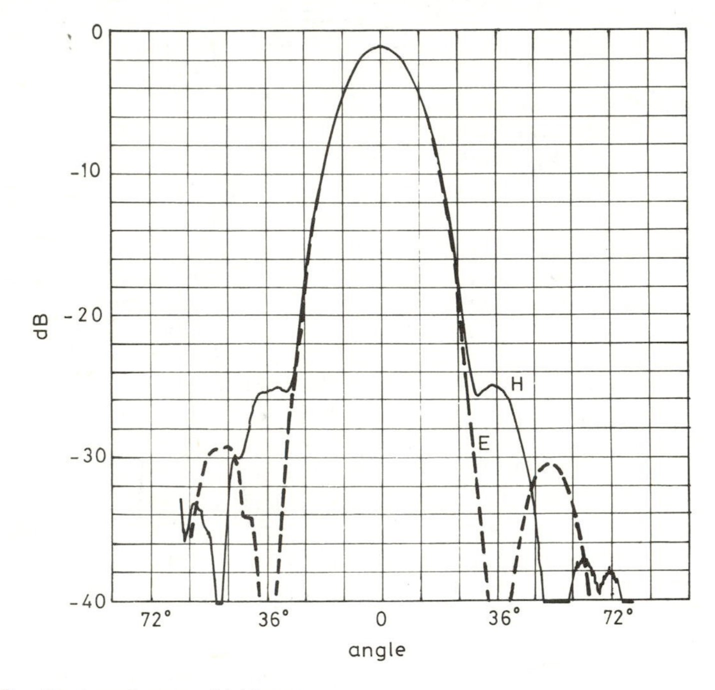

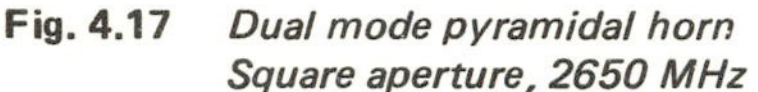

Fig. 4.17 *Dual mode pyramidal horn*
Square aperture, 2650 MHz

For large reflector antennas the use of the Potter horn has been largely superseded by the wider-bandwidth hybrid-mode corrugated horn but when weight is critical and required bandwidths small, the dual mode smooth-walled horn still finds applications. The bandwidth limitation remains the most critical limitation of this device and, although work has been reported in this area from time to time no major breakthrough has been effected. However, the use of a dual-mode conical horn loaded with a dielectric ring, as demonstrated by Satoh[84], appears promising.

A somewhat different form of dual mode radiator, which can make a simple but elegant linearly-polarised front feed for a reflector with $F/D = 0{\cdot}3$ to $0{\cdot}4$, is due to Clavin[27]. As shown in Fig. 4.18, the slot in the base of the cylindrical cup excites a TE_{11} mode in the open cavity while, simultaneously, it parasitically excites the two monopoles which then give rise to a TM_{11} mode in the cup. The radiation pattern is said to be satisfactory over the X-band waveguide range (8·2 – 12·4 GHz) but the impedance bandwidth is limited to about 10% by the resonant length of the slot. It is possible that this could be improved by making the slot simply the open end of a reduced height waveguide, thereby avoiding resonance limitations. The diameter of the monopoles is approximately 0·05 λ.

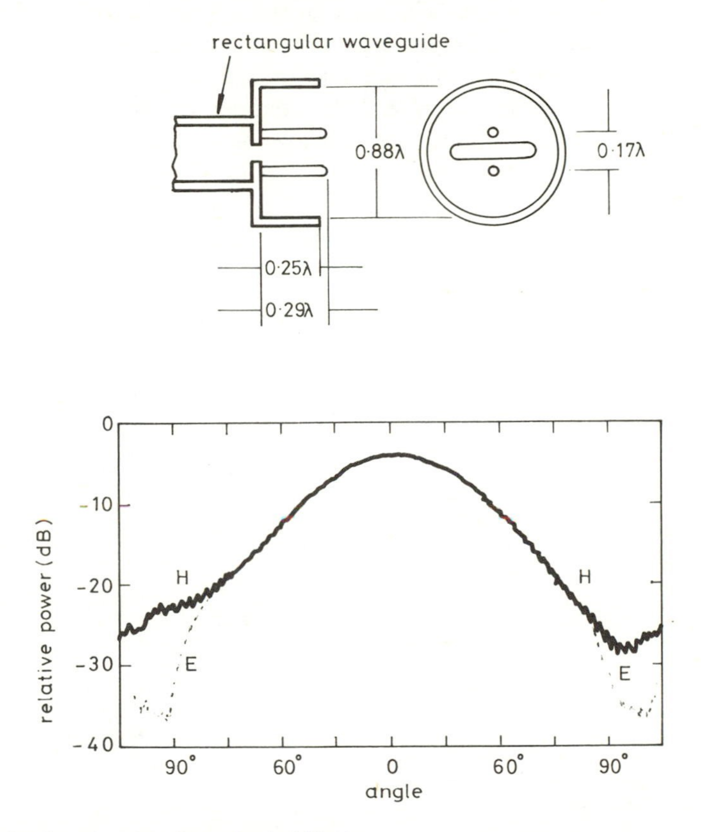

ig. 4.18 *Dual-mode prime focus feed of Clavin*
IEEE Trans., Sept. 1975, T-AP

.7 Hybrid mode horn feeds

'he development of the hybrid-mode horn was inspired in the early 1960s by the eed for improvements in the performance of large reflector antennas employed n satellite tracking, radar and radio astronomy. Improved antenna efficiency and educed spillover were the principal objectives, rather than polarisation purity which has come to the fore in recent times. The principal example of a hybrid-mode horn is the corrugated horn, to which the rest of this section is devoted, but similar properties are present in dielectric cone feeds[73].

The desirable radiating properties of a horn that carries a mixture of TE_{11} and TM_{11} modes are realised in the corrugated horn, along with two additional, important advantages. The effect of the circumferential slots is to force H_ϕ to vanish, along with E_ϕ at the horn wall. As a result, the boundary conditions for TE and TM waves are identical and the natural propagating modes are linear combinations of the two, and are referred to as hybrid modes. In particular, the dominant mode in a corrugated round waveguide or conical horn comprises a hybrid mixture of TE and TM components which can be readily excited without the need for a mode convertor. This fundamental hybrid mode is usually termed HE_{11}. The second advantage comes from the fact that the TE and TM components in the HE_{11} mode have the same cut-off and the same phase velocity. They therefore remain in correct phase relationship everywhere along the guide, independent of frequency, so that the bandwidth limitation of the dual-mode horn is removed. A detailed theoretical analysis has been given by Clarricoats and Saha[28], while simplified treatment will be found in Minnett and Thomas[29], Knop and Wiesenfarth[30] and Narasimhan and Rao[31,32].

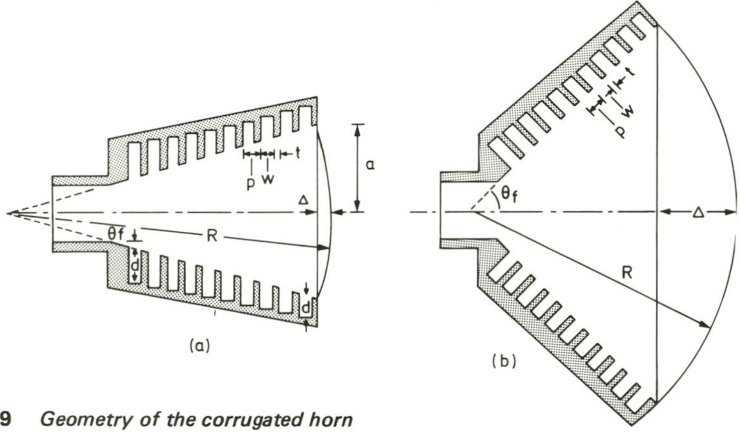

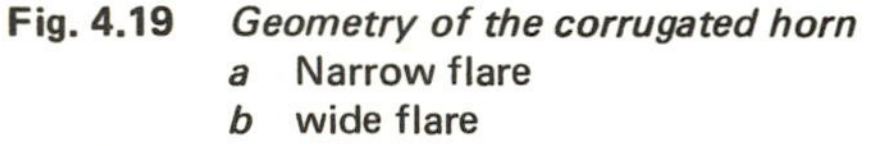
Fig. 4.19 *Geometry of the corrugated horn*
a Narrow flare
b wide flare

Two distinct versions of the hybrid mode horn are common: one in which th flare angle is large and the horn relatively short, the other in which the flare i very gradual so that the horn may have to be long, Fig. 4.19. The first of thes was introduced by Kay[33,34] and termed a scalar horn. Its distinguishing feature i that the large flare angle causes the spherical wave in the horn aperture to depar from a plane wave by a half wavelength or more. Because of this the shape of it radiation pattern depends very little on frequency, being determined almos entirely by the angle of flare. Over at least a 1·5:1 frequency range the width c the pattern at the −15 dB level is about equal to the total flare angle in the horn. The E- and H-plane phase centres are coincident, being located at the apex of the flare, and there is virtually no sidelobe structure over the same range. These desirable attributes hold for flare semi-angles up to about 70°, and for horn apertures exceeding about 2·5 λ in diameter. The scalar horn is thus an excellent prime focus feed in reflectors having *F/D* in the range 0·33 to 0·5, but blocking by the feed can become large in small diameter reflectors. For smaller *F/D* reflectors (e.g. 0·25) scalar horns have been built in which the flare semi-angle equals or exceeds 90°. These tend to resemble a simple round waveguide feed with a flanged aperture in which quarter-wave annular grooves are cut[35-38].

The other version is exemplified by a corrugated round waveguide with open end, or by a conical horn with a flare angle sufficiently small that the spherical wave in the aperture is within λ/5 or less of a plane wave., Fig. 4.19*a*. The bandwidth is thus limited by this condition and by the additional circumstance that the phase centre is now located between the apex and the aperture, its exact position depending on frequency.

In all cases, purely balanced HE_{11} mode operation occurs strictly only at one frequency, where the slots are of resonant depth and the mode content factor $\bar{\Lambda}$, is nearly equal to unity. The latter is defined as the ratio of the longitudinal magnetic field of the TE mode (multiplied by 377 Ω) to the longitudinal electric field of the TM mode. For positive values of $\bar{\Lambda}$, the TE and TM components are in phase and the mode is designated HE. Negative values of $\bar{\Lambda}$, are possible, leading to an antiphase condition, and the mode is then designated EH. The generation of this mode is generally undesirable and negative values of $\bar{\Lambda}$ are best avoided.

Narrow flare-angle corrugated horns

In the design of corrugated horns three problem areas must be considered:

(i) the junction between the smooth-walled waveguide feeder and corrugated horn
(ii) the flare section which expands the corrugated waveguide diameter to the required dimensions at the aperture
(iii) the radiation conditions established at the aperture to ensure that the radiated field exhibits the desired degree of symmetry and cross-polar suppression.

The last of these three areas has probably received the greatest attention in the literature, particularly in the treatment of small flare-angle horns. However, the increasing need for exceptionally low cross-polar radiation levels has made all three areas equally critical. A paper by Dragone[83] provides an excellent analysis of the large-aperture, small flare-angle corrugated horn with a number of useful and simple equations for modelling the performance of this complex device.

The junction between the feeder waveguide and the corrugated horn normally occurs in the throat region of the horn and the requirements here are at least two-fold. First the fundamental TE_{11} mode in the feeder must be matched into the HE_{11} mode in the corrugated structure. Secondly, considerable care must be taken to avoid mode conversion to higher-order modes in this region. For example, to satisfy specifications which demand cross-polar peak levels below −40dB over the main lobe of the radiation pattern clearly implies that mode conversion in the throat of the horn must be suppressed to very low levels.

The input reflection of a smooth waveguide connected to a corrugated feed of the same diameter is simply given by the coefficient

$$\rho_1 = \frac{\beta_1' - \beta_1}{\beta_1' + \beta_1} \tag{4.21}$$

where β_1' and β_1 are the propagation constants in the two waveguides.

For a smooth waveguide there is a simple relationship between the propagation constant β of a mode and the waveguide diameter, but no such simple relationship exists for a corrugated guide unless its radius a is much greater than a wavelength. This will often be true of the aperture of the horn but not necessarily true in the throat region. When the radius is large in terms of the wavelength, then for some modes using Dragones notation[83]

$$\beta = \sqrt{(ka)^2 - u_{0m}^2} \tag{4.22}$$

where u_{0m} is the mth zero of the Bessel function J_0 of order zero, i.e. $J_0(u_{0m}) = 0$.

For all other modes, except the surface-wave mode,

$$\beta = \sqrt{(ka)^2 - u_{2m}^2} \tag{4.23}$$

where u_{2m} is the mth root of the Bessel function of order two, i.e. $J_2(u_{2m}) = 0$.

For corrugated waveguides which do not satisfy the large normalised radius condition it is necessary to compute the appropriate propagation constant from the boundary condition for the infinite corrugated waveguide. If the horn throat must pass higher-order modes (e.g. in a multimode tracking system) then the match at the throat must be examined for both the fundamental and higher-order modes of interest.

A good impedance match at the throat is obtained when the slot depth of the first corrugation, d, is made a half wavelength at the highest frequency of operation. Succeeding slots should taper down from the half wavelength depth to the depth d required for resonance.

In the practical design of a high-performance corrugated horn the mode conversion along the flared section can be a serious problem. For example Dragone[83] has derived an equation for the mode conversion coefficient M from the HE_{11} mode to the HE'_{i1} mode* in small flare-angle horns which can be expressed in the form

$$M \leqslant \frac{1{\cdot}357 \times 10^{-2} \tan^2 \theta f}{(1 - t/\rho)^2 \tan^2 kd} \tag{4.24}$$

To suppress this mode conversion to a low level it is necessary either to keep a very small, or alternatively to maintain a high, wall reactance (i.e. resonant slots) such that $\tan^2 kd \to \infty$.

Normalised co-polar radiation patterns for the small flare horn are shown in Fig. 4.20, taken from Thomas[39], with spherical wave error Δ as parameter. From the geometry of Fig. 4.19 it is seen that Δ is given by

$$\Delta = a \tan \frac{\theta_f}{2} = R \sin \theta_f \tan \frac{\theta_f}{2} \tag{4.25}$$

When ka is less than 12 the abscissa in Fig. 4.20 must be truncated at the value corresponding to $\theta = 90°$. If ka is reduced below about 6 (corresponding to a half power beamwidth greater than about 37°) then the flange surrounding the aperture begins to affect the patterns and to destroy the pattern symmetry; the curves are then no longer valid.

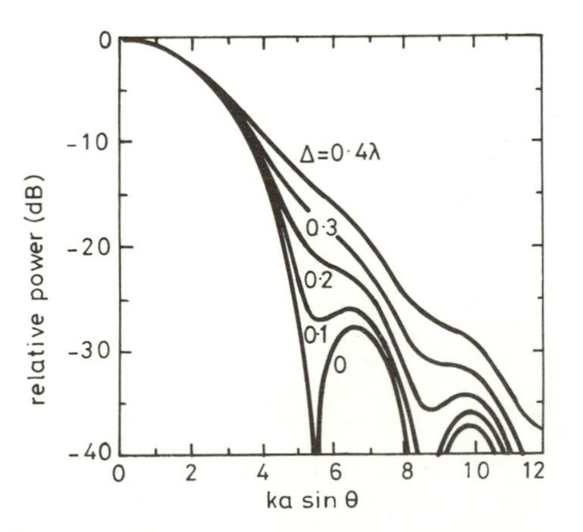

Fig. 4.20 *Normalised radiation patterns of small flare horns with Δ/λ as parameter*

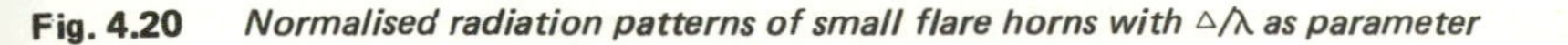

The half beamwidths at the −3dB, −10dB and −20dB power levels relative to the boresight power are shown in Fig. 4.21 as a function of the aperture diameter, for the case $\theta_f = 0$, i.e. an open-ended waveguide. They are also useful for the initial design of any small flare angle horn.

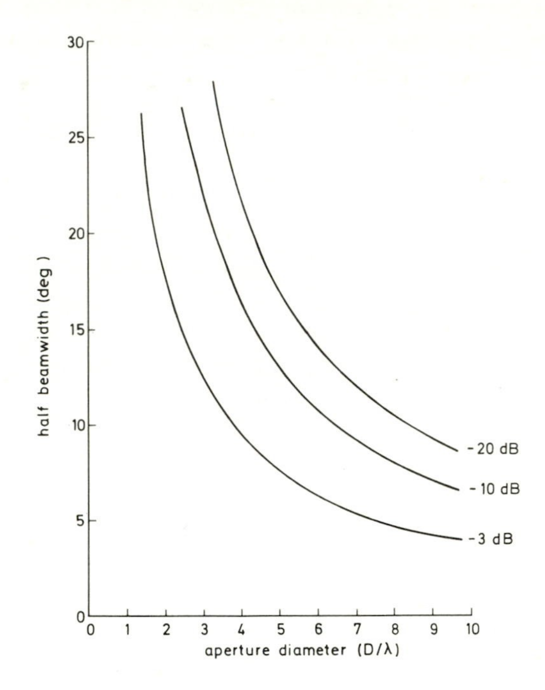

Fig. 4.21 *Half beamwidth of −3 dB, −10 dB and −20 dB copolar power levels against normalised aperture diameter for open-ended corrugated waveguide*

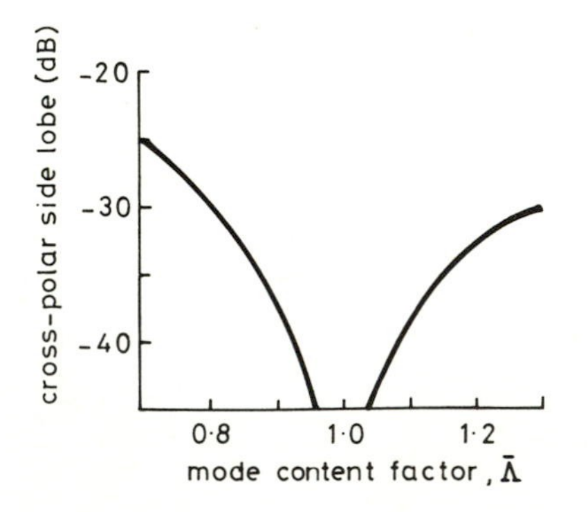

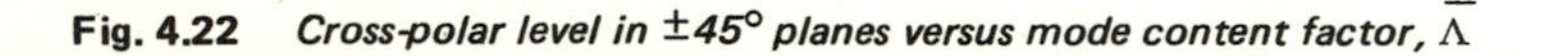

Fig. 4.22 *Cross-polar level in ±45° planes versus mode content factor,* $\bar{\Lambda}$

The HE_{11} cross-polar radiation pattern in the 45° plane has a null on boresight and peaks on either side of boresight. The peak occurs at approximately the −9dB power level of the co-polar pattern and therefore normally falls within the part of the beam which will illuminate a reflector. The variation of the peak cross-polar power with mode content factor $\bar{\Lambda}$ is shown in Fig. 4.22. Measured values of cross-polar power for horns with $ka > 9$ agree well with this result. The way in which $\bar{\Lambda}$ varies with frequency is shown in Fig. 4.23, where f_0 and k_0 refer to the frequency at which the slots are of resonant length. Figs. 4.22 and 4.23 are from Thomas[39]. The parameter is $k_0 R\theta_f$, or approximately $k_0 a$ for cylindrical guide. The figure shows that $\bar{\Lambda}$ is nearly linearly related to frequency over the range of frequencies normally used in a corrugated horn, so that Fig. 4.22 also shows the peak cross-polar behaviour with frequency. The frequency bandwidth at a given cross-polar level increases as the flare angle and aperture increases.

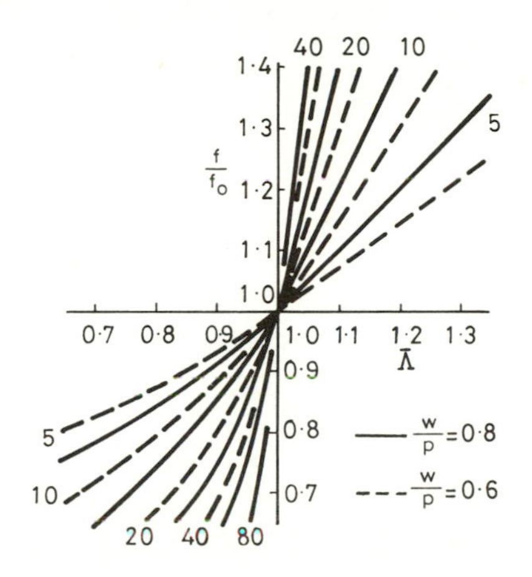

Fig. 4.23 *Variation of $\bar{\Lambda}$ against frequency with $K_0 R\theta_f$ as parameter*
IEEE Trans., March 1978, T-AP 26

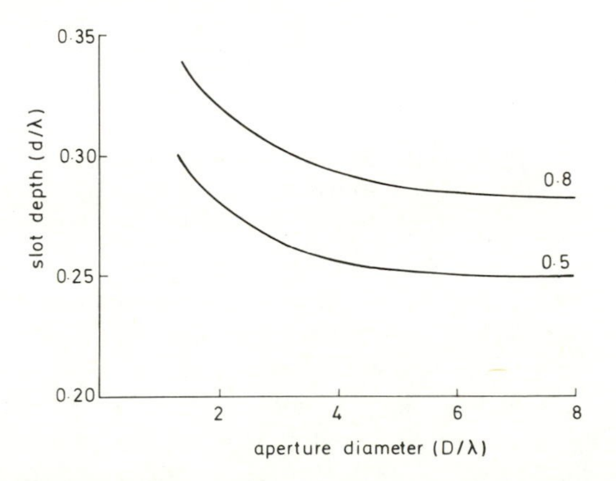

Fig. 4.24 *Normalised slot depth to produce minimum cross-polarisation. Parameter slot width/pitch ratio*

The slot resonant frequency is determined by the slot depth, the aperture diameter and the slot width/pitch ratio. The general assumption is often used that the slot depth be $\lambda/4$. This is adequate for horns where the cross-polarisation is of little interest. However to obtain low cross-polarisation the slot depth must be 'tuned' to resonance. Fig. 4.24 shows the slot depth for minimum cross-polarisation as a function of aperture diameter. For small apertures or high slot width/pitch ratios the slot must be considerably deeper than $\lambda/4$.

The number of slots per wavelength, i.e. the pitch, has only a marginal effect on the radiation characteristics. It is usual to have at least four slots per wavelength, but the number can be chosen for mechanical convenience.

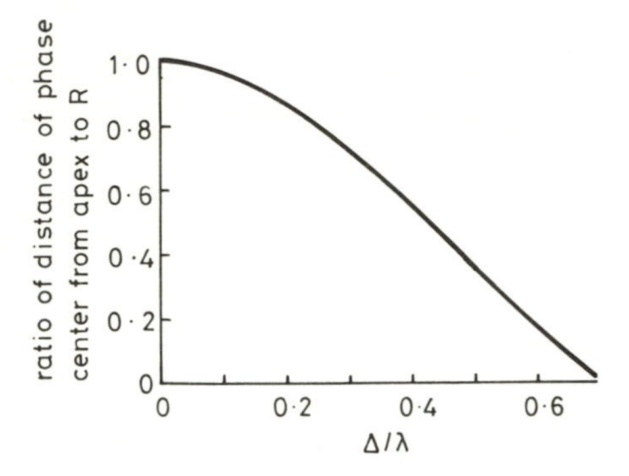

Fig. 4.25 *Normalised distance of horn phase centre from apex versus* Δ/λ

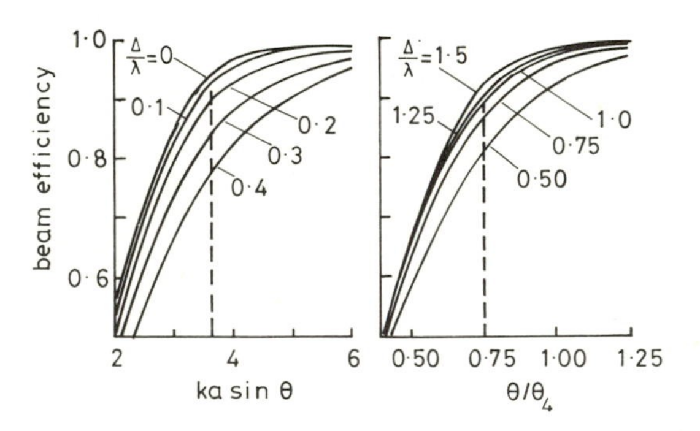

Fig. 4.26 *Beam efficiencies for small flare (left) and scalar (right) horns*
IEEE Trans., March 1978, **T-AP 26**

The location of the phase centre in the narrow flare horn is a function of the parameter Δ, as shown in Fig. 4.25[39]. If there were no phase error then the phase centre would lie in the plane of the horn aperture. For scalar horns in which Δ exceeds about 0·7 λ the phase centre is effectively at the apex. Beam efficiency curves, from which spill-over past the reflector may be estimated, are given in Fig. 4.26[39]. The dashed line approximately indicates the efficiency at the angle θ corresponding to the -10dB power level. It is clear that values of Δ less than about 0·2 λ are required if spill-over loss is to be kept below about 10%.

As an example, suppose a feed is required for a Cassegrain system in which the effective F/D ratio is 1·0, corresponding to an opening angle $\theta^* = 28°$ from Fig. 4.1. The −13dB pattern level occurs at about $ka \sin \theta = 4{\cdot}0$ (for small Δ) as shown by Fig. 4.20. Putting $\theta = \theta^* = 28{\cdot}0°$ then yields $ka = 8{\cdot}52$, so that the aperture radius is $a = 1{\cdot}36\ \lambda$. If Δ is taken to be 0·1 λ then Fig. 4.26 shows that spill-over beyond the subreflector will be approximately 5%. In this case, from eqn. 4.25, θ_f is found to be 8·46° and the slant length is $R = 9{\cdot}22\ \lambda$. Fig. 4.25 shows that the phase centre will lie at a distance 0·95 R from the apex; this is about 0·4 λ behind the aperture plane. If the frequency of operation is increased to $1{\cdot}5 f_0$ then both $ka \sin \theta^*$ and Δ/λ will increase by 1·5:1. Fig. 4.20 then indicates that the illumination in the direction θ^* will fall from −13dB at f_0 to about −22dB at $1{\cdot}5\ f_0$, leading to underillumination at the edge of the reflector aperture. The phase centre will move farther behind the aperture, to about 0·9 R, while spill-over loss will decrease because the effect of the increase in $ka \sin \theta^*$ from 4·0 to 6·0 more than compensates for the increase in Δ/λ.

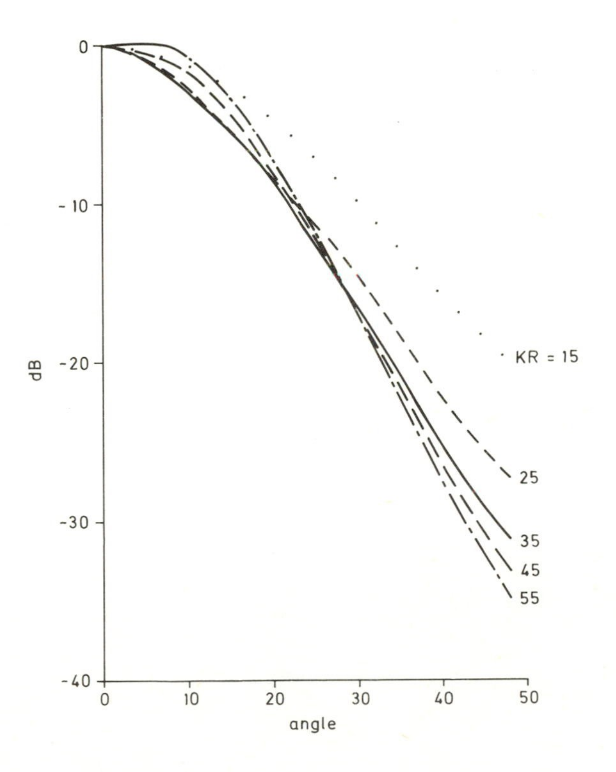

Fig. 4.27 *Copolar patterns of 30° semi flare angle scalar horn. Parameter normalised slant length = kR*

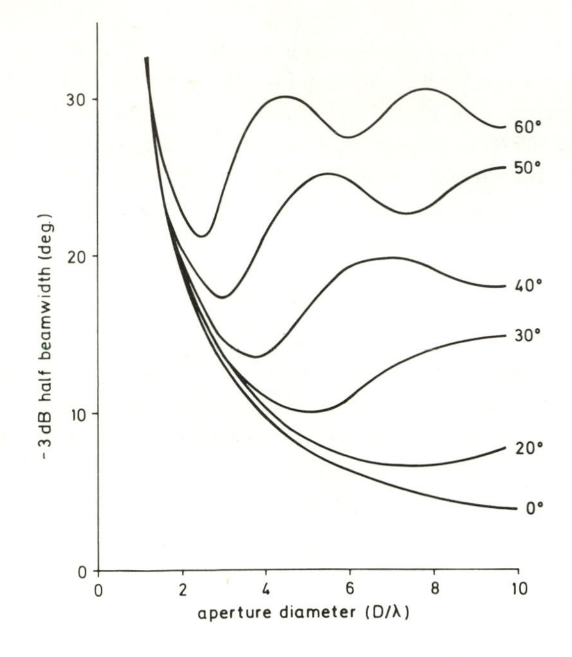

Fig. 4.28 *−3dB half beamwidth against normalised aperture diameter. Parameter: Horn semi-flare angle θ_f, deg*

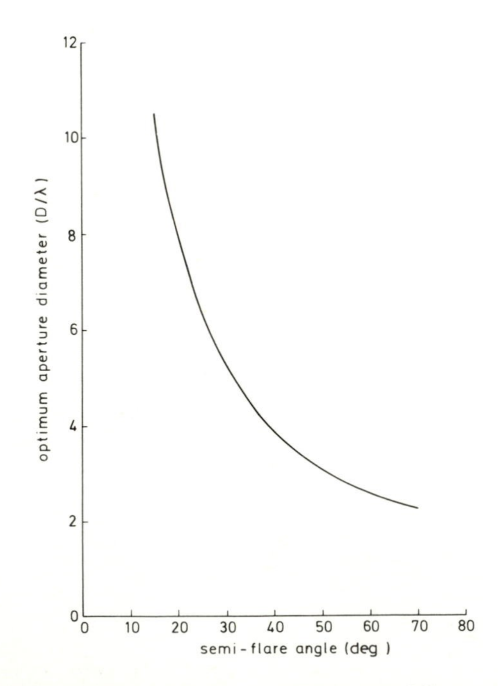

Fig. 4.29 *Optimum normalised aperture diameter against semi-flare angle for scalar horn*

Scalar horns

The copolar radiation characteristics of a scalar horn with $\theta_f = 30°$ and various normalised lengths are shown in Fig. 4.27. Comparison with Fig. 4.20 for the narrow flare-angle horn shows that the radiation behaviour is quite different. The shape of the scalar horn pattern near boresight changes with frequency but the beamwidth at the −15dB level is nearly constant for the larger normalised lengths. This result indicates the general form of the radiation pattern for all scalar horns. The −3dB half beamwidth is plotted against aperture diameter in Fig. 4.28 for a number of semi-flare angles. The half beamwidth for lower power levels has a similar form except that the ripple amplitude is reduced and consequently the lower power beamwidth is nearly constant as frequency changes. A very wide band horn can therefore be constructed. However the boresight gain is a maximum at the minimum −3dB half beamwidth, and so an optimum size can be specified for any semi-flare angle. This is shown in Fig. 4.29 and the corresponding −3dB, −10dB and −20dB half beamwidths in Fig. 4.30. The −10dB and −20dB curves are approximately true for any aperture size. As a guide, the half beamwidth at the −15dB level is about equal to the semi-flare angle of the scalar horn.

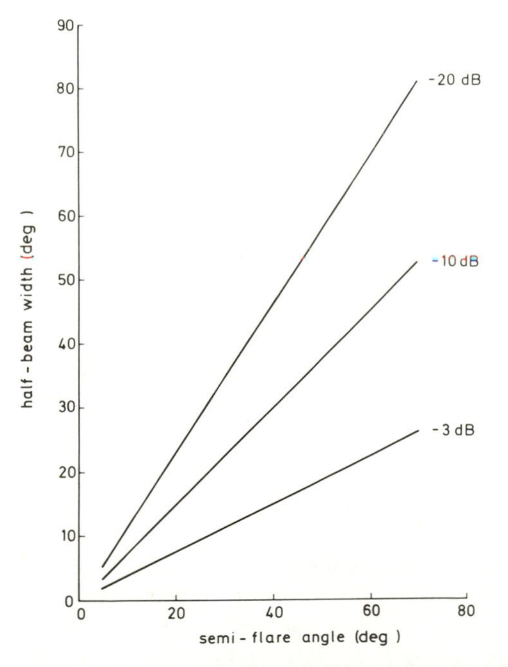

Fig. 4.30 ***−3 dB, −10 dB and −20 dB half beamwidth against semi-flare angle for optimum scalar horns***

The phase centre of the scalar horn will remain essentially fixed at the apex. If it is desired to design a scalar horn feed for the Cassegrain system in the previous example, then $\theta^* = 28°$ and Fig. 4.30 indicates that a −13dB edge illumination will be given by a horn with a semi-flare angle $\theta_f = 30°$. The optimum aperture diameter will be $5{\cdot}2\lambda$ and the corresponding spherical phase error Δ will be $1{\cdot}4\lambda$. Reference to Fig. 4.26 shows that the spill-over loss for this feed will be somewhat higher than for the narrow flare angle horn, but its phase centre remains stationary so that operation will improve as frequency is increased above the design frequency.

The radiation patterns of the balanced hybrid mode horns discussed above have a main beam which is approximately Gaussian in shape. If, in addition to the HE_{11} mode, the next higher HE_{12} mode is introduced, the pattern can be made to approximate a sector beam having a flatter, or even a rippled top, and steeper sides than the Gaussian beam.[40,41] For reflector feed purposes this clearly leads to a more nearly uniform aperture illumination: the resulting increase in gain can be expected to be about 10%.

The characteristic equation for all HE_{ln} modes in a corrugated guide is given by[28].

$$F_1\,(k_c a) - \frac{\beta^2}{k_2\,F_1\,(k_c\,a)} = \left(\frac{k_c}{k}\right)^2 S_1\,(ka,\,ka') \tag{4.26}$$

where $F_1\,(k_c a) = k_c a\,\dfrac{J_1'\,(k_c a)}{J_1\,(k_c a)}$

and $S_1\,(ka_1\,ka') = ka\,\dfrac{J_1'\,(ka)\,Y_1\,(ka') - J_1\,(ka')\,Y_1'\,(ka)}{J_1\,(ka)\,Y_1(ka') - J_1\,(ka')\,Y_1\,(ka)}$

$a' = a + d$, $\beta = 2\pi/\lambda_g$ is the mode propagation constant,
$k_c = 2\pi/\lambda_c$ is the cut-off wavenumer and $\beta^2 + k_c^{\,2} = k^2$

The mode content factor, $\bar{\Lambda}$, is positive for balanced HE_{ln}modes and negative for the EH_{ln} modes. Dispersion curves calculated from these equations are shown in Fig. 4.31, for the first two HE and EH modes. Although the EH_{11} mode has the same cutoff as the corresponding TE_{11} modes in a smooth guide, its dispersion characteristics are quite different, as the Figure shows. In a single hybrid mode guide the EH_{11} mode will have passed to high frequency cutoff when the HE_{11} propagates and so it can be disregarded. In a two mode guide, however, the EH_{12} and HE_{12} modes have nearly the same propagation characteristics. Thus, if the former mode is generated it will propagate to the aperture where it will radiate a purely cross-polarised pattern with an axial null.

The usual technique for generating the desired HE_{12} mode is by means of a large step discontinuity in diameter near the throat of the horn,[43] as shown in Fig. 4.32. As the ratio of diameters a/a_0 increases the power in the HE_{12} mode increases relative to that in the HE_{11} and approaches the optimum value. A large input guide diameter, $2a_0$, improves the match and reduces the level of the unwanted EH_{12} mode at the expense of less than optimum power in the HE_{12} mode. The corrugated section diameter, $2a$, should be less than $2{\cdot}72\ \lambda$ in order to ensure that the HE_{13} mode is below cutoff. The length l of the constant-diameter section must be chosen so that the HE_{11} and HE_{12} modes arrive in phase at the aperture. Theoretical and measured patterns for a feed of this type designed to illuminate a reflector with $F/D = 0{\cdot}41$ have been derived by Vu and Vu[40] and are reproduced here in Fig. 4.33. The aperture diameter of the horn is $2{\cdot}6\ \lambda$. The patterns ***a*** represent the case in which the length l is such as to cause the two modes to be in phase at the aperture; for the other patterns l has been changed to cause the modes to be in antiphase. No information as to the actual value of l has been given for either case.

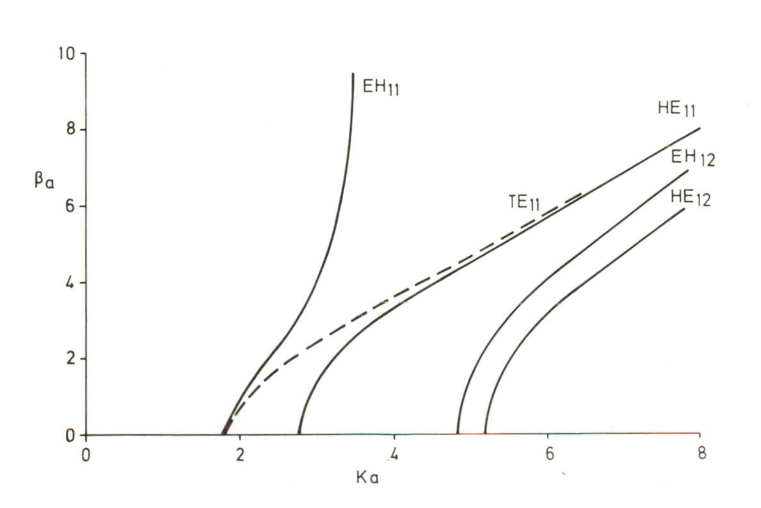

Fig. 4.31 *Propagation characteristics of corrugated waveguide for a (a + d) = 0·7 Dotted curve shows TE_n mode of smooth-wall circular waveguide*

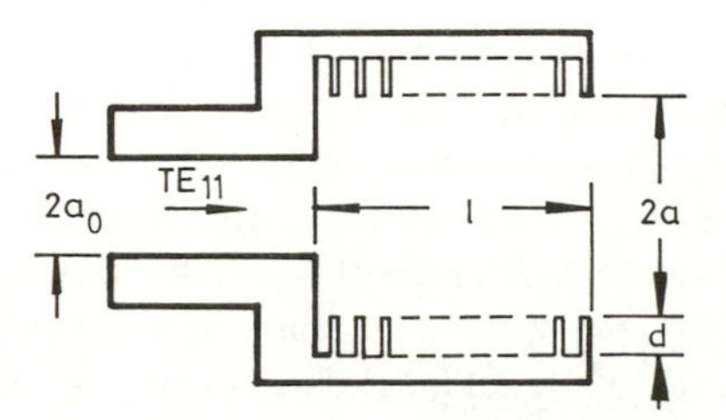

Fig. 4.32 *Two-hybrid mode feed for sector beam approximation*

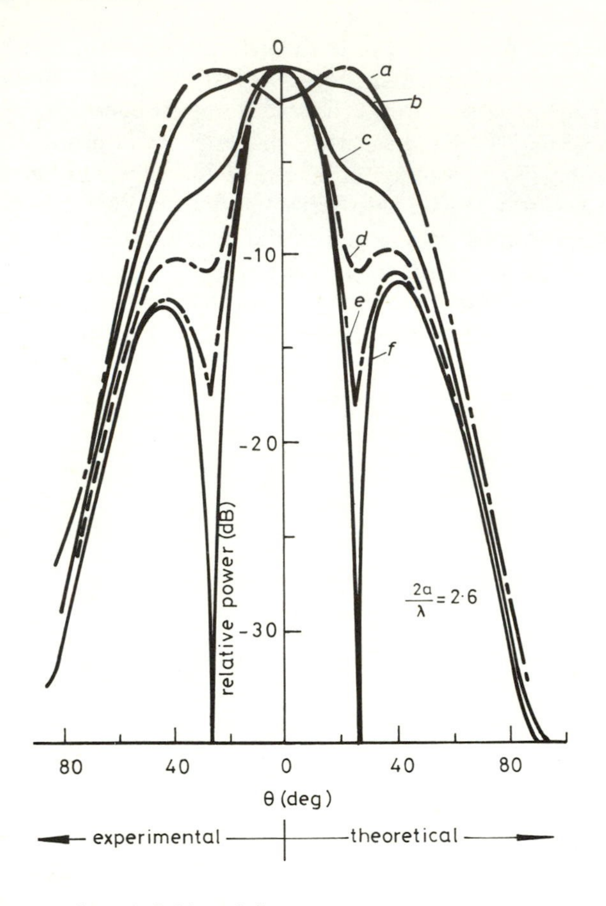

Fig. 4.33 *Patterns of two-hybrid mode horn*
Relative phase of modes depends on l;
a: 0°, *b*: 36°, *c*: 72°, *d*: 108°, *e*: 144°, *f*: 180°
Electron. Lett., 19 Mar. 1970, **6**

4.8 Tracking feeds

Antennas that must track moving targets, such as aircraft or satellites, need direction-of-arrival information from which signals may be derived for pointing control. A reflector antenna with the usual altazimuth pedestal requires two such signals: one to control the azimuth drive motor, the other for elevation drive control. The specialised feed arrays that perform these functions are often called two channel monopulse tracking feed systems because, in principle, they are capable of yielding both range and angle-of-arrival information from the flight of a single radar pulse.

Perhaps the simplest system for two-channel tracking is that shown in Fig. 4.34. Four identical horns are closely arrayed and are interconnected in the manner shown to produce a sum signal Σ and two difference signals Δ_{az} and Δ_{el}. The interconnecting network is termed a monopulse bridge circuit comparator, and utilises hybrid junctions to perform the summing and differencing functions. The sum channel signal Σ results from in-phase addition of the signals in all four horns A, B, C and D. The difference channel signals arise by anti-phase addition of the signals from appropriate pairs of honrs, i.e. (A + B) – (C + D) and (A + C) – (B + D), as indicated schematically in the figure.

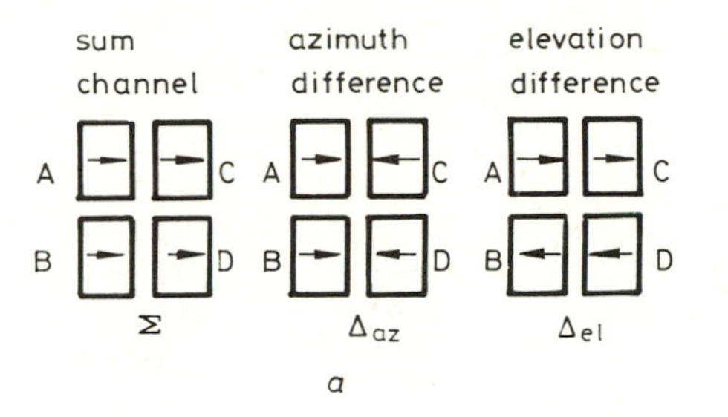

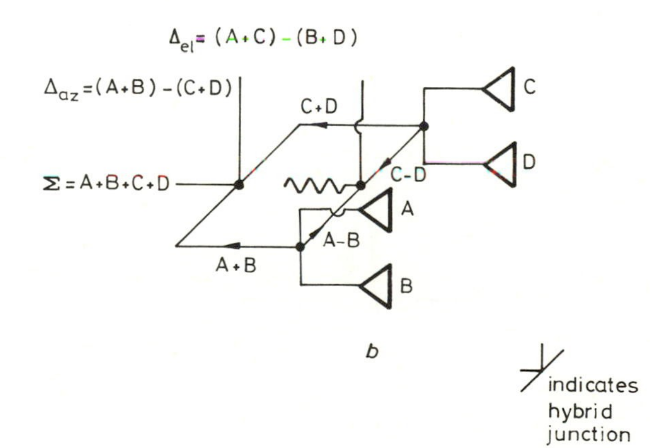

Fig. 4.34 *Simple two-channel tracking feed*
a Modes in horn apertures
b Comparator bridge network

The radiation pattern of this cluster of four horns is described by the product of an array factor and an element factor. The latter is simply the pattern of an individual horn and is fixed. The nature of the array factor depends upon which channel is under consideration. The sum channel signal clearly derives from an in-phase array of 4 elements so that the array factor has the shape of a cosine function.

Thus, the array illuminates the reflector in the usual way, tapering from a maximum at the centre of the aperture to a low value at the edge as shown in Fig. 4.35. In the azimuth difference channel the pair A and B taken together and the pair C and D together form an array of two elements in phase opposition so far as the azimuth plane is concerned. Therefore, in this plane the array factor is a sine function whose period is the same as that of the cosine array factor of the sum channel. In the difference-channel mode the reflector illumination is such that the left half of the reflector is out of phase with the right half as indicated by the dashed line in Fig. 4.35.

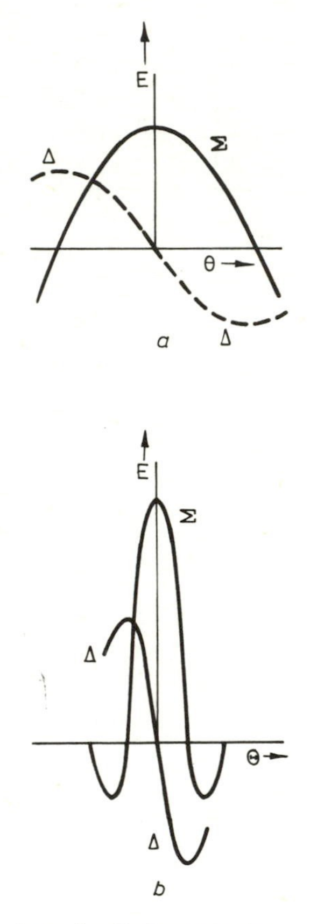

Fig. 4.35 *Sum and difference channel radiation patterns*
a Feed illumination patterns
b Reflector far-field patterns

The resulting far field voltage patterns of the reflector are as indicated schema tically in Fig. 4.35*b*. It can be seen that if the target is not on the boresight ($\theta = 0$) a signal will be received in the difference channel Δ whose magnitude and sign depend upon angle off boresight. This error signal can be amplified and applied to a servo-controlled drive system which then reduces the pointing error to a very small value.

With the simple system discussed above it is not possible to optimise the sum and difference channel reflector illuminations simultaneously. For example, suppose the reflector opening angle θ^* is chosen to correspond to the null in the sum channel pattern Σ in Fig. 4.35*a*. Such a highly tapered illumination pattern will yield a low sidelobe secondary pattern with moderate aperture efficiency. However, the same Figure shows that only half of each difference pattern lobe falls on the reflector. The high level of edge illumination creates high sidelobes in the secondary difference pattern and aperture efficiency will be poor owing to the large amount of spill-over. If the reflector opening angle is increased in order to improve the difference pattern performance then sum pattern performance deteriorates rapidly because sidelobes in the sum channel feed pattern now illuminate the reflector. This and other problems have been thoroughly discussed by Hannan,[44,45] who also suggests techniques for obtaining more nearly optimised sum and difference channel patterns. Implementation of these sophisticated techniques introduces considerable complexity and usually requires a multiplicity of horns (up to 12) or the generation of higher modes in the horn apertures.[46,47]

Less compromise and better performance in both the sum and difference channels may be obtained from a five-horn feed in which a central horn, used for the sum channel, is surrounded by four other horns whose functions are to provide the azimuth and elevation channel difference signals.[48,49,50] A clever arrangement for maximising sum mode gain in a five-horn system, while permitting operation with circular polarisation, is fully described in Ref. 51. By means of a directional coupler a small amount of sum mode power in the four outer horns is coupled into the transmission line feeding the central horn. By proper phasing, the two signals are anti-phased with the result that the sum mode pattern has a flattened top and steeper skirts. Since this pattern is a closer approximation to uniform illumination than that of the central horn alone, aperture efficiency is improved.

If the benefits of the scalar horn (polarisation purity, low sidelobes and wide bandwidth) are to be introduced into a five-horn monopulse feed system,[52] complications arise. These occur chiefly because the larger aperture of a scalar horn forces a greater separation between the four horns in the difference mode array. A thorough investigation has been carried out by Viggh.[53]

4.9 Multi-octave bandwidth primary feeds

A paraboloidal reflector is inherently a broadband device with a low frequency limit corresponding to the wavelength at which its diameter is, perhaps, two wavelengths and a high frequency limit set by the accuracy of its surface profile. This range may span several decades in frequency and is generally far greater than can be provided in any single feed. So-called frequency-independent antennas exist, however,[54,55] which can function reasonably efficiently over a 10:1 or even a 20:1 range in frequency. Some of these can be adapted for use as prime focus feeds, thereby permitting a single reflector to operate over a decade or more in frequency.

The beamwidth of the reflector antennas will, of course, vary approximately proportionally to wavelengths over these broadbands.

One of a class of log-periodic antennas, called a trapezoidal tooth structure, is shown in Fig. 4.36. It is defined by three angles α, β and ψ and the dimensionless ratio τ which is defined in the Figure. This antenna radiates essentially a undirectional, linearly polarised beam in which the electric field is parallel to the teeth. Its electrical properties repeat periodically with the logarithm of frequency, one period being defined as the range from τf to f. Since variations within one period are generally small they will similarly be small over many periods, leading to a very wide band antenna. The two halves of the structure are fed against one another by means of a balanced transmission line connected to the apices.

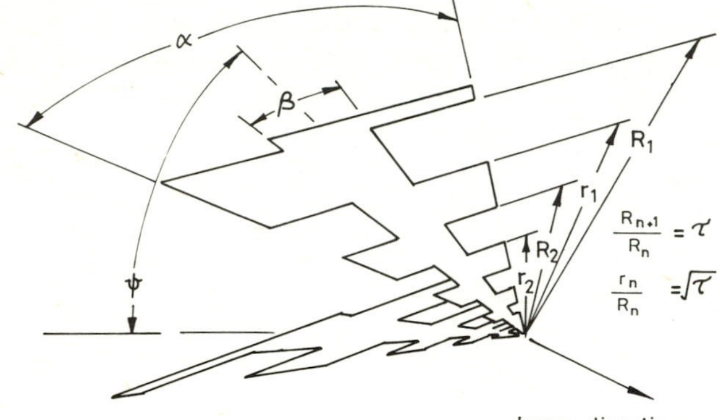

Fig. 4.36 *Log-periodic trapezoidal tooth structure*
IRE Nat. Conv. Record I, 1958

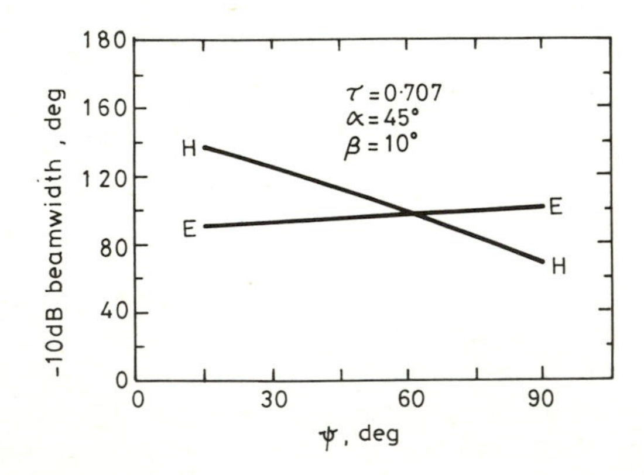

Fig. 4.37 *E- and H-plane −10 dB beamwidths*
IRE Nat. Conv. Record I, 1959

The properties of this structure, considered as a primary feed, have been experimentally measured[56] as functions of the four parameters α, β, ψ & τ. The experimental data is too extensive to be given here in its entirety, but by fixing the values of α, β, and τ it is possible to present useful feed design data in terms of the single parameter ψ. Thus, Figure 4.37 displays the variation in -10dB beamwidth as a function of ψ for the case $\alpha = 45°$, $\beta = 10°$ and $\tau = 0{\cdot}707$ while Fig. 4.38 shows the distance in wavelengths of both the E and H plane phase centres from the apex of the structure. These graphs are representative of conditions at a fixed frequency, but the period τf to f is not large (the range is 0·707 to 1) and variations within the period are reasonably small. Thus the curves may be regarded as essentially independent of frequency.

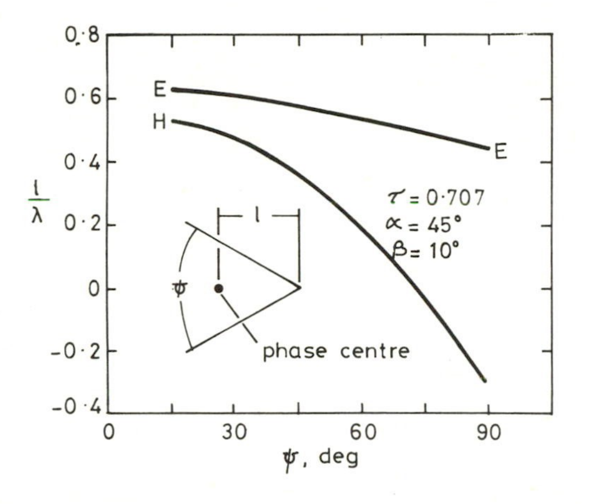

Fig. 4.38 *Location of phase centre versus Ψ*
IRE Nat. Conv. Record I, 1959

For the value $\psi = 60°$ the E and H plane beamwidths are each equal to about 100° and such a tooth structure would therefore provide circularly symmetrical illumination for a suitable paraboloidal reflector. Figs. 4.1 and 4.2 show that a focal ratio of about 0·5 would be a good choice, leading to an edge taper of perhaps -14dB. Reflectors with smaller *F*/*D* ratios can be accommodated with some sacrifice in efficiency and loss of circular symmetry in the illumination pattern. For example, the subtended angle ($2\theta^*$) for a reflector with $F/D = 0{\cdot}4$ is 128°. A tooth structure with $\psi = 45°$ has -10dB widths of 90° in the E plane and 112° in the H plane. It will therefore under-illuminate the reflector (more so in the E than in the H plane) leading to decreased aperture efficiency but, at the same time, to a low sidelobe pattern. The input impedance of these toothed structures will vary, over one period in frequency, about a nominal value called the characteristic impedance Z_0. For $\psi = 45°$ Z_0 is about 170Ω, increasing to about 190 Ω when ψ becomes 60°. The range of variation in impedance over one period can be expected to be about 1·5:1. At low frequencies VHF) a simple two-wire line may be connected to the feed points at the apex. At higher frequencies, where the use of coaxial cable is desirable, some kind of wide-band balun is needed.[54,55]

The non-coincidence of the E and H plane phase centres in these structures creates astigmatism when they are used in conjunction with a paraboloidal reflector. Added to this is the fact that the phase centre locations move with frequency, making it difficult or impossible to achieve perfect focusing, except perhaps at one frequency. Nevertheless, the loss in gain and increase in sidelobe levels due to these effects can be rendered acceptably small. A displacement Δ of the phase centre from the geometrical focus of the reflector causes a quadratic phase error which reaches a maximum value of δ at the aperture edge given by

$$\delta = \frac{2\pi\Delta}{\lambda}(1 - \cos\theta^*) \tag{4.27}$$

where θ^* is the reflector subtended semi-angle. The loss in gain due to the quadratic error δ has an upper bound given by[57]

$$L = 20\log_{10}\left[\frac{\sin\frac{\delta}{2}}{\frac{\delta}{2}}\right] \text{dB} \tag{4.28}$$

Reference to Figure 4.38 shows that the vertex of the feed structure should be moved closer to the reflector rather than being placed at the focus. Suppose, for example we define a mean phase centre location as the average values of l for the E- and H-planes; for the structure with $\psi = 45°$ the Figure shows that $l = 0{\cdot}46\,\lambda$. If the vertex is now placed at $0{\cdot}46\,\lambda$ away from the focus the mean phase error will vanish at a wavelength λ which lies somewhere within the desired operating range λ_1 to λ_2. A reasonable choice for λ is then the one that makes the mean error at λ_2 equal, but of opposite sign, to the error at λ_1. A little arithmetic shows that λ is the harmonic mean between λ_1 and λ_2. Although the mean phase error vanishes at wavelength λ the actual errors in the E and H planes amount to $\pm 22°$ and are acceptably small.

By combining two of the above structures to form a four-sided pyramid, and by exciting the orthogonal pairs in phase quadrature, a circularly polarised feed is obtained. A simple way to achieve the quadrature condition is to scale one of the pairs by the factor $\tau^{1/4}$ in relation to the other. Reference to Fig. 4.39 shows that there is a geometrical constraint to be observed; the structure angle α must be less than the value obtained from the equation

$$\tan\frac{\alpha}{2} = \sin\frac{\psi}{2}\,. \tag{4.29}$$

The aperture blocking created by these log-periodic feeds appears to be approximately equal to the physical cross-section of the structure, regardless of the frequency.

There are other forms of log-periodic antennas that can also be used as wideband feeds. These include the wire trapezoidal and zig-zag types, and the log-periodic dipole array. The latter is also well suited as a feed for reflectors having F/D ratios from 0·4 to 0·5 and has the advantage of having coincident E- and H-plane phase centres.

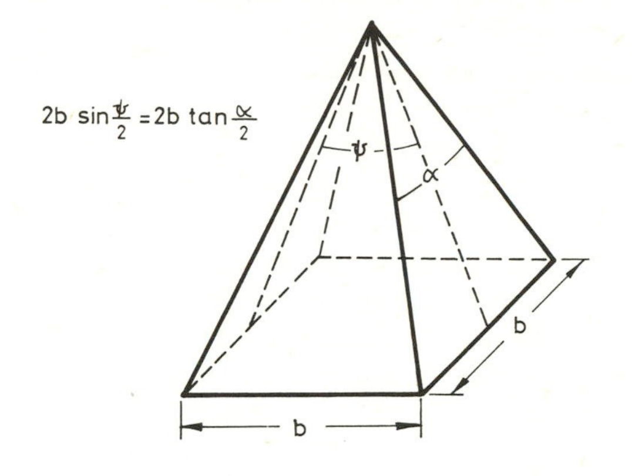

Fig. 4.39 *Pyramidal structure*

Another kind of frequency independent antenna which can be used as a feed for a paraboloidal reflector is the two-arm equiangular spiral wound on a cone.[58] When the half-angle θ_0 is less than about 15°, the conical spiral radiates an essentially unidirectional beam off the apex of the cone along the axis. The polarisation is nearly circular over the major portion of the pattern; the axial ratio is typically no worse than 3dB. The −10dB beamwidth may vary from 120° to 150° for wrap angles in the range 70° to 85°, making the spiral suitable as a feed for F/D ratios from 0·3 to nearly 0·5. The pattern exhibits good axial symmetry with a variation of perhaps 3° to 9° in the −10dB beamwidth between orthogonal cuts. The characteristic impedance for the two-arm self-complementary spiral on a cone with θ_0 = 15° is about 150 Ω. Front-to-back ratio can be expected to be about 15dB for a wrap angle of 70°, increasing to about 18dB for a wrap angle of 80°. The phase centre of the radiation pattern is on the cone axis at the location corresponding to a cone circumference of approximately $3\lambda/4$. Since the phase centre location moves with frequency, the same considerations that apply to the log-periodic structures will govern the placement of the cone apex relative to the reflector focal point.

The problem of the moving phase centre is eliminated in the planar spiral in which the cone angle is increased to θ_0 = 90°. In this case, the spiral radiates bidirectional lobes of opposite senses of polarisation. When installed over a cavity, the pattern becomes unidirectional, as required for a feed, but the bandwidth is sensitive to cavity design. The pattern of the cavity-backed planar spiral is somewhat broader than that of the conical spirals discussed above, so that it is better suited to a reflector with a smaller F/D ratio, say about 0·3 to 0·35.

4.10 Mathematical modelling of primary feeds

The increasing demands for improved electrical performance from reflector antennas has stimulated significant advances on the application of mathematical techniques for their design and optimisation. Considerable progress has been made in the analysis and synthesis of single and dual-reflector systems, both axisymmetric and asymmetric.[72,74-77]

More demanding performance specifications have also had a very significant impact upon primary-feed design, particularly in connection with improved polarisation purity. Extensive research and development has been performed on the properties of horn antennas[78] and our understanding of their polarisation properties in particular has advanced significantly over the last decade.

The use of mathematical techniques based upon physical optics has become very common in reflector antenna analysis and design. In essence these methods provide a prediction of the vector radiated fields from the overall antenna based upon an approximate knowledge of either the surface current distribution on the reflector, or the reflected vector fields in a plane immediately in front of the reflector, when the reflector is illuminated by a primary-feed[2,79]

Physical optics methods have been found to provide good predictions of the radiated fields provided that the surface dimensions are large relative to the operating wavelength and provided also that the fields incident upon the reflector from the primary-feed are accurately specified. For design purposes it is very desirable therefore to employ a mathematical model of the primary feed horn which will reproduce the principal co-polar and cross-polar characteristics of the practical device. To integrate effectively with the mathematical model for the reflector system it is also necessary that the feed model be mathematically simple.

The primary-feed model described here can provide useful predictions of the co-polar and cross-polar radiation fields from a variety of waveguide horns. It can also be used to model the illumination characteristics of fundamental radiators such as electric and magnetic current elements, and the combination of these to form a Huygens source. The analytical model has other attributes in that it is sufficiently accurate to provide a valuable insight into the radiating properties of waveguide horns, and can thereby aid in the design of these components. Examples are discussed in the following Sections.

Field-equivalence principles

Field-equivalence principles can be applied to determine the radiated fields from an electromagnetic source. These methods are treated in a number of standard texts[80,81] and the basic concepts are discussed in Section 1.11.

To apply the field-equivalence principle to the primary-feed horn problem we can enclose the horn with a surface as shown in Fig. 4.40*a*. For convenience the surface includes the aperture of the horn. If we can now postulate an electric and magnetic current distribution over the Huygen surface, the horn can be removed and the complete radiated fields outside of S calculated using the methods outlined in Section 1.11.

We could postulate for example that the surface currents were given by

$$\left.\begin{aligned} \boldsymbol{J} &= \boldsymbol{a}_n \times \boldsymbol{H}_a \\ \boldsymbol{J}_m &= \boldsymbol{E}_a \times \boldsymbol{a}_n \end{aligned}\right\} \text{ over the aperture of the horn} \tag{4.30}$$

and $\boldsymbol{J} = \boldsymbol{J}_m = 0$ elsewhere on the surface

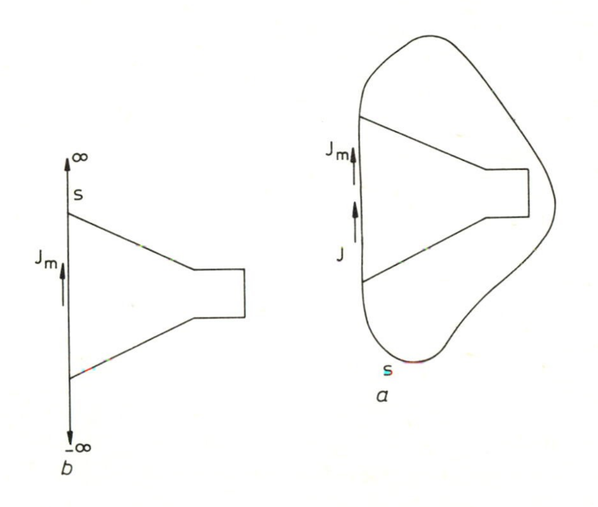

Fig. 4.40 *a* Electric and magnetic current model (complete sphere)
b Magnetic current model (forward hemisphere only)

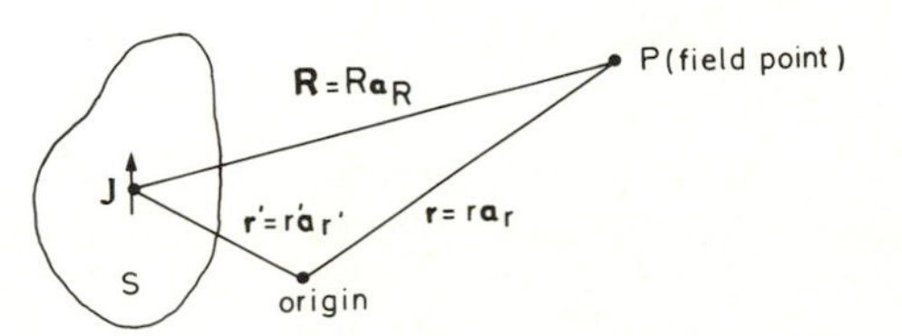

Fig. 4.41 *Vector nomenclature*

where E_a, H_a are the electromagnetic field distributions in the mouth of the horn. Approximations for E_a, H_a can be obtained by employing the solutions to Maxwell's equations in infinite waveguide structures.

This approach produces the equations originally derived by Chu and which can be found in Silver's text[2] or Love's selected reprints.[78] The method provides useful predictions of the co-polar radiated fields over the complete far-field sphere but the cross-polar predictions have been found to be very poor.[65,66]

An alternative approach is illustrated in Fig. 4.40*b*. Here the Huygens surface is continued through the aperture of the horn to form an infinite plane, and an infinite electric conducting sheet is postulated on one side of the surface. In this case only the tangential electric fields need be specified on the surface. The magnetic current is then given by

$$J_m = \begin{cases} 2(\boldsymbol{E}_a \times \boldsymbol{a}_n) & \text{over the horn aperture} \\ 0 & \text{elsewhere} \end{cases} \tag{4.31}$$

However, the radiated fields must now be determined in the presence of the conducting sheet and solutions can only be obtained for the radiation fields in the forward hemisphere. This configuration represents an accurate model of a waveguide horn with an infinite conducting flange. Its validity for finite-flange horns remains to be seen but its important advantage is that it is dependent only upon a knowledge of the tangential electric field distribution in the aperture of the horn.

Vector potential theory

With a knowledge, albeit approximate, of either $\boldsymbol{J}$ and/or $\boldsymbol{J}_m$ then *vector potential theory* can be applied in a surrounding unbounded homogeneous region with permittivity ϵ, *permeability* μ. With the nomenclature of Fig. 4.41 the *magnetic vector potential A* is defined as

$$A = \frac{\mu}{4\pi} \int_s \frac{J}{R} \exp\left[-jkR\right] ds \tag{4.32}$$

where R is the distance from a general point on the surface S to a general field point and k is the wave number ($k = 2\pi/\lambda$). The fields due to $\boldsymbol{J}$ are then given by

$$E^e = -j/\omega\mu\epsilon \left(\nabla \times \nabla \times A\right) \tag{4.33}$$

$$H^e = 1/\mu \left(\nabla \times A\right) \tag{4.34}$$

where

$$\bar{\nabla} = \frac{\partial}{\partial_x} a_x + \frac{\partial}{\partial_y} a_y + \frac{\partial}{\partial_z} a_z \tag{4.35}$$

Similarly the *electric vector potential* ***F*** can be defined as

$$F = \frac{\epsilon}{4\pi} \int_s \frac{J_m}{R} \exp\,[-jkR]\; ds \tag{4.36}$$

and the associated fields are

$$E^m = -\,1/\epsilon\,(\nabla \times F) \tag{4.37}$$

$$H^m = -j/\omega\mu\epsilon\,(\nabla \times \nabla \times F) \tag{4.38}$$

If the source comprises both electric and magnetic current distributions then the total fields outside S are given by

$$E = E^e + E^{m'} \tag{4.39}$$

$$H = H^e + H^m \tag{4.40}$$

Electric-field model

To construct the radiation model it is convenient to adopt the hybrid coordinate system shown in Fig. 4.42, in which we employ Cartesian or polar coordinates to

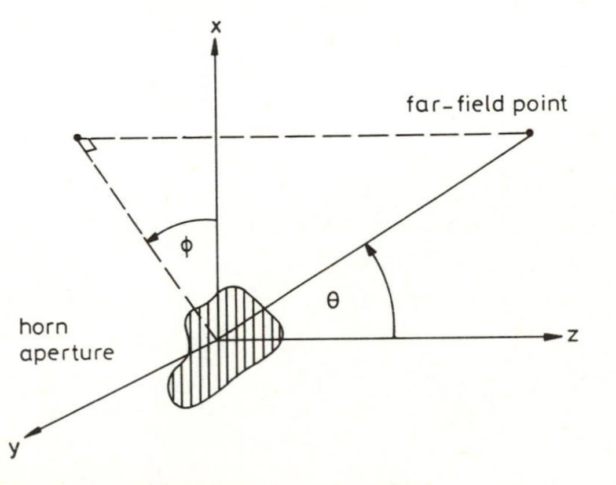

Fig. 4.42 *Coordinate system for horn radiation model*

define a source point in the aperture plane and spherical coordinates to define the far-field point. Since the model is to be based on the tangential electric field in the horn aperture plane we can make use of the electric potential to determine the radiated fields. For distant radiated fields the process is greatly simplified by use of the far-field approximation for R, i.e. if $R \gg r'$ then

$$\exp[-jkR]/R \simeq \exp[-jkr] \exp[jk(\boldsymbol{a}_r \cdot \boldsymbol{r}')]/r \tag{4.41}$$

Employing the hybrid coordinates of Fig. 4.42 and 4.41 and noting that

$\boldsymbol{r}' = x\boldsymbol{a}_x + y\boldsymbol{a}_y + z\boldsymbol{a}_z$ and $\boldsymbol{a}_r = \sin\theta\cos\phi\boldsymbol{a}_x + \sin\theta\sin\phi\boldsymbol{a}_y + \cos\theta\boldsymbol{a}_z$

we obtain the familiar expression

$$R \cong r - (x\sin\theta\cos\phi + y\sin\theta\sin\phi) \tag{4.42}$$

for the phase term.

If the aperture plane of the horn is located in the $z = 0$ plane of Fig. 4.42 then from eqns. 4.36, 4.41 and 4.42 the electric vector potential is given by

$$F = \frac{\epsilon}{4\pi r}\exp[-jkr]\int_x\int_y 2(E_a \times a_z)\exp[jk(x\sin\theta\cos\phi + y\sin\theta\sin\phi)]\,dxdy \tag{4.43}$$

Introducing a Fourier transform notation where the vector Fourier transform of the horn aperture tangential electric field is termed $\boldsymbol{F}(\theta,\phi)$, the vector potential can be more conveniently written as

$$\boldsymbol{F}(\theta,\phi) = \frac{\epsilon}{2\pi r}\exp[-jkr]\left\{f_x(\theta,\phi)\,\boldsymbol{a}_y - f_y(\theta,\phi)\,\boldsymbol{a}_x\right\} \tag{4.44}$$

where

$$\boldsymbol{F}(\theta,\phi) = \int_x\int_y \boldsymbol{E}_a(x,y)\exp[jk(x\sin\theta\cos\phi + y\sin\theta\sin\phi)]\,dxdy \tag{4.45}$$

and $f_x = \boldsymbol{f}\cdot\boldsymbol{a}_x$, $f_y = \boldsymbol{f}\cdot\boldsymbol{a}_y$

Making use of eqns. 4.37 and 4.44, the radiated fields can then be determined in conventional spherical-coordinate components (E_r, E_θ, E_ϕ) as

$$\begin{aligned} E_\theta &= -j\omega f \cdot a_\phi \\ E_\phi &= j\omega f \cdot a_\theta \\ E_r &= 0 \end{aligned} \tag{4.46}$$

giving

$$E_\theta = j\,\frac{\exp\,[-jkr]}{\lambda r}\,(f_x\,(\theta,\phi)\cos\phi + f_y\,(\theta,\phi)\sin\phi) \tag{4.47}$$

$$E_\phi = j\,\frac{\exp\,[-jkr]}{\lambda r}\,\cos\theta\,(f_y\,(\theta,\phi)\cos\phi - f_x\,(\theta,\phi)\sin\phi) \tag{4.48}$$

Since the predictions of the primary feed model will be compared directly with measured data it will be advantageous to express the radiated fields in a form which correlates conveniently with standard measurement techniques. Employing 'measured field components' as defined by Ludwig's third definition (see Section 1.7) and by substitution of eqns. 4.47, 4.48 into eqn. 4.9 the co-polar E_p and cross-polar E_q radiated fields from the primary feed can be expressed as

$$\begin{bmatrix} E_p\,(\theta,\phi) \\ \\ E_q\,(\theta,\phi) \end{bmatrix} = K\cos^2\theta/2 \begin{bmatrix} 1 - t^2\cos 2\theta & t^2\sin 2\phi \\ \\ t^2\sin 2\phi & 1 + t^2\cos 2\phi \end{bmatrix} \begin{bmatrix} f_y\,(\theta,\phi) \\ \\ f_x(\theta,\phi) \end{bmatrix} \tag{4.49}$$

where

$$K = j\exp\,[-jkr]/\lambda r \quad \text{and} \quad t = \tan\theta/2$$

Eqn. 4.49 is an exact expression for the distant radiated fields in the forward half-plane of any aperture antenna, providing the tangential electric field distribution is known exactly over the complete planar Huygen surface. Clearly this will not be the case in practice and the validity of the prediction is dependent upon the quality of the initial approximation for the tangential electric fields. Because of its dependence upon the tangential electric field this approach has been termed the E-field model[66]. Since the expressions can be usefully applied to a wide range of antenna types it is worth noting that, for high-gain antennas, where the radiation is largely concentrated into a small range of the angle, then as $\tan^2\theta/2 \to 0$ and $\cos^2\theta/2 \to 1$,

$$E_p(\theta, \phi) \simeq K f_y(\theta, \phi) = K\, FT < E_a \cdot a_y > \tag{4.50}$$

$$E_q(\theta, \phi) \simeq K f_x(\theta, \phi) = K\, FT < E_a \cdot a_x > \tag{4.51}$$

and it can be seen that a direct Fourier transform relationship exists between the co-polar and cross-polar measured field components and the tangential aperture fields. Even when this approximation is not completely valid, it is evident that the transform function will be a dominant factor in shaping the far-field characteristics of the antenna. This explains to a large extent the (sometimes surprising) effectiveness of design techniques based upon the scalar Fourier transform.

Applications in primary-feed design

The electrical design of primary-feed antennas involves a number of dependent factors. The vector characteristics of the feed horn radiation are of prime importance in reflector antenna design and the optimum horn characteristics are dependent upon the reflector to be illuminated and the overall performance objectives. The overall antenna requirements may involve pattern shape, efficiency (or gain) sidelobe and cross-polar specifications. There may also be mechanical, mass, or dimensional constraints which will influence the design of a practical feed antenna.

To achieve a desired radiation characteristic it is necessary to choose the type of horn; the dimensions and geometry of the radiating aperture; the geometry of the flared section coupling the waveguide feeder to the radiating section; and to design the throat of the horn to provide effective matching and to avoid the generation of unwanted higher order modes.

The E-field model can provide a means of examining the effects of the radiation characteristics of aperture dimensions and geometry and can be utilised (with a mathematical model for the reflector) to optimise the overall radiation characteristics of the antenna. The model can deal with higher-order waveguide modes but the magnitude and relative phase of the modes must be calculated separately. In the Sections which follow examples are provided of the application of the E-field model to a range of primary-feed antennas.

Fundamental-mode pyramidal and conical horns

Approximate expressions for the tangential aperture fields for horn antennas can be obtained from solutions to Maxwell's equations in infinite waveguides. These solutions are available in many standard texts for common geometries such as rectangular and circular cross-section pipes[2,81]. The solutions take the form of an infinite series of orthogonal field distributions (or modes) each of which propagates in the guide with a different phase velocity. By restricting the dimensions of the waveguide, higher-order modes can be excluded and design can be based upon a known fundamental-mode field distribution at the throat of the horn.

The effects on these solutions of flaring and truncating the pipe to form a horn radiator are more complex. In general for simple linear-flare horns the flaring will generate higher-order modes unless the flare semi-angle is kept small (typically less than 10°). Truncating the pipe produces relatively small effects for aperture dimensions greater than about one wavelength and small-angle flared-horns are traditionally designed on the assumption of only a fundamental mode in the mouth of the horn. Changes in flare-angle can be used deliberately in the design of horns which make use of higher order modes to modify their radiation characteristics.

The co-polar radiation characteristics and gain of pyramidal and conical horns has been the subject of extensive work which is well documented in the literature[2,78,81]. However, the cross-polarised performance of waveguide horns is less well documented and is currently of considerable interest. For rectangular or circular cross-section guides the Fourier transformations necessary for the evaluation of eqn. 4.49 can be performed analytically and closed-form expressions for the fundamental waveguide modes are available in many standard texts,[2,81] while the transforms for higher order modes can be readily derived.

For example, the electric-field distribution in the aperture of a small flare-angle fundamental-mode pyramidal horn with dimensions $2a \times 2b$ can be expressed approximately as[2]

$$E_a\,(x, y) = E_0 \cos \frac{\pi x}{2a} \mathbf{a}_y \tag{4.52}$$

where the horn's principal polarisation is aligned with the y axis and E_0 is a constant defining the maximum field strength in the horn aperture.

The Fourier transform of this aperture field distribution can be derived analytically as

$$f_y\,(\theta,\phi) = \frac{\pi E_0}{4} \left[\frac{\cos\,(ka \sin\theta \cos\phi)}{1 - \dfrac{(2ka \sin\theta \cos\phi)^2}{\pi^2}} \right] \left[\frac{\sin\,(kb \sin\theta \sin\phi)}{kb \sin\theta \sin\phi} \right] \tag{4.53}$$

and we note that for this case $f_x\,(\theta,\phi) = 0$

When these transform functions are substituted into eqn. 4.49 it is found that the fundamental mode (TE_{01}) pyramidal horn radiates a relatively high level of cross-polar radiation when its aperture dimensions are small[66]. Providing the horn flare-angle is sufficiently small to avoid over-moding, this level of cross-polarisation reduces significantly as the horn dimensions are increased. Employing the E-field model the predicted peak level of the cross-polar radiation from square pyramidal horns is shown in Fig. 4.43. The maxima occur in the intercardinal planes of the horn radiation pattern (i.e. $\phi = \pm 45°$ and $\pm 135°$).

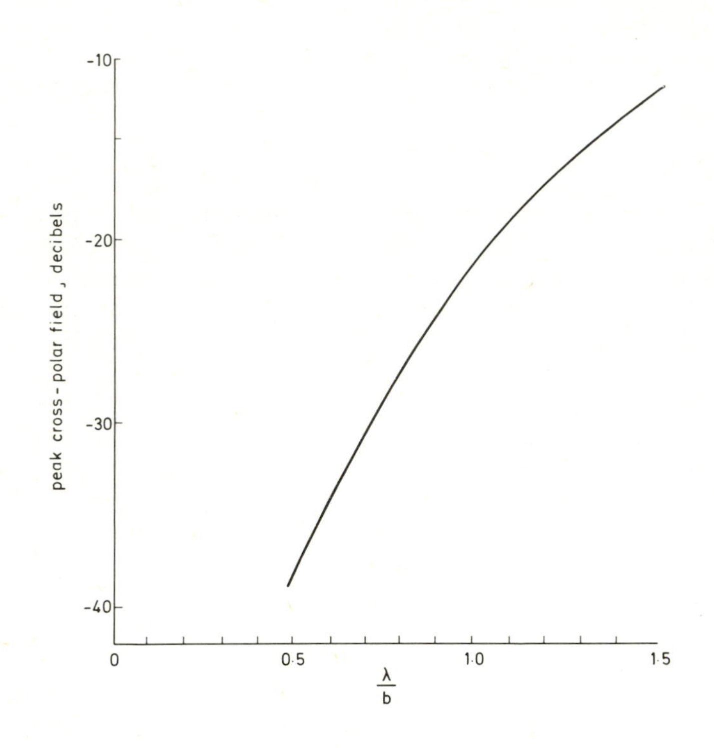

Fig. 4.43 *Peak cross-polar levels (in diagonal plane) for square pyramidal horns of side b*

Experience has shown that the predictions obtained from the E-field model are good for aperture dimensions exceeding about one wavelength. For smaller horns the radiation characteristics are increasingly flange dependent. For example Fig. 4.44 shows the radiation fields in a diagonal plane for a flangeless pyramidal horn with aperture dimensions $0{\cdot}9\lambda \times 1{\cdot}1\lambda$. In this case the predictions are still useful but the lack of a flange is resulting in obvious differences between the predicted and measured fields.

For large-aperture horns the radiation pattern predictions can be improved by incorporating a simple phase term in the expression for the aperture field, to account for the sphericity of the wavefront. In general the inclusion of this term implies that the Fourier transform function can no longer be derived analytically and must be computed. The quadratic phase error term is simply related to the

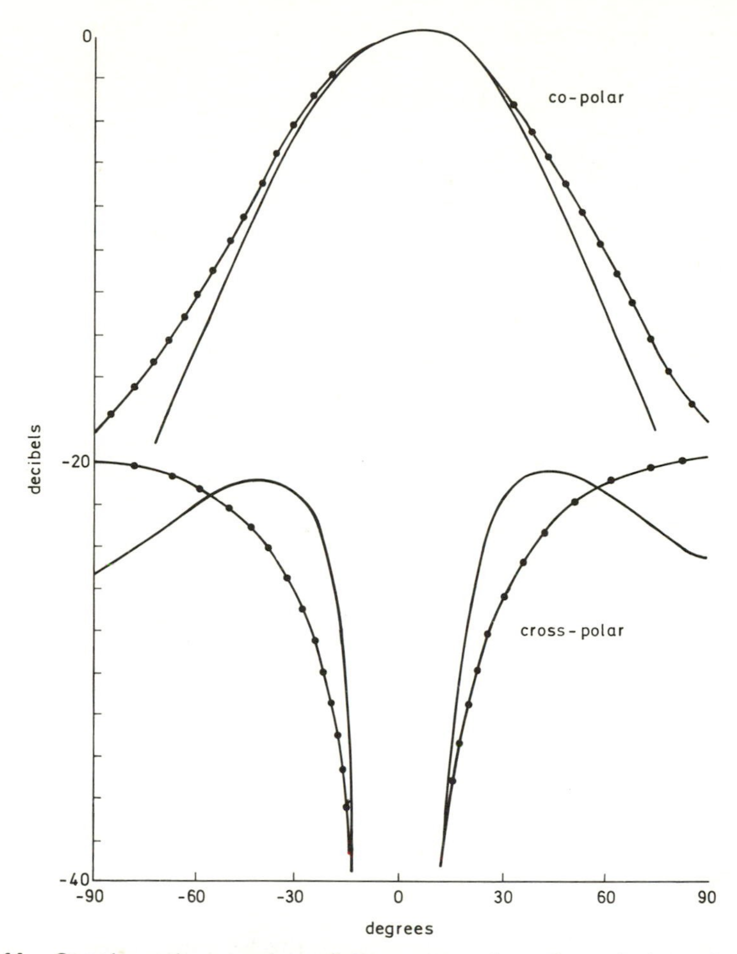

Fig. 4.44 *Co-polar and cross-polar radiation patterns in a diagonal plane of a pyramidal horn of dimensions 0·9 λ x 1·1λ*
predicted —•—•—•—
measured ————

length of the horn and the flare angle and can be modelled by multiplying the aperture field expression by

$$\exp\left[-j\,\frac{\pi}{\lambda}\left(\frac{x^2}{L_x}+\frac{y^2}{L_y}\right)\right] \tag{4.54}$$

where L_x L_y are the slant radii of the horn in the principal d,y planes. The slant radius is defined in Fig. 4.45.

The fundamental mode conical horn can be similarly treated and for small flare angle horns with radius a and principal polarisation aligned in the y-direction, the Fourier transform functions for the TE_{11} mode have the form

$$f_y(\theta, \phi) = A_1(\theta) \sin^2 \phi + A_2(\theta) \cos^2 \phi \tag{4.55}$$

$$f_x(\theta, \phi) = \tfrac{1}{2}(A_1(\theta) - A_2(\theta)) \sin 2\phi \tag{4.56}$$

The functions A_1, A_2 are given by

$$A_1(\theta) = 2J_1(ka \sin \theta)/(ka \sin \theta) \tag{4.57}$$

$$A_2(\theta) = \cos \theta \, J_1{}'(ka \sin \theta)/(1 - (ka \sin \theta/u')^2) \tag{4.58}$$

where $J_1{}'(u) = \dfrac{d}{du} J_1(u)$ and u' is the first root of $J_1{}'(u) = 0$ i.e. $u' = 1{\cdot}841$.

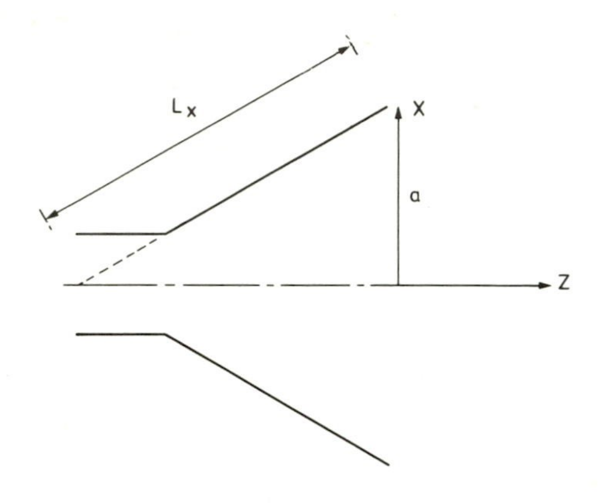

Fig. 4.45 *Pyramidal horn dimensions*

The fundamental mode (TE_{11}) conical horn provides a useful demonstration of the application of the E-field model to small waveguide horns. Fig. 4.46 from a paper by Adatia *et al.*[65] shows a comparison of the predicted cross-polar peaks employing the E-field model and the equations derived by Chu[2] which essentially employ the combined electric and magnetic tangential aperture field approach. For small conical horns the E-field model predicts a range of aperture dimensions about the value $d = 1{\cdot}15\lambda$ where the cross-polar levels are reduced to low levels. The measured data, which was obtained from a number of conical horn antennas with a variety of planar flanges, shows that a region of cross-polar cancellation does exist in practice. Fig. 4.47 shows the predicted and measured radiation fields in the inter-cardinal plane of a conical horn with a diameter of $1{\cdot}25\lambda$. It is noteworthy that for rotationally symmetric horns of this type low levels of cross-polarised radiation must also imply a high degree of radiation-pattern symmetry.

For conical horn diameters of greater than about one wavelength the measured data follows the E-field model, and to a first order is relatively independent of the flange dimensions. For diameters of less than one wavelength the horn cross-polar radiation is very flange dependent and will follow the general characteristics of the E-field model only when a planar flange is located around the horn aperture. It is in this range of small apertures (i.e. $\lambda/d > 1{\cdot}0$) where the presence of chokes and other

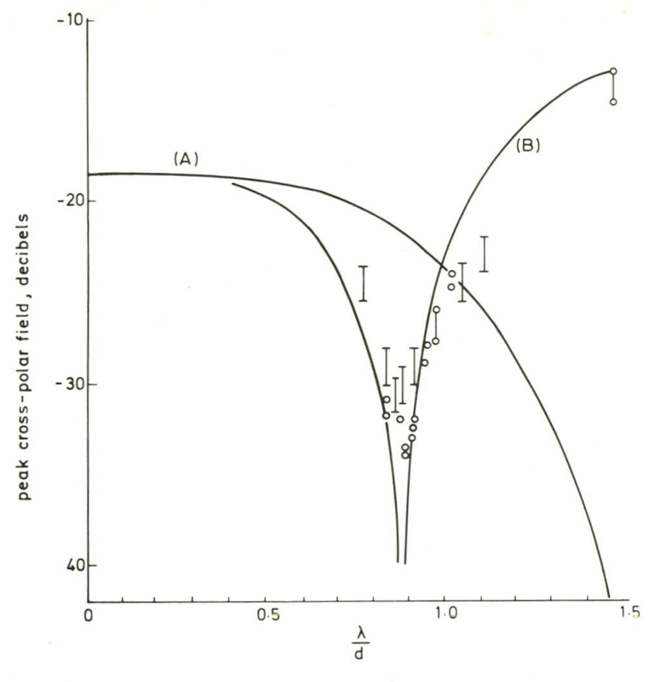

Fig. 4.46 *Peak cross-polar radiation in a diagonal plane for TE_{11} mode conical horns of diameter d*
A Chu models
B E field model and various measured data

obstacles around the aperture can result in very significant improvements in the cross-polar performance of conical horns[36,38]. Thin-walled horns without flanges can also provide good cross-polar performance in this region, but the horn performance is dependent on the currents on the outer walls of the waveguide, which must be considered as an inherent part of the radiating structure[13,67]

It is sometimes incorrectly assumed that for a primary-feed horn to produce low cross-polarisation it is necessary for the aperture of the horn itself to be purely linearly polarised. The small pyramidal horn provides a purely linearly polarised aperture but it is apparent from Fig. 4.43 that it will radiate a significant level of cross-polarisation. Conversely the TE_{11} mode field lines in the aperture of a small fundamental-mode conical horn are noticeably curved, but Fig. 4.46 shows that low cross-polarisation can be achieved. The E-field model can offer some insight in this respect. From eqn. 4.49 the cross-polarised radiated field from a horn is given by

$$E_q(\theta,\phi) = K\cos^2\theta/2\ [f_x(\theta,\phi)(1+t^2\cos 2\phi) + f_y(\theta,\phi)\,t^2\sin 2\phi] \tag{4.59}$$

If the horn aperture is purely linearly-polarised in the y-direction then $f_x = 0$ and the horn radiates a cross-polarised field due to the second term in the equation. This field component will have maxima in the intercardinal or diagonal planes which

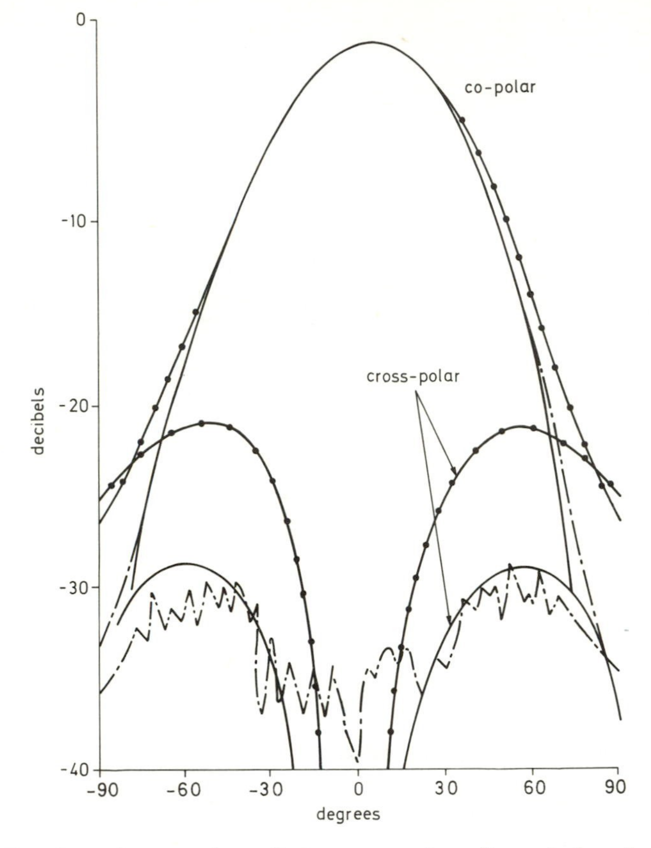

Fig. 4.47 *Copolar and cross-polar radiation patterns in a diagonal plane for a TE_{11} mod[e] conical horn of diameter 1·25 λ. Predicted data from:*
Chu model ——•——•——
E field model ————

will become increasingly significant as the horn dimensions reduce. This account[s] for the observed poor cross-polar performance of the small pyramidal horn. Wit[h] decreasing aperture dimensions the radiation from the pyramidal horn approache[s] that of the infinitesimal magnetic dipole, and hence the second term in eqn. 4.5[9] can be identified as the magnetic dipole component of the cross-polarised fiel[d].

For large pyramidal horns the radiated field will be concentrated into a sma[ll] range of θ angles where the magnetic dipole component will be suppressed by th[e] $\tan^2 \theta/2$ factor. Hence for large horns a purely linearly polarised aperture wi[ll] result in good cross-polar suppression.

It is evident from eqn. 4.59 that the condition of zero cross-polarised radiation a[...]

all field points cannot be satisfied with a small purely linearly-polarised radiating aperture, since to compensate the magnetic dipole component it is necessary that f_x be non-zero. With regard to the tangential electric fields in the horn aperture this implies that, with both f_x and f_y present, the field lines must exhibit curvature for optimum cross-polar suppression. The first term in eqn. 4.59 thus represents the field-curvature effects which are essential for small horns if their cross-polar radiation is to be suppressed to low levels. As the horn aperture dimensions are increased the required curvature component decreases and for large horns the optimum condition will correspond to a pure polarisation with linear field lines in the mouth of the horn.

Shaped aperture horns

The point has been made that for small waveguide horns the zero cross-polar radiation condition does not correspond to a purely linear polarisation in the horn aperture. By application of inverse Fourier transform methods the general form of the field-line curvature in the horn mouth which satisfies this condition can be obtained from eqn. 4.59. This is illustrated in Fig. 4.48 for small waveguide horns. For minimum cross-polar radiation the field lines must become progressively more parallel as the horn dimensions increase, and thus for any horn dimension there will be a unique degree of field line curvature to provide optimum cross-polar suppression.

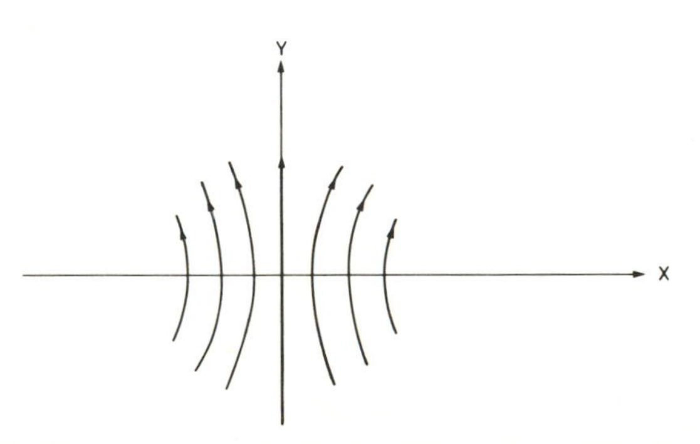

Fig. 4.48 *General form of electric-field lines in aperture plane of small horns to achieve zero cross-polar radiation*

This concept has been applied to a more general range of aperture cross-sectional geometries by Rudge and Philippou.[82] The tangential electric field in the aperture of the horn has been derived by use of finite-element methods. With this technique the horn aperture (which is taken as the cross-section of an infinite waveguide) is divided into triangular sub-regions and Helmholtz's equation is then solved using a variational expression together with a Rayleigh-Ritz procedure. The computed aperture fields are then Fourier transformed and used in eqn. 4.59 to predict the radiated cross-polar level.

These techniques have been applied to the design of a variety of horn aperture geometries including 6- and 8-sided polygons[82]. Geometries of these general categories appear attractive for use in multi-element feed configurations where compact feed clusters are necessary. In every case optimum dimensions have been determined which provide minimum cross-polar radiation. Note that the cross-polar suppression in these cases is not achieved by multi-moding the horn but by finding a structure with a fundamental mode which provides the desired degree of field curvature.

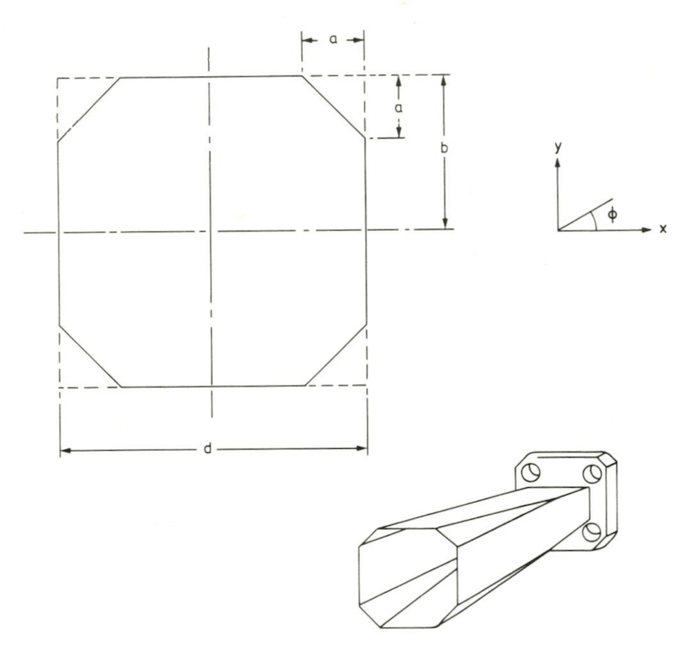

Fig. 4.49 *The 8-sided primary-feed horn*

An example is shown in Fig. 4.49. In Fig. 4.50 the computed peak cross-polar radiation (occurring in the diagonal planes) is shown against the geometric aperture parameters a/b and d/λ. Significant cross-polar suppressions are clearly apparent and the relatively wideband properties of the case with $a/b = 0{\cdot}4$ is of particular interest. A prototype feed horn with these dimensions has been evaluated in K-band. Unfortunately in this experimental model the cross-polar bandwidth of the prototype was reduced as a result of overmoding in the flared section, but this is not fundamental to the shaped aperture concept and the general cross-polar suppression properties were confirmed. The measured data for the 8-sided horn is shown in Fig. 4.51.

The 8-sided horn is not a dual polarised device since its co-polar radiation pattern has an elliptical contour with a major axis, which is always aligned with the H-plane of the horn. When the direction of the input polarisation vector is turned

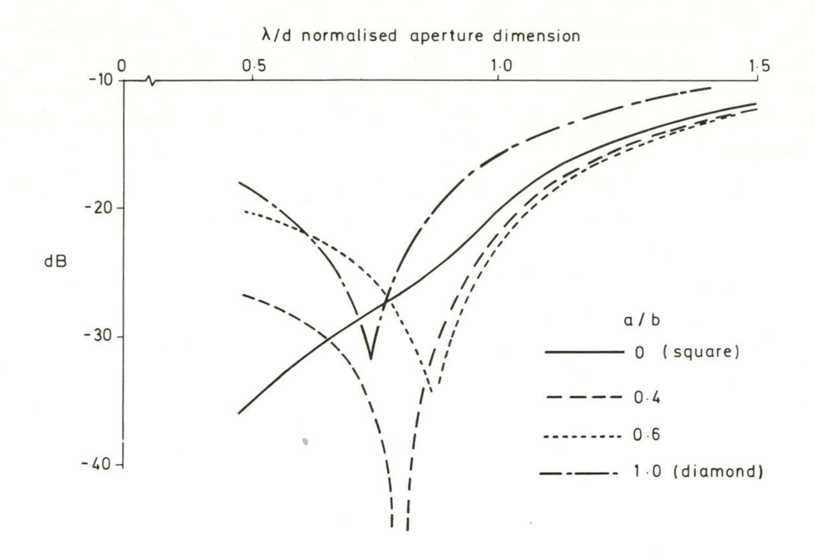

Fig. 4.50 *Predicted cross-polar characteristics of 8-sided feed horn for various parameter combinations*

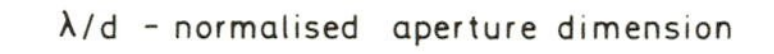

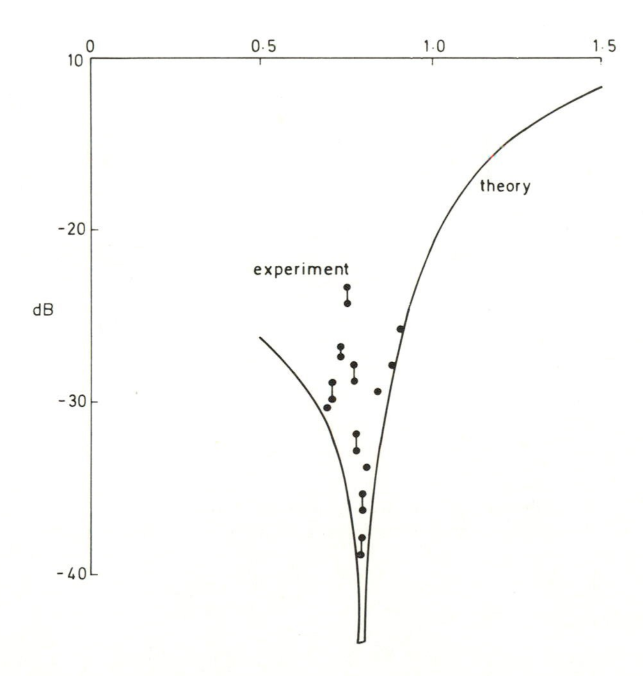

Fig. 4.51 *Predicted and measured values of peak cross-polar radiation from the 8-sided prototype horn*

through 90° the orientation of the ellipse follows. However, the horn does provide a simple linearly-polarised primary-feed with good polarisation purity. Hexagonal shaped apertures can provide good polarisation performance for dual-polarised applications and these devices offer interesting properties for use in multi-element feed clusters.

The diagonal horn antenna which is illustrated in Fig. 4.52 falls into this general category of shaped aperture horns. This device was investigated in the 1960s by Love[78] in the interests of providing a symmetrical radiation pattern. With regard to improved pattern symmetry the work was successful although the cross-polar performance of the diagonal antenna was found to be poor, with peaks of the order of −16dB. From Fig. 4.50 it can be seen that there is an optimum dimension for small diagonal horns which will reduce the cross-polar peaks to below −25dB over a small band but the cross-polar performance is evidently limited. The principal plane sidelobes of the diagonal horn antenna are at least 30dB down but higher lobes (−20dB) occur in the diagonal planes.

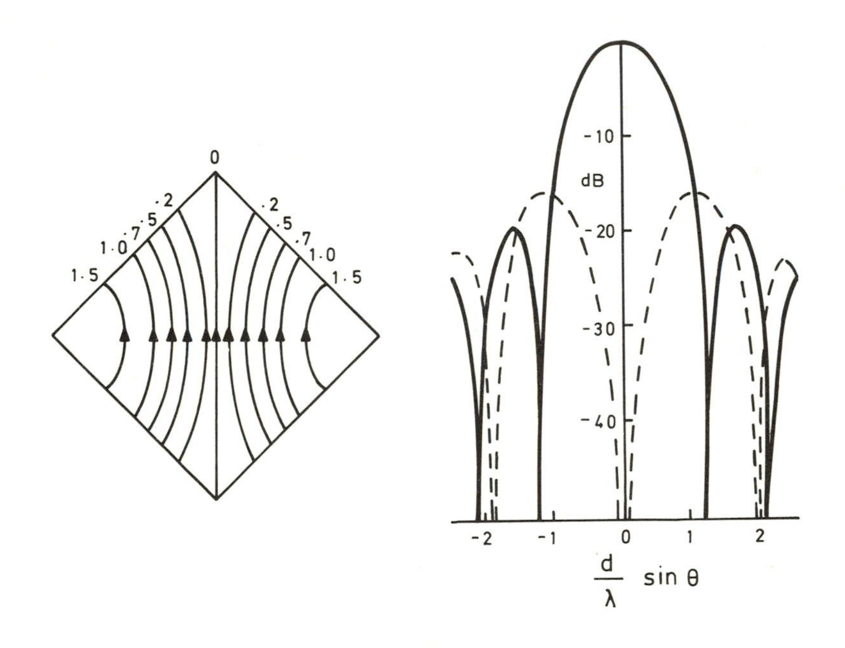

Fig. 4.52 *a* Field lines in the aperture of a diagonal horn
b Co-polar and cross-polar radiation fields in a diagonal plane

Multi-mode horns

The E-field model can be employed to model the dual mode (Potter) horn by adding in a component of the TM_{11} mode in the expression for the horn aperture field. The TM_{11} mode radiation field has only an E_θ component and the Fourier transforms for this mode take the form:

$$\begin{aligned} f_y(\theta,\phi) &= E_\theta \sin\phi \\ f_x(\theta,\phi) &= E_\theta \cos\phi \end{aligned} \tag{4.60}$$

and the radiation field component is given by

$$E_\theta(\theta,\phi) = \left\{ \frac{B_{11}\cos\theta}{1-(u_o/ka\sin\theta)^2} \quad \frac{J_1(ka\sin\theta)}{ka\sin\theta} \right\} \frac{e^{jkr}}{r} \sin\phi \tag{4.61}$$

Where B_{11} is a complex coefficient defining the relative amplitude and phase of the TM_{11} mode to the fundamental TE_{11} mode and u_0 is the first root of $J_1(u) = 0$, i.e. $u_0 = 3{\cdot}832$.

To model the Potter horn eqn. 4.60 is added to eqns. 4.55 and 4.56 respectively, and the sum substituted into eqn. 4.49. To minimise the cross-polar radiation the magnitude of B_{11} will be of the order of 0·7 and the phases of the two modes must be in quadrature in the aperture of the horn. To maintain at least a 10dB suppression of the fundamental-mode cross-polarisation this phase-quadrature condition must be maintained within approximately ±15°. While the cross-polar radiation remains low the co-polar radiation pattern will also exhibit good circular symmetry.

The limited bandwidth of the Potter horn arises largely as a consequence of the differing phase velocities of the TE_{11} and TM_{11} modes in smooth-walled circular waveguide. The bandwidth properties of the horn can be modelled about the phase-quadrature condition by use of an approximate phase differential function. This can be given the form

$$B_{11} \triangleq |B_{11}| \exp\left[j\left(\frac{\pi}{2} + 2\pi l\left(\frac{1}{\lambda_e} - \frac{1}{\lambda_m}\right)\right)\right] \tag{4.62}$$

where l is the distance from the moding step to the mouth of the horn and λ_e and λ_m are the average values of the respective guide wavelengths of the TE_{11} and TM_{11} modes in this region.

Corrugated horns

We have already seen that for large-aperture horns low cross-polar radiation will be achieved if the horn aperture is linearly polarised. To provide a relatively simple model for the corrugated horn, the aperture fields can be expressed in the form[83]

$$E_a(r,\phi) = J_0\left(\frac{r}{a}u_o\right)\mathbf{a}_x - \left(\frac{\bar{\Lambda}-1}{\bar{\Lambda}+1}\right) J_2\left(\frac{r}{a}u_2\right)(\cos 2\phi\,\mathbf{a}_x + \sin 2\phi\,\mathbf{a}_y) \tag{4.63}$$

where r, ϕ are polar co-ordinates in the horn aperture plane, a is the inner radius. u_n is given by the first roots of $J_n(u_n) = 0$ (i.e. $u_0 = 2{\cdot}4048$) $\bar{\Lambda}$ is the ratio of the TM and TE components of the hybrid mode and a_x, a_y are the cartesian unit vectors.

In the balanced hybrid conditions $\bar{\Lambda} = 1$ and the aperture of the horn is purely linearly polarised. This condition is synonymous with an infinite wall reactance and hence approximately quarterwave slot depths. The first term in eqn. 4.63 can then be used in the E-field formulae to compute the radiation field. A multiplying term of the form

$$\exp\left[-j\frac{\pi}{\lambda R}r^2\right] \tag{4.64}$$

can be included in the aperture field expression to account for the sphericity of the phase in the aperture of the horn. The dimension R is the slant length of the corrugated horn (see Fig. 4.19). It is also evident from the equation that at all other frequencies at which the slots are non-resonant there will be a cross-polar component in the horn aperture. For narrow flare angle horns with $ka \gg 1$ the far-field cross-polarisation will be totally dependent upon the aperture field conditions of the horn and, neglecting mode conversion components, the power ratio of the peak cross-polar component in the intercardinal planes to the peak co-polar level on boresight can be simply expressed with the notation of Fig. 4.19 as [83]

$$C^2 = \frac{0{\cdot}14}{(1 - t/p)^2\, k^2 a^2 \tan^2 kd} \tag{4.65}$$

Fig. 4.53 shows an example of the variation of cross-polar suppression as the frequency varies about the resonant condition.

For small aperture corrugated horns (i.e. diameter of less than about 3 wavelengths) the minimisation of cross-polar radiation demands some curvature of the horn aperture field lines to compensate for the 'magnetic dipole' component of eqn. 4.49. Inspection of eqn. 4.63 indicates that a component having the necessary $\sin 2\phi$ characteristic can be obtained by deliberately introducing a small imbalance between the TM/TE components of the hybrid mode. Since this imbalance occurs naturally with frequency, the effect for small horns will be to shift the centre frequency of the optimum cross-polar response.

This effect is illustrated in Fig. 4.53 which is taken from a paper by Adatia, Rudge and Parini [65]. The measured cross-polar response of a corrugated horn is compared with three theoretical curves. Curves 1 and 2 neglect the magnetic dipole component and curve 3, which employs the E-field model, includes this component. The aperture field lines of the horn have been computed at various frequencies in the band. The minimum cross-polar condition corresponds to case C which clearly illustrates the significant degree of field line curvature necessary to compensate the magnetic dipole component.

For large corrugated horns the flange around the horn aperture has little effect on the antenna radiation characteristics. To quantify this statement, for horn diameters of approximately two wavelengths the effect of the flange is likely to be discernible when cross-polar levels of the order of −35dB are of interest. At three wavelengths diameter Thomas[39] has reported that flange effects on narrow angle horns are discernible, but relatively small, at the −40dB level. The aperture dimensions indicated above refer to the frequency at which the hybrid mode is balanced and thus the slots in the aperture region are resonant.

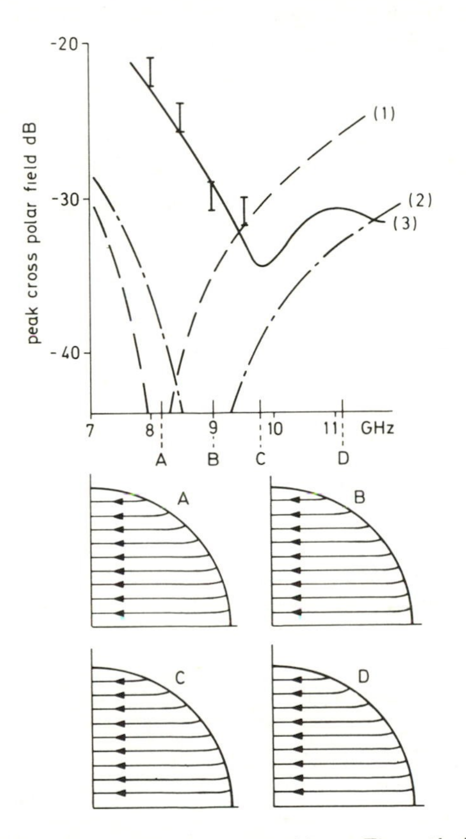

Fig. 4.53 *Cross-polarisation from a corrugated horn. Theoretical curves*
(1) Chu model
(2) Chu model with space harmonics
(3) E field model with space harmonics
Field lines in horn aperture shown at four points in the band

4.11 Primary feeds for offset-reflector antennas

Offset-reflector focal-plane fields

The offset parabolic reflector has many obvious advantages, (Section 3.3) but the depolarising beam-squinting phenomena represent a major limitation in many applications[69]. As an alternative to optimised dual reflector configurations[60,76,77] primary-feed antennas which can overcome this limitation are electrically feasible and practical devices have been demonstrated[75]. To understand the principles governing their design it will be useful to consider the form of the electromagnetic fields set up in the focal region of the offset reflector when a plane wave is normally incident upon the reflector aperture plane.

Bem[68] has performed an analysis of the focal-plane fields which, although limited to normally incident waves, has the particular merit that the transverse focal-plane fields E_x, E_y can be expressed approximately in a simple closed form. Valentino and Toulios[69] confirmed Bem's results by extending them to include incident waves making small angles with the reflector boresight and comparing their results with measured data. Ingerson and Wong[70] have also employed a focal-region field analysis to determine beam-deviation factors for offset-reflector antennas. For offset-reflectors with long focal length and polarised in the plane of symmetry, we have from Bem's analysis:

$$E_x(u, \phi_o) = \frac{2J_1(u)}{u} + \frac{jd \sin\theta_o}{F} \frac{J_2(u)}{u} \cos\phi_0 \tag{4.66a}$$

$$E_y(u, \phi_o) = \frac{-jd \sin\theta_o}{F} \frac{J_2(u)}{u} \sin\phi_o \tag{4.66b}$$

where r', ϕ_o are polar coordinates in the reflector focal plane with origin at the geometric focus, $u(r')$ is a normalised parameter representing the distance r' to a point in the focal plane, and all multiplying constants have been suppressed. $J_n(u)$

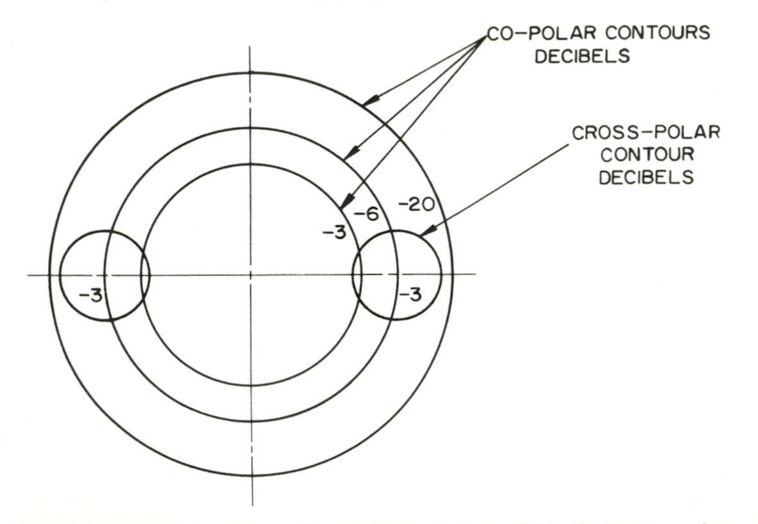

Fig. 4.54 *Approximate contour plot of typical focal-plane field distribution of an offset parabolic reflector uniformly illuminated from a distant linearly polarised source*

is the Bessel function of order n. The solutions for the wave polarised in the plane of asymmetry is achieved from eqn. 4.66 by interchanging x and y and replacing ϕ_0 by $2\pi - \phi_0$.

Inspection of eqn. 4.66 reveals that the cross-polar component E_x is an asymetric function with a magnitude increasing with the offset angle θ_0 and in phase quadrature with the principal axisymmetric copolar component. The axisymmetric copolar term is also modified by the presence of a quadrature component which is identical to the cross-polar term in all but its dependence upon ϕ_0. Figure 4.54 shows an idealised contour plot of the amplitude of the focal-plane field in the vicinity of the geometric focus.

Offset-reflector matched-feed concept

If the primary-feed is to provide an optimum conjugate match to the incoming field, then its aperture fields must exhibit similar polarisation characteristics.

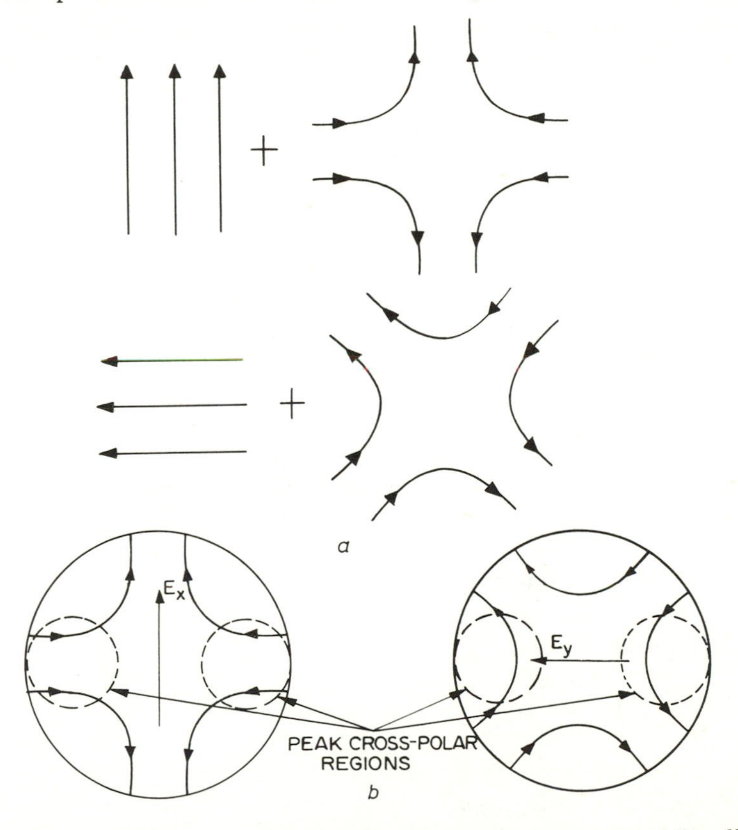

Fig. 4.55 *a* Field configurations of symmetric and asymmetric components of the offset-reflector focal-plane fields with the incident wave polarised in the plane of symmetry and in the plane of asymmetry
b Required field configuration in the aperture of a circular horn for focal-plane field matching in the orthonogal linear polarisations.

Conventional high-performance axisymmetric feeds (such as the corrugated horn) provide a conjugate match to the copolar component only, which results in the apparently poor polarisation properties of the offset reflector.

The focal-plane field distributions described by eqns. 4.66 and 4.67 can be matched very effectively by making use of higher-order asymmetric waveguide modes. Rudge and Adatia[61,62,63] have proposed a class of new primary-feed designs employing mode combinations in cylindrical, rectangular and corrugated waveguides. To illustrate the general principle of this approach, Figure 4.55 illustrates the nature of the symmetric and asymmetric components which make up the offset-reflector focal-plane fields for two linear polarisations. To adequately match this characteristic in a smooth-walled cylindrical waveguide, the required asymmetric mode is the TE_{21}. Fig. 4.55*b* shows the field distribution in the conical-horn mouth for the two orthogonal TE_{21} modes, which, for convenience, are designated TE^1_{21} and TE^2_{21}. The transverse field components of these modes (E^1_x, E^1_y, E^2_x, E^2_y) can be readily derived from the solution of the vector wave equation in the circular waveguide[2]. In terms of a normalised distance parameter u' and a polar angle ϕ_0 and omitting multiplying constants, these components can be expressed as:

$$E^1_x(u',\phi_0) = K'[J_1(u')\cos\phi_0 + J_3(u')\cos 3\phi_0] \tag{4.67a}$$

$$E^1_y(u',\phi_o) = -K'[J_1(u')\sin\phi_o - J_3(u')\sin 3\phi_o] \tag{4.67b}$$

$$E^2_x(u',\phi_0) = K''[J_1(u')\sin\phi_0 + J_3(u')\sin 3\phi_0] \tag{4.68a}$$

$$E^2_y(u',\phi_o) = K''[J_1(u')\cos\phi_o - J_3(u')\cos 3\phi_o] \tag{4.68b}$$

where K' and K'' are constant factors proportional to the complex coefficients of the two TE_{21} modes.

In the principal planes ($\phi_o = 0$ and $\pi/2$). eqn. 4.67 reduce to

$$E^1_x(u',0) = 4K'J_2(u')/u' \tag{4.69a}$$

$$E^1_y(u',0) = 0 \tag{4.69b}$$

$$E^1_x(u',\pi/2) = 0 \tag{4.69c}$$

$$E^1_y(u',\pi/2) = 4k'J_2(u')/u' \tag{4.69d}$$

Comparing eqn. 4.69 with the asymetric terms in eqn. 4.66, it can be seen that the coefficient of the TE^1_{21} mode can be selected to provide a perfect match to both the copolar and cross-polar asymmetric components when the reflector is polarised in the plane of symmetry ($-x$ axis). For operation with the principal polarisation along the y axis (plane of asymmetry) from eqn. 4.68 the TE^2_{21} mode gives in the principal planes:

$$E^2_x(u',0) = 0 \tag{4.70a}$$

$$E^2_y(u',0) = 2K''J'_2(u') \tag{4.70b}$$

$$E^2_x(u',\pi/2) = 2K''J'_2(u') \tag{4.70c}$$

$$E^2_y(u',\pi/2) = 0 \tag{4.70d}$$

where $J'_2(u') = dJ_2(u')/du'$.

Eqns. 4.70 can be compared with the asymmetric components of the reflector focal-plane field for a wave polarised in the plane of asymmetry. Making the necessary simple transformations in eqn. (4.66), the asymmetric components E^a are given by:

$$E^a_x(u,0) = 0 \tag{4.71a}$$

$$E^a_y(u,0) = \frac{jd\sin\theta_0}{2F}\;\frac{J_2(u)}{u} \tag{4.71b}$$

$$E^a_x(u,\pi/2) = \frac{jd\sin\theta_o}{2F}\;\frac{J_2(u)}{u} \tag{4.71c}$$

$$E^a_y(u,\pi/2) = 0 \tag{4.71d}$$

Comparison of 4.70 with 4.71 indicates a nonideal match. However, the Bessel differential functions $J'_2(u)$ have very similar general characteristics to the $J_2(u)/u$ distribution; and, by judicious choice of the constant K'', the two functions can be closely matched over the aperture of the feed horn.

The differences in the principle-plane distributions of the orthogonal TE_{21} modes is a consequence of the boundary conditions imposed by the smooth-walled cylindrical structure. The boundary conditions have similar implications with regard to the lack of axisymmetry in the copolarised radiation provided by the fundamental TE_{11} mode. To provide this axisymmetry it is necessary to add a component of the TM_{11} mode. This technique is well established and forms the basis of the well-known dual mode or Potter horn (Section 4.5). For cylindrical corrugated structures in which the fields satisfy anisotropic boundary conditions, it can be shown that the corresponding HE_{21} hybrid modes have identical distributions in the principal planes. Combined with the fundamental HE_{11} mode, these structures can provide a close to ideal match in the two hands of principal polarisation.

In Table 4.1 the modes required to provide focal-plane matching with three different feed structures are summarised. The cylindrical structures are also suitable for use with two hands of circular polarisation. The corrugated rectangular case, although equally feasible, is not shown since the mode designations for this structure are not standardised.

Table 4.1 Waveguide mode for offset reflector focal-plane matching

Feed structure	Waveguide modes	
	Principal polarisation	
	plane of symmetry (−x)	plane of asymmetry (y)
Smooth-walled cylindrical*	$TE^1_{11} + TM^1_{11} + TE^1_{21}$	$TE^2_{11} + TM^2_{11} + TE^2_{21}$
Corrugated cylindrical*	$HE^1_{11} + HE^1_{21}$	$HE^2_{11} + HE^2_{21}$
Smooth-walled rectangular	$TE_{10} + TE_{11}/TM_{11}$	$TE_{01} + TE_{20}$

Efficiency optimisation techniques, such as those described in a review paper by Clarricoats and Poulton[72] can be applied to optimise the performance of the offset antenna. If E_1, H_1 are the offset-reflector focal-plane fields and E_2, H_2 the fields created at the aperture of the primary feed when unit power is transmitted, then the efficiency can be obtained from an integration of these fields over the aperture plane of the feed(s).

$$\eta = \frac{1}{2} \int_s (E_1 \times H_2 + H_1 \times E_2) \cdot ds \qquad (4.72)$$

Optimisation of this η parameter will lead to the optimum values of the mode coefficients, but the optimum efficiency condition will not be the desired condition for all applications.

Alternatively and more simply in this case, the antenna copolar performance can be optimised independently to whatever criterion is relevant, and the cross-polar performance simply optimised by use of the higher-order mode coefficients. The interaction with the copolar characteristic will be small and generally favourable in that the higher-order modes act to compensate for the asymmetric space-attenuation factor introduced by the offset reflector.

The optimum values of the complex coefficients of the modes can be determined by using the mathematical models described in the early Sections, with the introduction of modified primary-feeds models which include the higher order modes. By examining the cross-polar radiation pattern of the offset antenna, for a range of values of the mode coefficients, the optimum characteristics can be readily determined. The close similarity between the value of the first cross-polar lobe peaks and the required level of the higher-order mode will be found to be an excellent guide in many cases.

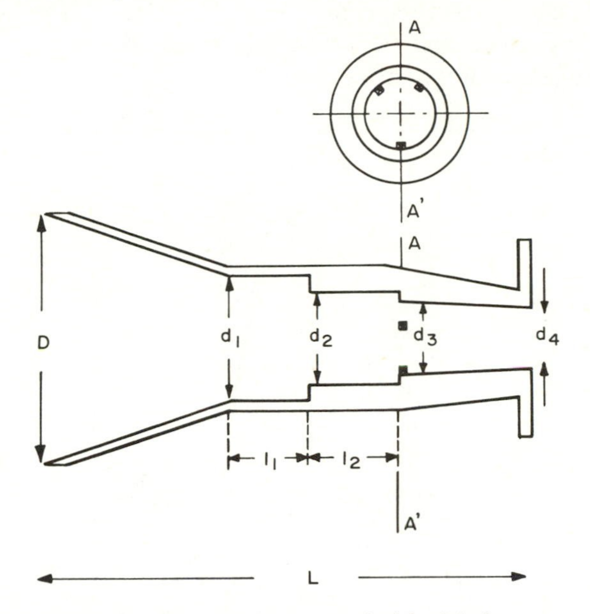

Fig. 4.56 *Prototype trimode offset-reflector matched-feed device*

Trimode devices based upon the matched-feed principle have been constructed in smooth-wall cylindrical guides, and dual-mode rectangular feeds have also been demonstrated. These prototype devices were constructed for operation with a single linear polarisation. A fully dual-polarised version of the trimode cylindrical structure has been developed[71]. The practical aspects of the matched-feed approach can best be realised by examining the design of the prototype trimode device described in 1975[61,75]. The basic configuration is illustrated in Fig. 4.56.

Prototype trimode matched feed

The trimode primary feed is essentially a small flare-angle conical horn with two steps or discontinuities. The first step region (d_3/d_2) is asymmetric and generates the TE_{21} mode. The diameter d_2 cuts off all higher modes. The second step (d_2/d_1) is axisymmetric, and the guide dimensions cut off all modes above the TM_{11}. The symmetry of this discontinuity avoids the further generation of the

TE_{21} mode. The amplitudes of the modes are governed by the ratios d_3/d_2 and d_2/d_1, and the relative phases of the modes are adjusted by the constant-diameter phasing section which follows each discontinuity. The mode amplitudes required are a function of the offset angle θ_o and the semiangle θ^* of the offset reflector. Typically, the required mode amplitudes lay in the range 20 – 30dB below the fundamental. The diameter of the primary-feed aperture is selected in the usual way to satisfy the illumination requirements of the reflector. The overall length of the feed is between 0·25 and 1·0 wavelengths greater than a conventional axi-symmetric dual-mode feed of the Potter type.

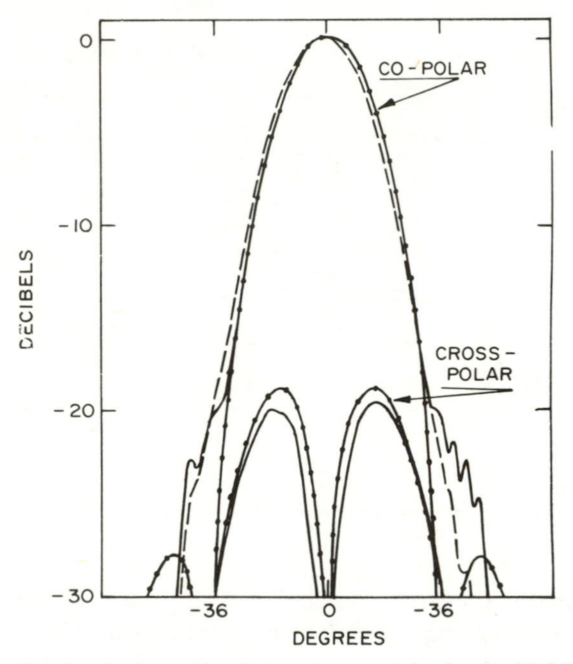

Fig. 4.57 ***Predicted and measured radiation characteristics for the 30 GHz prototype matched-feed. Predicted (H plane) —•——•— Measured (H plane) ——— (E plane) - - - - - - - - - - - -***

Predicted and measured radiation characteristics of this feed are shown in Fig. 4.57. The feed has an aperture diameter of 2·8λ, and the measurements were made at 30GHz. When used with a precision offset reflector with parameters F = 22·7λ, θ_0 = 44°, and θ^* = 30°, the matched feed provides a significant improvement in cross-polar suppression over a conventional feed. Typically, feeds of this particular design can provide a minimum of 10dB additional suppression of the reflector cross polarisation (relative to a conventional primary-feed) over a 4-5% bandwidth. At midband the additional cross-polar suppression can approach 20dB. Typical radiation pattern and bandwidth characteristics are shown in Figs. 4.58 and 4.59. These characteristics should not be interpreted as defining the funda-

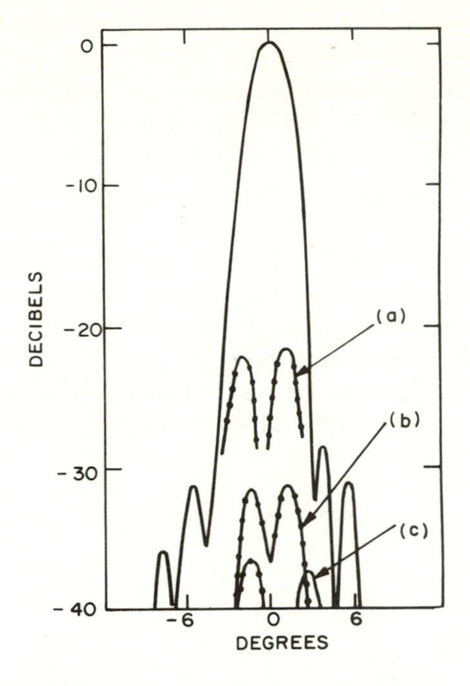

Fig. 4.58 *Measured radiation patterns in plane of asymmetry for offset reflector antenna ($F = 22{\cdot}7\lambda$, $\theta_o = 44°$, $\theta^* = 30°$)*
a Conventional Potter-horn feed.
b Matched feed (initial measurement)
c Optimised matched feed. Copolar ———. Cross-polar –•–•–

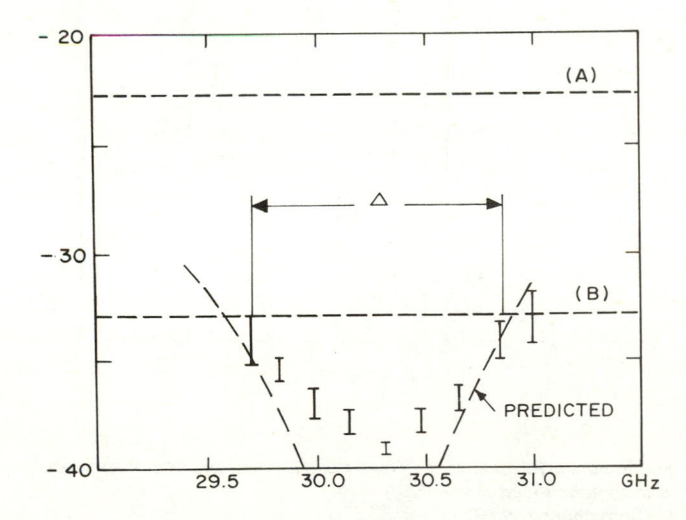

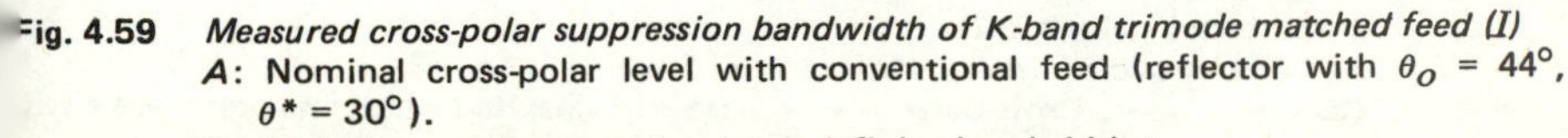

Fig. 4.59 *Measured cross-polar suppression bandwidth of K-band trimode matched feed (I)*
A: Nominal cross-polar level with conventional feed (reflector with $\theta_o = 44°$, $\theta^* = 30°$).
B: 10dB improved suppression level, defining bandwidth △

mental bandwidth limitations of multi-mode matched-feed devices. A number of techniques can be applied to improve the bandwidth of the feeds. For example, a dual-mode corrugated horn device, employing the HE_{11} and HE_{21} hybrid modes has been described which offers improved bandwidth potential.[71]

Matched feed for monopulse radars

The undesirable effect of offset-reflector depolarisation upon precision tracking, monopulse radars has been mentioned. Rudge and Adatia[62,63] proposed that the matched-feed principle could be applied to this problem by incorporating the necessary higher-order modes in a 4-horn static split feed. The questionable features of this application were identified as: (i) the level of cross-polar suppression which could be achieved by an essentially off-axis rectangular feed element, and

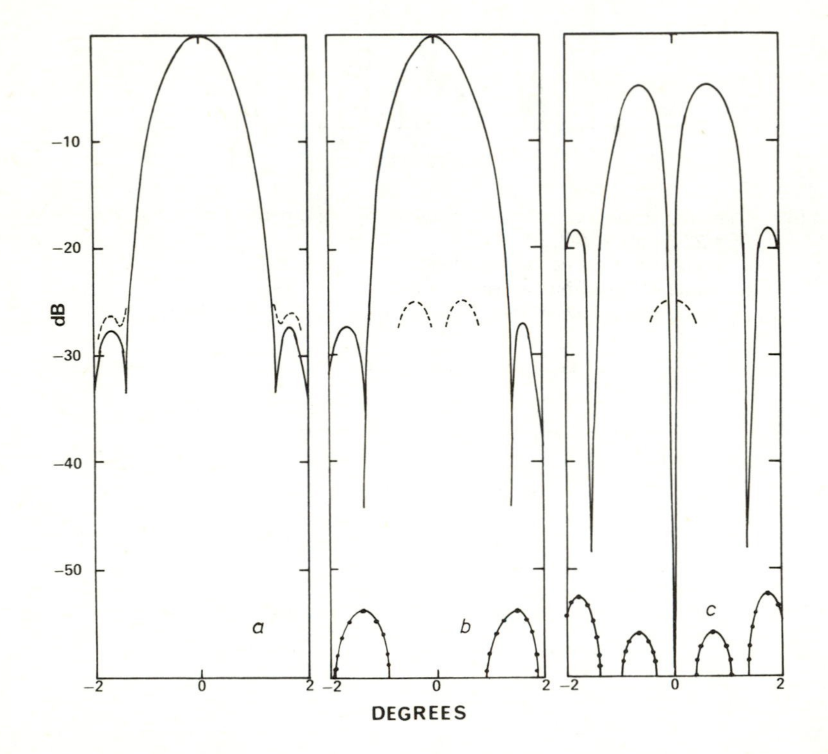

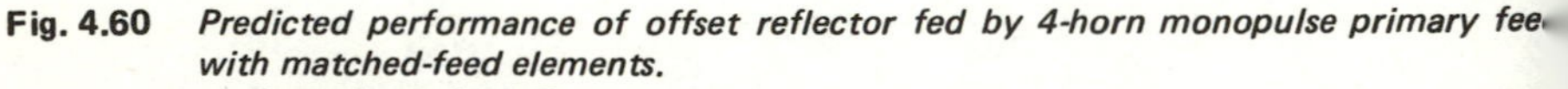

Fig. 4.60 *Predicted performance of offset reflector fed by 4-horn monopulse primary feed with matched-feed elements.*
a Sum channel $\phi = 0$.
b Sum channel $\phi = \pi/2$.
c Difference channel $\phi = \pi/2$.
Copolar ———. Cross-polar —•—•—. Dashed lines indicate cross-polar levels with conventional feed elements.

(ii) whether the matched-feed condition could be maintained for both sum and difference excitations of the monopulse feed.

The proposal was investigated by mathematical modelling of the offset reflector with its multimoded feed elements and the construction and evaluation of one ele-

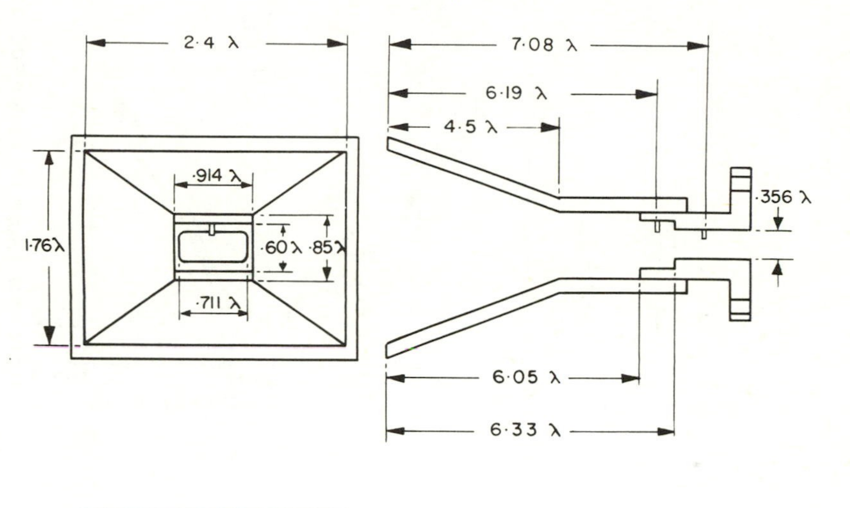

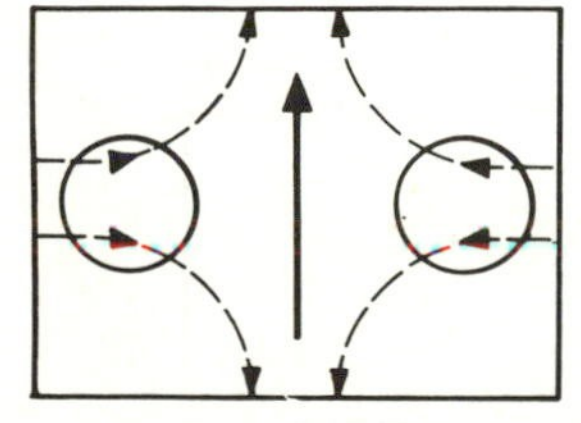

Fig. 4.61 Prototype rectangular-horn matched-feed configuraiton showing asymmetric mode aperture-plane-field distributions (TE_{11} mode) in mouth of rectangular waveguide horn with fundamental mode polarisation. Principal cross-polar regions are circled

ment of the 4-horn feed. Fig. 4.60 shows the computed cross-polar levels for a 4-horn matched feed and a conventional 4-horn device. The difference-channel radiation pattern in the principal plane ($\phi = 0$) has zero cross-polar radiation and is not shown in the Figure. Fig. 4.61 shows the rectangular feed element, and Fig. 4.62 summarises the measured radiation pattern and VSWR bandwidth performance of this K-band device. It was concluded that significant reductions in boresight jitter can be achieved by matched-feed techniques.

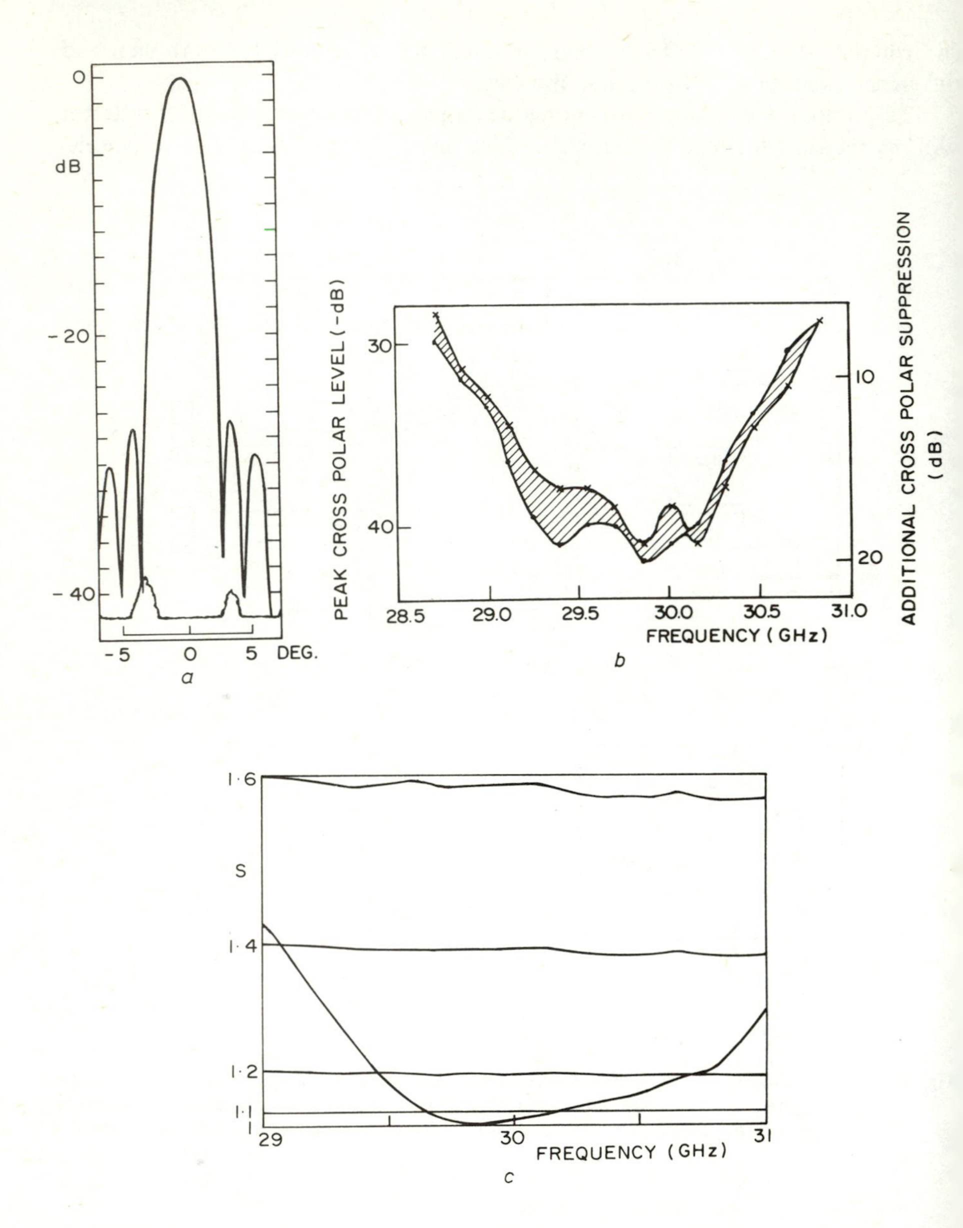

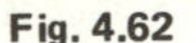

Fig. 4.62 *a* Measured radiation patterns of offset reflector fed by a rectangular matched feed element. Lower trace is cross-polar level which has been suppressed from its nominal level of −23 dB.

b Measured cross-polar suppression of prototype matched-feed element against frequency. Graph shows peak levels of lobes on either side of boresight.

c VSWR of prototype matched-feed with 2-step transformer

4.12 References

1 HANNAN, P.W.: 'Microwave antennas derived from the Cassegrain telescope', *IRE Trans.*, **AP-9,** March 1961, pp 140-153

2 SILVER, S.: 'Microwave antenna theory and design'. McGraw-Hill, N.Y., 1949.

3 BROWN, J.: 'Microwave lenses', Metheun, London, Wiley, New York, 1953

4 LUDWIG, A.C.: 'The definition of cross polarization', *IEEE Trans.* **AP-21,** January 1973, pp. 116-119

5 JONES, E.M.T.: 'Paraboloid reflector and hyperboloid lens antennas', *IRE Trans.* **AP-2,** July 1954, pp. 119-127

6 KOFFMAN, I.: 'Feed polarization for parallel currents in reflectors generated by conic sections', *IEEE Trans.* **AP-14,** January 1966, pp. 37-40

7 THOMAS, B. Mac A.: 'Cross-polarization characteristics of axially symmetric reflectors', *Electron. Lett.,* **12,** April 1976, pp. 218-219

8 SILBERBERG, R.W.: 'The paradisc antenna – A novel technique to improve the axial ratio of a circularly polarized high gain antenna system', *IEEE Trans.* **AP-21,** January 1973, pp. 108-110

9 CUTLER, C.C.: 'Parabolic-antenna design for microwaves', *Proc. IRE,* **35,** Nov. 1947, pp. 1284-1294

10 CLAVIN, A.: 'A new antenna having equal E- and H-plane patterns', *IRE Trans.* **AP-2,** July 1954, pp. 113-119

11 RUDGE, A.W. and WITHERS, M.J.: 'Design of flared-horn primary feeds for parabolic reflector antennas', *Proc. IEE* **117,** September 1970, pp. 1741-1749

12 TRUMAN, W.M. and BALANIS, C.A.: 'Optimum design of horn feeds for reflector antennas', *IEEE Trans.* **AP-22,** July 1974, pp. 585-586

13 JAMES, G.L. and GREENE, K.J.: 'Effect of wall thickness of radiation from circular guide', *Electron. Lett.,* **14,** February 1978, pp. 90-91

14 ADATIA, N.A., RUDGE, A.W. and PARINI, C.: 'Mathematical modeling of the radiation fields from microwave primary-feed antennas', Proc. 7th European Microwave Conference, Copenhagen, 1977

15 SHIMIZU, J.K.: 'Octave bandwidth feed horn for paraboloid', *IRE Trans.* **AP-9,** March 1961, pp. 223-224

16 POTTER, P.D.: 'A new horn antenna with suppressed sidelobes and equal beamwidths', *Microwave J.,* **6,** June 1963, pp. 71-76

17 TURRIN, R.H.: 'Dual mode small aperture antennas', *IEEE Trans.* **AP-15,** March 1967, pp. 307-308

18 JENSEN, P.A.: 'A low-noise multimode Cassegrain monopulse feed with polarization diversity', Northeast Electron. Res. and Eng. Meeting, November 1963, pp. 94-95

19 HARDY, W.M., GRAY, K.W. and LOVE, A.W.: 'An S-band radiometer design with high absolute precision', *IEEE Trans.* **MTT-22,** April 1974, pp. 382-390

20 COHN, S.B.: 'Flare angle changes in a horn as a means of pattern control', *Microwave J.,* **13,** October 1970, pp. 41-46

21 AJIOKA, J.S. and HARRY, Jr., H.E.: 'Shaped beam antenna for earth coverage from a stabilized satellite', *IEEE Trans.* **18,** May 1970, pp. 323-327

22 AGARWAL, K.K. and NAGELBERG, E.R.: 'Phase characteristics of a circularly symmetric dual-mode transducer', *IEEE Trans.,* **MTT-18,** Jan. 1970, pp. 69-71

23 ENGLISH, W.J.: 'The circular waveguide step-discontinuity mode transducer', *IEEE Trans.* **MTT-18,** Jan. 1970, pp. 69-71

24 TOMIYASU, K.: 'Conversion of TE_{11} mode by large diameter conical junction', *IEEE Trans.* **MTT-17,** May 1969, pp. 277-279

25 KOCH, G.F.: 'Coaxial feeds for high aperture efficiency and low spill-over of paraboloidal reflector antennas', *IEEE Trans.* **AP-21,** March 1973, pp. 164-169

26 KUMAR, A.: 'Reduce cross-polarization in reflector-type antennas', *Microwaves,* March 1978, pp. 48-51

27 CLAVIN, A.: 'A multimode antenna having equal E- and H-planes', *IEEE Trans. Antennas Propagat.,* **AP-23**, September 1975, pp. 753-757

28 CLARRICOATS, P.J.B. and SAHA, P.K.: 'Propagation and radiation behavior of corrugated feeds', *Proc.IEE,* **118**, Parts I and II, Sept. 1971, pp. 1167-1186

29 MINNETT, H.C. and THOMAS, B. MacA.: 'A method of synthesizing radiation patterns with axial symmetry', *IEEE Trans.* **AP-14**, Sept. 1966, pp. 654-656

30 KNOP, C.M. and WIESENFARTH, H.J.: 'On the radiation from an open-ended corrugated pipe carrying the HE_{11} Mode', *IEEE Trans.* **AP-20**, Sept. 1972, pp. 644-648

31 NARASIMHAN, M.S. and RAO, B.V.: 'Hybrid modes in corrugated conical horns', *Electron. Lett.,* **6**, Jan. 1970, pp. 32-34;

32 NARASIMHAM, M.S. and RAO, B.V.: 'Diffraction by wide-flare-angle corrugated conical horns', *Electron. Lett.,* **6**, July 1970, pp. 469-471

33 KAY, A.F.: 'The scalar feed', AFCRL Report 64-347, AD601609, March 1964

34 SIMMONS, A.J. and KAY, A.F.: 'The scalar feed – A high performance feed for large paraboloid reflectors', In Design and construction of large steerable aerials, IEE Conference Publ. 21, 1966, pp. 213-217

35 WOHLLEBEN, R., MATTES, H., and LOCHNER, O.: 'Simple small primary feed for large opening angles and high aperture efficiency', *Electron. Lett.,* **8**, Sept. 1972, pp. 474-476

36 GRUNER, R.W.: 'A 4- and 6-GHz, prime focus, CP feed with circular pattern symmetry', IEEE AP-S Symposium Digest, June 1974, pp. 72-74

37 SHAFAI, L.: 'Broadening of primary-feed patterns by small E-plane slots', *Electron. Lett.,* **13**, Feb. 1977, pp. 102-103

38 COWAN, J.H.: 'Dual-band reflector-feed element for frequency-reuse applications', *Electron. Lett.,* **9**, Dec. 1973, pp. 596-597

39 THOMAS, B. MacA.: 'Design of corrugated conical horns', *IEEE Trans.,* **AP-26**, March 1978, pp. 367-372

40 VU, T.B. and VU, Q.H.: 'Optimum feed for large radiotelescopes: Experimental results', *Electron. Lett.,* **6**, March 1970, pp. 159-160

41 THOMAS, B. MacA.: 'Prime-focus one-and-two-hybrid mode feeds', *Electron. Lett.,* **6**, July 1970, pp. 460-461

42 AL-HAKKAK, M.J. and LO, Y.T.: 'Circular waveguides with anisotropic walls', *Electron. Lett.,* **6**, Nov. 1970, pp. 786-789

43 COOPER, D.M.: 'Complex propagation coefficients and the step discontinuity in corrugated cylindrical waveguide', *Electron. Lett.,* **7**, March 1971, pp.135-136

44 HANNAN, P.W.: 'Optimum feeds for all three modes of a monopulse antenna I: Theory' *IRE Trans.* **AP-9**, Sept. 1961, pp. 444-454

45 HANNAN, P.W.: 'Optimum feeds for all three modes of a monopulse antenna II: Practice' *IRE Trans.* **AP-9**, Sept. 1961, pp.454-461

46 RICARDI, L.J., and NIRO, L.: 'Design of a twelve-horn monopulse feed', IRE International Convention Record, Part I, Antennas and Propagat., 1961, pp. 49-56

47 HANNAN, P.W. and LOTH, P.A.: 'A monopulse antenna having independent optimization of the sum and difference modes', IRE International Convention Record, Part I, Antenna and Propagat., 1961, pp. 57-60

48 MILNE, K. and RAAB, A.M.: 'Optimum illumination tapers for four-horn and five-horn monopulse aerial systems', IEE Conference Publication 21, 1966, pp. 12-16

49 SCIAMBI, A.F.: 'Five horn tracking feeds for large antennas', IEE Conference Publication 21, 1966, pp. 158-162

50 SCIAMBI, A.F.: 'Five-horn feed improves monopulse performance', *Microwaves,* **11** June 1972, pp 56-58

51 CLARK, R.T. and JENSEN, P.A.: 'Experimental Cassegrain-fed monopulse antenna system', NASA Contractor Report No. CR-720, May 1967

52 DAVIS, D.: 'Corrugations improve monopulse feed horns', *Microwaves*, **11**, April 1972, pp. 58-63

53 VIGGH, M.: 'Study of design procedures and limitations for monopulse scalar feeds', Rome Air Development Center Report No. RADC-TR-69-303, ASTIA Document 862515, Nov. 1969

54 DUHAMEL, R.H. and ISBELL, D.L.: 'Broadband logarithmically periodic antenna structures', IRE National Convention Record, Part I, 1957, pp. 119-128

55 DUHAMEL, R.H. and ORE, F.R.: 'Logarithmically periodic antenna designs', IRE National Convention Record Part I, 1958, pp. 139-151

56 DUHAME, R.H. and ORE, F.R.: 'Log periodic feeds for lens and reflectors', IRE National Convention Record, Part I, 1959, pp. 128-137

57 LOVE, A.W.: 'Quadratic phase error loss in circular apertures', *Electron. Lett.*, **15**, May 1979, pp. 276-277

58 DYSON, J.D.: 'The characteristics and design of the conical log-spiral antenna', *IEEE Trans.* **AP-13**, July 1965, pp. 488-499

59 CHU, T.S. and TURRIN, R.H.: 'Depolarization properties of offset reflector antennas', *IEEE Trans.* **AP-21**, May 1973, pp. 339-345

60 TANAKA, H. and MIZUSAWA, M.: 'Elimination of cross-polarization in offset dual-reflector antennas', *Electron. and Commun. in Japan,* **58-B**, 1975, pp. 71-78

61 RUDGE, A.W. and ADATIA, N.A.: 'New class of primary-feed antennas for use with offset parabolic reflector antennas', *Electron. Lett.,* **11**, Nov. 1975, pp. 597-599

62 RUDGE, A.W. and ADATIA, N.A.: 'Matched feeds for offset parabolic reflector antennas', Proc. 6th European Microwave Conference, Rome, Italy, Sept. 1976, pp. 1-5

63 RUDGE, A.W. and ADATIA, N.A.: 'Primary feeds for boresight jitter compensation of offset reflector radar antennas', IEE Conference Publication 155, Oct. 1977, pp. 409-413

64 RUDGE, A.W. and SHIRAZI, M.: 'Multiple beam antennas: Offset reflectors with offset feeds', Univ. Birmingham, England, July 1973, Final Report ESA Contract 1725/72PP

65 ADATIA, N.A., RUDGE, A.W. and PARINI, C.: 'Mathematical model of the radiation field from microwave primary feed antennas), Proc. 7th European Microwave Conference (Copenhagen) Sept. 1977, pp. 329-333

66 RUDGE, A.W. and PRATT, T., and FER, A.: 'Cross-polarised radiation from satellite reflector antennas', Proc. AGARD Conf. on Antennas for Avionics (Munich, Germany), Nov. 1973, **16**, p. 1-6

67 HANSEN, J.E., and SHAFAI, L.: 'Cross-polarised radiation from waveguides and narrow-angle horns', *Electron. Lett.* May 1977, pp. 313-315

68 BEM, D.J.: 'Electric field distribution in the focal region of an offset paraboloid', *Proc. IEE,* **116**, 1974, pp. 579-584

69 VALENTINO, A.R. and TOULIOS, P.P., 'Fields in the focal region of offset parabolic reflector antennas', *IEEE Trans.* **AP-24**, Nov. 1976, pp. 859-865

70 INGERSON, P.G. and WONG, W.C.: 'Focal region characteristics of offset fed reflectors' IEEE Int. Symp. AP-S (Georgia) 1974, pp. 121-123

71 WATSON, B.K., RUDGE, A.W. and ADATIA, N.: 'Dual-polarised mode generator for cross-polar compensation in offset parabolic reflector antennas'. 8th European Microwave Conf., Paris, France, Sept. 1978

72 CLARRICOATS, P.J.B., and POULTON, G.T.: 'High efficiency microwave reflector antennas – A review', *Proc. IEEE,* **65**, Oct. 1977, pp. 1470-1504

73 CLARRICOATS, P.J.B. and SALEMA, C.E.R.C.: 'Antennas employing conical dielectric horns, Parts I and II, *Proc. IEE*, **120**, 1973, pp. 741-749

74 LOVE, A.W.: 'Reflector antennas'. Selected Report Series, IEEE Press, 1978

75 RUDGE, A.W. and ADATIA, N.A.: Offset parabolic reflector antennas. A Review, *Proc. IEEE* **66**, Dec. 1978, pp. 1592-1618

76 GALINDO-ISRAEL, V., *et al*: 'Aperture amplitude and phase control of offset dual reflectors', *IEEE Trans.* **AP-27**, March 1979, pp. 154-164

77 WESTCOTT, B.S. *et al*: 'Exact synthesis of offset dual reflectors' *Electron. Lett.* **16**, Feb. 1980, pp. 168-169
78 LOVE, A.W.: 'Electromagnetic horn antennas'. Selected Report Series, IEEE Press, 1976
79 RUSCH, W.V.T. and POTTER, P.D.: 'Analysis of reflector antennas'. Academic Press, New York, 1970
80 HARRINGTON, R.F.: 'Time harmonic electromagnetic fields'. McGraw Hill, New York, 1961
81 COLLIN, R.E. and ZUCKER, R.: 'Antenna theory Parts I and II'. McGraw Hill, New York, 1968
82 RUDGE, A.W. *et al*: 'Low cross-polar waveguide horns for multiple-feed reflector antennas'. IEE Conf. Publ. 169, Antennas and Prop., Nov. 1978, pp. 360-363
83 DRAGONE, G.: 'Reflection, transmission and mode conversion in a corrugated feed'. *BST J.*, **56**, July-Aug. 1977, pp. 835-867
84 SATOH, T.: 'Dielectric-loaded horn antenna'. *IEEE Trans.* **AP-20**, March 1972, pp. 199-201

Chapter 5

Hybrid antennas

R. J. Mailloux

5.1 Introduction

The hybrid antenna will be defined as an antenna system that consists of an electronically scanned array combined with some radiating aperture. Hybrid antennas would be unnecessary if phased arrays could be made very inexpensively. If the systems designers' dream of a low-cost array with thousands of little elements, each costing a few dollars and controlled by some central processor, had happened or would soon happen, there would be little need to expend much time or effort in the development of hybrid antennas. The opposite is true. Phased arrays are extremely expensive and for that reason hybrid antennas are among the most important, most promising and most rapidly developing items of antenna technology.

The basic hybrid system, consisting of phased array and radiating aperture which will be referred to as the objective aperture, may also include another structure, perhaps a second lens, multiple-beam matrix or subreflector. Excluded from the basic definition are antennas that scan a beam because of some mechanical motion (conventional scanners) or those which are switched multiple-beam systems; these topics are considered in other chapters. The basic feed system for the antennas discussed in this chapter is the phase scanned array.

Hybrid antennas perform a variety of tasks for military and civilian systems. These tasks are more often related to radar than communication applications because rapid electronic beam agility is demanded of many modern radars. Hybrid antennas with limited sector coverage are used in fire-control systems, weapon locators and air traffic control systems, while those with extremely wide angular coverage find use in strategic radar applications. Hybrid antennas are used in certain non-scanning applications as well. These include systems with special feed structures for tolerance control and adaptive jammer cancellation. In each case, however, the justification is the same; to control the large objective aperture by changing the excitation of relatively few antennas in the array feed.

This Chapter is divided into five Sections. Section 5.2 introduces the basic concepts used in hybrid scanning systems and outlines general principles common to all of the antenna techniques described later in the chapter. Section 5.3 describes some of the more pertinent characteristics of reflector antennas and explores the development of hybrid scan systems using a single reflector or a reflector in combination with another reflector, lens or constrained matrix. Section 5.4 reviews the technology of lens-array, dual lens-array hybrids and hybrid systems consisting of a lens and a constrained multiple beam matrix. Section 5.5 comments on the use of these kinds of feed systems in error tolerance control, pattern null steering and sidelobe control, and in addition discusses the sidelobe suppression achievable with spatial filtering techniques.

In each case the levels of technical detail has been chosen to introduce the reader to the fundamental principles, delineate the geometrical constraints, and provide the engineer with tools to select an appropriate antenna for a given application.

5.2 Aperture scanning fundamentals of hybrid systems

5.2.1 Introduction

This Section groups the hybrid scanning antennas into several categories for convenient study. The categories and principles developed here admittedly do not include all of the existing systems, but they share common features with so many that the grouping is a useful one.

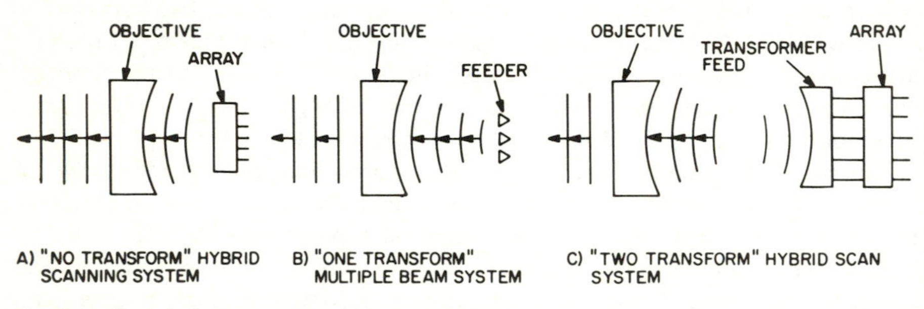

Fig. 5.1 *Basic hybrid scanning and multiple beam systems*

The analysis and properties of scanned apertures are treated in this Section, which also discusses the importance of orthogonal beams, the minimum number of control elements for a given aperture and scan sector and design criteria that are common to most of the hybrid scanning systems. The process of interrelating various hybrid systems is here carried out by identifying three categories of systems, each distinguished from the other by the number of Fourier transforms performed within the system, Fig. 5.1. Focusing objectives are well known to perform the Fourier transform operation in converting the aperture distribution to a spot at the focal plane; so reflectors, lenses and Butler matrices can all be thought of and analysed as Fourier transformers. In addition, the relationship between the aperture illumination and the far-field pattern is a Fourier transform, and one can make use of these relationships to gain insight into required antenna parameters. For example, a system that has one Fourier transformer (lens, reflector or matrix) focuses a pencil beam to a single input port, Fig. 5.1*b*. The same pencil beam, received by a two transform system results in a relatively constant amplitude distribution across an array feed, Fig. 5.1*c*. Tapering the array amplitude of the two transform system results in the same amplitude taper at the radiating aperture, and this additional aperture control is the key to the excellent low side-lobe potential of such dual transform systems.

Some scanning systems do not have a Fourier transform. For example, a focusing objective with an array feed well inside the focus has a converging field at the array but little amplitude variation; the array is a transverse feed, and the system takes no Fourier transform, Fig. 5.1*a*. The transform perspective also helps to explain the formation of overlapped subarrays; a principle of vast import for antennas with limited sector coverage, as well as for tolerance control and adaptive beam forming and null steering.

5.2.2 The plane aperture: transforms, orthogonal beams and scanning limitations

For the purposes of this Section, it is convenient to neglect constant factors and write the radiated far field of a linearly polarised antenna at a point $p(x, y, z)$ as a simple aperture integration:

$$g(\boldsymbol{k}) = \int f(\boldsymbol{r})\, e^{j\boldsymbol{k}\cdot\boldsymbol{r}}\, da \tag{5.1}$$

carried out over the area, a, of the aperture of Fig. 5.2 with aperture coordinates (x', y', z'). The parameters $\boldsymbol{r}$ and $\boldsymbol{k}$ are defined below.

$$\begin{aligned} \boldsymbol{r} &= \boldsymbol{a}_x(x-x') + \boldsymbol{a}_y(y-y') + \boldsymbol{a}_z(z-z') \\ \boldsymbol{k} &= \frac{2\pi}{\lambda}(\boldsymbol{a}_x u + \boldsymbol{a}_y v + \boldsymbol{a}_z \cos\theta) \end{aligned} \tag{5.2}$$

where $u = \sin\theta \cos\phi$ and $v = \sin\theta \sin\phi$.

The aperture distribution $f(\boldsymbol{r})$ can be reconstructed by taking the inverse transform of the far-field distribution

$$f(\boldsymbol{r}) = \int g(\boldsymbol{k})\, e^{-j\boldsymbol{k}\cdot\boldsymbol{r}}\, d\sigma \tag{5.3}$$

integrated over the surface of a unit sphere with element of solid angle $d\sigma$. Transform constants have been omitted in this expression and through the Section.

The analysis of this Section will be limited to two dimensions, and so the transform pair becomes:

$$g(u) = \int f(x)\, e^{j(2\pi/\lambda)ux}\, dx$$

and

$$f(x) = \int g(u)\, e^{-j(2\pi/\lambda)ux}\, du \tag{5.4}$$

where $u = \sin\theta$ for θ measured in the x, z plane of Fig. 5.2.

In this case the inverse transform for $f(x)$ is taken assuming $g(u)$ a finite real parameter bound within the region $-1 \leqslant u \leqslant 1$.

Consider a one-dimensional aperture of length D. If the aperture were excited by a uniform amplitude, progressive phase distribution of the form

$$f_i(x) = e^{-j(2\pi/\lambda)u_i x} \tag{5.5}$$

the resulting far field pattern is given by:

$$\begin{aligned} g_i(u) &= \int_{-D/2}^{D/2} e^{j(2\pi x/\lambda)(u-u_i)}\,dx \\ &= D\,\frac{\sin[(\pi D/\lambda)(u-u_i)]}{(\pi D/\lambda)(u-u_i)} \end{aligned} \tag{5.6}$$

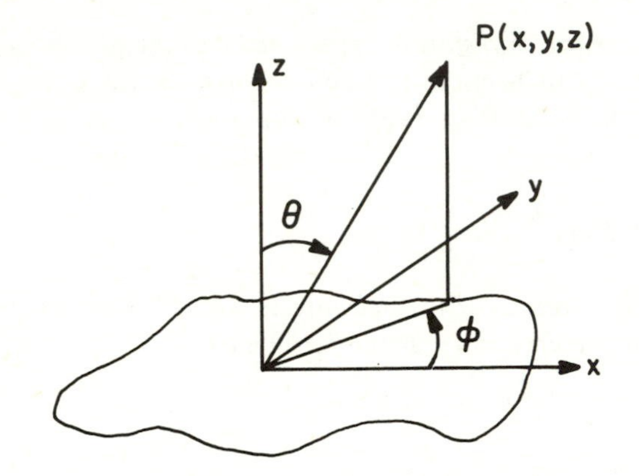

Fig. 5.2 *Radiating aperture coordinates*

This radiation pattern has the special characteristics that it is one of an orthogonal set of pencil beams if the inter-beam spacing u_i is chosen

$$u_i = \frac{\lambda}{D} i \quad \text{for } i = \pm\frac{1}{2}, \pm\frac{3}{2} \ldots$$

In addition to orthogonality, the beams have the characteristic that the beam with index i has its maximum at u_i, and has zero at any other value u_j for j within the set defined above. Thus it is possible to synthesise a shaped beam pattern

$$g(u) = \sum_i g_i(u_i) \tag{5.8}$$

by matching the value of the ith beam to the desired pattern at the angle denoted by $u_i = \sin\theta_i$. Since only one beam has a non-zero value for a particular choice of i, the synthesis is performed by exactly matching the required pattern at N-points, with N constituent beams of the orthogonal set. This type of synthesis was developed by Woodward and Lawson,[48] and is described in some detail in a number of texts (see Hansen[15] Vol. 1, p. 80). For an aperture of electrical length D/λ, N beams will fill a sector of width $(N-1)\lambda/D$ in $\sin\theta$ space. A given shaped pattern within this sector can thus be matched at N points by the use of N separate controls, assuming a network that can switch between the various beams. Thus the minimum number of controls required to form or scan a beam within the given sector is equal to the number of pencil beams within the sector.

A similar set of patterns results if the aperture is made up of discrete elements instead of a continuous aperture. Consider an array of length D, consisting of N elements spaced d_x apart. The elements are excited by progressive phase distribution

$$a_n = e^{-j(2\pi/\lambda)d_x u_i n} \tag{5.9}$$

where again $u_i = (\lambda/D)i = (1/N)(\lambda/d_x)i$ and $i, n = \pm\frac{1}{2}, \ldots \pm(N-1)/2$. Here, and throughout this Section, the index n is chosen to be an odd multiple of $\frac{1}{2}$ for convenience in analysing structures with even numbers of beams and elements. Eqn. 5.4, integrated over the elements to give the element patterns $g_e(u)$ (which are assumed to be the same for all elements), leaves the following summation:

$$g_i(u) = g_e(u) \sum_{n=-(N-1)/2}^{(N-1)/2} e^{j(2\pi/\lambda)x_n(u-u_i)} \tag{5.10}$$

$$= Ng_e(u)\frac{\sin[(N\pi d_x/\lambda)(u-u_i)]}{N\sin[(\pi d_x/\lambda)(u-u_i)]}$$

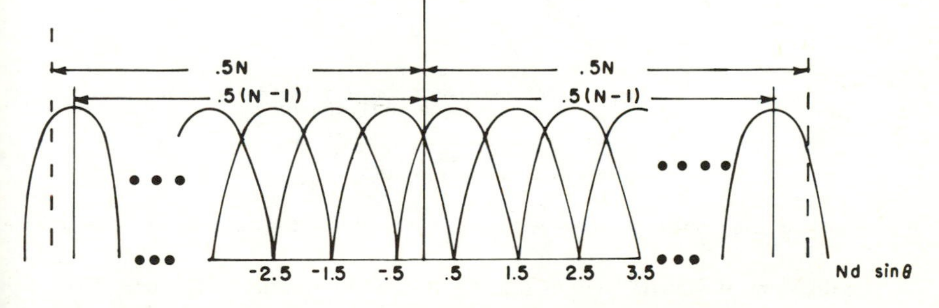

Fig. 5.3 *Orthogonal beam positions for an array of N-elements (N even)*

The set of beams $g_i(u)$ is also orthogonal and occupies the beam positions shown in Fig. 5.3. The outermost beam of this set has its peak value at

$$u_{(N-1)/2} = \frac{\lambda}{2d_x}\frac{(N-1)}{N} = \sin\theta_{max} \tag{5.11}$$

and the phase progression between elements for this beam is

$$\Delta = \frac{2\pi}{N}\frac{N-1}{2} = \pi\left(1 - \frac{1}{N}\right) \tag{5.12}$$

There can be no beam with $\sin\theta$ larger than this, because, if there were, it would have a grating lobe at

$$\sin\theta - \frac{\lambda}{d_x}$$

which would be of significant size and would contribute to a loss of gain and pattern ambiguity. Eqn. 5.12 again implies a minimum number of control elements necessary for an array by stating that the array can be scanned to one half a beamwidth of the angle of grating lobe onset. For N large, the array has the scan limit

$$\frac{d_x}{\lambda}\sin\theta_{max} = 0.5 \tag{5.13}$$

These measures of the minimum number of controls for scanning a given volume have been generalised by several authors to apply to other than orthogonal constituent beams.

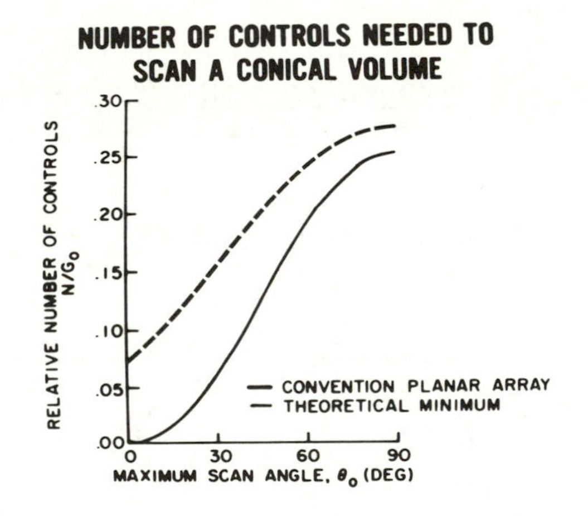

Fig. 15.4 *Minimum number of control elements for a conical sector* (*courtesy of J. Stangel*)

The theorem of Stangel[41] states that the minimum number of elements is:

$$N = \frac{1}{4\pi}\oint G_0(\theta,\phi)\,d\Omega$$

where $G_0(\theta,\phi)$ is the maximum gain achievable by the antenna in the (θ,ϕ) direction, and $d\Omega$ is the increment of solid angle. Fig. 5.4 shows how this number varies for an array scanning a conical sector. This curve is based upon an array scanning according to the gain envelope $G_0(\theta,\phi) = G_0\cos\theta$, and so in this general form includes beam broadening as a function of scan. The equation assumes uniform illumination for a planar array, and so does not account for beam broadening due to aperture taper.

Another measure of the minimum number of array elements is contained in the definition of a parameter introduced by Patton[24] and called the 'element use factor'. This parameter is N/N_{min}, where N is the actual number of phase shifters in the control array, and N_{min} is a reasonable number of control elements as defined below:

$$N_{min} = \frac{\sin\theta^1_{max}}{\sin(\theta^1_3/2)} \quad \frac{\sin\theta^2_{max}}{\sin(\theta^2_3/2)} \tag{5.14}$$

θ^1_{max} and θ^2_{max} are the maximum scan angles in the two planes measured at the peak of each beam and θ^1_3 and θ^2_3 are the half power beamwidths in these planes. This equation is clearly equivalent to the condition given in eqn. 5.13 upon assuming an approximate beamwidth λ/Nd_x.

For narrow beams and a rectangular scan sector, the element use factor can be written in the form

$$\frac{N}{N_{min}} = \frac{0.25\,\lambda^2}{(d_1 \sin\theta^1_{max})(d_2 \sin\theta^2_{max})} \tag{5.15}$$

Although more general than Patton's formula, Stangel's is equivalent for the case of small scan angle. Borgiotti[4] and others have obtained similar criteria, but for the purposes of this Chapter and most engineering work it is most convenient to count the number of beamwidths required to fill the scan sector (with -3 dB at the scan limit), and consider this number to be the theoretical minimum.

5.2.3 Scanning systems with no Fourier transform: Transverse feeds

One must either use no Fourier transforms, or two Fourier transforms (Fig. 5.1) to excite a scanned aperture with a phased array. Systems with one transform (focusing lens, reflector, or constrained matrix) convert the tilted received wavefront into an off-axis focal spot and are multiple beam systems, not scanning systems. There are a number of antenna systems in which the objective serves to project the incoming wavefront onto the array face, but does not focus at the array. The array serves simply as an extended transverse feed. The system does not perform a Fourier transform because the planar incident wavefront is converted to another nearly planar wavefront at the array, and a tapered array distribution transforms to a more or less equally tapered aperture illumination. Systems of this type include many of the reflector-array and lens-array hybrid antennas used for limited sector scanning applications and the wide-angle scanning Dome antenna Schwartzman and Stangel.[33] These techniques and others are described in more detail in Sections 5.3 and 5.4.

The key feature of such systems is that the objective must be large because the scanned array illuminates a spot that moves across the main aperture as a function of scan. This is a direct result of the need to use the whole array at all times, so that instead of requiring an amplitude weighted illumination to move around on the array face, the economical choice is to use the array efficiently and the main aperture inefficiently. Design is usually based upon the criteria that the array-aperture illumination be the complex conjugate of the received field distribution for an incident plane wave (see Assaly and Ricardi[1] and Winter[47]). This places a minimum limit on the size of the array because the usual requirement to scan with phase only requires that the array must be outside of the region of nonuniform fields near the focus.

5.2.4 *Apertures followed by one Fourier transform: Multiple beam systems*

A focusing lens or reflector objective operates on the aperture distribution as a Fourier transformer to focus, for example, a uniform incoming wavefront into a type of $(\sin x)/x$ distribution at the focal plane. A detailed description of this transform relationship for a reflector is included in the work of Rudge and Withers[27] and will be described in more detail in Section 5.3. Except for the fundamental description given in this Section, these systems are not the subject of this Chapter for they are multiple beam, not hybrid scanning antennas. The reflector or lens aperture produces a spot focus and a number of independent feeds are located in the focal region to receive energy from off-axis beams. Often it is difficult to produce low sidelobe multiple beams, but this problem is most usually avoided in practice by synthesising with clusters of these constituent beams, each with relatively high sidelobes (see Galindo Isreal *et al.*[13]).

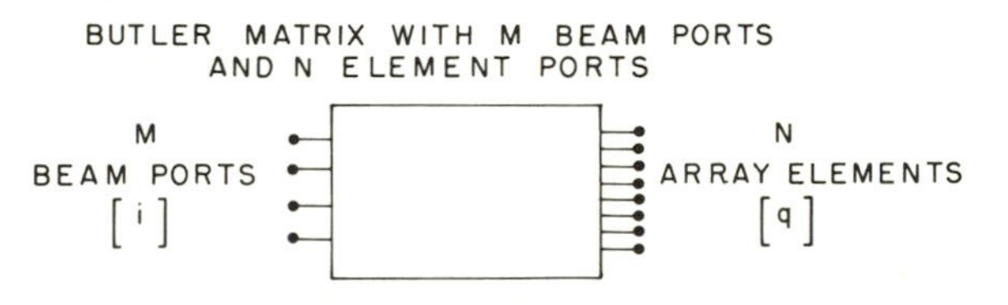

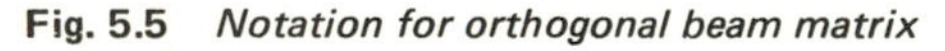

Fig. 5.5 *Notation for orthogonal beam matrix*

The Butler matrix is an ideal feed for exciting orthogonal multiple beams of the type described by eqn. 5.10, and for synthesis of complex shaped beam sector coverage using the Woodward–Lawson technique. Shown schematically in Fig. 15.5, the Butler matrix is an orthogonal beam network that forms a discrete Fourier transform using a fast Fourier transform algorithm. The matrix shown has N array element ports and M beam ports (with $N \geqslant M$), and has the property that a signal I_i applied at the ith beam port produces signals at each of the element ports according to the relation

$$i_{ni} = \frac{I_i}{\sqrt{N}} e^{-j2\pi(n/N)i} \tag{5.16}$$

which is the proper phasing to excite the ith orthogonal beam of eqn. 5.9. Thus the array excitation i_n is the Fourier transform of the input beam signals I_i, in that

$$i_n = \sum i_{ni} = \frac{1}{\sqrt{N}} \sum_{i=-(M-1)/2}^{i=(M-1)/2} I_i e^{-j2\pi(n/N)i} \quad (5.17)$$

and the matrix excites M beams out of the set of N available orthogonal beams.

The Butler matrix therefore is the ideal network for Woodward-Lawson synthesis because it makes each constituent beam available at a separate port. The matrix is orthogonal and no power is lost because of interbeam coupling. Unfortunately, such matrices are complex and often lossy, so although much larger matrices have been built it is uncommon to find Butler matrices used in applications that require more than 8 or 16 elements in a plane.

Larger multiple beam systems use lens objectives, which can form excellent wide-angle multiple beams. Still larger systems use reflector objectives that are limited in the number of relatively undistorted beams available, but which are much less expensive than large lenses.

Multiple-beam systems can grow into scanning antennas by the addition of a power-divider network to select the proper combination of constituent beams when excited by a set of signals with progressive phase. The network is again a Fourier transformer and the end result is a scanning two-transform system; which is the subject of the next Section.

5.2.5 Scanning systems with two Fourier transforms: The Principle of Overlapped subarrays

Two transform scanning antennas consist of a focusing objective, a subreflector, lens or constrained beam, matrix and a phased array. Antennas of this type have significant advantages in terms of scanning characteristics, main aperture size, element use factor (array size) and sidelobe levels.

The superior sidelobe characteristics result from two factors. First, the taper imposed upon the array is translated to the main aperture and so the aperture illumination can be carefully controlled. Secondly, the action of the several transforms is to form a subarray at the main aperture, and the subarray radiation pattern provides further suppression of certain sidelobes. In addition, two transform systems possess convenient properties that allow correction for surface tolerance errors in the objective, broadbanding, null steering and adaptive pattern control at the subarray level.

The concept of subarrays is fundamental to the design of systems with two Fourier transforms. Consider the infinite aperture of Fig. 5.6. The aperture is continuous but the dotted lines indicate a periodic grid where the points x are projected back onto an array feed. The radiated field is given by:

$$g(u) = \int_{-\infty}^{\infty} f(x) e^{j(2\pi ux/\lambda)} dx \quad (5.18)$$

The aperture distribution can be written as a sum of coefficients multiplying subarray excitations $\phi(|x - x_p|)$ centered at x_p for $x_p = pd_x$ and $p = \pm(\frac{1}{2}, \frac{3}{2}$ etc.)

$$f(x) = \sum a_p \phi(|x - x_p|) \tag{5.19}$$

and so

$$g(u) = \sum_p a_p \, e^{j(2\pi/\lambda)ux_p} \int_{-\infty}^{\infty} \phi(|x - x_p|) \, e^{ju(x-x_p)(2\pi/\lambda)}$$

$$= \sum_p a_p \, e^{(2\pi/\lambda)ux_p} g_0(u) \tag{5.20}$$

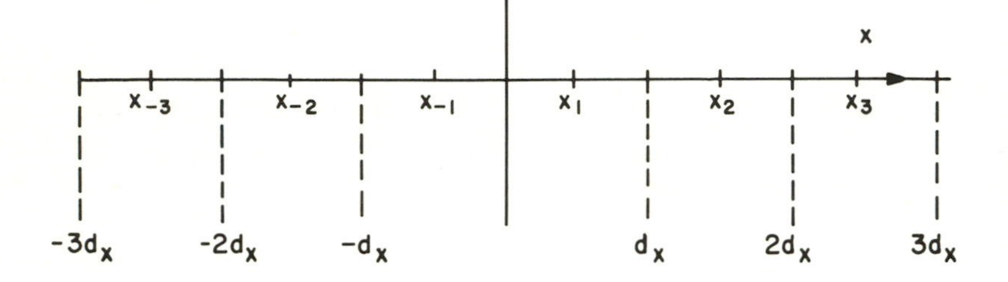

Fig. 5.6 *Infinite periodic phased array*

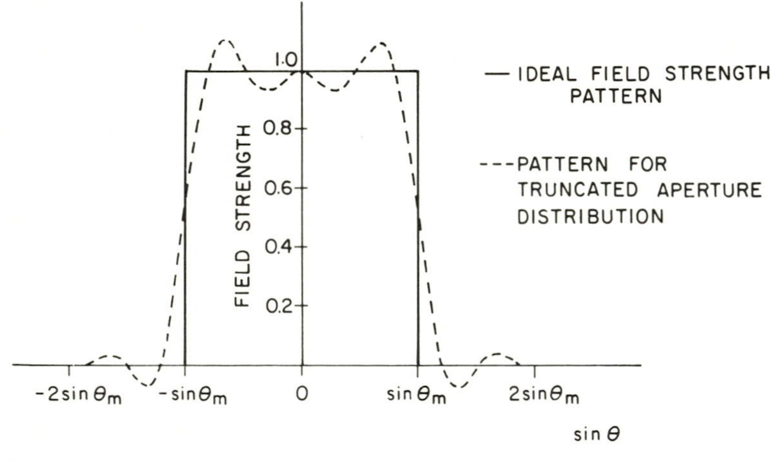

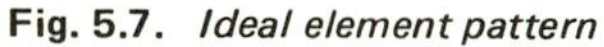
Fig. 5.7. *Ideal element pattern*

This expression shows the radiation pattern written as the sum of contributions from an infinite number of subarrays each with the same subarray pattern $g_0(u)$. For a number of applications involving limited sector scanning, it is desirable to have a subarray pattern $g_0(u)$ like the solid line of Fig. 5.7, where the radiation is uniform from $u = u_m = -\sin\theta_m$ to $\sin\theta_m$, and zero elsewhere. The subarray illumination for this pattern is the Fourier transform of the pulse shown in the figure:

$$\phi(|x-x_p|) = \int_{-\infty}^{\infty} g_0(u)\, e^{-ju(x-x_p)(2\pi/\lambda)}\, du$$
$$= u_m \frac{\sin[(2\pi/\lambda)u_m(x-x_p)]}{(2\pi/\lambda)u_m(x-x_p)} \tag{5.21}$$

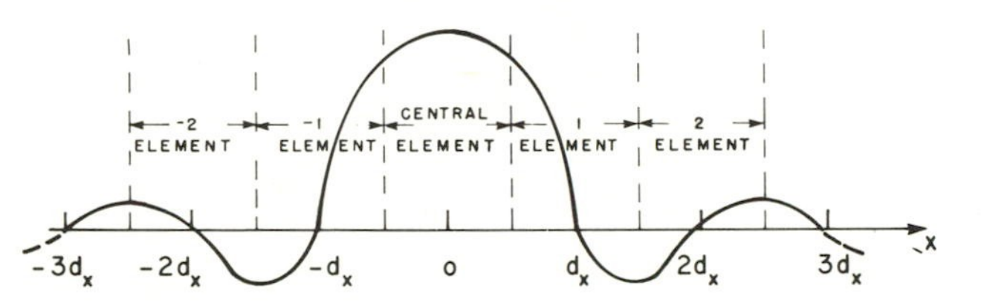

Fig. 5.8 *Overlapped subarray distribution to form ideal element pattern*

Fig. 5.8 shows this subarray distribution $\phi(|x-x_p|)/u_m$, which is called an overlapped subarray because it extends over an infinite number of periods 'd_x' of the array lattice. Thus the subarray formed at each x_n overlaps with every other subarray. Since the array has a grating lobe at

$$\sin\theta = \sin\theta_m - \frac{\lambda}{d_x}$$

the ideal subarrays can be spaced d_x apart, where

$$\frac{d_x}{\lambda}\sin\theta_m = 0.5$$

in order that they can allow scan to the maximum θ_m with total grating lobe suppression. This case yields the element use factor unity as explained in Section 5.2.2. The dashed curve of Fig. 5.7 shows the flat-topped subarray pattern achievable if the $\phi(|x-x_p|)$ is truncated at $|x-x_p| = 3d_x$. In this case the subarray pattern presents an added 20 dB suppression for grating lobes and sidelobes that do not fall within the region $-u_m \leqslant u \leqslant u_m$.

Several two-transform systems will be described in Sections 5.3 and 5.4, but for the purposes of explaining the formation of subarrays it is convenient to describe an idealised antenna feed system consisting of a set of back-to-back orthogonal beam (Butler) matrices as shown in Fig. 5.9. This network was proposed by Shelton,[36] who explained its use to form subarrays. Shelton used the network as a feed for an objective, and so was addressing its use in a three-transform multiple-beam system where it was shown to provide closely spaced beams with high cross-over but without excessive spillover. In a two-transform scanning system the network forms a flat-topped subarray as described below. The matrix that excites the arrays is the same as that of Fig. 5.5. Eqn. 5.16 gives the output signals i_{ni} in response to a given input I_i at the ith beam port.

Thus, apart from the factor $\sqrt{N}$, eqn. 5.10 gives the field pattern $g_i(u)$ that results when the ith port of the $M \times N$ Butler matrix is excited. Now, however, this matrix is excited by the $M \times M$ matrix with the result that the signal at the ith beam port due to an applied signal of J_m is given by:

$$I_{im} = \frac{J_m}{\sqrt{M}} e^{j2\pi(m/M)i} \tag{5.23}$$

BUTLER MATRICES FOR OVERLAPPING SUBARRAY PORTS

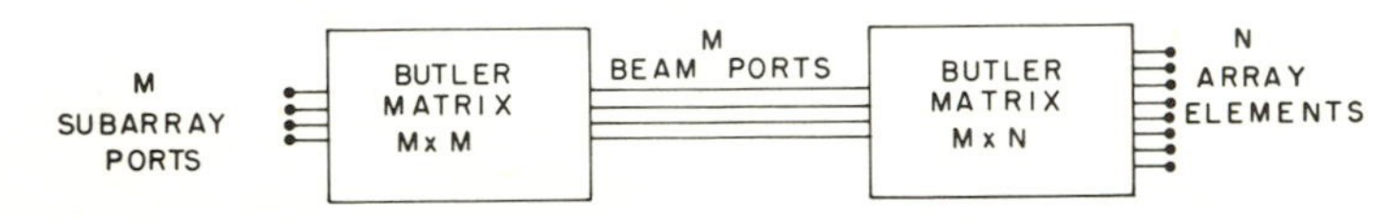

Fig. 5.9 *Butler matrices for overlapping subarray ports*

Thus the signal J_m excites all of the M beam ports of the matrix that feeds the array. The subarray pattern for this mth subarray:

$$g_m(u) = g_e(u) \sum I_{im} g_i(u) \tag{5.24}$$

$$= \frac{NJ_m g_e(u)}{\sqrt{MN}} \sum_{i=-(M-1)/2}^{(M-1)/2} e^{j2\pi(m/M)i} \left(\frac{\sin\left[(N\pi d_x/\lambda)(u-u_i)\right]}{N \sin\left[(\pi d_x/\lambda)(u-u_i)\right]} \right)$$

This expression is a sum of M orthogonal pencil beams arranged to fill the sector and, taken together, to form a flat-topped pattern for the mth subarray.

The aperture illumination corresponding to this subarray is obtained from eqns. 5.17 and 5.23, after performing the summation over the i beam ports.

$$i_n = \frac{1}{\sqrt{N}} \sum_{i=-(M-1)/2}^{(M-1)/2} I_{im}\, e^{-j2\pi(n/N)i}$$

$$= \frac{J_m}{\sqrt{NM}} \sum_{i=-(M-1)/2}^{(M-1)/2} \exp\left(j2\pi \left[\frac{m}{M} - \frac{n}{N} \right] i \right) \tag{5.25}$$

$$= N \frac{J_m}{\sqrt{NM}} \frac{\sin M\pi\left[(m/M)-(n/N)\right]}{M \sin \pi\left[(m/M)-(n/N)\right]}$$

This expression shows that each input I_m excites an i_n distribution that peaks at $n/N = m/M$, and so the amplitude distribution at the feed array is transferred directly to a proportionate distribution at the main aperture. Thus systems of this type can offer excellent control of near sidelobes, and can scan to the theoretical limit without grating lobes.

The scanning characteristics of this system can be inferred from eqn. 5.24, for if a set of progressively phased signals

$$\{\exp[-2\pi(i/M)m]\} = \{J_m\}$$

corresponding to the ith term of the orthogonal set were applied at the M input subarray terminals, the array elements will radiate the ith beam only since

$$g(u) = \sum_{m=-(M-1)/2}^{(M-1)/2} e^{-j2\pi(i/M)m} g_m(u) \tag{5.26}$$

$$= g_e(u) \frac{NM}{\sqrt{MN}} \frac{\sin[(N\pi d_x/\lambda)(u-u_i)]}{N \sin[(\pi d_x/\lambda)(u-u_i)]}$$

Thus the circuit radiates one of the orthogonal beams when the input progressive phase is an orthogonal set, and when the input signals do not exactly correspond to the orthogonal set, the circuit combines adjacent beams in an appropriate manner so that the resultant is the product of the array pattern times an element factor which has a nearly flat top.

One final conclusion can be drawn from the above equations. The circuit uses M-phase steered controls to scan across the set of M-orthogonal beams, and so its element use factor is unity. The important conclusion is that in order to access the furthest beam ($i = (M-1)/2$), the phase difference between adjacent input ports is

$$\pi\left(1 - \frac{1}{M}\right) \tag{5.27}$$

which, as previously recognised, is the phase control required to scan to one half a beamwidth of the theoretical scan limit. Failure to scan the array to this limit would necessarily result in an antenna system with element use factor greater than unity.

Similarly, if the circuit were replaced by one in which the Fourier transforms were performed by a lens or reflector, and if the elements of the feed array were spaced a distance d_x apart, then this array would necessarily scan to θ_{max}, such that

$$d_x \sin\theta_{max} = 0.5\lambda\left[1 - \frac{1}{M}\right] \tag{5.28}$$

This is precisely the condition for the feed-array maximum scan angle, and so if a scanning system is made up of a scanned feed array, spatial Fourier transformers, and a final aperture, then the minimum number of control elements as specified by the maximum scan angle and the size of the final aperture can only be achieved if the feed array is made to scan to its limit. Some examples of the implications of this condition are given in Sections 5.3 and 5.4.

5.3 Hybrid reflector systems for limited sector scanning

5.3.1 Introduction

The development of reflector-array hybrid systems is one of the most promising new directions in antenna technology. Reflectors have long been recognised as the least expensive antenna types for large apertures, and for this reason array reflector hybrid systems have attracted more attention than any other technique for application to high-gain limited sector scanning. Typical applications for such antennas include fire control systems, weapons locators, air traffic control radars and synchronous satellite communications. A major requirement for each of these applications is the need for gain in excess of 40 dBi coupled with rapid electronic scanning. The chief advantage of reflector or lens limited scan systems is that they make it possible to reduce the number of control elements to somewhere near the number of total beamwidths in the scan sector, and so to strive for an element use factor near unity. Unfortunately this goal is extremely difficult to obtain, for unlike the Butler matrix described in the second Section, reflector antennas are not ideal Fourier transformers. Attempts to form an off-axis collimated beam with a focusing reflector system results in a number of aberrations that are well known from classical optics. These include defocusing, coma and astigmatism, as well as higher-order aberrations (see Born and Wolf[5]).

As described in Section 5.2, there are two basic concepts employed in the design of limited scan reflector systems. The first is called transverse field matching, and consists of providing a conjugate match to the squinted beams of the reflector. Examples of this type of matching are the studies of Assely and Ricardi[1] and Winter.[47] This field matching takes place in a region where the field is fairly uniform and so the system involves no Fourier transform. The second technique involves the formation of overlapped subarrays on the main reflector and requires two Fourier transforms instead of the single one provided by the reflector. The second Fourier transformer is a constrained feed matrix, a lens or a second reflector. Design concepts for either type of system must fit the constraints imposed by the off-axis focal field distributions of the reflector geometry, and so this Section begins with a discussion of these field characteristics and the related geometrical considerations. Later in the Section the impact of these geometrical factors is evaluated for the several reflector-hybrid antenna techniques.

5.3.2 Scanning characteristics of an off-axis parabola

A plane wave incident upon a parabolic reflector is transformed into a converging field upon reflection. Except for diffraction effects a wave of normal incidence is perfectly focused, but when the incident angle departs from the normal there are a number of aberrations that distort the focal region fields. A detailed study of the focal fields of the receiving parabola is beyond the scope of this Chapter, and so this Section will highlight some of the properties of this converging field that most influence antenna design.

Fig. 5.10 shows a reflecting parabola and several rays for zero (solid) and an off-axis (dotted) angle incidence. The intersection of these edge rays does not define a point of best focus (except for the normal wave incidence), but it does indicate the direction of focal spot motion, and suggests that feed displacement presents an effective means of beam scanning. Lo,[20] Sandler,[31] Ruze[30] and more recently Imbriale *et al.*[17] and Rusch and Ludwig[28] have investigated the effect of feed translation, and collectively present substantial data on the amount of available scan for given feed displacements. Apart from these aberrations, lateral displacement of the feed by an amount 'd' results in scanning the reflector beam to an angle θ_B that is less than the feed offset angle ($\tan^{-1}(d/f)$) by a factor called the beam deviation factor (BDF)

$$\text{BDF} = \frac{\theta_B}{\tan^{-1}(d/f)} \tag{5.29}$$

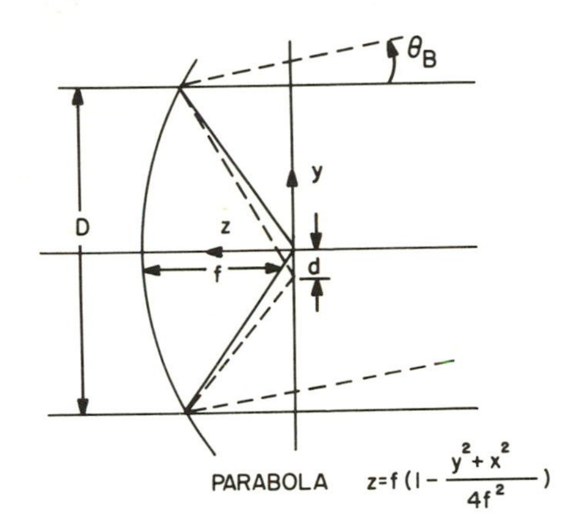

Fig. 5.10 *Off axis focusing of paraboloid*

This parameter is generally between 0.7 and 0.9, and increases with f/D. Lo[20] gives typical curves of BDF as a function of f/D. This factor can be used to estimate feed motion or size, for if the 3 dB beamwidth is given approximately by $\theta_3 \sim \lambda/D$, the required displacement is given in terms of the number of beamwidths scanned (θ_B/θ_3) as:

$$\frac{d}{\lambda} \doteqdot \frac{1}{\text{BDF}}\left(\frac{\theta_B}{\theta_3}\right)\left(\frac{f}{D}\right) \tag{5.30}$$

Increasing f/D thus requires a larger feed displacement, and hence a larger array or multiple feed network to scan the beam.

Ruze[30] shows that within the limits of geometrical optics, and when astigmatism is neglected, the feed locus for sharpest nulls is given by the equation:

$$z = -\frac{y^2}{2f} \tag{5.31}$$

in the coordinate system of Fig. 5.2. This equation defines the Petzval surface; a paraboloid of focal length one half that of the main reflecting surface. Ruze shows that the number of beamwidths that can be scanned (with a -10.5 dB coma lobe and 1 dB loss in gain) is given by:

$$\frac{\theta_B}{\theta_3} = 0.44 + 22\left(\frac{f}{D}\right)^2 \qquad (5.32)$$

Rusch and Ludwig[28] derive the surfaces of maximum gain using physical optics, and show that for an offset fed parabola and up to 10 beamwidths of scan, this surface is close to but not at the Petzval surface. In addition they prove that unless the f/D is very large or spillover is excessive a higher gain is achieved by pointing the displaced directional feed parallel to the axis of the reflector instead of redirecting it toward the vertex. Although useful in many multiple beam systems where sidelobe criteria are not severe, the literature indicates that it is generally not possible to scan more than four or five beamwidths by feed displacement without causing sidelobes on the order of −15 dB, and this severely restricts the application of feed displacement for low sidelobe scanning systems.

5.3.3 Transverse feeds for off-axis reflectors

Displaced feed systems cannot provide low sidelobes at wide scan angles because they cannot match the complex focal region fields of the scanned reflector, and it is for this reason that most limited scan systems use an array located less than the focal distance from the reflector to match the converging field. The use of such a transverse feed has been investigated by a number of authors, and has proven to be one of the more effective means of providing limited sector scanning of reflector antennas. In the early 1960s, the study of White and DeSize[46] showed that an array of feeds placed on a spherical surface concentric with the focal point of a parabolic reflector can be used to obtain limited beam scanning. The technique achieved about 10 beamwidths of scan in one plane.

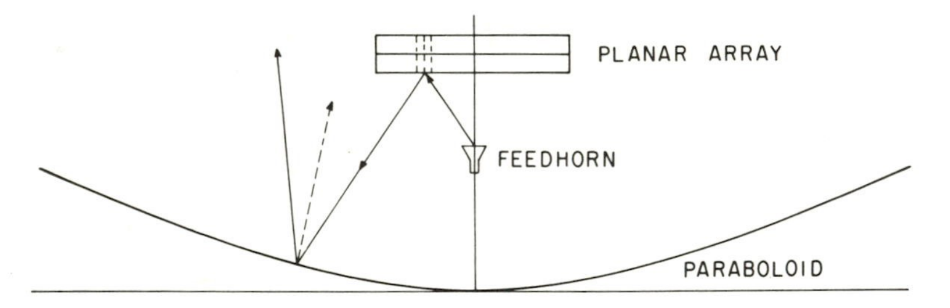

Fig. 5.11 *Planar array and paraboloid*

Winter[47] described a flat array feed for a parabola similar to that shown in Fig. 5.11 and achieved 7 beamwidths of scan with E-plane sidelobes approximately −15 dB relative to the main beam in the E-plane, and approximately −10 dB in the H-plane. The phase shifts required for achieving these scanned conditions were computed from geometrical optics. These were not progressive in either plane of scan and so row and column steering could not be used. The array consisted of 980 elements and so has an element use factor of approximately 5. The aperture efficiency is low because the illuminated spot is allowed to move about on the main reflector to achieve scanning. This study is of great importance because of its timeliness and because it showed the reflectors could be scanned by a phased array without the accompanying coma lobes observed for offset feeds.

A more recent study of a planar array scanning parabola is described by Tang[42] and Howell.[16] This structure, the AGILTRAC antenna, is a C-band array of 499 elements that scan an 18 ft parabola over a 7° half angle conical sector. The antenna has sidelobes of about − 18 dB at broadside, and maintains a monopulse null depth of 20 dB out to 5°. The gain is nearly constant at 41 dB out to the scan limit, and the element use factor is about 2.5. Losses throughout the system reduce the gain about 3.8 dB below the area gain for the chosen taper. System bandwidth is 9%.

Some general design principles for the development of such systems are given by Tang[42] and Howell.[16] Array size, location and approximate phasing are usually computed by geometrical optics. Tang gives an equation for the percentage of blockage as a function of reflector size, illuminated diameter scan angle and focal length, and points out that sidelobe ratios on the order of − 18 dB for a design of this type can be achieved with reasonable size reflectors, but if one insists on sidelobes lower than − 20 dB a much larger main reflector must be chosen.

The following analysis, based upon the geometric constraint given by Howell, results in an equation for array size equivalent to that derived by Tang. The analysis is then extended to evaluate the element use factor and to explore its relationship to the feed-array scan limits.

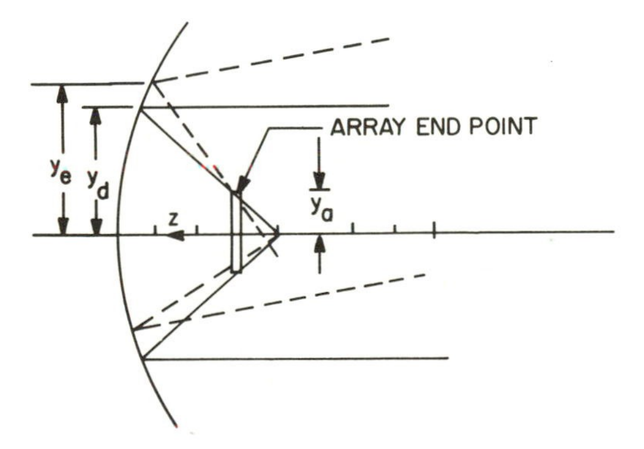

Fig. 5.12 *Array size determination for symmetric parabola*

Howell's geometric criterion is shown in Fig. 5.12. This Figure shows a symmetrical (parabolic) reflector with an array feed of size $2y_a$ and two rays that cross at the top of the array. The lower ray at an incident angle 0° hits the reflector at y_d, and reflects to pass through the focus. The second ray is incident at an angle θ, and after reflection crosses the first ray path at the array edge. Howell's condition is to choose the array size so that it intercepts all the reflected rays that come from the active region on the reflector for all up to the maximum scan angle. The Figure shows that the illuminated region must be allowed to move, for if it did not, then the array would not be fully utilised except at broadside. Choice of a larger y_e tends to make the array smaller. Howell presents no analysis and so the following results are included because they have significant implications for design. Snells law

$$\boldsymbol{a}_s = \boldsymbol{a}_{s_0} - 2(\boldsymbol{a}_\eta \cdot \boldsymbol{a}_{s_0})\boldsymbol{a}_\eta \tag{5.33}$$

gives the unit vector of the reflected ray for any incident unit vector $\boldsymbol{a}_s$ intercepting the reflector with surface normal $\hat{\eta}$. For the parabola of Fig. 5.12 along the centre line ($x = 0$)

$$z = f - \frac{y^2}{4f}$$

the resultant reflected ray unit vector for any ray is:

$$\begin{aligned}\boldsymbol{a}_s &= \frac{1}{(y/2f)^2+1}\{\boldsymbol{a}_y\{[(y/2f)^2-1]\sin\theta - y\cos\theta\} \\ &\quad + \boldsymbol{a}_z\{[(y/2f)^2-1]\cos\theta + y\sin\theta\}\} \\ &= \frac{1}{(y/2f)^2+1}\{\boldsymbol{a}_y g_1(y,\theta) + \boldsymbol{a}_z g_2(y,\theta)\}\end{aligned} \tag{5.34}$$

After some manipulations one can solve for the point where the rays cross and so obtain the array size $2y_a$ in terms of the other reflector parameters and the given scan angles:

$$\frac{y_a}{y_d} = 1 + \kappa_1 \tag{5.35}$$

where

$$\kappa_1 = \left\{\frac{g_2(y_e,\theta)[(y_e - y_d)/f] + g_1(y_e,\theta)[(y_e/2f)^2 - (y_d/2f)^2]}{g_2(y_e,\theta)y_d + g_1(y_e,\theta)[(y_d/2f)^2 - 1]}\right\}$$

Fig. 5.13 shows the resulting normalised array size as a function of maximum scan angle θ for several effective focal length ratios $f/2y_d = 0.5$ and 1.0 and several values of the allowable spot motion ratio $R = y_e/y_d$.

This result, equivalent to an expression derived by Tang,[42] can be used to estimate reflector and array size and location for a given coverage sector and hence to evaluate gain reduction and sidelobes due to blockage. In addition, it leads directly to an estimate of the element use factor. A reflector scanning a rectangular angular sector $\theta_1 \times \theta_2$ radians, using effective aperture sizes y_{d_1} and y_{d_2} and array element spacings a_1 and a_2 has an element use factor (eqn. 5.14) of:

$$\frac{N}{N_{min}} = \frac{0.25}{\theta_1\theta_2}\left(\frac{y_{a_1}}{y_{d_1}}\right)\left(\frac{y_{a_2}}{y_{d_2}}\right)\left(\frac{\lambda^2}{a_1 a_2}\right) \tag{5.36}$$

For example, a square 10° scan sector and a square section of a parabola with $f/2y_d = 1$ is seen from Fig. 5.13 to have $y_a/y_d = 0.44$ (for $R = 1.6$), and with

0.7λ array element spacing in both planes the resulting element use factor is 3.28. Clearly this design would have been improved by increasing the value of R to obtain a smaller array. This cursary analysis is only a rough guide to determining array size and position and a far more detailed analysis is required for design. However, in addition to its utility as an engineering guide, it demonstrates that single reflector and array hybrid systems require a relatively large array and an oversize reflector, and hence suffer blockage that ultimately limits their sidelobe ratio.

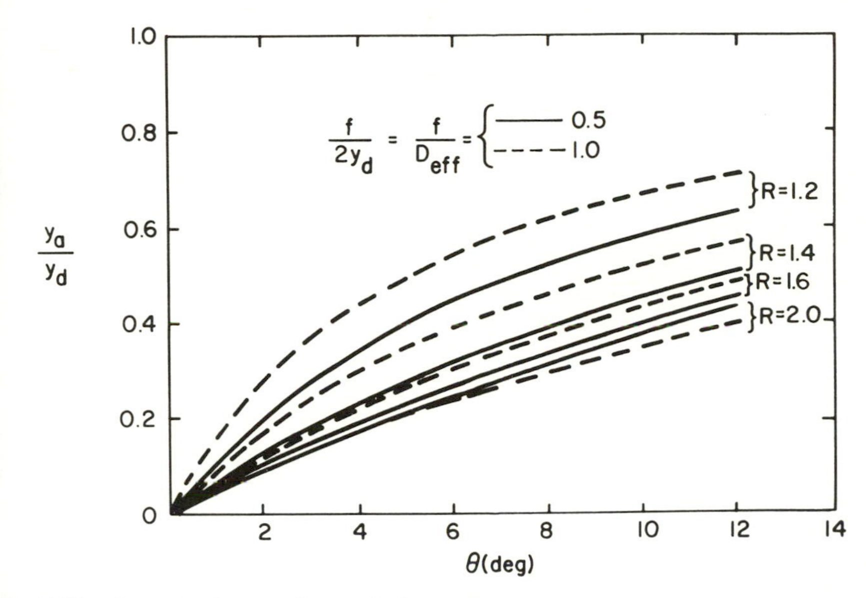

Fig. 5.13 *Array size for scanning parabola*

The need for improved pattern performance has led to the development of offset array structures with low aperture blockage and low sidelobes. Notable among these is the PAR antenna of the TPN-19 GCA system, which was developed for the US Air Force by Raytheon Corporation. This antenna, shown in Fig. 5.14 uses a hyperbolic main reflector and an offset phased array feed with a modified row and column steering algorithm for off axis scanning. The antenna has 1.4° azimuthal beamwidth and 0.75° elevation beamwidth, and scans ± 9.5 beamwidths in elevation and ± 6.65 beamwidths in the azimuthal plane. The array used 824 phase shifters, and so has an element use factor of approximately 3.25. The first sidelobes are at the − 20 dB level throughout most of the scan sector, although rising to − 18 dB at the limits. The system operates over a 2% bandwidth at X-band, and has a realised gain of approximately 39 dB at the scan limits. This realised gain is about 3.5 dB below the directive gain on boresight due to horn feed spill-over, array face matching, and phase-shifter losses. The main reflector size is about 3.0 by 3.8 m corresponding to an aperture efficiency of approximately 30%, for the illuminated spot moves across the oversize reflector as a function of scan.

In addition to the above, there is one more conclusion to be drawn about the operation of such limited scan reflectors. Eqn. 5.15 gives the element use factor for a generalised aperture in terms of the interelement distance, scan angles and beamwidths. That expression reveals that if one uses an objective lens or reflector for magnification, then the element factor can still be unity if the array elements scan to their limit $d \sin\theta = 0.5$ in both planes. Applied to the reflector-array geometry of Fig. 5.13, this says that for an array with 0.7λ spacing, the element use factor could be unity if the array scanned to $\pm 45°$. However, the maximum scan angle of the array in the configuration of Fig. 5.13 is approximately:

$$\theta_{max} \doteqdot \tan^{-1}\left[\left(\frac{y_d}{f}\right)\frac{[(y_e/y_d)-1]}{[1-(y_a/y_d)]}\right] \qquad (5.37)$$

Fig. 5.14 *Precision approach radar antenna of AN/TPN-19*

which is 24.5° for the case considered. Assuming a rectangular scan sector and rectangular reflector, the element use factor is computed by eqn. 5.15 and eqn 5.37 to be 2.98, which is relatively close to the exact number evaluated from eqn. 5.36. This result is obtained without considering any details of the reflector surface, and is based only upon the array scan limits, but it illustrates how dominant a part is played by these purely geometric constraints.

Clearly the reflector-array hybrid system needs a relatively large number of phase controls because one cannot scan over the extremely wide sector (± 45° for 0.7λ array spacing) to obtain a low element use factor. Attempts to scan that wide a range would necessarily mean inefficient use of the main objective.

Further advances with reflectors of this type could result from the use of spatial limited scan array feed techniques designed to have an element use factor nearer unity for the narrow scan ranges required to steer a reflector. This could include the use of constrained feed overlapped subarray techniques or the angular filter techniques described in Section 5.5.

5.3.4 Fourier transform feed for off-axis parabola

Recent work by Rudge and Withers[27] addresses the focal spot distortion for a parabola with an off-axis incident wave, and demonstrates the Fourier transform relationship between the aperture fields and focal plane fields. Specifically, Rudge gives the focal plane field for the linearly polarized rectangular parabolic section of Fig. 5.15 as:

$$E(x, y) = jk \int_{-\infty}^{\infty} \int_{-\infty}^{\infty} G(p, q) \exp\{jk(xp + yq)\} \tag{5.38}$$

where

$$\begin{aligned} p &= u \cos\phi \qquad & x &= t \cos\phi' \\ q &= u \sin\phi \qquad & y &= t \sin\phi' \end{aligned} \tag{5.39}$$

with

$$G = \begin{cases} (1 - p^2 - q^2)^{-1/2} F(p, q) & |p| \leqslant p^*, \quad |q| \leqslant q^* \\ 0 & |p| > p^*, \quad |q| > q^* \end{cases}$$

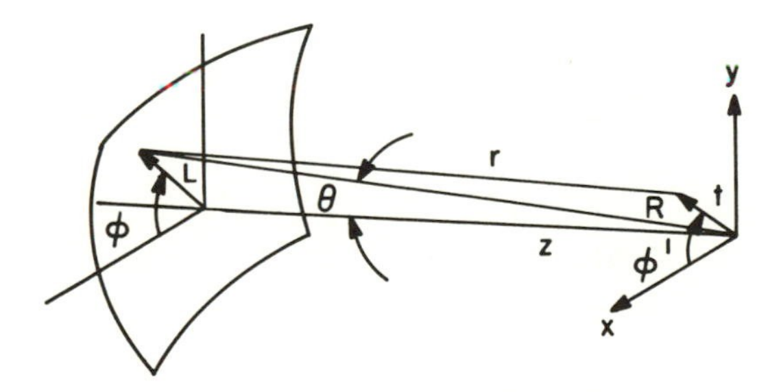

Fig. 5.15 *Parabolic section and coordinates*

and $p^* = u^* \cos\phi$ and $q^* = u^* \sin\phi$. Here $u = \sin\theta$ and $u^* = \sin\theta^*$, for θ^* the maximum half angle subtended by the reflector from the geometric focus. For a normally incident linearly polarised plane wave the focal plane field has the variation (for $y = 0$)

$$E(x) \propto \begin{cases} \dfrac{2p^* \sin kxp^*}{kxp^*} & p^* < 0.5 \\ \pi J_0(kx) & p^* = 1.0 \end{cases}$$

These functions as shown in Fig. 5.16 illustrate the relationship between the uniform aperture field and its Fourier transform, and Rudge shows that for an off-axis wave the same condition is maintained providing that the focal point is moved off axis so that the total angle subtended by the reflector as measured at the focal point is a constant equal to $2\theta^*$. With this choice of a new 'focal point' the new focal plane is defined to be normal to the centre line of the focal cone of width $2\theta^*$. The locus of feed motion follows the circle passing through the true focus and has diameter

$$\frac{d}{\sin 2\theta^*} = f\left[1 - \left(\frac{d}{4f}\right)^2\right]\sec^2\theta^* \tag{5.41}$$

as shown in Fig. 5.17, for an aperture of length 'd' in the plane of scan.

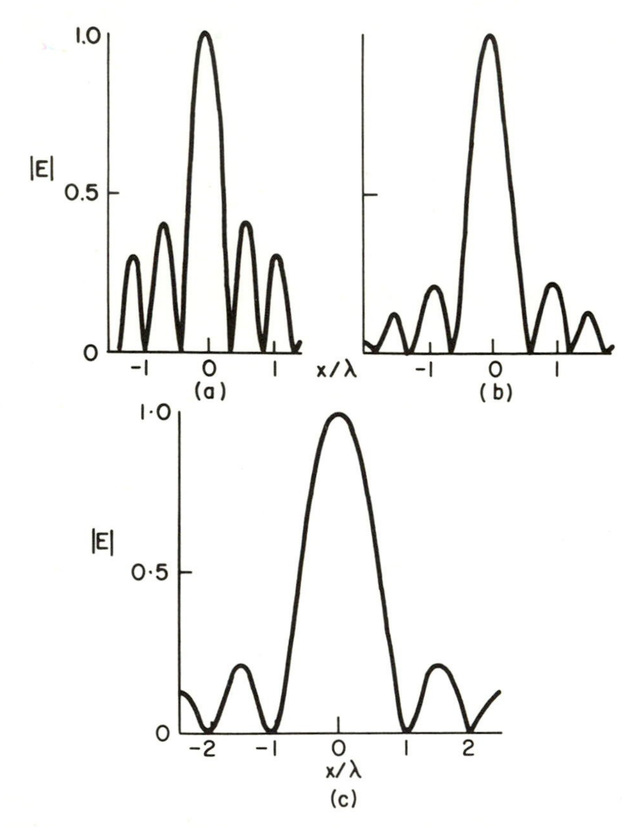

Fig. 5.16 *Focal plane electric field distributions*: (*a*) $f/D = 0.25$, (*b*) $f/D = 0.5$, (*c*) $f/D = 1.0$ (courtesy of A. Rudge)

Having thus found a plane in which the focal field variation with angle is always the Fourier transform of the aperture plane, it is clear that a second Fourier transformer placed at the new focal plane will have a uniform amplitude output and a phase variation which is the complex conjugate of the aperture plane field. Stated another way, the parabola acts as an ideal Fourier transformer of the off-axis incident wave for a feed at the new focal plane. Since the new focal plane field is the Fourier transform of a tilted wavefront, the second transform output duplicates the aperture field and phase shifters alone are required to refocus the aperture. In an experiment to test the validity of the concept for a single plane of scan this Fourier transformer was implemented as shown in Fig. 5.18 by an 8-element Butler matrix with element spacing $\lambda/2 \sin\theta^*$. By moving this device along the locus of constant $2\theta^*$, Rudge has demonstrated excellent scanning characteristics out to $\pm$ 15 beamwidths. Extension to dual plane scanning will probably require a quasi-optical transformer in the form of a second reflector or lens.

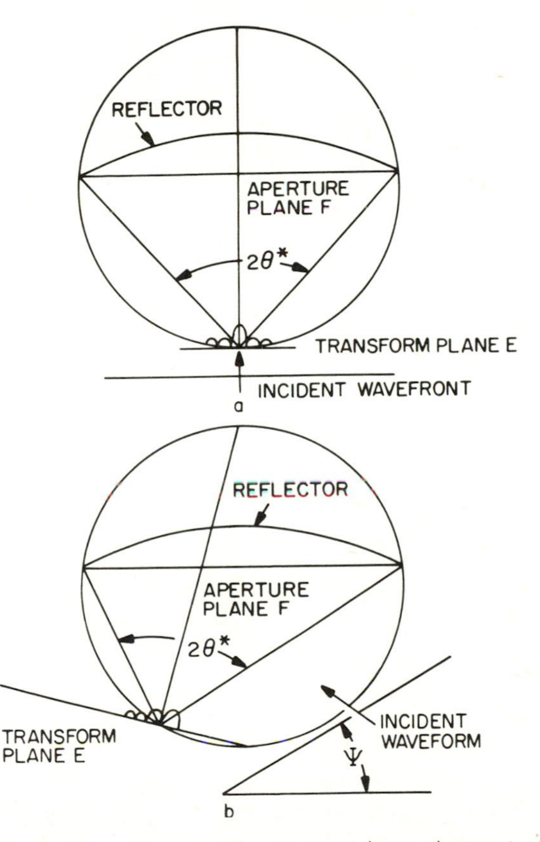

Fig. 5.17 *Locus of transform planes with associated (typical) electric field distributions* (courtesy of A. Rudge)

Although basically an embodiment of a two-transform system, this technique is conceptually new because it utilises the moving Fourier transform feed and an optimised feed locus. Not a true hybrid system in the strictest sense, its description is included here because of its influence on the dual transform design to be described in Section 5.3.6 and because it explains the fundamental physics of other dual transform systems. The antenna system of Section 5.3.6 substitutes a second focusing surface to provide the required feed motion but its operation is basically very similar to the above.

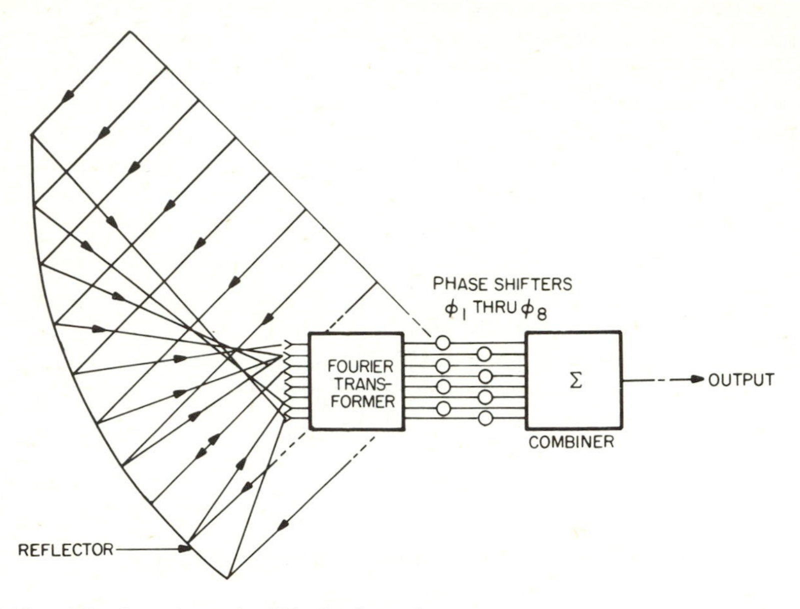

Fig. 5.18 *Adaptive-primary-feed block schematic*

5.3.5 Off-axis characteristics of paraboloid sections

A different perspective is gained by considering the focal properties of segments of the reflector in order to determine their positions of best focus. This point of view was investigated by Sletten *et al.*[39] for the purpose of developing fixed-beam corrective line sources for paraboloids, but the work is suggestive of the subarraying concepts of Section 5.2. That aspect is of direct concern to the development of dual transform systems with a parabolic objective. Particularly important for the purposes of this text is that Sletten's results lead directly to control of overlapped subarrays in the reflector aperture and to deriving means for exciting these sub arrays with a second Fourier transformer.

Sletten shows that for a strip across the reflector at height y_m there is a line of best azimuth focus, and that within the limits of geometrical optics this line is given by

$$x = 0; \qquad \frac{z_0}{y_0} = -\frac{y_m}{2f} \tag{5.42}$$

for the parabola

$$z = f\left(1 - \frac{y^2 + x^2}{4f^2}\right)$$

This is a straight line parallel to the slope of the reflector at $y = y_m$.

Fig. 5.19 shows one of these lines for the slice $y_m = 0.5$ (normalised to $f = 1$). A feed located at the position $(0, y_0, z_0)$ with therefore give a nearly perfect azimuth focus for the section at height y_m (for all elevation angles θ).

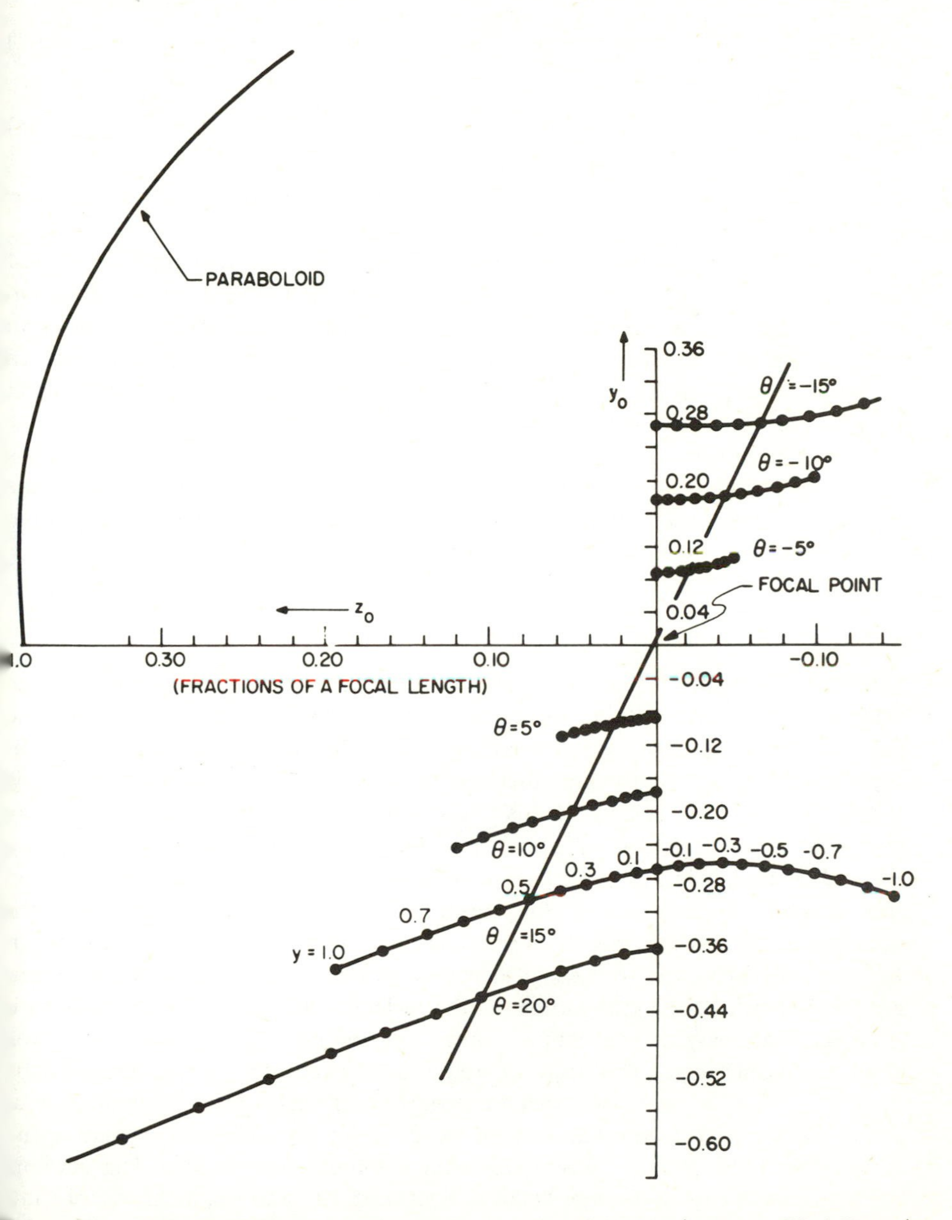

Fig. 5.19 *Ridge lines for rays incident at different angles of elevation* (courtesy of C. J. Sletten)

The second set of lines shown in Fig. 5.19 connecting the dots are called ridge lines. They are lines of best azimuth focus for a given elevation angle θ, and they are plotted parametrically for various strip heights 'y' across the reflector. Each line corresponds to a given θ, and each point on the line a different section (y) of the reflector. These lines are given by the parametric equations Sletten *et al.*:[39]

$$y_0 = -\frac{\sin\theta(x^2 + y^2 + 4f^2)}{4f\cos\theta - 2y\sin\theta} \tag{5.43}$$

$$z_0 = \frac{\cos\theta(x^2 + y^2 + 4f^2)}{4f\cos\theta - 2y\sin\theta} - 2f + z \tag{5.44}$$

A distributed feed located along a ridge line and phase corrected can thus produce a collimated beam at the angle θ.

Fig. 5.20 shows a third set of lines (and in addition shows the lines of best azimuth focus). This third set are circles that represent the locus of best elevation focus for the vertical strip in the centre of the reflector ($x = 0$). Sletten gives an equation for these surfaces, but the Figure shows that for any chosen point on the reflector y_m, the circle of best elevation focus is tangent to the reflector at y_m and passes through the true focus (0, 0, 0).

Each point on the reflector y_m ($x = 0$) has an azimuth focal locus given by a straight line and an elevation focal locus given by a circle. These intersect at the true focus (0, 0, 0) and at one other point, depending upon the value y_m (and hence the angle θ). This point of secondary focus is given by the angle

$$\tan\theta = \frac{y_m}{z} \tag{5.45}$$

with z given by the equation of the parabola. Sletten used these equations to describe two devices: a Ridge line corrector which is a line source properly phased and placed along a ridge line to coherently absorb the power from a given elevation angle θ, and a second corrector that is an array or group of sources placed along a given line of best azimuth focus. This second corrector directs all the array energy at a given strip $y = f/2$ and is called a mid-point corrector. It forms a focused beam in azimuth and a fanned or shaped beam in elevation. Sletten shows that by using an array placed along the line of best azimuth focus and near the intersection of azimuth and elevation focusing lines it is possible to have beams that are relatively well focused in elevation as well as in azimuth. These beams are focused using the same section of the reflector, and since the illumination is relatively stationary the technique tends to minimise reflector size for a given elevation beam shape. This type of corrector has recently been investigated by Tang *et al.*,[45] who have shown that the spectrum of beams formed by an array of sources along a corrector line can be sampled by a multiple beam matrix and scanning array to form an efficient technique for limited section scan. This development will be described in more detail in Section 5.3.6, and has its historical roots in the technology of overlapped subarrays as described in Section 5.2.5, in this work of Sletten, and in the Fourier transform feed system of Rudge,[27] but the analysis and design of the dual transform system of Tang *et al.* is carried out following the methods and practices outlined above.

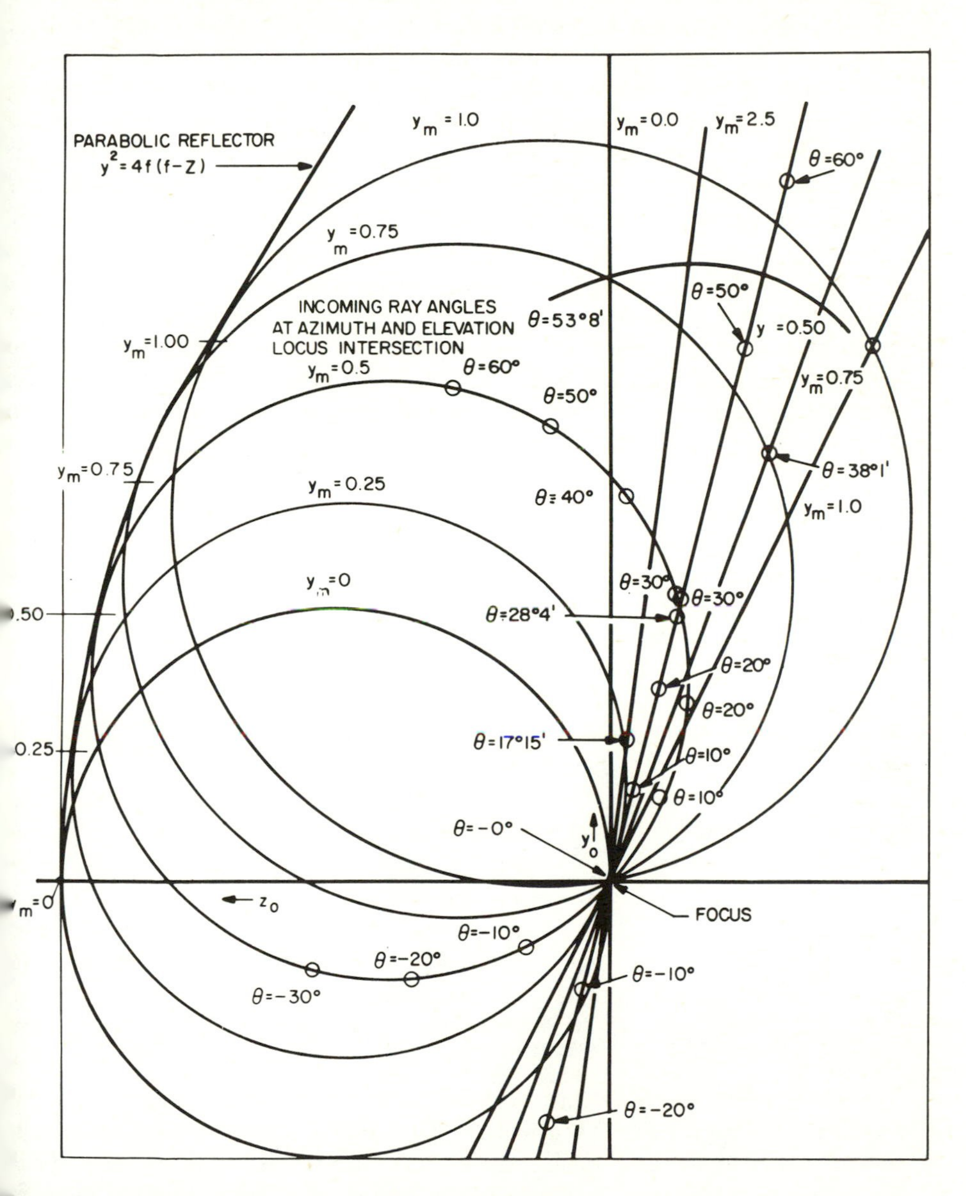

Fig. 5.20 *Azimuth and elevation focusing loci for various correction points* (courtesy of C. J. Sletten)

Key: Elevation focus is denoted by circles and azimuth focus by straight lines; ports of correction are indicated by y_m

5.3.6 Dual reflector and reflector-lens limited scan systems

Single reflector or lens structures with a phased array feed are simple but require a relatively large number of phase controls (element use factors of 2.5–3.25). All of these antennas require oversize main apertures because the illumination moves with scan (typical aperture efficiencies are 20–25%). Recent technological efforts have been concerned with developing systems for scanning over wider angles with pencil beams, and for providing these scanning capabilities with relatively small high-efficiency primary apertures. Two recent studies have dealt with the quest for larger scan multiples. One of these concerns a near field Cassegrain antenna shown in Fig. 5.21. This geometry is compact and has a length diameter ratio L/D of approximately 0.4, but the array diameter 'd' required was relatively large, with typical d/D ratios considered between 0.25 for a 400λ main reflector, and 0.35 for a 250λ main reflector. Computations show up to 17 beamwidths scan with element use factors between 6 and 7. These relatively large element use factors arise because the array is only required to scan over a narrow range of angles, and so itself is a limited scan array. For this reason, element spacings greater than 0.9λ can be used to maintain element efficiency (Fitzgerald[11]).

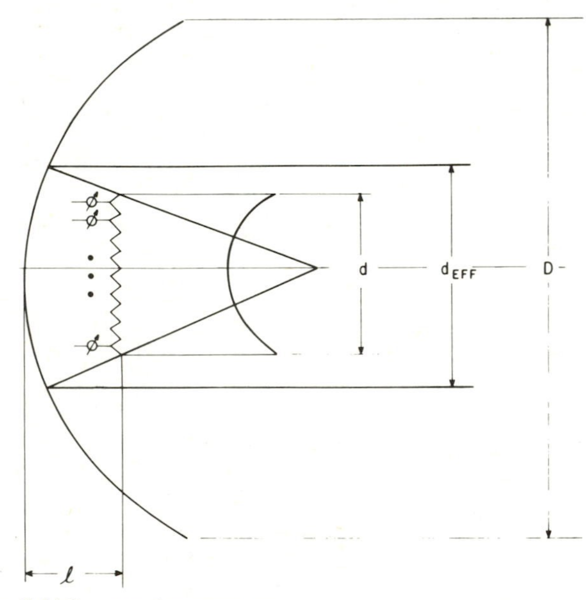

Fig. 5.21 *Near field Cassegranian geometry*

The use of 1.4λ oversized elements would reduce the element use factors for thi approach to somewhere between 3 and 4. This array scheme also has the advantag of using linear phase control and since its L/D ratio is small, can be mounted an moved conveniently. It has relatively high sidelobes, of the order of – 13 dB due t the inherent aperture blockage of the system. Another significant study concern the offset-feed Gregorian geometry of confocal coaxial parabolic sections show in Fig. 5.22. This geometry was first proposed by Dudkovsky[10] and proposed an investigated as a scanned geometry by Skahil *et al.*[37] Fitzgerald[12] describes a system with -15 to -17 dB sidelobes, $\frac{1}{2}°$ beamwidth which scans 14 beamwidths and, with an array 45 on a side, has an element use factor of about 2.5. The array itself scans only to about $\pm 20°$. The L/D ratio is approximately 1.5, and so the structure is awkward for pedestal mounting. More recent results by Miller *et al.*[23] have shown that the scanning properties of this system can be improved (and the element use factor reduced to about 2) by optimising the main and subreflector contours. The reports describing these two studies show substantial filling of the monopulse null from its broadside levels of -25 to -30 dB to values of about -10 dB for the plane of the offset reflector for the Gregorian structure, and about the same for the Cassegranian antenna.

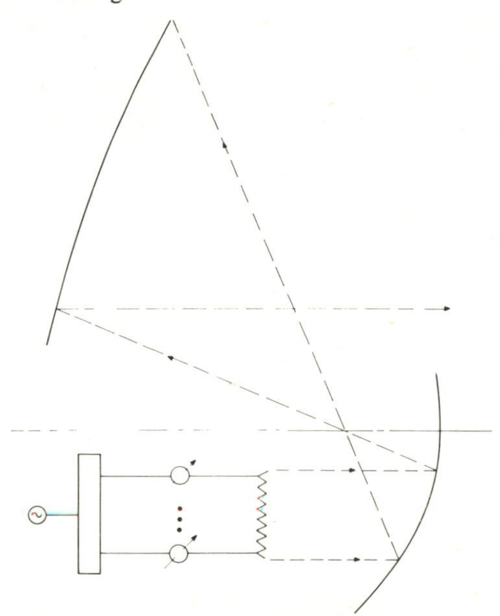

Fig. 5.22 *Offset-feed Gregorian geometry*

More recent studies have sought to increase the aperture efficiency by allowing movement of an illuminated zone on a sub-aperture and to decrease element use factors by maximising the phased array scan. Tang *et al.*[45] have performed a theoretical and experimental study of a limited scan system consisting of a main reflector, a multiple beam forming system and a phased array. Shown schematically in Fig. 5.23, this concept has its earliest roots in the subarraying systems described in Section 5.2.5, but the geometry is very similar to the Gregorian sub-reflector antenna of Fitzgerald. The element use factor is much smaller than that of Fitzgerald, indicating more efficient use of the array which is made to scan over a wider angle in accordance with the principle of eqn. 5.28. The additional scan is possible because the feed lens is made very large compared to the sub-reflector of Fitzgerald. The system is also suggestive of the technique studied by Rudge[27] and described in Section 5.3.4, but again the oversize lens is substituted for the moving feed.

The design procedure followed by Tang *et al.*[45] is based in part upon the studies of Sletten as described in Section 5.3.5. Considering a horizontal cut through the mid-point of a receiving reflector, as the incident angle θ is varied there is one line of best azimuth focus. Other sections of the reflector have a different line of best focus because of the spread of the ridge lines. If the ridge lines could be minimised the scanning could be accomplished by switching in a horn at the locus of mid-point focus, as originally recognised by Sletten. Like Sletten, Tang *et al.*[45] suggest that this be done by a focusing lens or reflector, but their design proceeds from that point to introduce the concept of designing the subarrays. If a perfect transforming matrix is used to switch between each of the feed points on the locus of mid-point focus, and if the feed points are chosen to correspond to separate (ideally orthogonal) beams formed by the reflector aperture, then each received beam will correspond to an orthogonal progressive set of input phases (eqn. 5.9) at the matrix input. The reflector and matrix is thus approximately equivalent to the ideal dual Fourier transform scanning system described in Section 5.2.

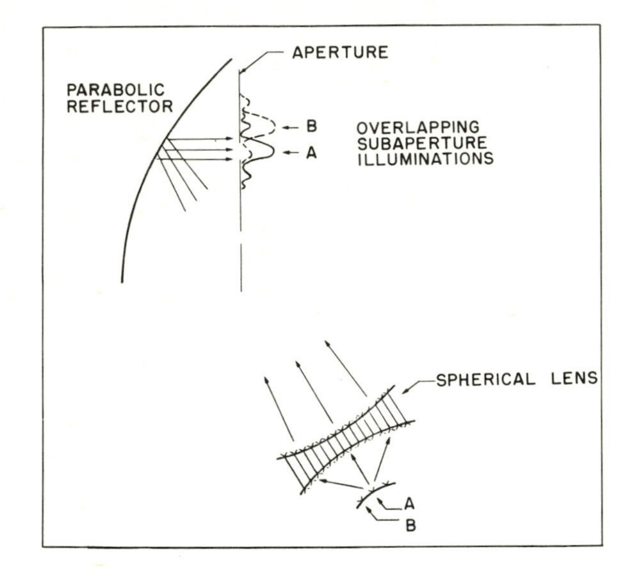

Fig. 5.23 *Reflector-lens limited scan concept*

A consequence of this relationship is that each input terminal of the matrix can thus be associated with a flat-topped subarray across the aperture. The antenna design is simplified because probing the main aperture can assure that the subarrays are equally spaced with low sidelobes as they would be in the ideal transformer system. Fig. 5.24 shows a typical subarray pattern for this system. Results obtained with a spherical lens feed system indicate that a 1° beam can be scanned over a ± 10° sector with sidelobes at − 20 dB and with an element use factor of approximately 1.4. Figs. 5.25 and 5.26 show the resulting radiation pattern at 0° and − 9° for the experimental model.

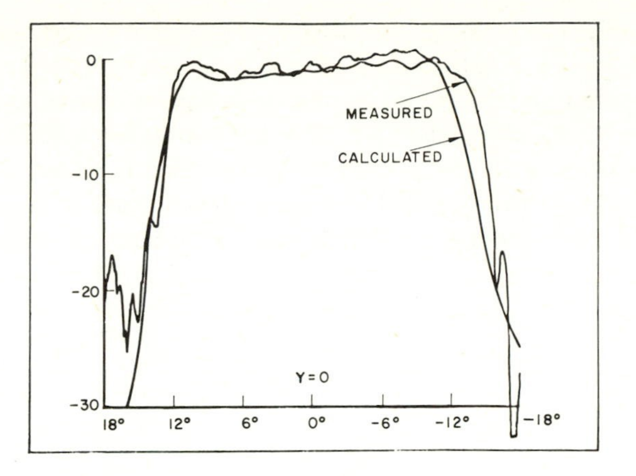

Fig. 5.24 *Subarray pattern of reflector-lens hybrid scanner* (Courtesy of R. Tang)

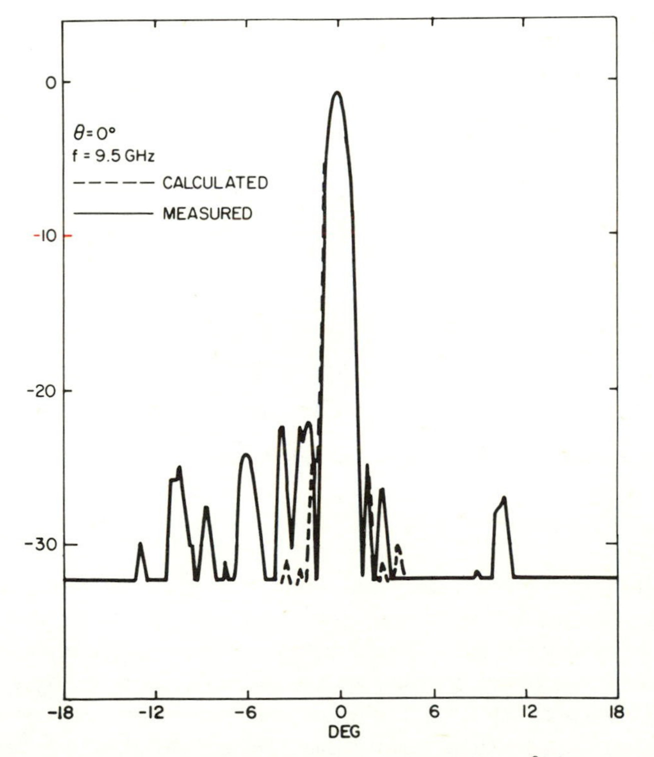

Fig. 5.25 *Radiation patterns of reflector-lens hybrid scanner* ($\theta = 0°$) (Courtesy of R. Tang)

The low element use factor of this system results from the wide angle of scan required of the array (as indicated in Section 5.2). Unfortunately this scanning forms a moving spot and results in the use of a very large feed (approximately 0.65 the size of the main reflector), which in turn contributes to the bulk of the structure. Comparison with the studies of Fitzgerald[11,12] reveals that the main distinction between the two approaches is the relatively small ratio of subreflector to main reflector size (about 0.25 to 0.35) and the small scan angle required of Fitzgerald's array.

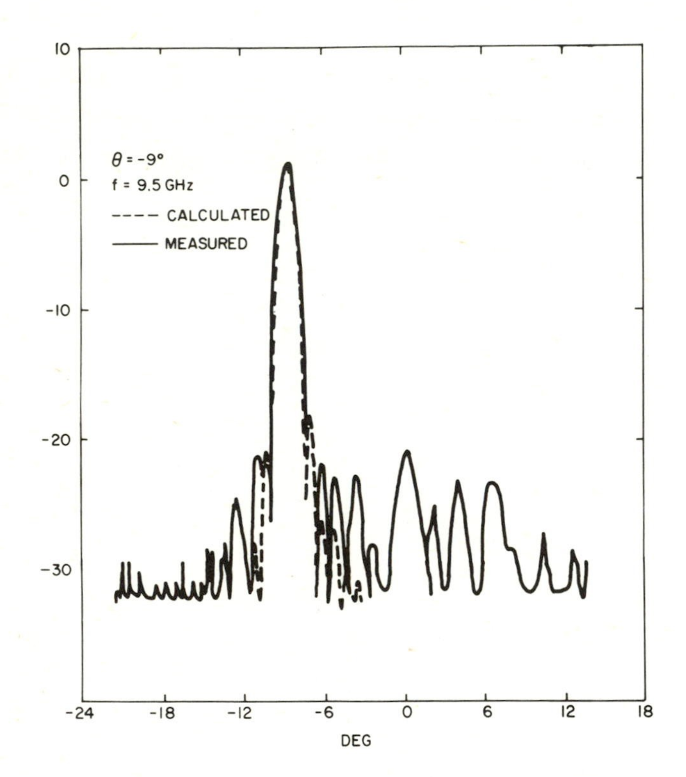

Fig. 5.26 *Radiation patterns of reflector-lens hybrid scanner* ($\theta = -9°$) (Courtesy of R. Tang)

Although not strictly an exercise in limited scan, the work of Chen and Tsandoulas[7] has demonstrated the effectiveness of the subarraying concept in dual reflector systems. These authors have shown that a small array with time delay steering can be used in combination with a dual reflector geometry as a feed for a large array with phase shifters. The space feed system creates time-delayed subarrays that provide wide-angle performance over bandwidth of 10–20% with main array apertures of up to 100λ on a side.

5.4 Hybrid lens/array scanning systems

5.4.1 Introduction

Microwave lenses can provide much wider scan multiples than reflectors because they inherently offer more degrees of freedom in design. Given the feed point location and illumination, reflector pattern control is limited to choosing the reflector contour, but even lenses with constant dielectric parameters have the front and back lens surfaces as variables. More complex lenses have additional degrees of freedom, and these are used to optimise scanning performance.

Until recently, the term 'electronic scanning' of a lens was generally interpreted to mean switching to an off-axis beam. The lens was used to focus the incident plane wave to a specific feed corresponding to a given scan angle, and so functioned as a multiple beam matrix. The desired off-axis beam was then selected by means of a switching tree. In fact, any such multiple beam system can be converted to a scanned beam system through the use of an array and a multiple beam matrix or a second lens. This combination produces a two-transform scanner that is a practical realisation of the idealised circuit described in Fig. 5.7.

A number of different types of lenses could be used for this application. These include various Luneberg configurations and R–2R lenses as described in the texts by Jasik,[18] Hansen (Ref. 15, Vol. I, chap. 3) and Collin and Zucker,[8] but the discussion in this Section will be restricted to the use of microwave constrained lenses in combination with arrays for the purpose of providing a scanned beam, and will include several transverse feed scanning systems (no transform) as well as a two transform scanner.

In order of increasing complexity and improved performance, the most commonly used constrained microwave scanning lenses are those with constant index of refraction but satisfying the Abbé sine condition; next those with variable index of refraction and two points of perfect focus; and finally, the so called 'bootlace' lenses which can have an added point of perfect focus. Two-dimensional constrained lenses with a cylindrical back face and a single focal point are free of coma and satisfy the Abbé sine condition (Brown[6]). Two-dimensional lenses with three parameters are derived following the work of Ruze[29] and make use of the three available degrees of freedom, the front and back lens face shapes and the variable index of refraction provided by waveguide or metal-plate cross-section. These three degrees of freedom are used to maximise the scan angle, to add an additional third point of perfect focus or to make one of several other possible parametric choices which are summarised in the original work and in the text by Brown.[6] Ruze's study dealt with a two-dimensional lens, but this work has been generalised to three-dimensional waveguide lens structures as well. Dion[9] described a waveguide lens with a multiplicity of beams filling a restricted spatial sector. The inner surface of the lens was nearly spherical and the outer surface approximately a spheroid, but stepped to reduce weight and increase bandwidth.

The third major category, called the 'bootlace' lens, was developed by Gent[14] and later the specific case of a straight front face was extensively investigated by

Rotman and Turner.[25] This lens type differs from that of Ruze in that the corresponding points on the two lens contour are connected by flexible cables and so are not necessarily equidistant from the lens axis; in addition, the variable refractive index of the Ruze design is replaced by the ability to adjust compensating cable lengths because of the flexibility provided by the 'bootlace' geometry. This added degree of freedom allows formation of a third point of perfect focus while retaining the flat front face for coma minimisation as in the Ruze lens.

A recent 'bootlace' design (Scott *et al.*[34]) uses a spherical inner surface and planar outer surface. It has a single focus at the lens axis. The lens elements are joined by TEM meander delay lines to provide the proper delay for focusing and scanning. The antenna is circularly polarised and operates over the frequency range from 3.7 to 6.425 GHz. The lens diameter is 23 wavelengths at mid-band, and feed horns are available to access 61 constituent beams. The beams are combined in clusters and each cluster has sidelobes on the order of 31 dB. Each of the lens types forms a multiplicity of focused beams for use in a scanning system or shaped coverage antenna. Rapid electronic scanning is available through the use of a transverse feed or another lens or Butler matrix to produce a two-transform scanner. Several of these combinations are discussed in the next Sections.

5.4.2 Scanning lens/array combinations

5.4.2.1 Limited sector scanning: A new concept in limited sector scanned antennas has been proposed by Schell.[32] The technique has application to reflector or lens antennas, but the lens version is described here. The concept uses an array disposed around a cylinder to scan a reflector or lens surface that is contoured according to an optimum scan condition, rather than a focusing condition. The reflector is then stepped or the lens phase corrected to achieve focusing. Preliminary design results (McGahan[22]) show that in one plane of scan the technique achieves an element use factor of about unity, while using an oversize final aperture to again allow motion of the illuminated spot.

Fig. 5.27 shows a schematic view of the array-lens combination and Fig. 5.28 demonstrates its scanning properties. The array element currents are equal in amplitude and have a progressive phase given by $\beta n \Delta\theta$. The reflector surface (or lens back face) is chosen to transform this phase variation into a linear wavefront normal to the beam direction. The condition for determining the curvature of the reflector (or lens) is that a constant incremental phase change in θ along the circular arc tangent to the centre of the back face of the lens produces a constant incremental phase change in the 'y' coordinate along the aperture.

Thus

$$\frac{dy}{d\theta} = \text{constant} = R \tag{5.46}$$

and, since $y = \rho \sin \theta$

$$\rho = \frac{R\theta}{\sin\theta} \tag{5.47}$$

This curvature satisfies the scan requirement, but does not guarantee that the wave will focus. Focusing is achieved for the reflector through the use of confocal parabolic sections stepped so that their centres lie along the scan surface, and for the lens by adjusting the path lengths so that they are equal at some angle. The array represented by the data of Fig. 5.28 consists of 25 elements. The half angle θ_0 subtended by the array is 45°, $k_0R = 197$, $k_0a = 48$ and k_0 times the final aperture width is approximately 526 ($D/\lambda = 83.7$).

The far field beam angle is given by

$$\sin\theta_B = \beta/kR \tag{5.48}$$

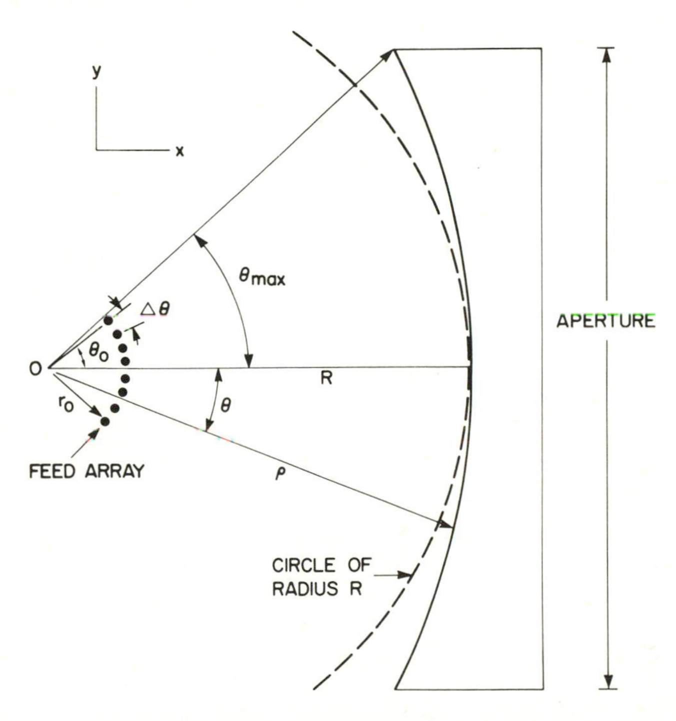

Fig. 5.27 *Scan corrected lens antenna*

and for the case shown in the Figure the maximum θ_B is about 11.5°. At this scan angle the gain is reduced about 2.5 dB with respect to broadside and a far-sidelobe has risen to the − 20 dB level. The aperture illumination is nearly uniform, and the − 13 dB near sidelobe ratio is maintained throughout the scan sector.

These computations have been confirmed experimentally by McGahan for the lens geometry scanning in one plane. Since the amplitude distribution on the array is transferred very simple onto the inner lens surface, it is possible to produce very low sidelobe patterns with this geometry. Preliminary theoretical data indicate that with perfect phase and amplitude control this structure can have sidelobes below −40 dB.

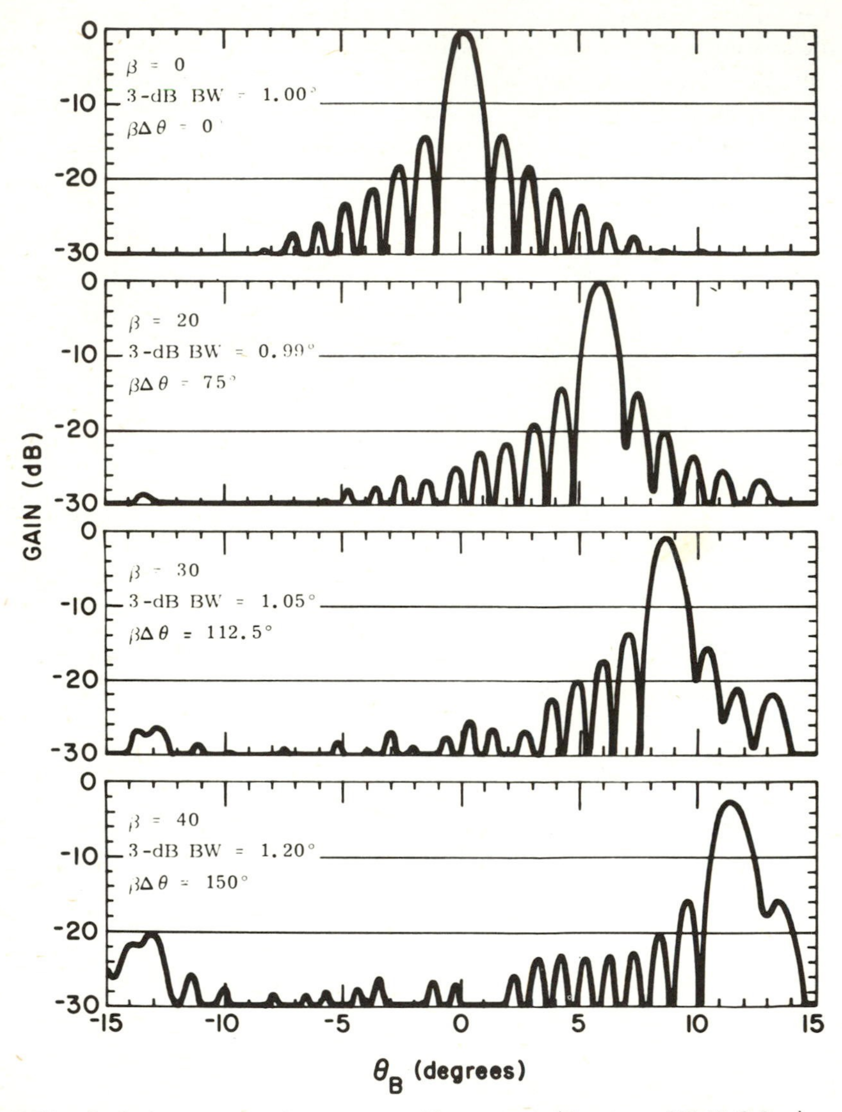

Fig. 5.28 *Radiation patterns of scan corrected lens antenna* (Courtesy of R. McGahan)

The technique studied by Schell and McGahan, like all other single-reflector or lens antennas, requires an oversize main aperture because the illumination moves with scan. Resulting typical aperture efficiencies are 20–25%. The main advantages of the technique described above are its extremely low element use factor and good pattern performance.

Several dual-lens configurations show promise of providing high-quality performance over limited angular sectors. The first is the structure shown in Fig. 5.29, which was investigated by Tang and Winter[43] using a computer simulation. The planar array feed consists of 437 elements arranged on a square grid. The array focuses a small spot on the elliptical rear face of a lens which transfers the spot to a region on the focal arc of a final lens with spherical back face. This main lens is an equal-path 'bootlace' lens with the front face chosen to provide corrections at two points $(0, \pm 10^\circ)$. Sidelobes are at approximately the -16 dB level for $\pm 10^\circ$ scan. At 65λ diameter, the antenna forms a beam of approximately 1.2° for all scan angles, and the aperture efficiency is about 60%. The element use factor is 1.5, and although the L/D ratio is at least 1.7, the main aperture diameter D is small because of its efficient illumination. The intermediate lens is about 0.7 the size of the final lens. The total system loss, including -2.1 dB aperture efficiency is about -6.5 dB. System bandwidth of $\pm 20\%$ is achieved in the computer study.

A recent investigation by Borgiotti[4] included an analytical and numerical evaluation of the antenna system of Fig. 5.30 for one-dimensional scan. This system is similar to the idealised two-transform limited scan system of Section 5.2, and to the completely overlapped subarraying system (HIPSAF) described by Tang.[44] The number of control elements used is approximately equal to the theoretical minimum. The system consists of a hybrid matrix and a 'bootlace' lens with linear outer profile and circular inner profile. The geometry is also similar to that studied by Tang *et al.*,[45] but with the lense objective instead of the reflector. Borgiotti presents a number of design details to facilitate parameter selection. Given the main aperture width $2a$, focal length F and array length $2b$, define ϵ as the end point of the subarray radiation pattern, then denote the 'magnification' μ as the ratio of lens aperture to array size:

$$\mu = \frac{\sin\theta_a}{\sin\epsilon} = \frac{a}{b} \tag{5.49}$$

where

$$\sin\theta_a = \frac{a}{F}$$

for a lens with θ_a as the half angle spanned by the lens as measured from the feed. Borgiotti introduces the phase delays ϕ_p for each pth element of the array feed, so that

$$\phi_p = \frac{2\pi}{\lambda} x_p \sin\theta_0 \tag{5.50}$$

for the selected subarray centres at

$$x_p = \frac{p\lambda}{2\sin\epsilon} \tag{5.51}$$

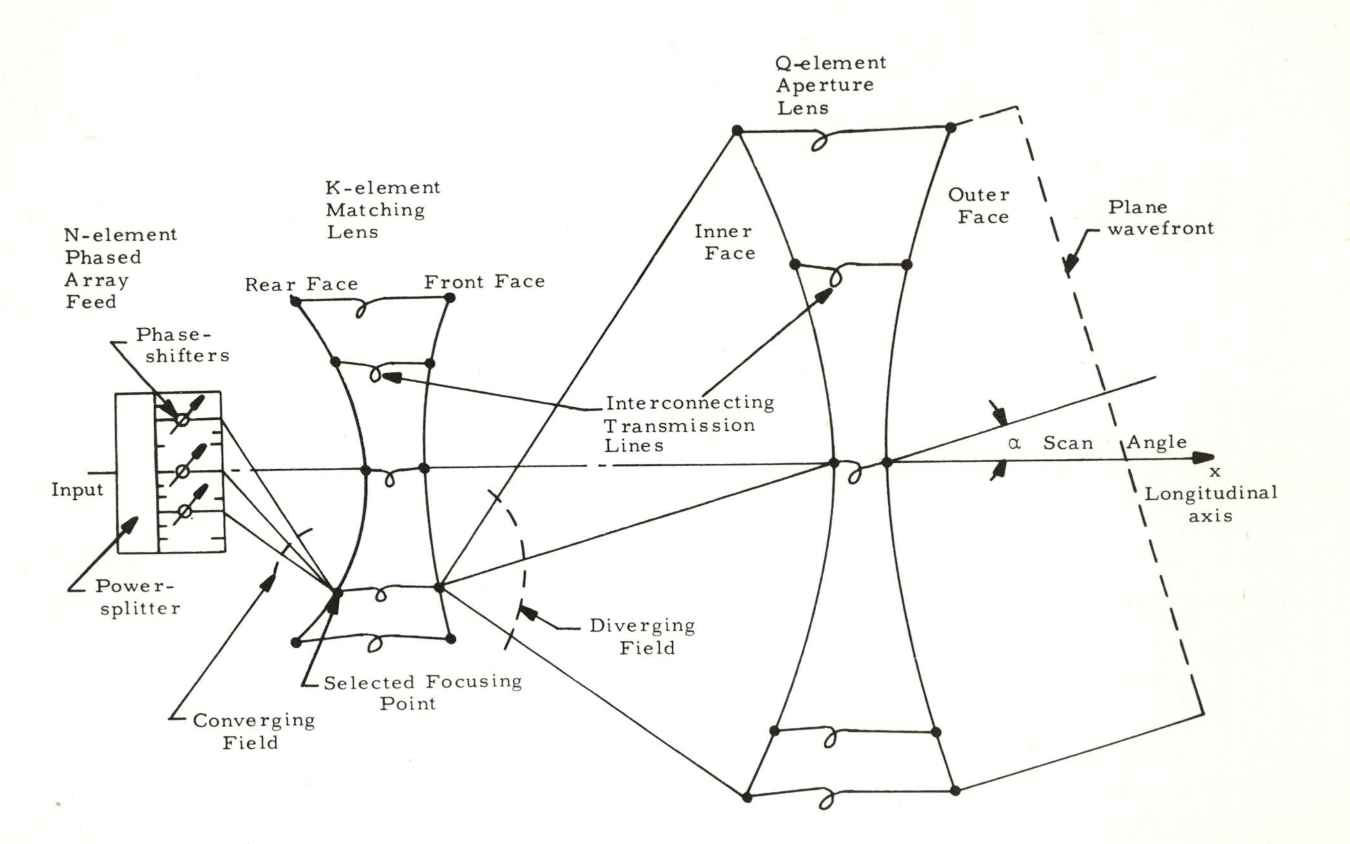

Fig. 5.29 *Dual lens limited scan concept*

on the main aperture. This choice of subarray spacings yields the theoretical minimum number of control elements in accordance with eqn. 5.13.

The progressively phased array distribution passes through the matrix array distribution to its output, where fixed time delays are added to focus the energy onto points on the inner surface of the lens.

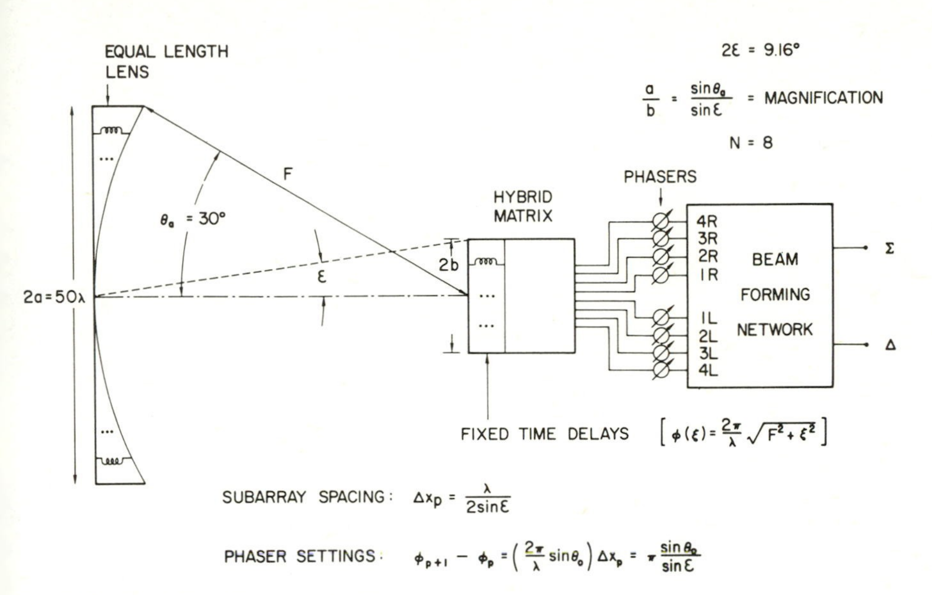

Fig. 5.30 *Dual transform scanning lens*

The design procedure consists of selecting an aperture illumination to meet required sidelobe level, choosing the lens size $2a$ for the desired beamwidth and a long focal ratio $(2a/F) = 1$. The desired subarray field of view (2ϵ) is taken to be about one beamwidth wider than the required scan angle to give good suppression of the lens grating lobes. The number of subarrays is the ratio of aperture length to subarray spacing

$$N = 2a\left(\frac{2\sin\epsilon}{\lambda}\right) \tag{5.52}$$

for N an integer.

Given these parameters it is possible to compute the magnification factor μ, the focal array size and number of elements, and the fixed time delay corrections. Fig. 5.31 shows a typical subarray pattern for the example studied by Borgiotti, and Fig. 5.32 shows the radiation pattern at broadside and scanned several degrees off axis. Notice that the near sidelobes are at the level given by the array taper and scan condition, but sidelobes beyond the subarray field of view are dramatically suppressed. This illustrates the obvious advantages of subarraying to suppress the far sidelobes associated with phase shifter tolerance or quantization errors.

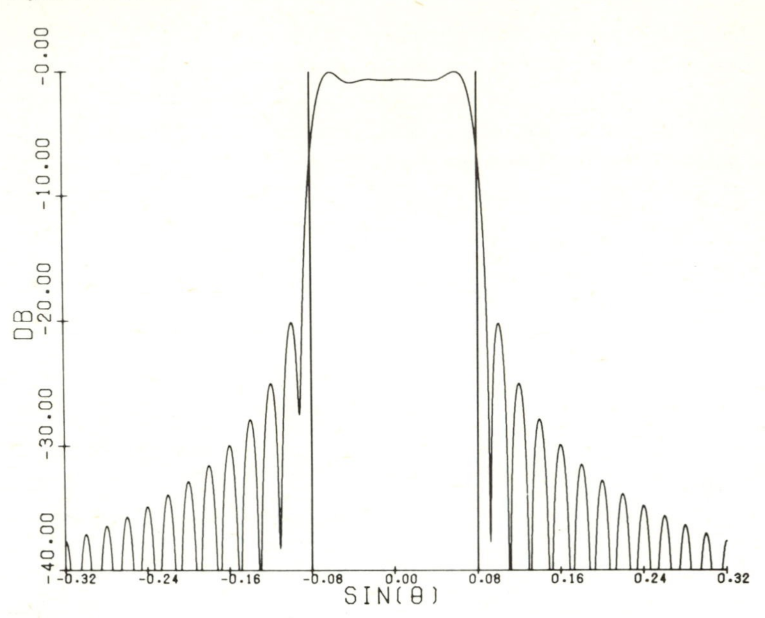

Fig. 5.31 *Subarray pattern of dual transform lens system* (courtesy of G. Borgiotti)

5.4.2.2 The Dome antenna: One of the more promising recent innovations in wide-angle scanning is the Dome antenna (Fig. 5.33) (Schwartzman and Stangel[33]); this novel structure uses the vertical projection of the Dome to achieve increased gain at low angles of elevation. The basic antenna has a passive spherical lens made up of fixed phase shifters, and a conventional planar phased array. As shown in Fig. 5.34 the array steers an illuminated spot to various portions of the lens, and the fixed phase delays of the lens are computed so as to convert the array scan direction φ' into a factor K times that angle, or a lens scan angle φ.

$$\varphi = K\varphi' \tag{5.53}$$

Analysis of the dome and phase shifter settings are done by ray tracing. The array phase-shifter settings are determined to form the non-linear phase progression required to scan the searchlight type beam to various spots on the lens. Although the radiated beamwidth varies with scan angle, the Dome can achieve scanning over sectors larger than a hemisphere, and in fact have achieved scan to ± 120° from zenith. Steyskal *et al.*[40] gives equations for the gain limits of a given circular cylindrical dome based upon allowable scan angles for the feed array, and shows that the ratio of the average gain to the broadside feed array gain is bounded by the limit:

$$\bar{g} \leqslant \frac{2}{\pi} \eta \sin \alpha_{max} \tag{5.54}$$

where η is the feed array aperture efficiency and α_{max} is the maximum permissible

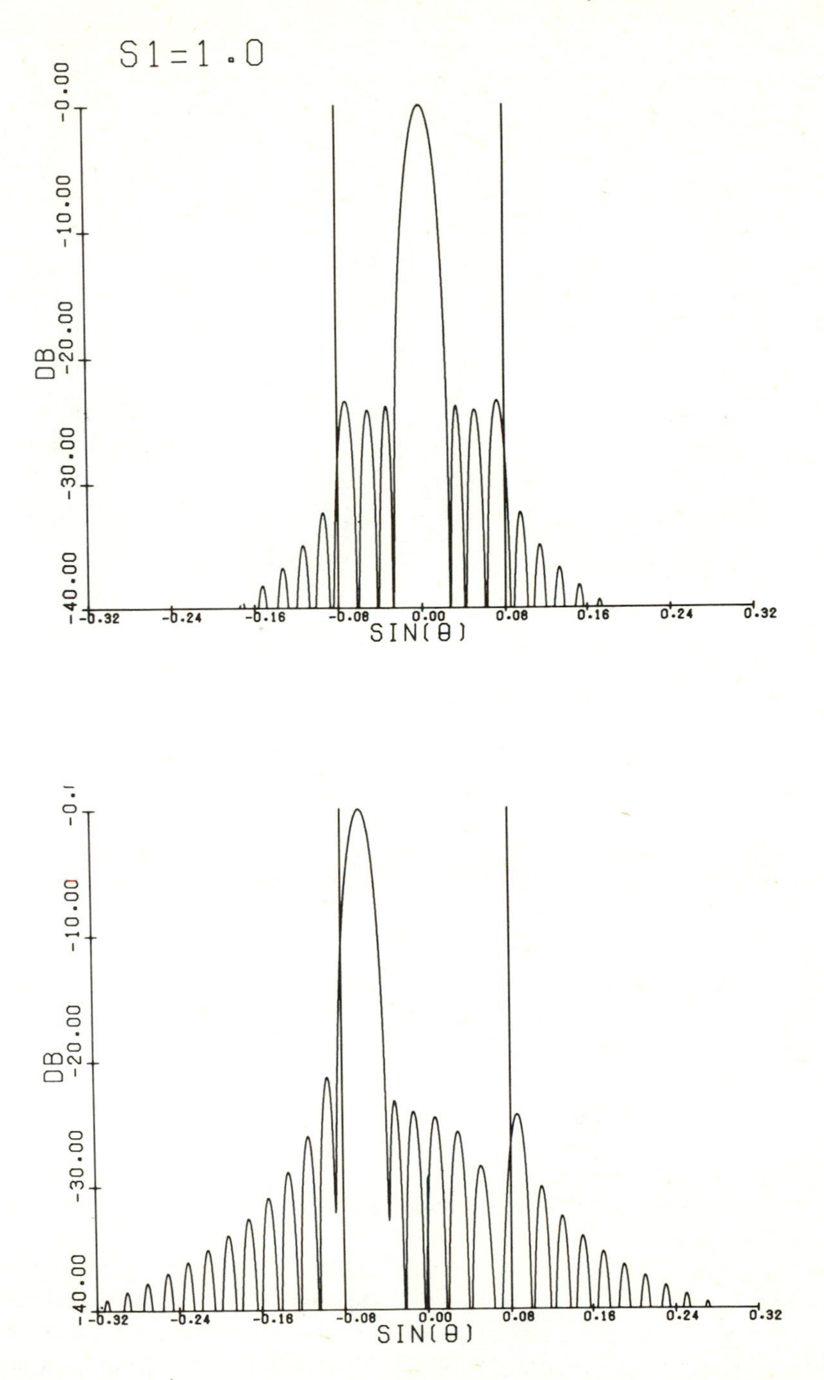

Fig. 5.32 *Radiation patterns of scanned dual transform lens system*: (*a*) at 0° (*b*) at −3.6° (courtesy of G. Borgiotti)

feed array scan angle. Steyskal also gives an expression for the refraction through the dome in terms of the angular parameters and dome constants:

$$\sin(\varphi - \varphi') = \sin(\alpha - \varphi') + \phi(\varphi') \tag{5.55}$$

Fig. 5.33 *The Dome antenna* (courtesy of J. Stangel)

where φ' is the angular dome coordinate denoting position on the radome, and th array scan angle measured from its centre φ is the scan direction, α the local sca angle of the feed array and

$$\phi(\varphi') = \left(\frac{\lambda}{2\pi a}\right)(d\psi/d\varphi') \tag{5.56}$$

is the normalised dome phase gradient for a dome with radius a and insertio phase ψ at position φ'.

Depending upon specific design criteria, it is possible to select a lens/arra configuration that emphasizes end-fire radiation. The reference (Schwartzma and Stangel[33]) illustrates several possible gain/scan profiles for a dome lens. On design, measured at 5.4 GHz, has about 20 dB gain at zenith, peak gain in exces of 26 dB at 65° from zenith, and 24 dB at the horizon. These data highlight th possibility of obtaining substantial gain at the array horizon. The array feed for thi lens measured 40 in in diameter, and so the array broadside directivity is approximately 35 dB. This figure thus indicates that there is a substantial penalty associated with the improvement of coverage at 90° and beyond, but the penalty is not nearly as severe as would be incurred for a conventional array scanned to end-fire.

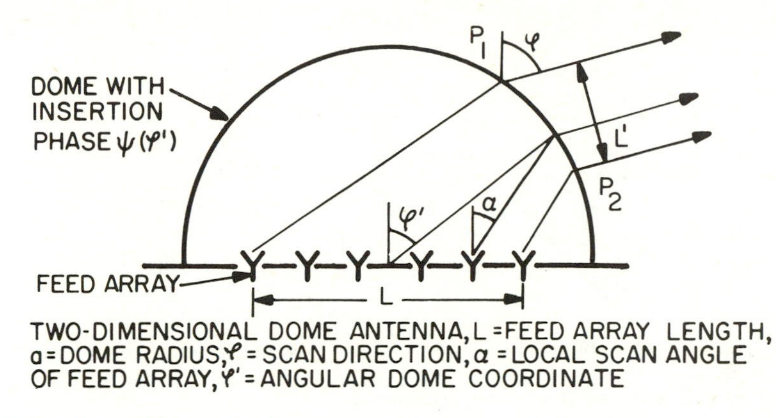

Fig. 5.34 *Ray paths through a Dome antenna* (courtesy of H. Steyskal)

5.4.3 Lens feeds and reflectarrays

The dome antenna consists of passive phase shifters inserted in a lattice throughout a dome radome, and so was indeed a lens, although not a true time-delay lens. The antenna structures described in this Section are more properly considered phased array techniques than lenses; for the constrained lens serves basically as an optical power divider. Optical systems have become the most common feeds for large arrays because of their excellent high-power behaviour, simplicity and low cost. In addition, since the array phase shifters are used to correct for the spherical incident wavefront as well as to provide the phase for scanning, the effects of phase quantisation are reduced because of the non-linear nature of the phase front. Optical techniques can also be used when it is desired to have separate transmit and receive feeds or different illumination under certain circumstances. In this case the array steers the beam and focuses at one of several feed horns.

Examples of lens fed arrays are given in many of the standard texts on antennas or radar (see Skolnik[38] pp. 11–50 and 11–58), and the technical literature contains references to a number of operational systems using optical feeds. Among the largest of these is described by Kahrilas.[19] Other significant arrays with optical feeds are discussed by Patton.[24]

The reflectarray concept as described by Berry[2] is another important optical array feed technique. Conceptually similar to the lens feed systems, the reflectarray face contains a number of antenna elements connected to shorted transmission lines with series phase shifters. The phase shifters vary the apparent line lengths so that the spherical wavefront of the feed horn shown in Fig. 5.35 is collimated and steered. The reflectarray can also include active transmit receive elements. Initially more common than lens-fed optical systems, the reflectarray is somewhat less flexible than the lens, and this, coupled with the sidelobe limitations imposed by feed blockage, has made lens systems more attractive for a number of applications.

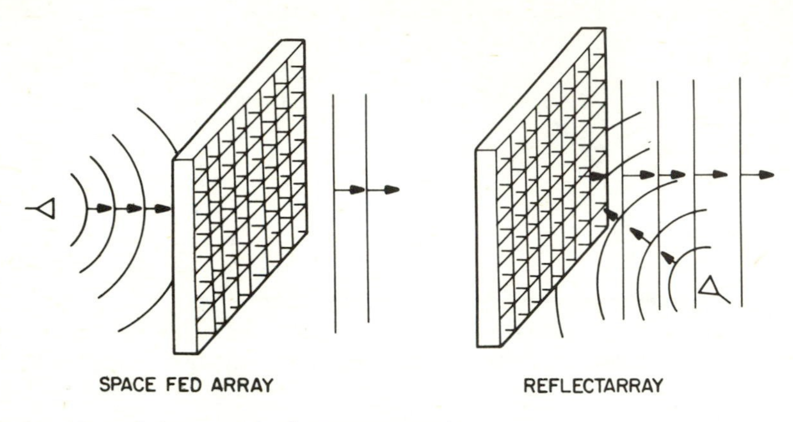

Fig. 5.35 *Space fed array and reflectarry geometries*

5.5 Radiation pattern control

5.5.1 Sidelobe control

The most common reason for combining a focusing objective with an array is to provide rapid electronic scan of a large aperture. However, the antenna types described in this Chapter have a number of special features that offer additional pattern control as compared with simpler focussed systems. This Section addresses the subject of low sidelobe feeds, tolerance control and null steering feeds to highlight these applications of hybrid systems.

The Chapter has described two basic feed systems; the transverse feed and the dual transform system. Both of these are useful for scanning, but they also allow direct control of the main objective illumination and so have great utility as low sidelobe antenna feeds. Transverse feeds have been used for this purpose for a number of years because of the added flexibility and precision control obtainable when an array illuminates the objective. Two transform systems offer an added dimension of control because of the formation of subarrays as described in Section 5.2. Since the radiation pattern of such a system is the product of the subarray pattern and the array factor, careful design of the subarray can provide substantial sidelobe reduction outside of the subarray pattern. The dual transform combinations of Tang *et al.*[45] and Borgiotti[4] provide in excess of 20 dB sidelobe suppression outside of the subarray field of view. Borgiotti has demonstrated that this suppression reduces phase shifter error or quantisation sidelobes to negligible levels, and so offers good control of these errors with short correlation lengths.

Coupled with error sidelobe control afforded by the subarrays, the excellent illumination transfer characteristics of dual transform systems gives them the potential of scanning beams over limited spatial sectors and extremely low sidelobes outside of the subarray field of view without the substantial expense of high-precision analogue phase shifters.

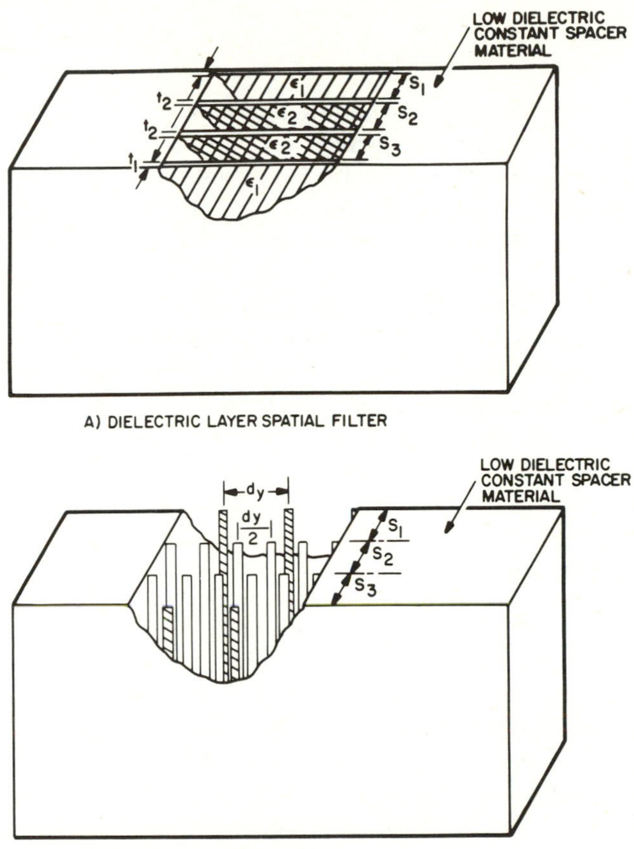

Fig. 5.36 *Dielectric and metallic grid spatial filters*

Additional sidelobe control is provided for reflector and lens systems through the use of a spatial filter that can be built in the form of a flat (or curved) radome as shown in Fig. 5.36, consisting of several layers of dielectric or metal grids. Spatial filters are designed to have an angular passband field of view and an angular stop band for sidelobe and/or grating-lobe suppression. The first filter application (Mailloux[21]) was to the problem of grating lobe suppression for an array with limited sector coverage. That filter used sandwiched dielectric layers and was designed to produce a Chebyshev pass-band for scan over a ± 12° sector centred at broadside. Design of a Chebyshev dielectric spatial filter consists of defining the parameter

$$\zeta = (2\pi/\lambda)S\cos\theta \tag{5.57}$$

for inter-layer spacing S, defining the ripple in the inverse of the power transmission factor to be Δ^2 and writing the power loss polynomial as:

$$A_{\|}A_{\|}^{*} = 1 + \Delta^2 T_m^2\left(\frac{\sin\zeta}{\sin\zeta_1}\right) \tag{5.58}$$

Here the $A_{\|}$ is the inverse of the filter transmission coefficient, T_m^2 is the square of the mth order Chebÿshev polynomial and ζ_1 defines the end of the pass band. Design curves are available for dielectric filters with up to four layers. Their use requires selection of the pass band width θ_1, the inter layer spacing S (usually λ or $\lambda/2$), the desired rejection at some out-of-band angle, and the allowable pass-band ripple.

Metal grid filters offer many advantages as compared with dielectric layer filters. Metallic gratings can be etched on printed-circuit boards and mounted on honeycomb or dielectric spacer material to make a lightweight, rigid and inexpensive radome structure. Recent studies indicate that metal strip filters of the type shown in Fig. 5.36 (in this case a linearly polarised version) have no second pass band or Brewster phenomena as is observed with the dielectric filter. Their design is exactly analogous to the procedure for shunt susceptance frequency bandpass filter design but with angle θ as parameter instead of frequency. Fig. 5.37 demonstrates the sidelobe suppression characteristics available with a metal grid filter in front of a small parabola. Fig. 5.38 shows the filter characteristics somewhat more dramatically by comparing the radiation patterns of a small horn element alone or placed behind an angular filter with an absorbing tunnel. In this case the filter transfer function is the difference between the two patterns, and the potential for 20–30 dB sidelobe suppression is clearly evident. The filter is used with an absorbing tunnel because energy that is not allowed to radiate through the pass band would contribute to a large end-fire sidelobe without the tunnel.

Angular filters are also frequency filters as indicated in eqn. 5.37, but the effect of varying frequency is to broaden or narrow the angular pass band. Maximum achievable bandwidth occurs for infinitely steep rejection skirts, and is

$$\frac{\cos\theta_m - \cos\theta_s}{\cos\theta_s}$$

for pass band maximum θ_m and minimum stop band angle θ_s.

A remaining factor in filter design is the determination of minimum filter size to produce a specified pass band sector. One can anticipate that to form a sector of given width $\Delta\theta$ the filter must be at least $1/\Delta\theta$ wavelengths across, but to determine the minimum filter size it is convenient to note that the filter is actually a subarray forming network. It does not perform a Fourier transform but merely spreads out the field to synthesise an efficient shaped subarray as shown in Fig. 5.38. The width of the subarray illumination is thus obtained by taking inverse transform of the synthesised subarray pattern (filter pattern), and then noting that

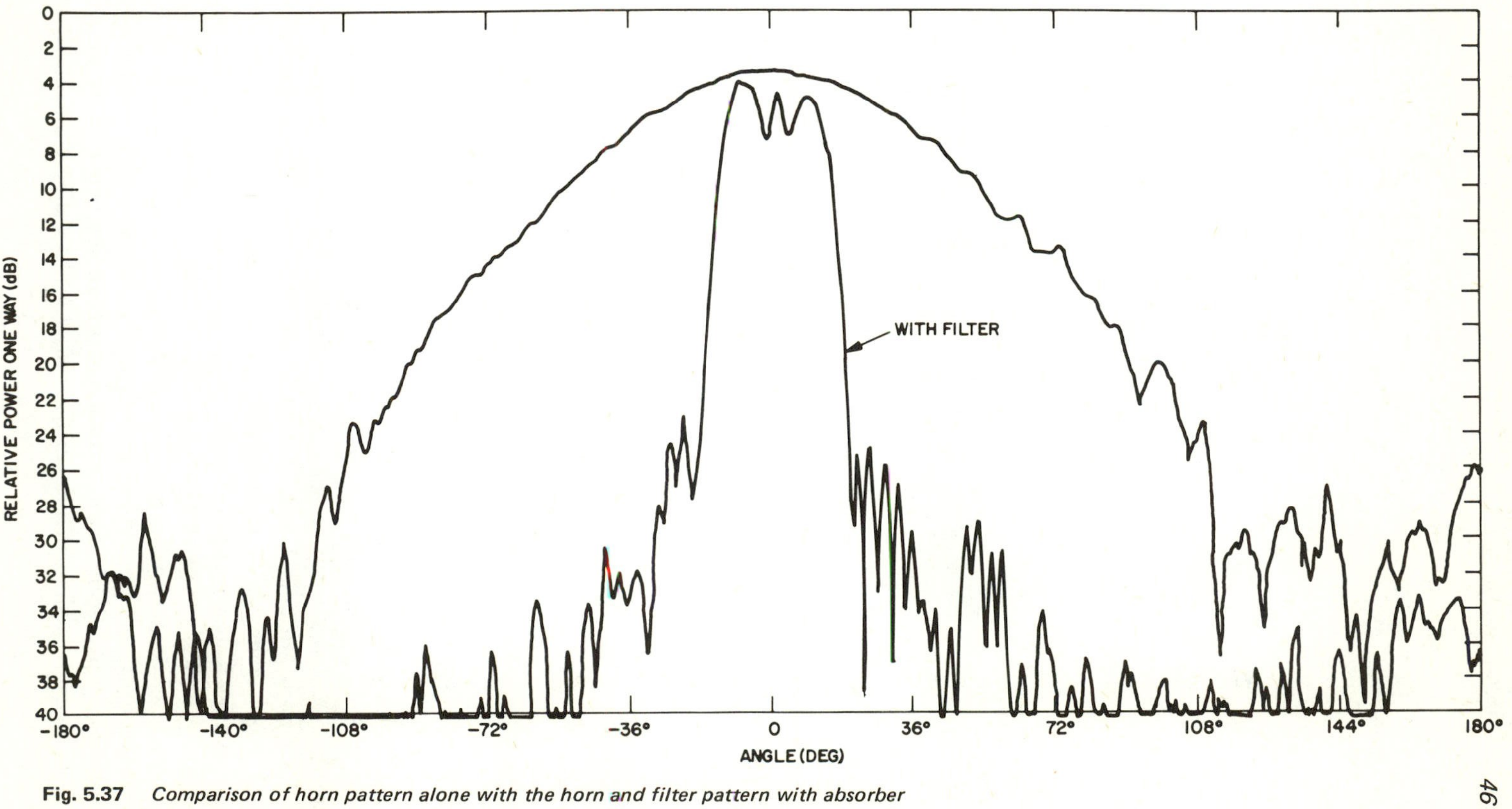

Fig. 5.37 *Comparison of horn pattern alone with the horn and filter pattern with absorber*

the filter must be approximately this amount larger than the aperture. For a flat-topped subarray pattern the filter should be at least $1/\Delta\theta$ wavelengths larger than the aperture.

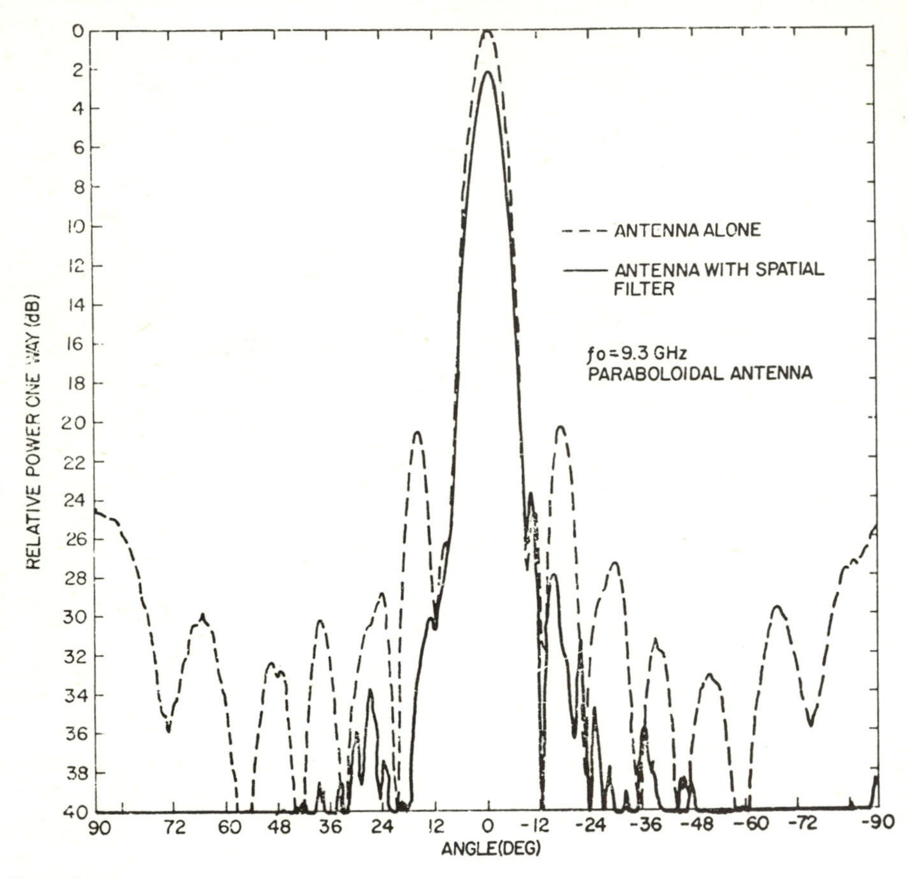

Fig. 5.38 *Sidelobe suppression by metallic grid filter*

5.5.2 Active pattern control in hybrid systems

Since hybrid systems use relatively few array elements to maintain a given field distribution across the large aperture of the objective, these systems are ideal in combination with null forming networks, adaptive optimisation circuits and real-time or programmed error compensation circuits. They also allow the instantaneous pattern modification features of the array for zooming or broadening the beam.

Compensation of reflector profile errors can be accomplished using systems with transverse feeds or with two transform feeds. Rudge *et al.*[26] present a detailed description of a two transform circuit using a Butler matrix to feed a reflector objective. Displacement errors with long correlation lengths across the objective can be compensated to first order by adding a phase correction to the terminal of the subarray illuminating a given section of the reflector. Eqn. 5.25 describes how the subarray phase and amplitude are transferred proportionally to the objective aperture and hence can conveniently remove these long correlation errors. Errors with short correlation length are minimised automatically by the subarray pattern.

5.6 References

1 ASSALY, R. N., and RICARDI, L. J.: 'A theoretical study of a multi-element scanning feed system for a parabolic cylinder', *IRE Trans.*, **PGAP,** 1966, pp. 601–605

2 BERRY, D. G., MALECH, R. G., and KENNEDY, W. A.: 'The reflectarray antenna', *IEEE Trans.*, **AP-11,** 1963, pp. 646–651

3 BORGIOTTI, G. V.: 'Degrees of freedom of an antenna scanned in a limited sector'. IEEE G-AP, International Symp., 1975, pp. 319–320

4 BORGIOTTI, G. V.: 'An antenna for limited scan in one plane: Design criteria and numerical simulation', *IEEE Trans.*, **AP-25,** 1977, pp. 232–243

5 BORN, M., and WOLF, E.: 'Principles of optics'. Pergamon Press, Oxford, England, 1959

6 BROWN, J.: 'Microwave lenses'. Methuen's Monographs on Physical Subjects, London, 1953

7 CHEN, M. H., and TSANDOULAS, G. N.: 'A dual-reflector optical feed for wide-band phased arrays', *IEEE Trans.*, **AP-22,** 1974, pp. 541–545

8 COLLIN, R. E., and ZUCKER, F. J.: 'Antenna theory'. McGraw Hill Book Co., 1969

9 DION, A. R., and RICARDI, L. J.: 'A variable-coverage satellite antenna system', *Proc. IEEE*, **59,** 1971, pp. 252–262

10 DUDKOVSKY, A.: 'A system for exciting large parabolic antennas'. Russian Patent No. 146365, 1962

11 FITZGERALD, W. D.: 'Limited electronic scanning with a near field cassegranian system'. ESD-TR-11-271, Technical Report 484, Lincoln Laboratory, 1971

12 FITZGERALD, W. D.: 'Limited electronic scanning with a near field gregorian system'. ESD-TR-71-272, Technical Report 486, Lincoln Laboratory, 1971

13 GALINDO-ISRAEL, V., LEE, S. W., and MITRA, R.: 'Synthesis of a laterally displaced cluster feed for a reflector antenna with application to multiple beams and contoured patterns', *IEEE Trans.*, **AP-26,** 1978, pp. 220–227

14 GENT, H.: 'The bootlace aerial', *Roy. Radar Establishment J.*, 1957, pp. 47–57

15 HANSEN, R. C. (Ed.): 'Microwave scanning antennas'. Academic Press Inc., Vol. 1, 1964; Vol. 2, 3, 1966

16 HOWELL, J. M.: 'Limited scan antennas'. IEEE AP/S, Int. Symp. Digest, 1974

17 IMBRIALE, W. A., *et al.*: 'A large lateral feed displacement in a parabolic reflector', *IEEE Trans.*, **AP-22,** 1974, pp. 742–745

18 JASIK, H. (Ed.): 'Antenna engineering handbook'. McGraw Hill Book Co., 1961, Chap. 15

19 KAHRILAS, P. J., and JAHN, D. M.: 'Handpoint demonstration array radar'. Supplement to *IEEE Trans.*, **AES-2,** 1966, pp. 286–299

20 LO, Y. T.: 'On the beam deviation factor of a parabolic reflector', *IRE Trans.* **AP-8,** 1960, pp. 347–349

21 MAILLOUX, R. J.: 'Synthesis of spatial filters with Chebysher characteristics', *IEEE Trans.*, **AP-24,** 1976, pp. 174–181

22 McGAHAN, R. V.: 'A limited-scan antenna comprised of a microwave lens and a phased array feed'. AFCRL-TR-75-0242, Air Force Cambridge Research Laboratories, Hanscom AFB, 1975

23 MILLER, C. J., and DAVIS, D.: 'LFOV optimization study'. Final Report No. 77-0231, Westinghouse Defense and Electronic systems Center, System Development Division, Baltimore MD., ESD-TR-72-102, 1972

24 PATTON, W. T.: 'Limited scan arrays' *in* 'Phased array antennas'. Proceedings of the 1970 Phased Array Antenna Symposium, Edited by A. A. Oliner and G. H. Knittel, Artech House, Inc., Dedham MA, 1972, pp. 332–343

25 ROTMAN, W., and TURNER, R. F.: 'Wide-angle microwave lens for line source applications', *IEEE Trans.*, **AP-11**, 1963, pp. 623–632

26 RUDGE, A. W., and DAVIS, D. E. N.: 'Electronically controllable primary feed for profile-error compensation of large parabolic reflectors', *Proc. IEE*, **117**, 1970, pp. 352–358

27 RUDGE, A. W., and WITHERS, M. J.: 'New techniques for beam steering with fixed parabolic reflectors', *Proc. IEE*, **118**, 1971, pp. 857–863

28 RUSCH, W. V. T., and LUDWIG, A. C.: 'Determination of the maximum scan-gain contours of a beam-scanning paraboloid and their relation to the Petzval surface', *IEEE Trans.*, **AP-21**, 1973, pp. 141–147

29 RUZE, J.: 'Wide angle metal plate optics', *Proc. IRE*, **38**, January 1950, p. 53

30 RUZE, J.: 'Lateral-feed displacement in a paraboloid', *IEEE Trans.*, **AP-13**, 1965, pp. 660–665

31 SANDLER, S.: 'Paraboloidal reflector patterns for off-axis feed', *IRE Trans.*, **AP-8,** 1960, pp. 368–379

32 SCHELL, A. C.: 'A limited sector scanning antenna'. IEEE G-AP International Symposium, Dec. 1972

33 SCHWARTZMAN, L., and Stangel, J.: 'The Dome antenna', *Microwave J.*, Oct. 1975 pp. 31–34

34 SCOTT, W. H., LUH, H. S., SMITH, T. M., and GRACE, R. H.: 'Testing of multiple-beam lens antennas for advanced communication satellites'. AIAA/CASI 6th Communications Satellite Systems Conference, 5–8 April 1976, Montreal, Canada

35 SHELTON, J. P., and KELLEHER, K. S.: 'Multiple beams from linear arrays', *IEEE Trans.*, **AP-9**, 1961, pp. 154–161

36 SHELTON, J. P.: 'Multiple-feed systems for objectives', *IEEE Trans.*, **AP-13**, 1965, pp. 992–994

37 SKAHILL, G. E., DESIZE, L. K., and WILSON, C. J.: 'Electronically steerable field reflector antenna techniques'. RADC-TR-66-354, 1966

38 SKOLNIK, M. I. (Ed.): 'Radar handbook'. McGraw Hill Book Co., 1970

39 SLETTEN, C. J., *et al.*: 'Corrective line sources for paraboloids', *IEEE Trans.*, **AP-6,** 1958, pp. 239–251

40 STEYSKAL, H., HESSEL, A., and SCHMAYS, J.: 'Limitations on gain-vs-scan for a dome antenna'. IEEE AP-S/URSI Int. Symp. Record, Univ. of Maryland, May 1978

41 STANGEL, J.: 'A basic theorem concerning the electronic scanning capabilities of antennas'. URSI Commission VI, Spring Meeting, 11 June 1974

42 TANG, C. H.: 'Application of limited scan design for the AGILTRAC-16 antenna'. 20th Annual USAF Antenna Research and Development Symposium, Univ. of Illinois, 1970

43 TANG, C. H., and WINTER, C. F.: 'Study of the use of a phased array to achieve pencil beam over limited sector scan'. AFCRL-TR-73-0482, ER 73-4292, Raytheon Co. Final Report Contract F19628-72-C-0213, 1973

44 TANG, R.: 'Survey of time-delay beam steering techniques' *in* 'Phased Array Antennas'. Proceedings of the 1970 Phased Array Antenna Symposium, Artech House Inc., Dedham MA, 1972, pp. 254–260

45 TANG, R., McNEE, F., JOE, D. M., and WONG, N. S.: Final Report, Limited Scan Antenna Technique Study. AFCRL-TR-75-0448, Contract F19628-73-C-0129, 14 August 1975

46 WHITE, W. D., and DESIZE, L. K.: 'Scanning characteristics of two-reflector antenna systems'. 1962 IRE International Conv. Record, Pt. 1, pp. 44–70

47 WINTER, C.: 'Phase scanning experiments with two reflector antenna systems', *Proc. IEEE*, **56,** 1968, pp. 1984–1999

48 WOODWARD, P. W., and LAWSON, J. D.: 'The theoretical precision with which an arbitrary radiation pattern may be obtained from a source of finite size', *J. AIEE*, **95,** Pt. 1, Sept. 1948, pp. 362–370

Chapter 6

Multiple beam antennas

L. J. Ricardi

6.1 Introduction

The potential of multiple beam antennas has probably been recognised since the design of the first reflector or lens antenna. Having invested in the reflector, its support, scanning mechanism and possibly control system, the relatively low cost and simple inclusion of more feed elements can result in an antenna system capable of producing more than one beam from the same aperture. Unfortunately the early arrays of beams had a fixed direction with respect to one another and the com- posite solid angle, defined by their beam cross-sections, was usually too small to be of general use. When radar systems were required to perform both surveillance and tracking functions at angular rates taxing or exceeding the physical capability of mechanically scanned systems, antennas using the wide-angle scanning properties of toroidal and spherical reflectors and an array of feedhorns, introduced the multiple beam antenna as we know it today. Specifically, tens, or even hundreds of beams are formed from a single radiating aperture by impressing signals on a corresponding number of input ports, or terminal pairs. Directional, or pattern shaping, characteristics can be carried out by distributing transmitter signals among the beam ports in a desired fashion or by appropriate weighting of the signal received at each port prior to summing them as an input to a receiver.

The development of satellite communication systems has opened up a majo area of use of multiple beam antennas. Their potential flexibility is used i advanced satellite and earth bound systems to provide variable pattern shaping[3] t accommodate changing surveillance or communication requirements and suppres interfering signals.

In this Chapter, the fundamental characteristics of the multiple beam antenn (MBA) are presented, with some of the associated physical limitations governing th synthesis of the desired radiation patterns. Since these antennas have a wide variet of performance capability, a 'figure of merit' is presented as an aid in assessing th performance of an MBA. These systemic features, a description of classical version of the MBA, the associated beam-forming networks, and some examples ar presented as an introduction to the multiple beam antenna.

6.2 Composition and performance characteristics of an MBA

6.2.1 Composition of an MBA system

An MBA system consists of three major components: an antenna, a beam-forming network (BFN), and the control circuits required to vary the power dividing or combining functions of the BFN. Any antenna with more than one input (or output) port is technically an MBA provided a different radiation pattern is obtained when signals are impressed on each of the input ports. It is generally accepted that a pencil-beam pattern results when signals are impressed on a single port and that $N > 5$ ports are available. The restriction that N exceeds 5 is needed to exclude five- or four-hour sequential lobing, or monopulse antennas from the general class of an MBA. Still further, a multiple-horn fed reflector, or lens, etc., using an organ pipe scanner, or its equivalent, is not considered an MBA as defined and discussed in this chapter.

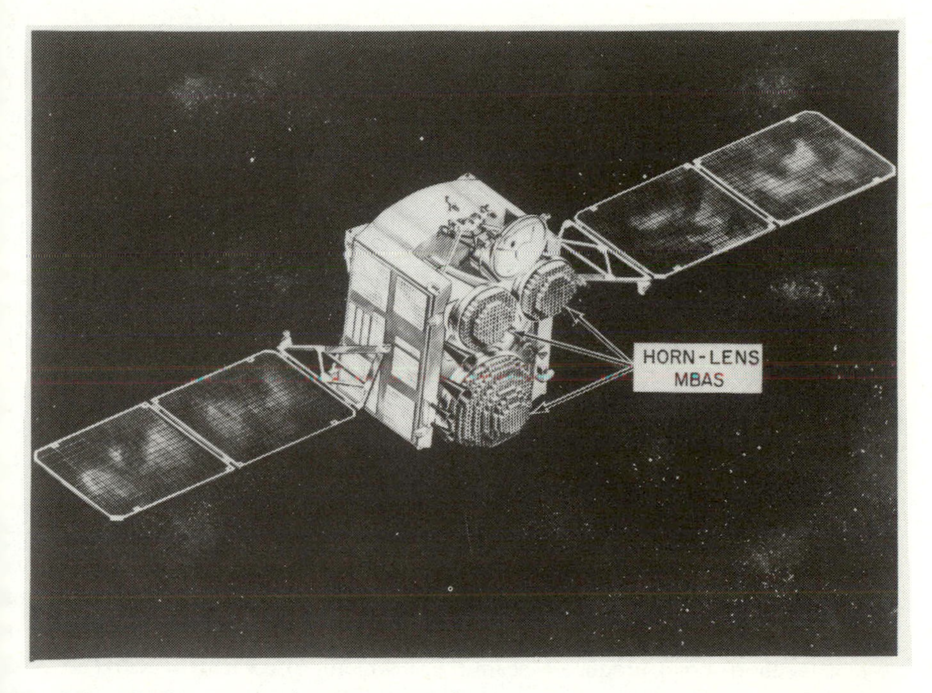

Fig. 6.1 *DSCS III Satellite (courtesy GE Co.)*

An MBA can be a lens antenna fed by an array of feedhorns (e.g., the DSCS III satellite antenna) as shown in Fig. 6.1, a reflector illuminated by an array of feedhorns (see Fig. 6.2), a 'planar' array excited by a Butler beam-forming network (BBFM), or some variation and/or combination of these. Impressing a signal on one of the N ports produces a beam pointing in a direction unique to that port. The N beams produced by exciting each port individually define, or span, the field of view (FOV) of the MBA. These beams form a fundamental basis for any radiation pattern produced by the antenna system. Consequently, they will be referred to as the fundamental beams of an MBA. Appropriate excitation of the N beam ports will produce the 'desired' antenna radiation pattern; similarly, appropriate weighting (i.e. adjustment of amplitude and phase) of the received signals, prior to summing them at the output terminal, will result in the 'desired' variation of the antenna receiving cross-section as a function of the direction to the received signal source.

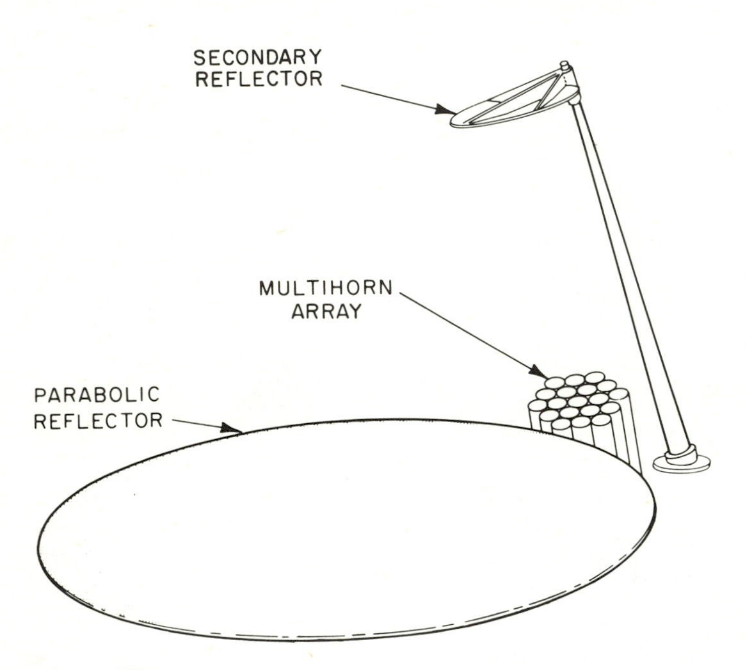

Fig. 6.2 *Reflector MBA*

In order to further clarify the fundamental characteristics of an MBA, consider the stepped waveguide lens schematically indicated in Fig. 6.3. (The lens could have been made of solid dielectric, or any of the devices discussed in Chapter 3; its precise character is not essential to this discussion. Signals impressed on the input port of feedhorn no. 4 produce a beam (i.e. beam no. 4) coincident with the focal axis of the lens. The beam direction is along the focal axis because the feedhorn is located on the focal point and the focal axis. Signals impressed on the input port of feedhorn no. 3 produce a beam squinted off the focal axis of the lens as indicated by beam no. 3. The angular displacement of beam no. 3 is related to displacement of feedhorn no. 3 as discussed in Chapter 3. Similarly, signals impressed on feedhorn no. 5 produce beam no. 5, etc., for all seven feedhorns. Spacing, location, and size of these feedhorns is critical to the MBA performance and will be discussed in detail later; the principal point to be derived here is that each port (feedhorn input) provides a place to inject signals and produce a beam pointing in a different direction. The composite spatial 'area' covered by the seven beams defines the field of view (FOV) of the MBA.

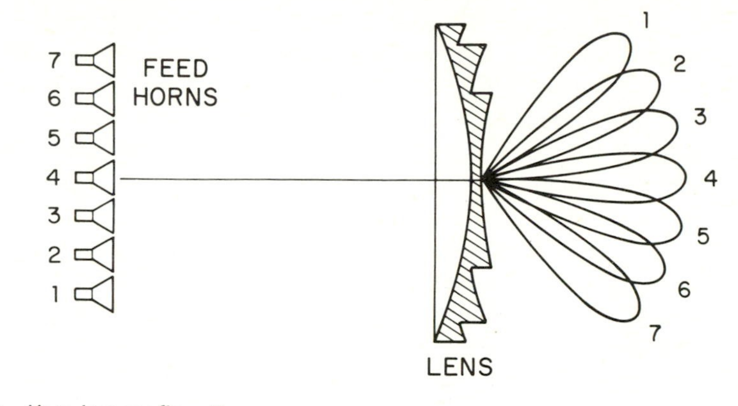

Fig. 6.3 *Horn-lens configuration*

The lens in Fig. 6.3 could have been a paraboloid reflector with a similar array of feedhorns illuminating it. Although the radiation patterns will be changed substantially by the introduction of aperture blockage, the feedhorn array would produce a corresponding 'pin cushion' of beam patterns, and hence the antenna would be an MBA. The use of beams and the previous discussion implies an antenna transmitting energy. However, the MBA is almost always a reciprocal device; hence the foregoing concept applies when the antenna is receiving signals. The concept of a transmitting rather than a receiving antenna is used because its performance characteristics are easier to visualise.

Although an MBA is essential to a variable coverage antenna system, the circuitry that divides the input power among the beam ports and/or combines the signals received by the beam ports, is of prime importance – in addition to being essential. In fact, the beam-forming network (BFN), as this RF circuit is called, is the 'heart' of MBA systems since it provides the desired flexibility and/or control of the antenna radiation pattern shape and/or direction. MBA systems use both variable and fixed power dividing or combining BFNs. With the latter, the radiation pattern shape is determined during the manufacture of the antenna; a variable BFN permits the radiation pattern shape to meet changing operational requirements for which the antenna is designed.

When the variable BFN is used, the MBA system must include control circuitry required to change the BFN power dividing or combining functions. Operational characteristics of the control circuitry determine the transient effects and reproducibility of the MBA radiation pattern. However, further discussion of this component of an MBA system is beyond the scope of this chapter.

6.2.2 *MBA configurations*

The reader should be aware that there are many classes of lens, reflector, and array type radiating devices. A popular version of a lens MBA is discussed in Section 3.7 and consists of an array of waveguides assembled to form a constrained lens. The waveguides have a square, or circular, cross-section and their lengths are adjusted to introduce the appropriate time delay in signals passing through them. Specifically, the lens introduces that delay required to convert a spherical wave, emanating from the focal point of the lens, into a plane wave propagating away from the radiating aperture of the lens. Since it is a reciprocal device, incident plane waves are focused onto the focal point. The lens indicated in Figs. 6.1 and 6.3 has been 'stepped' to make it lightweight and thinner. The stepping process permits sections of waveguide, L_D, to be removed when the differential phase shift introduced by L_D equals $n2\pi$ radians, where n is an integer. The number of steps and their location is a design parameter chosen to produce special characteristics. For example, if the lens is stepped such that the individual waveguide lengths are chosen such that they change the insertion phase of energy propagating through it less than 2π radians, the lens design is optimum in the sense that it is the thinnest lens possible and its weight varies approximately as the square of its diameter. This design is often called the 'minimum thickness' lens (see Section 3.7). Stepping a lens introduces a phase error that tends to compensate for the error introduced by the phase dispersion of the waveguide. These apparent degrading effects can be combined to produce a broadbanding effect that can be exploited by changing the number of steps and results in a thicker and heavier lens. The latter properties are often tolerable and to be considered worth the improved broadband performance.

It is also possible to reduce the phase error introduced by a broadband constrained lens by combining the foregoing design with what might be called a 'half wave-plate circularly polarised lens'. This compound lens has sufficient variables to result in perfect focusing at two frequencies. Appropriate choice of these two frequencies can increase the bandwidth of an uncompensated stepped waveguide lens from ~10% to ~40%! The reader is referred to Section 3.7 for further details of this design.

Reflector antennas illuminated by an array of feeds (see Fig. 6.2) also produce multiple beams and provide performance similar to that of a lens MBA. Howeve it is usually necessary to offset feed, as opposed to center feed, a reflector in orde to decrease or eliminate any degrading aperture blockage effects. Reflector antenna are usually preferred over all other antennas when the antenna aperture exceed say, 100λ, because of the ease with which the antenna can be constructed or, i necessary, deployed from a spacecraft. For example, a constrained lens, wit aperture diameter greater than 100λ, requires more than 30 000 individual cells c waveguides. Any attempt to separate this in sections which are deployed from spacecraft to form an integral assembly is usually more difficult than to deplo a reflector antenna with an equivalent radiating aperture.[7] Also, it is more difficu to manufacture the 30 000, or more, waveguide elements and assemble them t form a lens than it is to manufacture a reflector.

Finally, a planar array antenna in conjunction with a Butler beam-forming matrix (BBFM) possesses all the fundamental properties of an MBA. An array is usually the preferred configuration when the number of elements N_e ranges between a few to a few hundred. These antennas usually have N_e beams which span the FOV. All elements of the array are identical, and each element covers the entire FOV. Hence the N_e beams span that space approximately defined by the half-power beamwidth of the radiation pattern of an element of the array.

6.2.3 Pattern shaping

The facility with which the antenna radiation pattern can be shaped is undoubtedly the major advantage of an MBA system. Radar and communication system designers often need to know the pattern shaping limitations in order to realise either adequate or best overall system performance. Although particular applications may restrict their realisation, some performance characteristics that are fundamental to an MBA can be used as guidance in system design. This and the next four Sections address the variable radiation pattern characteristics that are potentially available with an MBA system. In this Section pattern shaping, in order to realise a particular 'coverage' over the field of view (FOV), is discussed with respect to general limitations.

The radiation pattern of an MBA is determined basically by the size of the MBA radiating aperture, the specific beam excitation distribution, and the number of beams. For example, excitation of all beams with in-phase signals of the same amplitude produces a radiation pattern that defines the FOV of the MBA system. With an N-beam MBA, the radiation pattern can be described uniquely at N points within the FOV. Assuming a classical spherical coordinate system where the z-axis is orthogonal to the aperture of the MBA (see Fig. 6.4), uniform illumination of the assumed circular aperture yields the well-known far zone radiation pattern $E(\theta, \phi)$ given by

$$E = 2\frac{J_1(u)}{u} \tag{6.1}$$

where $u = 2\pi a \sin\theta/\lambda$, a = radius of the aperture, λ is the wavelength, and θ is as shown in Fig. 6.4. When the phase of the aperture illumination is proportional to x, the radiation pattern scans off the z-axis. For example, the illumination I_n given by

$$I_1 = e^{j\alpha x} \tag{6.2}$$

causes the radiation pattern to scan in the plane $\phi = 0^\circ$. The resulting beam is of the form

$$E_n = 2\frac{J_1(u-u_n)}{u-u_n} \tag{6.3}$$

where $u_n = 2\pi a \sin\theta_n/\lambda$. A set of N such 'beams' span the FOV as described by

$$E = 2\sum_{n=1}^{N} a_n \frac{J_1(u-u_n)}{u-u_n} \tag{6.4}$$

The MBA radiation pattern E is determined by eqn. 6.4 and the complex excitation coefficients a_n. The radiation pattern 'fall off' E' is given by

$$E' = \frac{\partial E}{\partial \theta} = 2 \sum_{n=1}^{N} a_n \frac{(u-u_n)J_0(u-u_n) - 2J_1(u-u_n)}{(u-u_n)^2} \tag{6.5}$$

If $N = 1$ the pattern 'fall off' is exactly the same as that of a single beam as given by eqn. 6.1. This is also true at the edge of the FOV. However, adjacent beams can be spaced and excited to produce a larger radiation pattern slope as given by eqn. 6.5. For example, consider the two beam MBA radiation pattern given by

$$E_2' = 2\left[\frac{J_1(u-u_1)}{(u-u_1)} - \frac{J_1(u+u_1)}{(u+u_1)}\right] \tag{6.6}$$

where $a_1 = -a_2$ and $u_1 = u_2$.

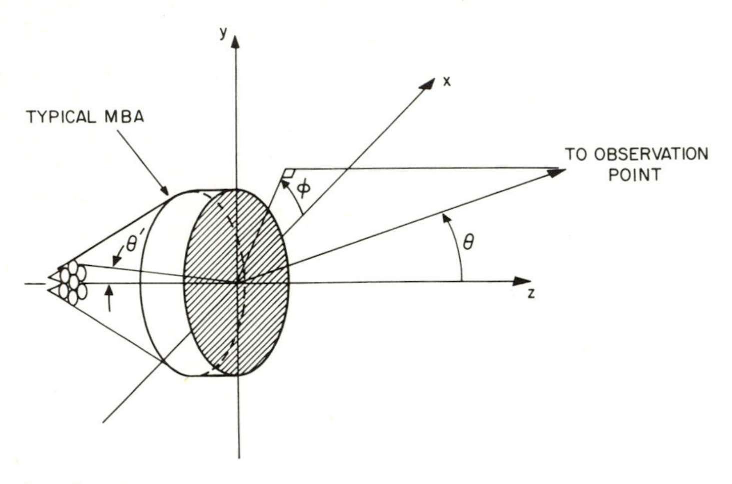

Fig. 6.4 *Coordinate system*

E_2' vanishes at $\theta = 0$ and it takes on a positive value for small positive values of u and it becomes negative for small negative values of u. Using the expression

$$\frac{J_1(u)}{u} \approx \frac{1}{2} - \frac{1.12}{2}\left(\frac{u}{3}\right)^2 \tag{6.7}$$

(i.e. for $-1 \leqslant u \leqslant 1$) eqn. 6.6 becomes

$$E_2' \approx \tfrac{4}{9} u_1 u \tag{6.8}$$

and

$$\frac{\partial E_2'}{\partial u} \approx \tfrac{4}{9} u_1 \tag{6.9}$$

This can be compared to the radiation pattern slope on the edge of a single beam by writing[1] E, from eqn. 6.2, with a maximum at u_1, the approximate location of a maximum of E_2'. That is

$$E \approx 1 - 0.1244(u - u_1)^2 - 0.0052(u - u_1)^4 \tag{6.10}$$

for $|u - u_1| \leqslant 3$. Therefore, at $u = 0$,

$$\frac{\partial E}{\partial u} \approx 0.24u_1 - 0.02u_1^3 \tag{6.11}$$

where $u_1 \leqslant 3$. The ratio of eqn. 6.9 to eqn. 6.11 gives the relative slope of the radiation pattern for approximately the same angular measure from the beam minimum. That is

$$\frac{\partial E_2'/\partial u}{\partial E/\partial u} \approx \frac{1.8}{1 - 0.08u_1^2} \tag{6.12}$$

The slope of the two-beam radiation pattern is greater than the slope of a single-beam pattern so long as $u_1 \leqslant 3$. The plot of E_2' shown in Fig. 6.5 demonstrates the change in pattern slope at $u = 0$ as u_1 is varied. Since the null of E occurs at $u \approx 4$, the curve $u_1 = 4$ in Fig. 6.5 clearly indicates the increased slope of E_2' when $u_1 \sim 2.5$.

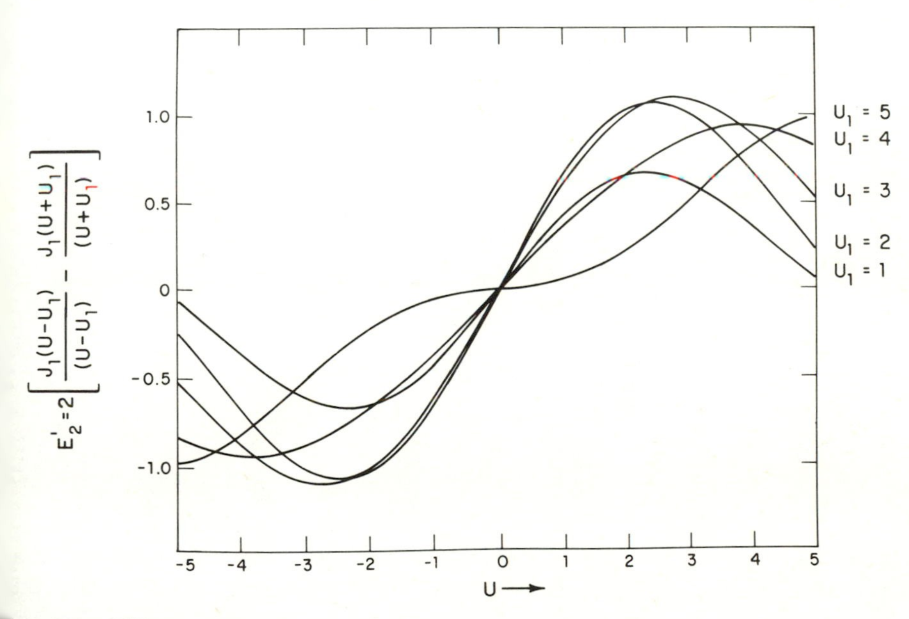

Fig. 6.5 *Difference pattern*

In summary the radiation pattern shape is limited, near the edge of the FOV, to that of a single beam of the MBA. Within the FOV, the rate of change of the radiation pattern with change in far zone angle can be approximately twice that obtained with a single beam. In both cases the radiating aperture size determines the maximum pattern 'roll off'; the beam spacing can affect the 'role off' obtained with two beams.

6.2.4 Beam scanning

Scanning a beam with an MBA is often misunderstood to consist of a series of steps from one beam position to the next. Actually the composite beam can be scanned smoothly over the entire FOV. To illustrate this, let us consider a scan of an MBA beam pattern from no. 1 beam direction to the direction of the adjacent beam no. 2 (see Fig. 6.3). In particular let us divide the total input power between feedhorns nos. 1 and 2. Referring to Fig. 6.6 notice as the power into feedhorn no. 2 increases from 0 to the case when it receives the entire input power, the beam direction varies smoothly and almost linearly from 0° to ~3.8°. Note that both experimental and analytical data are shown indicating the excellent agreement between measured and calculated data. Also note that the antenna gain varied ~1.0 dB. These results were obtained using a 30 in diameter stepped lens with a 30 in focal length, operating at 7.68 GHz and illuminated by an array of feedhorns spaced two inches on centre.

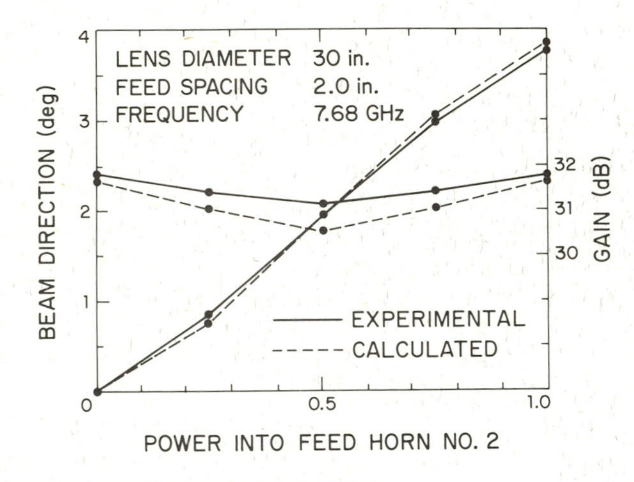

Fig. 6.6 *Beam-scanning performance characteristics*

Increasing the feedhorn spacing S would increase the variation in antenna gain a the beam is scanned from beam no. 1 to no. 2. Decreasing S will reduce thi variation in antenna gain. Still further, the foregoing considers beam scanning i only one plane. For the more general case of two-dimensional scanning the followin algorithm will aid in choosing a best aperture diameter and selecting an optimur value of S. This algorithm assumes that either one, two or three adjacent beams ar excited in phase and the variation in gain is calculated as the resulting beam is scanned over the FOV. It is important to note that the change in beam shape during 'scanning' results in a change in the instantaneous coverage, that is, the pattern formed by exciting two or more feeds simultaneously will be different from that obtained by exciting a single feed. The analytic procedure first assumes stepwise scanning, in which either one, two or three adjacent feeds are excited equally, and then the case of continuous scanning is considered.

Restricting the present discussion to this stepwise scanning also simplifies the concept of the optimisation procedure. Assuming an FOV = 18° (which is 0.6° larger than the solid angle subtended by the earth at synchronous altitude and allows for satellite attitude variation), the minimum directive gain is calculated for operation of the MBA with either one, two or three feeds excited in-phase and with equal amplitude. The diameter of the lens (keeping focal-length/diameter ratio constant) and the feedhorn spacing are varied and the directive gain at the design frequency is calculated. Since we are interested in determining the minimum value of directive gain anywhere in the FOV, it is necessary to examine only two situations. These are: the minimum directive gain obtained within, and at the edge of, the field of view.

Assuming a lens diameter in the range 20 to 34 in, a 19-beam MBA and a design frequency of 7.5 GHz, we start with a feedhorn spacing so small that the FOV is less than 18°. Thus increasing S increases the minimum directive gain G_e on the edge of the FOV to the maximum value, G_{emax}, at a spacing $S = S_{max}$. Further increases in S vary G_e between G_{emax} and $\sim 1/2\, G_{emax}$.

Considering the same initial conditions ($S < 1$), the minimum directive gain G_i, within the FOV, is larger than the gain at the edge of the FOV and decreases as S is increased. Hence, there exists a unique spacing, S_0, for a given lens diameter and frequency of operation, for which the minimum directive gain, G_{min}, over the entire FOV, is maximum. At this spacing $G_e = G_i$. In order to determine S_0, G_e and G_i were calculated and are plotted in Fig. 6.7 for various lens diameters between 20 and 34 in. Since S_0 must be less than S_{max}, S is varied over that range which demonstrates the equality of G_e and G_i.

It is reasonable to pass a curve through all points for which $G_e = G_i$ to obtain the solid curve that indicates the minimum directive gain versus spacing. However, it is necessary to recognise that for each feed spacing there exists only one lens diameter that will give the indicated G_{min}. Thus we see that, to a first order, a feedhorn spacing and antenna aperture size can be chosen which maximises the minimum gain over the FOV. Studies have shown that G_{min} can be increased ~ 0.5 dB if the excitation distribution is optimised. However, this has little effect on S_0 and the antenna aperture diameter.

The data, described in the previous paragraphs of this Section, were calculated for a planar array of conical feedhorns. With this feed configuration, the exterior beams have a slightly different phase with respect to the centre beam. Adjusting the relative phase of the signals exciting the feedhorns increases the directive gain when more than one feedhorn is excited simultaneously. The long dashed curve shown in Fig. 6.7 shows this increase in G_{min} and the corresponding value S_0. Still further increase in G_{min} can be obtained[11] by increasing the gain of the feedhorns. For example, adding an appropriately designed endfire element in the aperture of a feedhorn can increase the antenna gain by ~ 1 dB.[10]

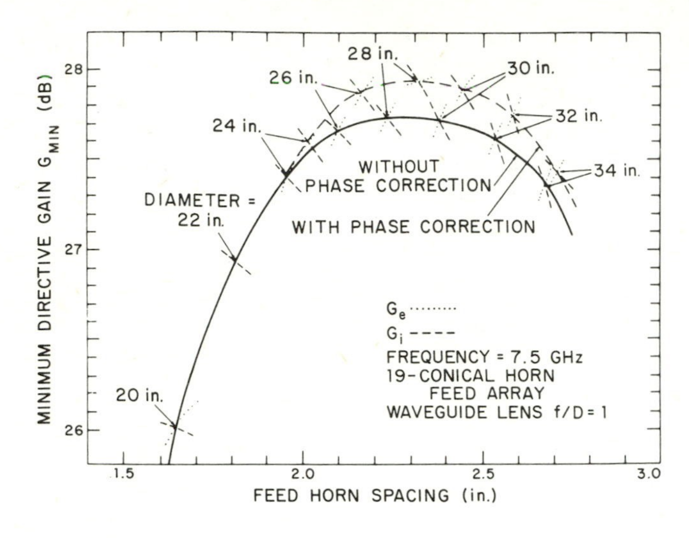

Fig. 6.7 *Minimum directive gain over field of view (FOV)*

In summary, Fig. 6.7 shows that a waveguide lens antenna illuminated by an array of 19 conical feedhorns excited in the TE_{11} mode will have a minimum directive gain of 27.9 dB over an 18° FOV if operation is limited to the excitation of a single feed or equal distribution of input power between two adjacent feedhorns or three feedhorns arranged in the form of a triangle. The device is reciprocal; hence the same is true when the antenna is receiving signals. Addition of an endfire element in the aperture of the feedhorn can increase the minimum directive gain G_{min} to ~ 29 dB. The calculated and measured maximum gain for a feedhorn spacing S_0 and a 28 in lens diameter is 32 dB; hence the antenna efficiency is 50% and the beam scanning loss is 3 dB. Studies[10] indicate that the latter may be reduced to ~ 2.5 dB by appropriate excitation of more than three horns.

It is important to note that potentially the scanning performance of an MBA is comparable to that of a planar phased array designed to scan over the FOV defined by one of its elements. A comparison of the losses in the beam-forming network of an MBA system to the losses in the power dividing and phase shifting network of a comparable phase array is necessary to determine which has the best performance characteristics.

The nature of the coverage obtained by exciting either one feedhorn, or two or three simultaneously, so as to maximise G_{min} is demonstrated in Fig. 6.8. The solid curves shown represent the directive gain contours corresponding to the excitation of a single feedhorn. For a 19-beam lens system described in the foregoing, the feedhorns are arranged in a hexagonal grid as indicated in Fig. 6.9. The dotted contour (Fig. 6.8) corresponds to the simultaneous and equal excitation of two adjacent feeds located at 31 and 32, 31 and 21, 32 and 21. The dashed contour is

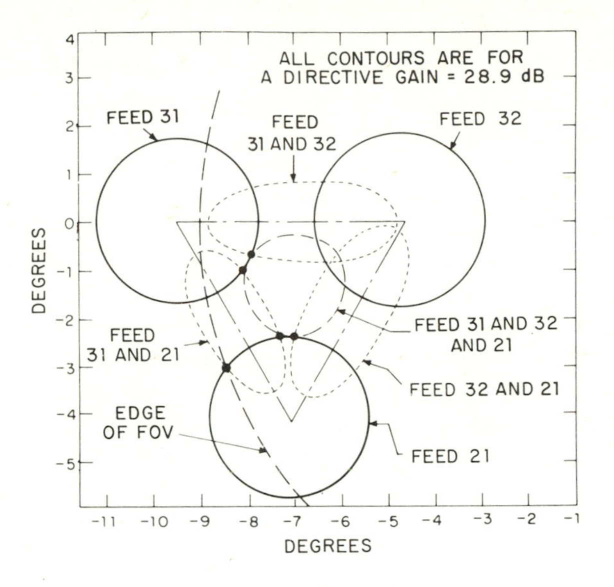

Fig. 6.8 *G_{min} contours for excitation of the indicated feeds*

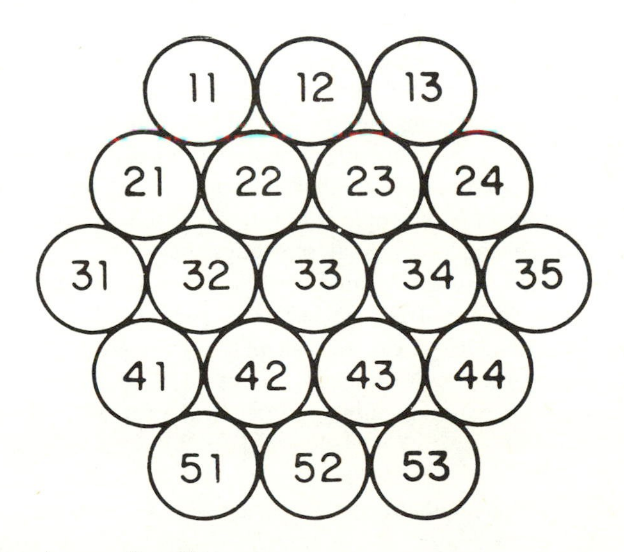

Fig. 6.9 *19-beam horn configuration*

obtained when all three feeds are excited simultaneously and with equal-amplitude signals. All contours represent a directive gain of 28.9 dB; the directive gain within any closed contour is $>$28.9 dB. The directive gain over most of the triangle (shown by the dash-dot lines) is > 30 dB. The MBA system has an 18° FOV, the edge of which is indicated by a dashed line. Since the triangle formed by these three feeds is representative of that formed by any three feeds, it follows that G_{min} exceeds 29.8 dB over the entire field of view. Note that the dots indicate those points where the gain of the MBA system equals G_{min}. Clearly unequal excitation (instead of equal excitation as assumed in this analysis) of two or three feeds would increase the antenna gain. In fact, further study of the foregoing configuration yielded ~ 0.6 dB increase in G_{min}.

In conclusion, continuous beam scanning is possible and may increase the minimum value of directive gain over that achievable as calculated in the foregoing by exciting one, two adjacent, or three adjacent feeds. A detailed presentation of the actual increase in directive gain is beyond the scope of this chapter. Initial calculations indicate that G_{min} is increased more than 0.6 dB, and the resulting radiated beam cross-section can be made circular even when the beam direction is midway between the beam directions corresponding to two adjacent fields.

6.2.5 Fundamental limitations

The facility with which the antenna radiation pattern can be shaped is undoubtedly the major advantage of an MBA system. Radar and communication system designers often need to know the pattern shaping limitations in order to realise either adequate or best overall system performance. Although particular applications may restrict their realisation, some performance characteristics that are fundamental to an MBA can be used as guidance in system design. For this reason, the concept of degrees of freedom (DOF) and resolution are introduced in this section.

Pattern shaping requires specification of the MBA system directive gain at a finite number P of points in the antenna field of view (FOV). Clearly, P cannot exceed M, the number of beams, and it may be less than M. The information can be expressed as specific values with acceptable bounds; however, it must be adjusted to result in pattern shapes that are within the fundamental limitations of the antenna system. For example, the rate of change of the antenna pattern with change in angle and the angular separation between a pattern 'null' and a pattern maximum are strongly related[8,9] to the MBA aperture size. These performance characteristics are often referred to as 'pattern fall-off' and interference-user angular separation. They can be calculated with great precision; however, reasonable estimates can be obtained as discussed in the following paragraphs. With a uniformly illuminated circular aperture, the half-power beamwidth (HPBW) is given by[14]

$$\text{HPBW} = 1.02 \frac{\lambda}{D} \text{ radians} \tag{6.13}$$

The relationship between HPBW, resolution and degrees of freedom is complex and a quantitative answer depends on the definition of the latter two performance characteristics. However, previous studies indicate that the number of circles M_D, with angular diameter equal to the HPBW, that encompass the FOV of the MBA is essentially equal to the degrees of freedom available with an aperture diameter, D. An MBA, using this size aperture, with $M = M_D$ beams takes full advantage of the DOF available. Increasing M will not significantly improve the MBA pattern shaping or nulling performance, over the FOV. It follows that the radiation pattern (amplitude and phase) can be defined at only one point within each circle defined by the HPBW, and the radiation pattern 'fall-off' near the edge of the FOV is very nearly equal to that of a beam with a HPBW $= \lambda/D$ radians. At points within the FOV, the change in directive gain[7] with either θ, or ϕ, (i.e. radiation pattern 'fall-off') can be substantially larger because the beams associated with several beam ports (i.e. more degrees of freedom) are potentially available to produce the antenna radiation pattern as was discussed in Section 6.2.2.

Since the radiation pattern produced by excitation of a single beam port has an approximate circular cross-section, arranging the beam direction (and the feed array) in a triangular grid provides the most uniform coverage of the FOV. When the FOV is circular, the beams are arranged on a hexagonal grid as indicated in Fig. 6.9. It follows that the number of beams M_D required to cover a circular FOV is given by

$$M_D = 1 + \sum_{i=1}^{(I-1)/2} 6i \tag{6.14}$$

where $I\,(= 2, 3, \ldots,$ etc.) equals the number of beams along a diameter of a circular FOV (e.g. in Fig. 6.9, $I = 5$). Expressing the diameter of the FOV in terms of the angular measure of the vertex angle ψ_o of a right circular cone that defines the FOV, it follows that

$$I \approx \text{integer value}\,[\psi_o/\text{HPBW}] \tag{6.15}$$

where HPBW $\sim \lambda/D$. For a geostationary satellite $\psi_o = 17.3° = 0.3$ radians and from eq. 6.15

$$I \approx \text{integer value}\,[0.3(D/\lambda)]. \tag{6.16}$$

Still further, the minimum gain G_{min} over the FOV is ~ 3 dB less than the antenna directivity[7] D_A when a single beam port is excited. Using the relationship

$$D_A = \eta \left(\frac{\pi D}{\lambda}\right)^2 \tag{6.17}$$

where η = antenna efficiency,[7] I can be expressed as

$$I \approx \text{integer value} \left(\frac{\psi_o}{\pi}\sqrt{\frac{D_A}{\eta}}\right) \tag{6.18}$$

where ψ_o is expressed in radians. Since a geostationary satellite is often used with satellite communication systems, it is interesting to note that from eqns. 6.16 and 6.17

$$I \approx \text{integer value} \left(0.1\sqrt{\frac{D_A}{\eta}}\right) \tag{6.19}$$

The foregoing expressions for I yield approximate values; it may be necessary to increase I by 2 in order to realise adequate coverage of the FOV. It is important to note that, from eqn. 6.14, increasing I by 2 increases the number of beams (i.e. feed ports) by $3(I+1)$! Consequently, the choice of I, and hence the number of beams M_D, should include an appropriate compromise between G_{min}, I and the spacing between feeds.

Interference suppression performance (i.e. nulling) depends principally on the aperture size D, the tolerable loss in gain to a desired signal source, and the number of interfering sources. For the present discussion, it is convenient to resolve the 'nulling' pattern into two patterns: (*a*) the desired pattern without nulls, and (*b*) the 'maximum directivity' pattern produced when the antenna aperture is uniformly illuminated and the phase of the aperture distribution forms a plane wave propagating in the direction of the desired 'null'. (For the case of a pattern with multiple nulls, a 'maximum directivity' pattern is pointed in the direction of each desired 'null'). An optimum 'nulling' pattern is formed by then setting the excitation amplitude and phase of the 'maximum directivity' pattern equal to, and 180° out of phase with, the amplitude and phase of the desired pattern, respectively. Simultaneous excitation of these radiation patterns via a single summing port will produce an optimum nulling pattern. It has been shown[8] that the resulting radiation pattern is optimum in the sense that it is the best root-mean-square fit to the desired radiation pattern with a null of infinitesimal angular width in the direction of the interfering source, or sources.

All interfering sources within a resolution cell (i.e. within a circle of angular diameter λ/D) are suppressed with a single DOF. Therefore, the number of interfering sources that can be suppressed by an MBA may exceed M_D. However, placement of one interfering source in each beam of an MBA will 'capture' all the antenna system DOF; this results in little, if any, antenna gain in the direction of desired signal sources.

Satisfactory performance (i.e. adequate directive gain in desired signal source(s) direction(s)) may be obtained when the angular separation between desired and interfering signal sources is less than an HPBW; however, directive gain in the direction of the desired signal will be reduced. This loss in directive gain L_g depends on the aperture size and field distribution, the number of interfering sources, etc. However, one can use a bound[8,9] in the form of the minimum L_g that can be obtained with reference to a scenario consisting of a single interfering source and a single desired signal source subtending an angle $\Delta\theta$ measured at the nulling antenna (see Fig. 6.10). Alternatively, one can use the rule

$$L_g \approx 3 + \left(\frac{\text{HPBW}}{\Delta\theta} - 2\right) 3\ \text{dB} \tag{6.20}$$

for $\Delta\theta < \text{HPBW}/2$. This rule is derived from rigorous analysis similar to that described in the following paragraphs, and it represents the minimum loss that can be achieved. For most applications, L_g will be larger than that given by eqn. 6.20.

Loss in gain due to nulling was accurately determined from several rigorous analyses[8, 9] and an assumed limiting scenario where the angular separation between the interfering signal source and the desired signal source is varied.

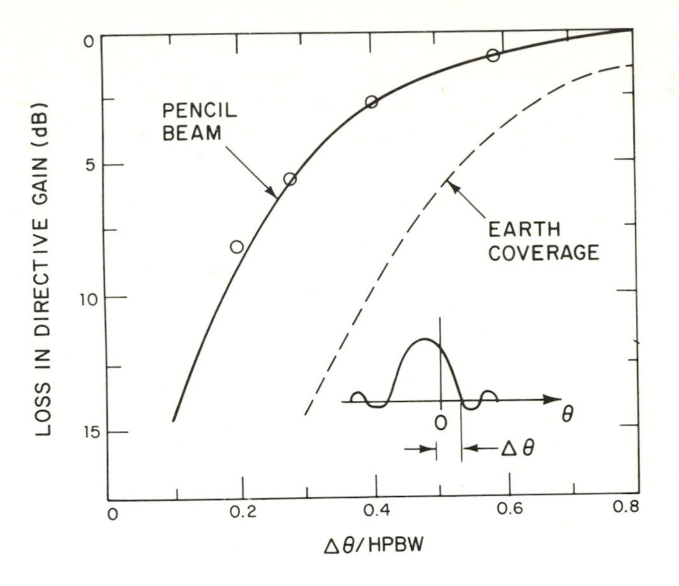

Fig. 6.10 *Loss in gain due to a single null*

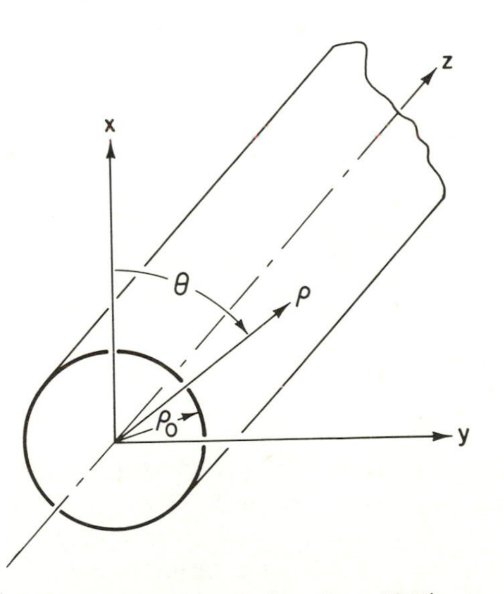

Fig. 6.11 *Cylindrical antenna* (*IEEE Trans. Ant. Prop.* Sept. 1975)

In one analysis,[8] a two dimensional problem is considered rigorously; the general results were extended to a more complicated three dimensional analysis.[9] For the two dimensional analysis, consider the cylindrical antenna indicated in Fig. 6.11. It extends from $z = -\infty$ and is excited to produce a radiated field which is a function of only ρ and θ, the radial distance to and angular location of the observation point, respectively. The excitation of the antenna can be in the form of currents on the conducting surface $\rho = \rho_0$; however, it is only necessary to define $\rho = \rho_0$ as a surface which encloses the antenna whatever the configuration of its cross-section might be. It is conceptually convenient to think of it as a cylindrical antenna with circular cross-section because, as will be explained, it provides an upper bound on the performance of the enclosed antenna. Without loss in rigour and only a slight decrease in generality, the antenna will be polarised in the z-direction; i.e. the electric field will be assumed to have only a z-component. In accordance with fundamental theory,[5] the electric field outside the antenna is given by

$$E = -j\omega\mu \sum_{n=-\infty}^{\infty} a_n H_n^{(2)}(k\rho)\, e^{jn\theta}\, \boldsymbol{a}_z \tag{6.21}$$

where $H_n^{(2)}(k\rho)$ is the cylindrical Hankel function of the second kind, $k = 2\pi/\lambda$, ω is the angular frequency, $\boldsymbol{a}_z$ is a unit vector in the z-direction, and μ is the permeability of free space. The $a_n = a_n^r + j a_n^i$ are complex quantities that will be determined later. The radiated power per unit length is given by

$$P_{rad} = 4\omega\mu \sum_{n=-\infty}^{\infty} |a_n|^2 \tag{6.22}$$

The directivity per unit length is defined as

$$D' = \left\{\frac{2\pi\rho|E|^2}{\zeta P_{rad}}\right\}_{\rho\to\infty} \tag{6.23}$$

where ζ is the impedance of free space. Since we are interested in maximising D' in the direction of the desired signal (i.e. $\theta_s = 0$), E in eqn. 6.23 will be evaluated at $\theta = 0$. Using eqn. 6.21 and the asymptotic form of the Hankel function as ρ approaches ∞, eqn. 6.23 reduces to

$$D' = \frac{\left|\sum_{n=-\infty}^{\infty} j^n a_n\right|^2}{\sum_{n=-\infty}^{\infty} |a_n|^2} \tag{6.24}$$

For physical reasons, it is desirable to restrict the allowable form for E in eqn. 6.21 so that the antenna does not supergain.[6] This is accomplished by allowing the sum in eqn. 6.21 to contain terms only of order $N < k\rho_0$ so as to eliminate those terms for which the wave impedance is highly reactive. These latter waves lead to antenna excitations which are impractical, if not impossible, to realise physically because a large amount of energy must be stored in the antenna's near-field in order to produce the desired far field. Truncating the series representation accordingly gives

$$D' = \frac{\left|\sum_{n=-N}^{N} j^n a_n\right|^2}{\sum_{n=-N}^{N} |a_n|^2} \tag{6.25}$$

where N = integer value of $k\rho$.

In accordance with the assumed scenario, it is necessary to maximise D' in eqn. 6.25 subject to the constraint that the directivity is zero in M specified directions $\theta = \theta_m$, $m = 1, \ldots, M$. The maximisation can be carried out by employing the techniques of variational analysis. Briefly, we form the function[15]

$$F[a_n] = \left|\sum_{n=-N}^{N} j^n a_n\right|^2 + \lambda_0 \sum_{n=-N}^{N} |a_n|^2 + \sum_{m=1}^{M} \lambda_m \sum_{n=-N}^{N} a_n j^n e^{+jn\theta_m} \tag{6.26}$$

subject to the constraints $E(\theta_m) = 0$, $m = 1, \ldots, M$, and maximise it with respect to the $\{a_n\}$. The $\lambda_0, \ldots, \lambda_M$ are Lagrange multipliers introduced to allow for the M constraints imposed on the solution. Making the appropriate substitutions and maximising eqn. 6.26, with respect to the α_n, one finds that the radiated electric field which maximises eqn. 6.25 takes the form

$$E(\theta) = A\left\{E_0(\theta) - \sum_{m=1}^{M} \alpha_m E_0(\theta - \theta_m)\mathbf{a}_z\right\} \tag{6.27}$$

where A is independent of θ and z, and $E_0(\theta)$ is the maximum directivity radiation pattern, given by

$$E_0(\theta) = \frac{1}{2N+1} \sum_{n=-N}^{N} e^{+jn\theta}; \tag{6.28}$$

that is the $\alpha_n = j^{-n}$.

The α_m are selected to satisfy the M constraints $E(\theta_m) = 0$, $m = 1, \ldots, M$. Using eqn. 6.28 in eqn. 6.27, the radiation pattern with prescribed nulls at $\theta = \theta_m$ can be expressed as

$$E(\theta) = \frac{A}{2N+1} \sum_{n=-N}^{N} \left(1 - \sum_{m=1}^{M} \alpha_m e^{-jn\theta_m}\right) e^{jn\theta}\, a_z \tag{6.29}$$

$$= \sum_{n=-N}^{N} a_n e^{jn\theta} \tag{6.29a}$$

and the α_m are given by the matrix equation

$$[\alpha_m] = [E_0(\theta_m - \theta_n)]^{-1}[E_0(\theta_n)] \tag{6.30}$$

where $[\alpha_m]$ and $[E_0(\theta_n)]$ are column matrices of the α_m and $E_0(\theta_n)$, respectively, and $[E_0(\theta_n - \theta_m)]$ is the square matrix obtained by letting n correspond to row and m to column in the matrix. M is the number of prescribed nulls. For $M = 1$,

$$\alpha_1 = E_0(\theta_1) \tag{6.31}$$

For $M = 2$,

$$\alpha_1 = \frac{E_0(\theta_1) - E_0(\theta_2)E_0(\theta_1 - \theta_2)}{1 - E_0^2(\theta_1 - \theta_2)} \tag{6.32}$$

$$\alpha_2 = \frac{E_0(\theta_2) - E_0(\theta_1)E_0(\theta_1 - \theta_2)}{1 - E_0^2(\theta_1 - \theta_2)} \tag{6.33}$$

It is possible to further simplify eqn. 6.29*a* by combining terms of positive and negative values of n. Since $E(\theta)$ is a real function, α'_n equals complex conjugate of α'_{-n} and

$$E(\theta) = \sum_{n=0}^{N} \epsilon_n |a'_n| \cos(n\theta + \Psi_n) \tag{6.34}$$

where $\Psi_n = \tan^{-1}(a_n^i / a_n^r)$ is the phase angle of a'_n. Antenna directivity is given by

$$D' = \frac{\left| \sum_{n=0}^{N} \epsilon_n |a'_n| \cos \Psi_n \right|^2}{\sum_{n=0}^{N} \epsilon |a'_n|^2} \tag{6.35}$$

where $\epsilon_n = 1$ when $n = 0$ and 2 for all other values of n.

Next it is important to develop a suitable angular normalization and calculate D' as the angular direction θ_1, to an interfering source, approaches $\theta_s = 0$. Specifically, the physical limitations imposed by truncating eqn. 6.24 results in $\rho_o = N\lambda/2\pi$. Since the antenna 'aperture' equals $2\rho_o$, the half-power beamwidth, corresponding to a uniform aperture illumination, can be approximated as

$$\text{HPBW} \approx \frac{\lambda}{2\rho_o} = \frac{\pi}{N} \tag{6.36}$$

Furthermore, when D' is maximised in the absence of any interfering signals the $a_n = j^{-n}$ and the first null in $E(\theta) = E_0(\theta)$ occurs at $\tau = 2\pi/(2N + 1) \approx$ HPBW for large N. Consequently, D' can be calculated using a reasonable value for N, say 20, and expressing the interfering source(s) location θ_m in terms of $\tau \approx$ HPBW.

Assuming $N = 20$, D' was calculated[8] for a single interfering source located at θ_1. When $\theta_1 = \tau$, D_1, is maximum because from eqn. 6.31 $\alpha_1 = 0$. Consequently the loss $\Delta D'$ (expressed in dB) caused by placing a nulling $E_0(\theta)$ at θ_1 is given by

$$\Delta D' = 10 \log D'(\theta_1 - \tau) - 10 \log D'(\theta_1) \tag{6.37}$$

The results obtained from eqn. 6.37 are shown as the pencil beam curve in Fig. 6.10.

Analysis[9] similar to the foregoing extended these results to include a finite-size circular antenna aperture. The circles in Fig. 6.10 indicate the loss in directivity[9] in the direction $\theta = 0$, when a null is placed $\theta = \theta_1$; the results are independent of the direction angle ϕ because of the assumed circular symmetry. The excellent agreement between the results obtained from these two analyses (solid curve and circles in Fig. 6.10) indicates that the loss in directivity of any antenna is equal to or greater than that given in Fig. 6.10 when the angular separation between a null and the maximum of radiation pattern equals θ_1 expressed in HPBWs.

The foregoing assumed that, in the absence of an interfering source, the radiation pattern is a maximum pencil-beam with HPBW $\sim \lambda/D$. For a communication satellite antenna, this 'null free' radiation pattern is often much broader than the maximum gain pencil beam. For example, the 'earth coverage' pattern from a geostationary satellite is uniform over the earth disc in order to provide the same antenna gain to all potential users in view of the satellite. In order to maximise the antenna directivity, the latter is reduced toward zero at all directions away from the earth. For this reason, Mayhan[9] calculated the loss in directivity when a null is placed in an earth coverage pattern. The results are shown in Fig. 6.10. It is important to note that by choosing the 'null free' radiation pattern appropriately, say with increased gain in the direction of the null; the loss in directivity of a broad coverage beam can approach that given for the maximum directivity beam shown in Fig. 6.10.

6.2.6 Degrees of freedom

The flexibility of a multibeam antenna often leads one to question the degree to which the radiation pattern can be shaped. In the previous Section fundamental limitations of pattern shaping were addressed in the form of pattern 'roll-off' and nulling resolution. As in any system where performance is controlled by a set of system parameters, it is common to express system capability in terms of the available degrees of freedom (DOF). Although F, the number of DOF of an MBA, indicates a relative potential of the system to shape its radiation pattern, the MBA with the largest F is not necessarily best for a given application. In this Section, a general interpretation of F will be discussed with respect to potential shaping performance.

Considering an M port MBA receiving signals from its field of view, each of the M complex weights, used to vary the phase and amplitude of the signals received at each port, represent a DOF. However, if the signals received at the mth port through the corresponding mth beam are much smaller than the signals received at the other $M-1$ ports, the mth weight will have little effect in shaping the pattern. Therefore, the M-port MBA has potentially M degrees of freedom, but, because of this specific scenario, only $M-1$ ports contribute significantly to the signals received by the MBA. Hence, for the foregoing scenario, the MBA has $M-1$, not M, available DOF.

Still further consider the transmitting MBA with M beams with radiation pattern $\Psi_m(\theta, \phi)$. The excitation α_m of these beams represent the M potential DOF. The radiation pattern $E(\theta, \phi)$ for the MBA can be expressed as

$$E(\theta, \phi) = \sum_{m=1}^{M} \alpha_m \Psi_m \tag{6.38}$$

It is possible to specify E at M points in the far zone of the antenna (i.e. E_1, $E_2, \ldots, E_m$) and determine the α_m that will produce the specified set of E_1 through E_m. For example writing eqn. 6.38 in matrix form, we have

$$E = \Psi A \tag{6.39}$$

where E is the column matrix of the $E(\theta_r, \phi_r)$; Ψ is the square matrix with elements

$$\Psi_{mr} = \Psi_m(\theta_r, \phi_r) \tag{6.40}$$

and A is a column matrix of the α_m. The α_m can be determined from

$$A = \Psi^{-1}E \tag{6.41}$$

The existence of a solution for the α_m depends on the character of Ψ^{-1}. Specifically Ψ^{-1} must have a non-zero determinant; this in turn requires that the Ψ_m are linearly independent. Since Ψ_m are the measured or calculated radiation patterns of an MBA, it is virtually impossible for Ψ^{-1} to be singular (e.g. that the determinant of $\Psi = 0$) so long as the θ_r, ϕ_r pairs define M different points in space. Clearly installation of the α_m will produce the specified set of E_r. Unfortunately this pattern synthesis method permits E to take on any value at points other than the set of θ_r, ϕ_r points. Furthermore, specification of a physically 'unrealisable' set of E_r will result in a corresponding set of α_m that are impractical, or virtually impossible to install.

The foregoing discussion is presented to demonstrate the inadequacy of relating potential DOF of an MBA to its expected performance. However, the available DOF can be used as a measure of the pattern-shaping potential of an MBA system. In other words, the number M of beam ports indicates the potential DOF and pattern shaping capability of an MBA; whereas the number of beams F that defines the FOV within which the specified radiation pattern is of interest, tends to indicate the number of available DOF. In most applications of an MBA, the number of available DOF tends to equal F. Unfortunately, availability of a DOF can vary qualitatively from unavailable to available, leaving the antenna designer unable to draw any quantitative conclusions that depend on M or F.

It is the author's experience that the use of DOF to describe an MBA performance represents, at best, the designers subjective view of the MBA system potential capability. The number of DOF usually portrays a conceptual view of the MBA performance and possibly the designer's confirmation that the beams encompass only the FOV of interest. That is, essentially all radiation patterns the MBA is designed to produce require non-zero excitation of F of the M beams. Even in those cases where a null in the "null free' radiation pattern is specified it is very unlikely that the excitation of a beam port will vanish. Rather, it is likely that at least a small but non-trivial excitation is required to produce a null in the radiation pattern.

6.3 Beam-forming network (BFN)

The true versatility and uniqueness of an MBA system manifests itself in the capability of the BFN. Some MBA systems use a fixed beam-forming network, and consequently have a fixed antenna pattern suited for a particular application. Still others use a BFN capable of weighting the signals received at each port over a wide range of relative amplitude and phase. Hence the associated antenna pattern can take on literally millions of different shapes; virtually any shape pattern the aperture of the MBA is capable of forming. In this Section, two BFNs are described, some operational characteristics are presented, and a special switching algorithm is outlined.

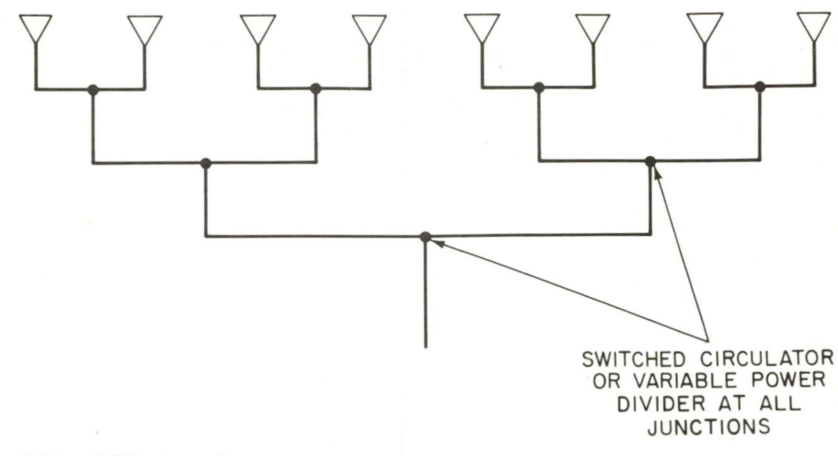

Fig. 6.12 *BFN schematic*

A commonly used BFN uses a corporate tree configuration similar to that shown in Fig. 6.12. Power division at each junction can be fixed (as with a power dividing waveguide 'Tee'), it can have a binary characteristic (as with a wideguide circulator switch), or it can have many controlled power division states (as with a variable power dividing junction). A BFN using fixed power-dividing junctions is comparatively straightforward, and the associated MBA has the potential to produce any pattern shape within the fundamental limits of its aperture size. Discussion here will be limited to the switchable and variable BFNs.

6.3.1 BFN with circulator switches

Waveguide circulator switches have been in the antenna designer's 'bag of tricks' for more than three decades. They are a three-port device (see Fig. 6.13*a*) with a symmetrically shaped ferrite rod placed at the centre of a symmetrical waveguide *H*-plane junction. For magnetisation of the ferrite in one direction, signals incident on port *A* arrive at port *B* with approximately 1/4 dB decrease in amplitude. Simultaneously these signals arrive at port *C* with ~ 20 dB decrease in amplitude.

Reversing the magnetisation of the ferrite interchanges the roles of port B and port C; that is, essentially all the signals arrive at port C. Appropriate setting of one switch in each level of the tree connects the input (or output) port to a single beam port of the MBA and in effect disconnects all other beam ports from the input port. The function of this device is often represented schematically as shown in Fig. 6.13*b*. The circular arrow indicates the flow of signals around the 'ring' connecting the ports. Reversing the magnetisation of the ferrite reverses the sense (direction) of the arrow.

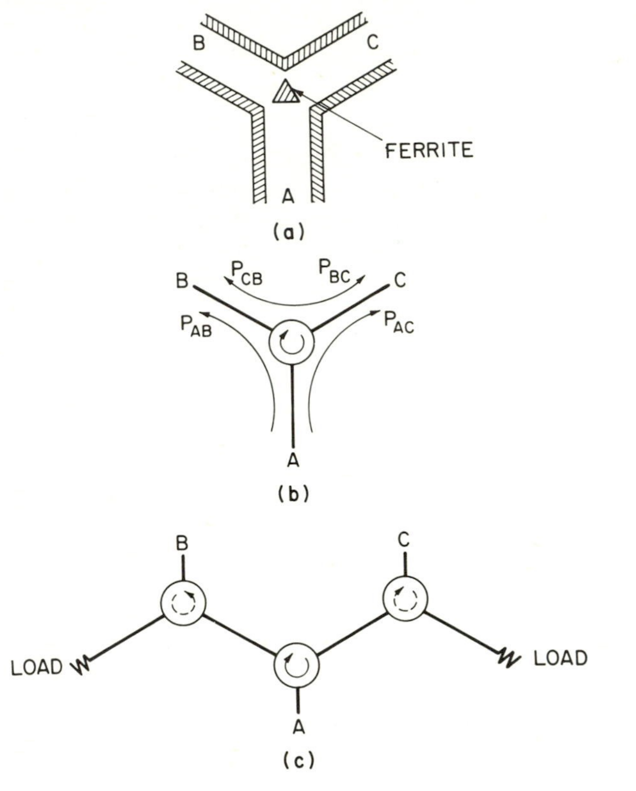

Fig. 6.13 *Three-port isolator switch*

A BFN using circular switches introduces a minimum insertion loss usually slightly greater than the combined insertion loss of those switches located in the path between the input and beam ports. Assuming all switches are identical, the insertion loss L_s of a BFN with M beam ports has the minimum value $L_s > \alpha p$ where α is the insertion loss of each switch and $p = \log_2 M$. For frequencies near 10 GHz $\alpha \sim 0.25$ dB and for a 16-port BFN, $L_s > 1$ dB.

Although low insertion loss is usually a design goal, isolation of 'off' beams often requires the most attention. Referring to Fig. 6.13*b*, note that power flowing from A to C, P_{AC}, combines with power P_{BC} reflected from the load on port B. Assuming the circulator has 20 dB isolation, and neglecting the insertion loss between port A and port B, the reflected power from a load at port B, with impedance Z_B, will cause $|P_{BC}| = |P_{AC}|$, if $|Z_B| = 1.2Z_0$, where Z_0 is the characteristic impedance of the waveguide forming the circulator. If the voltages associated with P_{AC} and P_{BC} are in phase, the isolation between port A and port C will decrease from 20 dB to 14 dB! If they are 180° out of phase, the isolation will increase theoretically to ∞ dB. Sometimes it is possible to locate the load, or successive switches, so as to produce the 'out of phase' performance and realise high isolation. However, when this is impossible, additional fixed circulators can be added as indicated in Fig. 6.13*c*. The fixed circulators are positioned to maximise the circulator switch isolation; then the load Z_B etc. can be located any distance from the junction because power reflected from Z_B is essentially absorbed in the load attached to the fixed circulators. Although this increases the number of circulator switches in the BFN from $M - 1$ to $3(M - 1)$, only one third of the switches requires control circuitry. Furthermore, the fixed circulator switches can have a lower insertion loss because the ferrite and its magnetising (or keeper) circuit have fewer design and manufacturing constraints than the circulator switches.

Isolation can be increased by adding two or more switching circulators in series. In particular, if all the isolator junctions shown in Fig. 6.13*c* are switchable (that is the sense of the arrows in all three switches can be controlled), simultaneous switching of three junctions can yield >40 dB isolation between the input and the output ports. For example, assume that signals are incident at port A (Fig. 6.13*c*) and that the switches at ports B and C are magnetised in the same direction; that is, the sense of both dashed arrows is as shown for the switch at port B. The input signals travel to port B with $\sim 1/2$ dB reduction in amplitude. The insertion loss between ports A and C will be greater than the minimum isolation of each switch; that is, for units of this type, the isolation will be greater than $14 + 14 = 28$ dB. Typically, this isolation will be greater than ~ 34 dB because the loads will have a VSWR ~ 1.01 and the corresponding reflection at port B will have a relatively small effect on signals arriving at port C. Reversing the magnetisation of all three switches connects port A to C with $\sim 1/2$ dB insertion loss and introduces ~ 34 dB insertion loss between port A and B.

It is also true that if the switches at ports B and C are magnetised in the opposite sense to that shown, the signal out of either port B or port C will be reduced more than 14 dB below the input signal! Consequently, a BFN (Fig. 6.12) will be able to 'turn off' any port by providing more than 34 dB insertion loss between the input port and the beam port if three circulator switches (as in Fig. 6.14*c*) are installed at those junctions, of the corporate tree, that are connected to the beam ports. Only a single circulator switch is required at the remaining junctions in the corporate tree of the BFN.

Accurate prediction of BFN performance is possible given a specific configuration using circulator switches with known transfer characteristics. Since many scenarios must be considered for even a four-port BFN, it is common to calculate the expected performance on a statistical basis. Considering only the worst cases, if they are known, may only yield a bound on the expected performance rather than an estimate of the most likely performance. In any case, further discussion of this type of BFN is beyond the scope of this chapter.

6.3.2 BFN with variable powered dividers

Variable power divider junctions (VPD) consist of a 3 dB hybrid coupler, two nonreciprocal latching ferrite phase shifters, and a 'magic tee' with a matched waveguide load on its 'difference' port. In the following, these components are discussed separately and then as a complete assembly.

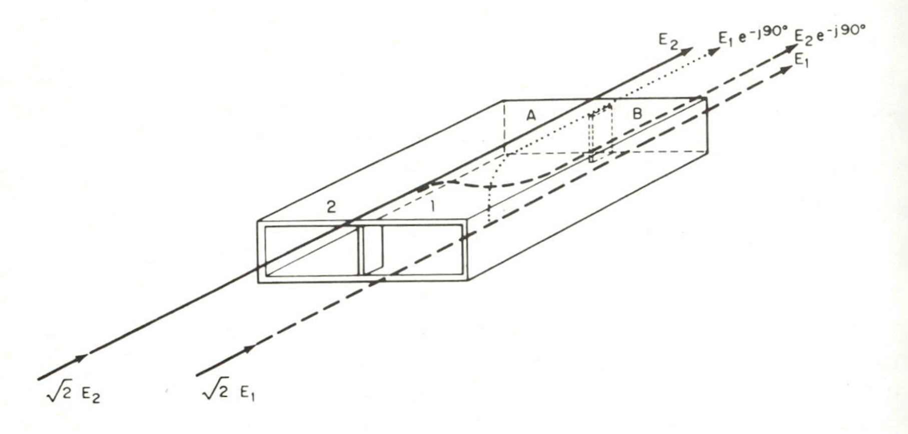

Fig. 6.14 *3 dB hybrid coupler*

The 3 dB hybrid coupler is shown in Fig. 6.14. Understanding its properties i the key to understanding the VPD. Basically, it is made of two standard rectangula waveguides which have a common 'narrow', or side, wall. Energy travelling in on guide is coupled to the other guide by removing the common wall between guide for an appropriate length along their axes. Small sections of the common wall ar shown (Fig. 6.14) at the input and output ends of the device.

Consider first a wave incident on port 1; energy contained in this wave i divided equally between ports A and B and essentially none appears at port 2 Signals arriving at ports A and B are in phase quadrature, with the signal at port lagging that at port B as indicated. Due to the symmetry of the junction, energ in a wave incident on port 2 will also divide equally between A and B; however, th signal at port B will lag that at port A by 90°. It is important to note that ports and B are isolated; hence the load on port A does not affect the power delivered t port B and vice versa.

Assume that simultaneously signals at the same frequency and of $\sqrt{2E_1}$ an $\sqrt{2E_2}$ amplitude are incident on ports 1 and 2, respectively. If the signals are phase and have equal amplitude, the signals at port A and B will also be in phas and have an amplitude of $\sqrt{2E_1}$. If the signal at port 1 leads that at port 2 b 90°, the two signals arriving at port B are 180° out of phase, and hence vanis at port A they add in phase to produce an amplitude $2E_1$. Similarly, if the signal at port 2 leads that at port 1 by 90°, there is no output at port A, and all the energy exits from port B. Suffice it to say that appropriate choice of the phase difference, in the range ± 90°, between signals at ports 1 and 2 will deliver any fraction of the total input power to port A or port B, and the sum of the output power will be equal to the total input power minus any power dissipated in the coupler or reflected from ports A and B. Typically, the power dissipated in the coupler (i.e. its insertion loss) is approximately 0.05 dB at 8 GHz.

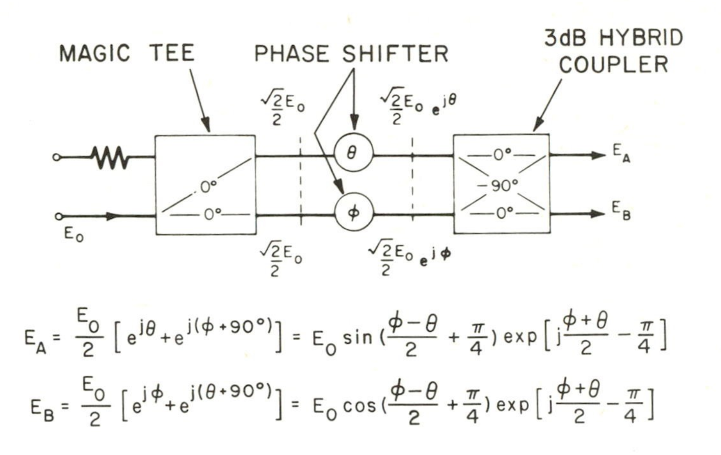

Fig. 6.15 *Variable power divider*

The 'magic tee' is a four-port device customarily referred to as having input 'sum' and 'difference' ports and two 'output' ports. Signals incident on the sum port divide equally between the output ports, and these output signals are in phase. Signals incident on the difference port also divide equally between output ports, but the output signals are 180° out of phase. This device is (as is the 3 dB hybrid coupler) reciprocal, linear and bilateral; consequently, equal-amplitude signals incident on the output ports appear only at the sum port if they are in phase and at the difference port only if they are 180° out of phase. If this device is connected to the 3 dB hybrid coupler, through identical phase shifters as indicated schematically in Fig. 6.15, signals E_0 incident on the sum port of the magic tee will divide between the output ports (i.e. ports A and B) in accordance with ϕ and θ the insertion phase shift of the phase shifters. The insertion phase of E_A and E_B depends on the sum of ϕ and θ the insertion phase of the phase shifters. Keeping the sum, $\phi + \theta$, constant, while varying the difference, $\phi - \theta$, varies E_A/E_B while holding the insertion phase constant. The following expressions indicate the relationship between power division and ϕ and θ:

$$P_A = P_0 \sin^2\left(\frac{\phi - \theta}{2} + \frac{\pi}{4}\right) \tag{6.42}$$

$$P_B = P_0 \cos^2\left(\frac{\phi - \theta}{2} + \frac{\pi}{4}\right) \tag{6.43}$$

Variations of θ and ϕ can be obtained by the use of latching ferrite phase shifters. These often consist of a ferrite toroid mounted in the centre of a rectangular waveguide as indicated in Fig. 6.16. The state of magnetisation of the ferrite controls the insertion phase of signals passing through it. A single-turn control wire determines the magnetic flux density in the ferrite. The wire enters the waveguide parallel to the broad wall, and hence has negligible interaction with the waveguide fields. To ensure that essentially all of the RF energy passes through the ferrite toroid, matching end sections are used to provide matched coupling to the unloaded waveguide; this reduces the insertion loss of the device and provides essentially reflection-free performance. The desired insertion phase is obtained by first driving the ferrite into saturation. This establishes a 'calibration' or 'set' condition; a second drive pulse of controlled width (i.e. length of time) and amplitude is applied to change the flux an amount $\Delta\Psi$ and establishes the desired remnant flux in the ferrite or 'reset' condition.

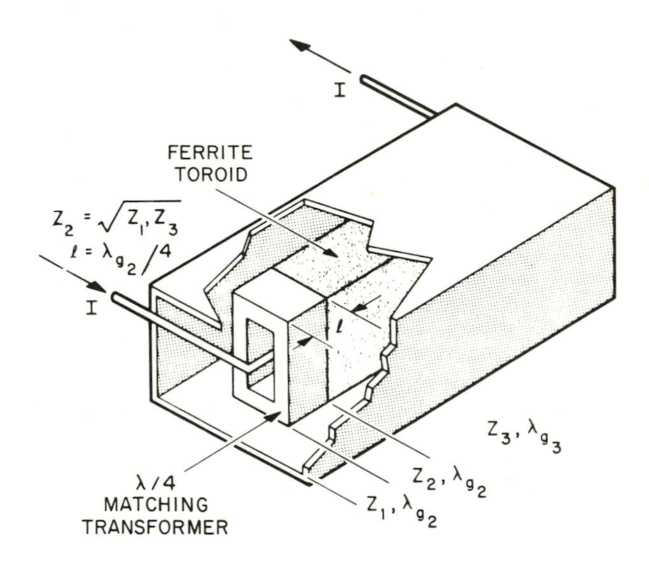

Fig. 6.16 *Latching ferrite phase shifter*

6.3.3 *Switching algorithm*

There is a definite convenience, simplicity and inherent accuracy associated with latching ferrite phase shifters whose phase is set by first placing the ferrite in a saturated remnant magnetisation state and then applying a controlled flux drive as described in the foregoing paragraph. However, use of this type of phase shifter in a variable power divider of a BFN changes the excitation (or sum of received signals) of the beam ports on an MBA. This change is, in general, different than either the initial, or final, excitation desired in an intended reconfiguration of an MBA. To illustrate this, consider the case of an MBA transmitting signals in accordance with the BFN schematically represented in Fig. 6.17. A four-port device is represented with the power (in watts) out of each port (i.e. A through D) represented by numbers shown immediately to the right of each port. The upper numbers represent the initial distribution; the lower numbers represent the final distribution of the total input power P_i. For convenience, P_i is set equal to one watt. Numbers adjacent to each variable power divider indicate the associated power division for the initial (upper number) and final (lower number) BFN configurations.

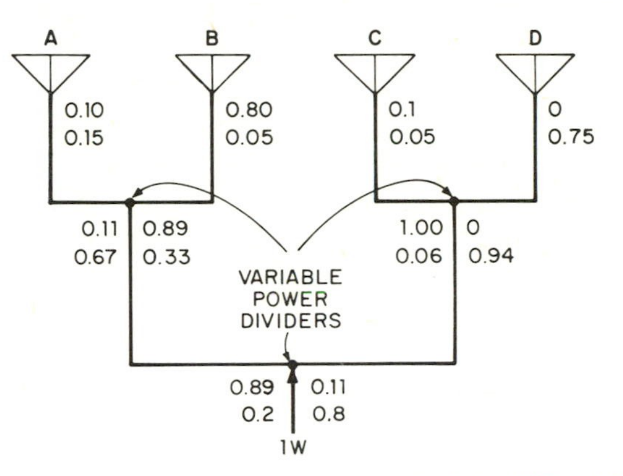

INITIAL DIST. — UPPER NO.
FINAL DIST. — LOWER NO.

		POWER (W)			
LEVEL	PORT	A	B	C	D
	INITIAL	0.10	0.80	0.1	0.0
2	SET	0.45	0.45	0.05	0.05
2	RESET	0.675	0.225	0.007	0.093
1	SET	0.375	0.125	0.031	0.469
	FINAL	0.15	0.05	0.05	0.75

ALL NUMBERS REPRESENT WATTS

Fig. 6.17 *Beam-forming network power distribution*

If the entire BFN was configured by simultaneously 'setting' and 'resetting' the VPDs, the power delivered to beam port B would decrease from 0.8 to 0.25 when the phase shifters in each VPD are 'set' prior to resetting them. This is because during the set operation, the VPD divides the power at its input port *equally* between its output ports (i.e. $\phi = \theta$). This 5 dB reduction in the power delivered to port B may introduce an undesirable systematic transient performance. Should this be the case, an alternative switching algorithm developed by DeSize and Simmons[2] may be used. Their algorithm guarantees that the power delivered to any beam port will always be greater than -3 dB referred to the power required at that port for either the final, or initial, power distribution. The -3 dB constraint is necessary to allow for the equal power distribution that results when the phase shifters are being set. Note that the set operation also changes the insertion phase $(\theta + \phi)$ of the VPD; this can be compensated by an external phase shifter that is switched in synchronism with the VPD.

Alternatively, one may wish to switch each level of the BFN in sequence beginning with that level closest to the beam ports. The power at the beam ports is indicated in Fig. 6.17 for each sequence in the procedure. That is, the VPDs in the 2nd level (i.e. the level closest to the beam ports) are set, then they are reset, etc. Note that the power delivered to port C drops 8.5 dB below its final value and 11 dB below the initial value when the 2nd level VPDs are reset. In order to avoid this decrease in power, the DeSize–Simmons algorithm may be used; proof of its validity can be obtained from their report.[2] The procedure is straightforward and is for a p-level BFN where $p = \log_2 M$ and M equals the number of beam ports. It requires that all VPDs in a level be set simultaneously and that the pth level VPDs are set first, then the $(p-1)$th are set, and so on until the VPD at the input to the BFN is set and reset. At the pth through 1st levels, each VPD is reset to equal power division unless the total power into the VPD is less for the initial than for the final distribution of power among the beam ports. After setting and resetting the first level VPD, the VPDs in levels 2 through p are again set and then reset to the value required by the final distribution. Note that this algorithm requires all but the first level VPD to be set and reset twice. Furthermore, for a VPD that can be switched directly from one state to another, the DeSize–Simmons algorithm guarantees that the power delivered to any beam will always be greater than or equal to the power delivered to that beam port for either the initial or final beam power distribution.

Using the DeSize–Simmons algorithm, the power distribution variation for the BFN considered in Fig. 6.17 is summarised in Table 6.1

Table 6.1

Level	Operation	A	B	C	D
	Initial	0.10	0.8	0.1	0.0
2	Set	0.45	0.45	0.05	0.05
2	Reset	0.675	0.225	0.05	0.05
1	Set	0.375	0.125	0.25	0.25
1	Reset	0.15	0.05	0.4	0.4
2	Set	0.1	0.1	0.4	0.4
2	Reset	0.15	0.05	0.05	0.75

Note that the power out of a port is never less than either the initial or the final values; whichever is the smaller. Note the power out of port A falls only 1.7 dB below its final value and is always greater than its initial value. The power at port B decreases smoothly from its initial value 0.8 to its final value 0.05 except for the 3 dB increase to 0.1 in the set operation preceding the final reset operation.

Similarly the power at port D increases smoothly from the initial value to the final value. Only the power at port C undergoes rather large variation from either its initial or final values, but it is always greater than either of the latter.

6.3.4 BFN with VPD and variable phase shifters

In the previous Section, the BFN described could vary principally the amplitude of the excitation (or received signal) at the beam ports. Any change in phase was due primarily to imperfections in the devices, and most likely not within operational control of the MBA system. Even so, exceptional good pattern shaping results. However, when it is desired to produce a good beam shape over a wide angle of scan, or produce a deep null in the radiation pattern, it may be necessary to vary both the amplitude and the phase of the excitation at the beam ports. Variable phase shifter (VPS) similar to those used in the VPD described in the previous Section and shown in Fig. 6.16 can be used if insertion phase shift less than 90° is required. When it is necessary, these devices can, with appropriate design, provide insertion phase shift in excess of 360°. However, a VPS with an insertion phase variation between 0° and 360° is perhaps adequate for most applications. Furthermore, ferrite phase shifters are discussed here primarily because at frequencies higher than, say, 5 GHz, their high power-handling capability and low insertion-loss properties make them potentially more desirable than the *p–i–n* diode or other type VPS devices. Clearly, other VPD devices could also be used; however, further discussion of these devices is beyond the scope of this chapter.

6.3.5 Other types of beam-forming networks

Often the MBA designer is asked to provide independent beam-forming operation for two or more transmitters or receivers. Given sufficient space, weight and cost allocation, this requirement can be satisfied by constructing a separate MBA for each transmitter. However, taking advantage of the isolation obtained with dual polarised antennas, the MBA can 'reuse' not only the same frequency band but also the same MBA. Specifically, if the feed corresponding to each beam has two ports, one for say vertical polarisation and one for horizontal polarisation, a BFN can be connected to each of these 'orthogonal' ports, resulting in two independent input ports. Two transmitters operating over the same frequency band could excite these BFNs and produce completely different radiation patterns while using the same MBA.

Placing the foregoing in sharper focus, consider the MBA schematically represented in Fig. 6.17. If the feedhorns are linearly polarized, with the plane of polarisation parallel to the page, essentially all of the radiated field is also linearly polarised in the same direction. Although they may be negligible, cross-polarised fields will exist as determined by the inherent polarisation purity of MBA.[4] Since polarisation of the feedhorn depends on the manner in which it is excited, a probe exciter parallel to the page and located in the throat of the feedhorn will result in the radiated field described in the foregoing. Orienting a second probe perpendicular to the page and locating it in the throat of the feedhorn will result in a radiated field linearly polarised perpendicular to page. These orthogonally polarised fields can be produced by independent excitation of the feedhorn probes. The probes can be excited by separate transmitters through a BFN attached to one set of probes and a BFN attached to the set of orthogonally polarised probes.

6.4 Pattern synthesis

Variable pattern shaping is undoubtedly the most useful characteristic of an MBA system. There are many techniques available for shaping the radiation pattern, but all must determine the beam excitation or weight coefficients a_m for a transmitting or receiving system, respectively. It is instructive to consider at least one class of beam shaping. Specifically, sidelobe level reduction will be demonstrated.

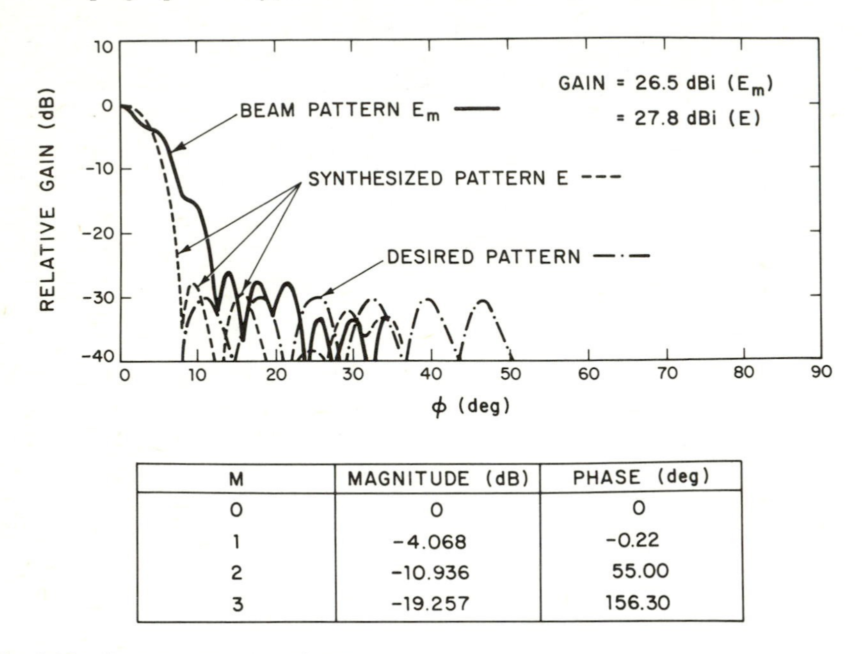

M	MAGNITUDE (dB)	PHASE (deg)
0	0	0
1	-4.068	-0.22
2	-10.936	55.00
3	-19.257	156.30

Fig. 6.18 *Pattern-synthesis performance*

Assume a wide-angle scanning lens antenna where the pattern $E_m(\phi)$ obtained by exciting the mth port of the MBA feedhorn array is as indicated in Fig. 6.18. Note the distorted beam peak, shoulders and high sidelobes. Assume that this beam shape is identical to that of the three adjacent beams on either side and in that plane in which the pattern shaping will be carried out. With an 8.5λ antenna aperture and an angular separation between beams ($\Delta\phi$) equal to $3.5°$, let us determine the excitation coefficients of these seven beams that will provide a single beam $E_0(\phi)$ with a $6°$ HPBW and a 30 dB sidelobe level. Specifically, it is desired that the MBA radiation pattern

$$E(\phi) = \sum_{m=-3}^{3} a_m E_m(\phi - m\Delta\phi) \tag{6.44}$$

approximates $E_0(\phi)$ within an acceptable error. The a_m are complex with real a_{m1} and a_{m2} parts. Toward this end, $E_0(\phi)$ and $E_m(\phi)$ are expanded in Fourier series as

$$E_0(\phi) = \sum_{i=0}^{I} \epsilon_i b_i \cos i\phi \tag{6.45}$$

and

$$E_m(\phi) = \sum_{j=0}^{J} \epsilon_j c_j \cos j\theta \tag{6.46}$$

where $\epsilon_i = 1$ when $i = 0$ and $\epsilon_1 = 2$ when $i > 0$. Next the error between the desired and MBA patterns is formed as

$$\epsilon = E_0 - E \tag{6.47}$$

the synthesis procedure is based on minimising

$$\sigma = \int_0^{2\pi} \epsilon \cdot \epsilon^* \, d\phi \tag{6.48}$$

where the asterisk indicates the complex conjugate. After making the appropriate substitutions, the differentials $\partial\sigma/\partial a_{m1}$, $\partial\sigma/\partial a_{m2}$ (for $m = -3, -2, \ldots, 2, 3$) are formed and set equal to zero. This results in a set of equations in the excitation coefficients a_m (i.e. seven equations in a_{m1} and seven equations in a_{m2}).

The foregoing set of equations can be written in the matrix form:

$$D = C * A \tag{6.49}$$

where C is a 7 x 7 matrix with elements

$$C_{1,m} = \sum_{j=0}^{J} \epsilon_j e_m |C_j|^2 \cos(j1\,\Delta\phi) \cos(jm\,\Delta\phi) \tag{6.50}$$

A is a column matrix of the a_m and D is a column matrix with elements

$$d_1 = \sum_{j=0}^{J} \epsilon_j b_j C_j^* \cos(j1\,\Delta\phi) \tag{6.51}$$

In the example under consideration, $J = 110$, and the desired pattern was chosen as

$$E_0 = \cos\left(\frac{\pi\phi}{2\xi}\right) \qquad \text{for } -\xi < \phi < \xi \tag{6.52a}$$

$$E_0 = E_x \sin\left(\frac{\pi\theta}{\xi}\right) \qquad \text{for } \left.\begin{matrix} \xi < \phi < \theta_s \\ -\theta_s < \phi < -\xi \end{matrix}\right\} \tag{6.53b}$$

$$E_0 = E_s\left(\frac{\phi_s}{\phi}\right) \sin\left(\frac{\pi\phi}{\xi}\right) \qquad \text{for } \left.\begin{matrix} \theta > \theta_s \\ \theta < -\theta_s \end{matrix}\right\} \tag{6.54c}$$

The calculated radiation pattern, obtained when the seven adjacent beams are excited as determined by the foregoing procedure, is shown in Fig. 6.18. Of particular note is the reduced sidelobe level and the improved beam shape. The desired radiation pattern $E_0(\theta)$ is also shown in Fig. 6.18 along with a tabulation of the a_m excitation coefficients.

The foregoing analysis demonstrates the estimated MBA performance when sidelobe reduction and beam shaping are emphasised. The case considered produced a pencil beam pointing along the axis of one of the individual beams of the MBA. Similar results can be obtained if the synthesised beam $E(\phi)$ is not coincident with the axis of a fundamental beam.

6.5 Performance evaluation

The commonly used parabolic reflector illuminated by a single feed has a radiation pattern resembling a classical pencil beam with sidelobe structure. Evaluation of its performance characteristics is relatively straightforward and often presented in terms of the radiation pattern half-power beamwidth and near-in sidelobe level, the antenna gain and input impedance. Multiple-beam antennas are substantially different and require a not so straightforward method of performance evaluation. Although a general performance evaluation procedure applicable to all MBA systems is probably not possible, a statistical procedure has been developed[12] for communication antennas employing adaptive nulling. An appropriately modified version of this method is probably suitable for almost any MBA or other antenna system. The following is a description of the method and an example of its use.

Performance of an adaptive antenna operating in a communication system is judged by the ratio of the directive gain D_d, in the direction of the desired signal S_d source (or sources), to the directive gain D_u in the direction of the undesired signal S_u sources. The system designer needs to know the ratio D_d/D_u so that he may set S_d large enough to guarantee uninterrupted communication even in the presence of undesired and/or unexpected interference. Unfortunately, it is not always possible to specify the location and strength of all S_u; consequently, the system designer might consider the 'worst case' performance characteristics in specifying the antenna performance. This could lead to an antenna system much too large, too complicated, or perhaps even impractical. Hence the system designer must compromise, or trade-off, the system performance. He might do this by specifying the antenna performance on a statistical basis; that is, $D_d/D_u > X$ for $Y\%$ of the communication system operating hours. In short, an intelligent quantitative means of trading off system performance characteristics can be very useful.

In response to the foregoing, consider a figure of merit (FOM), which consists of first dividing the angular field of view (FOV) into a grid of cells as indicated in Fig. 6.19. Next select a particular location and strength of desired and undesired signal sources and allow the adaptive antenna under test to adapt. Then the directive gain* is determined at each cell in the FOV and the cells are grouped according to their angular separation from the undesired signal sources. For example, zone 1 might include all cells within 1° of an undesired signal source S_u, zone 2 would include all cells between 1° and 3° from S_u, zone 3 would include all cells more than 3° from S_u. The values of directive gain would be sorted in accordance with their associated zone. In the case of more than one S_u, a cell may be in more than one zone. In this case, the cell would be assigned to the zone closest to an undesired signal source.

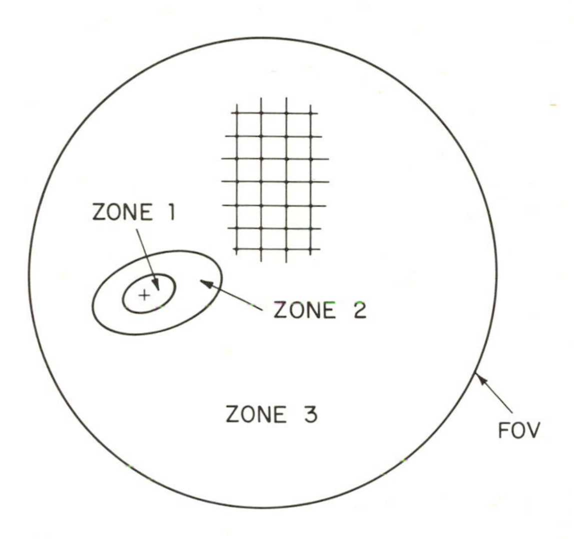

Fig. 6.19 *Figure of merit*

Next a new scenario is selected, the antenna is allowed to adapt, and the directive gain is determined and sorted as described in the foregoing paragraph. This procedure is repeated until all important scenarios have been considered, or a data base, sufficient for statistical analysis, has been accumulated. Using this data base, one can determine the probability of realising D_d/D_u, greater than a selected value, in any of the zones. Of course, the worst and best results could be presented in those cases where the designer wishes to demonstrate a guaranteed performance.

In those instances when the strength of the undesired signal strength is known, the ratio D_d/D_u can be used to determine the effective isotropic radiated power (EIPR) required by the desired signal source to overcome that signal radiated by the undesired signal source. The results of this calculation could be presented as a statistical distribution; that is, what EIRP is required for $S_u/S_d \geqslant A$ with probability X. In those cases where more than one interfering source of different intensity is present, one could determine an effective D_u (i.e. D_u^0) in accordance with

* Since the antenna is receiving signals, the antenna's receiving cross-section A is required. However, for a reciprocal antenna (as is almost always true), $D = 4\pi A/\lambda^2$; hence D will be used instead of receiving cross-section A

$$D_u^0 = \left(\sum_{i=1}^{I} D_i P_i\right) \Big/ \left(\sum_{i=1}^{I} P_i\right) \tag{6.53}$$

6.5.1 *Illustrative example of an FOM*

In order to demonstrate the utility of the proposed FOM, let us consider the following two adaptive antennas (Fig. 6.20):

(i) A hexagonal planar array of seven identical 8 dB gain antennas with an interelement spacing equal to 1 m.
(ii) Same array as in (i) but with an interelement spacing equal to 4 m.

Fig. 6.20 *Array antenna configurations*

We will assume each antenna has the same adaptive algorithm and must operate at 450 MHz, in the presence of two undesired signal sources each in any one of 25 different locations. The desired signal sources are assumed to be anywhere in the FOV. We are in fact attempting to determine the interelement spacing best suited to this particular set of 25 scenarios with all other antenna design parameters held constant. All values of directive gain will be determined analytically assuming phasing and matching errors appropriate to limit the depth of null to a credible value. Array antennas are used primarily for conceptual simplicity; an MBA is equally amenable to this method of analysis.

With the antennas fully adapted to a particular scenario, the directive gain at all cells in the FOV was calculated; the resulting antenna radiation pattern contour plots for array 1 and array 2 are shown in Figs. 6.21 and 6.22. The location of the interfering sources is indicated by a large solid dot on each plot. The FOV is centered on the antenna broadside direction and subtends an angle of 17.3°. The system designer (in our present analysis) assumes that the desired signal sources could be located anywhere in the FOV indicated by the edge of the contour plot.

These results are not too different than one would have expected; that is,

(i) The smaller array has a smooth radiation pattern with null centered on each interfering source.

(ii) The large array provides an irregular, uneven coverage of the FOV but the null on each undesired source is sharper and lesser in extent than that of the smaller array.
(iii) The larger array produces a large directivity (i.e. 12.8 dBi vs. 8.8 dBi).

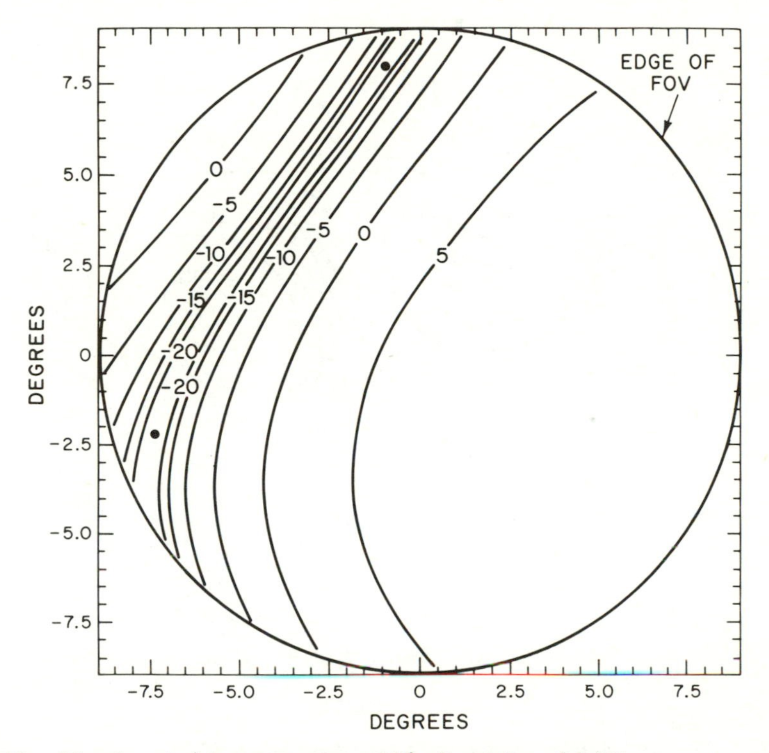

Fig. 6.21 *Directive gain (element spacing = 1 M); directivity = 8.8 dBi*

In short, the larger antenna aperture provides increased resolution of the desired nulls but introduces extra nulls because the interelement spacing is large enough to produce 'grating' lobes and nulls within the FOV. The point of this comparison, however, is that if 0 dBi directive gain is adequate, which antenna gives the best coverage of the FOV. Choice of the best antenna becomes further confused when we compare similar results obtained with a different location of the undesired sources.

The antenna performance associated with many scenarios can be assessed by comparing the FOM for the directive gain of these arrays. In particular, the statistical distribution of directive gain for each array with desired signal sources in three angular zones is shown in Fig. 6.23. Clearly the larger array has about 10 dB more directive gain than the smaller array when the desired and undesired signal sources have an angular separation between 0.5° and 2°. For all other locations of the desired signal source, the antenna directive gain is increased by about 5 dB when the spacing between the elements of the array is increased from 1 m to 4 m. Hence the larger array is best.

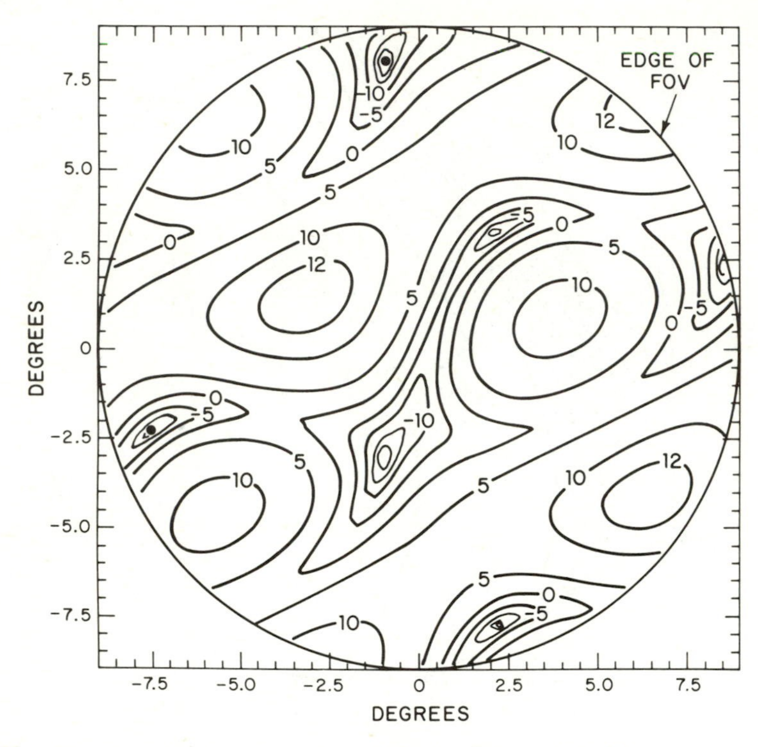

Fig. 6.22 *Directive gain (element spacing = 4M); directivity = 12.8 dBi*

It is also possible to determine the probability of realizing a gain in excess o *A* dB. For example, with the larger array, the directive gain (see Fig. 6.23), tc the desired signal sources located at least 2° from any undesired signal source will exceed 5 dBi (dB referred to as an isotropic radiator) with a probability o 0.56. The same Figure indicates that, for the larger array, the maximum directiv gain to a desired signal source is about 13.5 dBi, and the minimum directive gai is ~ − 30 dBi.

The ratio D_d/D_u may also be of prime importance in determining thos antenna parameters best suited for the desired system performance. Radiatio contour plots do not give this information directly. However, it can be deter mined from the same data base resulting from the foregoing calculations. Th probability of a desired signal source realising $D_d/D_u \geqslant A$ is plotted in Fig. 6.2 for both arrays. Notice that, for events with probability > 0.5, D_d/D_u is abou the same for each antenna array when the desired signal is between 0.5° and 2.0° from the undesired signal. It is also true that D_d/D_u is about 9 dB better for the *smaller* array than for the larger array, when the desired and undesired signal sources have an angular separation > 2.0° ! These conclusions are not intuitively obvious; however, they can be explained by the smooth as opposed to uneven radiation patterns of the small vs the large array.

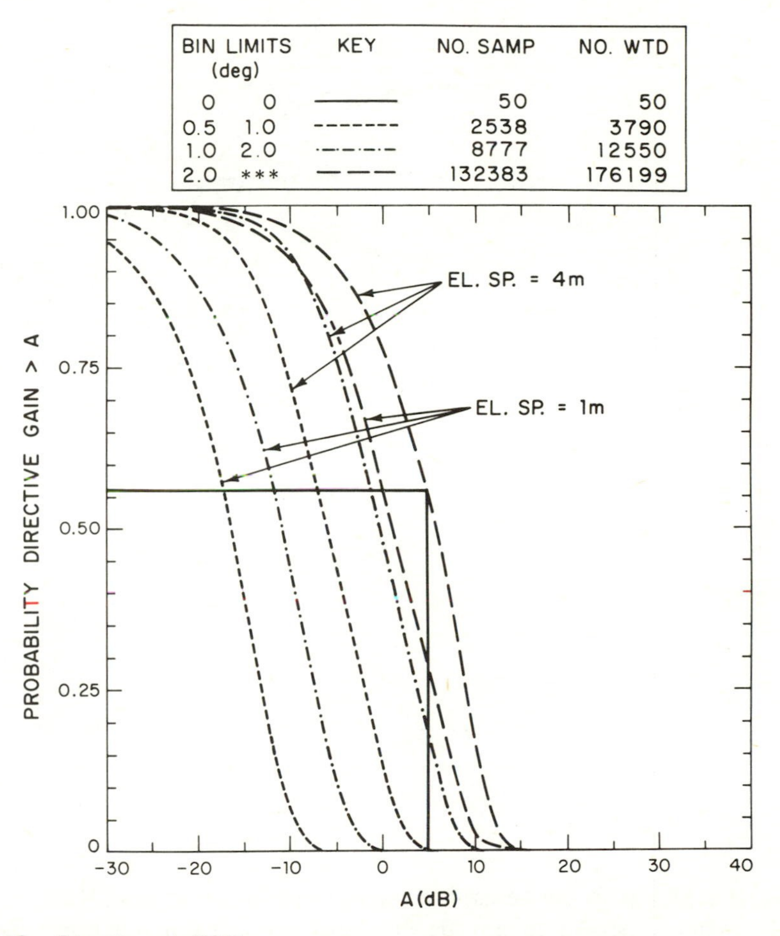

Fig. 6.23 *Directive gain FOM*

From the foregoing, we see that this FOM analysis indicates that the large array is preferred over the smaller array if maximising D_d is of prime importance. However, the smaller array is better if one wishes to maximise D_d/D_u. These results are not intuitively obvious, and it is doubtful that a *visual* inspection of the radiation contour plots for all 25 scenarios would yield the same conclusions.

The 'NO. SAMP' listed at the top of the Figures indicates the number of data points in each zone. The 'NO. WTD' listed indicates the weighting of these points in accordance with their individual representation of a 'cell' on the surface of the earth.

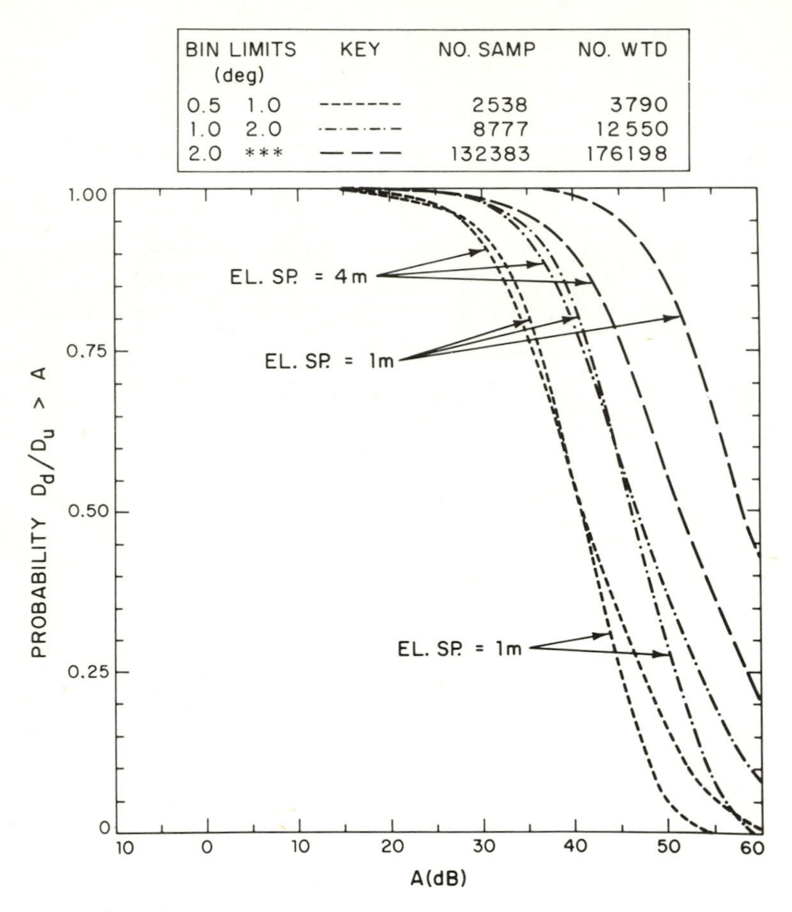

Fig. 6.24 *D_u/D_d FOM*

This same procedure can be used to assess the performance of an MBA, or other antenna, system designed to provide adequate directive gain for many 'typical' distributions of directive gain over the antenna system FOV. The evaluation method strives to examine many possible scenarios, calculate the observables of interest, and present the results with a measured probability of their occurrence. In short it examines all cases and presents the performance achievements in terms of a smooth continuous rating as opposed to the 'go' and 'no go' assessment of specific, or worst, cases. It has the added advantage of objectively assessing vast amounts of performance data as opposed to human 'judgement' of the same data.

6.6 References

1 ABRAMAWITZ, M. and STEGUN, I. A.: 'Handbook of mathematical functions'. Dover Publications, Inc., 1965, p. 370, Section 9.4.6

2 DESIZE, L. K. and SIMMONS, A. J.: 'How to siwtch a beam-forming network with minimum disturbance to existing communication channels'. MIT, Lincoln Laboratory, Technical Note 1976–20, 1976

3 DION, A. R. and RICARDI, L. J.: 'A variable-coverage satellite antenna system', *Proc. IEEE*, **59**, 1971, pp. 252–262

4 DUNCAN, J. W., HAMADA, S. J., and INGERSON, P. G.: 'Dual polarization multiple beam antenna for frequency reuse satellites'. AIAA/CASI 6th Communications Satellite Systems Conference, 1976, Montreal, Canada

5 HARRINGTON, R. F.: 'Time harmonic electric fields'. McGraw-Hill, 1961, p. 233

6 HARRINGTON, R. F.: 'Time-harmonic electromagnetic fields'. McGraw-Hill, New York, 1961, p. 309

7 IEEE Definitions of Terms for Antennas, *IEEE*, **45,** 1973

8 MAYHAN, J. T., and RICARDI, L. J.: 'Physical limitations of Interference reduction by antenna pattern shaping', *IEEE Trans.,* **AP-23,** 1975

9 MAYHAN, J. T.: 'Nulling limitations for a multiple-beam antenna', *IEEE Trans.,* **AP-24,** 1976

10 POTTS, B. M.. 'Radiation pattern calculations for a waveguide lens multiple-beam antenna operating in the AJ mode'. MIT, Lincoln Laboratory, Technical Note 1975–25, 1976

11 RICARDI, L. J., *et al.*: 'Some characteristics of a communication satellite multiple-beam antenna'. Defense Tech. Inf. Center AD/A-006405, 1975

12 RICARDI, L. J.: 'Methodology of assessing antenna performance'. Lincoln Laboratory Technical Note 1978–24, 1978

13 SCOTT, W. G., LUH, H. S., and MATTHEWS, W. E.: 1976, 'Design tradeoffs for multibeam antennas in communication satellites'. Conference Record, *I*, presented at 1976 International Conference on Communications, Philadelphia, PA

14 SILVER, S.: 'Microwave antenna theory and design'. McGraw-Hill, 1949, Table 6.2, p. 195

15 WEINSTOCK, R.: 'Calculus of variations'. McGraw-Hill, New York, 1952, p. 6

Chapter 7

Low- and medium-gain microwave antennas

J. Bach Andersen

7.1 Introduction

The isolated and relatively small microwave antenna is the topic in this chapter. Array elements and large antennas are covered in other chapters. The basis for the selection has been novelty, either in design or in theory of operation. Most of the material in this chapter has not appeared in previous antenna design books, which date back 10 to 15 years, although of course some overlap is unavoidable.

An understanding of the physical principles, in network theory or wave propagation, is the underlying philosophy, since this seems to be the basis for new, innovative design. The mathematics has been kept to a minimum, but the references will carry the reader on to greater details.

7.2 Helical and spiral antennas

7.2.1 Introduction

Helical and spiral (wire or flat tape) antennas are very useful when broadband, circularly polarised antennas are needed. The physical principles are well understood, but exact theories are scarce. Detailed design information does not exist in general, so the reader must rely on modelling or computer simulation. These broadband structures are not very sensitive to small changes, so it should be possible to design an antenna on the basis of the present section and the references given.

7.2.2 Helical antenna (Unifilar helix)

General properties and radiation mechanism: The helix or helical antenna consists of a single wire (the unifilar helix) or narrow tape wound like a left-hand or right-hand screw, self supporting or wound on a dielectric cylinder. Typically it is excited by a coaxial line over a small groundplane with diameter G (Fig. 7.1) and the other end of the wire is left as an open circuit. The helix may be characterised by some or all of the following parameters: n = number of turns; D = diameter of helix; d = diameter of wire or tape; S = spacing between turns; C = circumference, $C = \pi D$; L = unfolded length of single turn, Ψ = pitch angle, $\tan \Psi = S/C$.

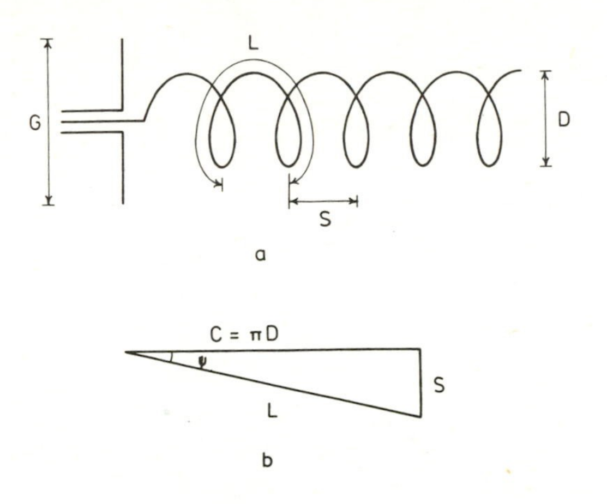

Fig. 7.1 *(a) Helix antenna (b) unfolded single turn*

Two different modes of operation are important for the helix: (*a*) the normal mode, (*b*) the axial mode.

(*a*) *The normal mode.* The normal mode is a low-frequency mode and exists for $C/\lambda < 0.5$. In this mode the helix operates more or less as a shortened monopole, and the current distribution along the wire is approximately the same as if the wire was stretched out ($v_W \sim c$). On this basis the far-field pattern may be computed fairly accurately, and the result does not deviate too much from that of a small monopole, i.e. a null along the axis of the helix and maximum radiation normal to the axis, provided the total length is much less than a wavelength. Due to the shortening the bandwidth is less than for the stretched-out monopole, but it may be made resonant for $nL \sim \lambda/4$. Since the application of the normal mode in antennas is rather limited, we shall not give any details.

(*b*) *The axial mode.* The axial mode exists in a limited frequency range

$$0.75 < C/\lambda < 1.25$$

for a limited range of pitch angles. Properly designed, the antenna will have circular polarisation and close to optimum length for maximum endfire gain over most of the frequency range, and, furthermore, the impedance will be resistive and almost constant. These properties make the helix a very popular and useful antenna.

The radiation mechanism and some of the design criteria may best be understood by considering the possible modes on the infinite, periodic structure. A single turn of the helix is considered as one element of the periodic structure; thus the constant spacing between the elements is S. The current carrying wire in the helix serves two functions; it constitutes the radiating elements and at the same time functions as a transmission line connecting the elements, where the velocity along the wire is of the order of the velocity of light.

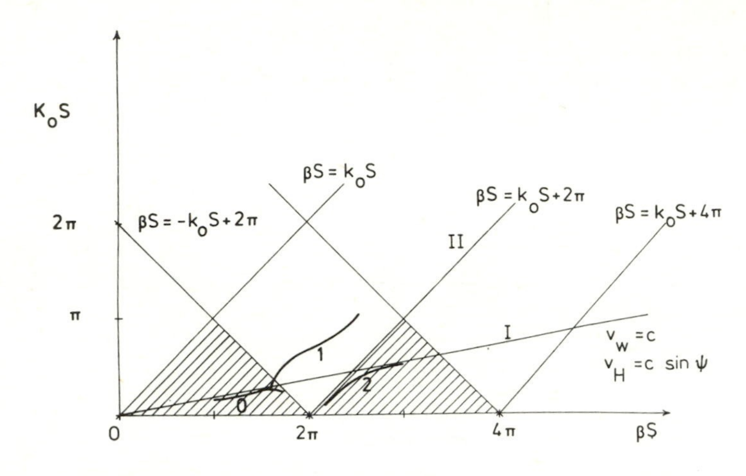

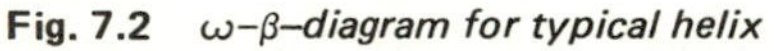

Fig. 7.2 *ω–β–diagram for typical helix*

- I: Dispersion curve for uncoupled wave along wire with $v_{wire} = c$, $v_{helix} = c \sin \psi$
- II: Uncoupled free-space wave with $v = c$ in forward direction. First space-harmonic ($m = 1$) of $\beta S = k_0 S + m2\pi$
- 0: Basic slow-wave mode, corresponds to normal mode
- 1: Fast, radiating mode
- 2: Slow wave, $v \lesssim c$, corresponds to axial mode

Fig. 7.2 shows an example of the ω–β diagram for a helix. The ordinate is the normalised frequency, $k_0 S$, and the abscissa is the normalised axial wavenumber, βS. For a detailed discussion the reader is referred to Hessel.[38] All the straight lines represent the uncoupled waves; one set

$$\beta_n S = k_0 S + m2\pi \tag{7.1}$$

represent space-harmonics of a wave in free space with velocity equal to c, the velocity of light. One interpretation of eqn. 7.1 would be that when the wave phase is sampled at intervals spaced S, the phase may only be determined within a multiplum of 2π. The curve marked II corresponds to $m = 1$ and is of special interest for the axial mode.

The curve marked I is related to the transmission-line function mentioned above. It corresponds to a current wave with velocity c, $v_w = c$ (w for wire), which gives an axial velocity or helix velocity of $v_H = c \sin \Psi$ or

$$\beta S = \frac{k_0 S}{\sin \Psi} \tag{7.2}$$

Whenever the two wave species have the same β, i.e. a crossing of the lines, we have an interaction since they follow each other with the same velocity. This interaction is also called mode coupling and has the effect of splitting the curves, so the combined wave with wavenumber $\beta_H S$ is now given by the curves 0, 1, 2. More curves follow for higher frequencies, but they will be neglected here. The strength of the coupling is determined by the thickness of the wires.

Inside the first triangle we have curve 0, which is only a small perturbation of I, so we have an essentially slow, non-radiating wave, propagating with the speed of light along the wires, axially $v_H = c \sin \Psi$. Mode 0 corresponds to the normal mode mentioned in the beginning of this Section. The first mode-coupling occurs at the boundary of the first triangle

$$\beta S = -k_0 S + 2\pi \tag{7.3}$$

which line after subtraction of 2π gives $\beta = -k_0$, a radiating wave in the backward direction. When the frequency is increased, mode 1 can propagate (Klock[55]), a radiating fast wave with a corresponding radiation damping. In the present context with the forward radiating helix this mode is of no interest.

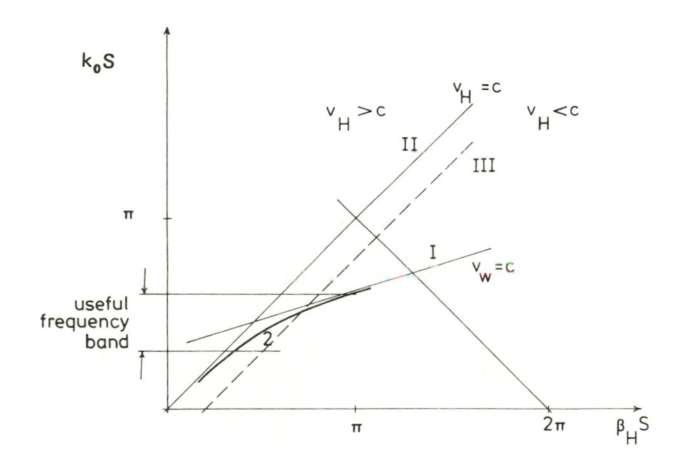

Fig. 7.3 *Dispersion curve for axial mode shifted into first triangle*
I, II, 2: same as in Fig. 7.2
III: Optimum dispersion curve for maximum gain according to Hansen–Woodyard condition.

$$\beta_H S = k_0 S + (\pi/n). \quad n = 5.$$

Mode 2, the result of the coupling between I and II, corresponds to the axial mode. In order to study the mode in greater detail we subtract 2π from βS and redraw the second triangle in Fig. 7.3. We note that the helix velocity v_H is slightly less than c:

$$v_H = \frac{\omega}{\beta_H} < c \tag{7.4}$$

and the wave gets slower and slower as frequency increases. This should not be confused with the velocity along the wire, v_w, which increases and approaches c when frequency increases. It is well known that the optimum velocity on a slow wave structure for maximum endfire gain is less than c, and according to the Hansen–Woodyard condition the following is approximately true:

$$\beta_H nS = k_0 nS + \pi \tag{7.5}$$

which states that the total phaseshift along the total length nS is π greater than the corresponding free space phaseshift. Eqn. 7.5 may also be written

$$\beta_H S = k_0 S + \frac{\pi}{n} \tag{7.6}$$

which is given as III in Fig. 7.3 for 5 elements. It is noted that curve 2 passes this line in the useful frequency band, which explains some of the properties of the helix.

From this periodic-structure analysis we may get some design guidelines. Line I must cut line II well within the second triangle, which means $\sin \Psi < \pi/3\pi$ or $\Psi < 19.5°$. In practice Kraus[57] recommends $12° < \Psi < 14°$. The frequency range may be determined approximately from the figure as $0.3\pi < k_0 S < 0.6\pi$ or

$$0.15 < S/\lambda < 0.3 \tag{7.7}$$

which gives

$$0.71 < C/\lambda < 1.20 \tag{7.8}$$

corresponding well with the empirical results of Kraus,[57] quoted previously. The upper frequency limit is determined by a safe distance from the right side of the triangle to avoid the backward wave and to avoid too slow a helix velocity v_H.

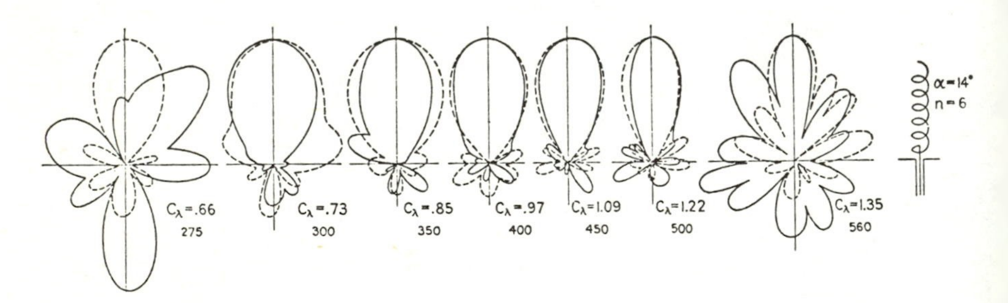

Fig. 7.4 *Measured field patterns of 14°, six-turn helix*
[Solid patterns are E_ϕ and dashed patterns E_θ, adjusted to same maximum. (From Kraus[57])].

The lower frequency limit is partly determined by the presence of the normal mode, curve 0, and partly by avoiding too fast a helix velocity v_H. In general the experience is that the helix antenna is a very uncritical antenna to design. Fig. 7.4 shows some experimental patterns over the useful frequency range.

The finite helix with finite groundplane is very difficult to analyse; some numerical results simulating an infinite ground-plane has been given by Hugh.[42] His results seem to indicate that groundplane effects are important for accurate numerical analysis. Near field measurements of phase and magnitude substantiate the phase velocity results discussed above (Cha[12]).

Design information. The following design information taken from Kraus[57] is of an empirical nature, but agrees well with later theoretical results. Fed over a small groundplane the impedance is nearly resistive and close to

$$R = 140C/\lambda\,\Omega \tag{7.9}$$

for the axial mode. The 3 dB beamwidth in degrees is

$$\Delta\theta = \frac{52}{C/\lambda\sqrt{nS/\lambda}} \tag{7.10}$$

and the corresponding directivity is

$$D = 15(C/\lambda)^2 nS/\lambda \tag{7.11}$$

Typical gain values in practice are between 10 and 15 dB over isotropic. The side-lobes are typically below −10 dB. Wide bandwidth gain characteristics have recently been studied by King and Wong[53] and some of their results are shown in Fig. 7.5. Note that the peak gain occurs at lower and lower frequencies as the number of turns increases in agreement with the discussion of Fig. 7.3.

7.2.3 Multiwire helix, bifilar and quadrifilar

The multiwire helix consists of a number of single-wire helices equally spaced circumferentially. For N wires this means N feedpoints, when we work without a groundplane. N feedpoints means $N-1$ different, independent modes, where it is customary to choose the model such that they have a progressive phase variation circumferentially. Arm k will thus have an excitation of

$$\exp\left(-2\pi j\frac{km}{N}\right) \tag{7.12}$$

for mode m. As with usual discrete Fourier transforms any excitation may be expressed as a linear combination of these orthogonal excitations. A discussion of the effect of this N-fold symmetry may be found in Rumsey[79] and Deschamps and Dyson.[21]

The angular variation of the radiated fields will depend on the mode and its higher harmonics; thus mode m will radiate fields with angular variation as

$$e^{j(m+iN)\varphi} \tag{7.13}$$

where $i = 0, \pm 1, \pm 2, \ldots,$ the higher-order variations being excited by the structure. $m = 1$ is the only mode with non-zero radiation in the axial direction and since we are mainly interested in this case the discussion will be limited to $m = 1$.

If we want a very pure excitation of the $m = 1$ mode the phase should be enforced at many points which calls for a large N. The situation is illustrated in Fig. 7.6, which shows the phase variation around the circumference assuming that phase equals zero for $\varphi = 0$. Since the phase is only determined within a multiplum of 2π we may add or subtract $j2\pi$. Assume that $e^{j\varphi}(m = 1)$ is the wanted mode.

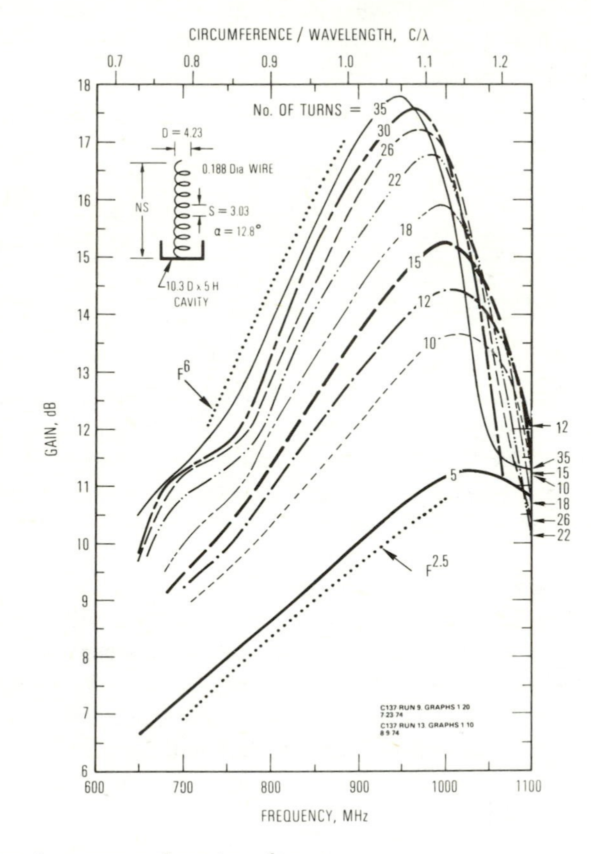

Fig. 7.5 *Gain v. frequency and number of turns*
$\psi = 12.8°$. (King and Wong[53]).

For the unifilar helix the phase is only enforced at $\varphi = 2\pi$, but by adding 2πs (the small circles) we see that the e^{j0} mode (the normal mode in eqn. 7.2) and $e^{-j\varphi}$, $e^{j2\varphi}$, etc. all may be excited in principle, since the circles fall on the lines of the modes. Whether a mode is strongly excited or not or whether it contributes in the far field is more difficult to determine; some discussion will be given later.

The balanced bifilar helix is phase enforced at $\varphi = \pi$ as well as at $\varphi = 2\pi$ (small squares); we note that only the odd-numbered modes are excited. Similarly, the quadrifilar helix excited for the $m = 1$ mode has the phase enforced at four points, and only, $e^{j\varphi}$, $e^{j5\varphi}$, $e^{-j3\varphi}$ etc. are excited. This agrees of course exactly with eqn. 7.13. When N tends to infinity the sheath helix is approached, and only the $e^{j\varphi}$ is excited.

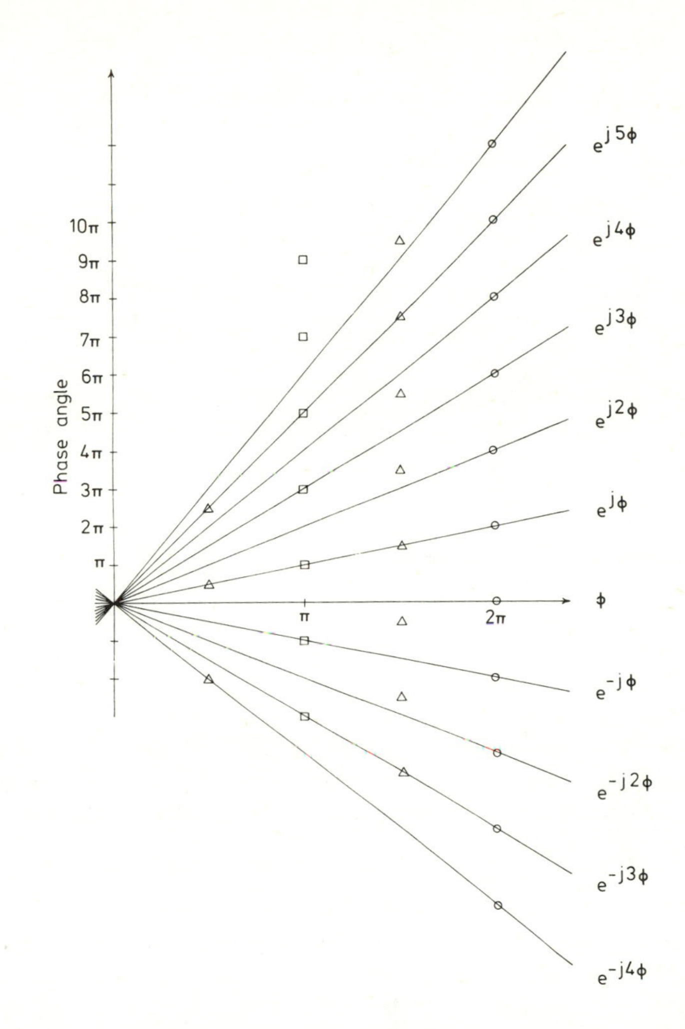

Fig. 7.6 *Circumferential phase variation and mode chart*
Phase enforced for:
○ unifilar helix
□ bifilar helix
△ quadrifilar helix

The enforcement of the phase at N points has interesting implications for the ω–β diagram. The special screw symmetry of the helix dictates a relationship between the axial and angular symmetry, such that the fields must be written like

$$F(z, \varphi) = e^{-j\beta_0 z} H(v) \tag{7.14}$$

where $H(v)$ is periodic with the period S, and

$$v = z - \frac{S}{2\pi}\varphi \tag{7.15}$$

$v =$ constant is the equation of the helix. Expanding $H(v)$ in its Fourier series gives

$$H(v) = \sum_{i=-\infty}^{+\infty} c_i e^{-ji(v/S)2\pi} \tag{7.16}$$

or by insertion of eqn. 7.15

$$F(z, \varphi) = e^{-j\beta_0 z}\{c_0 + c_1 e^{-j(z/S)2\pi + j\varphi} + c_2 e^{-j(z/S)4\pi + j2\varphi} + \ldots + c_{-1} e^{+j(z/S)2\pi - j\varphi} + c_{-2} e^{+j(z/S)4\pi - j2\varphi} + \ldots\} \tag{7.17}$$

Thus the angular variation of $e^{jn\varphi}$ is followed by the spatial variation of $e^{-j(z/S)n2\pi}$. For the quadrifilar helix this means, according to Fig. 7.6 or eqn. 7.13, that only

$$\ldots, c_{-7}, c_{-3}, c_1, c_5, c_9, \ldots$$

are different from zero. This gives a ω–β diagram like that in Fig. 7.7, where a pitch angle of $\Psi = 40°$ is chosen as an example. Comparing with Fig. 7.2 we note that all the coupling points with the lines emanating from $\beta S = 4\pi$, 6π and 8π have been eliminated due to the phase enforcement. The result is a much larger bandwidth. The resulting modes are sketched on the Figure, where (*a*) is the backfire mode and (*b*) the forward endfire mode.

The forward endfire mode was first described by Gerst and Worden;[32] a comprehensive paper was later written by Adams *et al.*,[1] giving both experimental and numerical results. The forward axial mode exists approximately in the range

$$0.4 < C/\lambda < 2$$

a 5:1 frequency range. The range is in good agreement with the sketch in Fig. 7.7 and should be compared with the unifilar range of $0.75 < C/\lambda < 1.25$.

Note that the normal mode of the unifilar helix cannot be excited, since it corresponds to $m = 0$, or c_0 in eqn. 7.17. The large bandwidth must be paid for with a greater complexity, since the excitation is somewhat complicated with four signals of equal amplitudes and relative phases of 0°, 90°, 180° and 270°, and these phase differences must be kept over the whole bandwidth. The results of Adams *et al.*[1] also indicate that the quadrifilar helix is more sensitive to ground-plane size than the unifilar helix.

The backfire mode (Fig. 7.7*a*) is a fast mode when the dispersion curve lies outside the triangle; so for an antenna of finite length the resulting pattern will be of a conical shape with a minimum along the axis. This pattern is well suited for satellite communications, giving a uniform signal at the receiver throughout the satellite pass. Design data including experimental results have been given by Kilgus.[51]

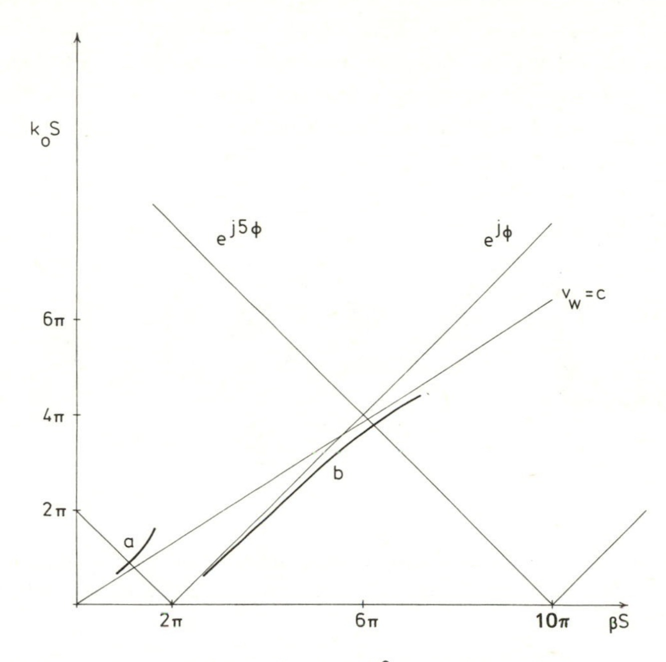

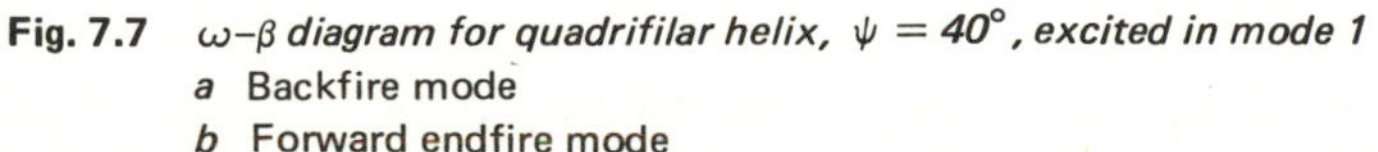

Fig. 7.7 *ω–β diagram for quadrifilar helix, $\psi = 40°$, excited in mode 1*
a Backfire mode
b Forward endfire mode

7.2.4 Multiwire conical spirals

It is well established now that almost all properties of the log-spiral conical antennas may be understood by reference to the local properties of the corresponding helix. The main advantage of turning the cylinder into a cone is a large increase in bandwidth. This is due to the fact that the cone is defined by an angle only (θ_0 in Fig. 7.8), whereas the cylinder is defined by the diameter, which necessarily must be related to a wavelength.

The balance two-arm or bifilar antenna has been discussed in great detail by Dyson,[24] the more general multiwire case by Deschamps and Dyson[21] and Atia and Mei,[7] where the latter give some numerical results for four-arm spirals. These references must be consulted for accurate design data; the following discussion will be limited to the physical mechanisms and some representative results.

A single arm of the spiral is defined by

$$\theta = \theta_0 \qquad r = e^{a\varphi}$$

where a determines the tightness of the structure. For an N-arm antenna the N wires are symmetrically spaced around the cone, similar to the multiwire helix. The structure is fed at the apex, usually in a balanced way without a groundplane, so the backfire modes can radiate without reflections. Applying the local helix principle and the ω–β diagrams discussed in the previous Sections, the first part of the structure may be denoted as the slow-wave region (Fig. 7.8), since the period is so small compared with wavelength that the dispersion curve lies inside the first triangle. The structure behaves as a transmission line and very little is radiated.

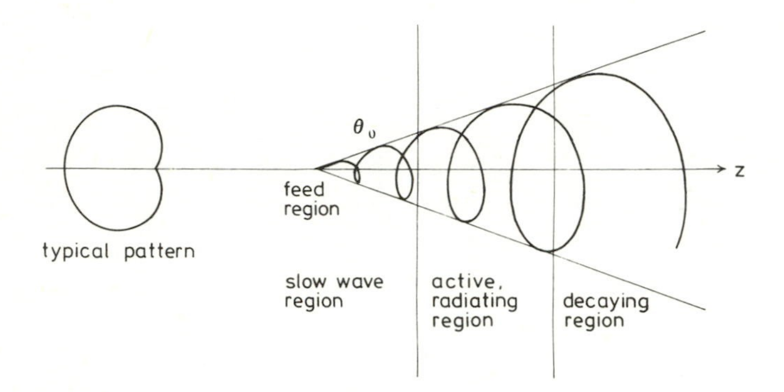

Fig. 7.8 *Single arm of log-periodic conical spiral antenna. Feed region is at apex and the main radiation is in backward direction and circularly polarized. Typical pattern shown is for M_1-mode*

When the boundary of the triangle is passed, as the backfire mode in Fig. 7.7, the active, radiating region is entered and almost all the energy is lost by radiation. After this region the current rapidly decays, so very little is left at the truncated ends. When the frequency varies, the active region moves, but the relative current distribution is the same, so impedance, pattern etc. will be practically unchanged.

The spirals have some mode-selective properties in normal operation, which are intuitively clear from the mode charts. The sense of polarisation is determined by the sense of the spiral and the main direction of radiation. If the conical spiral axis is oriented in the direction of the apex, the spiral is called right-hand if placing the thumb of the right hand along the axis, the fingers indicate the direction of the arms as they spiral out away from the axis (Deschamps and Dyson[21]). A right-hand spiral produces a predominantly right-hand polarised field.

As an example consider the left-hand spiral given by

$$v = z - \frac{S}{2\pi}\varphi$$

with two arms. According to the discussion in the previous Section all the odd-numbered space harmonics may be excited, but the dominant one will be C_{-1} in eqn. 7.17 for a current wave in the positive z-direction (see also Fig. 7.9). C_{-1} has a phase-variation as exp $(-j\varphi)$, which corresponds to left-circular polarisation in the negative z-direction. This can be seen simply by including the time variation, $e^{-j\varphi+j\omega t}$, an increase in time must be followed by an increase in φ in order to have an invariant field picture, and an increase in φ looking in the negative z-direction is counterclockwise or left-handed. The mode-coupling chart of Fig. 7.9 contains the same information as the conventional ω–β diagram, but it is more convenient for the discussion of polarisation. It is seen very clearly that there is no coupling to the right-hand circular polarisation.

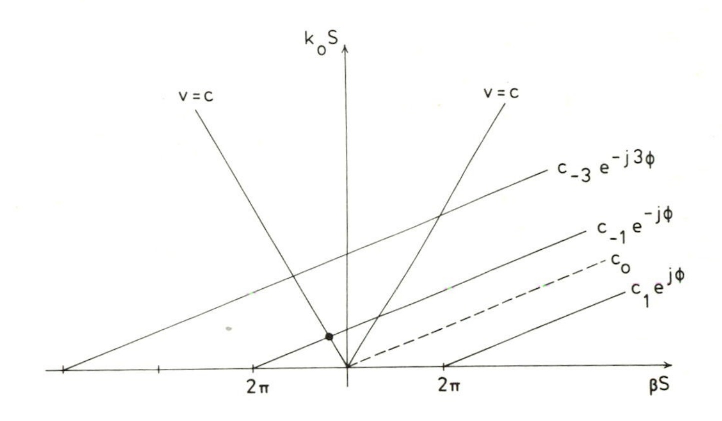

Fig. 7.9 *Mode coupling chart for left-handed bifilar conical spiral. Coupling is to the $e^{-j\varphi}$-harmonic, which gives left-handed circular polarisation in the negative z-direction*

These considerations are of course based on the assumption of an essentially infinite structure. If the active region is so large (or the structure so small) that a considerable fraction of the energy is reflected from the end, a wave of opposite polarisation may be radiated. This effect may sometimes be useful over a restricted frequency band. An example of this is given by Kim and Dyson,[54] where both circular polarisations are obtained for a four-arm log-spiral conical antenna with a simple hybrid matrix connected to the apex terminals. This effect is successful over a 1.5 : 1 bandwidth.

Some illustrative results for a four-arm spiral, excited in the M_1-mode, are shown in Fig. 7.10 (Atia and Mei[7]). Typically we have gain value between 4 and 8 dB, the highest gains for the largest values of Q, the slowness factor. Q is defined by $Q = (1 + \sin^2\theta_0/a^2)^{1/2}$ where θ_0 and a are defined previously. This effect may be understood as the result of a larger active region for a tightly wrapped structure. If the bandwidth is the most important factor, the antenna should be designed with a very narrow active region, corresponding to a smaller gain. The parameter δ is related to the thickness of the wires or tapes and is simply the angular width of the exponentially expanding arms.

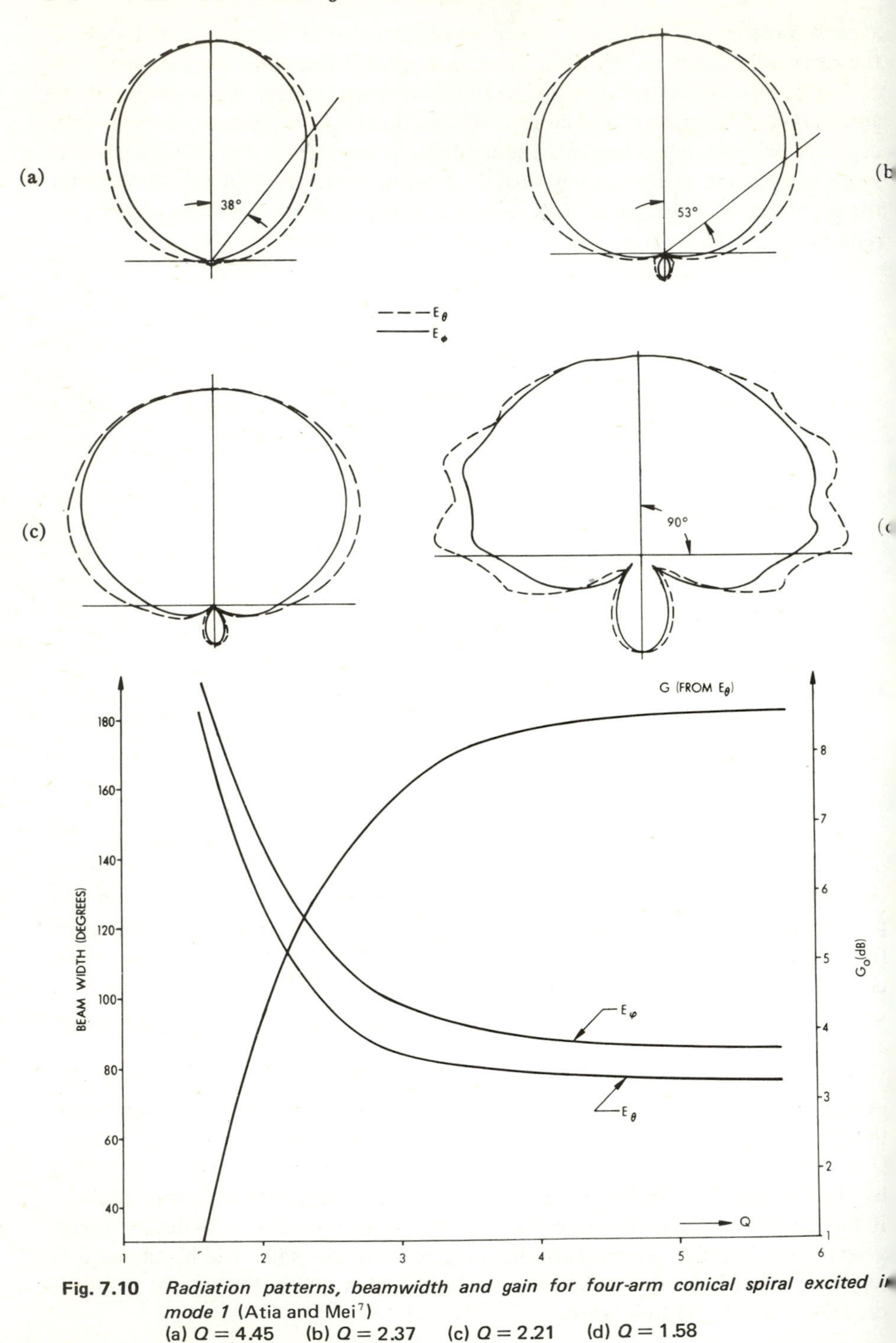

Fig. 7.10 *Radiation patterns, beamwidth and gain for four-arm conical spiral excited in mode 1* (Atia and Mei[7])
(a) $Q = 4.45$ (b) $Q = 2.37$ (c) $Q = 2.21$ (d) $Q = 1.58$

7.2.5 Planar spirals

The evolution of conical antennas from cylindrical helices may be continued, so that when $\theta_0 = 90^\circ$ a completely flat spiral results. It is remarkable that most of the principles discussed previously continue to be valid in this limit. The inner part of the spirals still serves as a nonradiating transmission line, carrying the current out to the active region, which is now a narrow circumferential band. Instead of identifying the active region as the region of a space-harmonic of a quasi-periodic structure, it is easier to view the two-arm spiral as a slot antenna, where the two sides of the slot are the two spiral arms (Fig. 7.11). The resonant place for $e^{jm\varphi}$ mode is the radius ρ_m, where the current and electric field vary approximately as $e^{jm\varphi}$. The viewpoint has been tested successfully by Cubley and Hayre,[19] who find the following far-field pattern for mode m

$$E_m = \cos\theta\,\{(\cos\theta + 1)J_{m-1}(K\rho_m \sin\theta) + (\cos\theta - 1)J_{m+1}(K\rho_m \sin\theta)\} \tag{7.18}$$

The sense of polarisation may be determined in the same way as for the conical spiral.

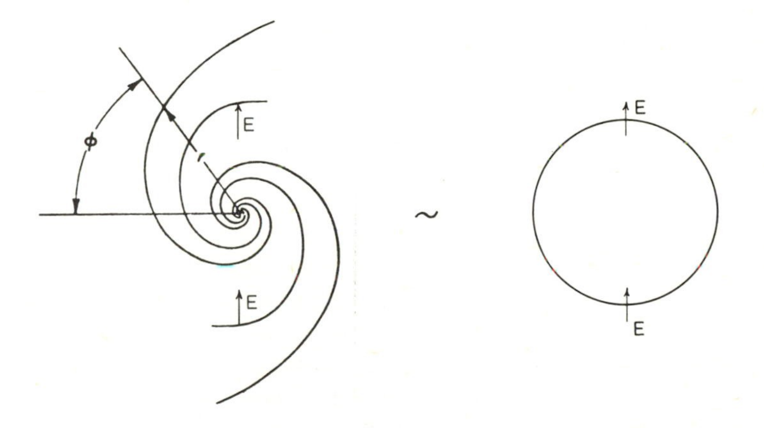

Fig. 7.11 *Planar log-spiral viewed as an approximate ring slot antenna*

The exact current distribution and radiation pattern are impossible to find analytically except for one case, the sheath spiral equivalent to infinitely many arms (Rumsey[79]). The result for the pattern of the $m = 1$ mode is

$$|E| = \frac{\cot(\theta)\tan(\theta/2)\exp(a^{-1}\tan^{-1}(a\cos\theta))}{\sqrt{1 + a^2\cos^2\theta}} \tag{7.19}$$

where a is the spiral constant in

$$\rho = \rho_0 e^{a\varphi} \tag{7.20}$$

Patterns for various values of a are shown in Fig. 7.12, and it is seen that they are almost independent of a for $a < 1$. The pattern tends to zero along the plane of the spiral, indicative of attenuation of the current along the spiral arms. The polarisation is circular everywhere. It turns out in practice that the sheath spiral is a good approximation for the spiral with a finite number of arms, even down to two arms.

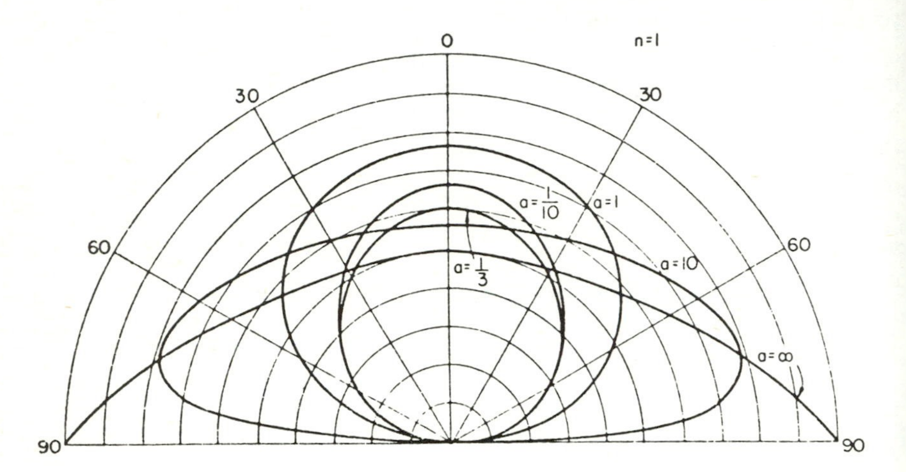

Fig. 7.12 *Radiation patterns for a log-spiral with an infinite number of arms* (Rumsey[79])

As for most travelling-wave antennas the resistive part of the input impedance depends on the wire or tape thickness; thin wires lead to high impedance values. For the spirals of increasing width of the arms the impedance depends on the angular extent δ of the arms, as for the conical spiral. The planar antennas have some additional, unique properties as discussed by Rumsey.[79] A self-complementary antenna is one which looks the same, when the metal part is exchanged with the non-metal part; for spirals this means that two-arm spirals with $\delta = 90^\circ$ are self complementary. All two-terminal self-complementary structures have a constant impedance of $60\pi \sim 189\,\Omega$. The impedance of multiterminal spirals depends on the interconnection of the arms; one particularly simple result is

$$Z = \frac{120\pi}{n} \text{ ohms} \tag{7.21}$$

for an n-arm spiral, where the alternate arms are connected together (Deschamps quoted in Rumsey[79]).

Here we have discussed the spirals as if they radiate to both sides of their plane but it is, of course, possible to use them as aperture antennas, for example with a cavity backing.

7.3 Slot antennas

7.3.1 Basic concepts

Any opening in an extended metallic surface may be considered a slot antenna, when electromagnetic energy is radiated through the opening or aperture. The radiated far and near fields may be determined from the currents running on the conducting surface, but in general an alternative, equivalent viewpoint is simpler. A slot in an infinite ground plane is equivalent to a distribution of magnetic currents confined to the slot area, so the radiated fields may be determined simply by integration over a small surface. Even more important, the complementary radiator of electric currents is often a well known antenna, so by simply interchanging electric and magnetic fields, the complete solution may be obtained directly. It is a condition for this equivalence that the source distribution is the same, e.g. the variation of electric field in the slot should be the same as the variation of magnetic field (or electric current) along the corresponding strip. This equivalence is never exact in practice, but it is a sufficiently good approximation. It is customary for narrow slots to make the same approximation of sinusoidal variation of magnetic current as is standard for electric currents for thin wires.

The equivalence is valid for the fields everywhere, also in the near field, which means that there is a specially simple relationship between impedances,

$$Y_{slot} = \frac{4Z_c}{Z_0^2} \qquad (7.22)$$

where $Z_0 = 120\pi$ ohms and Z_c is the impedance of the complementary conducting structure fed at the corresponding point. The complementary structure of a narrow slot is a flat wire, which is approximately equivalent to a thin, cylindrical wire. Thus we can conclude immediately for the narrow slot radiating to both sides of the ground-plane, that

(*a*) the conductance is independent of slot height, since the resistance of a wire is independent of thickness, as long as it is small
(*b*) the slot is inductive, where the wire is capacitive, and vice versa
(*c*) the resonance conductance equals $4 \times 73/(120\pi)^2 \sim 2.06$ m mhos ($R_{slot} \sim 500\,\Omega$)
(*d*) the radiation pattern is identical with that of the wire, with $\bar{E}$ and $\bar{H}$ interchanged.

Although the above statements are extremely important theoretically, they must be modified to take into account the more practical case, where the slot is a one-sided radiator, including the effect of feeding network. In the following Section we shall treat in detail the case of a cavity-backed slot.

The metallic surface in which the slot is embedded is not necessarily plane and infinite. The effect of a finite groundplane may be computed by means of the geometrical theory of diffraction (Section 2.2), if the distances to the edge are sufficiently large. The effect of curvature of the groundplane may be found approximately by various asymptotic methods, and for separable structures like cylinders and spheres the solution may be found directly. For details of the theories and some results the reader is referred to Compton and Collin.[16]

7.3.2 Cavity-backed slot antenna

As an example of an admittance analysis we shall take a narrow slot backed by a rectangular cavity. By admittance is meant the aperture admittance as seen by a feeding line at the centre of the slot. This would not be the normal way of feeding the antenna; instead we would excite the cavity with a capacitive probe or inductive loop, which would 'see' quite a different admittance. The value of finding the aperture admittance would lie in finding the antenna Q and the resonance frequency, both of which would only be changed slightly by different feeding systems. Thus we might as well think of an impressed aperture distribution as the source, knowing that it is only one out of many, but with the nice property that the solution for the fields may be written down immediately. A variational solution has been found by Galejs,[31] but such a solution, although rather accurate, tends to obscure the physics. The following is a slightly modified version of the work of Cockrell.[14]

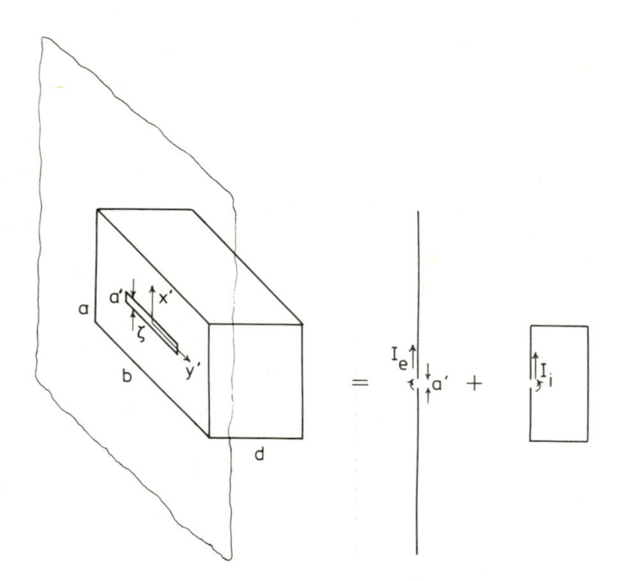

Fig. 7.13 *Cavity-backed slot in an infinite groundplane. Separated in an external and internal problem coupled via the slot*

The antenna is shown in Fig. 7.13 where the groundplane is assumed to be infinite. The slot is placed at the centre of the cavity, such that it couples to the fundamental mode of the corresponding waveguide; thus $b > a$. The length of the slot is l, the height a'; a' is considered to be small. The electric field is forced to be

$$E_{x'} = \frac{V_0}{a'} \sin k_0\left(\frac{l}{2} - |y'|\right) \tag{7.23}$$

where k_0 is the wavenumber $2\pi/\lambda$, and x', y' a local coordinate system at the slot. This distribution is the same as is used for thin, centre-fed wire antennas, which means that the external problem is known. When the aperture distribution is stiff, the internal and external problems are uncoupled (seen from the outside, the cavity may be changed arbitrarily, even removed, without changing the external fields). This means that the total current may be split into two parts, an external and an internal:

$$I = I_e + I_i \tag{7.24}$$

and the admittance

$$Y = Y_e + Y_i \tag{7.25}$$

The external admittance must be one half of Y_{slot} in eqn. 7.22, since in that case radiation was to both sides ($Y_i = Y_e$). The admittance has been given by Rhodes,[76] who showed that the conductance is independent of slot height.

For the internal part G_i will be zero when absorption is neglected. For the susceptance it is advantageous to split it in two parts, one connected to the propagating mode $B_{i,\,cavity}$ and one connected to the modes under cut-off, $B_{i,\,slot}$. The first one is proportional to the input of a shorted line (Fig. 7.14). The remaining part of the stored energy is connected with the storage in the immediate vicinity of the slot and is independent of d, if it is not too small. In fact, to a rough approximation, $B_{i,\,slot}$ would be independent of a and b as well, so $B_{i,\,slot} \sim B_e$.

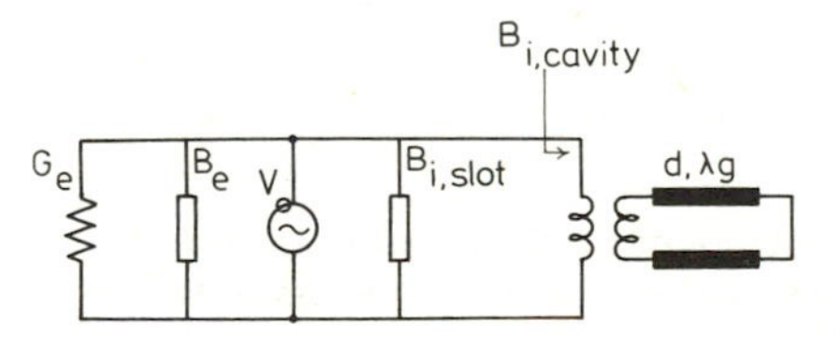

Fig. 7.14 *Equivalent circuit for slot and rectangular cavity*

The aperture admittance is then given by

$$Y = G_e(l) + j\,\{B_e(l, a') + B_{i,\,slot}(l, a', a, b) + B_{i,\,cavity}(l, a, b, d)\}, \tag{7.26}$$

where the dependence of the various dimensions is indicated. An example of the dependence of l and a' is shown in Fig. 7.15 (Cockrell[14]). Experimental results may be found in Long,[59] who also give some results for dielectric-filled cavities.

The interplay between the various parameters may be judged from the discussion above. In general, an increase in cavity depth lowers the resonance frequency and increases the Q-value. The same effect may be achieved by a dielectric filling.

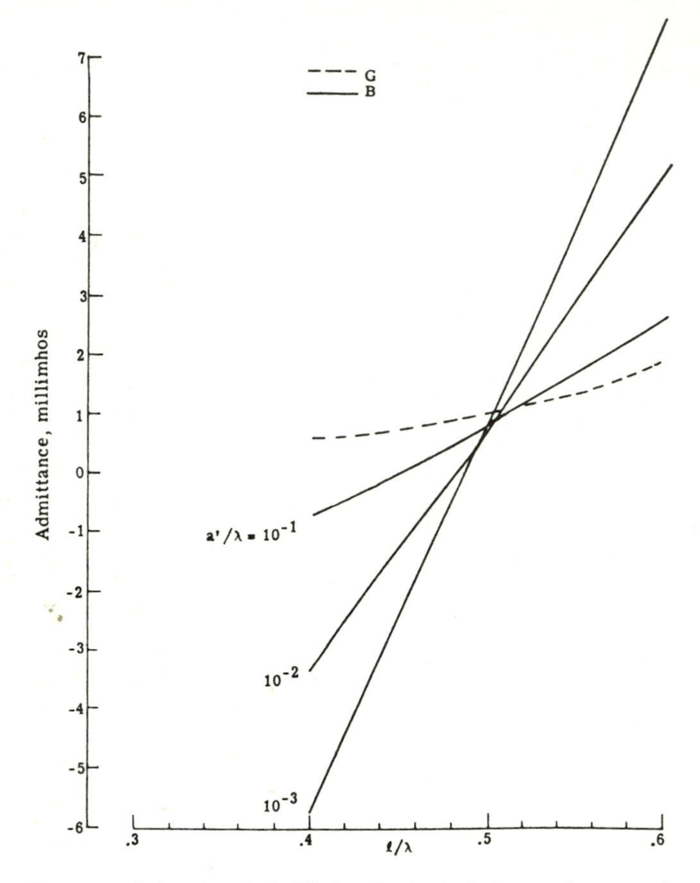

Fig. 7.15 *Aperture admittance of air-filled cavity-backed slot as a function of slot height and length* (Cockrell[14]) $a/\lambda = 0.3, b/\lambda = 0.6, d/\lambda = 0.5$

7.3.3 Broadband slots: T-bar fed and monopole slot

It is clear from the preceding discussion that the standard cavity-backed slot antenna with a narrow slot is essentially a narrow-band device in the same way as a thin-wire antenna, and the backing cavity will reduce the bandwidth further. We shall discuss two recent developments for slot antennas, which have a much larger bandwidth.

(*a*) *T-bar fed slot antenna.* A large slot gives large bandwidth and the extreme size is the cross-section of the cavity; care should then be taken to make a proper impedance transmission from a feed point and ensure that the right modes are excited. One such design is the T-bar fed slot antenna, first described by Jasik[46] and later refined by Newman and Thiele[71] and Crews and Thiele.[17] With reference to Fig. 7.16 the cavity is excited coaxially at the top wall, and the centre conductor is then flared out to a flat T-bar extending to contact the side walls.

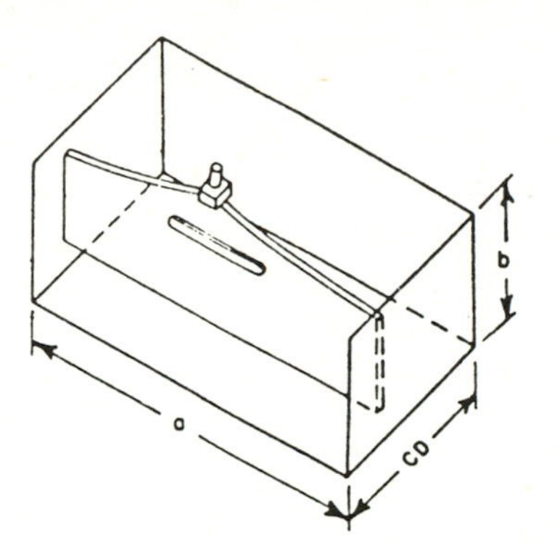

Fig. 7.16 *Cavity-backed T-bar fed slot antenna with top removed* (Crews and Thiele[17])

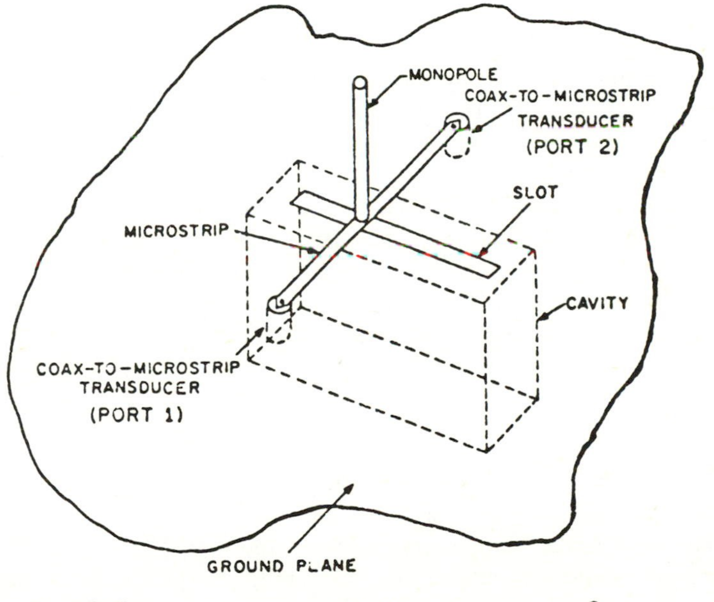

NOTE: PORT 2 is terminated in 50-Ohms when S_{11} and S_{21} are measured.

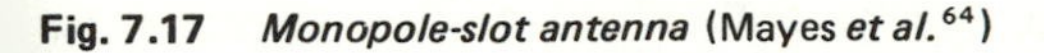

Fig. 7.17 *Monopole-slot antenna* (Mayes *et al.*[64])

No theories are available; the optimised designs are obtained experimentally. In one such design (Newman and Thiele[71]), a relative bandwidth of 2.3:1 was obtained where the bandlimit is defined by a standing wave ratio of 2.

(*b*) *Monopole slot antenna.* Bandwidth may be defined with respect to many parameters such as directivity, gain, impedance match etc., and it depends on the application which parameter is the most important. In receiver applications the efficiency is not always so important; the monopole slot antenna (Mayes *et al.*[64]) is an example of an antenna which has reasonably constant impedance and pattern characteristics over a 10:1 frequency band at the expense of the efficiency at the lower part of the band. The idea is to combine on the same feedline two structures with impedances which have dual properties relative to the feeder impedance; such two structures are the slot in a groundplane and a monopole mounted over the same groundplane (Fig. 7.17). The slot, backed by a cavity, is fed via a microstrip transmission line normal to the slot, and the monopole is placed on top of the microstrip. In the horizontal plane the pattern is a unidirectional cardioid pattern, the direction of which may be switched 180° depending on which port is excited.

7.4 Microstrip Antennas

7.4.1 Introduction

The microstrip antennas are important candidates when low profile, low weight and small size are required. They can be made conformal to a metallic surface, are of very rugged construction, and can be produced at a low cost with the same photo-etch technique as is standard with other microwave integrated circuits. Both linear and circular polarisation may easily be made and the single elements may be combined into arrays without many problems (Munson[68]). This seems almost like the perfect antenna, but there is one main drawback, the bandwidth, which typically is from a fraction of a percent to a few percent depending on the substrate dielectric constant and thickness. The gains are typically in the 4–7 dB range.

7.4.2 Basic properties of microstrip transmission line

A microstrip antenna is a transmission-line antenna, based on the microstrip transmission line (Fig. 7.18*a*, *b*). It is important to understand the basic properties of the microstrip in order to understand and apply the principles of the antenna design. The microstrip consists of a metallic groundplane, a thin dielectric sheet of permittivity ϵ_r and a 'centre conductor' or strip of width w on top of the dielectric. The fundamental mode of propagation is a quasi-TEM mode, where the transverse field distribution is well described by the static distribution (Fig. 7.18*b*). The mode is quasi-TEM because the medium between the conductors is inhomogeneous, which means that the phase velocity is different from the velocity in free space and the velocity in the dielectric. It has been found convenient to introduce an effective dielectric constant ϵ_{eff}, (Wheeler[90]) which takes the inhomogeneity into account and can be used for finding the effective velocity along the line and the characteristic impedance.

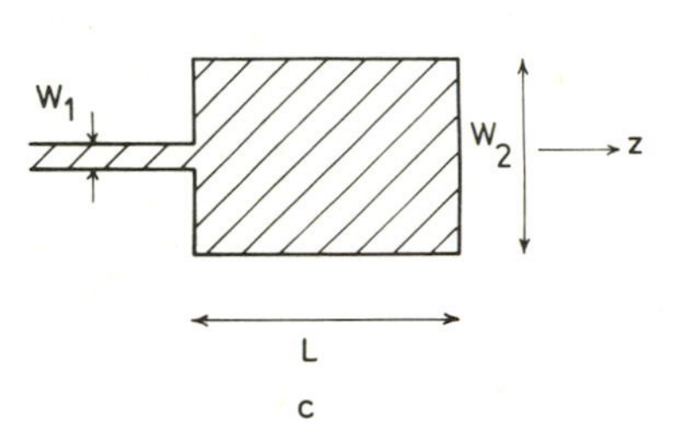

Fig. 7.18 *a* Cross-section of microstrip
b Electric-field distribution in cross-section of microstrip
c Top view of microstrip antenna consisting of a broad radiating element and a narrow feeding line

Having determined ϵ_{eff} we find the wavenumber β

$$\beta = k_0\sqrt{\epsilon_{eff}} = \frac{2\pi}{\lambda_g} \tag{7.27}$$

where λ_g is the guide waveguide, and the characteristic impedance Z,

$$Z = \frac{Z_0}{\sqrt{\epsilon_{eff}}} \tag{7.28}$$

where Z_0 is the free-space line impedance. The values may be corrected for the finite thickness of the strip. Due to the above-mentioned inhomogeneity some dispersion is present (ϵ_{eff} is a function of frequency), and this factor may be included by means of the equations by Getsinger.[33] In practice most of the field is concentrated in the dielectric, so ϵ_{eff} is close to ϵ_r, which may range from 2.5 to 10 for typical materials.

Thus the microstrip is an open, slow-wave structure. When viewed as an infinite line, the radiation loss is zero, as it is for all infinite slow-wave structures. A finite structure, on the other hand, will always have some radiation losses, and the art of antenna design consists of arranging the lines such that the radiation pattern, polarisation, efficiency and impedance satisfy the requirements. The exact solution of the radiation from a microstrip structure is extremely difficult and time consuming to find, but it is usually also unnecessary since most properties may be found and understood from a simplified theory.

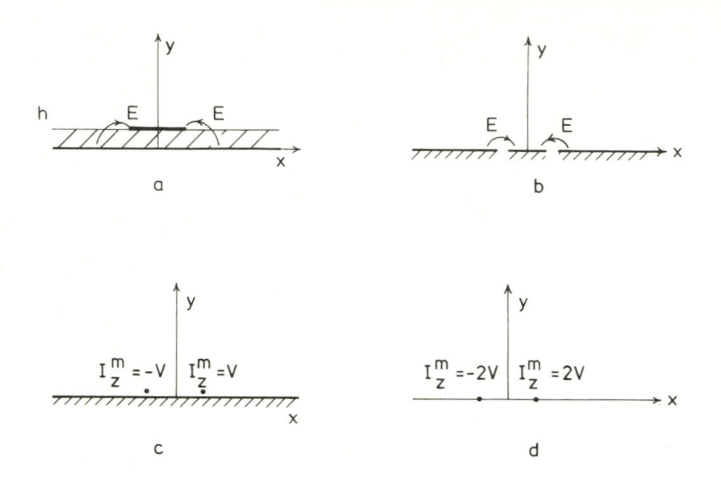

Fig. 7.19 *Equivalent versions of the microstrip line*
a Strip above groundplane
b Equivalent slots in groundplane
c Equivalent magnetic line currents above groundplane
d Equivalent magnetic line currents in free space (valid in upper halfspace, $y > 0$)

There are two, equally accurate, ways to analyse the antenna; one by means of the electric currents on the strips and groundplane, and one by means of the magnetic currents. The latter is the most convenient here. Since h, the thickness of the substrate, is very small compared to the wavelength, we may conceive the structure as one groundplane (Fig. 7.19*b*) with two slots of width approximately equal to h. The electric field distribution may be replaced by a magnetic line current which is z-directed and whose magnitude is given by

$$I^m = \int_{slot} \bar{E} \cdot \bar{ds} = V \tag{7.29}$$

Since the electric field in the slot to the left has the opposite direction, $I_z^m = -V$. Finally, we can use image theory and replace the groundplane by a doubling of the sources (Fig. 7.19*d*). This latter version is of course only valid in the upper half space.

This formulation is very convenient since only the voltage appears, the thickness h has dropped out. Thus we may conceive the microstrip antenna as a thin wire antenna, where the magnetic current in the wire is proportional to the voltage between the edge of the strip and the groundplane. Since this voltage is mainly determined by the transmission-line equations (usually little external coupling between the wire segments), we end up with a fairly simple antenna problem. Once the transmission-line problems have been solved all the radiated field components may be determined by integration over the wire currents.

7.4.3 Radiation mechanism of microstrip resonator

Consider the rectangular microstrip antenna in Fig. 7.18*c*. We have a wide section of width W_2 and length L and a narrow section, which we consider as the feed line. Forgetting for a moment the radiation, the wide line is terminated in an open circuit at $z = L$, where we have maximum voltage and minimum current.

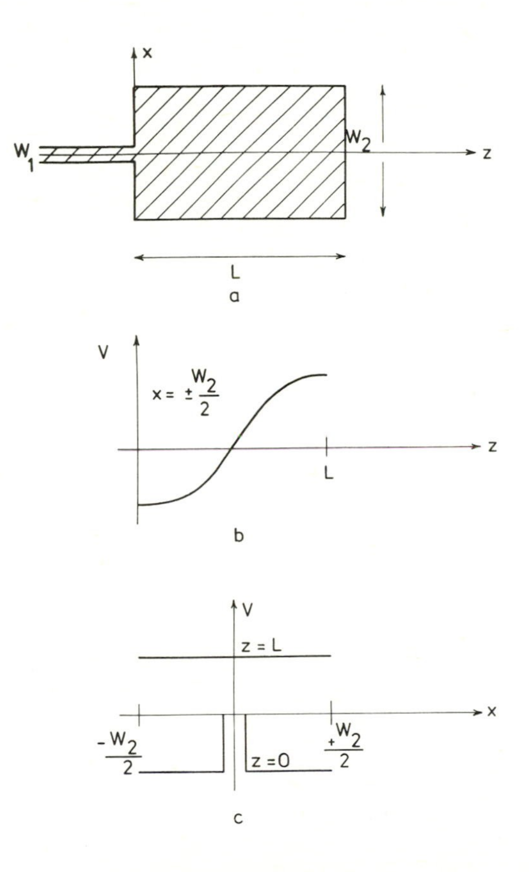

Fig. 7.20 *a* Resonant rectangular microstrip antenna
b Voltage distribution along side edges
c Voltage distribution along end edges

The structure is resonant, when $L \sim \lambda_g/2$ (lowest resonance); thus the voltage is maximum again at $z = 0$, but with a relative phase shift of 180° (Fig. 7.20*b*). Along the end edges the voltage is assumed to be constant, which is not true near the side edges. At the feedpoint a disturbing effect of the feedline is assumed, but it would be more correct to include the voltage along the feedline. Transforming the voltages to magnetic wire currents we get the situation of Fig. 7.21, where the sign reversals are due to the different directions of electric field as mentioned above.

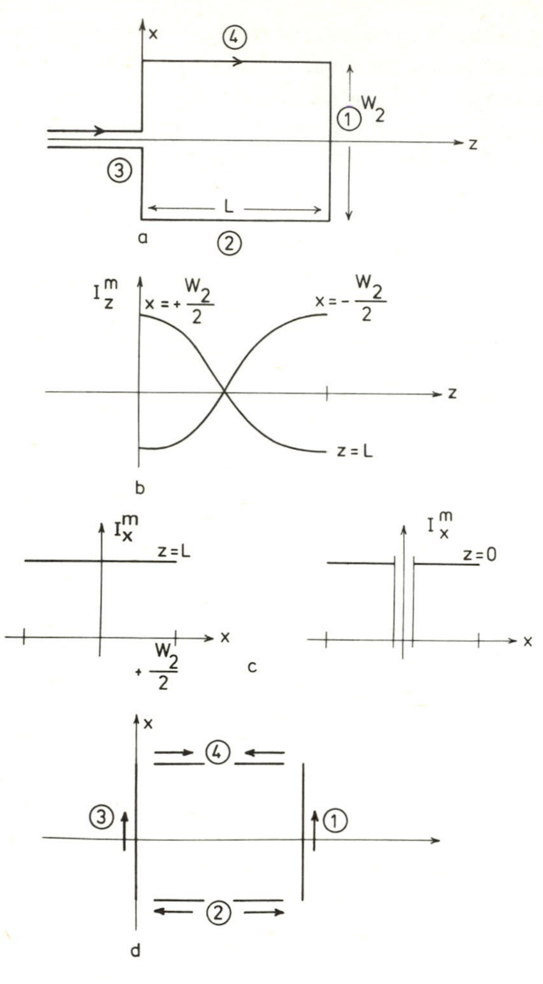

Fig. 7.21 *a* Equivalent wire antenna
b Magnetic current distribution along side edges
c Magnetic current distribution along end edges
d In-phase values of wire currents

The situation may be further visualised by Fig. 7.21*d*, where the arrows indicate in-phase values of the currents. We see immediately that ① and ③ will dominate since they are in phase, and constant along their length. The transmission-line currents ④ and ② tend to cancel in the far field; exact cancellation will take place in the planes of symmetry ($\theta = \pi/2$ and $\varphi = \pi/2$ in Fig. 7.22), but in other directions they will contribute to the total field.

Thus the microstrip antenna may be understood as a simple resonant slot antenna, where the total edge circumference acts as the source of radiation, and the volume between the strip or patch and the groundplane acts as the resonator.

7.4.4 Radiation pattern of rectangular microstrip antenna

Assume that a rectangular strip of length L and width W is situated in the xz plane as shown in Fig. 7.21. The magnetic current distribution is the following:

$$\begin{aligned} I_z^m &= 2V \cos\left(\pi \frac{z}{L}\right) & x &= W/2 \\ I_z^m &= -2V \cos\left(\pi \frac{z}{L}\right) & x &= -W/2 \\ I_x^m &= 2V & z &= 0 \quad \text{and} \quad z = L \end{aligned} \tag{7.30}$$

Using the far field expressions in Chapter 1 for radiation from wire antennas we readily find the following normalised far fields (normalised so $E_\theta = 1$ for $\theta = \pi/2, \varphi = \pi/2$).

$$E_\theta = \frac{\sin u}{u} \cos v \sin \varphi \tag{7.31}$$

$$\begin{aligned} E_\varphi &= \cos v \frac{\sin u}{u} \cos\theta \cos\varphi \left[1 - \frac{(k_0 L \sin\theta)^2}{\pi^2 - (k_0 L \cos\theta)^2}\right] \\ &= \cos v \frac{\sin u}{u} \cos\theta \cos\varphi \frac{\pi^2 - (k_0 L)^2}{\pi^2 - (k_0 L \cos\theta)^2} \end{aligned} \tag{7.32}$$

where $u = \frac{1}{2} k_0 W \sin\theta \cos\varphi$ and $v = \frac{1}{2} k_0 L \cos\theta$.

Of the two terms in the parenthesis in eqn. 7.32 the first one is due to the two end slots and the second one is due to the side slots. An example for a square microstrip is shown in Fig. 7.22 for $\epsilon_r = 2.5$, assuming that $k_0 L = k_0 W = \pi/\sqrt{\epsilon_r}$, which is true when fringing fields are neglected.

The E-plane pattern ($E_\theta = f(\theta)$ for $\varphi = \pi/2$) is very broad, corresponding to the closely spaced end slots. For $\epsilon_r \rightarrow \infty$ the E-plane pattern will approach a constant value, corresponding to the pattern from a single slot.

The horizontal H-plane pattern ($E_\varphi = f(\theta)$ for $\varphi = 0$) is also dominated by the end slots, the effect of the side slots being a narrowing of the pattern.

It may easily be seen from eqn. 7.32 that the relative contribution of the side slots is independent of W and depends only on $k_0 L$. For $k_0 L = \pi (\epsilon_r = 1)$ the two contributions are equal in magnitude, and E_φ vanishes identically.

Concluding, we can say that the rectangular microstrip antenna radiates a linearly polarised pattern, which is rather broad and with the maximum direction normal to the plane of the antenna.

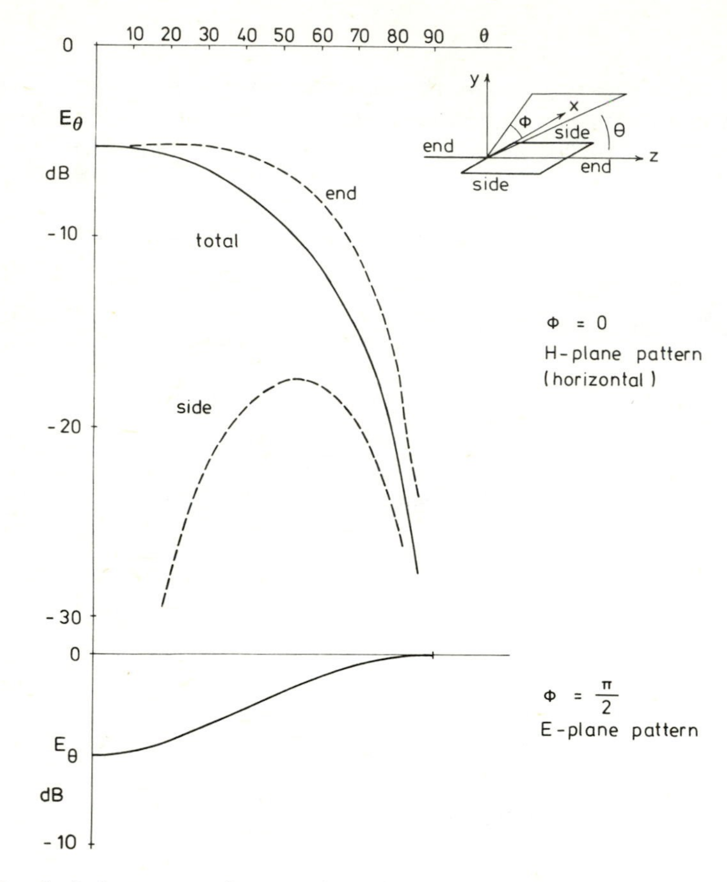

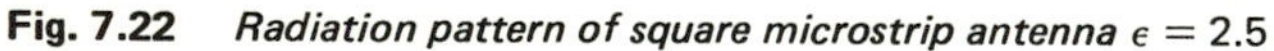
Fig. 7.22 *Radiation pattern of square microstrip antenna* $\epsilon = 2.5$

7.4.5 Directivity and gain

Having determined the far-field patterns we can determine the directivity by integration through the formula

$$D(\theta, \varphi) = \frac{4\pi(|E_\theta(\theta, \varphi)|^2 + |E_\varphi(\theta, \varphi)|^2)}{P} \tag{7.33}$$

where

$$P = 4 \int_{\varphi=0}^{\pi/2} \int_{\theta=0}^{\pi/2} (|E_\theta|^2 + |E_\varphi|^2) \sin\theta \, d\theta \, d\varphi \tag{7.34}$$

The maximum value of D is found for $\theta = \varphi = \pi/2$. In the general case, numerical integration is the easiest way to find the directivity. For square patches the directivity depends only on ϵ_r, see Fig. 7.23. The directivity varies between 9.8 dB and 4.8 dB, where the latter limit is the directivity of a single, small slot ($D = 3 = 4.77$ dB).

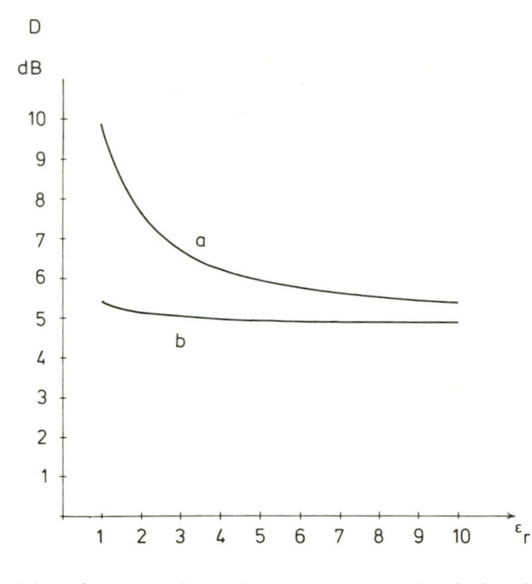

Fig. 7.23 *a* Directivity of square microstrip antenna, assuming $k_0 L = k_0 W = \pi/\sqrt{\epsilon_r}$
b Directivity of single slot of same width

It is customary to treat the single slots that make up the microstrip antenna as independent radiators; also in the sense that the total directivity is the sum of the directivities of the individual parts. This is only true, of course, when the radiation patterns are orthogonal or, equivalently, the mutual impedances or conductances are zero (Andersen and Rasmussen[6]). In Fig. 7.23*b* the directivity of a single slot of the same width as the square antenna is shown. Only for $\epsilon_r \sim 1.5$ is the directivity of the microstrip antenna 3 dB higher than that of the single slot.

The gain is less than the directivity due to the ohmic losses in the conductor material and the dielectric losses in the substrate material. These losses may be determined by standard transmission-line methods. Experimental results are reported at X-band with losses less than 0.5 dB (Munson,[68] Derneryd[20]).

It should be stressed that the square microstrip is a degenerate structure in the sense that side 2 and 4 might be the end slots instead of the side slots. Whether this orthogonal mode is excited or not depends on the feeding, in fact any polarisation may be achieved by a proper excitation of the two modes. In the foregoing discussions we have assumed that only one mode was excited.

It is possible to obtain circular polarisation by a simple method where only one feedline is necessary (Sanford and Munson[83]), see Fig. 7.24*c*. The feed point is at a corner, so the two linearly polarised modes are excited simultaneously. The element is not square, but rectangular in such a way that the two modes are excited with equal magnitudes but 90° out of phase, one having an impedance of + 45° and the other an impedance of − 45°.

7.4.6 Admittance and equivalent network

Feed network at the edges. The easiest way to feed a rectangular microstrip is to connect a thin (high-impedance) microstrip line to an edge (Fig. 7.24*a*).

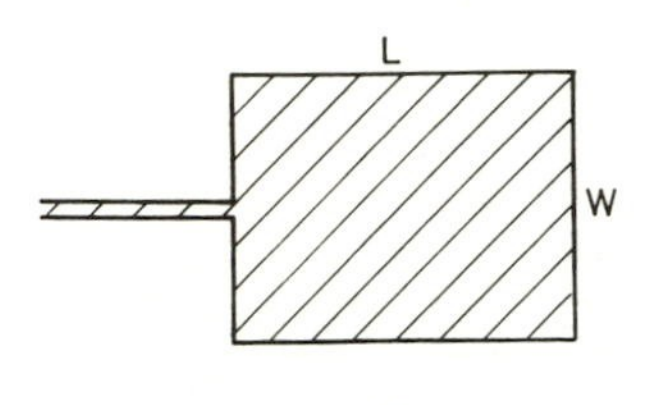

a

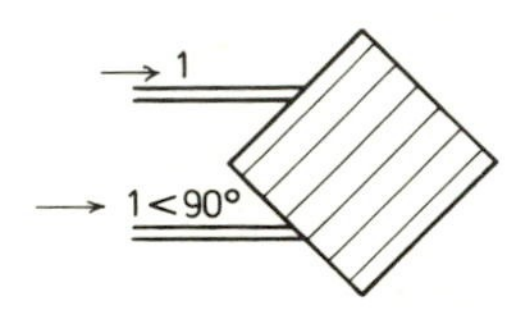

b

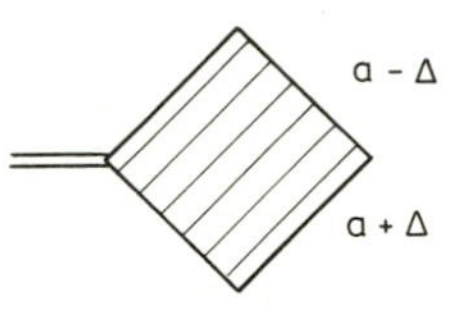

c

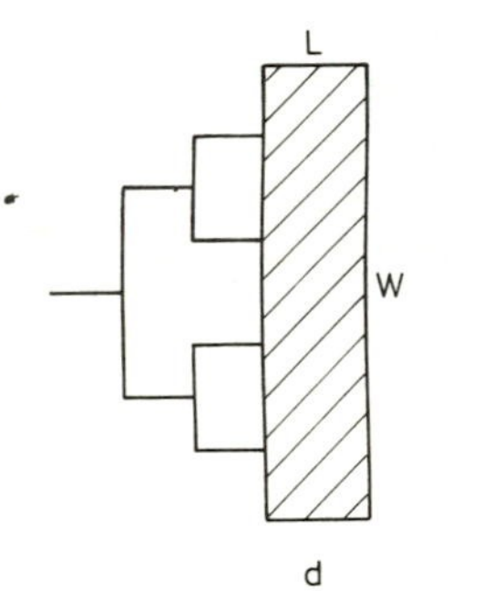

d

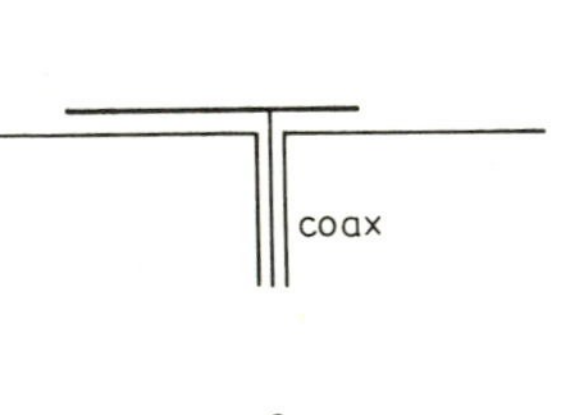

e

Fig. 7.24 *Various feeding mechanisms*
a Microstrip connected to edge of resonator
b Same principle, excitation of two modes to give circular polarisation
c As *b*, but only one feed. Unequal side lengths
d Parallel feed network
e Backfed through groundplane

As will be shown, this impedance is rather high so various transformers may be needed, but they are easily manufactured on the same substrate. By feeding the two sides with a 90° phase shift, circular polarisation may be achieved (Fig. 7.24*b*).

When the rectangular strip is very wide it may be necessary to support the mode at several places as indicated in Fig. 7.24*c* (Munson[68]).

Since a microstrip antenna in the present theory is voltage excited the conductance is well defined:

$$P_{rad} = G|V|^2 \tag{7.35}$$

where V is the voltage (effective value) at the edge, G is the conductance, and P_{rad} is the total radiated power. P_{rad} may be found through integrating over the far field, and since this quantity also appears in the formula for the directivity, we readily find for the rectangular microstrip antenna

$$G = \frac{2}{15D_{max}}\left(\frac{W}{\lambda}\right)^2 \tag{7.36}$$

which for square microstrip reduces to

$$G_{sq} = \frac{1}{30\epsilon_r D_{max}} \tag{7.37}$$

D_{max} is the maximum value of the directivity. G_{sq} is shown in Fig. 7.25 for a practical range of ϵ_r.

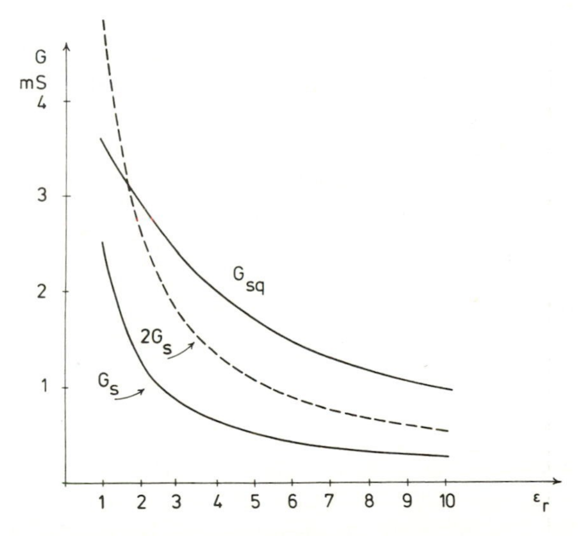

Fig. 7.25 G_{sq}: conductance of square microstrip
G_s : conductance of single edge of same width
$2G_s$: approximate value of conductance of square when external interaction is neglected

It has been found useful to treat each edge as an independent slot (Munson,[68] Howell,[41] Derneryd[20]) even though this is not strictly correct as mentioned in the previous Section. The conductance of a single slot, G_s, is given by

$$G_s = \frac{1}{30D_{max}}\left(\frac{W}{\lambda}\right)^2 \tag{7.38}$$

This may be found from eqn. 7.36 by letting the two slots coalesce and then divide by four to reduce to the value appropriate for one slot.

Eqn. 7.38 is valid for all values of W/λ, but it may be worthwhile to quote two limiting cases:

$$G_s = \frac{1}{90}\left(\frac{W}{\lambda}\right)^2 \qquad W \ll \lambda \tag{7.39}$$

and

$$G_s = \frac{1}{120}\frac{W}{\lambda} \qquad W \gg \lambda \tag{7.40}$$

A range of accurate values of G_s, based on eqn. 7.38, are given in Fig. 7.25, where the corresponding value of W/λ is given by $W/\lambda = 1/2\sqrt{\epsilon_r}$.

A simple transmission line model as shown in Fig. 7.26*a* may then be used to find the total input admittance, neglecting external interaction between the slots. Neglecting for a moment the slot susceptance B_s, the input admittance at resonance ($L = \lambda/2\sqrt{\epsilon_r}$) is

$$Y_{in} = 2G_s \tag{7.41}$$

This value is also indicated in Fig. 7.25 and it is seen that it is only equal to G_{sq} for $\epsilon_r \sim 1.5$, for larger ϵ_r it is too small. The conclusion is that the equivalent network in Fig. 7.26*a* should be used with care and should definitely not be used for large values of ϵ_r, where the two slots are closer to each other.

The equivalent network may be improved by including the mutual conductances between the slots (Fig. 7.26*b*). The input conductance in this situation at resonance is found to be

$$G_{in} = G_{11} + G_{22} + 2G_{12} \tag{7.42}$$

or

$$G_{in} = 2G_s + 2G_{12} \tag{7.43}$$

This explains the difference between G_{sq} and $2G_s$ in Fig. 7.25. In the limit when $\epsilon_r \to \infty, L/\lambda \to 0, G_{12} \to G_{11}$ and

$$G_{in} \to 4G_s \tag{7.44}$$

a doubling of the value found from the simple equivalent network.

In practice it is necessary to consider the slot susceptance or the effect of the fringing fields. The following treatment is due to Derneryd.[20] The effect of the fringing fields is equivalent to a line extension Δl, which is given by Hammerstad,[35]

$$\frac{\Delta l}{h} = 0.412\left[\frac{\epsilon_{eff} + 0.300}{\epsilon_{eff} - 0.258}\right]\left[\frac{W/h + 0.262}{W/h + 0.813}\right] \tag{7.45}$$

The corresponding end capacitance is given by

$$C = \frac{\Delta l}{vZ} \tag{7.46}$$

where v is the phase velocity and Z the characteristic impedance of the line. Due to this end susceptance, which is present at both ends of the line, the length should not be exactly half a wavelength in the line but somewhat shorter, given by

$$\tan \beta L = \frac{2YB}{G_s^2 + B^2 - Y^2} \tag{7.47}$$

where B is the susceptance ωC, and Y the line admittance. Eqn. 7.47 is of course only approximately true since it is based on the equivalent network of Fig. 7.26*a*. In practice, lengths between 0.46 and 0.49 of a wavelength in the dielectric have been reported.

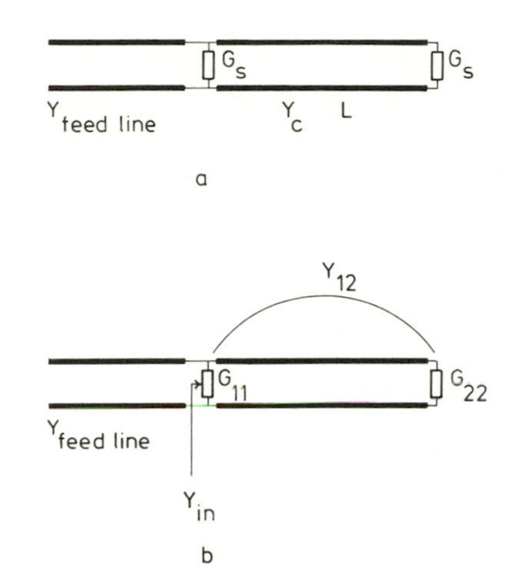

Fig. 7.26 *Equivalent networks for rectangular microstrip antenna*
a Approximate model neglecting interaction between slots
b More accurate model including interaction between end-slots

Backfed through groundplane. For some applications it may be useful to excite the resonator through the groundplane (Wiesbeck[91]) as indicated in Fig. 7.24*e*. This has the further advantage that a proper impedance level may be found through a proper position of the feed point. The variation of the impedance may be found by the following simple argument. Since the radiated power (the radiation pattern) is independent of the feed position,

$$G(z)\,|V(z)|^2 = G_0|V_0|^2 \tag{7.48}$$

where $G(z)$ is the conductance at position z and G_0 the value at the edge, and

$$V(z) = V_0 \cos\left(\pi \frac{z}{L}\right) \tag{7.49}$$

we find that

$$G(z) = \frac{G_0}{\left(\cos \pi \frac{z}{L}\right)^2} \tag{7.50}$$

or

$$R(z) = R_0 \left(\cos \pi \frac{z}{L}\right)^2 \tag{7.51}$$

This agrees well with the experimental results of Wiesbeck. At the centre of the resonator ($z = L/2$) $R = 0$, but at a position off the centre it is possible to match to a 50 Ω line.

7.4.7 Resonator Q

As already mentioned the bandwidth of a microstrip antenna is rather small, in practice of the order 1% or less. The easiest way to find a measure of the relative bandwidth is to determine the resonator Q (Ramo and Whinnery[75]) of the microwave resonator, considering that the only losses are the radiation losses:

$$Q = \frac{\omega_0 U}{P_{rad}} \tag{7.52}$$

where

$$U = \epsilon_r \epsilon_0 \iiint |E|^2 \, dx \, dy \, dz.$$

All the stored energy is assumed to exist in the interior of the antenna, i.e. between the strip and the groundplane.

Using eqn. 7.35 we find

$$Q = \frac{\sqrt{\epsilon_r} W}{240 h G} \tag{7.53}$$

or if we introduce the conductance G (eqn. 7.36)

$$Q = \frac{\sqrt{\epsilon_r}}{32} \frac{\lambda^2}{hW} D_{max} \tag{7.54}$$

where D_{max} is the maximum directivity. Since D_{max} tends to a constant when W tends to zero, the Q is inversely proportional to the cross-sectional area (hW) in this limit. For large values of W, D is proportional to W, so in this limit Q depends on h only. For square microstrip we find

$$Q_{sq} = \frac{\epsilon_r}{16} \frac{\lambda}{h} D_{max} \tag{7.55}$$

The product $Q_{sq}(h/\lambda)$ is independent of h and depends only on ϵ_r. Some values are given in Table 7.1.

Table 7.1 *Q for square microstrip antenna as a function of ϵ_r. Only radiation losses included*

ϵ_r	$Q_{sq}\dfrac{h}{\lambda}$
1	0.59
2.5	0.80
5	1.23
7	1.59
9	1.96
10.5	2.23

Example: A square microstrip antenna at 3 GHz with $h = 1$ mm, $\epsilon_r = 2.5$.

$$Q_{sq} = 0.80 \times \frac{10}{0.1} = 80$$

7.4.8 Non-rectangular resonator shapes

The main reason for the thorough discussion of the rectangular microstrip antennas was the simplicity of the theory, but it should be clear that there is a wide variety of possibilities, where the various resonators have various resonance frequencies and polarisation properties.

Within the same theoretical frame we may neglect the radiation and fringing fields and consider the resonators as dielectric-loaded resonators confined by electric walls at top and bottom and magnetic walls at the edges. Subsequently the radiated fields may be determined from the resulting magnetic wire currents as before. Consider as an example the circular disc antenna (Howell[41]). The fields in the resonator are:

$$E_z = AJ_n(k\rho)\, e^{jn\varphi} \tag{7.56}$$

$$H_\varphi = \frac{1}{j\eta} AJ'_n(k\rho)\, e^{jn\varphi} \tag{7.57}$$

where η is the impedance of the dielectric medium ($\eta = \sqrt{\mu/\epsilon}$), k the wavenumber $\omega\sqrt{\mu\epsilon}$ in the dielectric and J_n the Bessel function of the first kind. $|n| = 1$ corresponds to circular polarisation; linear polarisation may be obtained by combining $n = +1$ and $n = -1$. The resonance condition is $H_\varphi(\rho = a) = 0$ or

$$J'_n(k_c a) = 0 \tag{7.58}$$

The first roots are:

n	$k_c a$
0	3.83
1	1.84
2	3.05

Note that the roots are identical to those of TE-waves in a circular waveguide. Howell found that the measured frequencies were a few percent lower than predicted.

A ring current with radius a and a variation of $e^{jn\varphi}$ has a far field pattern of

$$E_\theta(\theta) = J_n'(k_0 a \sin\theta)$$

$$E_\varphi(\theta) = \cos\theta \frac{J_n(k_0 a \sin\theta)}{k_0 a \sin\theta} \tag{7.59}$$

where θ is measured from the normal.

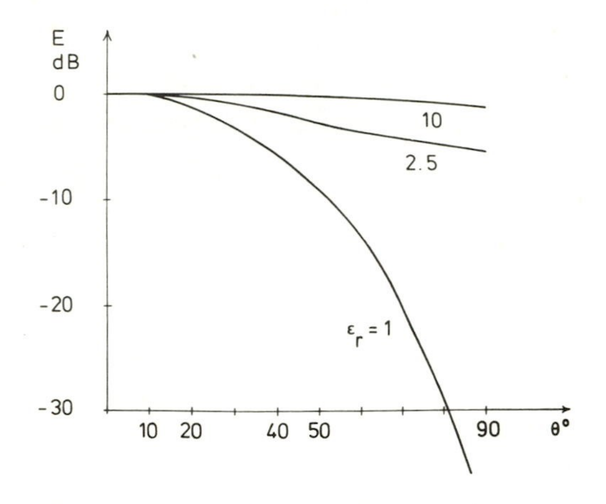

Fig. 7.27 *Vertical radiation pattern of circular disc antenna.* $n = 1$

We note from eqns. 7.58 and 7.59 that for $\epsilon_r = 1$ we have a pattern null along the groundplane. Some representative patterns for E_θ and $n = 1$ are shown in Fig. 7.27. Note the similarity with the E-plane pattern for the square microstrip in Fig. 7.22. Finite groundplane effects will of course change the pattern near $\theta \sim 90°$.

The antennas may be excited from the edge or from the back as in the rectangular case. Gains reported (Howell[41]) are between 6 and 7 dB over isotropic.

7.4.9 *Arrays of microstrip antennas*

Arraying of microstrip elements is especially attractive since the interconnecting feed network may be constructed in the same photo-etch technique as the antenna itself. For general design criteria for arrays, feed networks etc. the reader is referred to Chapters 9 and 10 of Volume II on arrays; so here we shall only mention some typical examples.

By arraying four of the wide elements in Fig. 7.24*d* Munson[68] obtained a gain of 21 dB at X-band with total outer dimensions of 7.6 cm × 12.7 cm × 0.79 cm. This corresponds to 90% aperture efficiency.

Linear arrays of square patches fed in series by narrow, high impedance lines have been designed by Derneryd.[20] He used the equivalent network of Fig. 7.26*a* and optimised the length of the lines to achieve maximum gain at broadside at the design frequency.

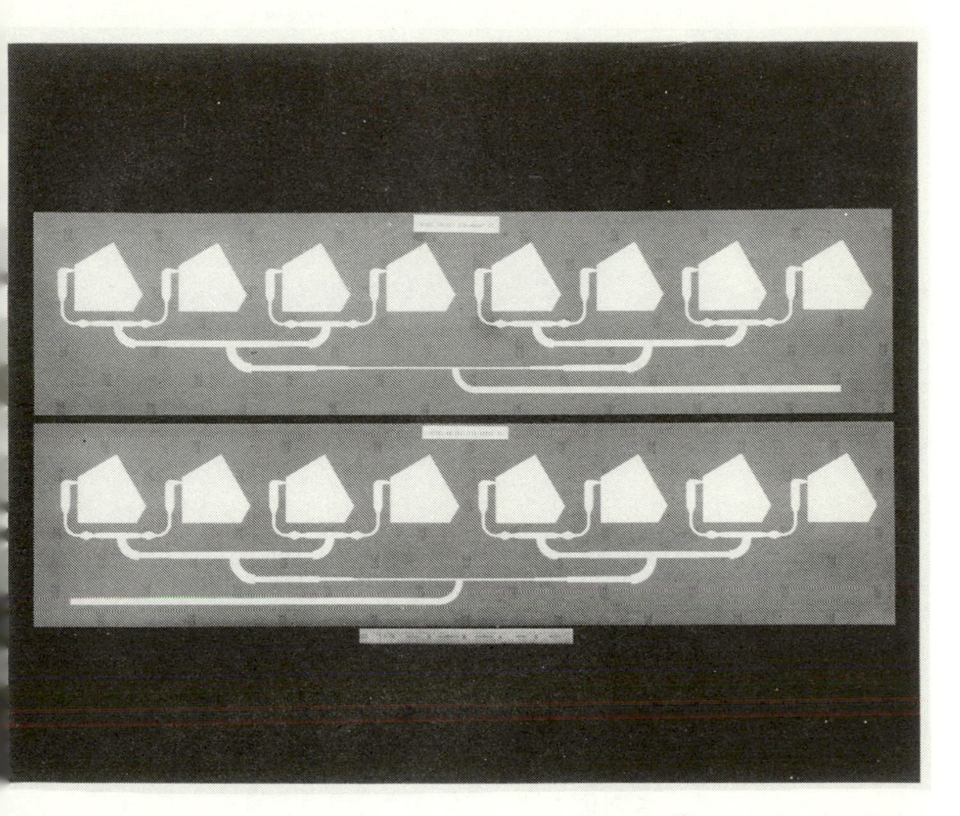

Fig. 7.28 *Circularly polarised S-band microstrip array* (Weinschel[89]). Reproduced with the permission of NASA

Fig. 7.28 shows an example of a circularly polarised S-band microstrip array, indicating some of the freedom the designer has to optimise the design (Weinschel[89]). The feed network is a so-called corporate feed network, which was also used by Munson[68] and Sanford and Klein[82] in microstrip phased arrays. The latter was an L-shaped 8-element array for use on aircraft for communication with satellite. Fig. 7.29 shows the phased array including the phase shifters. The basic element is square and circularly polarised. The whole array is not planar but conformal to the aircraft structure, another attractive feature of microstrip antennas.

A planar resonant array designed by James and Hall[44] is shown in Fig. 7.30. Here the ends of the stubs act as the radiating elements and by controlling the width of the stubs the amount of radiation leakage may be controlled (cf. eqn. 7.38,

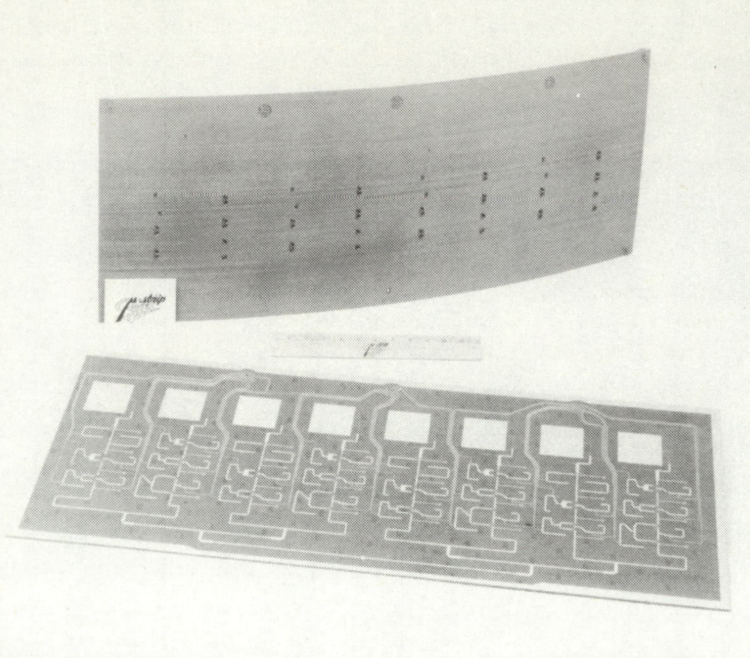

Fig. 7.29 *Conformal microstrip L-band phased array circularly polarised* (Sanford and Klein[82]). Reproduced with the permission of Ball Aerospace Systems Division

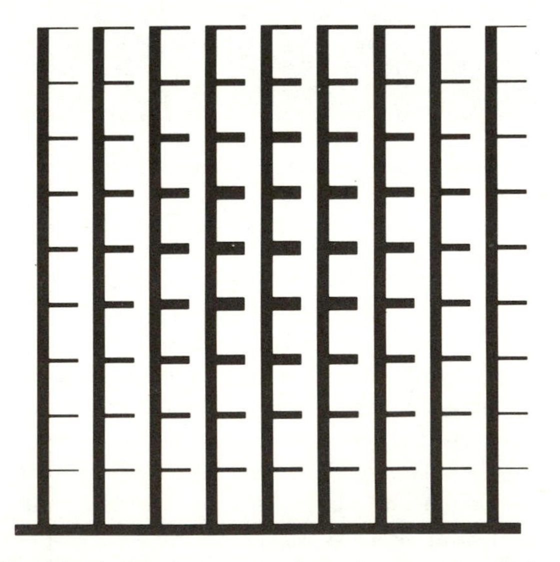

Fig. 7.30 *Two-dimensional J-band resonant microstrip array of radiating stubs* (James and Hall[44])

which gives the conductance of a single slot) in order to achieve a tapering of the aperture distribution. The bandwidth is only around a half percent, which is not surprising for a resonant array of resonant elements.

7.5 Backfire antennas

The backfire antenna is a compact, easy to construct, resonant-type antenna with excellent pattern characteristics. The gain is in an intermediate region from 15 to 25 dB, where slow-wave endfire antennas are too long and paraboloids too small. They compare favorably with horn antennas, but they are much shorter. The limiting factor will often be the bandwidth, which is surprisingly high though, from 10 to 100%.

Since the antenna operates in the resonance region, no simple theory is available and in fact all the design information has been obtained experimentally. The backfire antenna as invented by Ehrenspeck[25] was a long backfire ($L \sim 3$–4λ), essentially a slow-wave structure between reflecting plates. Later (Ehrenspeck[26]) it was found that a very short antenna ($L \sim \frac{1}{2}\lambda$) without a slow-wave structure would function as well, and this antenna was named short backfire. Although the short backfire was the last historically, we shall treat it first owing to its greater simplicity.

7.5.1 Short backfire antenna

Before embarking on the details consider the development in Fig. 7.31. In (*a*) we have a cavity-backed narrow slot in a groundplane; the cavity is excited by a dipole. The radiating currents are those on the external surface; the internal currents belong to the cavity modes, which leak out through the slot. In order to increase the bandwidth the slot size is increased; the 'internal' currents may now radiate directly, so we dispense with the groundplane (*b*). The current distribution is still to a high degree determined by the cavity resonances, to a lower degree by the feeding dipole. This antenna is close to the short-backfire and may be analysed as an open-end cavity (Hong *et al.*[40]). By reducing the cavity side-wall to a rim of height 0.25λ the cavity is transformed into two planar circular reflectors, one much larger than the other (*c*). This is the short backfire. The spacings between the plates is close to 0.5λ, the diameters of the two reflectors, being of the order 2 and 0.5λ, respectively. At higher frequencies the optical solution is relevant; the shapes must be transformed to satisfy the phase conditions of optical rays, and the open backfire cavity turns into a paraboloid with subreflector (*d* and *e*). The main advantage over the paraboloid is the fact that the reflectors are planar, making it much easier to construct. The position between an almost closed structure and an open optical antenna indicates the theoretical difficulties, and only numerical computations seem to be feasible. Nielsen and Pontoppidan[97] have obtained good agreement with experiment for a short backfire without the rim; the current distribution on the large reflector was assumed to be the same as for an infinite plane, while the small reflector was replaced by a number of short-circuited wires simulating a reflector of circular shape. The computed gain was 12.4 dB above isotropic. The rim is important in bringing down the side-lobes and leads to a further gain increase; Ehrenspeck[26] reports gains around 15 dB.

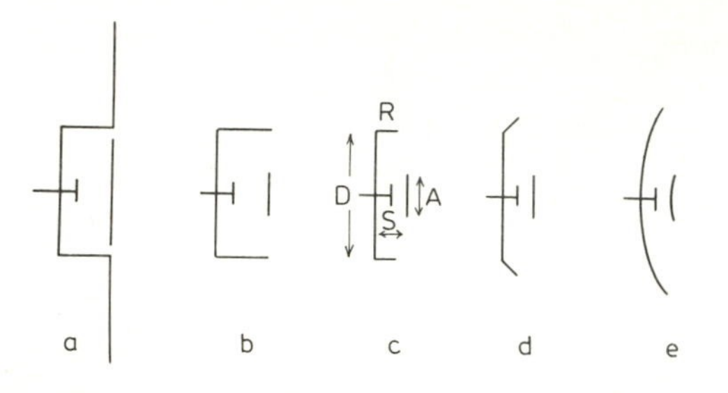

Fig. 7.31 The short backfire (*c*) viewed as a transition from a cavity-backed slot (*a*) and a paraboloid with sub-reflector (*e*)

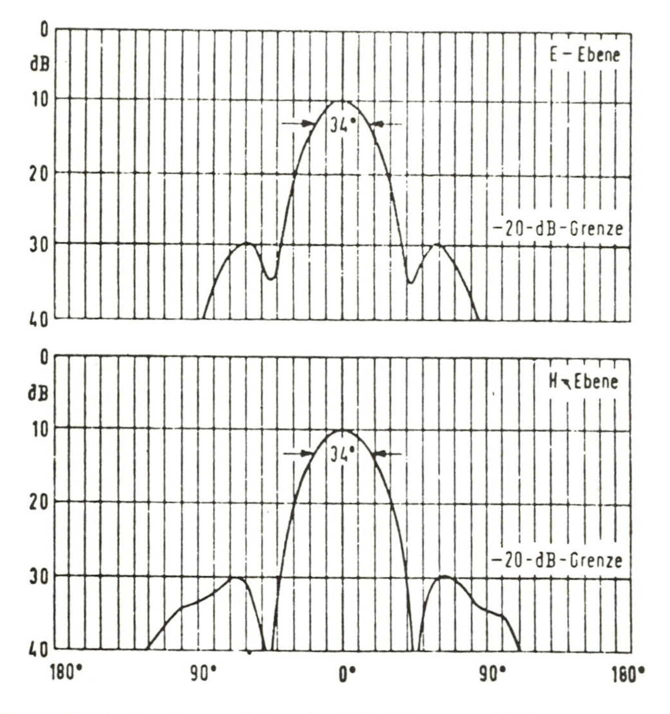

Fig. 7.32 *E- and H-plane patterns of short backfire* (Ehrenspeck[28])

Fig. 7.32 shows the radiation pattern in E- and H-plane for a short-backfire. The dimensions are $D = 2\lambda$, $A = 0.4\lambda$, $S = 0.5\lambda$, $R = 0.25\lambda$. The sidelobes are all below 20 dB and the backlobe is below 30 dB. Rather remarkably, considering the simple dipole feed, the E- and H-plane patterns are close to each other, which means that the short backfire is an excellent feed antenna for a parabolic antenna (Ehrenspeck[28]). The antenna is of course easily adapted to circular polarisation by a simple change of the feed.

The bandwidth is to a large extent determined by the exciting dipole, although a lower fundamental limit is set by the 'cavity' mode. By using a broadband dipole Ehrenspeck[28] obtained a VSWR less than 2 over a 2:1 frequency range. If the bandwidth requirements are less than 10% a waveguide feed may be used, which is somewhat easier at the higher microwave frequencies (Large[58]). The sidelobes are slightly higher in this configuration (15 dB in the E-plane).

7.5.2 Long backfire antenna

A gain increase may be obtained by stretching the antenna so that we have a greater length in wavelengths between the subreflector R and the main reflector M, which also would have to increase. If no further changes were made the antenna characteristics would rapidly deteriorate since the planar reflector would not satisfy the phase conditions of geometrical optics. Instead, a slow-wave structure is inserted between the two reflectors such that the wave energy is bound to a circular channel surrounding the structure. Since a bound surface wave has no phase variation in the radial direction a planar reflector has the optimum shape for in-phase excitation of the reflector. A row of short-circuited dipole elements is used in Fig. 7.33(*a*) similarly to a Yagi antenna, but other slow-wave structures could have been used as well. Ordinary endfire antennas are discussed in Section 7.6.

Radiation mechanism. The radiation mechanism of the long backfire antennas is closely related to that of ordinary surface wave antennas. It suffices to mention here that there is an optimum phase velocity for a given length of antenna, and that the gain is proportional to length when this optimised condition is satisfied. If we take a Yagi array as an example,

$$G_{iso} = 7\frac{L}{\lambda} \tag{7.60}$$

is a good approximation for the gain over isotropic (Zucker,[94] p. 16.32) for the optimised case. Let us for a start replace the finite reflector M in Fig. 7.33(*a*) by an infinite reflector (*b*). For the halfspace to the right we can replace the reflector by an image of the structure (*c*), such that the complete antenna has two feed points exciting a structure of double length. A sketch of the patterns is indicated below the antennas. Since the reflector R is effective in reducing the backlobe for the equivalent endfire antenna, the two patterns from F_1 and F_2 are almost independent, such that the total pattern in (*c*) consists of a narrow lobe to the left, excited by F_1, and a narrow lobe to the right, excited by F_2. Each lobe corresponds to the length $2L$, but in the real case of (*b*), where the reflector is present, of course only the lobe to the right is present. The gain in case (*b*) is then given by

$$G_{iso} = 14\frac{L}{\lambda} \tag{7.61}$$

Thus we have a 3 dB gain increase due to the infinite reflector and not 6 dB as is sometimes stated. We see now also the reason behind the name backfire since it is the image of the source, which gives the main contribution. The structure in (*b*) without the reflector would radiate to the left.

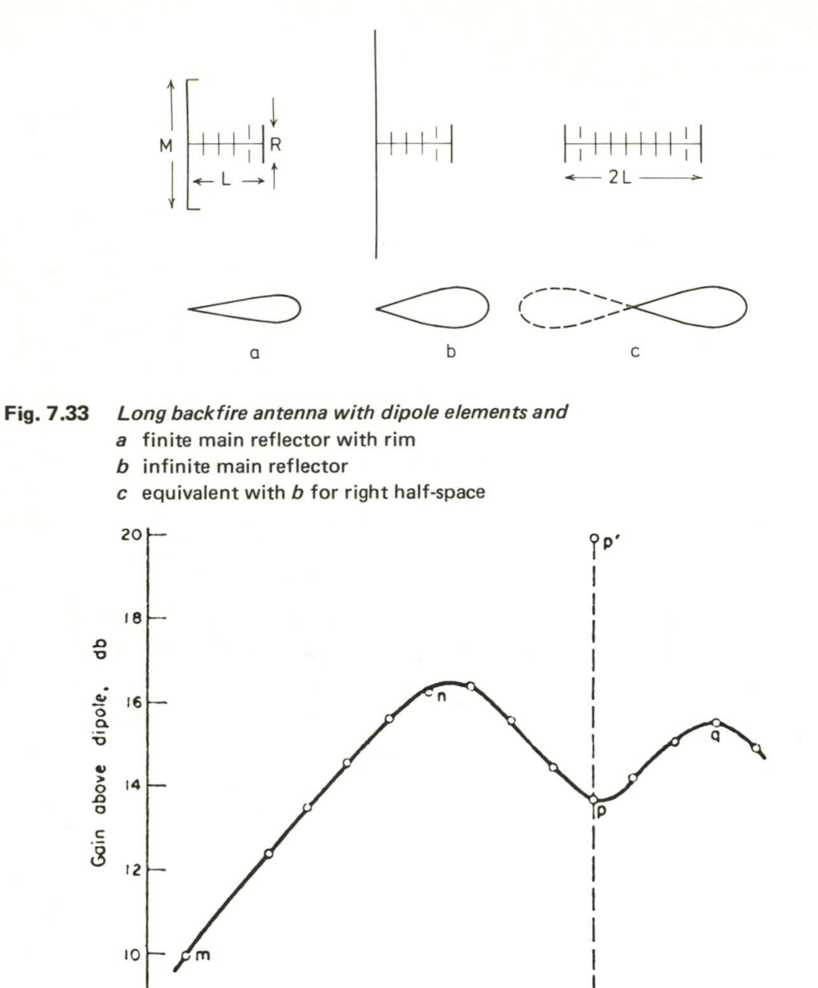

Fig. 7.33 *Long backfire antenna with dipole elements and*
a finite main reflector with rim
b infinite main reflector
c equivalent with *b* for right half-space

Fig. 7.34 *Gain for 4λ backfire antenna as a function of planar reflector M* (Ehrenspeck[27])

The size of the planar reflector is important for optimising the gain. Fig. 7.34 (Ehrenspeck[27]) shows the gain of a 4λ long backfire antenna with dipole elements as a function of the diameter of *M*. It is noted that the gain dB is a linear function of size for small reflectors.

When M increases further the gain starts to oscillate around the asymptotic value it would have for an infinite reflector. The gain in dB above a half-wave dipole as calculated from eqn. 7.61 is 15.3 dB, which is in good agreement with the experimental results of about 15.0 dB.

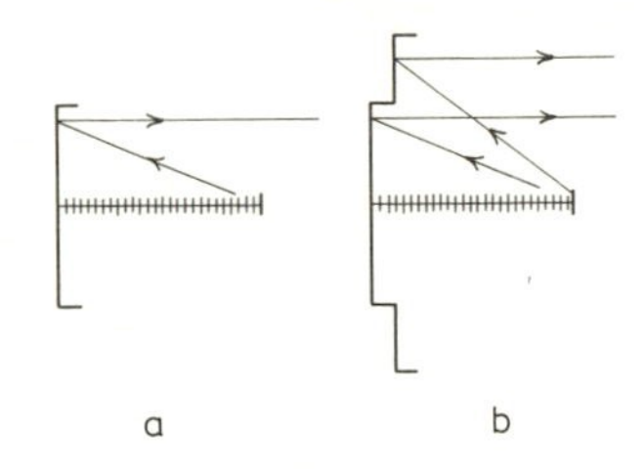

Fig. 7.35 *Influence of reflection from edge of reflector by direct feed radiation. Additional step in reflector creates the proper in-phase conditions in the far field*

The reason for the oscillations is the direct radiation from the feed which will dominate when the radius is larger than a certain value (Zucker[95]). In the beginning for $D \sim 4\lambda$ this reflected feed radiation will add an additional 1.5 dB to the gain due to proper phase relationships (Fig. 7.35*a*), but for $D \sim 6\lambda$ the next Fresnel zone is entered and the gain is less than that of the infinite reflector. This situation can be corrected by shifting the outer part as in a stepped zone reflector antenna (Fig. 7.35*b*), (Zucker[95]), and in this way a gain of 20 dB above a dipole may be obtained for $D \sim 6\lambda$. We note ther we are now approaching the situation in Fig. 7.31*e*, where the reflector must be shaped.

7.5.3 Design methods

As is evident from the preceeding Section a backfire antenna is a rather complicated structure and much of the design information available has been found through experiment.

A short backfire may be made with the dimensions mentioned earlier at a band-centre frequency, $D = 2\lambda$, $S = 0.5\lambda$, $R = 0.25\lambda$, $A = 0.4\lambda$. A broad band dipole with suitable balun and transformer to 50 Ω may be used as a feed antenna (Ehrenspeck[28]) with a rather surprising bandwidth of 2:1 (defined as VSWR less than 2) with low sidelobes over the whole band. Circular reflectors are natural for circular polarisation; for linear polarisation quadratic or rectangular shapes work equally well, and in this case it is only necessary to include the rim on the E-plane sides. The result is constructionally a very simple antenna with a gain over isotropic from 10.5 to 15 dB over the band.

Since the long backfire is equivalent to two oppositely directed endfire antennas of equal strength, when the main reflector is large, we can use the theory of endfire antennas in Section 7.6 to give a more accurate design. With reference to Fig. 7.36 the symmetric structure of Fig. 7.33*c* is excited independently by + 1 at the left-hand feed and − 1 at the right-hand feed. The minus sign is necessary for satisfying the boundary conditions on the reflector. We may consider the two sources one at a time, and owing to the linearity of the system the feed not in use should be short-circuited.

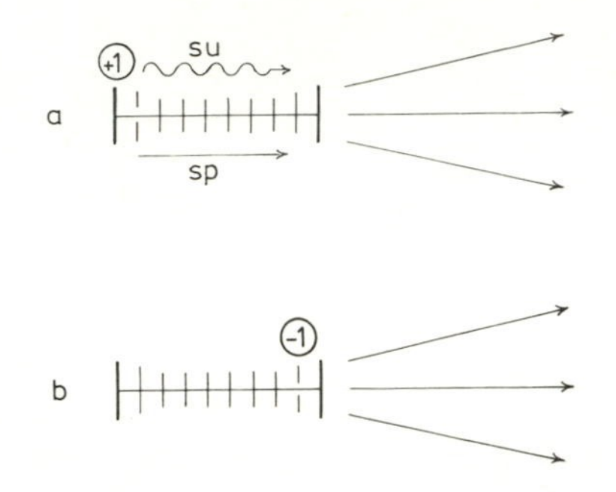

Fig. 7.36 *The far field to the right is given by two sources: + 1 at the left feed (a), and − 1 at the right feed (b). The feed at the left excites a surface wave, su, and a space wave, sp, which are both diffracted at the end. The feed at the right radiates a small backlobe*

The source at the left excites a surface wave with wavenumber $k_0\tau$ ($v = c/\tau$), the surface wave propagates without loss a distance $2L$ and is diffracted into a radiating space wave with relative amplitude and phase $A_{su} = |A_{su}|e^{j\varphi_{su}}$. At the same time a modified space wave travels the same distance with a velocity close to the velocity of light and is also diffracted into a space wave with an amplitude A_{sp}. The diffraction around the end is slightly different from the normal endfire case due to the small reflector, but the principles are the same. For the other excitation (*b*) we have a small back radiation F_b, where we assume F_b real.

Collecting the terms, we get for the total far field in the axial direction to the right

$$F_e \simeq \frac{e^{-jk_0R}}{R}\,[(+1)\{A_{sp}e^{-j2k_0L} + A_{su}e^{-j2k_0\tau L}\} + (-1)F_b] \tag{7.62}$$

where the feed to the right is used as phase centre. As far as the phases are concerned we may assume A_{sp} to be negative, real (see Section 7.6) and F_b positive, real. Assuming that we want the three terms to be in phase for maximum gain (which is not strictly true), we get

$$2k_0L = n2\pi \tag{7.63}$$

or

$$L = n\frac{\lambda}{2} \tag{7.64}$$

which brings the two space waves in phase.

$$2k_0L\tau = 2\pi n + \pi + \varphi_{su} \tag{7.65}$$

or

$$\tau = \frac{n + 0.5 + \varphi_{su}/2\pi}{n} \tag{7.66}$$

which then defines the slowness factor (or axial index of refraction). Experience shows that φ_{su} is around $-\pi/3$, so that

$$\tau = \frac{n + 0.33}{n} \tag{7.67}$$

in agreement with Ehrenspeck.[27] Eqn. 7.65 is equivalent to the Hansen–Woodyard condition, and eqn. 7.64 has been found experimentally by Ehrenspeck. Once τ is determined a slow-wave structure may be chosen and designed by standard techniques.

Eqn. 7.64 does not mean that the long backfire is a narrow band antenna. A 4λ antenna optimised for 3 GHz has a gain of 23.5 dB and a gain of 20 dB at 2 GHz, corresponding to a constant aperture area. The aperture efficiency turns out to be 60%, which is a standard value for a paraboloid. The pattern deteriorates at the lower end of the band with sidelobe level around 10 dB (Ehrenspeck[27]).

The main reflector should have such a size that the edge reflection for the ray from the feed adds in phase with the surface wave contribution. Zucker[95] has found that

$$\frac{D}{\lambda} \sim 2.2\sqrt{\frac{L}{\lambda}} \tag{7.68}$$

If the stepped reflector is used the ratio of the outer to inner diameter should be about 1.5, and the steps, including the rim, 0.25λ.

If higher gains than 25 dB are wanted, it may be advantageous to form a broadside group of elements on the same reflector.

7.6 Dielectric antennas

The use of dielectric materials in antennas adds an additional degree of freedom because both the shape and dielectric permittivity are important parameters. We shall concentrate on the dielectric rod antenna, since it presents a solution to a much wider class of antennas, namely surface-wave antennas. This means that, in most of the following, artificial dielectrics may be used if preferable; the general principles will be the same. Surface-wave antennas are of moderate gain, usually less than 20 dB, and broad band because reflections from the end are minimised. The main advantage is a relatively large effective aperture compared with the physical aperture, which makes surface-wave antennas well suited for flush mounting along a metallic surface. Among artificial-dielectric antennas are Yagi antennas and helical antennas.

Dielectric antennas have not been used to a very large extent as feed radiators. This is surprising because a shaping of the dielectric can change the illumination and cross-polarisation properties. The reason is probably the lack of design methods and easy computational methods. An increase in applications is to be expected in the future, when the techniques are more advanced. Some progress has been made concerning spherical and more general non-cylindrical shapes, whiwh will be covered at the end of the Section.

7.6.1 Long cylindrical rod

The finite dielectric rod of uniform cross-section as shown in Fig. 7.37 is assumed to be excited by a waveguide. The complete boundary-value problem is very complicated, and various approximate theories have in the past led to various misunderstandings or disagreements with experiment, but they all contain part of the truth.

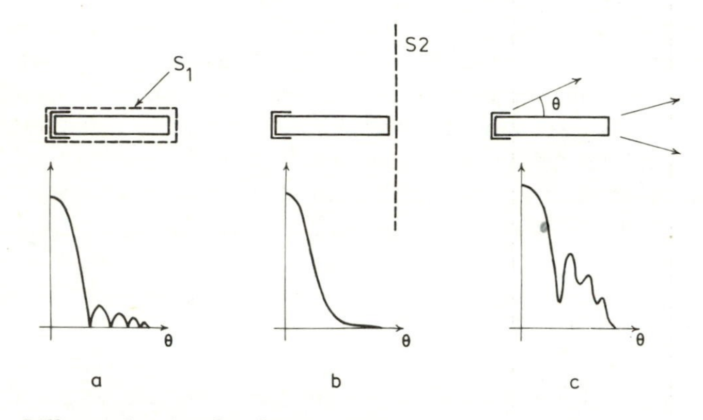

Fig. 7.37 *Different theories of surface wave antennas*
a Line-source with surface wave
b Aperture illuminated with surface wave
c Direct feed radiation plus scattering of surface wave and space wave around end

As is well known the far field may be determined when the near-field is known over a closed surface, as S_1 in Fig. 7.37*a*. This is valid exactly, the only problem being that the fields are not known to a high accuracy. In the early theories (Mallach[62] and Mueller and Tyrrell[67]) a simple line source was assumed with a propagating slow wave of constant strength

$$I \sim e^{-jk_0 \tau z}, \qquad 0 < z < L \tag{7.69}$$

which gives the pattern

$$F_a(\theta) = \frac{\sin u/2}{u/2} \tag{7.70}$$

where $u = k_0 L(\tau - \cos\theta)$. $k_0 \tau$ is the wavenumber of the surface wave and θ is the angle measured off axis. The pattern is also indicated in Fig. 7.37*a* and shows deep nulls contrary to the experimental results. As an example a 10-wavelength antenna with $\tau = 1.1$ should have a null in the endfire direction, since $u = k_0 L(\tau - 1) = 2\pi \times 10 \times 0.1 = 2\pi$, but a well-defined mainlobe usually results experimentally. The explanation is, of course, that the assumed aperture distribution is not the correct one.

Zucker[93] was the first to realise that it was better to consider a surface wave antenna as radiating from the discontinuities, that is from the feed and the termination. Instead of the surface of the antennas, the terminal plane, S_2, normal to the axis, is viewed as the effective aperture illuminated by the incident surface wave. Again, the formulation is exact, but the aperture illumination is only known approximately. Assuming a 100% surface wave content, the resulting pattern is approximately

$$F_b(\theta) \sim \frac{1}{u} \sim \frac{1}{\tau - \cos\theta} \tag{7.71}$$

as indicated in Fig. 7.37*b*.

The differences between the two theories are striking; the new model has an endfire gain independent of length, no sidelobes and a theoretical possibility of infinite gain. The terminal plane theory was applied to the dielectric rod by Brown and Spector,[10] but in order to get partial agreement with experiment they had to assume some feed radiation. Two simplifying assumptions were made about this radiation, first it was considered to be equal to the radiation from a horn in free space neglecting the dielectric, secondly its magnitude was unknown. It is the precise way of taking the source into account which is the crucial point in surface-wave antenna theory. In the following Sections the radiation is found as originating from the feed in the presence of the infinite rod, supplemented by the scattering at the end of the incident surface wave and the incident space wave (Fig. 7.37*c*). More details than covered here may be found in Andersen.[4]

The infinite cylinder. In order to fully understand the radiation mechanism of the finite cylinder it is important first to study the various waves excited on the infinite cylinder by a finite source, and the easiest source to treat is a magnetic ring current

$$\bar{K} = K\, e^{j\varphi}\delta(r - r_s)\delta(z)\hat{\varphi} \tag{7.72}$$

where r_s is the radius of the ring (see Fig. 7.38). Only the dipole-mode, $e^{j\varphi}$, is considered here, since it is the only one which has a non-zero field in the axial direction.

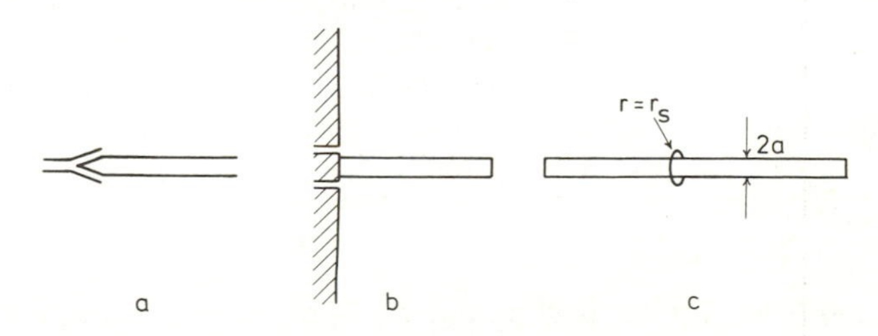

Fig. 7.38 *a* dielectric rod excited by waveguide
b dielectric rod excited by slot in groundplane
c dielectric rod excited by magnetic ring current

(*a*) *The surface waves.* The surface waves propagate without decaying (in the ideal case of a lossless dielectric) along the cylinder with a phase velocity less than the velocity of light by a factor τ, $v = c/\tau$.

Physically they are related to the trapping of the energy inside the cylinder due to total reflection from the boundaries, but nevertheless a large part of the energy (axial Poynting vector) may be situated outside the cylinder when the wave is loosely bound, $\tau \sim 1$. τ depends on ϵ and k_0a and may be determined numerically from a rather complicated transcendental equation

$$\left[\frac{\epsilon}{\sqrt{\epsilon\mu-\tau^2}}J_1'(u_1)K_1(u_2)+\frac{1}{\sqrt{\tau^2-1}}J_1(u_1)K_1'(u_2)\right]\cdot\left[\frac{\mu}{\sqrt{\epsilon\mu-\tau^2}}J_1'(u_1)K_1(u_2)\right.$$
$$\left.+\frac{1}{\sqrt{\tau^2-1}}J_1(u_1)K_1'(u_2)\right]-\frac{\tau^2(\epsilon\mu-1)^2J_1^2(u_1)K_1^2(u_2)}{u_1^2(\tau^2-1)^2(\epsilon\mu-\tau^2)} = 0 \qquad (7.73)$$

where $u_1 = k_0a\sqrt{\epsilon\mu-\tau^2}$ and $u_2 = k_0a\sqrt{\tau^2-1}$ are the transverse propagation constants in the dielectric and in the air multiplied by the radius. The dispersion equation, eqn. 7.73, is complicated because all field components are present, a so called hybrid mode, where it is not possible to separate into TE or TM-waves. The mode under discussion is usually called the HE_{11} mode, the first number refers to $e^{jn\varphi}$-variation ($n = 1$) and the second to the lowest root. If we define cut off by $\tau = 1$, the cut off frequencies are given by

$$J_1(k_ca\sqrt{\epsilon\mu-1}) = 0 \qquad (7.74)$$

or

$$k_ca\sqrt{\epsilon\mu-1} = 0, 3.8317, \ldots \qquad (7.75)$$

Thus we see that in theory the HE_{11} mode has no lower cut off frequency, in practice

$$k_0a\sqrt{\epsilon\mu-1} \geqslant 0.4 \qquad (7.76)$$

before the dielectric has any appreciable effect. Eqn. 7.76 should not be interpreted such that very thin rods may be used with high values of ϵ and μ, it may be shown that k_0a under all circumstances should be larger than about 0.3. Some numerical values for $\epsilon = 2.56$ are shown in Fig. 7.39*a* (Zucker[96]). The higher order modes shown are symmetric modes ($n = 0$), which may be avoided by proper excitation, the next dipole mode appears at $2a/\lambda_0 = 0.977$. Below is the launching efficiency, the relative amount of power launched into the surface wave by the ring source, the remaining power radiates directly. It is clearly seen that there is an optimum ring radius for a given τ (given k_0a), and that the launching efficiency rapidly goes to zero when τ approaches unity. In practice τ lies between 1.01 and 1.1 (see later).

(*b*) *Radiated field.* The far field pattern $F_\infty(\theta)$ of the ring in the presence of the infinite cylinder may be found by standard techniques, but the expression is too complicated to be given here. Instead we shall discuss briefly an approximate value valid in the limit where $k_0a, k_0r_s, k_0a\sqrt{\epsilon\mu-\cos^2\theta} < 1$,

$$F_\infty(\theta) = -j\frac{k_0a}{4}\,\frac{2(1+\epsilon)+(k_0a)^2(\epsilon-1)\sin^2\theta\,\ln j(k_0a\sin\theta\Gamma)/2}{(1+\epsilon)(1+\mu)+(k_0a)^2(\epsilon\mu-1)(1+\cos^2\theta)\ln j(k_0a\sin\theta\Gamma)/2} \qquad (7.77)$$

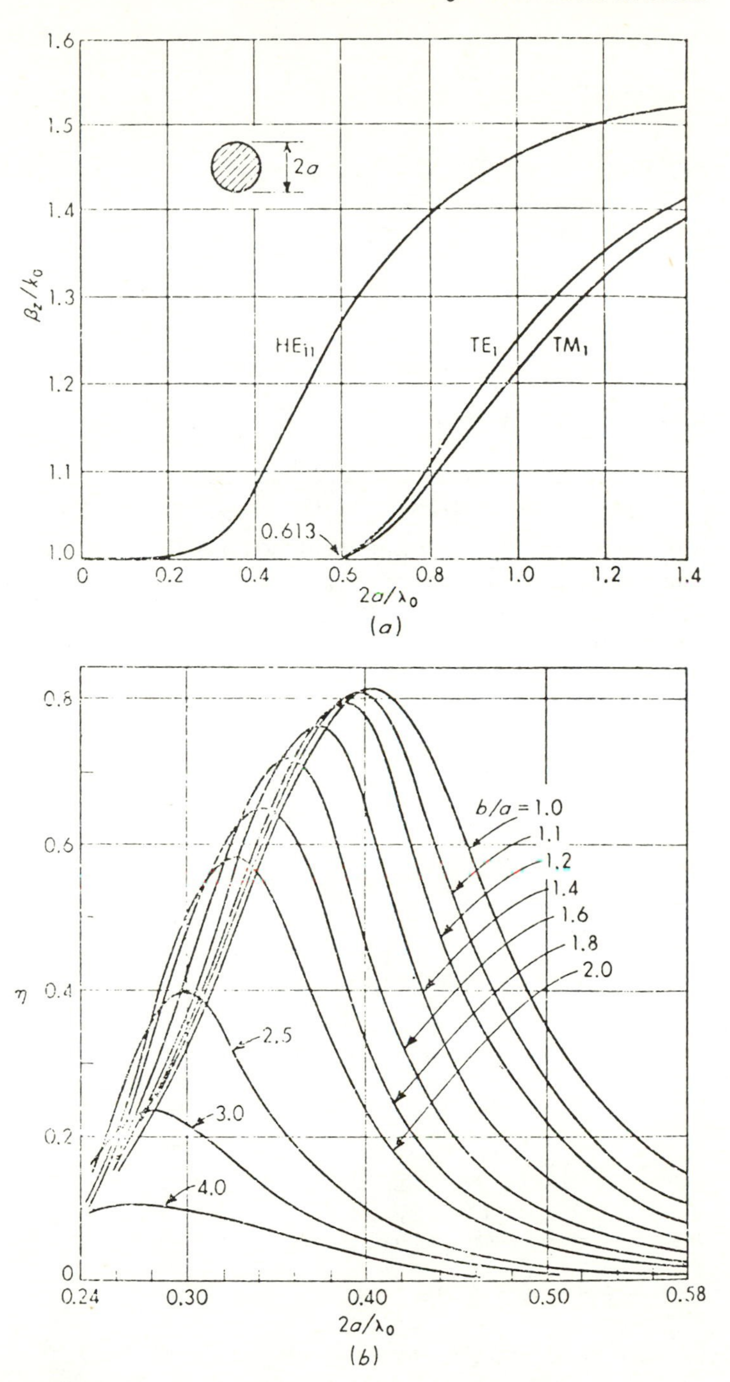

Fig. 7.39 *Dielectric cylinder with $\epsilon = 2.56$*

a relative axial wavenumber τ

b launching efficiency (power in surface wave over total power). *b* equals radius of magnetic ring current

(Zucker[96])

where $F_\infty(\theta)$ is the factor in

$$\zeta H_\varphi = F_\infty(\theta)\frac{e^{-jk_0R}}{R}K \tag{7.78}$$

and Γ equals 1.781 ($= e^\gamma, \gamma = 0.577\,21$, Euler's constant).

In eqn. 7.77 it is assumed that $r_s = a$.

It is noted that for $\epsilon = \mu = 1, F_\infty(\theta)$ reduces to

$$F_\infty(\theta) = -j\frac{k_0a}{4} \tag{7.79}$$

an omnidirectional pattern for a small ring. For any $\epsilon \neq 1$

$$F_\infty(\theta) \to 0 \quad \text{for} \quad \theta \to 0$$

since

$$F_\infty(\theta) \simeq -j\frac{k_0a}{4}\,\frac{1}{1+(k_0a)^2(\epsilon-1)/(\epsilon+1)\ln j(k_0a\sin\theta\Gamma)/2} \tag{7.80}$$

when θ is sufficiently small. This is an important fact, valid for all surface wave structures, that the feed pattern tends to zero along the structure. This means that to a certain approximation only the surface wave is present near the surface far from the source, and that in this approximation the far field pattern equals a combination of the feed pattern and the scattered surface wave. An example of $F_\infty(\theta)$ is shown in Fig. 7.40.

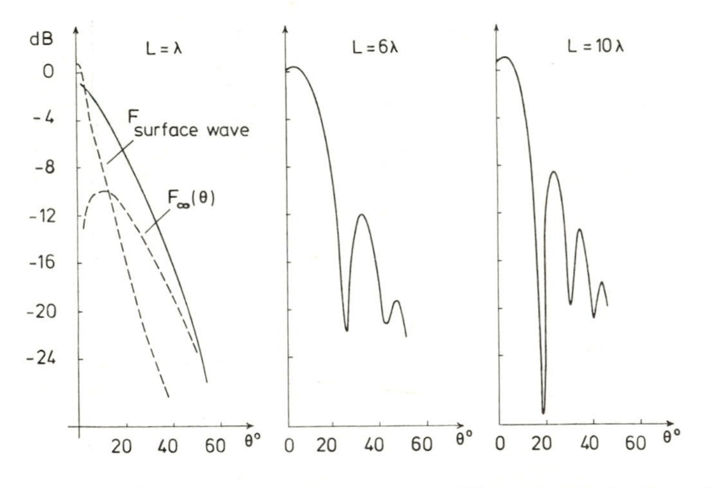

Fig. 7.40 *Radiation pattern for dielectric rod for three different lengths for* $F_{sp} = 0$
$\epsilon = 2.6$
$k_0a = 0.995$, $k_0r_s = 1.69$

(*c*) *Space wave along cylinder.* Although we just found that $F_\infty(\theta)$ was zero there is a remainder field along the cylinder which decays faster than R^{-1}, the characteristic spreading factor of a spherical wave. This remainder field is important for explaining the gain variations with length, and we call it a modified space wave or simply the space wave, since its characteristics are close to those of a true space wave.

For thin rods it may be found that

$$\zeta_0 H_{\varphi, sp} \simeq \frac{-j(\epsilon+1)^2}{(k_0 a)^3(\epsilon\mu-1)^2}\frac{e^{-jk_0 z}}{z}\frac{K}{\left(\ln\frac{j2z}{k_0 a^2\Gamma^2}-\frac{(\epsilon+1)(\mu+1)}{(k_0 a)^2(\epsilon\mu-1)}\right)^2+\pi^2} \tag{7.81}$$

which is the modified space wave on the surface of the cylinder for $r_s = a$. Apart from the spherical spreading $(1/z)$ it has a logarithmic decay asymptotically. In the limit $k_0 a\sqrt{\epsilon\mu-1} \to 0, \mu = 1$, the modified space wave approaches

$$\zeta_0 H_{\varphi, sp} \simeq -j\frac{k_0 a}{4}\frac{e^{-jk_0 z}}{z}K \tag{7.92}$$

in agreement with eqn. 7.78.

The finite cylinder. When the waves on the infinite cylinder meet the end they are scattered in a complicated way, partly into a backward-running surface wave, partly into a backward-running space wave, and partly into forward-running free-space spherical waves. When the waves are loosely bound the latter is the dominant one and we shall neglect the others completely. In Andersen[4] an approximate method based on the compensation theorem is derived, and the results will be given here. The method gives the far field pattern as a correction to the pattern of the source over the infinite cylinder; all we need to know are the fields on the infinite cylinder tangential to the surface. Referring to Fig. 7.37*a* we can say that the surface of integration is the cylindrical surface complementary to S_1, lying on the antenna extension. In this way we do not need to know the complicated fields near the source, only the asymptotic fields far away from the source.

The result is the following far field pattern for H_φ or E_θ, which is independent of φ (except for an $e^{j\varphi}$-variation for all fields) for this excitation.

$$\begin{aligned} F(\theta) = {} & F_\infty(\theta) \\ & -\frac{a}{4}\frac{\exp[-jk_0 L(\tau-\cos\theta)]}{\tau-\cos\theta}F_{su} \\ & -j\frac{k_0 a}{4}\int_L^\infty \exp[-jk_0(1-\cos\theta)z]f_1\left(\frac{z}{l}\right)dz\, F_{sp} \end{aligned} \tag{7.83}$$

where

$$F_{su} = [E^0_{z,su}(J_2-J_0)+j\zeta_0 H^0_{z,su}\cos\theta\,(J_2+J_0)+j\zeta_0 H^0_{\varphi,su}2\sin\theta J_1]$$

and similarly for F_{sp}, and the following abbreviations have been introduced:

$$J_n = J_n(k_0 a\sin\theta)$$

$$f_1(x) = \int_0^\infty \frac{e^{-xt}\,dt}{(\ln jt)^2+\pi^2}$$

$$l \simeq \frac{k_0 a^2 \Gamma^2}{2} \exp \frac{(\epsilon + 1)(\mu + 1)}{(k_0 a)^2 (\epsilon\mu - 1)}$$

$E_z^0, H_z^0, H_\varphi^0$ are amplitude factors for the surface fields for the surface wave and space wave. For thin cylinders the following relations are valid for the amplitude factors both for the surface wave and space wave:

$$\frac{E_z^0}{\zeta_0 H_\varphi^0} = jk_0 a \frac{\epsilon\mu - 1}{\epsilon + 1} \tag{7.84}$$

$$E_z^0 = -j\zeta_0 H_z^0 \tag{7.85}$$

The latter equation shows that the fields satisfy the so-called balanced hybrid condition which we shall mention later in Section 7.9 as a condition for rotationally symmetric patterns for linearly polarised fields.

The first term in eqn. 7.83 is the previously discussed pattern of the source over the infinite cylinder. The second term is the radiation pattern (E-plane) of the incident surface wave, where we note the similarity with eqn. 7.71. The third term may be interpreted as the end pattern due to the incident space wave, or alternatively as a correction to $F_\infty(\theta)$ due to the finite length.

(*a*) *Space wave along cylinder neglected.* The main effect of the space wave is in the endfire direction, so the general pattern shape, especially the sidelobe structure is well described by the first two terms in eqn. 7.83. The length enters only in the relative phase between the two terms. Fig. 7.40 gives examples of the pattern for three different lengths, where the surface wave end pattern and $F_\infty(\theta)$ are also indicated. When the length increases the number of sidelobes increases, their level increases and they occur closer to endfire. Deep minima occur for long antennas where an out of phase condition occurs together with an equal amplitude condition. The far-out sidelobes are determined by $F_\infty(\theta)$.

It is noted that in general the sidelobes for dielectric antennas are rather high, which is related to the difficulty of achieving a launching efficiency close to 100%. The power not launched into the surface wave appears in the sidelobes. As we shall see in Section 7.6.3 the situation may be somewhat improved by tapering.

The peculiar dip in the patterns near endfire in Fig. 7.40 vanishes when the space wave is included.

(*b*) *Space wave along cylinder included.* As mentioned, the space wave term has only a minor effect on the shape of the radiation pattern, so we shall concentrate on the field in the endfire direction or the endfire gain. For $\theta = 0$ eqn. 7.83 reduces to

$$F(0) = -\frac{a}{4} \frac{e^{-jk_0 L(\tau - 1)}}{\tau - 1} F_{su} - jl \frac{k_0 a}{4} f_2 \left(\frac{L}{l}\right) F_{sp} \tag{7.86}$$

since $F_\infty(0) = 0$, where

$$F_{su} = -(E^0_{z,su} - j\zeta_0 H^0_{z,su})$$

and similarly for F_{sp}

$$f_2(x) = \int_0^\infty \frac{e^{-xt}\,dt}{t[(\ln jt)^2 + \pi^2]} \tag{7.87}$$

$f_2(x)$ is a slowly decaying function of the argument. Asymptotically

$$f_2(x) \sim \frac{1}{\ln(-j\Gamma x)} \tag{7.88}$$

The maximum phase of $f_2(x)$ is about 20°. The function $f_2(x)$ is well known in antenna theory, since it describes the decay of current along an infinite, thin metallic cylinder.

Thus we see from eqn. 7.86 that asymptotically for large lengths the endfire gain will depend only on the surface wave, since the second term tends to zero. However, in practice the decay is so slow that both terms are of importance, and a relevant question is: What is the optimum length? From eqn. 7.86 we see that this happens when the two terms are in phase. It is of interest to trace back the origin of the various phases, which can be done by considering H_φ as the variable to follow:

$$\begin{aligned}
&\text{Excitation:} && \varphi_{sp} = 0 \quad \text{(by definition)}\\
& && \varphi_{su} = -\frac{\pi}{2}\\
&\text{Propagation:} && \varphi_{sp} = -k_0 z + ph(f_1)\\
& && \varphi_{su} = -k_0 \tau z\\
&\text{Forward scattering:} && \varphi_{sp} = \pi + ph(f_2)\\
& && \varphi_{su} = +\frac{\pi}{2}
\end{aligned} \tag{7.89}$$

Collecting the terms we find

$$\pi + ph(f_2) - k_0 L = -k_0 \tau L$$

or

$$k_0 L(\tau - 1) = \pi - ph(f_2) \tag{7.90}$$

This is in good agreement with the Hansen–Woodyard condition, derived on an entirely different basis. In words we can state it approximately as follows: In the endfire direction both the scattered surface wave and space wave contribute. Since the space wave has a scattering phase shift of π, the antenna should be so long that the additional propagation phase shift of the surface wave compensates. A sketch of the phase relationships is given in Fig. 7.41.

In practice, it turns out that the sum of the excitation and scattering phase shift for the surface wave is not zero, so we must add an additional $\Delta\varphi_{su}$; then eqn. 7.90 is replaced by

$$k_0L(\tau-1) = \pi - ph(f_2) + \Delta\varphi_{su} \tag{7.91}$$

$\Delta\varphi_{su}$ accounts for the fact that part of the surface wave energy is inside the cylinder, which was neglected in eqn. 7.90. Experimental results for a number of surface wave structures (Zucker[94]) gives

$$\Delta\varphi_{su} - ph(f_2) \sim -\frac{\pi}{3} \tag{7.92}$$

It should be stressed that this physical picture of dividing the process up into local processes is only strictly valid for the surface wave. The phase shift in $f_2(L/l)$ is an integrated effect since it depends on the length. The gain variation with length is shown in Fig. 7.42, where the details of course depend on the actual structure, launching efficiency, etc.

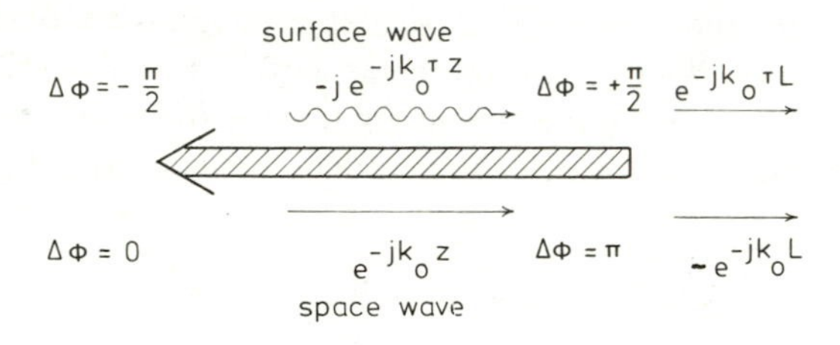

Fig. 7.41 *Phase relationships for excitation, propagation and scattering for the surface wave and the space wave along the rod*

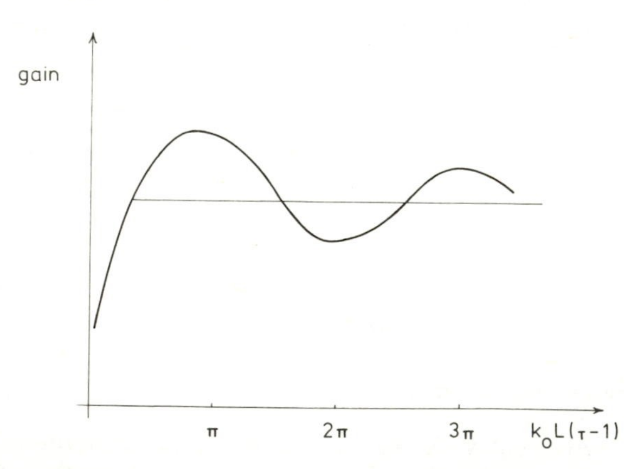

Fig. 7.42 *Gain variation for a surface-wave antenna due to interference between the surface wave and the space wave*

7.6.2 Short cylindrical rod

The short, cylindrical rod with length less than a wavelength (Fig. 7.43) cannot be analysed and understood as a surface wave antenna, since the various approximate asymptotic expressions of the previous Section break down. This does not prevent the antenna from being useful, it turns out in fact that a short rod excited by a metallic circular waveguide has very good properties as a feed antenna for a reflector antenna as shown by Dombek.[23]

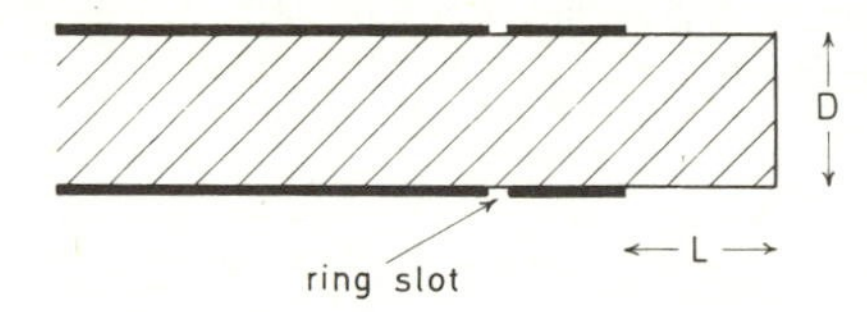

Fig. 7.43 *Short cylindrical rod excited by cylindrical waveguide. The ring slot is used for suppression of back-radiation* (Dombek[23])

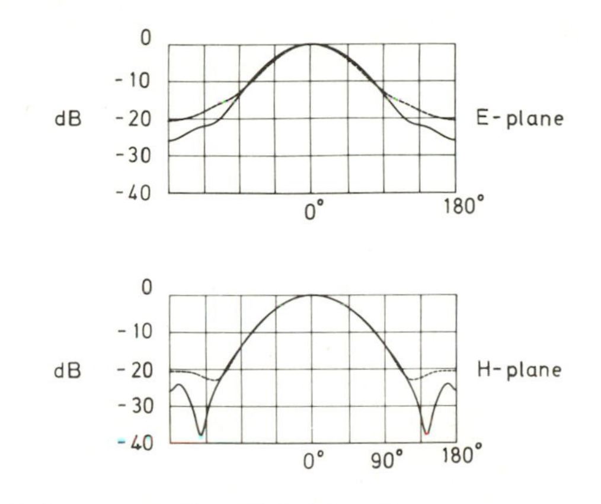

Fig. 7.44 *E- and H-plane pattern for a dielectric rod* $L = 0.4\lambda, D = 0.5\lambda, \epsilon = 2.5$ (Dombek[23])

In its simple form the dielectric simply protrudes from the end of a circular waveguide as shown in Fig. 7.43. In order to suppress the backlobe a ring slot at a suitable location may be used. Dombek analysed the antenna approximately by neglecting internal reflections from the dielectric interface (these are the ones that create the surface waves so the theory breaks down for longer lengths). The far field is found via a twofold application of the Kirchhoff-integrals; first the waveguide aperture is assumed to be illuminated by the dominant H_{11}-wave in the circular waveguide, and then the fields on the dielectric boundary are found assuming an infinite dielectric. These near-field fields are then used as a new aperture illumination over the dielectric boundary to find the far fields. Since the fields have a known φ variation the area integration may be reduced to a line integration along the boundary. The technique is described in detail in Dombek.[22] The results are rather good for the mainbeam as may be seen from Fig. 7.44. An important property as a feed is the equality of the H- and E-plane beamwidths, a property which is kept for other lengths, as is apparent from Fig. 7.45, which shows the 3, 10 and 20 dB beamwidths as a function of antenna length. The antenna may be matched by means of the previously mentioned ring slot, and an additional increase in bandwidth may be achieved by using a dielectric tube at the end instead of the solid cylinder. A reflection coefficient less than 5% may be achieved over a 15% bandwidth almost independent of antenna length.

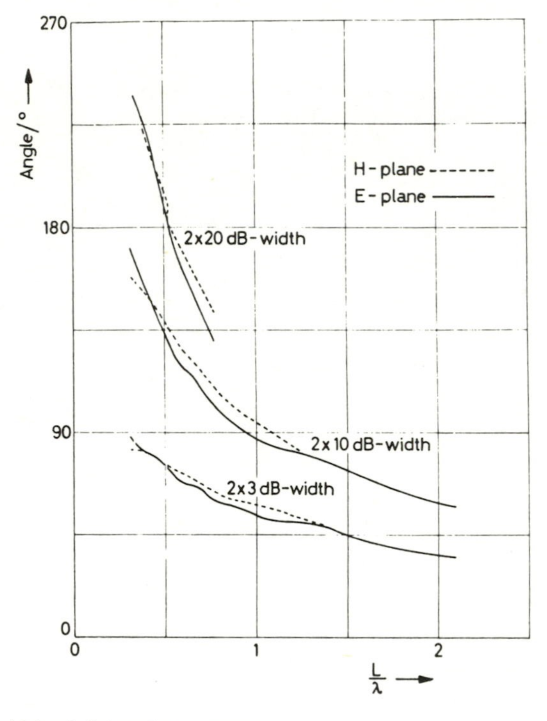

Fig. 7.45 *Beamwidth of dielectric antennas* (diameter $0.5\lambda_0$; $\epsilon_r = 2.5$) (Dombek[23])

7.6.3 Non-cylindrical shapes of dielectric

It is clear that some improvement for the long antennas may be achieved by tapering the cross section of the dielectric, such that the launching efficiency is highest near the launcher. In this way the rather high sidelobes may be lowered. Some rules of thumb are given in Zucker,[94] since no theoretical (or even numerical) results are available for tapered dielectrics. Over the first 20% of the length τ should be between 1.2 and 1.3, and then in a smooth transition approach the appropriate value for the length in question. It is advantangeous to have an end taper over the last half wavelength in order to reduce reflections from the end.

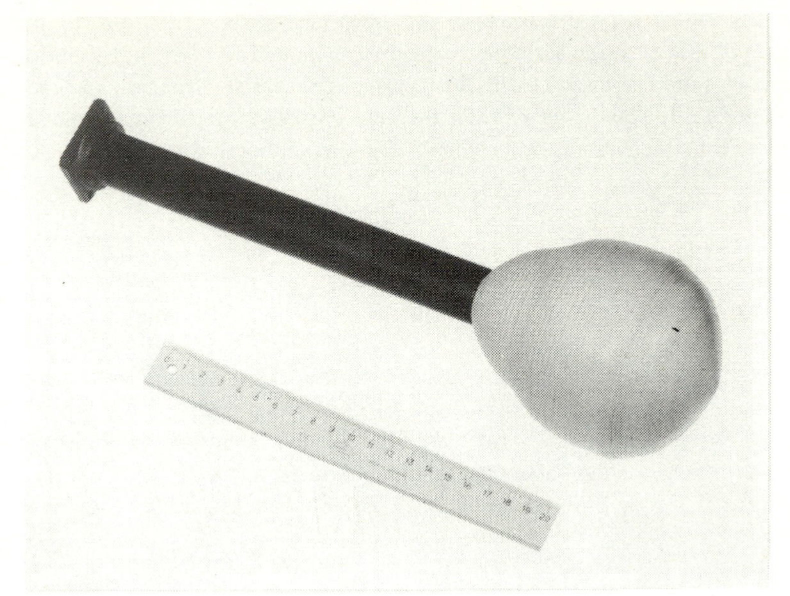

Fig. 7.46 *a*

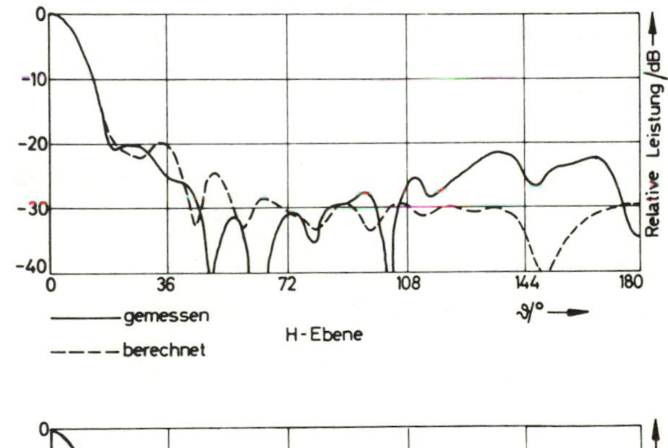

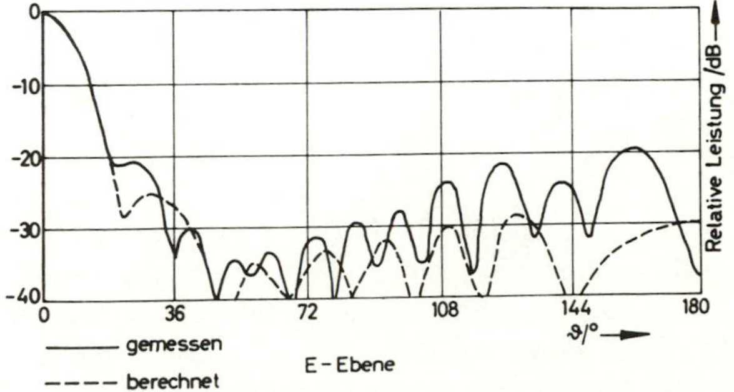

Fig. 7.46 *b Shape and patterns of dielectric antenna* (Dombek[23])

For shorter and fatter antennas the optimum shapes turn out to be quite different (Dombek[22]), the diameter increases towards the end as shown in Fig. 7.46. The sidelobes are less than 20 dB, the pattern is rather symmetrical and the gain is more than 21 dB. This is higher than the gain corresponding to the cross-sectional area, an effect which also appears when a dielectric sphere is inserted in the opening of a waveguide (Croswell *et al.*[18]).

7.7 Simple horn antennas

The electromagnetic horn antenna is one of the oldest microwave antennas, its history dating back to the 1930s. Most of the classic papers are contained in a recent reprint collection of Love[60] and the most relevant design information is contained in many antenna books. In this Chapter we shall therefore stress the more recent findings, although some basic material is necessary for a self-contained treatment.

7.7.1 Sectoral horns

Sectoral horns are horns where two of the side surfaces are parallel, the other two sides forming an angle, called the flare angle (Fig. 7.47). It is a limiting case of the more general pyramidal horn, where one of the two flare angles is zero. Since several polarisations are possible depending on the excitation at the throat, it is customary to name the sectoral horn after the polarisation of the dominant mode. Thus an E-plane sectoral horn is one where the horn flares out in the plane of the electric field of the dominant mode, and similarly for a H-plane horn. The name relates to the excitation and not to the horn itself.

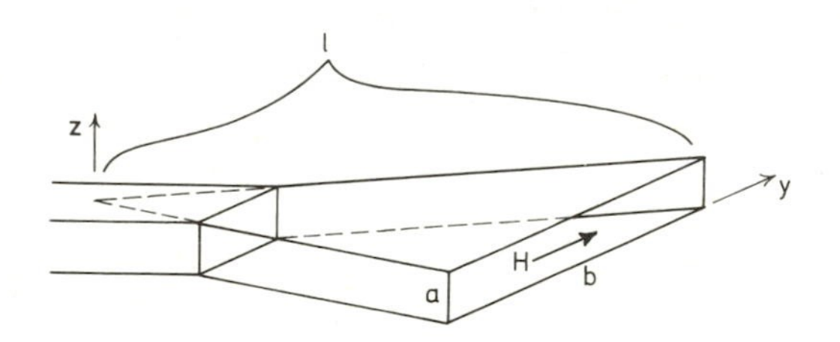

Fig. 7.47 *H-plane sectoral horn*

An antenna is always a transition between a closed region and an open region. It is a characteristic of horn antennas that this transition is slow and gradual, a property which leads to broad-band characteristics but also to rather long structures. Referring to Fig. 7.48 we may think of the waveguide end (the throat region) as the source and the sloping sides as guiding structures concentrating the field in the forward region. This viewpoint leads automatically to an edge diffraction theory to which we shall return later. Alternatively we may think of the horn opening as the true aperture and use aperture theory to find the far field. This is the classical approach which we shall consider initially.

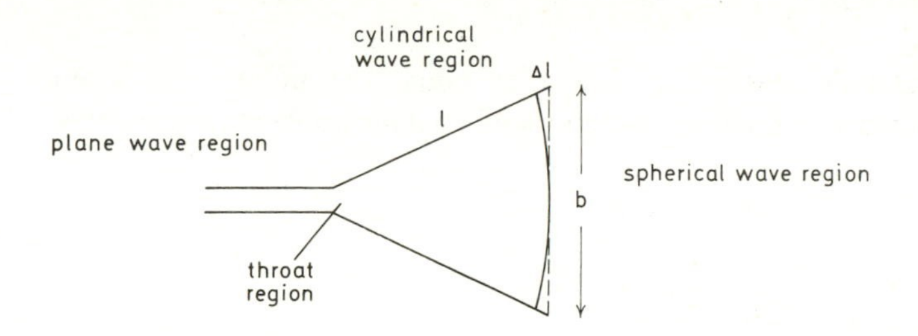

Fig. 7.48 *Top view of sectoral horn showing deviation from cylindrical wavefront at aperture*

Aperture theory. In standard aperture theory the field in the aperture is assumed to equal the incident field and to be zero outside. Thus we must know the incident field which in this case is known, because the sectoral horn admits separation of the variables (which the pyramidal horn does not). For the E-plane sector the following may be found for the basic mode:

$$
\begin{aligned}
E_\varphi &= AH_1(\gamma\rho)\cos\frac{\pi z}{a} \\
H_z &= -A\frac{\gamma}{j\omega\mu_0}H_0(\gamma\rho)\cos\frac{\pi z}{a} \qquad (7.93) \\
H_\rho &= -A\frac{\pi}{j\omega\mu_0 a\rho}H_1(\gamma\rho)\sin\frac{\pi z}{a}
\end{aligned}
$$

The z-axis is normal to the plane of the sector, $z = 0$ being the symmetry plane

$$\gamma = \sqrt{k_0^2 - (\pi/a)^2} \qquad (7.94)$$

is the propagation constant of the corresponding parallel-plate waveguide. No φ-dependent modes are assumed to be excited. In the usual horn theory, asymptotic expressions for the fields are used, with the result that the tangential aperture fields are

$$E_\varphi, H_z \sim \frac{1}{\sqrt{\gamma\rho}}e^{-j\gamma\rho}\cos\frac{\pi z}{a} \qquad (7.95)$$

the phase is constant along a circular arc. Introducing a y-coordinate in the aperture with $y = 0$ at the centre the phase deviation along the aperture is

$$\Delta\varphi = \gamma(l - \sqrt{l^2 - y^2}) \sim \frac{\gamma y^2}{2l} \sim \frac{\pi y^2}{\lambda l} \qquad (7.96)$$

The amplitude variation due to $\sqrt{\rho}$ may be neglected, so in the y-direction the aperture fields have constant magnitude and quadratic phase. It is customary to replace γ by k_0 in eqn. 7.96. Following Schelkunoff,[85] the on-axis far field is given by

$$E \sim j\frac{e^{-jk_0 r}}{\lambda r}\int_{-b/2}^{+b/2}\int_{-a/2}^{+a/2}\cos\frac{\pi z}{a}\, e^{-jk(y^2/2l)}\, dy\, dz \tag{7.97}$$

The radiated power is assumed to equal the power in the outward-going cylindrical wave in the horn. Schelkunoff's gain-formula may be written

$$G = \frac{32ab}{\pi\lambda^2}R_E = G_0 R_E \tag{7.98}$$

where

$$R_E = \frac{C^2(w)+S^2(w)}{w^2} \tag{7.99}$$

and

$$w = \frac{b}{\sqrt{2\lambda l}} \quad \text{and} \quad C(w)-jS(w) = \int_0^w e^{-j(\pi/2)t^2}\, dt$$

We note that w is proportional to the square root of the maximum phase-deviation over the aperture. G_0 is the in-phase aperture-gain for the area ab. Defining the aperture efficiency η as

$$G_0 = \eta 4\pi\frac{A}{\lambda^2}$$

we find

$$\eta = \frac{32}{4\pi^2} = 0.81 \tag{7.100}$$

The function R_E (plotted in Fig. 7.49) takes care of the gain reduction due to the quadratic phase variation; the argument is

$$s = \frac{2b^2}{\lambda l} \tag{7.101}$$

Let us put the question: What is the value of b which for a given a, l and λ gives maximum gain? Clearly a very small b gives a small aperture and a small G_0, a very large b gives a large gain reduction. By differentiating eqn. 7.98 we find $dG/db = 0$ for $-2sR_E'(s) = R_E(s)$ or $s_{opt} = 4.25$ corresponding to a $\Delta\varphi \sim 90°$ over the aperture and $R_E = -1.04$ dB. Thus the efficiency of the optimum E-sectoral horn is $\eta = 0.64$.

Horn antennas may conveniently be used as gain standards for which reason the accuracy of the gain formulas are very important. According to Jull and Allan,[50] the standard gain formula is only valid within about 1 dB, which is clearly inadequate. The reason for the discrepancy is the closely spaced edges which render the usual Kirchhoff approximations inapplicable, but this may be repaired by using the exact open-ended parallel plate waveguide results (Jull and Allan[50]). The reason for this is that the three-dimensional result may approximately be found as a product of two two-dimensional results, of which the parallel-plate waveguide is one. The improved gain formula is

$$G = \frac{16ab}{\lambda^2(1+\lambda_g/\lambda)} R_E(w) \exp\left[\frac{\pi a}{\lambda}\left(1-\frac{\lambda}{\lambda_g}\right)\right] \tag{7.102}$$

where

$$w = \frac{b}{\sqrt{2\lambda_g l}\,\cos(\varphi_0/2)}$$

and φ_0 is the flare angle of the horn.

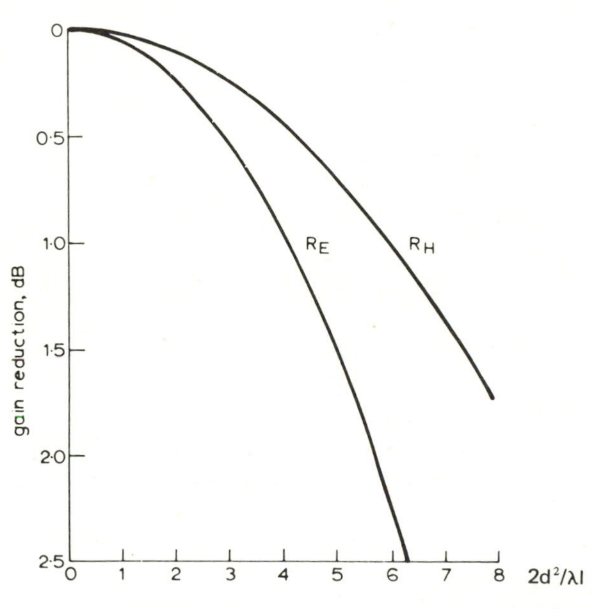

Fig. 7.49 *Gain reduction factors due to spherical wavefronts in aperture* (Jull[48])

For the H-plane sectoral horn the results are quite similar, the only difference being that the cosine variation of the aperture field is along the large dimension instead of the short. Eqn. 7.98 is replaced by

$$G = G_0 R_H \tag{7.103}$$

where R_H is plotted in Fig. 7.49 with $s = 2a^2/\lambda l$ as the argument. We note that for given dimensions the H-plane horn has a larger gain than the E-plane horn. Again there is an optimum value of s which by the same method as before may be found to be $s_{opt} = 6.35$, corresponding to a $\Delta\varphi \sim 145°$ and $R_H = -1.12$ dB. The H-plane horn can tolerate a larger phase deviation near the edges since we have a tapered aperture illumination. A correction for the closely spaced edges is not necessary in this case since the Kirchhoff method happens to yield the exact results for the gain of the open-ended parallel-plate waveguide of this polarisation (Jull and Allan[50]).

As far as radiation patterns are concerned the complete expressions based on the aperture theory are rather complicated, so they will not be given here. Furthermore, they are by their very nature only approximately valid in the half-space in front of the antenna. There are, however, distinct differences between the patterns of the E-plane and H-plane horn which should be mentioned. In the following we only consider radiation patterns in the plane containing the flare:

E-plane horn (optimum):

$$\theta_{3\,\mathrm{dB}} \simeq 2 \sin^{-1}\left(0.4 \frac{\lambda}{b}\right)$$

First sidelobe ~ − 10 dB

H-plane horn (optimum):

$$\theta_{3\,\mathrm{dB}} \simeq 2 \sin^{-1}\left(0.7 \frac{\lambda}{a}\right)$$

First sidelobe ~ − 17 dB (7.104)

The high E-plane sidelobes may of course be related to the untapered amplitude distribution.

We note that the sidelobes for the optimum gain horns are somewhat higher than those with zero phase deviation (− 13 dB and − 23 dB, respectively); another effect is the filling in of the nulls in the pattern. Thus the sidelobe in the H-plane case is hardly recognisable as a sidelobe. General patterns for various amplitude distributions with quadratic phase variation are given in Jasik,[46] Section 2.

The far-out sidelobes and the back radiation cannot of course be given by aperture theory.

Edge diffraction theory. Some of the limitations of aperture theory may be overcome by applying half-plane diffraction theory. The results are particularly useful for the sidelobe region, apart from the fact that alternative theories always cast some light and are helpful in providing a physical insight into the problem. The treatment follows the work of Russo, Rudduck and Peters,[81] Yu, Rudduck and Peters[92] and Jull.[49] For a more general treatment of diffraction theory the reader is referred to Chapter 2.

A magnetic line current radiates a cylindrical wave similar to the mode in the E-plane sector, and by using the exact solution for a magnetic line source near a conducting half-plane we can find an approximate expression for the far field from the horn, neglecting the interaction between the edges.

As indicated in Fig. 7.50 we may consider the line current to be situated at the apex radiating a direct cylindrical wave

$$H = \begin{cases} \dfrac{1}{\sqrt{k\rho}} e^{-jk\rho} & |\varphi| < \varphi_0 \\ 0 & |\varphi| > \varphi_0 \end{cases} \tag{7.105}$$

The diffraction solution for the upper edge alone is

$$H = \frac{e^{-jkr}}{\sqrt{kr}}\left\{\begin{bmatrix}\exp[jkl\cos(\pi-\varphi_0+\varphi)]\\0\end{bmatrix} + v(l,\pi-\varphi_0+\varphi)\right\} \qquad (7.106)$$

where

$$v(l,\alpha) = -\frac{e^{jkl\cos\alpha}}{2}\,\mathrm{sign}\,(u)\{1+(1+j)(C(u)-jS(u))\}$$

and

$$u = \sqrt{\frac{4kl}{\pi}}\cos\frac{\alpha}{2}$$

$C(u)$ and $S(u)$ are defined in eqn. 7.99.

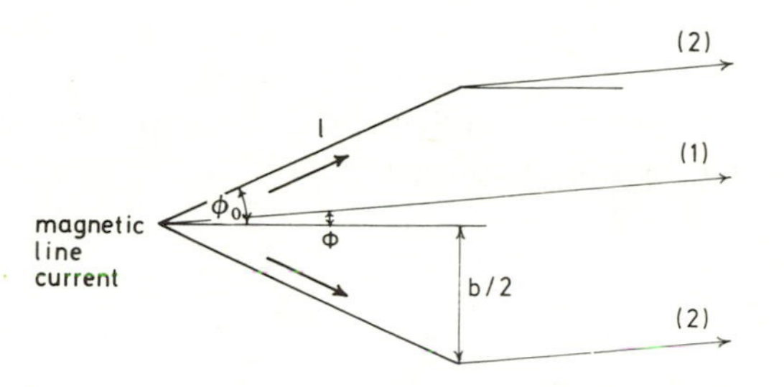

Fig. 7.50 *Two-dimensional E-plane horn viewed as a diffraction antenna with a direct ray (1) and two first-order difracted rays (2)*

The diffraction function $v(l, \alpha)$ is discontinuous at the shadow boundary ($\alpha = \pi$) to compensate for the jump in the first term, so that the complete expression is continuous. Asymptotic expansions for large u form the basis of the geometrical theory of diffraction, but u will in general not be very large in the horn case.

Assuming that the two edges are independent the total pattern is given by

$$F(\varphi) = \begin{Bmatrix}\exp[jkl\cos(\pi-\varphi_0+\varphi)]\\0\end{Bmatrix} + v(l,\pi-\varphi_0+\varphi)$$
$$+\exp[-j2kl\sin\varphi_0\sin\varphi]\,v(l,\pi-\varphi_0-\varphi) \qquad (7.107)$$

This solution is valid for $|\varphi| < \pi/2$. For $|\varphi| > \pi/2$ one of the diffracted waves are shadowed by the other edge, so the present solution is discontinuous at $|\varphi| = \pi/2$. Already at this stage, though the agreement with experiment is very good, further refinements may be achieved by considering the images of the edges in the other wall and second-order diffracted waves (Yu, Rudduck and Peters[92]).

It is also apparent from this diffraction theory that for electrically very large horns ($kl \gg 1$), the diffracted waves may be neglected and the beam width tends to equal the flare angle. For optimum horns this is of course not the case as shown previously.

It is interesting to compare the on-axis gain from a diffraction point of view (Jull[49]). The argument u in the Fresnel integral is the determining argument, and putting $\varphi = 0$ in eqns. 7.106 and 7.107 gives

$$u = \sqrt{\frac{4kl}{\pi}} \cos\left(\frac{\pi - \varphi_0}{2}\right) = \frac{b}{\sqrt{2\lambda l}} \frac{1}{\cos(\varphi_0/2)} \tag{7.108}$$

to be compared with $u = b/\sqrt{2\lambda l}$ in eqn. 7.99. Noting that we assumed a quadratic phase variation in the previous derivation the small difference in the arguments is explainable. Thus we see that there is essentially no difference between the two solutions, or stated differently, the aperture theory includes the geometrical optics field of the horn and the singly diffracted fields from the aperture edges. The important thing is that there is no systematic way of improving the aperture theory whereas the diffraction solution may systematically be improved by including double, triple etc. diffraction. It turns out that only doubly diffracted rays and singly diffracted rays reflected from the interior are important, and they result in minor oscillations in the gain vs. frequency curve.

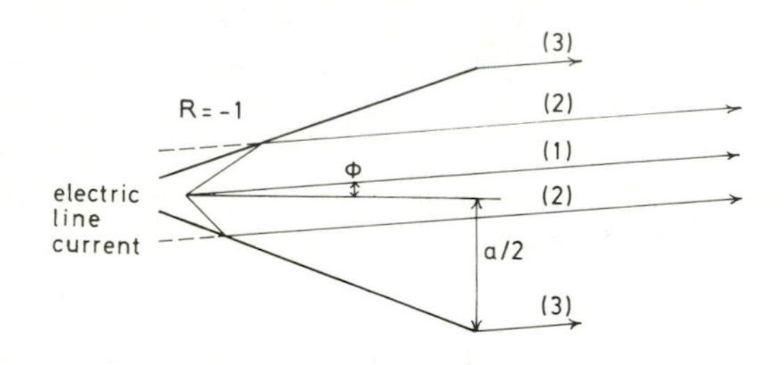

Fig. 7.51 *Two-dimensional H-plane horn viewed as a diffraction antenna with a direct ray (1) reflected rays (2) and diffracted rays (3)*

The H-plane horn may also be treated as a diffraction antenna with good results (Mentzner *et al.*[65]), although the diffracted edge waves play a minor role in this case. Instead of a magnetic line source we now need an electric line source, which of course has to be displaced slightly from the conducting wall (Fig. 7.51). The rays reflected from the side walls are reflected with a reflection coefficient of -1, which means that the optical pattern of the source in the presence of an infinite horn has a null in the direction of the walls ($|\varphi| = \varphi_0$). Thus the optical field is continuous on the shadow boundary, so the first-order diffraction field is zero. However, the slope of the field is discontinuous, so there is a small diffracted field (slope diffraction) to take care of this discontinuity. Put differently, the field incident on the edge is the radial magnetic field, which decays as $(k\rho)^{-3/2}$, much faster than the free-space cylindrical wave $(k\rho)^{-1/2}$.

The solution for the two-dimensional H-plane horn is as follows (Mentzner *et al.*[65]):

$$F(\varphi) = \begin{Bmatrix} E_0 \cos\left(\frac{\pi}{a} l \cos\varphi_0 \tan\varphi\right) e^{-jkl\cos(\varphi_0-\varphi)} \\ 0 \end{Bmatrix} \tag{7.109}$$
$$+ jkz_0 H_\rho^i e^{jkl} l^{3/2}$$
$$\times \{v_s(l, \pi-\varphi_0+\varphi) + v_s(l, \pi-\varphi_0-\varphi)\, e^{-jkl\sin\varphi_0\sin\varphi}\}$$

where

$$v_s(l,\alpha) = \frac{\sin\alpha}{\sqrt{\pi}} \operatorname{sign}(-u) \exp\left[-j\left(kl-\frac{\pi}{4}\right)\right] \times \left[e^{ju^2}\int_{|u|}^{\infty} e^{-j\tau^2}\,d\tau + \frac{j}{2|u|}\right]$$

and $u = \sqrt{2kl}\cos\alpha/2$. (7.110)

In eqn. 7.109 the first term is the optical radiation pattern ((1) and (2) in Fig. 7.51), which is zero at the walls and outside, $|\varphi| \geqq \varphi_0$. H_ρ^i is incident radial magnetic field and v_s is the slope diffraction function, continuous and non-zero at the shadow boundary but with a discontinuous slope.

7.7.2 Pyramidal horns

No exact solution is possible for the pyramidal horn, not even for the incident field; however, for all practical purposes we can consider the pyramidal horn as two independent sectoral horns, and use either the aperture methods or the diffraction methods.

As far as gain is concerned we might as well find it at a finite range, since this corresponds effectively to a different quadratic phase variation over the aperture. Following Jull,[48] we find

$$G = G_0 R_E(s_E) R_H(s_H) \tag{7.111}$$

where

$$G_0 = \frac{32ab}{\pi\lambda^2} \tag{7.112}$$

and $R_E(s_E)$, $R_H(s_H)$ are given by Fig. 7.49

$$s_E = \frac{2b^2}{\lambda l'_E}, \qquad l'_E = \frac{rl_E}{r+l_E} \tag{7.113}$$

and

$$s_H = \frac{2a^2}{\lambda l'_H}, \qquad l'_H = \frac{rl_H}{r+l_H} \tag{7.114}$$

r is the range, l_E and l_H the slant lengths in the E- and H-plane, respectively.

The diffraction patterns given in the previous Section were two-dimensional. The pyramidal horn is three-dimensional, and thus the cylindrical rays should be transformed into spherical. This is done simply by multiplying with divergence factors

$$f_E = \sqrt{\left(\frac{l_E}{l_E+r}\right)} \quad \text{in the E-plane}$$

and

$$f_H = \sqrt{\left(\frac{l_H}{l_H+r}\right)} \quad \text{in the H-plane}$$

This is not all though, because the rays diffracted at the E-plane edges will also contribute to the H-field pattern, but it turns out that the shape of the pattern is not changed. The solution may be refined even further by introducing equivalent edge currents, which are useful when the phase variation over the edge is not sufficiently rapid (Mentzer *et al.*[65]).

7.7.3 Conical horns

The conical horn excited by the dominant TE_{11} mode in a circular waveguide has properties very similar to pyramidal horns, as shown for example in the gain curves in Fig. 7.52 obtained by King.[52] The modes in the infinite cone may be expressed

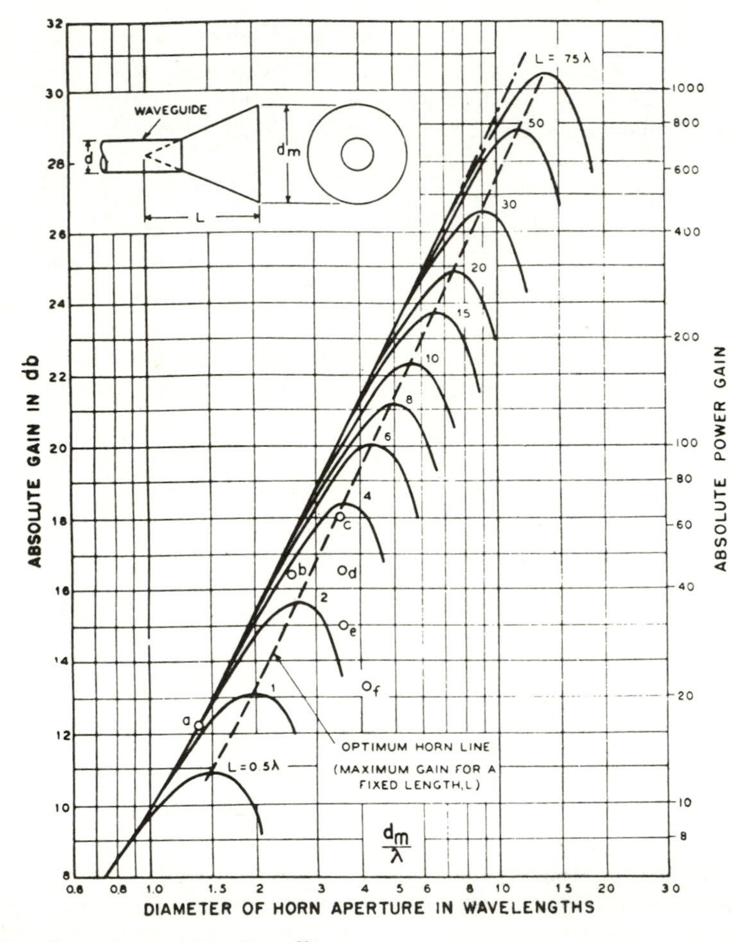

Fig. 7.52 *Gain of conical horn* (King[52])

exactly in terms of spherical Hankel functions and Legendre functions giving rather complicated results. A good, asymptotic solution has been obtained by Narasimhan and Rao[69] and Narasimhan and Sheshadri[70] with the following results for the fundamental mode:

$$
\begin{aligned}
E_\theta &= \frac{A}{\sin\theta} J_1\left(1.841\frac{\theta}{\theta_0}\right)\frac{e^{-jkr}}{r}\sin\varphi \\
E_\varphi &= A\,\frac{1.841}{\theta_0} J_1'\left(1.841\frac{\theta}{\theta_0}\right)\frac{e^{-jkr}}{r}\cos\varphi \qquad (7.115)\\
H_r &= \frac{A}{j\omega\mu}\left(\frac{1.841}{\theta_0}\right)^2 J_1\left(1.841\frac{\theta}{\theta_0}\right)\frac{e^{-jkr}}{r^2}\cos\varphi
\end{aligned}
$$

where $2\theta_0$ is the cone angle.

We note that there is some tapering of E_θ over the aperture, such that the side-lobes in the E-plane are slightly better than for the pyramidal horn. The optimum horn corresponds to

$$
s = \frac{2d^2}{\lambda l} = 4.8 \qquad (7.116)
$$

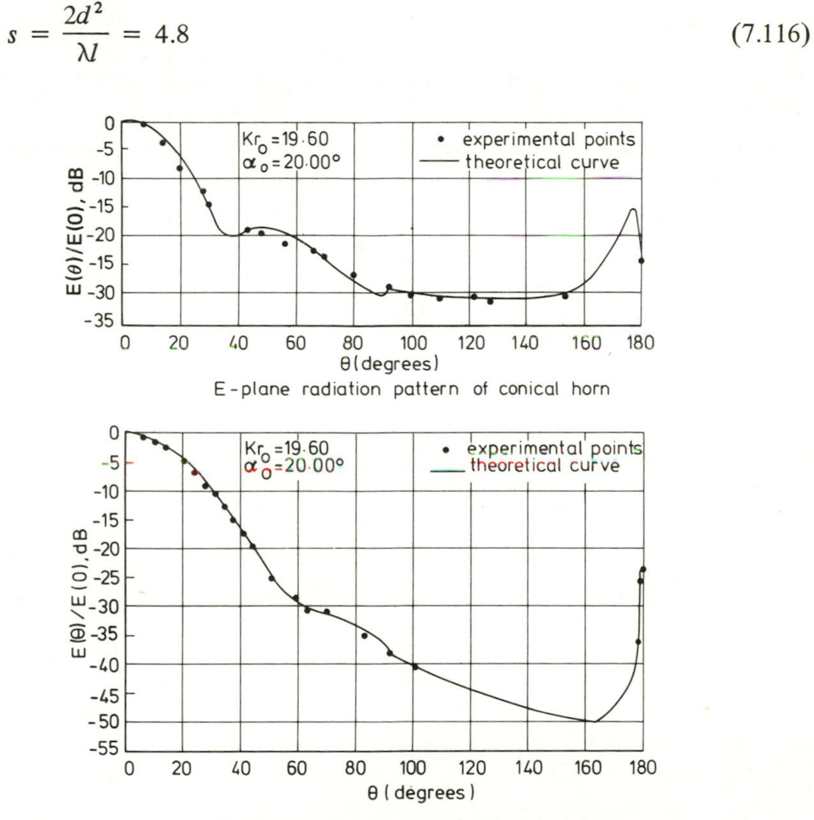

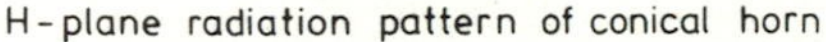

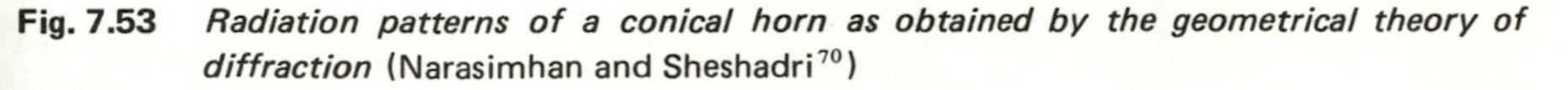

Fig. 7.53 ***Radiation patterns of a conical horn as obtained by the geometrical theory of diffraction*** (Narasimhan and Sheshadri[70])

equivalent to $\Delta\varphi \simeq 108°$ over the aperture. The efficiency of the optimum conical horn is 0.52.

The radiation pattern may with very good accuracy be obtained by a GTD (geometrical theory of diffraction) analysis very similar to what was mentioned under the sectoral and pyramidal horns (Narasimhan and Sheshadri[70]). The main difference is that the axis forms a caustic region since all points on the curved edge of the horn contributes, but by integrating over equivalent edge currents the situation is repaired. Fig. 7.53 shows some theoretical and experimental E- and H-plane patterns obtained by this method.

7.8 Multimode horns

For many applications the single-mode horn is not satisfactory, especially the rather high sidelobes and the difference between E-plane and H-plane patterns. Seen from an aperture-theory point of view the obvious solution is to modify the aperture illumination in such a manner that the desired properties are obtained. In doing this it is customary to neglect the sphericity of the horn modes and discuss the corresponding waveguides only, where the waveguide cross-section corresponds to the horn aperture. Sphericity can then be added afterwards to complete the picture. In this Section we shall treat smooth-walled waveguides and various ideas of how to create the proper ratio between the modes.

7.8.1 Multimode rectangular horns

The dominant mode TE_{10} in rectangular waveguide has an electric field distribution

$$E_y = A_0 \sin \frac{\pi x}{a}, \qquad E_x = 0 \tag{7.117}$$

with the tapered distribution in the H-plane, and the constant distribution in the E-plane gives the unwanted patterns mentioned above. The y-distribution may be changed by adding TE_{1m} and TM_{1m}-modes, since for these modes

$$E_y = A_m \sin \frac{\pi x}{a} \cos \frac{m y \pi}{b} \tag{7.118}$$

which will yield an unchanged x-distribution (H-plane pattern). For symmetric patterns we need a symmetric distribution, which only allows even values of m. The lowest higher-order modes are the TE_{12} and TM_{12}-modes, which taken alone also have an E_x field. By using proper symmetries they may be excited as a pair (TE/TM_{12}) such that the total E_x is identically zero (Jensen[47]). The total distribution may thus be written

$$E_y = \sin \frac{\pi x}{a} \left(1 + A \cos \frac{2\pi y}{b} \right) \tag{7.119}$$

and A may be chosen appropriately for a given purpose, for example for cancellation of the E-plane sidelobe. Such a cancellation requires a real A, which means that the two modes must be in phase in the aperture.

The far field pattern is found here assuming that the waveguide is situated in an infinite, metallic flange, whereby only the electric field in the aperture is required. Such an E-field model has been shown to provide good prediction of the radiation both for the co-polarised and cross-polarised fields[78]. (See Section 4.10).

The general result is

$$E_\theta = \frac{jk_0\, e^{-jk_0 r}}{2\pi r}(f_x \cos\varphi + f_y \sin\varphi) \tag{7.120}$$

$$E_\varphi = \frac{jk_0\, e^{-jk_0 r}}{2\pi r}\cos\theta(f_y \cos\varphi - f_x \sin\varphi) \tag{7.121}$$

where

$$f_y(\theta,\varphi) = \iint_{apert.} E_y(x,y)\exp[jk_0(x\sin\theta\cos\varphi + y\sin\theta\sin\varphi)]\,dx\,dy$$

and similarly for f_x. In the present case f_x equals zero. Performing the integration we find

$$f_y = ab\,\frac{\cos u}{(\pi/2)^2 - u^2}\left[\frac{\sin v}{v} + A\,\frac{v\sin v}{v^2 - \pi^2}\right] \tag{7.122}$$

where

$$u = \frac{k_0 a}{2}\sin\theta\cos\varphi \qquad v = \frac{k_0 b}{2}\sin\theta\sin\varphi$$

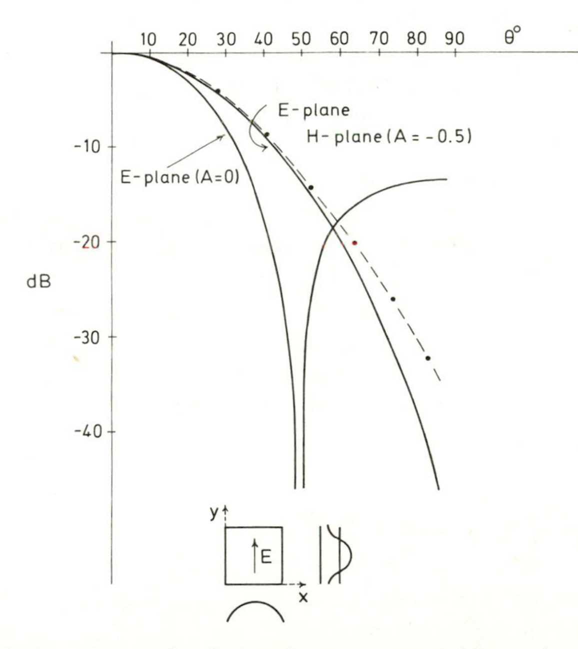

Fig. 7.54 *Radiation patterns for dual-mode square waveguide, a = b = 1.35λ. Content of TE/TM_{12} mode in aperture equalises E- and H-plane patterns*

The E-plane pattern is found for $\varphi = \pi/2$ and the H-plane pattern is found for $\varphi = 0$. An example for this dual mode antenna is shown in Fig. 7.54 for $A = -0.5$; it is seen that the E- and H-plane patterns are almost identical. The quadratic phase distribution associated with the pyramidal horn changes the pattern; some examples have been given by Profera.[74]

Small horns are often used as feed antennas for reflector antennas and these are dealt with in Chapter 4.

7.8.2 Multimode conical horn

Potter[73] was the first to use multimode techniques in a conical horn to achieve pattern control in connection with the design of cassegrainian feed systems. The technique is essentially the same as for the rectangular case with excitation of a higher-order mode of proper relative magnitude and a proper propagation phase shift to create in-phase aperture fields. The design of these horns is discussed in Chapter 4.

7.9 Corrugated horns

Corrugated horns (also called hybrid mode horns and scalar feeds) enjoy a great popularity as feed antennas due to their low sidelobes, rotationally symmetric patterns and bandwidth. The first two properties were achieved also by the multimode horns of the previous Section, which used smooth-walled guides. In a sense the corrugated guide or horn is also a multimode horn, but with a continuous coupling along the structure, such that the in-phase condition at the aperture is achieved over a large bandwidth. Design information for corrugated horns used as feeds for reflector antennas may be found in Chapter 4.

7.9.1 Propagation properties of a corrugated surface

Most of the properties of a corrugated surface are well described by assuming that there are many slots or teeths per wavelength, although in practice one might end up with as few as two slots per wavelength. In the limit of high corrugation density a simple network model leads to an equivalent surface reactance, which we shall derive. Consider first the case where the electric surface field is in the z-direction, normal to the slots (Fig. 7.55). The equivalent network is a series connection of the surface of the teeth (a short circuit) and the input impedance of the shorted transmission line, so the average surface impedance equals

$$Z_1 = j\frac{g}{g+t}Z_0 \tan(k_0 w) = jX_1 \tag{7.123}$$

In this approximation the edge fields have been neglected; in practice this means a slightly different effective value of w.

For the other polarisation where E is parallel to the teeth, we have a parallel connection of the short circuit and the input impedance of the guide. The result is, of course, zero reactance, so

$$Z_2 = 0 \tag{7.124}$$

Thus a corrugated surface is essentially an anisotropic reactive surface, where in one direction the surface reactance may be positive or negative depending on w. We note specifically that the surface reactance is infinite (an open-circuit) for $w = \lambda/4$, leading to zero magnetic fields in the x-direction.

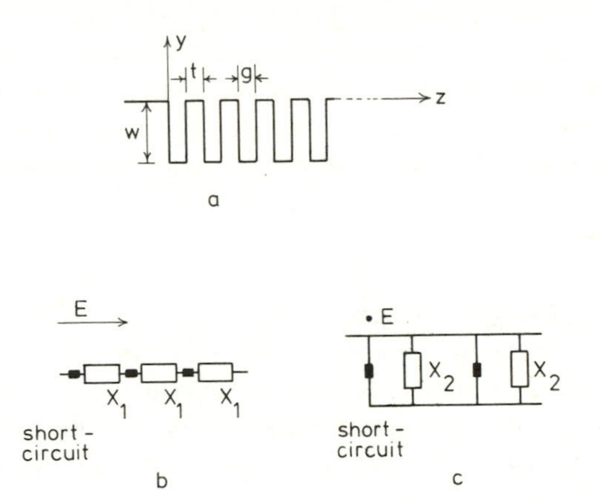

Fig. 7.55 *a* Rectangular teeth of a two-dimensional surface
b Equivalent network for electric field normal to teeth (series network)
c Equivalent network for electric field parallel to teeth (parallel network)

A transverse resonance calculation will show that a TM-surface wave may exist for

$$0 < w < \lambda/4$$

where the surface reactance is inductive. This situation must be avoided for the corrugated horns, since this would lead to a strong illumination of the horn edges. Thus, the design is usually made such that

$$\frac{\lambda}{4} < w < \frac{\lambda}{2}$$

where the surface reactance is capacitive.

Consider the reflection of a plane wave from such a surface. A 'horizontally' polarised wave, E-field in x-direction, will see the impedance Z_2 and be totally reflected with a reflection coefficient

$$\rho_E = -1 \tag{7.125}$$

independent of angle of incidence.

A 'vertically'-polarised wave, H-field in x-direction, will see the impedance Z_1 and be totally reflected with reflection coefficient

$$\rho_H = \frac{Z_0 \cos\theta - jX_1}{Z_0 \cos\theta + jX_1} \tag{7.126}$$

where θ is the angle of incidence, measured from the normal. We note that for grazing incidence, $\theta = \pi/2$, $\rho_H = -1$ independent of X_1, which means that the direct and reflected wave tend to cancel along the corrugated surface. Most efficiently of course for $X_1 \to \infty$, $w = \lambda/4$. A smooth surface with $X_1 = 0$ gives $\rho_H = +1$ and a doubling of the incident field.

Thus we see that the incident energy is forced off the surface for both polarisations, and that the two polarisations are treated almost equally. This property is the basis for all developments in corrugated horns, especially for the equality of E- and H-plane patterns, which is a condition for low cross-polarisation. The fact that X_1 need not be infinite for a grazing reflection of -1 explains the broadband properties of the surface.

7.9.2 Corrugated cylindrical waveguides

For low flare angle horns it is useful to understand the propagation in the corresponding waveguide. For simplicity we shall consider a rectangular guide with transverse corrugations on top and bottom walls only and use the surface impedance as explained in the previous Section. The presentation is close to that of Baldwin and McInnes,[8] which again is related to that of Bryant.[11]

A hybrid mode satisfying the boundary conditions on the vertical walls and one of the two impedance conditions ($Z_2 = E_x/H_z = 0$ on top and bottom) is described by the following equations, where we have chosen a mode where the transverse electric field (E_y) is symmetric in y. Asymmetric modes exist as well:

$$
\begin{aligned}
E_x &= 0 \\
E_y &= -\frac{j\beta K}{K_y} \cos\left(\frac{\pi}{a}x\right) \cos(K_y y) \\
E_z &= K \cos\left(\frac{\pi}{a}x\right) \sin(K_y y) \\
H_x &= \frac{j\beta_1^2 K}{\omega\mu_0 K_y} \cos\left(\frac{\pi}{a}x\right) \cos(K_y y) \\
H_y &= \frac{j\pi K}{\omega\mu_0 a} \sin\left(\frac{\pi x}{a}\right) \sin(K_y y) \\
H_z &= -\frac{\beta\pi K}{\omega\mu_0 K_y a} \sin\left(\frac{\pi x}{a}\right) \cos(K_y y)
\end{aligned}
\tag{7.127}
$$

Here $\beta_1^2 = k_0^2 - (\pi/a)^2$ assuming we have the lowest-order mode in the x-direction. K_y is the transverse wavenumber in the y-direction, as yet unknown. For slot depth equal to zero the mode reduces to the dominant TE_{10}-mode of the smooth guide. β equals the axial wavenumber and the wave-equation gives the following equation

$$\beta^2 + K_y^2 + (\pi/a)^2 = k_0^2 \qquad (7.128)$$

The surface impedance on the corrugated surface must be modified to take into account the variation of the field in the x-direction so instead of eqn. 7.123, we now get

$$Z_s = j\frac{g}{g+t} Z_0 \frac{k_0}{\beta_1} \tan \beta_1 h \qquad (7.129)$$

corresponding to the assumption of TE_{10}-mode in the slots. The same impedance as found from the mode fields gives

$$Z = -\frac{E_z}{H_x}\bigg|_{y=-b/2} = -jZ_0 \frac{k_0 K_y}{\beta_1^2} \tan \frac{K_y b}{2} \qquad (7.130)$$

and the two impedances must be equal (transverse resonance); which yields the following transcendental equation for K_y:

$$K_y \tan \frac{K_y b}{2} = -\frac{g}{g+t} \beta_1 \tan \beta_1 h \qquad (7.131)$$

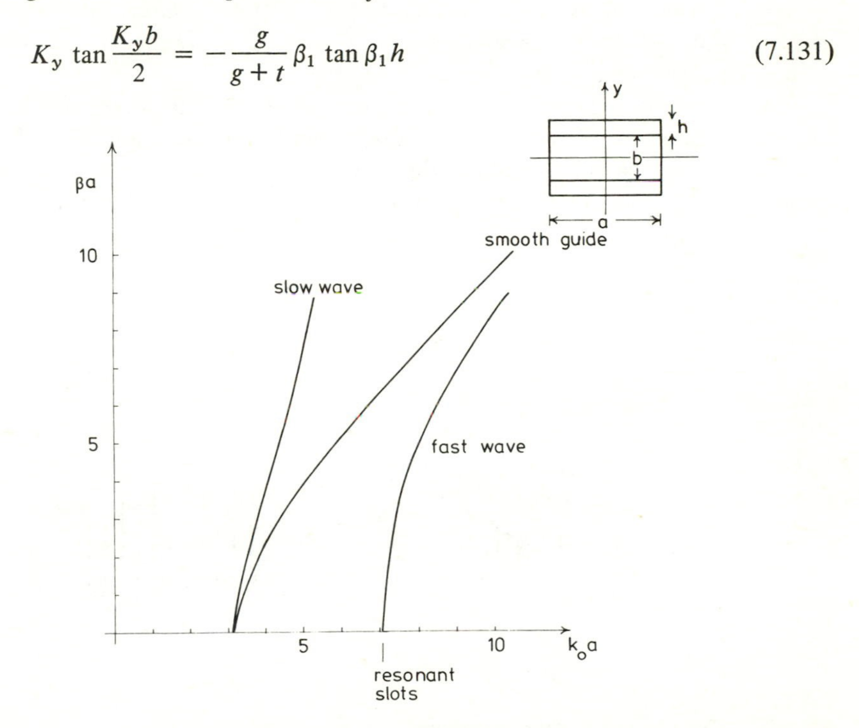

Fig. 7.56 *TE_{10} mode in rectangular corrugated waveguide with corrugations on top and bottom wall only.*
$a = 2b = 4h$. $t = 0$

The equation may easily be solved numerically or graphically, an example is shown in Fig. 7.56.

For $\beta_1 h < \pi/2$ (inductive slots) the lowest-order solution to eqn. 7.131 gives imaginary values of K_y, so the equation may instead be written

$$|K_y| \tan h \frac{|K_y b|}{2} = \frac{g}{g+t} \beta_1 \tan \beta_1 h \qquad (7.132)$$

Physically this means surface waves bound to the corrugations and a resulting high edge illumination at the aperture (the E-plane edges).

For $\beta_1 h > \pi/2$ (capacitive slots) real values of K_y are possible, leading to a cosine-distribution of E_y and a fast wave. Note that it is only at the resonance frequency for the slots ($\beta_1 h = \pi/2$) that $K_y b = \pi$, giving zero electric field at aperture edges. At the higher frequencies the tapering of the fields will not be so complete.

The case of a rectangular waveguide with all the walls transversely corrugated is theoretically difficult, since an infinity of transverse modes is needed to satisfy the boundary conditions (Andersen[5]). However, in practice good results are obtained by neglecting the corrugations on the vertical walls for vertical polarisation, as evidenced by the experimental results of Manwarren and Farrar[63] and Baldwin and McInnes.[9] The advantage of rectangular guides lies in the possibility of shaping a beam with an elliptical cross-section.

The corresponding problem in the corrugated circular waveguide has been studied in detail by Clarricoats and Saha,[13] part 1. The equations are, of course, somewhat more complicated, but the physical picture for the hybrid modes is almost the same. A sketch of a typical dispersion diagram is shown in Fig. 7.57,

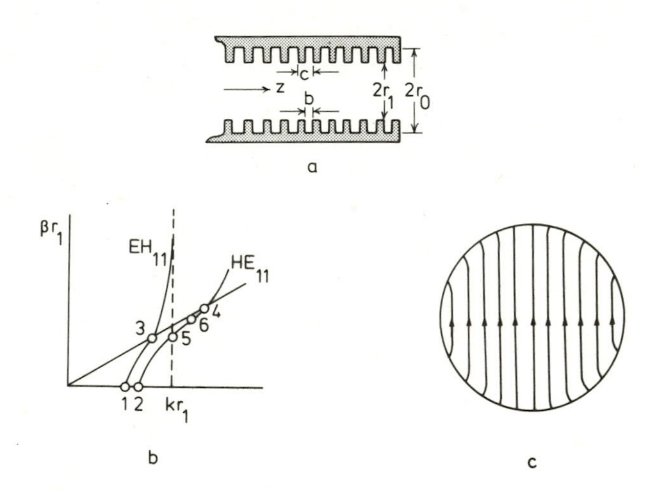

Fig. 7.57 *Cylindrical corrugated waveguide*
a Geometry
b Dispersion curve for lowest order modes, $r_1/r_0 = 0.8$
c Transverse field distribution under balanced hybrid conditions, point 5 in *b* (Clarricoats and Saha[13])

showing the split of the TE_{11}-mode in two branches, one of them having a slow-wave resonance ($\beta \to \infty$), where the slots have an open-circuit input impedance. Due to the radial waveguide character of the slots this does not happen at $\Delta r = \lambda/4$.

At the same frequency the other branch, which is fast, experiences the so-called balanced hybrid condition

$$\frac{H_z}{E_z} = -jY_0 \tag{7.133}$$

where Y_0 is the free space admittance. It is worthwhile to write down the complete fields at this frequency

$$\begin{aligned}
E_z &= J_1(x) \\
H_z &= -jY_0 \cdot E_z \\
E_r &= -j\frac{k}{K}\frac{J_1(x)}{x}\{\bar{\beta}F_1(x)+1\} \\
H_r &= -jY_0 \cdot E_r \\
E_\varphi &= \frac{k}{K}\frac{J_1(x)}{x}\{\bar{\beta}+F_1(x)\} \\
H_\varphi &= -jY_0 \cdot E_\varphi
\end{aligned} \tag{7.134}$$

where $x = Kr$, $K^2 = k^2 - \beta^2$, $\bar{\beta} = \beta/k$ and $F_1(x) = xJ_1'(x)/J_1(x)$. A factor $e^{j\varphi}$ has been suppressed.

We note a symmetry between the magnetic and electric fields

$$\bar{H} = -jY_0\bar{E} \tag{7.135}$$

which has been shown by Rumsey[79] to be a general condition to be satisfied in order that the power pattern radiated by an antenna should be φ-independent.

The mode we have discussed is usually labelled the HE_{11}-mode due to its hybrid nature of having both E_z and H_z components, but there is not a unique definition available.

7.9.3 Corrugated, conical horns

Analysis of propagation in a conical corrugated waveguide is considerably more complicated than the corresponding cylindrical one; in general, a multimode description is necessary. However, for the special case of infinite slot impedance and balanced hybrid conditions, some simple approximations for long horns are the following (Narasimhan and Sheshadri[70]), which are analogous to eqn. 7.115 for the smooth horn:

$$\begin{aligned}
E_\theta &= A\frac{2.405}{\theta_0}J_0\left(2.405\frac{\theta}{\theta_0}\right)\frac{e^{-jkr}}{r}\sin\varphi \\
E_\varphi &= A\frac{2.405}{\theta_0}J_0\left(2.405\frac{\theta}{\theta_0}\right)\frac{e^{-jkr}}{r}\cos\varphi \\
H_r &= \frac{A}{j\omega\mu}\left(\frac{2.405}{\theta_0}\right)^2 J_1\left(2.405\frac{\theta}{\theta_0}\right)\frac{e^{-jkr}}{r^2}\cos\varphi
\end{aligned} \tag{7.136}$$

A more complete analysis is given by Clarricoats and Saha,[13] part 2. The aperture distribution on the conical guide may then be used in a Kirchhoff–Huygens integration to find the far field, alternatively, expansion of the fields over a spherical cap in spherical modes of free space (Clarricoats and Saha[13]) or for wide flare angles a GTD-analysis based on the incident H_r-field above (Narasimhan and Sheshadri[70]).

7.10 References

1 ADAMS, A. T., GREENOUGH, R. K., WALLENBERG, R. F., MENDELOVICZ, A. and LUMJIAK, C.: 'The quadrifilar helix antenna', *IEEE Trans.*, **AP-22,** March 1974, pp. 173–178

2 AGARWAL, K. K., and NAGELBERG, E. R.: 'Phase characteristics of a circularly symmetric dual-mode transducer', *IEEE Trans.*, **MTT-18,** Jan. 1970, pp. 69–71

3 AGRAWAL, P. K., and BAILEY, M. C.: 'An analysis technique for microstrip antennas', *IEEE Trans.*, **AP-25,** Nov. 1977, pp. 756–759

4 ANDERSEN, J. B.: 'Metallic and dielectric antennas'. Polyteknisk Forlag, Lyngby, Denmark, 1971

5 ANDERSEN, J. B.: 'Propagation in the general rectangular waveguide'. Proc. 5th European Microwave Conference, Hamburg, 1975, pp. 595–597

6 ANDERSEN, J. B., and RASMUSSEN, H. H.: 'Decoupling and descattering networks for antennas', *IEEE Trans.*, **AP-24,** Nov. 1976, pp. 841–846

7 ATIA, A. E., and MEI, K. K.: 'Analysis of multiple-arm conical log-spiral antennas', *IEEE Trans.*, **AP-19,** May 1971, pp. 320–331

8 BALDWIN, R., and McINNES, P. A.: 'Corrugated rectangular horns for use as microwave feeds', *Proc. IEE,* **122,** 1975, pp. 465–469

9 BALDWIN, R., and McINNES, P. A.: 'A rectangular corrugated feed horn', *IEEE Trans.*, **AP-23,** Nov. 1975, pp. 814–817

10 BROWN, J., and SPECTOR, J. O.: 'The radiating properties of end-fire aerials', *Proc. IEE,* **104,** Pt. B, 1957, pp. 27–34

11 BRYANT, G. H.: 'Propagation in corrugated waveguides', *Proc. IEE,* **116,** Feb. 1969, pp. 203–213

12 CHA, A. G.: 'Wave propagation on helical antennas', *IEEE Trans.*, **AP-20,** Sept. 1972, pp. 556–560

13 CLARRICOATS, P. J. B., and SAHA, P. K.: 'Propagation and radiation behaviour of corrugated feeds: Pt. 1 – corrugated-waveguide feed, Pt. 2 – corrugated-conical-horn feed', *Proc. IEE,* **118,** 1971, pp. 1167–1186

14 COCKRELL, C. R.: 'The input admittance of the rectangular cavity-backed slot antenna', *IEEE Trans.*, **AP-24,** May 1976, pp. 288–294

15 COHN, S. B.: 'Flare-angle changes in a horn as a means of pattern control', *Microwave J.*, **13,** Oct. 1970, pp. 41, 42, 44, 46

16 COMPTON, R. T., and COLLIN, R. E.: 'Slot antennas' *in* 'Antenna theory' Pt. I, eds. Collin and Zucker. McGraw-Hill Book Co., 1969

17 CREWS, M. R., and THIELE, G. A.: 'On the design of shallow depth T-bar fed slot antennas', *IEEE Trans.*, **AP-25,** Nov. 1977, pp. 833–836

18 CROSSWELL, W. F., *et al.*: 'Radiation from a homogeneous sphere mounted on a waveguide aperture', *IEEE Trans.*, **AP-23,** Sept. 1975, pp. 647–656

19 CUBLEY, H. D., and HAYRE, H. S.: 'Radiation field of spiral antennas employing multimode slow wave techniques', *IEEE Trans.*, **AP-19,** Jan. 1971, pp. 126–128

20 DERNERYD, A. G.: 'Linearly polarised microstrip antennas', *IEEE Trans.*, **AP-24,** Nov. 1976, pp. 846–851

21 DESCHAMPS, G. A., and DYSON, J. D.: 'The logarithmic spiral in a single-aperture multimode antenna system', *IEEE Trans.*, **AP-19**, Jan. 1971, pp. 90–95
22 DOMBEK, K. P.: 'Dielektrische Antennen nicht-cylindrischer Form', *Nachrichtent. Z.*, **26**, 1973, pp. 529–535
23 DOMBEK, P. K.: 'Dielektrische Antennen geringer Querabmessungen als Erreger für Spiegelantennen', *Nachrichtent. Z.*, **28**, 1975, pp. 311-315
24 DYSON, J. D.: 'The characteristics and design of the conical spiral antenna', *IEEE Trans.*, **AP-13**, July 1965, pp. 488–499
25 EHRENSPECK, H. W.: 'The backfire antenna, a new type of directional line source', *Proc. IRE.*, **48**, Jan. 1960, pp. 109–110
26 EHRENSPECK, H. W.: 'The short-backfire antenna', *Proc. IEEE*, **53**, Aug. 1965, pp. 1138–1140
27 EHRENSPECK, H. W.: 'The backfire – A high gain cavity antenna', *in* 'Electromagnetic Wave Theory' (ed. J. Brown). Pergamon Press, 1967, pp. 739–749
28 EHRENSPECK, H. W.: '"Backfire" – Antennen', *NTZ*, **5**, 1969, pp. 286–292. (In German)
29 ENGLISH, W. J.: 'The circular waveguide step-discontinuity mode transducer', *IEEE Trans.*, **MTT-21**, Oct. 1973, pp. 633–636
30 FRANK, Z.: 'Very wideband corrugated horns', *Electron. Lett.*, **11**, 1975, pp. 131–133
31 GALEJS, J.: Admittance of a rectangular slot which is backed by a rectangular cavity', *IEEE Trans.*, **AP-11**, March 1963, pp. 119–126
32 GERST, C., and WORDEN, R. A.: 'Helix antennas take turn for better', *Electronics*, Aug. 1966, pp. 100–110
33 GETSINGER, W. J.: 'Microstrip dispersion model', *IEEE Trans.*, **MTT-21**, 1973, pp. 34–39
34 GUY, R. F. E., and ASHTON, R. W.: 'Crosspolar performance of an elliptical corrugated horn antenna', *Electron. Lett.*, **15**, 1979, pp. 400–402
35 HAMMARSTAD, E. O.: 'Equations for microstrip circuit design'. 5th European Microwave Conf., 1975, pp. 268–272
36 HAN, C. C., and WICKERT, A. N.: 'A new multimode rectangular horn antenna generating a circularly polarised elliptical beam', *IEEE Trans.*, **AP-22**, Nov. 1974, pp. 746–751
37 HARRINGTON, R. F.: 'Field computation by moment methods'. MacMillan, 1968
38 HESSEL, A.: 'General characteristics of traveling-wave antennas', *in* 'Antenna theory' eds. Collin and Zucker. McGraw-Hill Book Co., 1969
39 HOCHHAM, G. A., and OLVER, A. D.: 'Cross polarised performance of small corrugated feeds'. IEEE Symposium. Antennas and propagation, 1978, pp. 431–434
40 HONG, M. H., NYQUIST, D. P., and CHAN, K. M.: 'Radiation fields of open-cavity radiators and a backfire antenna', *IEEE Trans.*, **AP-18**, Nov. 1970, pp. 813–815
41 HOWELL, J. Q.: 'Microstrip antennas', *IEEE Trans.*, **AP-23**, Jan. 1975, pp. 90–93
42 HUGH, J. A. C.: 'Numerical computation of the current distribution and far-field-radiation pattern of the axial-mode helical aerial', *Electron. Lett.*, **9**, May 1973, pp. 257–258
43 JACOBSEN, J.: 'On the cross polarisation of asymmetric reflector antennas for satellite applications', *IEEE Trans.*, **AP-25**, March 1977, pp. 276–283
44 JAMES, J. R., and HALL, P. S.: 'Microstrip antennas and arrays Pt. 2 – New array-design technique', *Microwaves, Optics and Acoustics*, **1**, Sept. 1977, pp. 175–181
45 JAMES, J. R., and WILSON, G. J.: 'Microstrip antennas and arrays. Pt. 1 – Fundamental action and limitations', *Microwaves, Optics and Acoustics*, **1**, Sept. 1977, pp. 165–174
46 JASIK, H. (Ed.): 'Antenna engineering handbook', Sec. 8.9. McGraw-Hill, New York, 1961
47 JENSEN, P. A.: 'A low-noise multimode Cassegrain monopulse feed with polarisation diversity'. Northeast Electronics Research and Engineering Meeting, Nov. 1963, *Nerem record* (also in Ref. 60)
48 JULL, E. V.: 'Finite-range gain of sectoral and pyramidal horns', *Electron. Lett.*, **6**, 1970, pp. 680–681

49 JULL, E. V.: 'Errors in the predicted gain of pyramidal horns', *IEEE Trans.*, **AP-21**, 1973, pp. 25–31

50 JULL, E. V., and ALLAN, L. E.: 'Gain of an E-plane sectoral horn – A failure of the Kirchhoff theory and a new proposal', *IEEE Trans.*, **AP-22**, 1974, pp. 221–226

51 KILGUS, C. C.: 'Shaped-conical radiation pattern performance of the backfire quadrifilar helix', *IEEE Trans.*, **AP-23**, May 1975, pp. 392–397

52 KING, A. P.: 'The radiation characteristics of conical horn antennas', *Proc. IRE*, **38**, March 1950, pp. 249–251

53 KING, H. E., and WONG, J. L.: 'Gain and pattern characteristics of 1 to 8 wavelength uniform helical antennas'. 1978 Intern. Symp. on Antennas and Propagation, Washington, D.C., 15–19 May 1978, pp. 69–72

54 KIM, O. K., and DYSON, J. D.: 'A log-spiral antenna with selectable polarisation', *IEEE Trans.*, **AP-19**, Sept. 1971, pp. 675–677

55 KLOCK, P. W.: 'A study of wave propagation on helices'. Antenna Lab. Tech. Rep. 68, University of Illinois, 1963

56 KNOP, C. M., WIESENFARTH, H. J.: 'On the radiation from an open-ended pipe carrying the HE_{11} mode', *IEEE Trans.*, **AP-20**, Sept. 1972, pp. 644–648

57 KRAUS, J. D.: 'Antennas'. McGraw-Hill Book Co., 1950

58 LARGE, A. C.: 'Short backfire antennas with waveguide and linear fields', *Microwave J.*, August 1976, pp. 49–52

59 LONG, S. A.: 'Experimental study of the impedance of cavity-backed slot antennas', *IEEE Trans.*, **AP-23**, Jan. 1975, pp. 1–7

60 LOVE, A. W. (Ed.): 'Electromagnetic Horn Antennas'. IEEE Press, New York, 1976

61 LUDWIG, A. C.: 'The definition of cross-polarisation', *IEEE Trans.*, **AP-21**, Jan. 1973, pp. 116–119

62 MALLACH, P.: 'Dielektrische Richtstrahler', *Fernmeldetechn. Z.*, **2**, 1949, pp. 33–39

63 MANWARREN, T., and FARRAR, A.: 'Pattern shaping with hybrid mode corrugated horns', *IEEE Trans.*, **AP-22**, May 1974, pp. 484–487

64 MAYES, P. E. *et al.*: 'The monopole slot: a small broadband unidirectional antenna', *IEEE Trans.*, **AP-20**, July 1972, pp. 489–493

65 MENTZNER, C. A., PETERS, L. Jr., and RUDDUCK, R. C.: 'Slope diffraction and its application to horns', *IEEE Trans.*, **AP-23**, 1975, pp. 153–159

66 MITTRA, R. (Ed.): 'Numerical and asymptotic techniques in electromagnetics'. Springer-Verlag, 1975

67 MUELLER, G. E., and TYRRELL, W. A.: 'Polyrod Antennas', *Bell Syst. Tech. J.*, **26**, 1947, pp. 837–851

68 MUNSON, R. E.: 'Conformal microstrip antennas and microstrip phased arrays', *IEEE Trans.*, **AP-22**, Jan. 1974, pp. 74–78

69 NARASIMHAN, M. S., and RAO, B. V.: 'Modes in a conical horn: new approach', *Proc. IEE*, **118**, Feb. 1971, pp. 287–292

70 NARASIMHAN, M. S., and SHESHADRI, M. S.: 'GTD analysis of the radiation patterns of conical horns', *IEEE Trans.*, **AP-26**, 1978, pp. 774–778

71 NEWMAN, E. H., and THIELE, G. A.: 'Some important parameters in the design of T-bar fed slot antennas', *IEEE Trans.*, **AP-23**, Jan. 1975, pp. 97–100

72 PETTERSON, L.: 'Dielectric ring mode-generators for dual-mode radiators'. Research Laboratory of Electronics, Chalmars University, Gothenburg, Sweden, 1979

73 POTTER, P. D.: 'A new horn antenna with suppressed sidelobes and equal beamwidths', *Microwave J.*, **6**, June 1963, pp. 71–78

74 PROFERA, C. E.: 'Complex radiation patterns of dual mode pyramidal horns', *IEEE Trans.*, **AP-25**, May 1977, pp. 436–438

75 RAMO, S., WHINNERY, J. R., and VAN DUZER, T.: 'Fields and waves in communication electronics'. John Wiley and Sons, Inc., 1965

76 RHODES, D. R.: 'On the stored energy of planar apertures', *IEEE Trans.*, **AP-14**, Nov. 1966

77 RUDGE, A. W.: 'Multiple-beam antennas, offset reflectors with offset feeds', *IEEE Trans.*, **AP-23,** May 1975, pp. 317–322

78 RUDGE, A. W., and ADATIA, N. A.: 'Offset-parabolic-reflector antennas, a review', *Proc. IEEE,* **66,** Dec. 1978, pp. 1592–1618

79 RUMSEY, V.: 'Frequency independent antennas'. Academic Press, 1966

80 RUMSEY, V. H.: 'Horn antennas with uniform power patterns around their axes', *IEEE Trans.,* **AP-14,** Sept. 1966, pp. 656–658

81 RUSSO, P. M., RUDDUCK, R. C., and PETERS, L. Jr.: 'A method for computing E-plane patterns of horn antennas', *IEEE Trans.,* **AP-13,** 1965, pp. 219–224

82 SANFORD, G. G., and KLEIN, L.: 'Development and test of a conformal microstrip airborne phased array for use with the ATS-6 satellite'. Proc. IEE Conf. on Antennas for Aircraft and Spacecraft, 1975, pp. 115–122

83 SANFORD, G. G. and MUNSON, R. E.: 'Conformal VHF antenna for the Apollo-Soyuz test project'. Proc. IEE Conf. on Antennas for Aircraft and Spacecraft, 1975, pp. 130–135

84 SATOH, T.: 'Dielectric-loaded horn antennas', *IEEE Trans.,* **AP-20,** March 1972, pp. 199–201

85 SCHELKUNOFF, S. A., and FRIIS, H. T.: 'Antennas-theory and practice'. Wiley, New York, 1952

86 SILVER, S.: 'Microwave antenna theory and design'. MIT Radiation Laboratory Series, vol. 12. McGraw-Hill, 1949

87 TAKEDA, F., and HASHIMOTO, T.: 'Broadbanding of corrugated conical horns by means of the ring-loaded waveguide structure', *IEEE Trans.,* **AP-24,** Nov. 1976, pp. 786–792

88 TSANDOULAS, G. N., and FITZGERALD, W. D.: 'Aperture efficiency enhancement in dielectric loaded horns', *IEEE Trans.,* **AP-20,** Jan. 1972, pp. 69–74

89 WEINSCHEL, H. D.: 'A cylindrical array of circularly polarised microstrip antenna'. IEEE AP–S Intern Symp. Digest, 1975, pp. 177–180

90 WHEELER, H. A.: 'Transmission-line properties of parallel strips separated by a dielectric sheet', *IEEE Trans.,* **MTT-13,** 1965, pp. 172–185

91 WIESBECK, W.: 'Miniaturisierte antenne in mikrowellen streifenleitungstechnik', *Nachrichtent. Zeitschrift,* **28,** May 1975, pp. 156–159

92 YU, J. S., RUDDOCK, R. C., and PETERS, L. Jr.: 'Comprehensive analysis for E-plane of horn antennas by edge diffraction theory', *IEEE Trans.,* **AP-14,** 1966, pp. 138–149

93 ZUCKER, F. J.: 'Theory and application of surface waves', *Nuovo Cimento,* Supplemento, **9,** 1952, pp. 451–473

94 ZUCKER, F. J.: 'Surface and leaky-wave antennas', Chap. 16 *in* 'Antenna engineering handbook'. H. Jasik (Ed.). McGraw-Hill, 1961

95 ZUCKER, F. J.: 'The backfire antenna: a qualitative approach to its design', *Proc. IEEE,* **53,** 1965, pp. 639–641

96 ZUCKER, F. J.: 'Surface-wave antennas', *in* 'Antenna Theory', Pt. II, eds. Collin and Zucker. McGraw-Hill, 1969

97 NIELSEN, E. D. and PONTOPPIDAN, K.: 'Backfire antennas with dipole elements', *IEEE Trans.,* **AP-18,** May 1970, pp. 367–374

Chapter 8

Antenna measurements

J. Appel-Hansen, E. S. Gillespie, T. G. Hickman, J. D. Dyson

8.1 Introduction

In recent years the variety of antenna measurement techniques has expande considerably and these are dealt with in the following Sections. The topics covere commence with a study of the antenna as an unknown radiating structure in space which is to be investigated by using a source and a detector. A discussion on impe dance measurements is followed by current distribution measurements where probe is moved close to the antenna structure. This leads on to an extensive explan ation of near-field techniques where the field of the test antenna is sampled clos to the antenna and transformed to obtain the far-field.

Several techniques carried out in the intermediate range between near-fiel scanning and the conventional far-field distance are described, including compac ranges and defocusing methods. The traditional far-field methods for measurin directivity, gain, phase centre, boresight and scattering are then considered. Furthe details of these techniques may be found in IEEE Standard Test Procedures fo Antennas.[1] Lastly a description is given of both outdoor and indoor test ranges an the methods used to evaluate their performance.

8.2 Siting considerations

8.2.1 General concepts

The antenna to be investigated experimentally may be defined as an unknow structure in space. Usually the antenna has terminals and an antenna gap as show schematically in Fig. 8.1a. Generally, the ability of the antenna to take part i distributing electromagnetic fields in space is investigated by using a source and detector, and antenna measurements can be considered as scattering measurements. Suppose that a closed region in space is chosen appropriately around the antenn terminals. Then, four modes of operation can be characterised as shown in Fig. 8.1

The position of the source and the detector can be interchanged if the antenn is reciprocal which will be assumed in the following unless otherwise stated. Ther fore, in describing a measurement technique, we will use positions of the sourc and the detector which simplify the description although they may not be the best ones to choose in practice.

By positioning the source and the detector in smaller or larger parts of the closed and the open region, one is free to predict the field existing in a smaller or a larger part of space. In the prediction, the influence of the source and the detector on the test antenna should be taken into account together with other discrepancies between the experimental set up and the in-situ situation. It should be noted that in practice, it is often satisfactory to be able to predict the signal received at one point when a signal is impressed at another point, e.g. in point-to-point communication. For such purposes, special antenna parameters such as gain, receiving cross section, and mismatch factors are defined. The use of these characteristics gives accurate results for communication only over large distances. At short distances an ambiguity problem arises in the assessment of distance because the radiating and receiving structures are not point objects. Also, due to this fact a systematic measurement error is always associated with antenna measurements.[2] In this Reference, problems associated with correction factors and near-field parameters are also discussed.

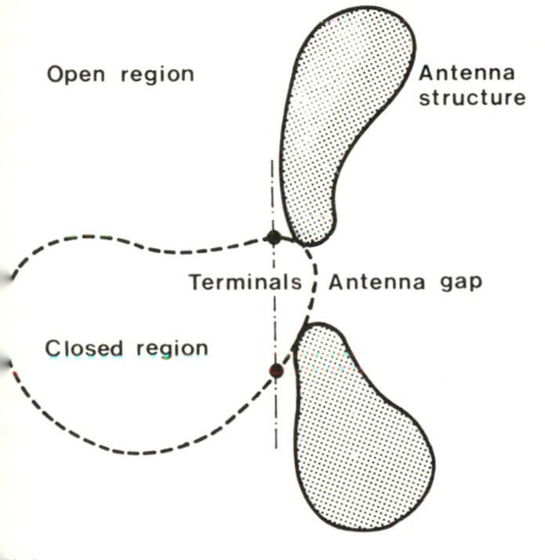

(a) Antenna structure

Source \ Detector	Closed region	Open region
Closed region	Reflecting	Receiving
Open region	Transmitting	Scattering

(b) Modes of operation

Fig. 8.1 *The antenna concepts*

In general, it is characteristic of antenna measurements that special sources of systematic errors have to be taken into account. Besides the ambiguity problem mentioned above, there is a systematic error associated with the fact that a perfect plane wave cannot be generated. Also, error signals due to such imperfections as multipath reflections from the test range and multiple reflections between source, detector and test antenna may give considerable systematic errors. Special problems are associated with identifying, evaluating and minimising the systematic errors. This may be done by averaging a number of measurement results. It should be noted that in order to decrease systematic errors in this manner some test parameters have to be changed from one measurement to the next.

8.2.2 *Distance criteria*

Long transmission distances usually exist between antennas, so it is usual to specify antenna behaviour in the limit of infinite distance. To emphasise the limiting process the characteristics may be called far-field characteristics. These characteristics, e.g. gain, radiation pattern and polarisation, may be measured with sufficient accuracy by using a large test distance between the test antenna and an illuminating source. This means that the antenna under test is illuminated with approximately a plane-wave field. However, for testing purposes it is more convenient to carry out measurements on test ranges of limited size. This creates a particular problem because the measured data change strongly with change in distance as the test range approximates to laboratory size. To understand this problem, it is noted that the field around a radiating antennna may be decomposed into reactive and radiating field components. In the region close to the antenna, within a distance of about $\lambda/2\pi$, the reactive field predominates. Therefore, this region is referred to as the reactive near-field region. Outside this region the reactive field decays rapidly and it can be neglected at a distance of a few wavelengths from the antenna structure. It is in general the distribution of the radiating component which is measured. As shown in Fig. 8.2, the region outside the reactive near-field region is subdivided into two major regions, viz. the radiating near-field region and the radiating far-field region. The boundary between the two regions is defined to be at that test distance for which measured data represent with sufficient accuracy the far-field parameters. This means that every test problem has its own boundary. It also means that, in the radiating near-field region, the measured data depends strongly upon the test distance.

The dependence is understood by considering the field at a point P at a distance R from the antenna, see Fig. 8.2. This field is a sum of field contributions from different parts of the antenna. The contributions propagate along different path lengths, e.g. r_1 and r_2 are the path lengths for the contributions from the area dS_1 and dS_2, respectively. As R is increased the path lengths r_1 and r_2 and their difference change. Therefore, the amplitudes and phases of the contributions from dS_1 and dS_2 change. But, as R gets sufficiently large, the path lengths become nearly parallel and the ratio between the amplitudes and the phase difference change insignificantly. This means that far-field characteristics are measured with sufficient accuracy for distances where P is in the radiating far-field region.

For gain measurements on electrically large aperture antennas, the boundary between the radiating near-field region and the radiating far-field region is usually accepted to be at a distance R, referred to as the far-field distance, $R = 2D^2/\lambda$ where D is the largest aperture dimension and λ is the wavelength. From a single geometrical consideration, it is found that this corresponds to a maximum path length difference of about $\lambda/16$, equivalent to a phase difference of $22.5°$, between the field contributions from the centre of the aperture and the edge of the aperture. Thus, if measurements are carried out at test distances larger than $2D^2/\lambda$, the gain and often also other characteristics may be measured with sufficient accuracy. However, high-precision measurements may require a larger distance, e.g. $50D^2/\lambda$

In some cases at lower frequencies and dependent upon the value of D, it might be necessary to carry out measurements at a distance which is larger than 10λ in order to avoid induction coupling. Further details on the siting considerations related to far-field measurements can be found in reference 1.

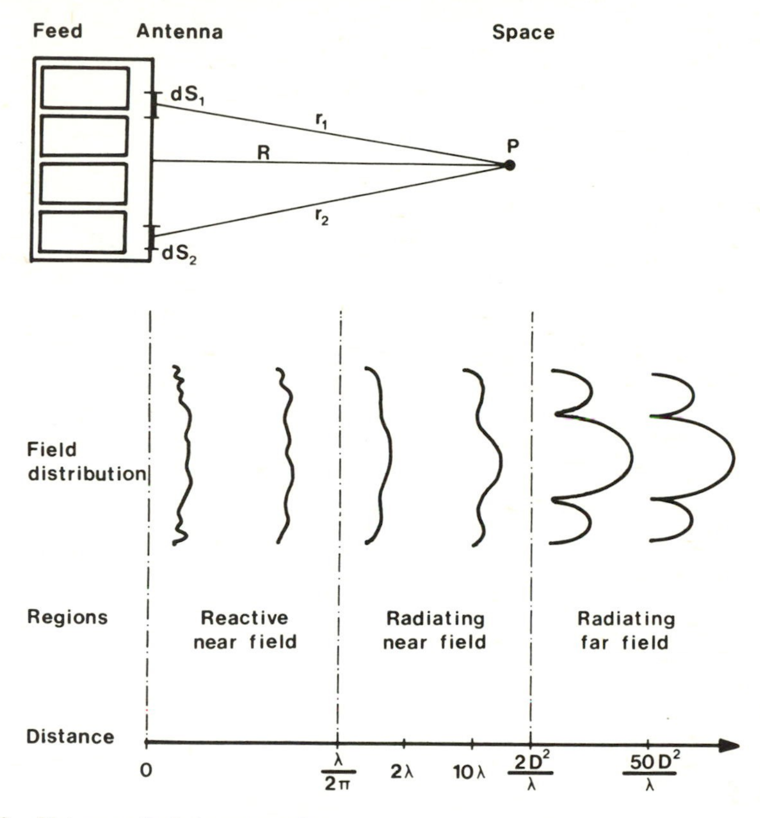

Fig. 8.2 *Distance criteria in open region*

It is seen that as antenna dimension and frequency increase, the far-field distance may become larger than the maximum test distance of the test range. Therefore, several techniques have been developed to measure within the range of $2D^2/\lambda$.

Let it be assumed that the space is divided into a closed and an open region as discussed in Section 8.2.1. For example, let a natural boundary between the two regions be chosen in the form of a surface consisting of waveguide walls and an antenna gap as shown in Fig. 8.3. Then, in accordance with the criteria given above the open region is divided into the three major regions, viz. the reactive near-field region, the radiating near-field region, and the radiating far-field region. Furthermore, the closed region can also be divided into two regions. In the characterisation of these two regions, use is made of the fact that in waveguides the total field can be conveniently expanded in terms of waveguide modes. Furthermore, the dimensions of the waveguide can be chosen such that behind a certain reference plane only a single mode can propagate. Therefore, the closed region is divided into two regions referred to as the higher-order mode region and the single-mode region as illustrated in Fig. 8.3. The measurement space is now divided into five major regions within which we can place the source and the detector.

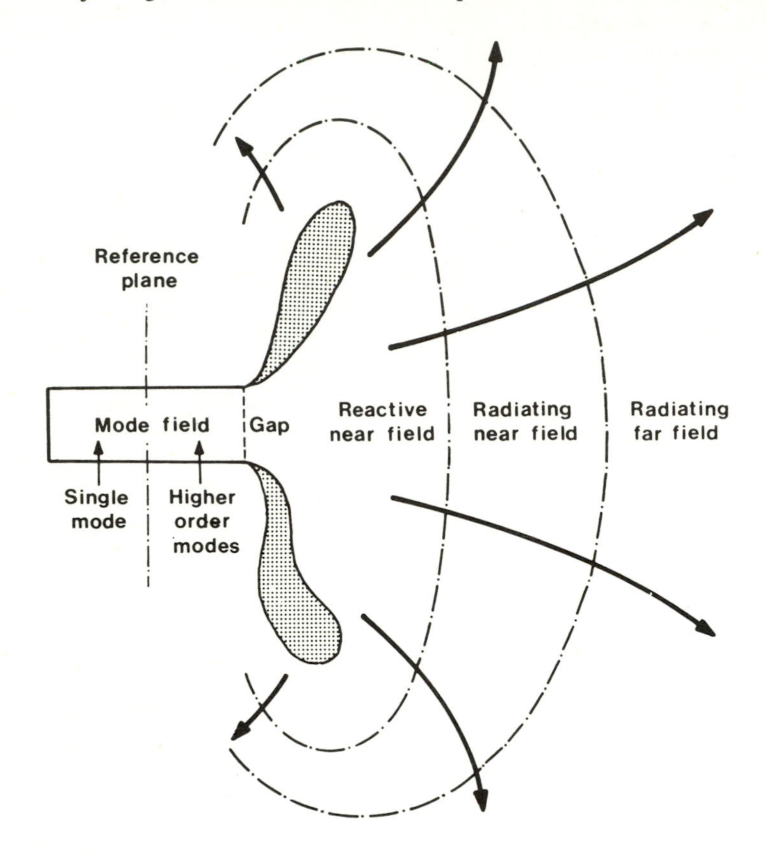

Fig. 8.3 *Field regions and measurement techniques*

A random choice of the positions of the source and the detector in the five regions provides 25 different measurement situations. In analogy with Fig. 8.1*b*, the situations may be described systematically. Not all the 25 modes of operation have been tried; new measurement procedures may be developed. An analysis shows that the procedures which are usually followed fall into five major categories:

Impedance measurements
Current distribution measurements
Near-field and intermediate range techniques
Conventional far-field measurements
Scattering measurements

The five major categories may be considered from a siting point of view, see Fig. 8.4. Impedance measurements may be said to be carried out with both the source and detector placed behind the antenna gap. Current distribution measurements are carried out with the source in the closed region and a small probe as close as possible to the antenna structure. In near-field scanning techniques, the test antenna is in the transmitting mode of operation and a probe scans a smaller or a larger part of a complex surface enclosing it. The smallest distance between the surface and the antenna structure is chosen to be so that the probe is scanning outside the reactive near-field region. Often the smallest distance is chosen larger than a few wavelengths.

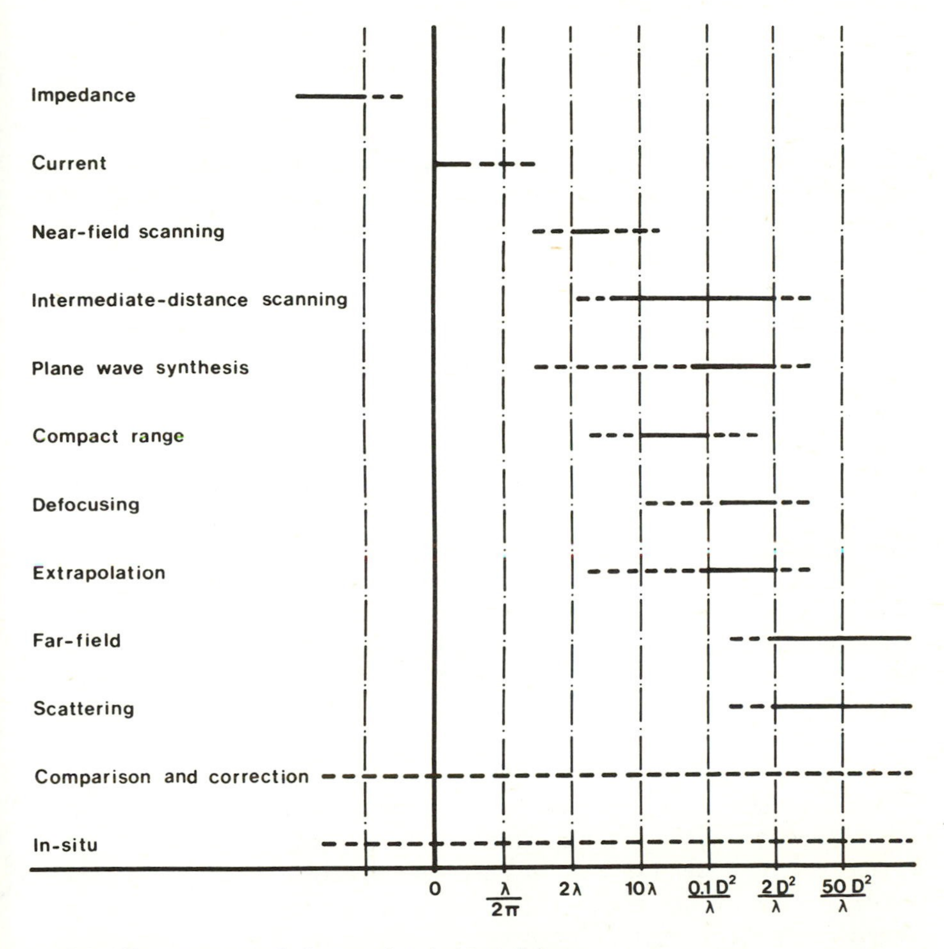

Fig. 8.4 *Measurement techniques and typical test distances*

In the radiating near field but at distances larger than 10λ, the intermediate range techniques are carried out. Depending on the actual measurement set up, these techniques are, for example, plane-wave synthesis, compact range, defocusing and extrapolation techniques. Additional types of intermediate range techniques are described in Reference 4. Conventional far-field techniques and scattering measurements are carried out beyond the far-field distance.

Two additional techniques are included in the Fig. 8.4. Both techniques are based on an appropriate theoretical model of the test antenna. The first technique is called the comparison technique. In this technique measured data are compared with theoretical data. If there is sufficient agreement between the two sets of data the theoretical model may be used to derive the proper antenna characteristics. The second technique is called the correction technique. Here, also, the two sets of data mentioned above are compared, but the experimental results are corrected with factors found from the theoretical model (or experience).

8.3 Impedance measurements

The measurement of the impedance of an antenna, can be performed with a wide range of laboratory equipment, from conventional bridges to sophisticated automatic systems. Furthermore, direct use can be made of the various techniques developed for measurement of impedance in network circuits at various frequencies. This is due to the fact that the antenna is a component which, together with other components such as oscillators, detectors and transmission lines, makes up a complete network circuit.

It is interesting to note that from a fundamental point of view, any network element is an antenna. However, antennas are special components which are designed so that they can interact in a satisfactory manner with each other without being connected to each other, i.e. wireless. It is this design which causes special error signals which have to be taken into account in antenna impedance measurements but which usually can be neglected in other types of impedance measurements. A complete description of the error signals is now given. For a description of equipment and measurement procedures the reader is referred to standard text books.

The error signals will be discussed by considering the space divided into a closed region and an open region as discussed in Section 8.2. The antenna is operating in the reflecting mode of operation; i.e. the source and detector is in the closed region and connected to the antenna reference plane by means of some network as illustrated in Fig. 8.5.

The input impedance is defined as the ratio between the voltage and current at the reference plane. Optimum operation of the antenna and the network usually requires that all power associated with a signal propagating in the closed region and incident on the reference plane is radiated into the open region where it is absorbed. However, often some mismatch between the transmission line and antenna causes a reflected signal to propagate back to the network. It is by measurement of the reflected signal in phase and amplitude relative to the signal incident on the reference plane that the input impedance of the antenna can be determined.

The influence of several error signals have to be taken into account. In order to outline these, suppose that the network in Fig. 8.5 is some properly chosen equipment for measuring the input impedance. Basically, the value of the measured impedance depends upon the interaction between the closed region and the open region. It is desirable that the open region is empty and that the interaction between the closed region and the open region only occurs through the reference plane. However, the open region is not empty and the interaction does not occur only through the reference plane. This is due to the fact that, in practice, any infinitesimal part of any body interacts with any other infinitesimal part of the same body or any other body. The total interaction effect can be integrated into a total detected signal. As illustrated in the Figure, it is convenient to consider the total integration as a sum of signals. The first signal is the one we wish to measure and the remaining signals are disturbing error signals. The measurement errors will be classified into five types.

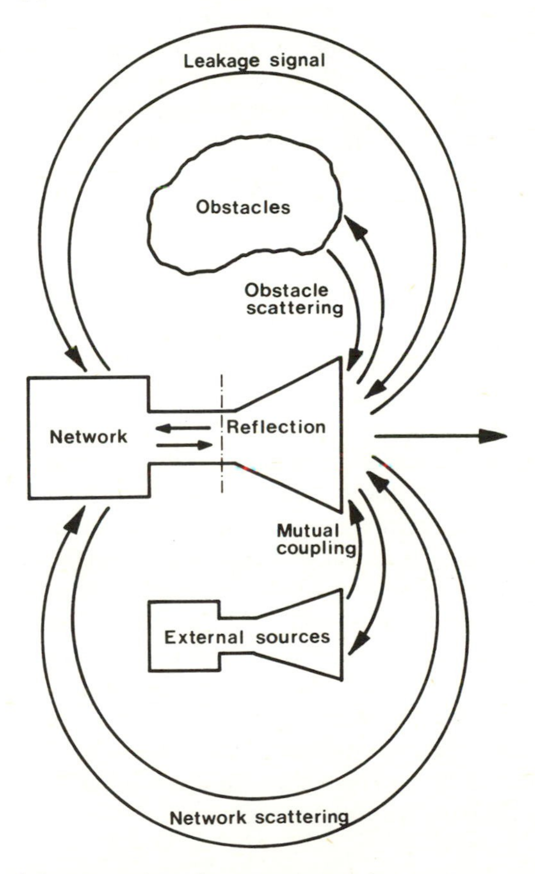

Fig. 8.5 *Error signals in antenna impedance measurements*

The first type of error signal is due to the fact that the radiated field of the antenna generates a current distribution on the outside of the closed region. This gives an error signal which is called network scattering. If the antenna has a ground plane, the signal can be minimised by placing the measuring network behind the ground plane. The second type of error signal is due to current distributions generated on all obstacles in the open region. This is called obstacle scattering. If the antenna is directive, the effect of obstacle scattering can be minimised by letting the antenna point in an obstacle-free direction or at least in such a direction that specular reflections are not detected through the antenna main lobe. If the antenna is omnidirectional it may be necessary to carry out the measurements in an anechoic chamber. A chamber that is not shielded has a third type of error signal which is due to external sources. This type of error signals is called mutual coupling. The effect of mutual coupling can be minimised by proper selective detection methods. The fourth type of error signal is called leakage. This is due to interaction between the two regions not only through the reference plane but also through the remaining boundary between the two regions. The effect of leakage can be minimised by proper shielding. Sometimes leakage causes a situation in which touching the cables causes a change in the detected signal. This situation is often characterised as 'hot'. In a fifth type of error signal, the remaining interactions referred to as scattering are included. The reader can imagine that there is also scattering between the obstacles and the external sources which may cause measurement errors. Sometimes the error signals are called extraneous signals or multipath signals.

Radio anechoic chambers for impedance measurements need not be so large as required for conventional pattern measurements. Often small caps are constructed for impedance measurements or impedance calibration purposes. If high-precision measurements are needed, it may be possible to correct for some if not all of the inevitable error signals by proper averaging of the measured data. For example, the influence of obstacle scattering may be eliminated by making measurements at various positions of the test antenna relative to the surroundings.

Sometimes, in order to measure the expected interaction between the antenna and its network, it may be necessary to measure the input impedance when the antenna is placed in its operational environment. In particular, this is necessary if the active impedance is needed for an antenna element in an array. This is defined as the input impedance of the element with all other elements excited as in an actual array operation. As an alternative to measure the active impedance, this may be calculated from the array impedance matrix consisting of measured mutual impedances between the elements and self impedances of the elements, IEEE Standard.[1] It is interesting to note that, if the antenna is placed in its operational environment, some of the extraneous signals discussed above are not error signals but included into the signal we wish to measure.

8.4 Current distribution measurements

From a knowledge of the amplitude and the phase of the current distribution on the surface of an antenna or scatterer, the field can be calculated in amplitude and phase in the entire space surrounding the antenna. Therefore, it can be claimed that measurement of the current distribution will give the most complete picture of the field distribution.

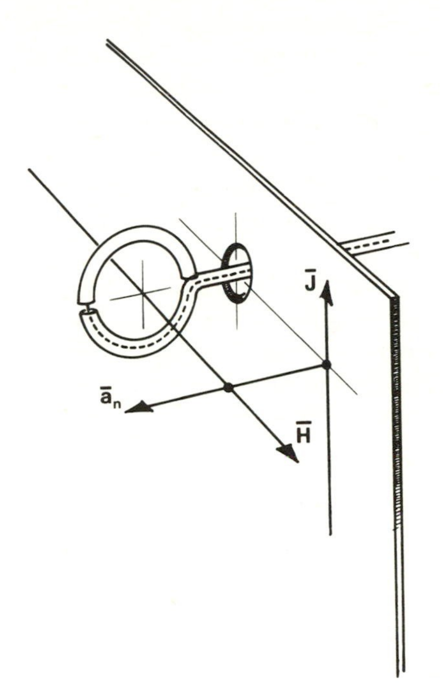

Fig. 8.6 *Measurement of current*

The most often used probe for measuring the current distribution is the small loop surrounding part of the tangential magnetic field $\bar{H}$ which is related to the surface current density $\bar{J}$ through the equation

$$\bar{a}_n \times \bar{H} = \bar{J} \tag{8.1}$$

where $\bar{a}_n$ is a unit vector pointing in the outward direction from the surface. Eqn. 8.1 is valid at the surface for a perfect conductor. Because the loop has a dimension and conductors are not perfect, the measured magnetic field at some distance from the surface will not be exactly proportional to the current, see Fig. 8.6. Also, the positioning of the probe close to the surface will change the original current distribution and cause measurement errors. It is understood that in order to keep measurement errors at a minimum the loop should be so small that it does not seriously disturb the current and that it measures the field essentially at a point as close to the surface as possible. Furthermore, the probe should only measure a linear component of the field.

A current distribution measurement may be considered as a near-field measurement with the requirement of a minimum distance to the antenna structure. This is in contrast to conventional near-field measurements, as described in Section 8.5, which are carried out at a distance of a few wavelengths from the antenna structure in order to keep out of the reactive near-field region.

In order to avoid disturbance from the mechanical means of supporting the loop and other instruments of the set up, the loop may protrude through holes or slots cut parallel to the direction of the flow of the expected current distribution. For a description of the design of probes and cable arrangements the reader may consult the tutorial paper by Dyson[5] and its list of references. Current distribution measurements have also been described by Kraus,[9] in his chapter on antenna measurements, and by King[6] in relation to current measurement on linear antennas. A test facility for measurements of surface fields on three-dimensional objects has also been developed as a diagnostic tool in scattering studies, Knott, Liepa and Senior.[8] Several examples of microwave homodyne systems making use of small modulated scatterers for measuring the complex $\bar{E}$- and/or $\bar{H}$-field distribution near either radiating or non-radiating structures can be found in King.[7]

8.5 Near-field scanning techniques

8.5.1 Introduction

Since the early 1960s there has been an increasing interest in implementing test ranges for predicting far-field antenna properties from near-field measurements. As surveyed by Johnson, Ecker and Hollis,[26] the techniques can be classified into three major classes: the focusing techniques, the compact ranges and the scanning techniques. At present, major emphasis is on the scanning techniques which are the subject of the present Section. In these techniques, the field of the test antenna is detected in phase and amplitude by a probe antenna scanning a surface which may often be only a few wavelengths from parts of the antenna structure. The surfaces, that can be realised easily in practice, are a planar surface in front of the antenna, a circular cylindrical surface surrounding the antenna and a spherical surface enclosing the antenna as shown in Fig. 8.7.

Fortunately, well-established theories for expanding the electromagnetic field in planar, cylindrical and spherical elementary waves or modes exist. By using the expansion coefficients and far-field terms for the modes, the intensities of the far field radiated by the test antenna can be computed. It is said that near-field measurement data is *transformed* into far-field data. Thus, usual antenna parameters, which are referred to as predicted parameters, can be found.

Due to the transformation of the near-field data to far-field, the conventional far-field criterion is set aside, i.e. the error due to the fact that measurements cannot be made at infinity is eliminated. Besides this obvious advantage, which means that long test ranges simulating far-field conditions are not needed, the near-field scanning techniques have several other advantages. These often mean that scanning techniques can compete with or even be better in several respects than conventional techniques.

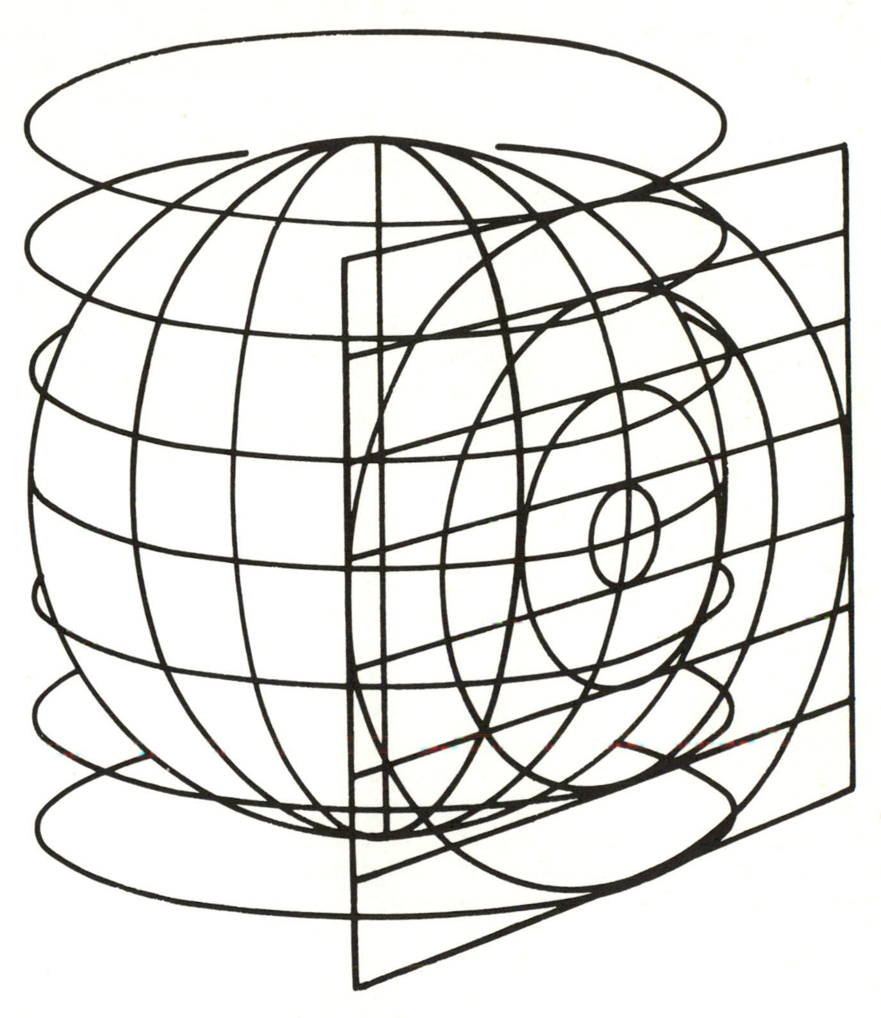

Fig. 8.7 *Planar, cylindrical and spherical scanning techniques*

With less effort, near-field scanning techniques can give much more information than ordinary far-field measurements. As soon as a scanning is carried out and the near-field data stored, a computer can be asked to calculate many details which would have required much conventional measurement work. For example, besides eliminating the proximity effects, it is possible to compute the radiated field in phase and amplitude at any point outside the scanning surface. Due to this fact, co-site interference may be evaluated. The requirement to measure phase and amplitude means that the data has to satisfy more stringent theoretical constraints than conventional far-field amplitude measurements which are made usually only for a limited number of cuts.

Since the probe is placed in the vicinity of the test antenna a large dynamic range can be obtained with standard laboratory equipment. Near-field measurements may also be used as a diagnostic tool to disclose possible antenna faults which may be difficult to find from conventional far-field measurements. Thus, production testing is possible.

The implementation of a near-field test range in a laboratory environment has several advantages in comparison with outdoor test ranges. Measurements can be carried out continuously without schedule disruptions and other disturbances due to changing weather conditions (rain, snow, wind, sun-shadow, day-night, summer-winter, absorption). Measurements can be carried out freed from usual outdoor reflections, e.g. from ground, trees, and buildings. An indoor facility also means that measurements can be carried out under controlled conditions, i.e. a certain degree of cleanliness, humidity, temperature, etc. Indoor ranges have the advantages that they can be used for other purposes when they are not operated for near-field measurements. Moreover, they can be placed close to other facilities. Thus, possible costs of large outdoor test range areas and transportation to and from the ranges of equipment and personnel can be saved.

It should also be mentioned that indoor ranges can be shielded. This means that measurements are not disturbed by unwanted signals (electric perturbation, broadcasting, radar etc.) and vice versa they do not disturb other experiments and can be kept secret if desired. A shielded environment means that multiple reflections from the test range can be kept constant and, by judicious use of absorbers, low. If the walls of the laboratory are sufficiently far away from the test setup, only a small amount of absorbing material may be needed in critical areas. Then expensive installation of an anechoic chamber can be avoided.

All the advantages mentioned above may not justify near-field measurements in some cases. Near-field techniques have an unnecessary complexity when conventional techniques can be applied easily. In particular, this will be the case if the antenna is small in terms of wavelengths or only a limited amount of data is required. The complexity of near-field techniques is due to the fact that the techniques have some special features. They always require a large amount of data even if only a single far-field value is needed. The detected signal has to be measured in phase and amplitude for many positions of the probe. The phase and amplitude data has to be digitised and transformed using computer programs. While many conventional measurements can be carried out without using a computer and by using a simple rotator and an arbitrary illuminating antenna, near-field techniques in general require special scanning apparatus. In addition, the probe has to be characterised in order to correct for its directive effect during the computations of predicted parameters. A certain degree of inaccuracy will always be associated with the assessment of a reference point and the corresponding characteristics.

A special problem is those parts of the scanning system, such as mounts, carriages and rotators, which are close to both antennas. In order to reduce multipath reflections, the scanning system and supports have to be covered with absorbers. Multiple reflections between the test antenna and probe may give appreciable contributions to the extraneous reflections. These contributions usually have negligible effect in conventional techniques, but often have to be included in the error budget for near-field techniques.

At the lower frequencies, the techniques may be limited by the cost of expensive ferrite absorbers. At the higher frequencies, precise measurements of phase at given positions for the probe may be difficult. Furthermore, antennas large in terms of wavelengths will require the use of large computers.

In order to do the measurements properly, each antenna often requires its own test parameters such as sampling spacing, necessary scan area, measurement distance, probe characterisation. Also the influence of extraneous reflections has to be estimated or kept under control.

Before describing the scanning techniques, an outline of their development will be given. This has mainly had its background in two approaches for finding the signal received in phase and amplitude by the probe. In the first approach, Kerns and Dayhoff[29] used a scattering matrix formulation known from microwave circuit theory. A detailed description of this approach is given by Kerns.[28] The matrices relate amplitudes of waveguide modes and expansion coefficients by linear matrix transformations. Besides using scattering matrices in the planar case, they have been used in the spherical case by Wacker[40] and Holm Larsen.[32] In the cylindrical case Yaghjian[44] introduced a source scattering matrix. In the present section the source scattering matrix is adopted in the cylindrical and spherical techniques. Hence, these techniques can be developed by analogy with the planar case.

In the second approach Brown and Jull[17] used a Lorentz reciprocity theorem formulation for the two-dimensional cylindrical case. The test antenna and the probe were enclosed in proper surfaces so that a source free volume was obtained. Then, using the Lorentz reciprocity theorem for the source-free volume an expression for the received signal was found. This formulation was used by Jensen[25] in the spherical case and by Leach and Paris[33] for scanning on a cylinder. As demonstrated by Appel-Hansen,[12] there is agreement between the two approaches. It should be mentioned that other approaches have been presented, e.g. Martsafey[53] considered plane wave synthesis in case of planar scanning, Bennett and Schoessow[14] adopted a plane wave synthesis technique for spherical coverage, Borgiotti[16] used an integral equation formulation in cylindrical scanning, and Rahmat-Sahmii, Galindo-Israel and Mittra[36] applied a Bessel–Jacobi series expansion for the polar planar scanning. This scanning in the plane is an alternative to the rectangular planar scanning. In fact, further alternatives exist; e.g. Wacker[40] has pointed out that there are five measurement lattices which have natural orthogonalities with respect to summation on the lattices. The deficiencies and merits of the different approaches have yet to be analysed in detail, but they will probably depend upon the antenna to be tested, the available apparatus and particularly the experience and background of the test people. Advantages and disadvantages of rectangular planar, circular cylindrical and spherical scanning have been discussed by Wacker and Newell.[41] A rectangular planar scanning for production testing is described by Harmening.[24]

Using the scattering matrix formulation, which is directly applicable also to non-reciprocal antennas, the scanning techniques are introduced in the Sections below. The three expositions, which do not include the polar planar scanning, follow the same guidelines. This shows the similarities and differences between the three techniques. The terminology is introduced in the section on the planar scanning. This section also has a special discussion on measurement accuracy. In the Section on spherical scanning, a general computer program is described. Its approach can also be adopted by the other scanning techniques. After the description of each of the techniques they are compared and some test ranges are described.

8.5.2 The planar technique

Geometrical configuration: In Fig. 8.8 the transmission system setup for the planar scanning technique is shown. The antenna under test is operated as a transmitting antenna and the probe is operated as a receiving antenna. The probe is moved on an imaginary planar surface referred to as the scan plane. The properties of the transmitting antenna is described in a rectangular xyz-coordinate system with its origin at O chosen conveniently in or close to the antenna structure.

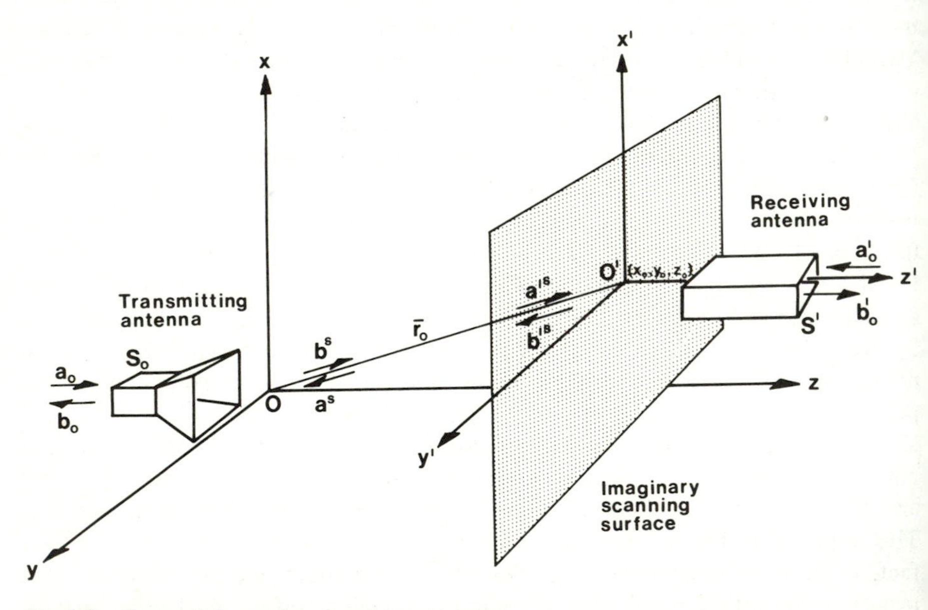

Fig. 8.8 *Transmission system for planar scanning*

The scan plane is placed parallel to the xy-plane at a distance $z_o > 0$ in the positive z-direction. The position of the probe is characterised by the point O', which has the position vector $\bar{r}_o$ with coordinates (x_o, y_o, z_o), where x_o and y_o can take any real values in the range $(-\infty, \infty)$ while z_o is kept constant. The properties of the probe are described in a rectangular $x'y'z'$ coordinate system with its origin at O' and its x', y' and z' axes parallel to the x, y and z axes, respectively. When the probe is moved, the orientation of the probe and z_o is kept constant while x_o and y_o are varied. For example, for discrete values of x_o, a scan is made by varying y_o. The area in which x_o and y_o vary is referred to as the scan area. The choice of the size of the scan area will be discussed in the section on measurement accuracy.

The antennas are considered as two-port transducers. For the transmitting antenna one port is placed in the feed line at S_o while the other port is chosen to be at the xy-plane. The quantities a_o and b_o are phasor wave amplitudes for incident and emergent travelling waves of a single waveguide mode at S_o. The quantities a^s and b^s are spectrum density functions for incoming and outgoing waves as defined in the following. As it appears from Fig. 8.8, primes are used to associate symbols with the receiving antenna.

Expansion in planar modes: The planar scanning technique is developed by expanding fields in a spectrum of plane waves. The vector function used in the expansion can be derived from the elementary planar scalar wave function, Stratton[39]

$$\psi = e^{-j\bar{k}\cdot\bar{r}} \tag{8.2}$$

where $\bar{r}$ is the position vector of the point (x, y, z) in a rectangular xyz-coordinate system and $\bar{k}$ is the propagation vector pointing in the direction of propagation of the plane wave. Let $\bar{k}$ have the components k_x, k_y, k_z, so that

$$k^2 = k_x^2 + k_y^2 + k_z^2 \tag{8.3}$$

At a fixed frequency, k takes the value $\omega\sqrt{\mu\epsilon}$, where ω, μ, and ϵ are, respectively, the angular frequency, permeability and permittivity of the medium which is supposed to have no conductivity. Thus, at a single frequency, only two components of $\bar{k}$ can be considered as independent. Let these components be k_x and k_y which are chosen to take values in the range $(-\infty, +\infty)$. For the third component k_z, it is convenient to relate it to a parameter γ given by

$$\gamma = \pm\sqrt{k^2 - k_x^2 - k_y^2} \tag{8.4}$$

and due to eqn. 8.3

$$k_z = \pm\gamma \tag{8.5}$$

The plus or minus sign in eqn. 8.4 is chosen so that γ is positive real for $k_x^2 + k_y^2 < k^2$ and negative imaginary for $k_x^2 + k_y^2 > k^2$. The plus or minus sign in eqn. 8.5 is chosen dependent on the required behaviour of the plane wave with variation in the z-direction. To indicate the choice, we use a superscript index (i) on $\bar{k}$. Thus

$$\bar{k}^{(i)} = k_x\bar{a}_x + k_y\bar{a}_y \pm \gamma\bar{a}_z \tag{8.6}$$

where the plus sign and the minus sign are chosen when i takes the values 1 and 2, respectively, and $\bar{a}_x, \bar{a}_y, \bar{a}_z$ are usual unit vectors parallel to the coordinate axes.

Due to the conventions given above, we use the expression

$$\psi^{(i)}_{k_x k_y}(\bar{r}) = e^{-j\bar{k}^{(i)}\cdot\bar{r}} \quad (8.7)$$

for the scalar wave function. To facilitate the reading, some of the indices i, k_x, k_y and the argument $\bar{r}$ may be omitted in some cases. It is noted that depending on the value of γ, $\psi^{(i)}$ represents propagating or evanescent waves in the positive and negative z-direction for i equal to 1 and 2, respectively. This is illustrated in Fig. 8.9. Here, a test antenna is placed relative to the xyz-coordinate system and its origin O. As indicated the antenna may be operated either as a receiving antenna or as a transmitting antenna.

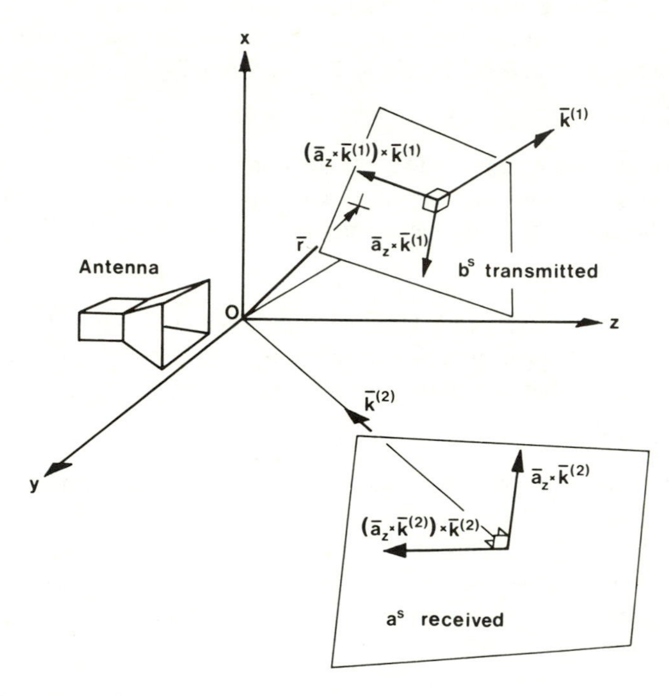

Fig. 8.9 *Resolution of waves in planar technique*

The transmitted or received field is constructed from the elementary planar vector wave functions $\bar{M}^{(i)}_{k_x k_y}(\bar{r})$ and $\bar{N}^{(i)}_{k_x k_y}(\bar{r})$ which are derived from $\psi^{(i)}_{k_x k_y}(\bar{r})$ through the relations

$$\bar{M}^{(i)}_{k_x k_y}(\bar{r}) = \frac{1}{|\bar{a}_z \times \bar{k}^{(i)}|} \nabla \times \bar{a}_z \psi^{(i)}_{k_x k_y}(\bar{r}) \quad (8.8)$$

and

$$\bar{N}^{(i)}_{k_x k_y}(\bar{r}) = \frac{1}{|\bar{a}_z \times \bar{k}^{(i)}|} \frac{1}{k} \nabla \times \nabla \times \bar{a}_z \psi^{(i)}_{k_x k_y}(\bar{r}) \quad (8.9)$$

It is found that

$$\bar{M}^{(i)}_{k_xk_y}(\bar{r}) = j\frac{\bar{a}_z \times \bar{k}^{(i)}}{|\bar{a}_z \times \bar{k}^{(i)}|} e^{-j\bar{k}^{(i)}\cdot\bar{r}} \tag{8.10}$$

$$\bar{N}^{(i)}_{k_xk_y}(\bar{r}) = -\frac{1}{k}\frac{\bar{a}_z \times \bar{k}^{(i)}}{|\bar{a}_z \times \bar{k}^{(i)}|} \times \bar{k}^{(i)} e^{-j\bar{k}^{(i)}\cdot\bar{r}} \tag{8.11}$$

In case $\bar{k}^{(i)}$ and $\bar{a}_z$ are parallel, we choose $\bar{a}_x$ instead of the vector $\bar{a}_z \times \bar{k}^{(i)} = \bar{0}$. Using $\bar{M}^{(i)}_{k_xk_y}(\bar{r})$ and $\bar{N}^{(i)}_{k_xk_y}(\bar{r})$ the plane wave spectrum expansion of the transmitted field $\bar{E}(\bar{r})$ in the region $z > 0$ is given by

$$\bar{E}(\bar{r}) = \int_{-\infty}^{\infty}\int_{-\infty}^{\infty} \{b^1(k_x, k_y)\bar{M}^{(1)}_{k_xk_y}(\bar{r}) + b^2(k_x, k_y)\bar{N}^{(1)}_{k_xk_y}(\bar{r})\}\, dk_x\, dk_y \tag{8.12}$$

The coefficients $b^s(k_x, k_y)$, where $s = 1, 2$, are referred to as spectrum-density functions of outgoing waves. Using Faraday's law and the symmetry relations

$$\bar{M}^{(i)} = \frac{1}{k}\nabla \times \bar{N}^{(i)} \tag{8.13}$$

$$\bar{N}^{(i)} = \frac{1}{k}\nabla \times \bar{M}^{(i)} \tag{8.14}$$

the magnetic field $\bar{H}(\bar{r})$ is found to be

$$\bar{H}(\bar{r}) = -\frac{k}{j\omega\mu}\int_{-\infty}^{\infty}\int_{-\infty}^{\infty} \{b^1(k_x, k_y)\bar{N}^{(1)}_{k_xk_y}(\bar{r}) + b^2(k_x, k_y)\bar{M}^{(1)}_{k_xk_y}(\bar{r})\}\, dk_x\, dk_y \tag{8.15}$$

Observing that $\bar{M}$ has no z-component, it appears from eqns. 8.12 and 8.15 that, with respect to the z-axis, the electromagnetic field is resolved into a TE-field with coefficients $b^1(k_x, k_y)$ and a TM-field with coefficients $b^2(k_x, k_y)$. Thus s may be referred to as a polarisation index. From a comparison of eqns. 8.10 and 8.11 with Fig. 8.9, it is seen that $\bar{M}$ and $\bar{N}$ are linearly polarised plane wave fields with polarisation perpendicular and parallel, respectively, to the plane of incidence determined by $\bar{a}_z$ and $\bar{k}^{(i)}$.

The plane-wave expansion of an incoming field $\bar{E}'(\bar{r})$ from a transmitting antenna placed somewhere in the region $z > 0$ is given by

$$\bar{E}'(\bar{r}) = \int_{-\infty}^{\infty}\int_{-\infty}^{\infty} \{a^1(k_x, k_y)\bar{M}^{(2)}_{k_xk_y}(\bar{r}) + a^2(k_x, k_y)\bar{N}^{(2)}_{k_xk_y}(\bar{r})\}\, dk_x\, dk_y \tag{8.16}$$

where $a^s(k_x, k_y)$ are used as spectrum density functions for the incoming waves. From the above it is seen that a^s and b^s are simple expansion coefficients of incoming and outgoing waves, respectively.

In order to find an expression for the signal b'_o received by the probe in Fig. 8.8, it is desirable to express the radiated field $\bar{E}(\bar{r})$ given by eqn. 8.12 in terms of

$\bar{M}^{(i)}_{k_x k_y}(\bar{r}')$ and $\bar{N}^{(i)}_{k_x k_y}(\bar{r}')$ associated with the $x'y'z'$-coordinate system. This may be done by inserting $i = 1$ and, see Fig. 8.10,

$$\bar{r} = \bar{r}_o + \bar{r}' \tag{8.17}$$

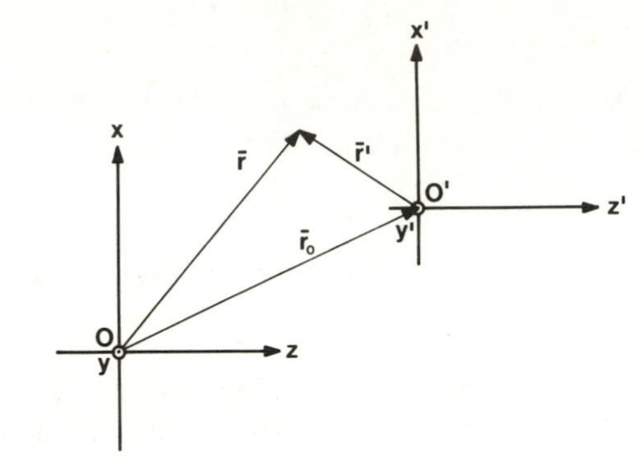

Fig. 8.10 *Derivation of translation theorem*

into eqns. 8.10, 8.11. Then the following translation theorems are obtained:

$$\bar{M}^{(1)}_{k_x k_y}(\bar{r}) = \bar{M}^{(1)}_{k_x k_y}(\bar{r}')\, e^{-j\bar{k}^{(1)}\cdot \bar{r}_o} \tag{8.18}$$

$$\bar{N}^{(1)}_{k_x k_y}(\bar{r}) = \bar{N}^{(1)}_{k_x k_y}(\bar{r}')\, e^{-j\bar{k}^{(1)}\cdot \bar{r}_o} \tag{8.19}$$

Insertion of eqns. 8.18, 8.19 into eqn. 8.12 gives for the field incident upon the probe

$$\begin{aligned}\bar{E}(\bar{r}') = \int_{-\infty}^{\infty}\int_{-\infty}^{\infty} \{(b^1(k_x, k_y)\, e^{-j\bar{k}^{(1)}\cdot \bar{r}_o}\, \bar{M}^{(1)}_{k_x k_y}(\bar{r}') \\ + b^2(k_x, k_y)\, e^{-j\bar{k}^{(1)}\cdot \bar{r}_o}\, \bar{N}^{(1)}_{k_x k_y}(\bar{r}')\}\, dk_x\, dk_y\end{aligned} \tag{8.20}$$

Thus $\bar{E}(\bar{r}')$ can be written as

$$\bar{E}(\bar{r}') = \int_{-\infty}^{\infty}\int_{-\infty}^{\infty} \{a'^1(k_x, k_y)\bar{M}^{(1)}_{k_x k_y}(\bar{r}') + a'^2(k_x, k_y)\bar{N}^{(1)}_{k_x k_y}(\bar{r}')\}\, dk_x\, dk_y \tag{8.21}$$

where

$$a'^s(k_x, k_y) = b^s(k_x, k_y)\, e^{-j\bar{k}^{(1)}\cdot \bar{r}_o} \tag{8.22}$$

are the spectrum density functions of the waves incoming on the probe. It should be noted that $a'^s(k_x, k_y)$ depends on the choice of both O and O' relative to the structures of the test antenna and the probe.

Coupling equation: For the transmitting antenna, there exists the following scattering matrix formulation relating input and output quantities defined above:

$$b_o = \Gamma_o a_o + \int_{-\infty}^{\infty}\int_{-\infty}^{\infty} \sum_{s=1}^{2} R^s(k_x, k_y) a^s(k_x, k_y)\, dk_x\, dk_y \tag{8.23}$$

$$b^s(k_x, k_y) = T^s(k_x, k_y)a_o + \int_{-\infty}^{\infty}\int_{-\infty}^{\infty} \sum_{\sigma=1}^{2} S^{s,\sigma}(k_x, k_y; \kappa_x, \kappa_y)a^\sigma(\kappa_x, \kappa_y)\, d\kappa_x\, d\kappa_y \tag{8.24}$$

where Γ_o, R^s, T^s, and $S^{s,\sigma}$ are the elements of the scattering matrix describing the antenna as a two-port transducer. From a consideration of output quantities, when different input quantities are zero, it is seen that Γ_o corresponds to the reflection coefficient at S_o looking toward the antenna, R^s and T^s are associated with the receiving and transmitting properties of the antenna, respectively, and $S^{s,\sigma}$ describes its scattering properties. When primed quantities are used in eqns. 8.23, 8.24, the scattering matrix formulation for the probe is obtained.

We are now able to derive a relation between the signal $b'_o(\bar{r}_o)$ received by the probe and the input signal a_o delivered to the test antenna. Let Z'_l be the impedance of the load connected to the probe. If Z'_l has the reflection coefficient Γ'_l, then

$$a'_o = \Gamma'_l b'_o \tag{8.25}$$

In addition to this, we neglect multiple reflections between the antennas. Hence, $a^\sigma(\kappa_x, \kappa_y)$ is neglected. Using eqn. 8.24 under this condition, we obtain

$$b^s(k_x, k_y) = a_o T^s(k_x, k_y) \tag{8.26}$$

Insertion of eqn. 8.26 into eqn. 8.22 gives for the spectrum density functions of the field incident on the probe

$$a'^s(k_x, k_y) = a_o T^s(k_x, k_y)\, e^{-j\bar{k}^{(1)}\cdot\bar{r}_o} \tag{8.27}$$

By taking the scattering matrix formulation for the probe and making use of eqns. 8.25, 8.27, the desired relation which is referred to as the coupling equation or transmission formula

$$b'_o(\bar{r}_o) = \frac{a_o}{1-\Gamma'_o\Gamma'_l} \int_{-\infty}^{\infty}\int_{-\infty}^{\infty} \sum_{s=1}^{2} R'^s(k_x, k_y)T^s(k_x, k_y)\, e^{-j\bar{k}^{(1)}\cdot\bar{r}_o}\, dk_x\, dk_y \tag{8.28}$$

is obtained.

By observing that

$$e^{-j\bar{k}^{(1)}\cdot\bar{r}_o} = e^{-j(k_x x_o + k_y y_o + \gamma z_o)} \tag{8.29}$$

and by noting that z_o is kept constant, eqn. 8.28 may be Fourier inverted to give

$$D(k_x, k_y) = \frac{(1-\Gamma'_o\Gamma'_l)\, e^{j\gamma z_o}}{4\pi^2 a_o} \int_{-\infty}^{\infty}\int_{-\infty}^{\infty} b'_o(\bar{r}_o)\, e^{j(k_x x_o + k_y y_o)}\, dx_o\, dy_o \tag{8.30}$$

where $D(k_x, k_y)$ is defined by

$$D(k_x, k_y) = \sum_{s=1}^{2} R'^s(k_x, k_y)T^s(k_x, k_y) \tag{8.31}$$

The integration in eqn. 8.30 can be made using standard FFT computer programs. The size of sample spacing will be discussed below. The quantity $D(k_x, k_y)$ is referred to as the coupling product of the receiving and transmitting properties of antennas. In fact, the coupling equation is a generalisation of Friis's transmission formula and the coupling product corresponds to the scalar product of the polarisation vectors of the antennas in the communication link. In the next Section the relationship between the transmission characteristic $T^s(k_x, k_y)$ and the far-field intensity will be stated.

The measurement procedure followed in the planar scanning technique can be described with reference to eqns. 8.30, 8.31. Suppose that the receiving antenna has known receiving characteristic $R'^s(k_x, k_y)$ and that we wish to measure the transmitting characteristic $T^s(k_x, k_y)$ of the transmitting antenna. Then by detection of $b'_o(\bar{r}_o)$ in a plane, eqn. 8.30 provides one equation in the two unknowns $T^s(k_x, k_y)$. An additional equation may be obtained by measuring with an additional known receiving antenna or simply by repeating the measurements for another aspect of the original receiving antenna. Usually, the antenna is simply rotated 90° about the z-axis. Further measurements may have to be carried out in the special cases where the two equations cannot be solved; e.g. the determinant is zero.

Due to the presence of the probe characteristic R'^s in eqns. 8.30, 8.31, it is sometimes stated that probe correction of the measured results is carried out. Measurements without probe correction are discussed in the next Section. It should be noted that multiple reflections are not related to probe correction. So far, it has not been possible in actual near-field measurements to take into account multiple reflections between the probe and the test antenna. These reflections are to some extent taken into account in the extrapolation technique which is also based on eqn. 8.28, Newell, Baird and Wacker.[34]

Far-field parameters: Above it was assumed that the characteristic R'^s of the probe is known. The characteristic can be determined from far-field amplitude and phase measurements on the probe itself. This is due to the fact that there is a simple relation between the far-field intensity $\bar{E}'(\bar{r}')$ and the transmitting characteristic T'^s. Without loss of generality, we will derive the relation with reference to an arbitrary antenna as shown in Fig. 8.9. When the antenna is transmitting, a combination of eqns. 8.10, 8.11, 8.12, 8.26 gives for the radiated field at a position $\bar{r}$ in the region $z > 0$ when $a_o = 1$

$$\bar{E}(\bar{r}) = \int_{-\infty}^{\infty}\int_{-\infty}^{\infty} \left\{ jT^1(k_x, k_y) \frac{\bar{a}_z \times \bar{k}^{(1)}}{|\bar{a}_z \times \bar{k}^{(1)}|} - \frac{T^2(k_x, k_y)}{k} \frac{\bar{a}_z \times \bar{k}^{(1)}}{|\bar{a}_z \times \bar{k}^{(1)}|} \times \bar{k}^{(1)} \right\} e^{-j\bar{k}^{(1)} \cdot \bar{r}} \, dk_x \, dk_y \tag{8.32}$$

This shows that the field at an arbitary point can be considered as an integration of plane waves propagating in all directions.

Before deriving the desired relation it should be noted that Fourier inversion of eqn. 8.32 gives

$$\left\{ jT^1(k_x, k_y) \frac{\bar{a}_z \times \bar{k}^{(1)}}{|\bar{a}_z \times \bar{k}^{(1)}|} - \frac{T^2(k_x, k_y)}{k} \frac{\bar{a}_z \times \bar{k}^{(1)}}{|\bar{a}_z \times \bar{k}^{(1)}|} \times \bar{k}^{(1)} \right\}$$

$$= \frac{e^{j\gamma z_o}}{4\pi^2} \int_{-\infty}^{\infty} \int_{-\infty}^{\infty} \bar{E}(\bar{r}_o)\, e^{j(k_x x_o + k_y y_o)}\, dx_o\, dy_o \tag{8.33}$$

where $\bar{E}(\bar{r}_o)$ is the radiated field intensity measured in a plane $z = z_o$ in front of the antenna. Thus, eqn. 8.33 is an alternative to eqn. 8.30 for determination of T^s. However, in the case of eqn. 8.33, the problem is to determine $\bar{E}(\bar{r}_o)$. The components of $\bar{E}(\bar{r}_o)$ can only be determined correctly by using a Hertzian dipole. If measured signals are interpreted as signals received by a Hertzian dipole, which can be approximately realised by small probes like a short dipole or an open-ended waveguide, it is said that T^s is determined without probe correction. If it is desired to obtain more accurate results, eqn. 8.30 should be used with probe correction, i.e. we need to determine R'^s as described in the following.

By the method of steepest descents, it can be shown that for r large, $\bar{E}(\bar{r})$ of eqn. 8.32 is given by

$$\bar{E}(\bar{r}) \simeq j2\pi k_{zo} \left\{ jT^1(k_{xo}, k_{yo}) \frac{\bar{a}_z \times \bar{a}_r}{|\bar{a}_z \times \bar{a}_r|} - T^2(k_{xo}, k_{yo}) \frac{\bar{a}_z \times \bar{a}_r}{|\bar{a}_z \times \bar{a}_r|} \times \bar{a}_r \right\} \frac{e^{-jkr}}{r} \tag{8.34}$$

where k_{xo}, k_{yo}, k_{zo} are the components of $\bar{k}_o^{(1)}$ defined by $\bar{k}_o^{(1)} = k\bar{a}_r$ where $\bar{a}_r$ is a unit vector in the considered direction of $\bar{r}$. It is seen that in the direction $\bar{r}$ the amplitude of the far field, expressed by eqn. 8.34, is proportional to the amplitude of that plane wave, integrated in eqn. 8.32, which propagates in the direction $\bar{r}$.

Introducing a spherical $r\theta\phi$-coordinate system with the usual orientation relative to the xyz-coordinate system, we obtain

$$\bar{E}(\bar{r}) \simeq j2\pi k \cos\theta \,\{jT^1(k_{xo}, k_{yo})\bar{a}_\phi - T^2(k_{xo}, k_{yo})\bar{a}_\theta\} \frac{e^{-jkr}}{r} \tag{8.35}$$

This is the desired relation by means of which T^s may be computed from usual linear $\bar{a}_\theta$- and $\bar{a}_\phi$-far-field components measured in phase and amplitude. Alternatively, eqn. 8.35 may be used to express the far-field from near-field measurements of T^s. This relation corresponds to the relation between angular spectrum and polar diagram as discussed by Booker and Clemnow.[15]

If the antenna is reciprocal, as will usually be the case for probes, R^s can be found from the reciprocity relation

$$\eta_o R^s(k_x, k_y) = -\eta T^s(-k_x, -k_y) \tag{8.36}$$

where η_o is the wave admittance of the single mode in the waveguide feed to the antenna and η is the free-space admittance. It is noted that in the directions corresponding to a null in the probe pattern, $R^s = 0$, probe correction as discussed in relation to eqn. 8.31 cannot be made.

From these considerations it can be concluded that the characteristics T^s and R^s express far-field transmitting and receiving parameters. Therefore, common far-field characteristics can be expressed by using these parameters, e.g. the gain $G(k_{xo}, k_{yo})$ is given by

$$G(k_{xo}, k_{yo}) = 16\pi^3 k^2 \cos^2\theta \frac{\{|T^1(k_{xo}, k_{yo})|^2 + |T^2(k_{xo}, k_{yo})|^2\}\eta}{\{1 - |\Gamma_o|^2\}\eta_o} \tag{8.37}$$

Measurement accuracy: The sources of errors in the determination of T^s can be derived from a consideration of the coupling product and the inverted coupling equation 8.30 and its background.

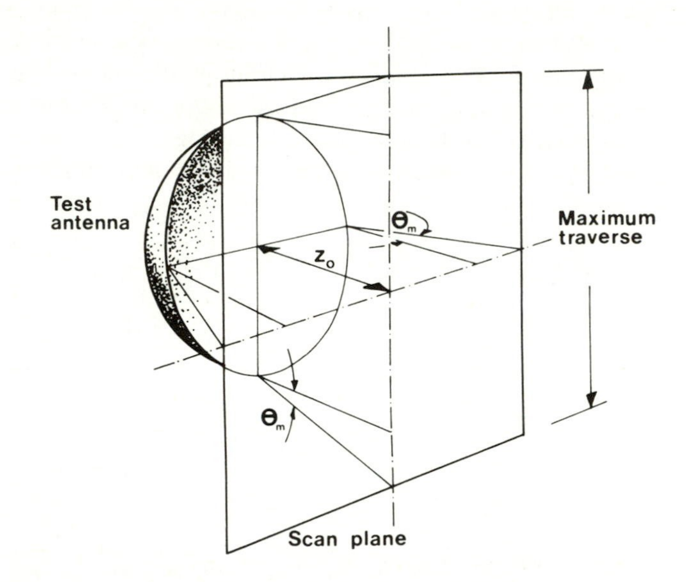

Fig. 8.11 *Scan area and expected range for accurate far-field data*

In near-field test ranges for rectangular scanning the signal $b'_o(\bar{r}_o)$ is sampled at discrete points on a rectangular lattice in the scan plane. The transverse extension of the scan area depends mainly upon the test antenna and the directions in which it is desired to determine the far field. Typically, the scan area will be a little more than twice the aperture area of the test antenna. Fig. 8.11 illustrates the maximum angle or truncation angle θ_m to which accurate far-field data can be expected. For angles larger than θ_m, computed far fields cannot be relied upon with any confidence, Yaghjian.[43] The truncation angle θ_{mx} for predicted far fields in the xz-plane is given by

$$\tan\theta_{mx} = \frac{L_x - D_x}{2z_o} \tag{8.38}$$

where L_x is the scan length and D_x is the diameter of the antenna. Both L_x and D_x are in the x direction. Thus, planar scanning has only conical coverage in front of the antenna. However spherical coverage is possible by placing at least four scan planes (tetrahedron) round the antenna. This can be arranged if the test antenna can be rotated around two axes. To control the measurements and increase the accuracy some overlapping can be used between the scans. Thus, this tetrahedron scanning may be a competitor to normal spherical scanning, in particular if a large accurate planar scanner is available.

The sampling spacing depends mainly upon the distance z_o between the test antenna and the scan plane. A criterion for the magnitude of the sampling spacing can be derived from a consideration of eqns. 8.28, 8.31. For $k_x^2 + k_y^2 > k^2$, γ is negative imaginary. Hence, the factor $e^{-j\gamma z_o}$ in eqn. 8.28 shows that evanescent waves contribute to $b_o'(\bar{r}_o)$. If z_o is chosen sufficiently large, the evanescent waves are attenuated so that they can be neglected. This means that $b_o'(\bar{r}_o)$ can be considered as a Fourier transform of a band-limited function of k_x and k_y. Then, $D(k_x, k_y)$ in eqn. 8.30 can be written as a summation where the sampling spacing is chosen in accordance with the two-dimensional sampling theorem. In order to determine the upper limit of the sampling spacing, suppose that the scan plane is placed in the far field. Then, the band limit of k_x and k_y is k. In this case, the required sampling spacings Δx and Δy are determined by

$$\Delta x = \Delta y = \frac{\pi}{k} = \frac{\lambda}{2} \tag{8.39}$$

Thus, an upper limit of $\lambda/2$ for the sampling spacing is approached when the scan plane is placed in the far field. In actual scanning, a sampling spacing of $\lambda/3$ for a measurement distance of two to three wavelengths is typical. This often causes several hundreds of dB attenuation of evanescent waves. The final choice of sampling spacing depends upon the dynamic range of the equipment, Joy and Paris.[27] The work in this Reference also shows that computational efforts may be reduced by filtering the measurement data in case only a portion of the far-field pattern is wanted. In some cases violation of the upper limit of $\lambda/2$ has been investigated. Sample spacings in the order of λ may not cause large errors in the boresight direction of directive antennas, but sidelobes turn out to be unreliable, Joy and Paris.[27]

Analyses of errors due to inaccuracies in x_o, y_o, and z_o have shown that precision of the order of 10^{-4} m in the positioning of the probe is required in order to measure gain with an accuracy of a few tenths of a dB at 10 GHz. This means that near-field techniques seem to require precision mechanical setups. In particular, the z-coordinate is critical at higher frequencies. Corey and Joy[21] have shown that it is possible to include compensation for probe positioning errors which may indicate that the mechanical requirements might be reduced.

Errors due to extraneous reflections may be severe. These can be resolved into two components. One component is due to reflections between the test antenna and the probe. The other component is due to reflections from the lining of the test range and the instrumentation. In order to consider each component separately, the first component is referred to as multiple reflections and the second as multipath reflections. The multiple reflections may be made small by not measuring too close to the test antenna and by using a properly matched probe with a low backscattering cross-section. The multipath reflections can be reduced by judicious use of absorbers. It should be noted that some compromise is necessary in reducing the two types of reflections. A highly directive probe may experience low multipath reflections but large multiple reflections. A measurement distance of several wavelengths may cause large multipath reflections and low multiple reflections.

In an actual measurement situation, it may be cost saving to carry out some preliminary tests to determine scan area, sampling spacing and the influence of extraneous reflections. The required scan area and sampling spacing may be determined by carrying out tests along several lines in scan planes for different values of z_o, scan length and sampling spacing. In particular, in case the probe is moved along several lines parallel to the z-axis, multiple reflections may be observed as a standing-wave oscillation with a period of $\lambda/2$. These regular variations will in general be superimposed on other variations which are due to near-field variations and multipath reflections.

It should be noted that near-field techniques have eliminated the systematic error which is present in conventional far-field measurements due to a finite measurement distance. However, near-field techniques have introduced the new systematic errors discussed above.

Some error studies will now be described. Rodrigue *et al.*[37] made computer simulations on hypothetical and measured near-field distributions. In this study and the work by Carver and Newell[19] measured and predicted far-field patterns and gain were compared. Yaghjian[43] used geometrical theory of diffraction and asymptotic evaluation of integral equations in a theoretical study. Here, in order to stipulate design criteria for setups, upper bound error expressions for gain, sidelobe level, beam width and polarisation ratio were obtained for aperture antennas taking into account finite scan area, probe position error, non-linearities in receiving system, error in determining mismatch factors, uncertainty in the receiving characteristic of the probe, and multiple reflections. In another study by Newell,[35] a Fourier transform analysis and error simulation on actual near-field data were made in order to find upper-bound errors for steerable beam antennas.

The error studies indicate that the major sources of errors are due to receiver nonlinearities and multiple reflections. The nonlinearities can be reduced by including calibration curves in the computer programs processing the measured data. The influence of extraneous reflections can be reduced by averaging predicted far-field values obtained for several scan planes which are separated, e.g. a tenth of a wavelength. Another important error source is due to the fact that far-field parameters cannot be predicted with an accuracy larger than the accuracy with which the probe is characterised. At frequencies above 10 GHz, the position error may be dominant. However, the relative magnitudes of the error types will depend upon the predicted parameter and the direction in space considered; e.g. area truncation may be negligible in the boresight direction, but dominant at sidelobe levels near the truncation angle. Nevertheless, error budgets for predicted parameters seem to demonstrate that the planar technique has come to a level of confidence where predicted parameters can be trusted without relying on checking with conventional far-field measurements.

8.5.3 The cylindrical technique

Geometrical configuration: In the cylindrical scanning technique the probe is moved on an imaginary circular cylinder surrounding the antenna under test. An outline of the transmission system setup for this technique is shown in Fig. 8.12.

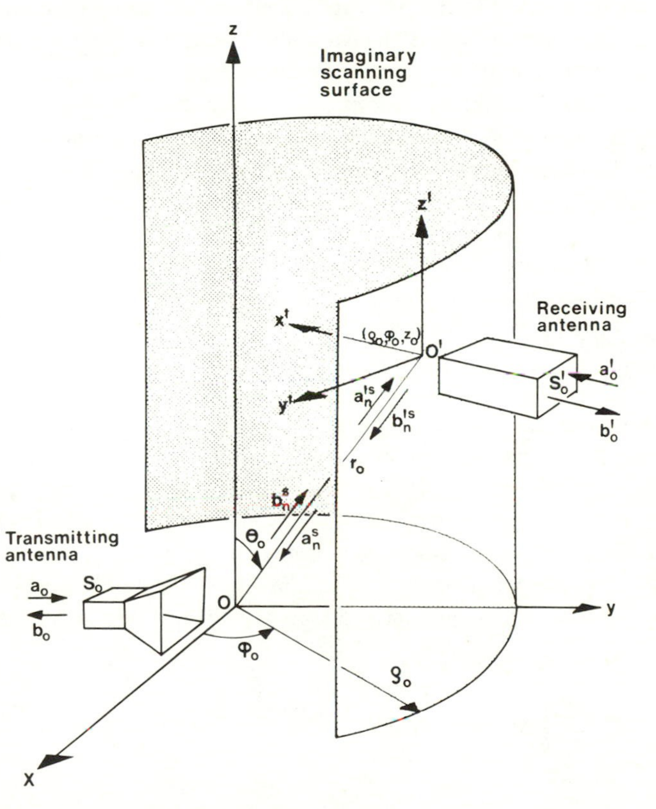

Fig. 8.12 *Transmission system for cylindrical scanning*

Let the probe be operated as a receiving antenna and the antenna under test as a transmitting antenna. Only part of the cylindrical scanning surface is shown. The properties of the transmitting antenna are described in a rectangular xyz-coordinate system with its origin at O which is a point chosen conveniently in or close to the antenna structure. Alternatively, properties may be described in cylindrical $\rho\phi z$- and spherical $r\theta\phi$-coordinate systems with their origin at O and conventional orientation with respect to the xyz-coordinate system. The cylindrical coordinate system is most convenient for cylindrical scanning and will be used here. Cylindrical unit vectors are denoted $\bar{a}_\rho$, $\bar{a}_\phi$ and $\bar{a}_z$. Also, as shown, a point O' characterising the position of the probe is designated by (ρ_o, ϕ_o, z_o). The position vector of O' is denoted $\bar{r}_o$. The properties of the probe may be described in a rectangular $x'y'z'$-coordinate system with its origin at O', its z'-axis parallel to the z-axis and the positive direction of its x'-axis intersecting the z-axis. When the probe is scanned, ρ_o is kept constant and ϕ_o and z_o are varied. For example, for discrete values of z_o, the test antenna is rotated 360° in azimuth. Whenever needed the properties of the probe may also be described in conventional cylindrical $\rho'\phi'z'$- and spherical $r'\theta'\phi'$-coordinate systems with their origin at O'.

As in the planar case the antennas are considered as two-port transducers. One port is located at a plane S_o inside the waveguide feed. As will be apparent from the following discussion, the other port may be considered as a cylindrical surface placed around the antenna. The amplitudes of the incoming and outgoing simple waveguide modes at S_o are a_o and b_o, respectively. a_n^s and b_n^s are spectrum density functions of the modes in which the electromagnetic field can be expanded in the region outside the antenna. Primes are used to characterise quantities related to the probe.

Expansion in cylindrical modes: It can be shown that in cylindrical coordinates every field can be derived from the so-called elementary cylindrical scalar wave function $\psi_{n\gamma}^{(i)}(\bar{r})$ given by, Stratton:[39]

$$\psi_{n\gamma}^{(i)}(\bar{r}) = Z_n^{(i)}(\kappa\rho)\, e^{jn\phi}\, e^{-j\gamma z} \tag{8.40}$$

where $\bar{r}$ is the position vector of the point (ρ, ϕ, z), n is any integer, γ is the propagation constant in the z-direction, a real number, $\kappa = \sqrt{k^2 - \gamma^2}$, where k is the wave number of free space, $Z_n^{(i)}(\kappa\rho)$ represents the circular cylinder functions $J_n(\kappa\rho)$, $N_n(\kappa\rho)$, $H_n^{(1)}(\kappa\rho)$, and $H_n^{(2)}(\kappa\rho)$ corresponding to the set of values 1, 2, 3, and 4, respectively, which the superscript index (i) can take.

In order to construct a vector field, the elementary cylindrical vector wave functions $\bar{M}_{n\gamma}^{(i)}(\bar{r})$ and $\bar{N}_{n\gamma}^{(i)}(\bar{r})$ are derived from $\psi_{n\gamma}^{(i)}(\bar{r})$ through the equations

$$\bar{M}_{n\gamma}^{(i)}(\bar{r}) = \nabla \times \bar{a}_z \psi_{n\gamma}^{(i)}(\bar{r}) \tag{8.41}$$

and

$$\bar{N}_{n\gamma}^{(i)}(\bar{r}) = \frac{1}{k} \nabla \times \nabla \times \bar{a}_z \psi_{n\gamma}^{(i)}(\bar{r}) \tag{8.42}$$

It is found that

$$\bar{M}_{n\gamma}^{(i)}(\bar{r}) = \left(\frac{jn}{\rho} Z_n^{(i)}(\kappa\rho)\bar{a}_\rho - \frac{\partial Z_n^{(i)}(\kappa\rho)}{\partial\rho}\, \bar{a}_\phi \right) e^{jn\phi}\, e^{-j\gamma z} \tag{8.43}$$

and

$$\bar{N}_{n\gamma}^{(i)}(\bar{r}) = \left(-\frac{j\gamma}{k} \frac{\partial Z_n^{(i)}(\kappa\rho)}{\partial\rho}\, \bar{a}_\rho + \frac{n\gamma}{k\rho} Z_n^{(i)}(\kappa\rho)\bar{a}_\phi + \frac{\kappa^2}{k} Z_n^{(i)}(\kappa\rho)\bar{a}_z \right) e^{jn\phi}\, e^{-j\gamma z} \tag{8.44}$$

The general expression for a radiated electric field $\bar{E}(\bar{r})$ is given by a linear combination of the functions $\bar{M}^{(4)}_{n\gamma}(\bar{r})$ and $\bar{N}^{(4)}_{n\gamma}(\bar{r})$

$$\bar{E}(\bar{r}) = \sum_{n=-\infty}^{\infty} \int_{-\infty}^{\infty} \{b_n^1(\gamma)\bar{M}^{(4)}_{n\gamma}(\bar{r}) + b_n^2(\gamma)\bar{N}^{(4)}_{n\gamma}(\bar{r})\}\, d\gamma \tag{8.45}$$

The coefficients $b_n^s(\gamma)$, where $s = 1, 2$, are referred to as spectrum density functions. It is seen that $b_n^s(\gamma)$ are the weights with which, for given values of n and γ, the elementary functions or modes $\bar{M}$ and $\bar{N}$ contribute to the total field.

From Faraday's law it is found that the magnetic field $\bar{H}(\bar{r})$ is given by

$$\bar{H}(\bar{r}) = -\frac{k}{j\omega\mu} \sum_{n=-\infty}^{\infty} \int_{-\infty}^{\infty} \{b_n^1(\gamma)\bar{N}^{(4)}_{n\gamma}(\bar{r}) + b_n^2(\gamma)\bar{M}^{(4)}_{n\gamma}(\bar{r})\}\, d\gamma \tag{8.46}$$

From the observation that $\bar{M}$ has no z-component, it is seen that with respect to the z-axis, the electromagnetic field is resolved into two partial fields, a TE field with coefficients $b_n^1(\gamma)$ and a TM field with coefficients $b_n^2(\gamma)$. Thus s may be referred to as a polarisation index.

In order to derive the coupling equation in cylindrical coordinates, it is desirable to express $\bar{E}(\bar{r})$ in the $\rho'\phi'z'$-coordinate system. This is done by making use of the vector translation theorem for the cylindrical vector wave function, Abramowitz and Stegun[10]

$$\bar{M}^{(4)}_{n\gamma}(\bar{r}) = \sum_{m=-\infty}^{\infty} (-1)^m H^{(2)}_{n+m}(\kappa\rho_o)\, e^{jn\phi_o}\, e^{-j\gamma z_o}\, \bar{M}^{(1)}_{-m\gamma}(\bar{r}') \tag{8.47}$$

$$\bar{N}^{(4)}_{n\gamma}(\bar{r}) = \sum_{m=-\infty}^{\infty} (-1)^m H^{(2)}_{n+m}(\kappa\rho_o)\, e^{jn\phi_o}\, e^{-j\gamma z_o}\, \bar{N}^{(1)}_{-m\gamma}(\bar{r}') \tag{8.48}$$

which is valid in the region $r' < r_o$. Insertion of eqns. 8.47 and 8.48 into eqn. 8.45 gives for the field in the primed coordinate system, after rearrangement

$$\bar{E}(\bar{r}') = \sum_{n=-\infty}^{\infty} \int_{-\infty}^{\infty} \{a_n'^1(\gamma)\bar{M}^{(1)}_{n\gamma}(\bar{r}') + a_n'^2(\gamma)\bar{N}^{(1)}_{n\gamma}(\bar{r}')\}\, d\gamma \tag{8.49}$$

where

$$a_n'^s(\gamma) = \sum_{m=-\infty}^{\infty} (-1)^n H^{(2)}_{m-n}(\kappa\rho_o)\, e^{jm\phi_o}\, e^{-j\gamma z_o}\, b_m^s(\gamma) \tag{8.50}$$

It is interesting to note that although the vectors $\bar{M}^{(1)}_{n\gamma}(\bar{r}')$ and $\bar{N}^{(1)}_{n\gamma}(\bar{r}')$ are derived by using the Bessel function, which represents standing waves, the expression for $\bar{E}(\bar{r}')$ gives the field incident on the probe. Therefore, we refer to $a_n'^s(\gamma)$ as the spectrum density functions of the modes in which the incident field is expanded.

Coupling equation: The relationship between the wave amplitudes a_o and b_o in the waveguide feed and the spectrum density functions $a_n^s(\gamma)$ and $b_n^s(\gamma)$ in the space surrounding the antenna is given by the source scattering matrix formulation

$$b_o = \Gamma_o a_o + \sum_{n=-\infty}^{\infty} \int_{-\infty}^{\infty} \sum_{s=1}^{2} R_n^s(\gamma) a_n^s(\gamma)\, d\gamma \tag{8.51}$$

$$b_n^s(\gamma) = T_n^s(\gamma) a_o + \sum_{\nu=-\infty}^{\infty} \int_{-\infty}^{\infty} \sum_{\sigma=1}^{2} S_{n,\nu}^{s,\sigma}(\gamma, h) a_\nu^\sigma(h)\, dh \tag{8.52}$$

where Γ_o is the reflection coefficient of the antenna, and its receiving, transmitting and scattering properties are characterised by the receiving spectrum $R_n^s(\gamma)$, the transmission spectrum $T_n^s(\gamma)$ and scattering spectrum $S_{n,\nu}^{s,\sigma}(\gamma, h)$, respectively.

The coupling equation for the transmission system in Fig. 8.12 may now be found. Using the source scattering matrix formulation, the spectrum density functions of the fields leaving the transmitting antenna are found to be

$$b_n^s(\gamma) = T_n^s(\gamma) a_o \tag{8.53}$$

when multiple reflections between the probe and test antenna are neglected, i.e. $a_\nu^\sigma(h) = 0$. Insertion of eqn. 8.53 with n replaced by m, into eqn. 8.50 gives for the spectrum density functions of the field incident on the probe

$$a_n'^s(\gamma) = a_o \sum_{m=-\infty}^{\infty} (-1)^n H_{m-n}^{(2)}(\kappa\rho_o)\, e^{jm\phi_o}\, e^{-j\gamma z_o}\, T_m^s(\gamma) \tag{8.54}$$

Let the probe be terminated with a load Z_l' with reflection coefficient Γ_l', then a_o' and b_o' are related by eqn. 8.25.

Insertion of eqn. 8.54 into the scattering matrix formulation for the probe gives

$$b_o'(\bar{r}_o) = \frac{a_o}{1 - \Gamma_o' \Gamma_l'} \sum_{n=-\infty}^{\infty} \int_{-\infty}^{\infty} \sum_{s=1}^{2} R_n'^s(\gamma) \times \left\{ \sum_{m=-\infty}^{\infty} (-1)^n H_{m-n}^{(2)}(\kappa\rho_o)\, e^{jm\phi_o}\, e^{-j\gamma z_o}\, T_m^s(\gamma) \right\} d\gamma \tag{8.55}$$

This is the desired coupling equation from which the unknown $T_m^s(\gamma)$ can be found from knowledge of $R_n'^s(\gamma)$ and $b_o'(\bar{r}_o)$ as explained in the next Section.

As in the case of planar scanning, it is convenient to introduce a coupling product $D(m, \gamma)$ given by

$$D(m, \gamma) = \sum_{s=1}^{2} \left\{ T_m^s(\gamma) \sum_{n=-\infty}^{\infty} (-1)^n H_{m-n}^{(2)}(\kappa\rho_o) R_n'^s(\gamma) \right\} \tag{8.56}$$

and which is determined by Fourier inversion of eqn. 8.55

$$D(m, \gamma) = \frac{1 - \Gamma_o' \Gamma_l'}{4\pi^2 a_o} \int_{-\infty}^{\infty} \int_{o}^{2\pi} b_o'(\bar{r}_o)\, e^{-jm\phi_o}\, e^{j\gamma z_o}\, d\phi_o\, dz_o \tag{8.57}$$

For a known probe receiving spectrum $R_n'^s(\gamma)$, eqn. 8.57 provides one equation in the two unknowns $T_m^s(\gamma)$ of the test antenna. Another equation may be obtained, by repeating the measurements for another aspect of the probe or with another known probe. After this, the solution of the two equations is straightforward. Cases where the determinant of the equations is zero may occur as in the planar case. Such cases may be solved by carrying out additional experiments.

In actual measurements $b_o(\rho_o, \phi_o, z_o)$ is sampled for the probe moved in a lattice with sample intervals $\Delta\phi_o$ and Δz_o. The angle ϕ_o is changed from 0 to 2π during a revolution in azimuth. The z_o traverse parallel to the z-axis has to be limited to the annular region in which radiation is appreciable. Thus, finite limits of z_o can be introduced in the integral of eqn. 8.57. Then the coupling product $D(m, \gamma)$ can be written as a summation which is easily computed by using the Fast Fourier Transform algorithm. The sampling spacings can be chosen in accordance with the two-dimensional sampling theorem. For most practical antennas the transmission spectrum $T_m^s(\gamma)$ becomes negligible for $|\gamma| \geqslant k$ and $|m| \geqslant ka$, where a is the radius of the smallest cylinder enclosing the antenna. From eqn. 8.56 it is seen, therefore, that $D(m, \gamma)$ is band limited provided the summation over n does not grow extraordinarily large for $|m| > ka$. This will be the case if ρ_o is chosen sufficiently large. If ρ_o is chosen sufficiently small, the summation can be made appreciable for arbitrary large values of m even when $R_n'^s(\gamma)$ exists only for a few values of n. This means that small probes can be used to measure antennas for which $T_m^s(\gamma)$ exists for large values of m. The maximum value of $|n|$ is determined for the probe in a similar manner as m for the test antenna. Practice seems to reveal that it is sufficient if the probe remains a few wavelengths away from the test antenna, i.e. the measurements are made outside the reactive near field. Then the sample spacings can be determined from

$$\Delta\phi_o = \frac{\pi}{ka} \tag{8.58}$$

and

$$\Delta z_o = \frac{\pi}{k} \tag{8.59}$$

i.e. $\Delta\phi_o$ must be less than or equal to $\lambda/2a$ and Δz_o must be less than or equal to $\lambda/2$.

Far-field parameters: The transmission spectrum $T_n^s(\gamma)$ determined above can be used to express the radiated far-field of the test antenna. For $a_o = 1$, we obtain from eqn. 8.52 $b_n^s(\gamma) = T_n^s(\gamma)$ when the test antenna is alone in space. Insertion of these values for the spectrum density functions into eqn. 8.45 gives

$$\bar{E}(\bar{r}) = \sum_{n=-\infty}^{\infty} \int_{-\infty}^{\infty} \{T_n^1(\gamma)\bar{M}_{n\gamma}^{(4)}(\bar{r}) + T_n^2(\gamma)\bar{N}_{n\gamma}^{(4)}(\bar{r})\}\, d\gamma \tag{8.60}$$

Using asymptotic expansion of the cylinder functions $\bar{M}_{n\gamma}^{(4)}(\bar{r})$ and $\bar{N}_{n\gamma}^{(4)}(\bar{r})$, the expressions

$$E_\phi(\bar{r}) = -\frac{2k \sin\theta\, e^{-jkr}}{r} \sum_{n=-\infty}^{\infty} j^n T_n^1(k\cos\theta) e^{jn\phi} \tag{8.61}$$

$$E_\theta(\bar{r}) = -\frac{2jk \sin\theta\, e^{-jkr}}{r} \sum_{n=-\infty}^{\infty} j^n T_n^2(k\cos\theta) e^{jn\phi} \tag{8.62}$$

are obtained for the ϕ- and θ-components of the far field, respectively. In order to find the field radiated in the direction (r, θ, ϕ) or $\bar{r}$, only $T_n^s(\gamma)$ with $\gamma = k\cos\theta$ need be known. Thus, it is sufficient to know $T_n^s(\gamma)$ for $-k \leqslant \gamma \leqslant k$. This means that values of γ outside this range represent evanescent waves in the vicinity of the antenna. Since γ is the propagation constant in the z-direction and $\kappa = k\sin\theta$ is the radial propagation constant, one may say that contribution to the far field in the direction $\bar{r}$ only stems from the spectrum density functions corresponding to cylindrical waves which propagate in the direction of $\bar{r}$. From eqns. 8.61, 8.62 relations between normal far-field parameters and transmission spectrum functions can be derived.

Above it was assumed that the receiving spectrum $R_n'^s(\gamma)$ of the probe was known. From equations analogous to eqns. 8.61, 8.62 it is seen that $T_n'^s(\gamma)$ may be determined by Fourier inversion of measured far-field components of the probe. Then $R_n'^s(\gamma)$ can be found from the reciprocity relation

$$\eta_o R_n'^s(\gamma) = (-1)^n \frac{4\pi\kappa^2}{k} \eta T_{-n}'^s(-\gamma) \tag{8.63}$$

where η_o is the wave admittance of the single mode in the antenna feedline and η is the free-space admittance. As an alternative to this procedure, $R_n'^s(\gamma)$ may be found by scanning of a known transmitting antenna and proper inversion of the coupling equation followed by solution of two equations with two unknowns. In the section on spherical scanning, it will be described how a single computer program may be used to do all the calculations.

8.5.4 The spherical technique

Geometrical configuration: In the spherical scanning technique the probe is moved on an imaginary sphere enclosing the antenna under test. An outline for the transmission system set-up is shown in Fig. 8.13. Let the probe be operated as a receiving antenna and the antenna under test as a transmitting antenna. Only part of the spherical scanning surface is shown.

The properties of the system are similar to the cylindrical coordinate system of Section 8.5.3 with r, θ, ϕ replacing ρ, ϕ, z.

The properties of the probe may be referred to a rectangular $x'y'z'$-coordinate system with its origin at O'. The orientation of the $x'y'z'$-coordinate system is such that its z'-axis is pointing in the radial direction characterised by the angles ϕ_o and θ_o, and its x'-axis is making an angle χ_o with respect to the unit vector $\bar{a}_\theta$ at O'

When the probe is scanned, r_o and χ_o are kept constant and ϕ_o and θ_o are varied. For example, conical cuts can be made for discrete values of θ_o when ϕ_o is changed 360°. This can be performed by mounting the antenna under test on a standard antenna rotator and keeping the probe fixed in space.

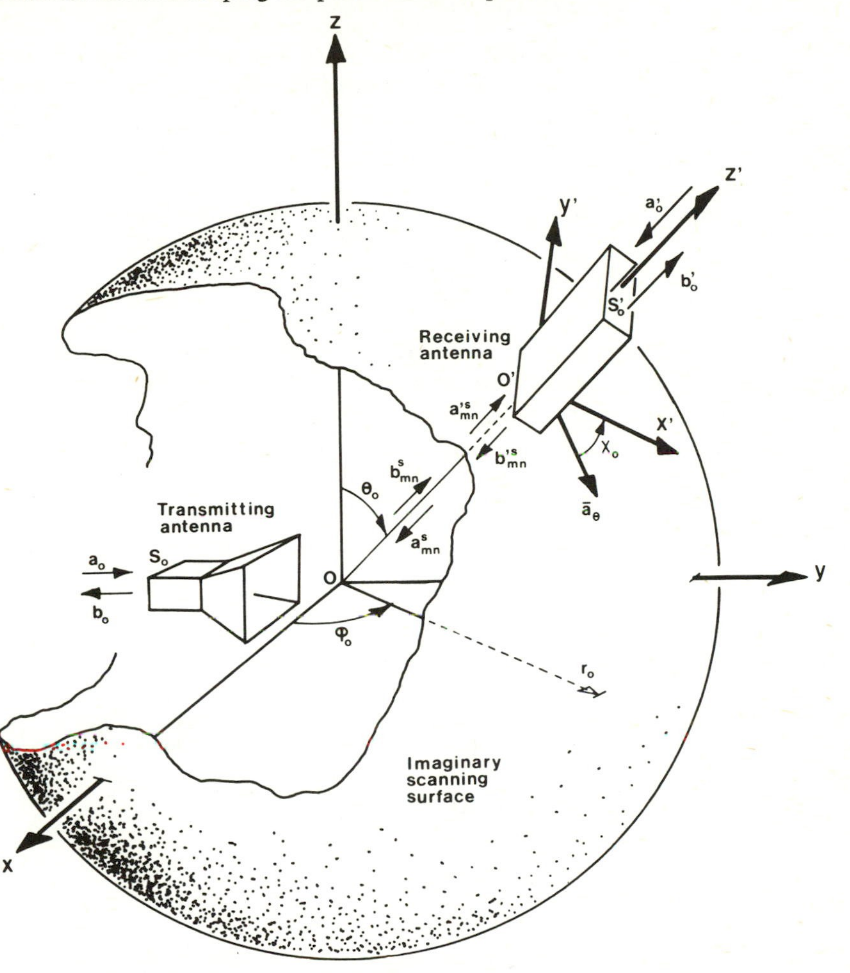

Fig. 8.13 *Transmission system for spherical scanning*

As in the other scanning techniques, the antennas are considered as two-port transducers. For the test antenna one port is located at a plane S_o inside the waveguide feed. The other port may be considered as a spherical surface placed around the antenna. The amplitudes of the incoming and outgoing single waveguide modes at S_o are a_o and b_o, respectively. From the following, it will be seen that a^s_{mn} and b^s_{mn} are spectrum density functions of incoming and outgoing spherical modes in which the electromagnetic field around the antenna can be expanded. Primed quantities are used to characterise quantities related to the probe.

Expansion in spherical modes: In spherical coordinates every field can be generated from the elementary scalar wave function $\psi_{mn}^{(i)}(\bar{r})$ given by Stein[38] and Bruning and Lo,[18]

$$\psi_{mn}^{(i)}(\bar{r}) = z_n^{(i)}(kr)P_n^m(\cos\theta)\,e^{jm\phi} \tag{8.64}$$

where $\bar{r}$ is the position vector of the point (r, θ, ϕ), n and m are integers for which $0 \leqslant n \leqslant \infty$, $-n \leqslant m \leqslant n$, k is the wavenumber of free space, $z_n^{(i)}(kr)$ represents the spherical Bessel and Hankel functions $j_n(kr)$, $n_n(kr)$, $h_n^{(1)}(kr)$, and $h_n^{(2)}(kr)$ corresponding to the set of values 1, 2, 3 and 4, respectively, which the superscript index (i) can take.

The functions $P_n^m(\cos\theta)$ are associated Legendre functions defined by

$$P_n^m(x) = \frac{(1-x^2)^{m/2}}{2^n n!}\frac{d^{n+m}(x^2-1)^n}{dx^{n+m}} \tag{8.65}$$

for $|m| \leqslant n$. It is seen that k can be interpreted as the propagation constant in the radial direction and the modal indices m and n are related to field variation with the azimuthal angle ϕ and the polar angle θ, respectively.

From $\psi_{mn}^{(i)}(\bar{r})$ the elementary spherical vector wave functions $\bar{m}_{mn}^{(i)}(\bar{r})$ and $\bar{n}_{mn}^{(i)}(\bar{r})$ are generated through the equations

$$\bar{m}_{mn}^{(i)}(\bar{r}) = \nabla\times\bar{r}\,\psi_{mn}^{(i)}(\bar{r}) \tag{8.66}$$

and

$$\bar{n}_{mn}^{(i)}(\bar{r}) = \frac{1}{k}\nabla\times\nabla\times\bar{r}\,\psi_{mn}^{(i)}(\bar{r}) \tag{8.67}$$

The explicit expressions are found to be

$$\bar{m}_{mn}^{(i)}(\bar{r}) = z_n^{(i)}(kr)\,e^{jm\phi}\times\left[\frac{jm}{\sin\theta}P_n^m(\cos\theta)\bar{a}_\theta - \frac{\partial P_n^m(\cos\theta)}{\partial\theta}\bar{a}_\phi\right] \tag{8.68}$$

and

$$\begin{aligned}\bar{n}_{mn}^{(i)}(\bar{r}) = {} & \frac{n(n+1)}{kr}z_n^{(i)}(kr)P_n^m(\cos\theta)\,e^{jm\phi}\,\bar{a}_r \\ & + \frac{1}{kr}\frac{\partial(r\,z_n^{(i)}(kr))}{\partial r}e^{jm\phi} \\ & \times\left[\frac{\partial P_n^m(\cos\theta)}{\partial\theta}\bar{a}_\theta + \frac{jm}{\sin\theta}P_n^m(\cos\theta)\bar{a}_\phi\right]\end{aligned} \tag{8.69}$$

The radiated field from the test antenna placed at O may be given by the expansion

$$\bar{E}(\bar{r}) = \sum_{n=1}^{\infty}\sum_{m=-n}^{n}\{b_{mn}^1\bar{m}_{mn}^{(4)}(\bar{r}) + b_{mn}^2\bar{n}_{mn}^{(4)}(\bar{r})\} \tag{8.70}$$

where b_{mn}^s, $s = 1, 2$, are the spectrum density functions illustrated in Fig. 8.13. It is seen that b_{mn}^s are in fact expansion coefficients or the weights with which the elementary functions $\bar{m}_{mn}^{(4)}(\bar{r})$ and $\bar{n}_{mn}^{(4)}(\bar{r})$ contribute to the field.

Using Faraday's law, the magnetic field is found to be

$$\bar{H}(\bar{r}) = -\frac{k}{j\omega\mu}\sum_{n=1}^{\infty}\sum_{m=-n}^{n}\{b_{mn}^{1}\bar{n}_{mn}^{(4)}(\bar{r}) + b_{mn}^{2}\bar{m}_{mn}^{(4)}(\bar{r})\} \tag{8.71}$$

Since $\bar{m}_{mn}^{(4)}(\bar{r})$ has no radial component, it is observed that with respect to $\bar{a}_r$, the field is resolved into two partial fields, a TE field with coefficients b_{mn}^1 and a TM field with coefficients b_{mn}^2. Thus s may be referred to as a polarisation index.

The coupling equation for the transmission system in Fig. 8.13 is derived by expressing $\bar{E}(\bar{r})$ in the $r'\theta'\phi'$-coordinate system. To do this, as can be seen from eqn. 8.70, we need to express $\bar{m}_{mn}^{(4)}(\bar{r})$ and $\bar{n}_{mn}^{(4)}(\bar{r})$ in the $r'\theta'\phi'$-coordinate system.

Fortunately, there exist so-called addition theorems which express the transformation of the vector wave functions from one coordinate system to another, Stein,[38] Bruning and Lo,[18] Edmonds,[22] and Appel-Hansen.[11] Since two right-handed coordinate systems can be brought into coincidence by rotations and translations, addition theorems exist for each type of movement. For the actual case of Fig. 8.13 the xyz-coordinate system may be brought to the position of the $x'y'z'$-coordinate system by rotating the xyz-coordinate system the angles ϕ_o, θ_o, χ_o and then translating it $\bar{r}_o$. The details can be explained from Fig. 8.14. First, the xyz-system is rotated ϕ_o about its z-axis to give the $x_1y_1z_1$-system. Next, the $x_1y_1z_1$-system is rotated θ_o about its y_1-axis to give the $x_2y_2z_2$-system. Then, the $x_2y_2z_2$-system is rotated χ_o about its z_2-axis to give the $x_3y_3z_3$-system. Finally, the $x_3y_3z_3$-system is translated r_o in the direction of the positive z_3-axis to give the $x'y'z'$-coordinate system associated with the probe.

For the rotation of the xyz-system the angles ϕ_o, θ_o, and χ_o to give the $x_3y_3z_3$ system, the addition theorem states

$$\bar{m}_{mn}^{(4)}(\bar{r}) = \sum_{\mu=-n}^{n} D_{\mu m}^{\mu n}(\phi_o, \theta_o, \chi_o)\,\bar{m}_{\mu n}^{(4)}(\bar{r}_3) \tag{8.72}$$

and

$$\bar{n}_{mn}^{(4)}(\bar{r}) = \sum_{\mu=-n}^{n} D_{\mu m}^{\mu n}(\phi_o, \theta_o, \chi_o)\,\bar{n}_{\mu n}^{(4)}(\bar{r}_3) \tag{8.73}$$

where $\bar{r}_3$ is the position vector in the $x_3y_3z_3$-system and the rotation coefficient $D_{\mu m}^{\mu n}(\phi_o, \theta_o, \chi_o)$ is given by

$$D_{\mu m}^{\mu n}(\phi_o, \theta_o, \chi_o) = (-1)^{m+\mu}\sqrt{\frac{(n-\mu)!\,(n+m)!}{(n+\mu)!\,(n-m)!}}\,e^{jm\phi_o}\,d_{\mu m}^{(n)}(\theta_o)\,e^{j\mu\chi_o} \tag{8.74}$$

in which $d_{\mu m}^{(n)}(\theta_o)$ is expressed by a summation involving the indices μ, m, n, $\sin(\theta_o/2)$ and $\cos(\theta_o/2)$.

For the translation of the $x_3y_3z_3$-system the distance r_o in the positive direction of the z_3-axis to give the $x'y'z'$-system, the addition theorem states

$$\bar{m}^{(4)}_{\mu n}(\bar{r}_3) = \sum_{\nu=(1,\mu)}^{\infty} \{(-1)^{n+\nu} A^{\mu n}_{\mu\nu} \bar{m}^{(1)}_{\mu\nu}(\bar{r}') + (-1)^{n+\nu+1} B^{\mu n}_{\mu\nu} \bar{n}^{(1)}_{\mu\nu}(\bar{r}')\} \tag{8.75}$$

$$\bar{n}^{(4)}_{\mu n}(\bar{r}_3) = \sum_{\nu=(1,\mu)}^{\infty} \{(-1)^{n+\nu} A^{\mu n}_{\mu\nu} \bar{n}^{(1)}_{\mu\nu}(\bar{r}') + (-1)^{n+\nu+1} B^{\mu n}_{\mu\nu} \bar{m}^{(1)}_{\mu\nu}(\bar{r}')\} \tag{8.76}$$

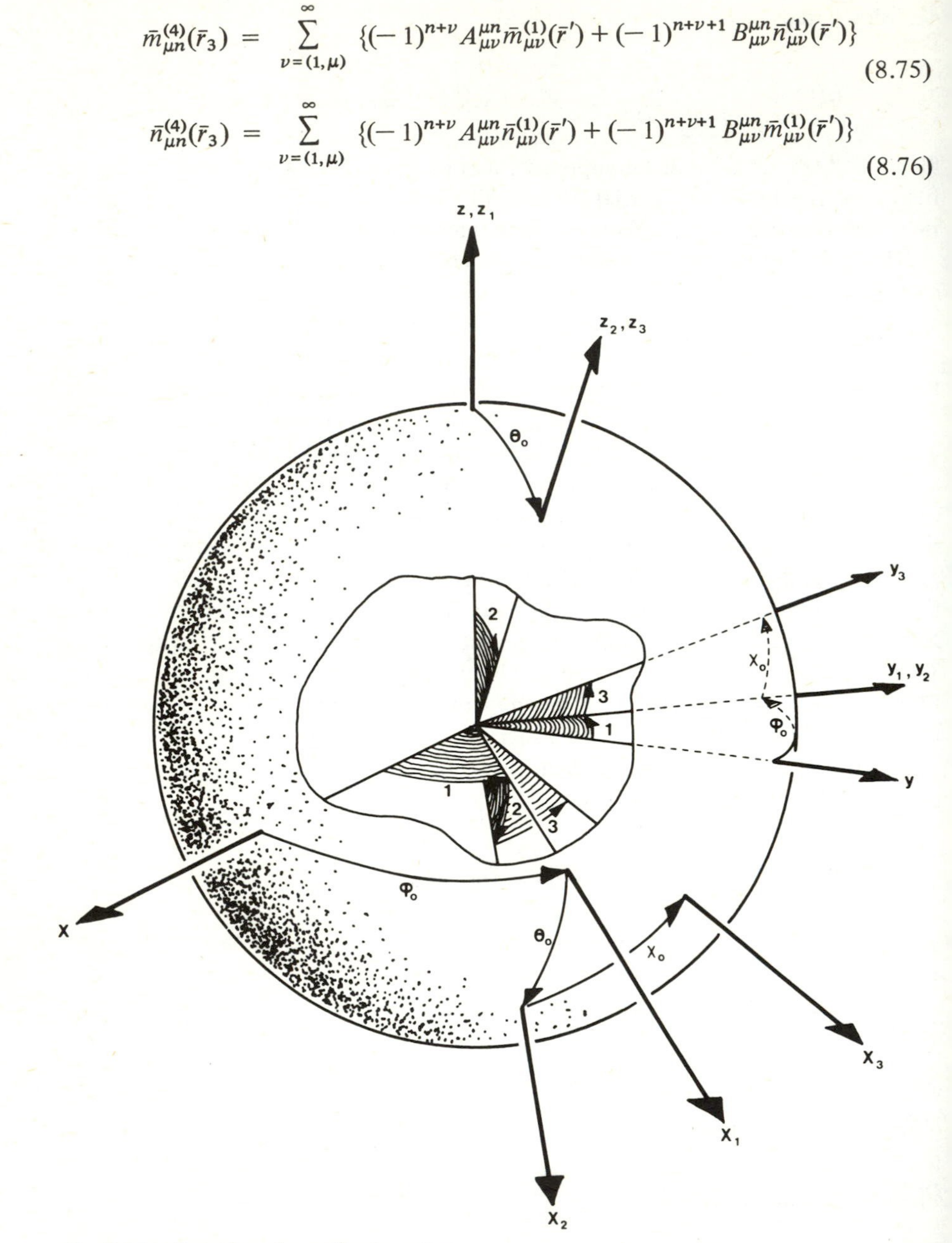

Fig. 8.14 *Rotation of coordinate system*

where the coefficients $A^{\mu n}_{\mu\nu}$ and $B^{\mu n}_{\mu\nu}$ are expressed by summations involving the modal indices μ, n, ν, $h^{(2)}_n(kr_o)$ and the Wigner 3-j symbols. By $\nu = (1, \mu)$ is meant that ν takes 1 or $|\mu|$ which ever is the largest. (In fact, $A^{\mu n}_{\mu\nu} = B^{\mu n}_{\mu\nu} = 0$ when $\nu < |\mu|$.

This is due to the definition of the associated Legendre function eqn. 8.65. Therefore, for the sake of brevity we might write $\Sigma_{\nu=1}^{\infty}$ instead of $\Sigma_{\nu=(1,\mu)}^{\infty}$).

Using $D_{\mu m}^{\mu n}$ for $D_{\mu m}^{\mu n}(\phi_o, \theta_o, \chi_o)$, eqns. 8.70, 8.72–8.76 give for the spectrum density functions in the primed coordinate system after rearrangement

$$\bar{E}(\bar{r}') = \sum_{\nu=1}^{\infty} \sum_{\mu=-\nu}^{\nu} \left\{ \left[\sum_{n=(1,\mu)}^{\infty} \sum_{m=-n}^{n} \right.\right. \times D_{\mu m}^{\mu n} \{ b_{mn}^{1} (-1)^{n+\nu} A_{\mu\nu}^{\mu n} + b_{mn}^{2} (-1)^{n+\nu+1} B_{\mu\nu}^{\mu n} \} \Bigg] \bar{m}_{\mu\nu}^{(1)}(\bar{r}') + \left[\sum_{n=(1,\mu)}^{\infty} \sum_{m=-n}^{n} \right. \times D_{\mu m}^{\mu n} \{ b_{mn}^{2} (-1)^{n+\nu} A_{\mu\nu}^{\mu n} + b_{mn}^{1} (-1)^{n+\nu+1} B_{\mu\nu}^{\mu n} \} \Bigg] \bar{n}_{\mu\nu}^{(1)}(\bar{r}') \Bigg\} \quad (8.77)$$

Analogous to the cylindrical case, eqns. 8.49–8.50, the expansion coefficients are denoted $a'^{\sigma}_{\mu\nu}$ and given by

$$a'^{\sigma}_{\mu\nu} = \sum_{n=(1,\mu)}^{\infty} \sum_{m=-n}^{n} D_{\mu m}^{\mu n} \{ b_{mn}^{\sigma} (-1)^{n+\nu} A_{\mu\nu}^{\mu n} + b_{mn}^{3-\sigma} (-1)^{n+\nu+1} B_{\mu\nu}^{\mu n} \} \quad (8.78)$$

Coupling equation: The source scattering matrix formulation for the spherical case is

$$b_o = \Gamma_o a_o + \sum_{n=1}^{\infty} \sum_{m=-n}^{n} \sum_{s=1}^{2} R_{mn}^{s} a_{mn}^{s} \quad (8.79)$$

$$b_{mn}^{s} = T_{mn}^{s} a_o + \sum_{\nu=1}^{\infty} \sum_{\mu=-\nu}^{\nu} \sum_{\sigma=1}^{2} S_{mn,\mu\nu}^{s,\sigma} a_{\mu\nu}^{\sigma} \quad (8.80)$$

where Γ_o is the reflection coefficient of the antenna, and its receiving, transmitting and scattering properties are characterised by the receiving spectrum R_{mn}^{s}, the transmission spectrum T_{mn}^{s} and scattering spectrum $S_{mn,\mu\nu}^{s,\sigma}$, respectively.

The coupling equation may now be found easily. When multiple reflections between the probe and test antenna are neglected, i.e. $a_{\mu\nu}^{\sigma} = 0$, we find from eqn. 8.80

$$b_{mn}^{\sigma} = T_{mn}^{\sigma} a_o \quad (8.81)$$

The desired coupling equation expressing the signal $b'_o(\bar{r}_o)$ received by the probe is now found as in the cylindrical case

$$b'_o(\bar{r}_o) = \frac{a_o}{1 - \Gamma'_o \Gamma'_l} \sum_{\nu=1}^{\infty} \sum_{\mu=-\nu}^{\nu} \sum_{\sigma=1}^{2} R'^{\sigma}_{\mu\nu} \times \left\{ \sum_{n=(1,\mu)}^{\infty} \sum_{m=-n}^{n} D_{\mu m}^{\mu n} \{ T_{mn}^{\sigma} (-1)^{n+\nu} A_{\mu\nu}^{\mu n} + T_{mn}^{3-\sigma} (-1)^{n+\nu+1} B_{\mu\nu}^{\mu n} \} \right\} \quad (8.82)$$

It is noted that $b'_o(\bar{r}_o)$ depends on χ_o. Thus, we may write $b'_o(\bar{r}_o) = b'_o(r_o, \phi_o, \theta_o, \chi_o)$.

This shows that the received signal is expressed as a multiplication of the receiving properties of the probe $R'^{\sigma}_{\mu\nu}$ and the transmitting properties of the test antenna T^{σ}_{mn} transformed to the probe coordinate system. As described in the following, it is inversion of this equation which gives us information on T^{σ}_{mn} from knowledge of $R'^{\sigma}_{\mu\nu}$ and $b'_o(\bar{r}_o)$.

From eqn. 8.74 it is apparent that Fourier inversion with respect to ϕ_o and χ_o may be done. To carry out inversion with respect to θ_o, we apply the orthogonality integral

$$\int_o^{\pi} d^{(n)}_{\mu m}(\theta_o) d^{(n')}_{\mu m}(\theta_o) \sin\theta_o \, d\theta_o = \frac{2}{2n+1}\delta_{nn'} \tag{8.83}$$

where $\delta_{nn'}$ is the Kronecker delta function.

Before the inversion it is convenient to change the sequence of summation in eqn. 8.82 so that we can write

$$b'_o(\bar{r}_o) = \frac{a_o}{1-\Gamma'_o\Gamma'_l} \sum_{n=1}^{\infty} \sum_{m=-n}^{n} \sum_{\mu=-n}^{n} D^{\mu n}_{\mu m}(\phi_o, \theta_o, \chi_o) D(m, n, \mu) \tag{8.84}$$

which is a short form of the coupling equation, where $D(m, n, \eta)$ is the coupling product defined by the relation

$$D(m, n, \mu) = \sum_{s=1}^{2} T^{s}_{mn} \sum_{\nu=(1,\mu)}^{\infty} \{R'^{s}_{\mu\nu}(-1)^{n+\nu} A^{\mu n}_{\mu\nu} + R'^{3-s}_{\mu\nu}(-1)^{n+\nu+1} B^{\mu n}_{\mu\nu}\} \tag{8.85}$$

Insertion of eqn. 8.74 into eqn. 8.84 followed by Fourier inversion gives

$$D(m, n, \mu) = \int_0^{\pi} b^{n}_{\mu m}(\theta_o) d^{(n)}_{\mu m}(\theta_o) \sin\theta_o \, d\theta_o \tag{8.86}$$

where

$$b^{n}_{\mu m}(\theta_o) = \frac{(1-\Gamma'_o\Gamma'_l)(-1)^{\mu+m}(2n+1)}{a_o 8\pi^2} \sqrt{\frac{(n+\mu)!\,(n-m)!}{(n-\mu)!\,(n+m)!}}$$
$$\times \int_0^{2\pi} \int_o^{2\pi} b'_o(r_o, \phi_o, \theta_o, \chi_o) e^{-jm\phi_o} e^{-j\mu\chi_o} \, d\phi_o \, d\chi_o \tag{8.87}$$

The summation with respect to ν in eqn. 8.85 is over known quantities. Then for specified values of m, n and μ, eqn. 8.86 is an equation in the two unknowns T^1_{mn} and T^2_{mn}. If the integral is computed for two values of μ, the two unknowns are simply found from the resolution of two equations. To facilitate the calculations, a probe is chosen with $R'^{s}_{\mu\nu} \neq 0$ for μ equal to $+1$ and -1 only, e.g. an open-ended circular waveguide. This has the advantage that the integration over χ_o in eqn. 8.87 can be made from measurements carried out with the probe in two polarisation orientations $\chi_o = 0°$ and $\chi_o = 90°$, for each set of values (θ_o, ϕ_o). The integration over ϕ_o in eqn. 8.87 can be made using standard Fast Fourier Transforms. The integral in eqn. 8.86 is also calculated using Fourier analysis in the following manner. First the functions $d_{\mu m}^{(n)}(\theta_o)$ and $b_{\mu m}^{n}(\theta_o)$ are Fourier expanded into the series.

$$d_{\mu m}^{(n)}(\theta_o) = j^{\mu-m} \sum_{m'=-n}^{n} \Delta_{m'\mu}^{n} \Delta_{m'm}^{n} e^{-jm'\theta_o} \tag{8.88}$$

and

$$b_{\mu m}^{n}(\theta_o) = \sum_{m''=-\frac{1}{2}N\theta_o}^{\frac{1}{2}N\theta_o} b_{\mu m}^{nm''} e^{jm''\theta_o} \tag{8.89}$$

where N_{θ_o} depends upon the size of the test antenna as discussed below. The quantities $\Delta_{m'\mu}^{n}$ and $\Delta_{m'm}^{n}$ can be computed from the definition

$$\Delta_{m'\mu}^{n} = d_{m'\mu}^{(n)}\left(\frac{\pi}{2}\right) \tag{8.90}$$

and the analogous for $\Delta_{m'm}^{n}$. The expansion coefficients $b_{\mu m}^{nm''}$ in eqn. 8.89 are found by defining $b_{\mu m}^{n}(\theta_o)$ in the interval $-\pi \leqslant \theta_o \leqslant \pi$. Although $b_o'(\bar{r}_o)$ is only measured in the interval $0 \leqslant \theta_o \leqslant \pi$ that is possible. This is due to the fact that $d_{\mu m}^{(n)}(\theta_o)$ is an odd function of θ_o if $m-\mu$ is odd and an even function of θ_o if $m-\mu$ is even. Then, from eqns. 8.74, 8.82, 8.87 it is seen that $b_o'(\bar{r})$ and $b_{\mu m}^{n}(\theta_o)$ have the same property.

Combining eqns. 8.86–8.89, we obtain

$$D(m,n,\mu) = j^{\mu-m} \sum_{m'=-n}^{n} \Delta_{m'\mu}^{n} \Delta_{m'm}^{n} \sum_{m''=-\frac{1}{2}N\theta_o}^{\frac{1}{2}N\theta_o} \left\{ b_{\mu m}^{nm''} \int_o^{\pi} e^{j(m''-m')\theta_o} \sin\theta_o \, d\theta_o \right\} \tag{8.91}$$

Let the integral in eqn. 8.91 be denoted $p_{m'm''}$ and we find

$$p_{m'm''} = \begin{cases} 0 & m'-m'' \quad \text{odd} \\ \dfrac{-2}{(m'-m'')^2} & m'-m'' \quad \text{even} \end{cases} \tag{8.92}$$

Thus, the inversion of the coupling equation has been changed to the following form, which can be efficiently implemented on a computer:

$$D(m,n,\mu) = j^{\mu-m} \sum_{m'=-n}^{n} \Delta_{m'\mu}^{n} \Delta_{m'm}^{n} \sum_{m''=-N\theta_o}^{N\theta_o} \{p_{m'm''} b_{\mu m}^{nm''}\} \tag{8.93}$$

where $b_{\mu m}^{nm''}$ and $p_{m'm''}$ from eqns. 8.89, 8.92 have been extended to be periodic with the period $2N_{\theta_o}$. This has been done, without changing the value of the right-hand side in eqn. 8.91 by defining

$$b_{\mu m}^{nm''} = 0 \tag{8.94}$$

in the intervals $-N_{\theta_o} \leqslant m'' < -\frac{1}{2}N_{\theta_o}$ and $\frac{1}{2}N_{\theta_o} < m'' \leqslant N_{\theta_o}$. It can be observed that the summation over m'' in eqn. 8.93 has the form of a convolution integral. This is used in the numerical calculations where $p_{m'm''}$ and $b_{\mu m}^{nm''}$ are Fourier transformed and the product of the transformed functions is calculated. Then this product is inverse Fourier transformed to give the summation over m''. This is only correct in the range $-\frac{1}{2}N_{\theta_o} \leqslant m' \leqslant \frac{1}{2}N_{\theta_o}$, but it is also only in this range that it is used in the final summation over m'. Here, some symmetry properties of the Δ's and $b_{\mu m}^{nm''}$'s are utilised, Holm Larsen.[31]

In order to do the Fourier transformations as described above, it has to be justified that $b_o(\bar{r}_o)$ is a band limited function of ϕ_o, θ_o and χ_o. The justification is made from the following observations of the properties of the coupling product $D(m, n, \mu)$ and the rotation coefficient $D_{\mu m}^{\mu n}(\phi_o, \theta_o, \chi_o)$ eqns. 8.74, 8.84–8.86. The coupling product can be neglected for large values of m, n, and μ. This is due to the fact that, for most practical antennas, expansions of the radiated field, eqn. 8.70, can be truncated at $n = ka$, where a is the radius of the smallest sphere enclosing the antenna. Thus, from eqn. 8.80, T_{mn}^s can be neglected for n and m larger than ka. Furthermore, the probe is conveniently chosen so that $R_{\mu\nu}'^{\sigma} = 0$ for $\mu \neq \pm 1$ as described above. Due to these limitations on m, n and μ, it is seen from eqns. 8.74, 8.88 that the largest frequencies in $D_{\mu m}^{\mu n}$ are $n = m = ka$ and $\mu = 1$, i.e. $b_o(\bar{r}_o)$, expressed as the summation in eqn. 8.84, is a band limited function. Then, the maximum value of the sampling spacings $\Delta\phi_o$ and $\Delta\theta_o$ is π/ka, i.e. the spacings must be less than or equal to $\lambda/2a$. When the choice for $\Delta\theta_o$ has been made, N_{θ_o} in eqn. 8.89 is given by

$$N_{\theta_o} = \frac{2\pi}{\Delta\theta_o} > 2ka \tag{8.95}$$

Far-field parameters: Using the asymptotic form for $h_n^{(2)}(kr)$, the ϕ- and θ-components of the far-field radiated by the test antenna into free space can be found. For $a_o = 1$, it is obtained from eqns. 8.68–8.70 and 8.80.

$$E_\phi(\bar{r}) \simeq j\frac{e^{-jkr}}{kr}\sum_{n=1}^{\infty}\sum_{m=-n}^{n} j^n e^{jm\phi}\left\{-\frac{\partial P_n^m(\cos\theta)}{\partial\theta}T_{mn}^1 + \frac{m}{\sin\theta}P_n^m(\cos\theta)T_{mn}^2\right\} \tag{8.96}$$

$$E_\theta(\bar{r}) \simeq \frac{e^{-jkr}}{kr}\sum_{n=1}^{\infty}\sum_{m=-n}^{n} j^n e^{jm\phi}\left\{-\frac{m}{\sin\theta}P_n^m(\cos\theta)T_{mn}^1 + \frac{\partial P_n^m(\cos\theta)}{\partial\theta}T_{mn}^2\right\} \tag{8.97}$$

Thus, from these equations, when T_{mn}^s is determined, the far-field characteristics of the test antenna may be computed.

In order to determine T_{mn}^s, it was assumed above that $R_{mn}'^s$ for the probe was known. From equations analogous to eqns. 8.96, 8.97 $T_{mn}'^s$ may be determined from measured far-field components of the probe. In this determination Fourier inversion and orthogonality properties of the Legendre functions are used. Then $R_{mn}'^s$ can be determined from the reciprocity relation

$$\eta_o R'^s_{mn} = (-1)^m \frac{\eta}{k^2} \frac{2\pi n(n+1)}{2n+1} T'^s_{-mn} \tag{8.98}$$

where η_o is the wave admittance of the single mode in the waveguide feed to the antenna and η is the free-space admittance. It is interesting to note that instead of using eqns. 8.96 and 8.97 to calculate the far field of the test antenna and the receiving spectrum of the probe, a single computer program based on the coupling equation may be used. Basically, the computer program shall be able to do two types of computations. In the first type of computation the program inverts the coupling equation as described above and finds a transmission spectrum from measured data with a known probe. In the second type of computation the program calculates a received signal when the transmission spectrum and the receiving spectrum of the antennas involved are known. In this second type of computation, Fourier inversions in θ, ϕ, and χ analogous to those used in the first type of computation are made. The program uses the fact that, when the probe is specified as a Hertzian dipole, the received signal is proportional to the component of the field intensity along the probe. Furthermore, when far-field data are used the distance can be specified as infinite and asymptotic forms used. Thus, a single program can be used to do all the calculations required in probe corrected near-field measurements.

To do the calculations two sets of measured data must be available. The first set of data is the radiation pattern of the probe itself. In these measurements it is assumed that a probe is used such that it can be approximated with a Hertzian dipole. Then, using the receiving spectrum of a Hertzian dipole, the coupling equation can be inverted to find T'^s_{mn}. After this, R'^s_{mn} is found by using the reciprocity relation. The second set of data is obtained from measurements of the test antenna with the probe. Using this set of data and R'^s_{mn}, the coupling equation is inverted to provide T^s_{mn} for the test antenna. Finally, using these values of T^s_{mn} and $R'^s_{\mu\nu}$ of a Hertzian dipole, the field radiated by the test antenna for any specified distance can be found from the coupling equation. It is understood that by proper use of electric and magnetic Hertzian dipoles the program can do general field transformations both in the near field and between far field and near field. Using such procedures, it is concluded that algorithms can be applied so that the limitations of spherical near-field measurements no longer lie in the numerical computations but in the speed and accuracy by which near-field data can be measured, Holm Larsen.[32]

8.5.5 *Comparison of scanning techniques*

Which scanning technique is to be used for testing a given antenna depends upon several factors. The major factors are the directions in which the near-field properties are to be predicted, the distribution of radiated field in the vicinity of the antenna structure, the physical form, size and movability of the antenna, the available instrumentation, the state-of-the-art of the three techniques and the know-how of the personnel on the test range.

The spherical technique has the advantage that the far field can be predicted in all directions; i.e. full spherical coverage is obtained, and directivity can be computed. Inherent in the cylindrical and planar techniques is that they have toroidal and conical coverage, respectively. Thus, the planar technique can be used for pencil-beam antennas and the cylindrical technique can be used for antennas mainly radiating in all directions in a plane. However, as discussed further below, it should be noted that the planar technique and the cylindrical technique might give full spherical coverage by proper combination of several scans, e.g. four planar scans and two cylindrical scans, respectively. Conversely, in order to cut the required number of measurement points, limited angular coverage may be arranged in the spherical case. One procedure would be to sample on a truncated part of the spherical surface. Then, proper use of zeros for the field outside the truncated surfaces might be used in the transformation programs. In the same manner the cylindrical surface may be truncated to give conical coverage. In this connection it should be mentioned that, from a stringent mathematical point of view, truncation means that predicted far-field values will not satisfy Maxwell's equations. However, there is experimental evidence of sufficiently accurate measurement results. It should also be noted that, without truncation, planar and cylindrical scanning have an infinite number of sample points while spherical scanning always has a finite number.

Often the angular coverage of the main beam is closely related to the near-field distribution. Therefore, a consideration of the field distribution in the vicinity of the antenna might indicate a possible technique; e.g. the planar or cylindrical technique might be used if the distribution is concentrated in a region which can be cut through by a plane or cylinder, respectively. However, depending upon angular coverage, spherical scanning may be required even in such cases.

It seems logical to use a scanning surface of the same form as the antenna structure or aperture. The argument is that one should try to have a measurement surface arranged so that the detected field is expected to have as small a variation as possible or a variation which might be predicted. This is likely to give the smallest errors. The scanning surface may be truncated where the field intensity is low, e.g. 40 dB below the maximum intensity.

The influence of movability of the test antenna on the selected technique is closely related to the available scanner. In all three techniques either the test antenna or the probe antenna or both antennas may be moved. Thus, there are a considerable number of more or less realistic movement possibilities. As an example, considering no translation of the test antenna but an increasing number of possible rotation axes for it, three cases may be discussed in accordance with Table 8.1. First, suppose that the test antenna cannot be rotated, i.e. there are no possible axes of rotation. This may be due to its size or in order to simulate a 0 *g* environment. Then, the movement of the probe can probably be carried out most accurately in the planar case. Two major planar techniques exist, viz. rectangular planar and polar planar. The table indicates rectangular scanning by 'linear/linear' for the probe movement. The cylindrical scanning may be realised by moving the probe along a line which can be moved around the test antenna along a circular rail as indicated by 'linear/circular'. The spherical scanning may be realised by moving the probe or a semicircular rail which can be rotated about an axis through its ends. This is indicated by 'semi-circular/circular'. Second, if the antenna can be rotated about one axis only, in order to keep gravity induced deformations constant, then a simple cylindrical scan where the probe is moved linearly might be appropriate to give toroidal coverage. However, this may also be obtained by planar scanning by making use of the rotation axis of the test antenna, e.g. rotating the test antenna in steps of 120° and after each rotation perform a planar scanning, in total three scans. Similarly, by making use of the rotation axis and performing two cylindrical scans, spherical coverage can be obtained. Spherical coverage may also be obtained by performing a spherical scanning where the test antenna is rotated and the probe moved along a fixed semi-circular rail.

Table 8.1 *Example of obtainable coverage as a function of a number of scans, probe and test antenna movements*

Axes \ Scanning	Planar	Cylindrical	Spherical
0	conical 1 surface linear/linear	toroidal 1 surface linear/circular	spherical 1 surface semi-circular/ circular
1	toroidal 3 surface linear/linear	spherical 2 surface linear/circular	spherical 1 surface semi-circular
2	spherical 4 surface linear/linear	spherical 2 curve linear	spherical 1 point fixed

Coverage: conical/toroidal/spherical
Number of scans: 1/2/3/4
Probe movement: surface/curve/point
linear/semi-circular/circular

Third, if the antenna can be rotated about two axes perpendicular to each other, the spherical scanning might be implemented easily with the probe in a fixed position. However, spherical coverage may also be obtained in this case by combining at least two cylindrical or four planar scans. The results, illustrated by Table 8.1, are obtained by making maximum use of possible rotation axes for the test antenna. The probe is constraint to be at a fixed point or move along a curve (line or circle) or a surface (made of lines or circles) fixed in space. The mode of probe measurements is chosen so that maximum coverage has priority over minimum movement of the probe. It should be noted that without constraints on the probe movement, spherical coverage can be obtained with any scanning technique independent of the movability of the test antenna; e.g., for an immovable antenna, four planar scans may give spherical coverage except for a conical region allowing some support mechanism for the antenna. The reader can imagine that tables similar to Table 8.1 might be created for different combinations of various translations and rotations of test antenna and probe.

If the movability of the test antenna is not limited, the different scanning techniques can be implemented as illustrated in Fig. 8.15. Here, the order of increasing mechanical complexity will in general be spherical, cylindrical, polar planar and rectangular planar technique. This is due to the fact that the first three techniques may use existing rotators. In particular, as discussed above, the spherical scanning can be implemented without the need for a probe carriage. Of course, this requires that the antenna can stand the rotations and that the available rotator can handle the antenna and is precise enough.

For large antennas, the planar technique has a particular advantage. This stems from the fact that by moving the whole scanner and test antenna relative to each other, a large scan area can be covered by several small scans and, in fact, arbitrary coverage may be obtained. An overlapping procedure may be used to survey proper measurements. It is surprising that this technique of combining several planar scans has not attracted more attention.

Also for large antennas, spherical scanning where the probe is moved on a semicircular rail which can be rotated about an axis through the end of the rail may be advantageous in some cases. This is due to the fact that the distance between the probe and the test antenna may be kept more constant for this type of probe movement than if the test antenna is rotated about two axes and the probe is fixed. Large antennas can cause varying bending moments on the rotator resulting in a varying distance between the probe and test antenna. Therefore, it may be worthwhile to provide an accurate probe carriage system which only has to be aligned once to carry probes of small weight.

Lining of the laboratory with absorbers is a special problem. If the antenna has conical coverage and the planar technique is used, only the part of the laboratory illuminated by the antenna may have to be lined with absorbers. Alternatively, the antenna may be pointing towards the sky through a roof which can be made like a radome or be removed during particularly accurate measurements. Similarly, cylindrical scanning may require only a limited lining of the laboratory with absorbers. But the spherical technique will in general require an anechoic chamber, and for large antennas this technique will often require the largest laboratory space.

From a data-processing point of view the planar technique is the simplest. The cylindrical is more complicated and the spherical technique is the most complex.

8.5.6 Near field scanning test ranges

A commercial planar scanner was first offered for sale in the early 1960s. This used an analogue Fourier integral computer described by Clayton and Hollis.[20] However, it is has taken 20 years plus the development of the digital computer and Fast Fourier Transform programs, before all three scanning techniques can be fully implemented.

Near-field scanning facilities can be considered as an integration of four parts: namely, scanner, controlling and data aquisition system, transmitter and receiver, and data-processing equipment with software. These four parts are also present in far-field ranges. Therefore, scanning can be implemented by using much of the conventional equipment. The additional equipment which is needed depends upon the type of scanning, the required accuracy and the present instrumentation. For a well equipped far-field range, it may only be necessary to implement a computer program for the transformation of data. However, generally, in addition to this, an accurate scanner, a position programmer which controls the scan and step axes, and some digital equipment have to be procured. As examples, two near-field test ranges will be outlined.

Wacker and Newell[41] have described a facility for making rectangular planar, circular cylindrical and spherical scanning. The planar system consists of a rectangular frame bolted to concrete pillars, and a vertical column which can run on rails in the horizontal direction. Along the vertical column a probe carriage runs on rails. Thus, a rectangular *XY* scanner which can move the probe over a quadratic scan area of the size $4.5 \times 4.5\ m^2$ is created.

In order to align the rails to an accuracy of ± 0.002 cm, use is made of a laser interferometer. This is also used for accurate measurement of x and y positions. In order to minimise amplitude and phase changes occurring during movement of the probe, this is connected to the receiver by a single semirigid cable properly supported and with loops. The scanner is covered with absorbing material and installed in a laboratory. For cylindrical and spherical scanning the probe is still mounted in the planar scanner, but the test antenna is placed on a standard antenna rotator as shown in Fig. 8.15. Multipath reflections are reduced by proper use of absorber panels.

The implementation of a spherical scanner in an anechoic chamber is described by Bach *et al.*[13] and Hansen and Larsen.[23] The test antenna is mounted in a rotator constructed as a standard rotator having two orthogonal axes of rotation which intersect each other in a center of rotation. The probe is mounted on a tower which can run on rails. Thus, it is possible to vary the measurement distance depending upon the test antenna. In order to obtain an accurate centre of rotation towards which the probe is pointing use is made of mirrors, levellers and a theodolite. The probe and different parts of the rotator are adjusted using various means such as translation shuttles and adjustable wedge blocks. The transmitter is connected to the test antenna through semirigid cables. To minimise phase and amplitude changes during rotations, the cables are not allowed to move and rotary joints are used. The receiving system is able to detect at the same time two orthogonal polarisations. This is obtained by using a three channel receiver and an orthomode transducer connected to a conical horn. Thus, instead of two scans with the probe rotated 90° between the scans, only a single scan is needed so that measurement time is halved. References to other test ranges using the scanning techniques can be found in the previous Sections.

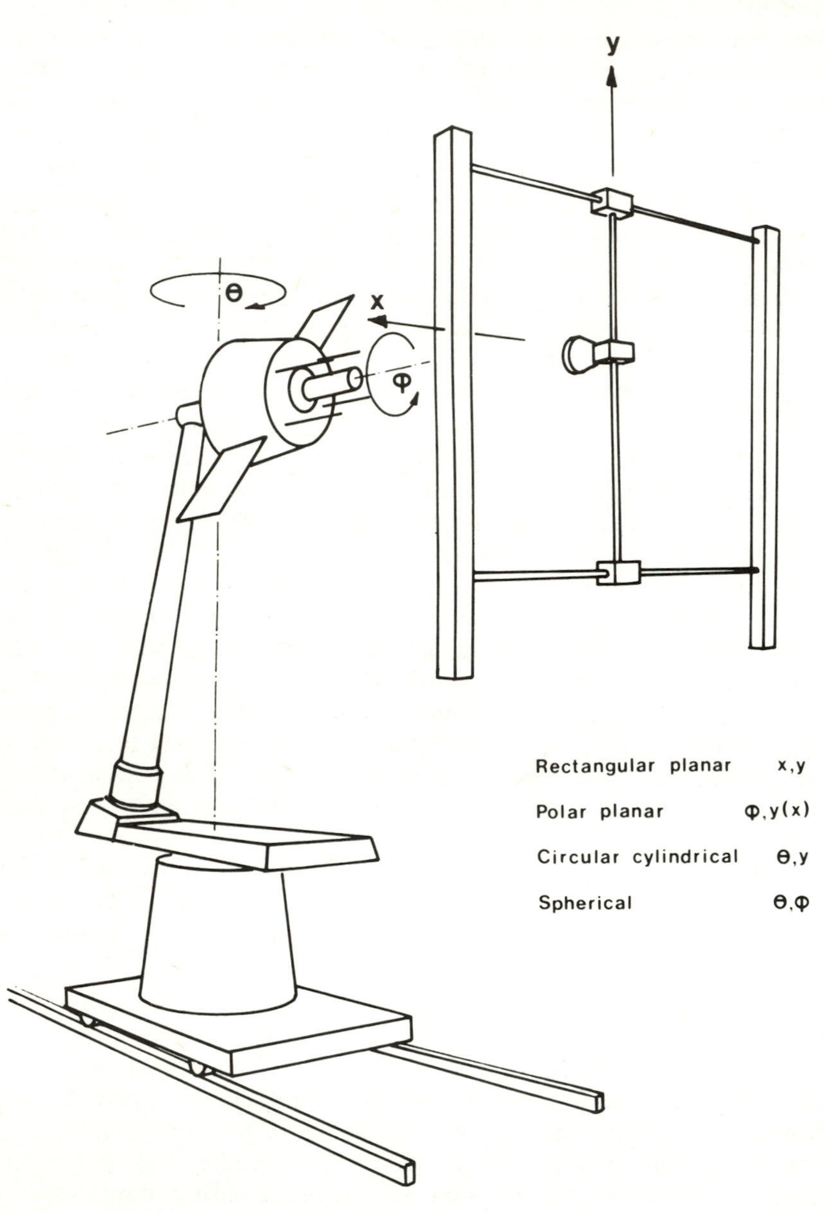

Fig. 8.15 *Implementation of scanning techniques*

8.5.7 Plane-wave-synthesis technique

A plane-wave-synthesis technique for near-field antenna measurements has been suggested by Bennett and Schoessow.[14] The idea is that a stepping radiator is scanned over a spherical surface. By proper feeding of the radiator, the addition of each field contribution arriving from the radiator in each of its positions synthesise a plane wave. The addition can be made with a computer and each contribution can be multiplied by a weighting function. In fact, this means that conventional feeding of the radiator is sufficient. This procedure is similar to that used in a work by Martsafey[53] who studied a 'mathematical collimator' with planar scanning.

That a perfect plane wave may be synthesised within a certain volume can be understood from the following consideration. Suppose that the volume is screened by a perfect conductor and that a plane wave is incident on the conductor. Then the currents generated on the conductor will radiate a perfect plane wave in the interior of the conductor in order to extinguish the incident plane wave. In the plane-wave-synthesis technique the stepping radiator approximates the continuous current distribution on a smaller or larger part of the surface enclosing the volume. Ludwig and Larsen[52] have analysed spherical near-field scanning and noted that it has the advantage of determining the size of the test volume within which the wave satisfies some specifications independent of the antenna under test.

A plane-wave-synthesis approach may be used to analyse any measurement procedure. The present introduction may provide the reader with a deeper insight into the plane-wave-synthesis technique. This insight may be used to understand and develop near-field scanning techniques and to obtain ideas for future measurement procedures for determining far-field antenna characteristics.

Infinite sensitivity probe: It is well known that at an arbitrary point in front of a transmitting antenna, the field may be considered as an integration of plane waves propagating in all directions away from the antenna. Furthermore, the amplitude of the wave propagating in a particular direction is proportional to the far-field amplitude in that direction. Therefore, if it were possible at an arbitrary point to place a probing antenna which were able to intercept only a single plane wave propagating in a certain direction, it would be possible to detect the far-field pattern simply by rotating the probe, or test antenna in order to obtain spherical coverage.

Such a hypothetical probe can be approximated by a probe having a high degree of directivity. Generally, this would require a large probe. The influence of the size of the probing antenna on the measured pattern has been studied by Brown[47] and Jull.[50] In particular, for short test distances, it is shown that measured side-lobe levels can be either above or below the far-field values. It is also shown that as the half-power beam width of the probe tends to zero the correct far-field pattern is measured as expected by the arguments stated above. In the following, to give a simple proof of this fact and as an introduction to the plane-wave-synthesis technique, the terminology introduced in the case of planar scanning is used, see Section 8.5.1.

Suppose that the probe in Fig. 8.8 is a linearly polarised antenna which has infinite sensitivity in a single direction. Let this hypothetical probe be orientated so that its receiving characteristic is given by

$$R'^s(k_x, k_y) = \delta_s \delta(k_x - k_{xo})\delta(k_y - k_{yo}) \tag{8.99}$$

where $\delta(k_x - k_{xo})$ and $\delta(k_y - k_{yo})$ are Dirac delta functions and δ_s is 1 for $s = 1$ and 0 for $s = 2$, i.e. the probe has infinite sensitivity for the $\bar{a}_\phi$-component of a plane wave propagating in the direction characterised by

$$\bar{k}_o^{(1)} = k_{xo}\bar{a}_x + k_{yo}\bar{a}_y + \gamma_o \bar{a}_z \tag{8.100}$$

Then, using the coupling eqn. 8.28 the following expression for the received signal is obtained:

$$b_o'^1(\bar{r}_o) = \frac{a_o}{1 - \Gamma_o' \Gamma_l'} T^1(k_{xo}, k_{yo}) e^{-j\bar{k}_o^{(1)} \cdot \bar{r}_o} \tag{8.101}$$

It is seen that $T^1(k_{xo}, k_{yo})$ may be determined from measured values of $b_o'^1(\bar{r}_o)$, a_o, Γ_o', Γ_l' and $e^{-j\bar{k}_o^{(1)} \cdot \bar{r}_o}$. By rotation of the probe, k_{xo} and k_{yo} may be varied so that the pattern of $T^1(k_{xo}, k_{yo})$ can be determined. By changing the polarisation of the probe, $T^2(k_{xo}, k_{yo})$ may be determined in a similar manner. Thus, the complete transmission characteristic can be determined.

Using the reciprocity relation 8.36 and eqn. 8.32, it can be shown that the hypothetical probe in the transmitting mode of operation radiates a plane wave back in the direction of infinite receiving sensitivity. As an alternative to the above proof, this readily explains that the correct far-field pattern is measured. However, the problem is that the hypothetical antenna generating a plane wave cannot be made. But in practice, it is possible within a certain test volume to generate approximately a plane-wave field. One method is to construct an antenna consisting of an array of elements.

Synthesis techniques: The idea of synthesising a plane-wave field within a test volume by using an array of elements is illustrated in Fig. 8.16. In order to implement the array, it is possible to follow one of two major approaches. In the first approach, the array of elements is really built and all elements radiate at the same time. This approach does not seem to have attracted much interest. But, the second approach has been made use of in implementing and evaluating near-field scanning. This approach will be described in some detail. Instead of having a radiator at all element positions, a single scanning radiator is stepped from one position to the next. Thus, a plane wave is not directly generated in the test volume. What is generated is a sequence of field contributions each of which is separated in time. By properly changing the feeding network of the radiator when stepping from one position to the next, the sum of the contributions can be made to approximate a plane wave. This imaginary wave is referred to as the synthesised plane wave.

Now, let a receiving antenna be inserted into the test volume. Furthermore, taking advantage of the time separation, let the antenna response to each field contribution be sampled. Then, the added samples will give a response which is equal to the response from the antenna if the synthesised plane wave were really generated. The sampling of the field contributions is a crucial point in the implementation and evaluation of the technique. Before the sampled responses are added, they can be multiplied by weighting functions. Therefore, the radiator can be fed as usual and the change in the feeding can be substituted by such a multiplication which can easily be made in a computer. This is equivalent to multiplication of the field contributions from a constantly fed radiator. The wave obtained by the addition of these field contributions is referred to as the equivalent synthesised plane wave. By proper change of the weighting functions, the direction of propagation of the equivalent synthesised plane wave can be changed. Thus, the antenna pattern can be obtained by using only sampled responses from a single scan.

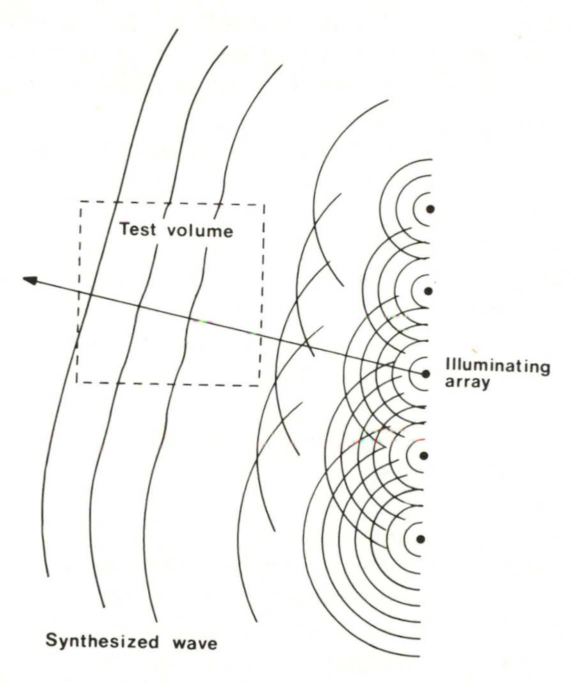

Fig. 8.16 *Synthesis of quasi-plane wave*

The approach has been studied by Martsafey[53] and Bennett and Schoessow[14] in the case of planar and spherical scanning, respectively. Ludwig and Larsen[52] have pointed out that synthesising a plane wave by a scanning radiator is the reciprocal case to predicting far-field parameters from near-field probing. Therefore, the same weighting functions can be used in the two cases. This means that conventional near-field scanning and the applied computer programs can be checked by examining the equivalent synthesised plane wave. By specifying a desired quality of the equivalent synthesised plane wave, the size of the test volume can be computed for a given scanning with given weighting functions. This also means that the synthesised wave may be optimised by varying the weighting functions. In fact, it is interesting to note that it is possible to try to generate approximately any specified field distribution.

As in conventional far-field measurements, where the quality is specified as an allowable phase error of 22.5°, the problem is to estimate the error of actually measured characteristics. Such an estimation may be made theoretically as done by Kinber and Tseytlin.[51] However, their evaluation, and probably any evaluation of the magnitude of error, requires knowledge of the test antenna. But, basically this is an a priori unknown. However, if some assumptions can be made, the error can be estimated. In particular, Kinber and Tseytlin[51] have analysed the measurement of the directive pattern of a round, uniformly excited disc. In some cases the major contribution to the error in measurement of main beam maximum was proportional to the square of the ratio between the diameter of the disc and the diameter of the cross-section of a circular test volume. In the cross-section the amplitude distribution of the synthesised wave decayed towards the edge according to a parabolic law.

Ludwig and Larsen[52] have investigated truncation of a spherical array to only nine element positions covering only part of the spherical surface. It is shown that for some cases the range distance may be reduced to one third of that distance required in conventional far-field techniques.

It should be mentioned that any attempt to make far-field measurements may be analysed from a plane-wave-synthesis point of view. For example, the compact range, to be described in the next Section may be considered as a continuous array of elements arranged on a parabolic surface.

8.6 Intermediate range techniques

8.6.1 Compact range

An antenna range consisting of the near field of a paraboloidal reflector antenna as shown in Fig. 8.17 is called a compact range. The compactness becomes apparent by noting that the test antenna is placed at a short distance in front of the reflector aperture. The compact range technique makes use of the collimated tube of rays existing in a region in front of a reflector antenna.

The merits and deficiencies of the compact range may be analysed by ray tracing as used in geometrical optics. Thus, we assume that we are in the limit of zero wavelength and that the feed is a point or a linesource located at the focal point or the focal line of the reflector, respectively. As illustrated in the Figure the field radiated with spherical or cylindrical fronts of constant phase from the feed is transformed into a field with planar fronts of constant phase in front of the reflector aperture. It is noted that the rays diverge from the feed toward the reflector. But no divergence is present after the rays have left the aperture. Thus, the power flux density in the collimated tube of rays incident on the test antenna is independent of the distance separation R between the aperture and the test antenna. Hence, the usual inverse distance variation in the illuminating field is absent in the compact range. With increase in frequency, conventional far-field techniques require increase in distance separation while compact range techniques do not require this. On the contrary, geometrical optics indicates that with increase in frequency the compact range performance may improve.

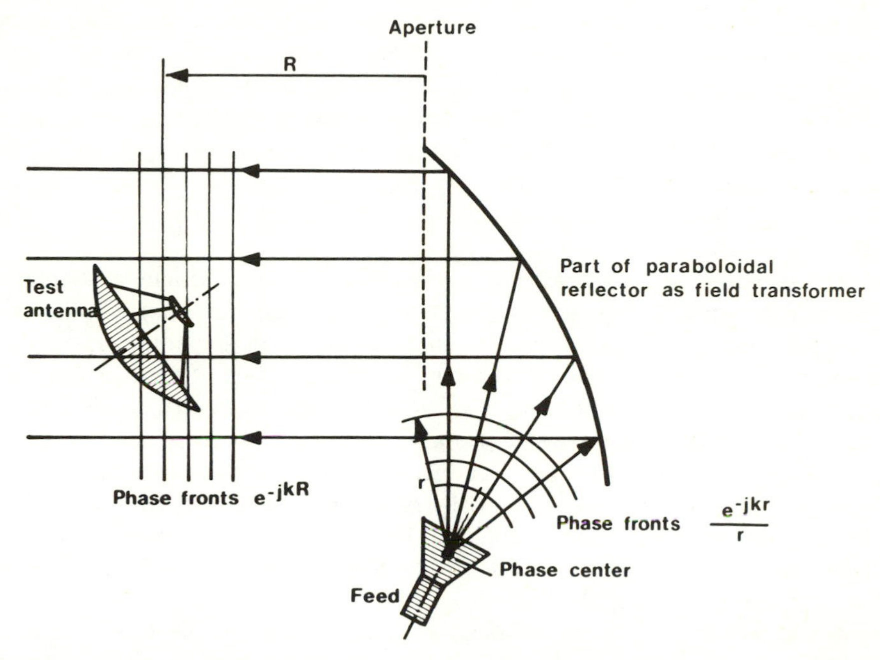

Fig. 8.17 *Compact range implemented by using an offset paraboloidal reflector*

In practice, several imperfections have to be coped with. This is mainly due to the facts that ray tracing, which gives the clue to the compact range concept, does not give exact results at finite wavelengths, the reflector surface is finite and cannot have a perfect focal point, the curved reflector causes some depolarisation of the incident field, the feed is not a point source, and extraneous signals exist.

Usually, it is required that the aperture of the reflector is three times that of the test antenna so that a sufficiently large region with plane wave illumination is ensured. The reflector surface deviations from that having a perfect surface are required to be a small fraction of a wavelength over any area of size λ^2. In order to avoid disturbing reflections from the edge of the reflector, this is serrated. A long focal length is required in order to obtain small depolarisation effects. Typically, a cylindrical test zone of diameter 1 m and length 1.5 m can be obtained in the frequency range 2 GHz to 40 GHz.

In the description above, the feed was assumed to be a point source radiating uniformly within a tube of rays incident on the reflector. This cannot be implemented.

Therefore, it is required to find a reference point for the feed, usually its phase centre. But, because no true phase centre exists for normal feeds, the positioning of the feed ends up as an optimization problem. In order to avoid aperture blockage, the feed may be placed asymmetrically outside the collimated tube of rays. However, this may cause direct sidelobe illumination of the test antenna. To minimise this extraneous signal, a screening plane with absorbers may be inserted between the feed and test antenna. Similarly, other extraneous signals may be avoided by installing the compact range in an anechoic chamber. In the region of the chamber where the compact range is placed, parts of the chamber surfaces may not require absorbing material. A special type of extraneous signal is multiple scattering between the test antenna and the reflector. However, by not carrying out measurements too close to the reflector, the effect of multiple scattering is low because the reflector directs back-scattering from the test antenna mainly toward the feed where it may be in part received by the feed and absorbed in a uniline and in part further scattered.

The imperfections described above may often not cause serious measurement errors. The compact range has the advantage that it can be evaluated using conventional probing for detection of such performance as amplitude taper, phase variation, depolarisation and extraneous scattering. When the range has been evaluated, measurement data can be recorded using conventional far-field equipment and procedures for antenna characteristics and radar cross-section measurements. The compact range may also create a plane wave field region for other studies such as radome and biological irradiation. The first in-depth analysis of the compact range was made by Johnson, Ecker and Moore.[49] However, optimum design data does not seem to be available. Some results have been reported by Olver,[109] Olver and Tong.[110]

A competitor to the compact range is a collimating lens made of a dielectric material. This has been used over the years in various situations. An improved version of a lens using loaded foam material has been investigated by Olver and Saleeb.[55] The analysis of the lens is usually based on ray tracing.

8.6.2 Defocusing technique

The defocusing technique was one of the first attempts to provide far-field antenna patterns at test distances shorter than normally required. The technique may be introduced by a geometrical-optics approach considering path lengths. The idea is that the path-length differences occurring at infinity for an actual antenna are established at a finite distance R which is usually less than the normal $2D^2/\lambda$ test distance. This may be obtained by modification of for example the physical structure of the antenna, the feeding network, or other components of the antenna. The antenna is modified to be focused at R or defocused from infinity. The antenna characteristics measured at R for the modified antenna are assumed to approximate the far-field characteristics of the original antenna. After the measurements the modified antenna has to be refocused at infinity. The defocusing technique will be illustrated with two cases.

As the first case, Fig. 8.18*a* shows a linear aperture for which the $2D^2/\lambda$ distance may exceed the available test range distance R. At infinity and in a direction normal to the aperture, there is no path-length differences between rays originating from different parts of the aperture. In order to establish this situation at the finite distance R at a point F', the aperture may be modified by mechanically bending it in a circle with radius R as shown in Fig. 8.18*b*. It should be noted that the identity in zero path length differences at infinity in Fig. 8.18*a* and at F' in Fig. 8.18*b* only occurs in the direction normal to the array. Therefore, when the antenna is rotated as in a pattern measurement, an accurate pattern can be expected only within a certain range of angles around the normal to the aperture. In the case of slotted waveguide arrays, Bickmore[46] has found this range for different test distances by using a criterion similar to the $\lambda/16$ path length criterion used in conventional far-field measurements. The defocusing technique in the present case has zero phase error normal to the array where the conventional technique has maximum phase error. In the direction parallel to the array the opposite situation exists.

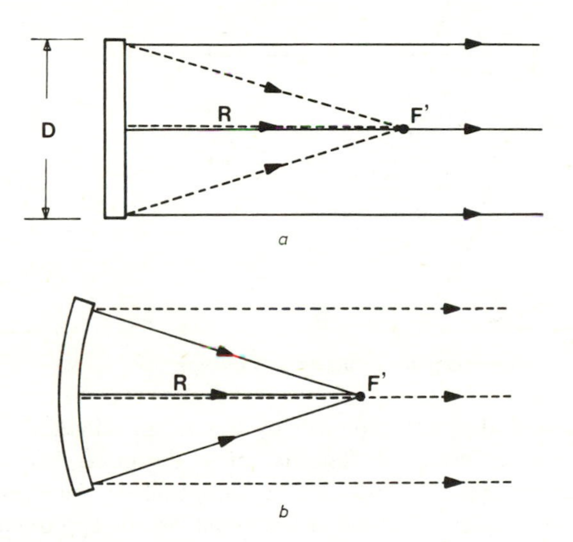

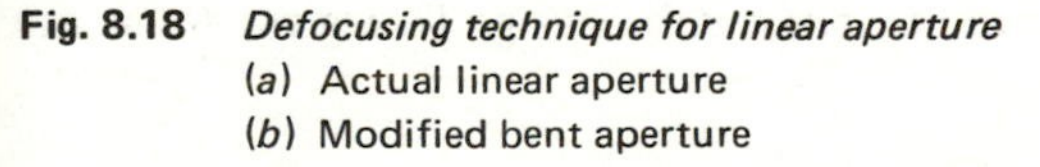

Fig. 8.18 ***Defocusing technique for linear aperture***
(*a*) Actual linear aperture
(*b*) Modified bent aperture

As the second case, Fig. 8.19 shows a paraboloidal reflector with its focus at F. This reflector has the property that in the aperture there is no path length differences between rays originating from the focus and reflected from the paraboloidal surface. Thus, at infinity in the direction of the axis of the paraboloid there are no path length differences either. To establish this situation at a point F' on the axis and at a distance R from the vertex would require that the reflector was modified to be ellipsoidal with foci at F and F'. Such a modification would probably be impractical or at least expensive. In practice, the original antenna is changed by displacing the feed a distance ϵ to a point F'' along the axis and away from the reflector. Then the different rays will tend to intersect the axis. A focusing at F' would now require an ellipsoidal reflector with foci at F' and F''. Therefore, some phase error will always exist at F'. This means that the problem is to optimise the value of ϵ so that a measurement of the antenna characteristics at F' represents with minimum error the far-field characteristics. Unfortunately, there does not seem to exist an approach which determines a unique value of ϵ. This is demonstrated by the numerous criteria which have been used in the assessment of the value of ϵ. In a simple geometrical optics approach by Moseley,[54] an expression for ϵ was found by requiring the two path lengths from F'' to F' via the vertex and the edge of the reflector be identical:

$$F''V + VF' = F''E + EF' \tag{8.102}$$

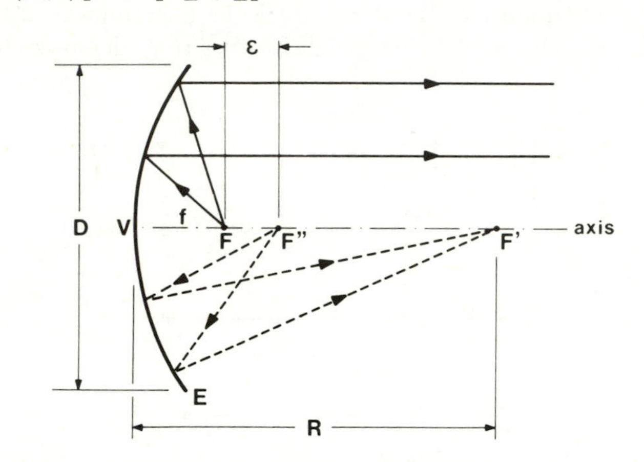

Fig. 8.19 *Defocusing technique for paraboloidal reflector*

In another geometrical optics approach by Cheng and Moseley,[48] another value for ϵ was found by approximate cancellation of the phase differences of all the reflected rays arriving at F'. This can be done because the Fresnel phase term in the diffraction integral of the original antenna can be approximately cancelled by a phase term introduced as a result of the defocusing. This gives for ϵ the expression

$$\epsilon = \frac{1}{R}\left[f^2 + \left(\frac{D}{4}\right)^2\right] \tag{8.103}$$

where D is the diameter of the aperture, f is the focal length, and it is assumed that the ratio f/R is small.

Instead of different geometrical optics approaches considering phase differences at F', the value of ϵ has also been found by computations of the variation of such antenna parameters as beam width, gain and null depth between the main lobe and the first sidelobe with variations in ϵ. Then, minimum beamwidth, maximum gain or deepest null have been used as criteria for finding the optimum value of ϵ. It is found that all approaches give in general slightly different values of ϵ. Johnson, Ecker and Hollis[26] have discussed several approaches. For $R = D^2/8\lambda$, the dashed pattern in Fig. 8.20 was found for a paraboloidal reflector by using maximum null depth as a criterion. The other pattern shown in the Figure was obtained by using approximate cancellation of the Fresnel phase term as described above, i.e. $\epsilon = \epsilon'$ is calculated by using eqn. 8.103. As indicated in the Figure, it turned out that the pattern with the deepest null has a value of ϵ which is 7.5% less than the value ϵ' obtained by eqn. 8.103.

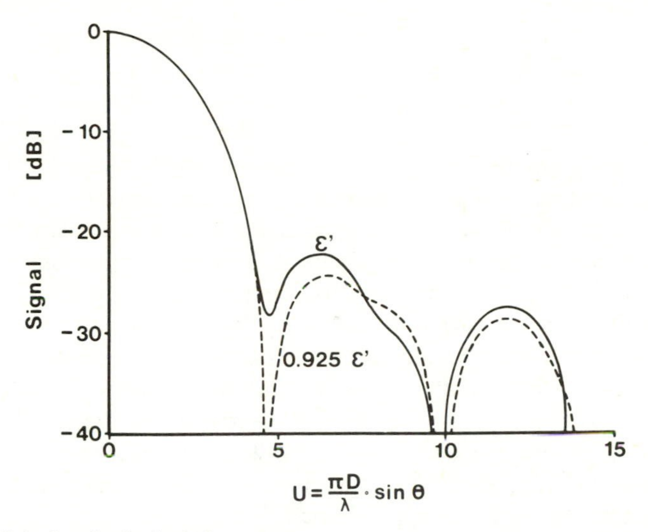

Fig. 8.20 *Calculated principal-plane patterns*

In general, it is believed that some value of ϵ which is 5% to 10% less than that obtained from eqn. 8.103 is a better value for antennas with typical aperture illuminations. It is also found that the distance of R should be chosen sufficiently large to ensure the required accuracy. In fact, for the same reasons as discussed related to the linear aperture, pattern measurements can only be trusted in a certain region around the axis. Range distances should in general be chosen from $D^2/4\lambda$ up to the normal far-field distance $2D^2/\lambda$ depending upon the required accuracy.

In the cases described above, measurements in the direction of the main beam have been considered. In this direction at infinity the path-length differences are usually zero, as indicated by saying that the antenna is focused at infinity. However, the defocusing technique may be used in an actual measurement situation for an arbitrary direction and any test distance. As above, proper modifications may be made by a path-length consideration. For example, in the case of dual reflector antennas, the focusing at short ranges is usually made by displacement of the subreflector but may also be made by a longer displacement of the feed. Antenna arrays may be defocused from infinity by introducing proper phase shifts in the feed network.

In general, in the modification of an antenna, it is difficult to take into account the fact that diffraction and mutual coupling between the different parts of the antenna vary with the modification. The technique usually has the disadvantage that it is only sufficiently accurate in a certain region around the direction in which it is defocused. Therefore, if the antenna is defocused in the main beam direction, accurate assessment of sidelobe level and detection of grating lobes may be impossible. Further details about the defocusing technique can be found in the cited literature and its references.

8.6.3 Extrapolation technique

In conventional far-field techniques antenna parameters are measured for a finite separation distance between test antenna and probe. The measured parameters are used as approximations to the far-field parameters which would be measured in case it was possible to carry out measurements for infinite distances between the antennas. In fact, this conventional approximation is the simplest form of the extrapolation technique. In its more advanced form, antenna parameters are measured when the distance between the antennas are varied within a certain range. The variation of the parameters with distance is then extrapolated to infinity. The mathematical details of the extrapolation technique are developed by combining two different expressions for the received signal as shown by Newell, Baird and Wacker.[34] One expression is a series expression derived by Wacker.[56] The other expression is the coupling equation used in near-field planar scanning technique, Kerns and Dayhoff[29] and Kerns.[28] In the present exposition the coupling equation is used in the form presented by Appel-Hansen.[45]

Geometrical configuration: The extrapolation technique can be used as a generalised three-antenna technique which permits absolute gain and polarisation measurements to be performed under the restriction only that no two antennas are circularly polarised. Therefore, suppose that three antennas numbered 1, 2 and 3 are available. Associated with each antenna is a rectangular xyz-coordinate system. The z-axis is supposed to be in the boresight direction and the x- and y-axes are chosen conveniently in accordance with the radiating properties or the physical appearance of the antennas. Similarly, the origins of the coordinate systems are placed conveniently and used as reference points from where distances to the antennas are measured. Fig. 8.21 shows a setup with the antennas 1 and 2 together with their coordinate systems $x_1y_1z_1$ and $x_2y_2z_2$, respectively. The relative aspects of the antennas are arranged so that the directions of their x- and z-axes are opposite and the boresight directions are colinear. The separation distance between the antennas is denoted by d.

Coupling equation and series expansion: Let antenna 1 be transmitting and antenna 2 receiving, then it is convenient to let the coupling equation, which expresses the received signal $b_o'(d)$, take the form, see eqn. 8.28

$$b'_o(d) = \frac{a_o}{1-\Gamma_2\Gamma_l} \int_{-\infty}^{\infty}\int_{-\infty}^{\infty} \sum_{s=1}^{2} {}^2R'^s(k_x, k_y)\, {}^1T^s(k_x, k_y)\, e^{-j\gamma d}\, dk_x\, dk_y \tag{8.104}$$

where a_o is the input signal to antenna 1, Γ_l the reflection coefficient of the receiving load, ${}^2R'^s(k_x, k_y)$ is the receiving characteristic of antenna 2, ${}^1T^s(k_x, k_y)$ is the transmitting characteristic of antenna 1, and k_x, k_y and γ are rectangular components of the propagation vector $\bar{k}$.

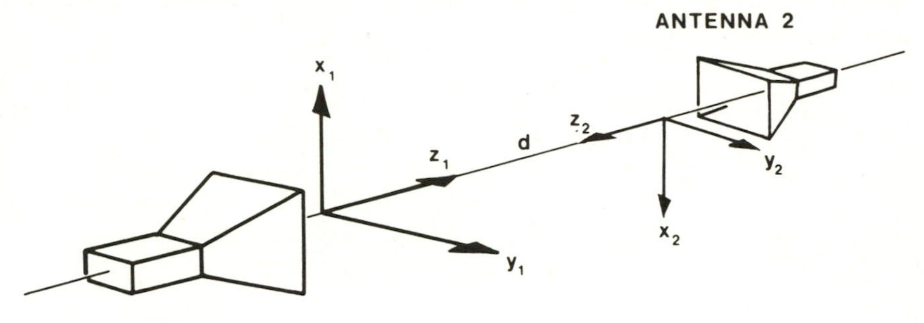

Fig. 8.21 *Geometrical configuration*

For large values of d, eqn. 8.104 takes the following asymptotic form:

$$b'_o(d) \simeq \frac{j2\pi k a_o}{1-\Gamma_2\Gamma_l} \frac{e^{-jkd}}{d} \sum_{s=1}^{2} {}^2R'^s\, {}^1T^s \tag{8.105}$$

where ${}^2R'^s$ and ${}^1T^s$ are ${}^2R'^s(0, 0)$ and ${}^1T^s(0, 0)$, respectively. Denoting the coupling product of ${}^2R'^s$ and ${}^1T^s$ by D'_{12}, we obtain

$$D'_{12} = \lim_{d\to\infty} \left\{ \frac{b'_o(d)\, d(1-\Gamma_l\Gamma_2)\, e^{jkd}}{j2\pi k a_o} \right\} \tag{8.106}$$

This is one equation in the four unknown quantities describing ${}^1T^s$ and ${}^2R'^s$. Rotating antenna 2 about its z_2-axis by 90° in the direction from x_2 to y_2, we obtain the additional equation

$$D''_{12} = \lim_{d\to\infty} \left\{ \frac{b''_o(d)\, d(1-\Gamma_l\Gamma_2)\, e^{jkd}}{j2\pi k a_o} \right\} \tag{8.107}$$

where $b''_o(d)$ is the received signal and D''_{12} is the coupling product of ${}^2R''^s$ and ${}^1T^s$ for the antennas in their new relative orientation. Equations analogous to eqns. 8.106 and 8.107 may be obtained for the antenna pair 1 and 3 and the pair 2 and 3. Thus, it is understood that six complex equations for six unknown parameters are obtained; e.g. when antenna n is transmitting and antenna m is receiving, we obtain by using the reciprocity relation

$$\eta_o{}^m R^s(k_x, k_y) = -\eta\,{}^m T^s(-k_x, -k_y) \tag{8.108}$$

where η_o is the admittance in the waveguide feed and η is the free-space admittance

$$\eta({}^nT^1\,{}^mT^1 - {}^nT^2\,{}^mT^2) = \eta_o D'_{nm} \tag{8.109}$$

$$\eta({}^nT^1\,{}^mT^2 + {}^nT^2\,{}^mT^1) = \eta_o D''_{nm} \tag{8.110}$$

where ${}^iT^s$, $i = 1, 2, 3$, is the transmitting property of antenna i in its own coordinate system.

Let us now consider the series expression for the signal received by antenna 2 in its first orientation

$$b'_o(d) = \frac{a_o}{1-\Gamma_2\Gamma_l} \sum_{p=0}^{\infty} \frac{e^{-j(2p+1)kd}}{d^{2p+1}} \sum_{q=0}^{\infty} \frac{A_{pq}}{d^q} \tag{8.111}$$

which can be written as

$$\begin{aligned} b'_o(d) = \frac{a_o}{1-\Gamma_2\Gamma_l} \cdot \Bigg\{ & \frac{e^{-jkd}}{d}\left(A_{00} + \frac{A_{01}}{d} + \frac{A_{02}}{d^2} + \ldots\right) \\ & + \frac{e^{-jk3d}}{d^3}\left(A_{10} + \frac{A_{11}}{d} + \frac{A_{12}}{d^2} + \ldots\right) \\ & + \frac{e^{-jk5d}}{d^5}\left(A_{20} + \frac{A_{21}}{d} + \ldots\right) + \ldots \Bigg\} \end{aligned} \tag{8.112}$$

where the first series represents the direct signal, the second series the first-order multiple reflection between the two antennas, and the other series the higher order multiple reflections.

Insertion of eqn. 8.112 into eqn. 8.106 gives

$$D'_{12} = \frac{A_{00}}{j2\pi k} \tag{8.113}$$

In the extrapolation technique measured values of $b'_o(d)$ over a certain range of distances may be fitted to a truncated part of the series (eqn. 8.112) and a certain number of the unknowns A_{pq} may be determined. Thus, A_{00} is determined for the different antenna pairs, and from eqns. 8.109, 8.110, 8.113 the far-field transmission characteristics are found. In this way correction for multiple reflections and proximity effects is carried out. In the actual solution of eqns. 8.109, 8.110 it is convenient to use circular components of the far-field antenna characteristics which are related to the linear components ${}^iT^s$ in the usual way.

Curve fitting: The fitting of $b'_o(d)$ to the series (eqn. 8.111) requires accurate measurement of amplitude and phase. Furthermore, due to the phase factor $e^{-j(2p+1)kd}$, the distance should be measured accurately, in particular at short wavelengths. The need for this accurate distance measurement can be avoided and the fitting procedure can be facilitated by measuring the ratio between transmitted and received powers. To illustrate this, let us define a function $v(d)$ as

$$v(d) = \left| \frac{b'_o(d) d\, e^{jkd}(1 - \Gamma_2 \Gamma_l)}{a_o} \right|^2 \tag{8.114}$$

Then, from eqn. 8.112 we obtain

$$v(d) = \left| A_0(d) + \frac{e^{-jk2d}}{d^2} A_1(d) + \frac{e^{-jk4d}}{d^4} A_2(d) + \ldots \right|^2 \tag{8.115}$$

or

$$v(d) = |A_o(d)|^2 + \frac{|A_1(d)\, e^{-jk2d}|^2}{d^4} + \frac{|A_2(d)\, e^{-jk4d}|^2}{d^8} + \ldots$$

$$+ 2\,\mathrm{Re} \sum_{n=0}^{\infty} \sum_{m=n+1}^{\infty} \frac{A_m(d) A_n^*(d)}{d^{2(m+n)}} e^{-2j(mk - nk^*)d} \tag{8.116}$$

where

$$A_n(d) = A_{no} + \frac{A_{n1}}{d} + \frac{A_{n2}}{d^2} + \ldots \tag{8.117}$$

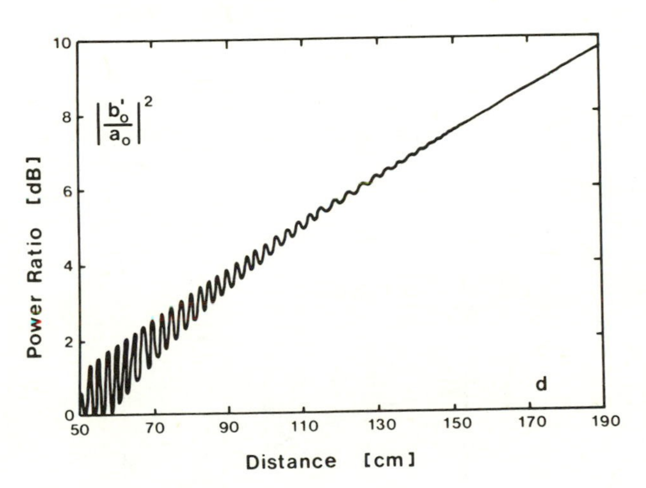

Fig. 8.22 *Effect of multiple reflections*

are the subseries in eqn. 8.112, $A_n^*(d)$ and k^* are complex conjugates of $A_n(d)$ and k. The summation in eqn. 8.116 represents a modulation in the received power due to multiple reflections. The modulation is illustrated in Fig. 8.22 and can be removed by averaging. The averaging can be done graphically, digitally or by filtering processes during the measurements. The averaged curve is fitted to a polynomial in the reciprocal distance given by

$$v'(d) = A'_{00} + \frac{A'_{01}}{d} + \frac{A'_{02}}{d^2} \tag{8.118}$$

where $v'(d)$ is $v(d)$ minus the summation term in eqn. 8.116. It is easily found that $A'_{00} = |A_{00}|^2$. Using eqn. 8.113 the magnitude of D'_{12} is found. In the solution of eqns. 8.109, 8.110 only the phase differences between the coupling products are needed. The phase differences are found by extrapolating measured phase differences between, e.g. $b'_o(d)$ and $b''_o(d)$. Thus,

$$\arg\left(\frac{D''_{12}}{D'_{12}}\right) = \lim_{d\to\infty} \arg\left(\frac{b''_o(d)}{b'_o(d)}\right) \tag{8.119}$$

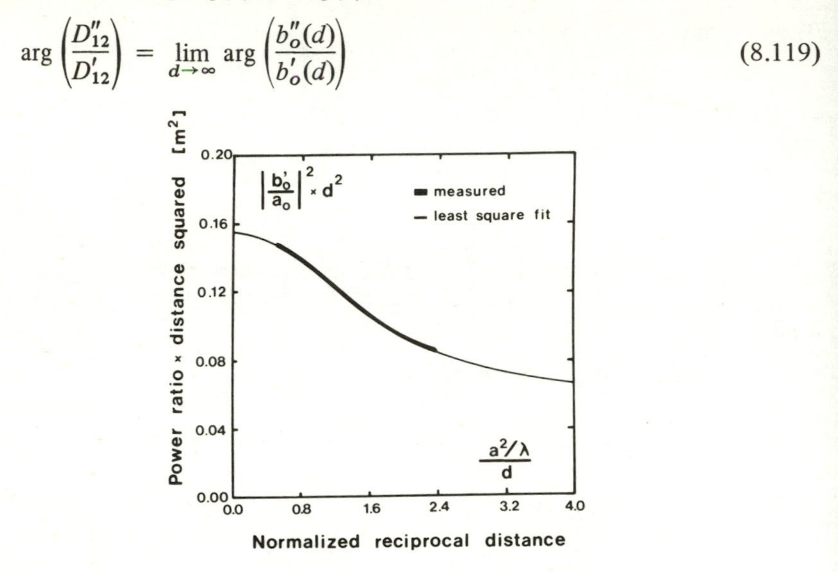

Fig. 8.23 *Extrapolation of power ratio*

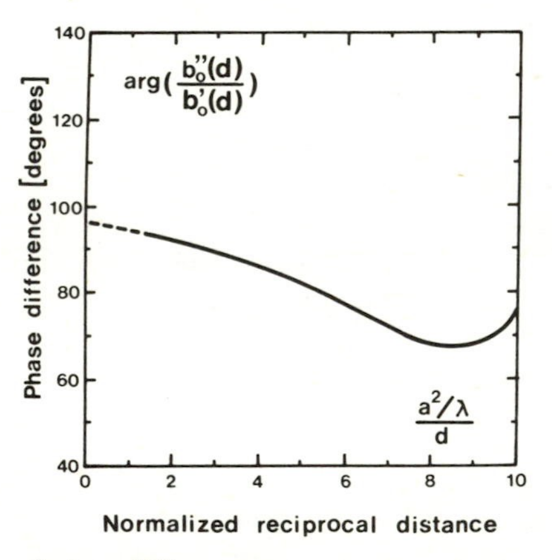

Fig. 8.24 *Extrapolation of phase difference*

In Figs. 8.23, 8.24 are shown examples of extrapolation data. Power ratio multiplied by distance squared and phase difference are shown as a function of the reciprocal distance normalised with a^2/λ, where a is the largest dimension of the antenna. This procedure has the advantage that the asymptotic data can be found at the intersection of the extrapolated curves with the ordinate axes. This is of particular value in the case for the phase difference where the functional variation with distance has not been derived yet. The normalisation implies that approximately the same range on the abscissa is used for different types of antennas. This is due to the fact that the distance interval for the extrapolation technique seems to be approximately between $0.2a^2/\lambda$ and $2a^2/\lambda$. For that distance range, ground reflections should not be large and it should be possible to determine 4 to 8 coefficients by curve fitting. In conventional far-field techniques a zero-order polynomial is used to determine D'_{12}.

Measurements: In the work by Newell, Baird and Wacker[34] several hints are given to the practical implementation of the extrapolation technique. Details of an outdoor extrapolation range and measurement procedures are described. It is mentioned that the influence of ground reflections can be averaged out like multiple reflections between the antennas. The ground reflections can be distinguished by a longer modulation period than the period of $\lambda/2$ presented by first-order multiple reflections. A statistical test of significance is used as a criterion to find a satisfactory number of terms in the polynomial fitting. An error analysis is carried out and actual errors are computed for gain and polarisation parameters determined for different types of antennas. It is concluded that accurate gain and polarisation measurements can be conveniently performed on ranges only 1/5 to 1/10 as large as required for conventional far-field measurements.

8.7 **Conventional far-field measurements** (Contribution by J. Appel-Hansen, J. D. Dyson, and E. S. Gillespie)

8.7.1 Directivity (ESG)

The directivity, $D(\theta, \phi)$, of an antenna is a measure of the spatial distribution of the radiation intensity transmitted by the antenna. It is defined as the radiation intensity in a given direction divided by the average radiation intensity. In this definition, the average radiation intensity is simply the total power radiated by the antenna divided by 4π steradians. Directivity, unlike gain, does not include any losses; therefore, it cannot be measured directly but must be computed from the radiation patterns for the antenna. The more complex that the radiation patterns are, the more difficult it is to obtain an accurate determination of directivity. Usually the directivity of a highly directional antenna with its narrow main beam and numerous sidelobes is not measured. For such antennas it is the gain that is measured; although their directivity can be estimated.

A notable example of a case where the measurement of directivity is required is that of antennas mounted on aircraft, spacecraft or missiles. Because the entire vehicle usually contributes to the shape of the radiation pattern, it is often necessary to use scale modelling techniques.[1] Usually, it is impractical to scale the finite conductivities and loss factors of the materials of which the antenna and vehicle are constructed; thus, gain measurements cannot be performed using scale models. The directivity can be measured, however. If the radiation efficiency of the full-scale antenna can be established by other means, then the gain can be determined since the directivity of a properly scaled model antenna and the corresponding full-scale antenna are the same.

Another case in which the directivity is measured is that of the determination of radiation efficiency. If both the gain and directivity are measured, then the efficiency can be determined since it is equal to the ratio of the gain to directivity.

From its definition, the directivity is a parameter associated with an antenna operating in the transmit mode. However, by the reciprocity theorem[1] it also applies to an antenna operating in the receive mode. The effective area, $A_e(\theta, \phi)$, is a more appropriate parameter for that case. The relationship between directivity and effective area is given by

$$D(\theta, \phi) = \eta_r \frac{4\pi}{\lambda^2} A_e(\theta, \phi) \tag{8.120}$$

where λ is the wavelength and η_r is the radiation efficiency.

Measurements: The directivity of an antenna needs only to be determined for a single direction since the directivity in any other direction can be obtained from the appropriate power patterns of the antenna. For directional antennas this direction is usually chosen to be that of the maximum radiation intensity, $\Phi_m(\theta_o, \phi_o)$, which is also the direction of peak directivity, $D_m(\theta_o, \phi_o)$.

From its definition, the peak directivity can be expressed as

$$\begin{aligned} D_m(\theta_o, \phi_o) &= \frac{\Phi_m(\theta_o, \phi_o)}{1/4\pi \int_0^{\pi} \left[\int_0^{2\pi} \Phi(\theta, \phi) \, d\phi \right] \sin\theta \, d\theta} \\ &= \frac{\Phi_m(\theta_o, \phi_o)}{P_t/4\pi} \end{aligned} \tag{8.121}$$

The double integral in the denominator represents the total power P_t radiated by the antenna. When P_t is divided by 4π steradians, the average radiation intensity is obtained.

The radiation intensity can be decomposed into two components which correspond to any two orthogonal polarisations.[1,57] Thus, eqn. 8.121 can be written as

$$\begin{aligned} D_m(\theta_o, \phi_o) &= D_1(\theta'_o, \phi_o) + D_2(\theta_o, \phi_o) \\ &= \frac{\Phi_{m1}(\theta_o, \phi_o)}{1/4\pi(P_{t1} + P_{t2})} + \frac{\Phi_{m2}(\theta_o, \phi_o)}{1/4\pi(P_{t1} + P_{t2})} \end{aligned} \tag{8.122}$$

in which the subscripts 1 and 2 denote the two orthogonal polarisations and P_{t1} and P_{t2} are given by

$$P_{t1} = \int_0^{\pi} \left[\int_0^{2\pi} \Phi_1(\theta, \phi)\, d\phi \right] \sin\theta\, d\theta$$

and (8.123)

$$P_{t2} = \int_0^{\pi} \left[\int_0^{2\pi} \Phi_2(\theta, \phi)\, d\phi \right] \sin\theta\, d\theta$$

$D_1(\theta_o, \phi_o)$ and $D_2(\theta_o, \phi_o)$ are called the partial directivities with respect to polarisations 1 and 2, respectively. It is important to note that *both* P_{t1} and P_{t2} must be evaluated in order to determine either $D_1(\theta_o, \phi_o)$ or $D_2(\theta_o, \phi_o)$.

This decomposition of the directivity into two components corresponding to two specified orthogonal polarisations is important from a practical point of view because, in general, the polarisation of an antenna changes with direction. The decomposition allows one to measure the partial directivities with respect to two specified polarisations and then add them to obtain the (total) directivity of an antenna. Usually the radiation patterns are measured on an antenna range which employs a spherical coordinate positioner; therefore, θ and ϕ linear polarisations are usually chosen. For some applications right and left hand circular polarisations are used.

A quantity that is generally sought is the partial directivity with respect to the design or intended polarisation of the antenna. Therefore, it is desirable to choose one of the orthogonal polarisations to be the intended polarisation of the antenna. In this way the partial directivities with respect to the intended polarisation and the cross polarisation will be obtained.

For example, if the antenna to be tested is designed to be essentially linearly polarised, then orthogonal linear polarisation should be chosen. This is accomplished by mounting the antenna to be tested on a spherical coordinate positioner in such a manner that its intended polarisation is co-polarised with either θ or ϕ polarisation. If right and left hand circular polarisations were chosen, the directivity of the antenna would be obtained, but the partial directivity with respect to the design polarisation would not be obtained.

Ideally, right and left hand circular polarisations should be chosen when testing a circularly polarised antenna; however, it is usually difficult to produce the circularly polarised fields with sufficient purity to make accurate measurements. Most often linear polarisations are used and a separate polarisation measurement is made to determine or to estimate the cross-polarisation component.

In order to simplify the following discussion, it will be assumed that the cross-polarised component is zero. The discussion will apply equally well to the measurements of the partial directivity with respect to the cross polarisation, which is usually required in practice. It should be kept in mind that even when one seeks only the partial directivity with respect to the design polarisation, the radiation patterns for both the co-polarisation and cross polarisation are required in order to compute the total power radiated by the antenna.

The radiation intensity is not measured directly but rather a normalised value, $\bar{\Phi}(\theta, \phi)$, is measured. The normalisation is usually with respect to the maximum value so that eqn. 8.121 becomes

$$D_m(\theta_o, \phi_o) = \frac{1}{1/4\pi \int_0^\pi \left[\int_0^{2\pi} \bar{\Phi}(\theta, \phi)\, d\phi \right] \sin\theta\, d\theta} \tag{8.124}$$

where

$$\bar{\Phi}(\theta, \phi) = \Phi(\theta, \phi)/\Phi_m(\theta_o, \phi_o)$$

The normalised radiation intensity may be measured by sampling the field over a large sphere centred on the antenna, called the radiation sphere of the antenna. It can be accomplished by making conical cuts, successive ϕ cuts over the radiation sphere at increments of θ, or by great-circle cuts, successive θ cuts at increments of ϕ. The number of increments required depends upon the complexity of the antenna's pattern structure. Generally, the number of increments required for a given accuracy increases as the pattern becomes less uniform.

If ϕ cuts are employed, then the θ interval from 0 to π rad. is divided into M equal spherical sectors, and the expression for the peak directivity becomes

$$D_m(\theta_o, \phi_o) = \frac{4M}{\sum_{i=1}^{M} \left[\int_0^{2\pi} \bar{\Phi}(\theta_i, \phi)\, d\phi \right] \sin\theta_i} \tag{8.125}$$

where $\theta_i = i\pi/M$ rad.

If θ cuts are employed, the ϕ interval from 0 to 2π is divided into N equal increments, and the expression for the peak directivity can be written as

$$D_m(\theta_o, \phi_o) = \frac{2N}{\sum_{j=1}^{N} \int_0^{\pi} \bar{\Phi}(\theta, \phi_j) \sin\theta\, d\theta} \tag{8.126}$$

Note that each $\bar{\Phi}(\theta, \phi_j)$ must be multiplied by $\sin\theta$ during the process of integration.

When a rectangular recorder is used in the measurement of the normalised radiation intensity, the linear or power scale must be used. If a polar recorder is employed, then the square root or voltage scale must be used. This is because the area of a differential wedge in polar coordinates is given by $1/2V^2\, d\phi$ in the case of ϕ cuts where V is the voltage recorded at the angle ϕ. Since the radiation intensity is proportional to V^2, the area in the polar curve will be proportional to the integral within the brackets of eqn. 8.125.

The directivity can be evaluated numerically by the use of a digital computer. The computer can be on-line, in which case real-time computations can be made. Alternatively, the data can be recorded digitally and entered into a computer at a later time.

Estimation of directivity: The directivity of an antenna is inversely proportional to the beam area or, as it is sometimes called, the areal beamwidth,[58] which is the product of the two principal half-power beamwidths.* Thus, the directivity can be expressed as

$$D_m(\theta_o, \phi_o) = \frac{K}{\theta_{HP}\phi_{HP}} \tag{8.127}$$

where θ_{HP} and ϕ_{HP} are the principal half-power beamwidths expressed in degrees and K is a constant of proportionality.

K has been given as being between 25 000 to 41 253. The upper value is of little practical importance since it corresponds to a conical beam with no sidelobes.[59] For most practical apertures, the value of 32 400 is reasonable.[58]

8.7.2 Gain (ESG)

The gain of an antenna, like the directivity, is a measure of the spatial distribution of radiation intensity; however, it differs from directivity in that it includes dissipative losses within the antenna. Gain can be defined as the ratio of the radiation intensity, in a specified direction, to the radiation intensity that would be obtained if the power accepted by the antenna were radiated isotropically.

There are two basic categories of gain measurements; they are: absolute-gain measurements and gain-transfer measurements.[57,60] When absolute-gain measurements are performed, no *a priori* knowledge of the gains of any of the antennas used in the measurements are required. The methods that fall into this category are usually used in the calibration of gain standards.

The most commonly employed method for the measurement of gain is the gain-transfer method, which is also called the gain-comparison method. This requires the use of a gain standard to which the gain of the antenna being tested is compared.

Absolute-gain measurements: The basis for absolute-gain measurements is the Friis transmission formula for two-antenna systems, such as depicted in Fig. 8.25. It can be written as

$$P_r = P_o G_A G_B \left(\frac{\lambda}{4\pi R}\right)^2 \tag{8.128}$$

where P_r is the power received, P_o is the power accepted by the transmitting antenna, G_A is the gain of the transmitting antenna, G_B is the gain of the receiving antenna and R is their separation.[57,60] For this formula, it is implicitly assumed

* For a directional antenna with a single, narrow main beam, the half-power contour on the radiation sphere is essentially elliptical in shape. The principal half-power beamwidths are obtained from the two pattern cuts which contain the major and minor axis of the ellipse, respectively. For example, if a rectangular phased array is linearly polarised with its mainbeam broadside-broadside, the major and minor axes are in the principal E and H planes. To measure the patterns, the antenna is usually mounted on a spherical coordinate positioner so that the appropriate pattern cuts are θ or ϕ cuts. If the beam is scanned in the principal E or H planes, the major and minor axes remain aligned with the θ or ϕ directions. However, if the beam is scanned in any other direction, the axes are tilted and no longer are aligned with the θ and ϕ directions. For these cases the principal half power beamwidths are difficult to determine.[58]

that the antennas are polarisation matched for their prescribed orientations and that the separation between them is such that far-field conditions prevail.

For convenience, eqn. 8.128 can be written in logarithmic form, from which the sum of the gains of the two antennas, in decibels, can be expressed as

$$(G_A)_{dB} + (G_B)_{dB} = 20 \log \left(\frac{4\pi R}{\lambda}\right) - 10 \log \left(\frac{P_o}{P_r}\right) \tag{8.129}$$

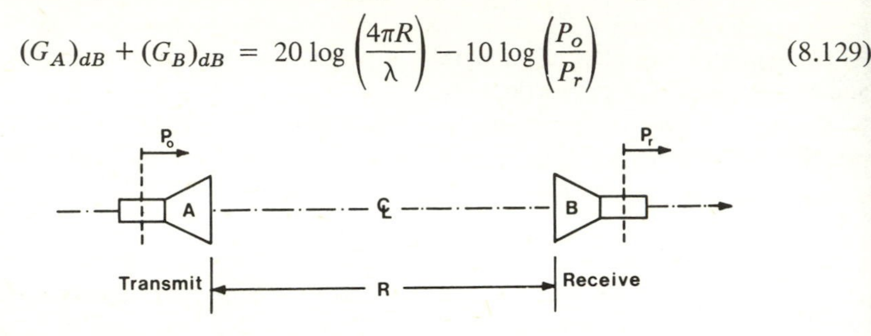

Fig. 8.25 *Two-antenna system*

If the two antennas are identical, then their gains are given by

$$(G_A)_{dB} = (G_B)_{dB} = \frac{1}{2}\left[20 \log \left(\frac{4\pi R}{\lambda}\right) - 10 \log \left(\frac{P_o}{P_r}\right)\right] \tag{8.130}$$

To determine the gain, one measures R, λ and $10 \log (P_o/P_r)$ and from this data, $(G_A)_{dB}$ is computed. This method is called the two-antenna method since two identical, or near identical, antennas are required. If two identical antennas are not available, then a third antenna is required in order to determine the gain of any one of the antennas.

When three antennas are used, three sets of measurements are performed using all of the possible combinations of the antennas. The result is a set of simultaneous equations given by

$$\begin{aligned}
(G_A)_{dB} + (G_B)_{dB} &= 20 \log \left(\frac{4\pi R}{\lambda}\right) - 10 \log \left(\frac{P_o}{P_r}\right)_{AB} \\
(G_A)_{dB} + (G_C)_{dB} &= 20 \log \left(\frac{4\pi R}{\lambda}\right) - 10 \log \left(\frac{P_o}{P_r}\right)_{AC} \\
(G_B)_{dB} + (G_C)_{dB} &= 20 \log \left(\frac{4\pi R}{\lambda}\right) - 10 \log \left(\frac{P_o}{P_r}\right)_{BC}
\end{aligned} \tag{8.131}$$

From these equations the gains of all three antennas can be determined. This method is called the three antenna method of gain measurement.

The accepted power P_o cannot be measured directly unless the transmitting antenna is impedance matched. If one measures the available power from the generator, P_A, then the accepted power is given by

$$P_o = P_A M_1 \tag{8.132}$$

where

$$M_1 = \frac{(1-|\Gamma_G|^2)(1-|\Gamma_T|^2)}{|1-\Gamma_G\Gamma_T|^2} \tag{8.133}$$

in which Γ_G and Γ_T are the complex reflection coefficients of the generator and the antenna, respectively, when viewed at the same reference terminal.[61] If the power meter had been matched to the transmission line, then one measures the line-matched power P_M and

$$P_o = P_M M_1' \tag{8.134}$$

where

$$M_1' = \frac{1-|\Gamma_T|^2}{|1-\Gamma_G\Gamma_T|^2} \tag{8.135}$$

A similar situation exists at the receiving system where the available received power is reduced by an impedance mismatch factor, M_2, which is given by

$$M_2 = \frac{(1-|\Gamma_R|^2)(1-|\Gamma_L|^2)}{|1-\Gamma_R\Gamma_L|^2} \tag{8.136}$$

In this equation Γ_R and Γ_L are the complex reflection coefficients of the receiving antenna and the receiver, respectively.

These factors can be evaluated by measuring Γ_G, Γ_T, Γ_R and Γ_L. With this data the measured gain obtained using eqns. 8.130 or 8.131 can be corrected. The correction term is given by

$$C_1 = 10\log M_1 + 10\log M_2 \tag{8.137}$$

This correction term is applied to each two antenna measurement. If P_M is measured, M_1' is used instead of M_1.

There is an alternative procedure for the measurement of absolute gain in which the impedance mismatch losses are included.[62] In this method the insertion loss between the terminals of the two antennas is measured. This is accomplished by connecting the transmitter directly to the load and measuring the received power, P_L^i. Then the antennas are inserted and properly spaced and the power absorbed in the load P_L^f is measured. From the Friis transmission formula, one can show that the sum of the gains, in decibels, of the two antennas is given by

$$(G_T)_{dB} + (G_R)_{dB} = 20\log\left(\frac{4\pi R}{\lambda}\right) + 10\log\left(\frac{P_L^f}{P_L^i}\right) + 10\log M \tag{8.138}$$

in which

$$M = \frac{|1-\Gamma_R\Gamma_L|^2\,|1-\Gamma_G\Gamma_T|^2}{(1-|\Gamma_R|^2)(1-|\Gamma_T|^2)\,|1-\Gamma_G\Gamma_L|^2} \tag{8.139}$$

If G_T and G_R are equal, then the gain of each of the antennas is equal to one-half of the right side of eqn. 8.138.

The effect of polarisation mismatch can be accounted for by multiplying eqn. 8.128 by the polarisation efficiency between the incident wave and the receiving antenna, p_{AB}. When carried through to the expressions for gain, this will result in a correction term for the measured gain given by

$$C_2 = |10 \log p_{AB}| \tag{8.140}$$

The polarisations of the incident wave and the receiving antenna must be known in order to determine p_{AB}, however.

In addition to impedance and polarisation mismatch errors, there are errors due to proximity effects and multipath interference. The proximity effects can only be estimated for the methods just described. These errors are of a systematic type and are negatively valued. The proximity effects are due to two causes:

(1) The angular distribution of the antenna's field actually varies with distance because of the presence of near-field components which may not have decreased to a negligible value.
(2) The amplitude taper of the incident wave at the receiving antenna.

In the first case the correction at a distance of $2D^2/\lambda$ for the gain of a typical aperture antenna is about 0.05 dB. It decreases with larger spacing. For the second case, if the amplitude taper is 0.25 dB across the aperture of a typical aperture type receiving antenna, then the correction will be about 0.1 dB. Thus, if the transmitting and receiving antennas are chosen such that the 0.25 dB taper occurs at a spacing of $2D^2/\lambda$, then a total correction of 0.15 dB may be expected.[1,57] The effects of finite spacing on the gain of electromagnetic horn antennas has been studied extensively[62–67] because of their use as a gain-standard antenna.

When multipath interference is present, the gain will exhibit an undulating characteristic as it is measured as a function of distance. This is due to the interaction of the direct wave and reflected wave at the receiving antenna. Where the spacing is close, less than D^2/λ, the multipath interference is primarily caused by multiple reflections between the antennas.

The errors due to proximity effects and multipath interference can be effectively removed by the use of an extrapolation technique.[34] This technique requires that the insertion loss be measured as a continuous function of the spacing between the transmitting and receiving antennas. The measurement is made at spacings less than $2D^2/\lambda$ and the far-field insertion loss is obtained by extrapolation. This technique requires a special range which maintains boresight between the antennas as their spacing is changed. When combined with a generalised three-antenna method, not only is the gain obtained, but also the polarisations of the antennas provided none of the antennas are circularly polarised. Accuracies of 0.05–0.08 dB can be obtained using this method.[60]

Gain-transfer method: The gain-transfer method is one in which the unknown gain of the antenna being measured is compared to that of a gain-standard antenna. In this way the gain of the standard is transferred to the antenna under test.[1,57,60]

Either a free-space or ground-reflection range may be employed for the measurement. It can also be performed on an antenna *in situ* provided there is adequate signal available.

Ideally, the illumination of the test antenna and the gain-standard antenna should be by an incident plane wave which is polarisation matched to both antennas. The antennas should be conjugately matched to the receiver. With these conditions met the gain of the test antenna in decibels, $(G_T)_{dB}$, is given by

$$(G_T)_{dB} = (G_S)_{dB} + 10 \log (P_T/P_S) \tag{8.141}$$

where $(G_S)_{dB}$ is the gain of the gain-standard antenna, P_T is the power received by the test antenna and P_S that received by the gain-standard antenna.

This method is less affected by multipath interference and proximity effects provided the antenna under test is not too dissimilar in physical size from the gain standard. The method is, however, sensitive to impedance and polarisation mismatches. The correction factor for impedance mismatch is given by

$$C_1 = 10 \log \left[\frac{(M_2)_T}{(M_2)_S}\right] \tag{8.142}$$

in which $(M_2)_T$ and $(M_2)_S$ are the M_2 correction factors for the test and gain-standard antennas, respectively. The correction factor for polarisation mismatches is given by

$$C_2 = 10 \log \left[\frac{p_S}{p_T}\right] \tag{8.143}$$

where p_S is the polarisation efficiency between the incident wave and the gain-standard antenna and p_T is that for the test antenna. Note that C_2 can be positive or negative depending upon the relative values of p_S and p_T.

The gains of circularly or elliptically polarised antennas are usually determined by measuring the partial gains with respect to two orthogonal polarisations. The (total) gain of an antenna measured in this manner is the sum of the two partial gains. Typically, a single linearly polarised gain-standard antenna is used. The partial gain is measured for one orientation of the antenna, say, horizontal polarisation; then the measurement is repeated with the transmitting and gain-standard antenna rotated 90° to vertical polarisation. The gain of the test antenna can be computed by the use of the equation

$$(G_T)_{dB} = 10 \log (G_{TV} + G_{TH}) \tag{8.144}$$

where G_{TV} and G_{TH} are the partial gains with respect to vertical-linear and horizontal-linear polarisation, respectively.

Use of extraterrestrial radio sources for gain measurement: Antennas that are electrically large generally cannot be tested on an antenna range because of the large distances required for far-field conditions to prevail. Furthermore, these antennas are usually physically large and, therefore, experience gravity-induced structural deviations when moved. Thus, it is necessary to measure their gains *in situ* as a function of the antenna's orientation. This can be accomplished by the use of extraterrestrial radio sources.[68–71] The sources that are most useful for gain measurements are the ones whose angular extents are less than those of the main beams of the antennas under test. A convenient rule of thumb is that the antenna beamwidth should be at least twice the angular extent of the source. Additionally, only the strongest of these sources are suitable for accurate gain measurements.

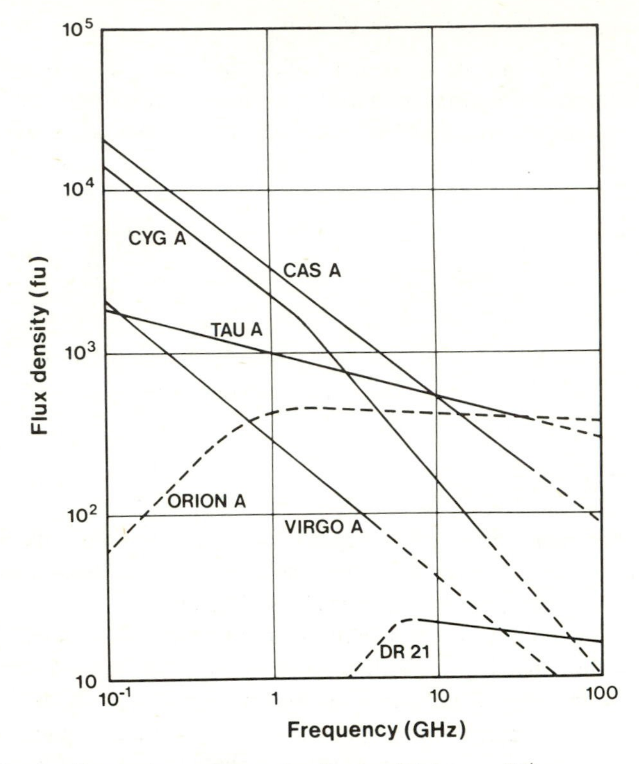

Fig. 8.26 *Flux density spectrum of several radio stars* (Reference 72)

The flux-density spectra of several radio stars are depicted in Fig. 8.26. The flux densities are expressed in flux units (f.u.) which are equivalent to 10^{-26} watts per meter squared per hertz. More details concerning these sources are given in Table 8.2. The Table includes the location, size, spectral index α, and type of phenomena believed responsible for the creation of the radio star. In addition, the visibility given refers to that interval of terrestrial latitudes from which the source's apparent daily path through the sky rises to at least 20° above the horizon. The 20° minimum elevation assures that the source is sufficiently high that it will be discernible above the atmospheric and ground noise.[72]

Of the sources listed in Table 8.2, Cassiopeia A, Cygnus A, and Taurus A are the most useful for gain measurements. They can be treated as point sources if the antenna under test has a beamwidth of 10 minutes or greater. The flux densities of these sources have been accurately measured over a frequency range of 30 MHz to 16 GHz.[73,74] These measurements were made at discrete frequencies; hence, the flux density may not be available for a frequency required for a particular gain measurement. The flux density, however, is proportional to the frequency raised to a power equal to the spectral density. If the flux density $S(f_k)$ is known, then the flux density $S(f)$ is given by

$$S(f) = \left(\frac{f}{f_k}\right)^{\alpha} S(f_k) \tag{8.145}$$

where f is the frequency at which the flux density is required and f_k is the frequency for which the flux density is known.

Table 8.2 *Information about several radio sources* (Reference 72)

Radio star	Type*	Position RA^h	Position Dec°	Size RA′ × Dec′	Visibility NL°	Visibility to SL°	Spectral index α
Cas A	SR	23.4	58.6	4 × 4	90	11	−0.787
Tau A	SR	5.5	22.0	3.3 × 4	90	48	−0.263
Orion A	EN	5.5	−5.4	3.5 × 3.5	65	75	0
Cyg A	RG	20.0	40.6	1.6 × 1	90	29	−1.205
Virgo A	RG	12.5	12.7	1 × 1.8	73	57	−0.853
DR 21	EN	20.6	42.2	<0.3	90	28	1.75, −0.13

* SR ≡ Supernova remnant
EN ≡ Emission nebula
RG ≡ Radio galaxy

There are certain characteristics of the radio stars that must be considered. Cassiopeia A has an annual decrease in flux density of 0.9 ± 0.1%, Cygnus A has a curvature in its spectrum and Taurus A exhibits a significant degree of linear polarisation.[69] These effects have been studied and the results well documented.

The power received by the antenna, P_R, when it points at a radio star is given by

$$P_R = \frac{SA_e}{2} = \frac{S\lambda^2 G_T}{8\pi} \text{ watts Hz}^{-1} \tag{8.146}$$

where S is the known power flux density, A_e is the effective area of the antenna, and G_T is its gain. The noise power radiated by the source has been assumed to be randomly polarised; hence, the polarisation efficiency is 1/2. The received power can be set equal to kT_a where k is Boltzmann's constant and T_a is the effective noise temperature of the radio star referred to the antenna's terminals. From this the gain of the antenna can be written as

$$G_T = \frac{8\pi k T_a}{S\lambda^2} \tag{8.147}$$

The gain can, therefore, be computed once the effective noise temperature T_a has been measured. This measurement requires the use of a radiometer.[75]

There are many factors which affect the accuracy of the measurement of gain using this method. Some of these factors are:

(1) K_1, the effect of atmospheric attenuation
(2) K_2, the angular extent of the source
(3) K_3, the antenna pointing error
(4) K_4, polarisation efficiency
(5) K_5, the instrumentation system response.

Additional factors are given in Reference 76. These factors can be included by writing eqn. 8.147 as

$$G_T = \frac{8\pi k T_a}{S\lambda^2} K_1 K_2 K_3 K_4 K_5 \tag{8.148}$$

Some of these factors can be computed, or determined by measurement, so that the measured value of gain can be corrected. In any event they all affect the uncertainty of the measurement.

8.7.3 Phase centres (JDD)

The radiation pattern of an antenna is completely described by the magnitude and phase of the radiated field components in two particular orthogonal polarisations. It is usual to measure the magnitude of the specified component of the field for which the antenna is designed; however for many applications the phase of this component may also be of importance.[77,78] For a complete description of the field the magnitude and phase of the cross-polarised component must also be measured.

A specified component of the far field produced by an antenna can be expressed in the form

$$E_u(r, \theta, \phi) = P(\theta, \phi)\, e^{j\psi(\theta,\phi)} \frac{e^{-jkr}}{r} \boldsymbol{a} \tag{8.149}$$

where P and ψ represent respectively the (θ, ϕ) dependence of the magnitude and phase of the specified component of the field and (r, θ, ϕ) are spherical coordinates of the observation point. The vector $\boldsymbol{a}$ is a vector of magnitude unity, ($\|\boldsymbol{a}\|^2 = \boldsymbol{a}^* \cdot \boldsymbol{a} = 1$) which indicates the polarisation of the specified component. For a linearly polarised component the vector $\boldsymbol{a}$ is taken as *real*, for instance it may be the *unit* vector $\boldsymbol{a}_\theta$ (or $\boldsymbol{a}_\phi$) in the direction of increasing θ (or ϕ) and the phase ψ is defined unambiguously. For an elliptically or circularly polarised component the vector $\boldsymbol{a}$ is complex and is *not* uniquely defined by its polarisation and the condition $\|\boldsymbol{a}\|^2 = 1$. For example, multiplying it by $e^{j\alpha}$ preserves these conditions, but this changes the phase ψ accordingly. Therefore, a definite convention must be used to specify $\boldsymbol{a}$ and, when discussing a pattern, this convention must be applicable to all directions (θ, ϕ) of interest.[1]

To describe completely the radiation of the antenna, the magnitude and phase patterns corresponding to two distinct vectors $\boldsymbol{u}$ and $\boldsymbol{v}$, preferably orthogonal, are needed. Alternatively the radiation can be described by means of a polarisation pattern, giving the polarisation as a function of (θ, ϕ), and by the magnitude and phase in that particular polarisation. To make this description unambiguous a convention must be used to choose a particular vector $\boldsymbol{u}$ of magnitude one and proper polarisation for every direction.

The exact position of the point 0, centre of the coordinate system, is important in defining the phase pattern and it must be clearly specified with respect to the antenna structure, within a fraction of a wavelength.

For many applications it is useful to simplify the form of the phase pattern by a judicious choice of the origin 0. In some cases it is possible by a shift of origin to reduce the phase ψ to a constant. When this is achieved 0, the centre of the resulting spherical equiphase surface, is said to be the *phase centre* of the antenna and the pattern is described for the specified polarisation by the single real function $P(\theta, \phi)$. This circumstance occurs for linear antennas and arrays that have odd or even crossing symmetry that is such that the excitation current satisfies the condition.[79]

$$I(-x) = \pm I^*(x) \tag{8.150}$$

For most antennas a unique point that can be described as the phase centre, valid for all directions of observations, does not exist. However, one may sometimes find a reference point such that ψ becomes constant over a range of directions of interest, for instance over a portion of the main beam of the antenna. This point has been also called a phase centre, or an 'apparent phase centre'.[80] It is particularly important to determine such a point for the primary feed of a reflector since it allows one to use geometrical optics to properly place the feed with respect to the reflector and to compute the radiation pattern.

Although the theoretical calculation of the apparent phase centre is sometimes possible, it is usually laborious and limited to those antennas for which the complete expression for the far field is known.[81–85] In doing so, the most straightforward approach is to calculate the radius of curvature of the equiphase surface at the point of interest. The rays from the phase reference point to the observation point (the radii of curvature) form a pencil of lines as one of the spherical coordinates, say θ, varies. The evolute of the far-field equiphase contour is traced by the envelope curve of the rays. The evolute is the locus of the phase reference points for varying θ. As θ and ϕ are both allowed to vary, the evolute of the equiphase surface generates a warped surface upon which the phase reference points or apparent phase centres lie. There are in general two principal phase centres on each normal. It is entirely possible for these points to lie outside the volume of the antenna or for the evolute surface to lie partly without the volume of the antenna.

For many applications, it becomes necessary to determine, or check by experimental methods, the location of the phase centre of the antenna. The experimental determination of a phase centre reduces to finding an equiphase surface in the far field. If this surface is a sphere its centre is the phase centre. If it is not, one can sometimes approximate the equiphase surface by a sphere over a limited range of directions and thus obtain the corresponding apparent phase centre.

Some antennas, such as the conical log-spiral, have a natural axis of symmetry. It is convenient to take this axis as the polar axis of the system of spherical coordinates. The phase pattern may then separate into a product

$$\psi(\theta, \phi) = \Theta(\theta)\Phi(\phi)$$

For a particular choice of O on the axis, over some limited angular range of interest, the phase ψ is independent of θ and reduces to $\Phi(\phi)$. This point O can be considered as a phase centre for the θ coordinate.[80] It can be found by the same methods as the ordinary phase centre by observing a cut of the phase pattern for constant value of ψ.

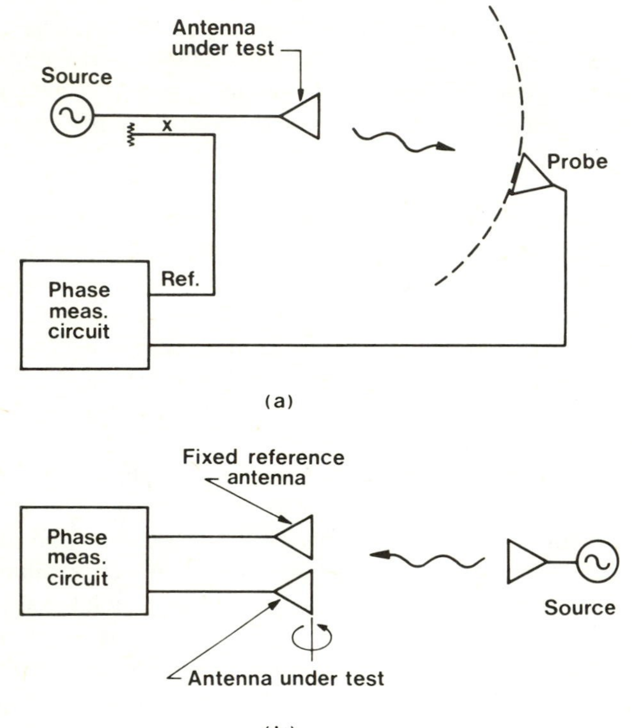

Fig. 8.27 *Measuring phase patterns*

For phase measurements made at short distances, the antenna under test may be used as the transmitting antenna and a simple receiving antenna or probe used to sample the radiated field,[5] as shown in Fig. 8.27*a*. A reference signal may be coupled out of the transmission line leading to the test antenna, and compared with the received signal in a suitable circuit. For measurements at distances too great to permit direct comparison between the sampled signal and this type of reference signal, the arrangement of Fig. 8.27*b* may be used wherein the signal from a distant source is received simultaneously by the antenna under test and by a fixed reference antenna. To measure a phase pattern, the antenna under test is rotated in the usual manner for measuring radiation patterns.

The most comprehensive information about the phase characteristics of an antenna can be gained by recording the complete phase patterns, preferably with an automatic system. From these phase patterns, contours of constant phase and the phase centre or reference point can be determined.

A consideration of the typical characteristics of the phase pattern of antennas with a well defined phase centre is useful in the interpretation of patterns from an antenna with unknown characteristics.[80] Assume that an antenna is given that has a phase centre on the axis of the structure and that the antenna is positioned in an r, θ, ϕ coordinate system with $\theta = 0^\circ$ as the antenna axis. If a phase pattern is recorded for this antenna about some origin, or centre of rotation, which coincides with the phase centre, this pattern over the main beam will by definition be a constant. If the amplitude pattern has sidelobes, there will be a 180° phase reversal at each null and the phase pattern will exhibit an abrupt transition. If now a phase pattern is recorded as this antenna is rotated about some origin which lies along the axis of the structure, but is displaced a distance d from the phase centre, the phase of the field of the antenna will be modified by a cosinusoidal function of θ. When d is very small compared to the distance to the point of observation, and the pattern is normalised to the phase of the signal received when the phase centre is at position A in Fig. 8.28*a*, resultant phase pattern for an antenna with only one radiated beam will be given by

$$\psi' \cong kd(1 - \cos\theta) \tag{8.151}$$

Normalisation to the phase of the signal received when the phase centre is at position B in Fig. 8.28*a* will produce the image of the first pattern.

In theory, the position of the phase centre can be calculated from one such phase pattern. Pattern and experimental anomalies will usually require that several patterns be recorded as the antenna is repositioned along its axis, in order to bracket the curve for $d = 0$.

If the apparent phase centre is displaced from the axis of the antenna by D, as indicated in Fig. 8.28 (*b*) and (*c*), phase change with rotation will be

$$\psi' = k(d^2 + D^2)\left[1 - \cos\left(\theta + \arctan\frac{D}{d}\right)\right] \tag{8.152}$$

If the minimum detectable phase change, that is to say the resolution of the phase detection system is 1/2° and a phase comparison is made over a range of θ of ± 10° from the axis of the antenna, the indeterminancy in the position of the measured phase centre may be as great as 0.1 wavelength. In practice, this uncertainty can be minimised by recording over a greater range of θ, if there is an apparent phase centre over this greater range, and/or by recording a second value for d_{min} when the phase centre approaches the centre of rotation from the opposite direction. In the ideal case, the geometry is such that the two values of d_{min} should indicate a bracket of the true apparent phase centre.

To make these measurements the antenna under test is placed in the far field of a source of the desired polarisation, on a positioner that is capable of precise rotation about a point and of precise translation along the axis of the antenna. To correct for and evaluate a transverse displacement of the phase centre from the axis, and to correct for minor mechanical tolerances in the positioner, it is highly desirable to also be able to precisely translate the antenna in a path which is orthogonal to its axis.

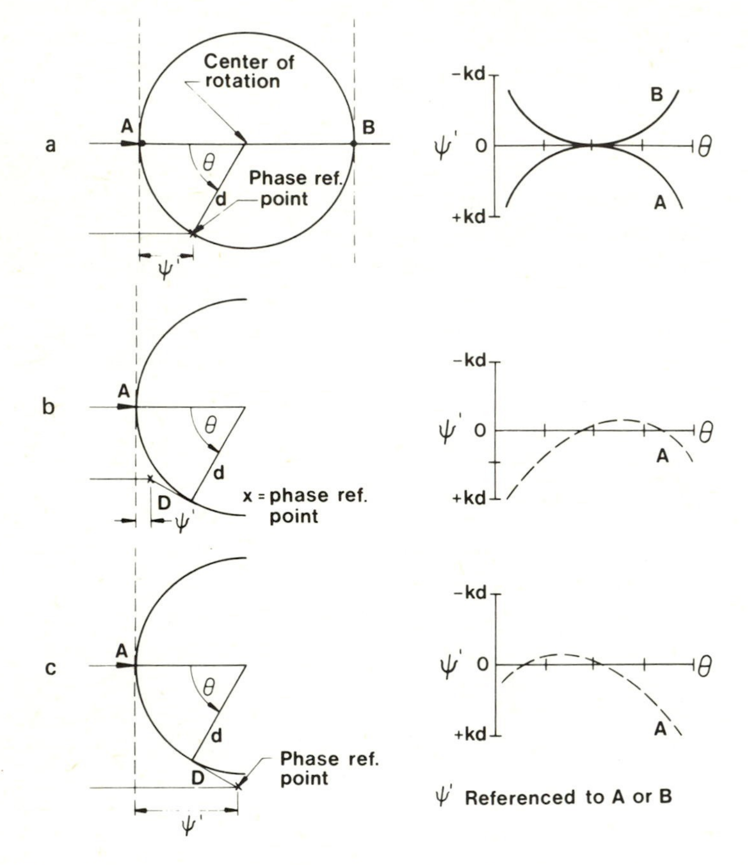

Fig. 8.28 *Phase-change resulting from the rotation of a displaced phase centre about a given origin*

Whether a commercial phase-measuring instrument or an assembled system[86,87] is used, considerable attention must be paid to numerous sources of error. A major source of error in the measurement of phase is due to the interaction of reflected waves from components which are not matched to the waveguide or transmission line in use.[88] The movement of rotary joints and the bending or flexing of coaxial cable can introduce non-negligible phase shifts at higher frequencies.

Measurement systems such as have been discussed, which involve the propagation of energy from the source to the phase measurement receiver through two paths, impose strict requirements on frequency stability of the source. If these paths are not exactly equal in electrical length, a shift in frequency of operation will cause a shift in the measured phase difference between the signals on the two paths. The requirement for equal length paths is, of course, a necessity for swept-frequency measurements.

Accurate phase measurements require careful attention to the details of the system, and assumed accuracies of say 5° or less must be viewed with caution unless all factors have been taken into account.

8.7.4 Boresight measurements

The electrical boresight of an antenna system is a direction determined by an electrical indication from the antenna system. Typically, the electrical boresight is the direction of a pattern maximum or minimum as shown in Fig. 8.29. In order to ensure proper operation of the antenna system, the electrical boresight is aligned within specified tolerances with a reference boresight which is a specified direction, e.g. the direction of a mechanical axis of the antenna or a stationary direction. An angular deviation of the electrical boresight is called the boresight error. It is understood that a boresight measurement is a measurement of direction. The boresight of a pattern maximum is usually determined as the bisector between, the directions to the -3 dB points of the pattern. But, possible asymmetries in the pattern have to be taken into account. In the measurement of the boresight of a pattern minimum, the manner in which the antenna system generates the minimum has to be taken into account in the test procedures.[1] The precision with which this boresight can be determined may be of the order of one hundredth of the beam width.

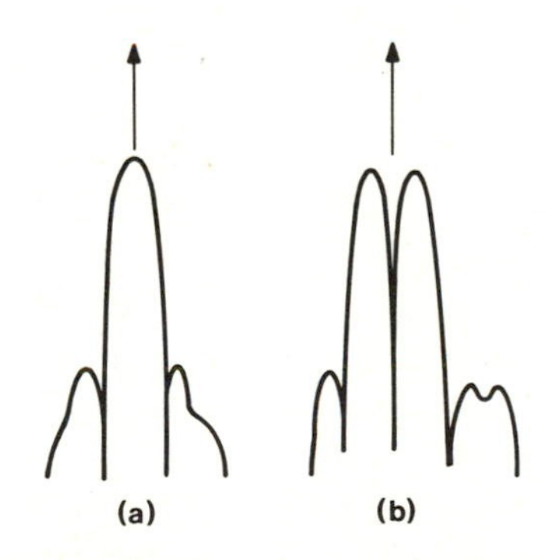

Fig. 8.29 *Electrical boresight directions*
(*a*) Pattern maximum
(*b*) Pattern minimum

A boresight measurement may be carried out by using similar equipment as in a conventional set up. In addition to this, such instruments as telescopes, cameras, mirrors, optical targets and special indicators may be added in order to obtain high precision in the measurement of variation of boresight with system parameters (Oh,[89]). Furthermore, appropriate signal processing equipment may be needed.

The major sources of errors in boresight measurements stem from positioning system, recording instrumentation and extraneous signals. Non-uniform illumination due to finite test distance may only cause minor errors as long as the illumination is symmetrical about the centre of the test antenna. However, it should be noted that, for finite test distances, the measured boresight may depend on the test distance. In particular, this may occur if there is a separation between the test antenna and the centre of rotation used in the experimental set up. Generally, however, tests are performed so that such parallax errors may be present only when testing asymmetrical antennas (Lyon and Fraley[57]). Boresight measurements are primarily made on monopulse antennas and conical-scanning angle-tracking antennas. Special boresight measurements are carried out in connection with measurement of the electrical properties of radomes,[1,90]

8.8 Scattering measurements

It is well known that associated with a receiving process is a scattering process. This is due to the fact that a current distribution is generated on a receiving antenna by the incident field. However, superimposed on this current distribution may be another current distribution due to some power which is reflected back to the antenna gap from an imperfectly matched load. It is based on this observation that techniques have been developed for measuring antenna parameters by radar cross-section measurements (Garbacz[92] and Appel-Hansen[91]). The idea is simply by intention to mismatch the antenna. Then some of the received power will be reradiated in accordance with the transmitting properties of the antenna.

Let us describe the radar cross-section technique with reference to the most simple procedure illustrated in Fig. 8.30. Here, the unknown antenna is connected through a transmission line of length l to a short circuit. When a wave is incident on the antenna a current distribution is generated on the antenna. This current distribution causes two effects. First, the current distribution radiates a scattered power P_S to which corresponds a scattered field $\sqrt{\sigma_s}\, e^{-j\phi_S}$, where σ_S is the scattering cross-section associated with P_S and ϕ_S is the phase of the scattered signal. Secondly, the current distribution causes a received power P_r propagating towards the short circuit. This reflects P_r which is reradiated and causes a scattered field $\sqrt{\sigma_r}\, e^{j\phi_r}$, where σ_r is the scattering cross-section associated with P_r, and ϕ_r is its phase. By making use of the definitions for the receiving cross-section A, the gain G, and the scattering cross-section, the surprisingly simple relation

$$\sigma_r = AG \tag{8.153}$$

can be found (Appel-Hansen[91]). By varying the length l, it is possible in a scattering measurement to determine σ_r and thus, by making use of the relation $A = G\lambda^2/4\pi$, G can be determined. It should be noted that the radar cross-section technique avoids a long disturbing feed line. In addition to this no reference standard gain antenna is needed, because the set up is calibrated using reference spheres for which the theoretical radar cross-section is trusted to a high degree of accuracy. During calibration a properly chosen reference point on the sphere should be used. A reasonably good choice will often be a reference point P near the phase centre of the sphere at a distance R_a from the radar given by

$$R_a = \sqrt{R(R+a)} \tag{8.154}$$

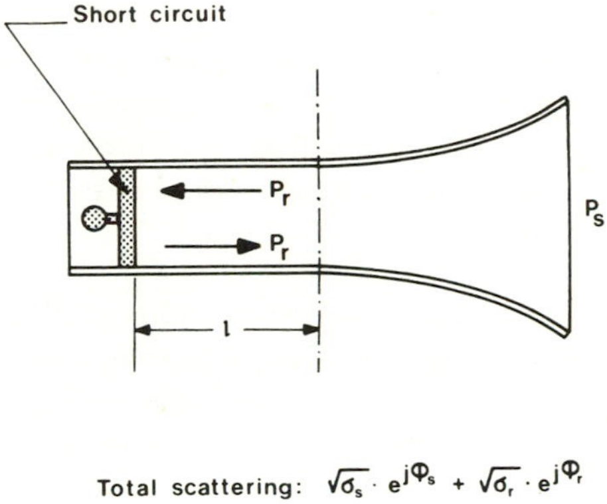

Short circuit

Total scattering: $\sqrt{\sigma_s} \cdot e^{j\Phi_s} + \sqrt{\sigma_r} \cdot e^{j\Phi_r}$

Fig. 8.30 *Radar cross-section technique*

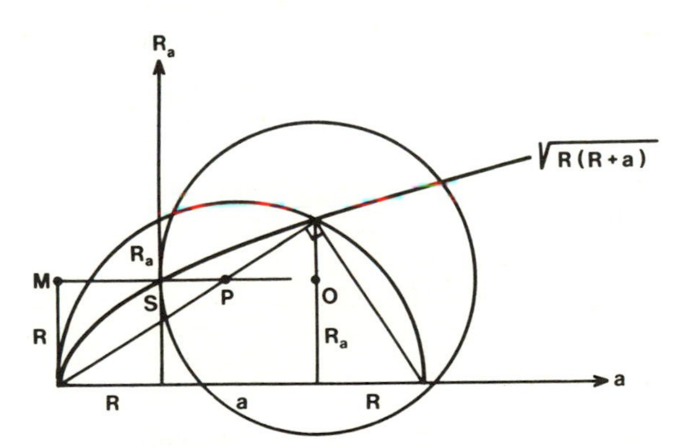

Fig. 8.31 *Construction of apparent radar reflection centre P*

where R and $R + a$ are the distances from the radar to the specular reflection point and to the centre of the sphere, respectively, and a is the radius of the sphere. This reference point can be considered as a good approximation to an apparent radar reflection centre with a radar cross-section corresponding to the true far-field radar cross-section of the sphere (Appel-Hansen[3]). An example construction of the apparent radar reflection centre is shown in Fig. 8.31. The radar technique also has the advantage that in properly designed experiments all antenna parameters can be determined including antenna scattering properties.

8.9 Outdoor test ranges (Contribution by T. G. Hickman)

8.9.1 Introduction

Most antenna measurements made today are on an outdoor test range. It is by far the most popular type of test facility and is frequently the only type that is practical. This Section will discuss the operation of an outdoor range and present the design criteria to which a range should comply in order to make accurate measurements. Rigorous derivations of the criteria are in other publications and will only be referenced here.

There are three basic kinds of outdoor antenna ranges; elevated or free space, ground reflection, and slant ranges. Before discussing these ranges individually, a brief discussion of sources of error will be presented. Each range will then be presented along with its method of controlling these errors.

8.9.2 Error sources

The ideal test environment is a plane wave of uniform amplitude. This environment is never attainable in practice and the design of a test facility is a process by which sources of error are reduced to a level compatible with the measurements to be made. Six forms of field irregularities are present on any antenna range. They are:

(1) inductive coupling between antennas
(2) radiative coupling between antennas
(3) phase curvature of the illuminating wavefront
(4) amplitude taper of the illuminating wavefront
(5) periodic variations caused by reflections
(6) interference from spurious radiating sources

Inductive coupling: Inductive coupling between antennas is rarely a problem except at low frequencies. From the field equations for an elemental dipole, it can be seen that the ratio of the induction field of an antenna to its radiation field is $\lambda/2\pi R$. At a separation of $R = 10\lambda$, this ratio is $1/20\pi$ or -36 decibels. The criterion

$$R \geqslant 10\lambda \tag{8.155}$$

is a commonly used rule of thumb for an antenna range. Such effects are usually considered negligible at greater separations.

Radiative coupling: Radiative coupling between antennas is the result of multiple reflections between the test antenna and the source antenna. The level of these multiple reflections increases as larger portions of the radiated energy illuminate the antenna under test. This effect is controlled primarily through the directivity of the source antenna. It can be shown that this coupling level is at least 45 dB below the original received signal if an amplitude taper of $\frac{1}{4}$ decibel or less illuminates the antenna under test.[57,93] This level is even less for smaller amplitude tapers. It is for this reason that amplitude tapers of less than $\frac{1}{4}$ decibel are usually recommended.

Phase taper: Phase curvature of the illuminating wavefront is controlled by the length of the antenna range. A rigorous development of this is developed in the previous Reference by Hollis. The source antenna is very closely approximated by a point source located at its phase centre. From this point source emanates a spherical wave which illuminates the antenna under test. As the radius of this spherical wave increases, the phase deviation across a fixed aperture decreases. The phase deviation across a test antenna is given by:

$$\Delta\phi = \pi D^2/4\lambda R \tag{8.156}$$

where D is the diameter of the antenna under test, λ is the wavelength of the signal, and R is the separation between the transmit antenna and the antenna under test. A commonly employed criterion is to limit this phase variation at a maximum to $\pi/8$ radians. With this restriction the range length R is limited to:

$$R \geqslant 2D^2/\lambda \tag{8.157}$$

It should be recognised that this separation is not adequate for all antenna measurements and is more restrictive than is required by some. It is, however, quite useful as a benchmark and should be recognised as such. Errors present in antenna patterns as a result of phase taper will be first observed in the nulls of the antenna pattern. A filling of the close nulls occurs with phase taper and a lack of definition in these nulls increases with increasing phase taper. Fig. 8.32 illustrates this effect.

Amplitude taper: The effect of amplitude taper on an antenna test facility is much less severe than for phase variations. Amplitude taper across the test antenna is controlled by the directivity of the source antenna. It was previously shown that the directivity of the source antenna also effects radiative coupling. The effect of amplitude taper on pattern distortion is shown in Fig. 8.33.

It is readily apparent that this effect is very small. This is reasonable when it is realised that an amplitude taper can be thought of as simply altering the aperture distribution of the test antenna, which is in itself typically 10 dB or more.

Reflections: The effect of reflections on pattern measurements can be quite severe. If the reflection is in phase with the direct path signal, the pattern level will appear too large and if it is out of phase with the direct path signal, it will appear too small. In Section 8.11 the magnitude of this error is discussed in detail.

External interference: Errors resulting from external radiating sources are less easily controlled by the physical design of the facility than for the previous sources of error. External sources are controlled primarily by the proper selection of test instruments. Receivers with very narrow bandwidths will discriminate against external emitters which do not fall within the bandwidth. Therefore, a proper selection of instrumentation is an important consideration in the design of any test facility.

8.9.3 Elevated antenna ranges

The elevated antenna range, sometimes known as a free space range, is the most commonly used outdoor antenna range. A diagram of an elevated range is shown in Fig. 8.34. It relies on directive source antennas and adequate tower heights to maintain reflections at a sufficiently low level for the measurements to be made. The parameters to be controlled are:

(1) range length
(2) height of the test tower
(3) height of the transmit tower
(4) beamwidth of the source antenna

In addition, diffraction fences can be utilised to provide additional suppression of reflected energy.

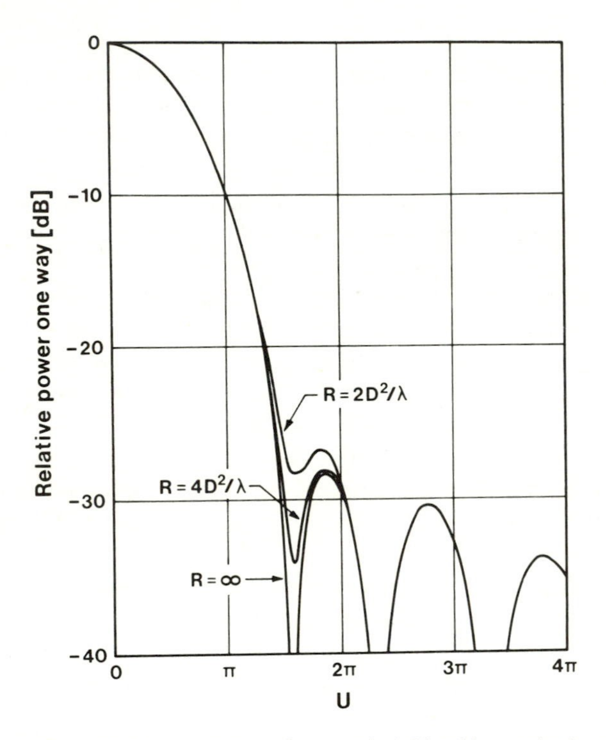

Fig. 8.32 *Calculated radiation patterns of a paraboloid with quadratic phase errors encountered in measuring at three ranges as indicated*

Range length: The range length, or separation between the transmit and test antennas, is determined either by inductive coupling or more generally by the phase taper which is acceptable for the test to be made. As was previously discussed, a commonly accepted rule of thumb is that the range length R is given by eqn. 8.157 This is the first parameter which must be determined.

Tower height: The height of the test antenna, and hence that of the test tower, is the second parameter which should be determined. The height h_r should be at least four times the diameter of the antenna to be tested. That is:

$$h_r \geqslant 4D \tag{8.158}$$

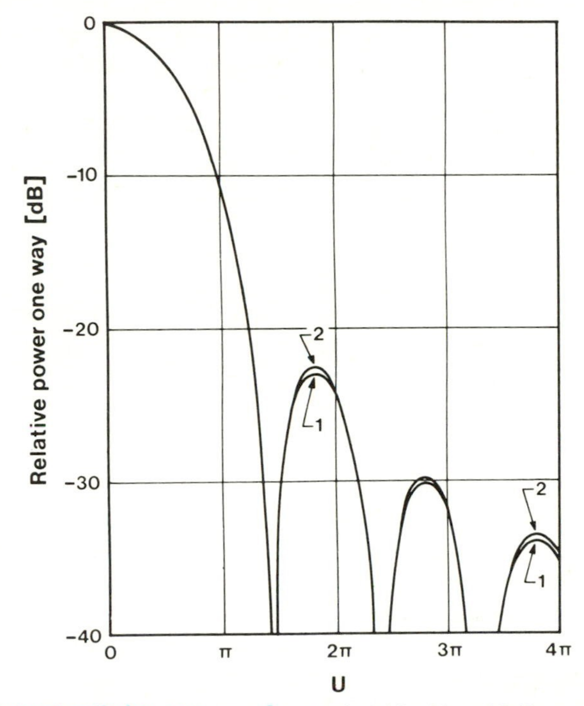

Fig. 8.33 *Calculated radiation patterns of a paraboloid with a 10 dB aperture illumination taper; (1) measured with a 0.5 dB (sin x)/x taper of the source antenna pattern, and (2) with no taper*

A rigorous derivation is presented by Hollis *et al.*[57] and will not be presented here. This restriction, however, results from the fact that the main lobe of the transmit antenna can be approximately characterised by a sin $(x)/x$ pattern and this pattern should illuminate the test antenna with an amplitude taper less than $\frac{1}{4}$ dB with no main lobe energy illuminating the range surface. In order for both conditions to hold simultaneously, the height of the test antenna must be at least four diameters. Sometimes it is not convenient to test with towers this tall. In such cases, compromises must be made and either main lobe energy illuminates the ground resulting in higher reflections or a more restrictive amplitude taper must be accepted, thereby resulting in higher radiative coupling.

Transmit-tower height: The height of the transmit antenna should nominally be equal to that of the test antenna. Variation from this can be made depending on the characteristics of the antenna being tested. If the pattern characteristics most critical to the antenna are near sidelobes, it is usually more advantageous to have the transmit antenna somewhat higher than the test antenna, thereby reducing the reflection levels when measuring these sidelobes. If on the other hand the backlobe region of the antenna, or front to back ratios, are of greater interest, it is somewhat more advantageous to have the test antenna at a greater height. In this case, measurement of the backlobes occur with the mainbeam looking skyward. This ensures that no reflections are received on the high-gain main lobe when measuring low backlobe levels.

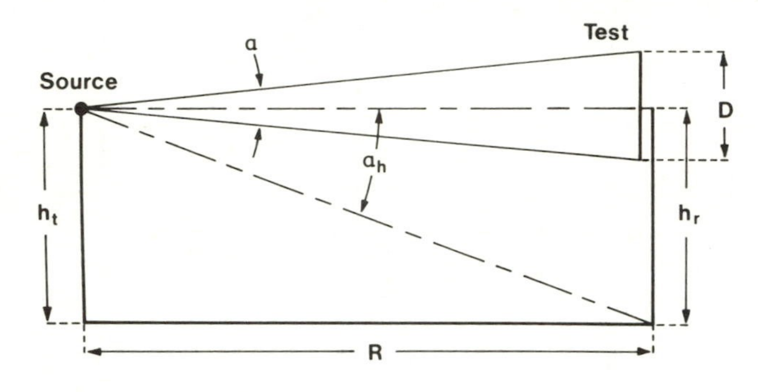

Fig. 8.34 *Elevated range geometry*

Source antenna beamwidth: The beamwidth of the source antenna should be selected so that the $\frac{1}{4}$ dB points fall just outside the antenna under test. In addition, the first null of the radiation pattern should fall above the base of the test tower. These two conditions can be satisfied only when eqn. 8.158 holds as was discussed previously. With only sidelobe energy illuminating the range surface, relatively low reflection levels should be expected.

Diffraction fences: Diffraction fences are not always required on an antenna range but can be used effectively to further reduce reflection levels when required. Extreme caution must be used with diffraction fences to ensure that main lobe energy of the transmit antenna does not illuminate the fence itself. In such cases, diffraction from the edge of these fences can easily be worse than the reflections they eliminate. Fences can be easily constructed with lumber frames and common aluminum window screen surface. Their purpose is to intercept energy normally reflected from the specular region of the range surface and reflect it skyward and away from the test antenna.

8.9.4 Slant ranges

The slant range is a special case of the elevated range. Consider an elevated range with a transmit antenna located essentially at ground level instead of on a tower. The objective is to select a transmit antenna which points at the test antenna and whose first null is along the range surface in the direction of the base of the receive tower. This is in practice difficult to obtain exactly. If the reflection levels obtained with such a configuration are adequate for the measurements to be made, this range can be used with satisfactory results. If, however, they are not sufficiently low, diffraction fences cannot be effectively used as on the elevated range because energy diffracted from the fence itself will probably be quite large.

Another special case of the elevated range is also sometimes called a slant range. That range utilises a tall transmit tower with the test antenna at a low level. This is used most frequently on satellite antennas when near sidelobes are of primary concern, and when it is highly desirable to test the satellite in a radome protected room adjacent to the fabrication facility. This is a practical means of testing antennas on a live satellite prior to launch.

8.9.5 Ground reflection ranges

The ground reflection range is primarily used for the testing of broadbeam antennas at lower frequencies. The elevated range discussed previously requires reasonably directive antennas in order to keep reflections at a low level. This directivity is impractical at low frequencies due to the size of antennas required and the expense of such antennas. Therefore, a different approach is taken for antennas at these low frequencies. This approach is called a ground reflection range.[57] The ground reflection range uses the specular reflection from the antenna range surface in conjunction with the direct path signal to create an interference pattern in which the test antenna is placed. In order for this interference pattern to be smooth and well defined, the antenna range surface must also be smooth. The height of the transmit antenna, h_t, is adjusted such that the distance between the transmit and receive antenna, R_D, is half a wavelength shorter than the path taken by the primary reflection in the range surface, R_R. This configuration is shown in Fig. 8.35. An additional 180° phase shift occurs on reflection, so the two signals arrive in phase at the centre of the test antenna. It can also be shown that this interference pattern is a cosine function with the peak at the centre of the test antenna and the null at the base of the test tower. In order for the antenna to have no more than a $\frac{1}{4}$ dB taper in amplitude, the height of the test antenna h_r must be at least $3.3D$, where D is the diameter of the test antenna. Since approximations are involved in the derivation, the recommended height is given by

$$h_r \geqslant 4D \tag{8.159}$$

just as was the case in elevated range.

This range also differs from the elevated range in that the height of the transmit antenna must be adjustable. This height is given by

$$h_t = \frac{\lambda R}{4h_r} \tag{8.160}$$

Each time the frequency is changed, this height must be readjusted so that the interference pattern is again peaked on the test antenna. The smoothness of the range surface must also be controlled. This smoothness is given by the Rayleigh criterion

$$\Delta h = \frac{\lambda}{m \sin \theta} \tag{8.161}$$

where Δh is the surface variation, λ is the wavelength, m is a smoothness constant nominally equal to 20, and θ is the angle of incidence of the reflected ray as measured from the range surface.

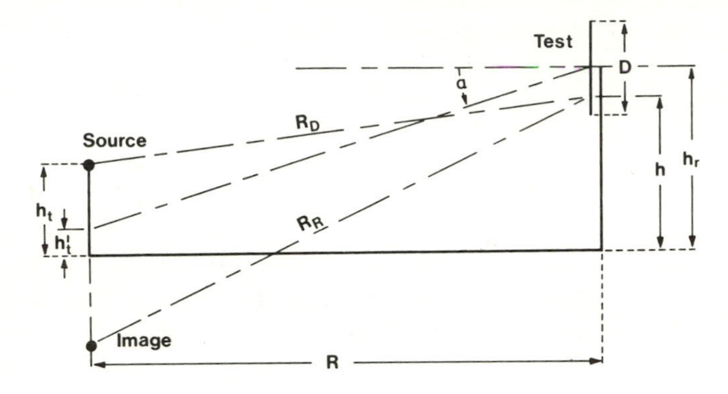

Fig. 8.35 *Ground reflection range geometry*

8.10 Indoor test ranges

8.10.1 Introduction

With the increasing number of applications of electromagnetic waves, the demand for various indoor test ranges has increased over the years. The purpose of a radio anechoic chamber is to simulate a reflectionless free-space environment in which the interaction between electromagnetic waves and many different devices can be investigated under controlled conditions. During the years this purpose has been met in many different ways, depending upon the desired type of investigations. In Section 8.2 indoor near-field techniques are described. The reader is also referred to this Section for a discussion of the advantages of indoor measurements. In Section 8.3 on intermediate range techniques, the compact range technique is described. The near-field and compact range techniques do not require in general a complete lining of the walls, floor and ceiling with absorbers as in anechoic chambers. However, the implementation of these techniques in properly designed anechoic chambers will in general improve the measurement capabilities.

The present Section outlines the construction and mode of operation of anechoic chambers. In particular, the rectangular and tapered chambers will be described. The description of the chambers is given in relation to conventional far-field measurements. A historical summary of the development of microwave absorbing materials and anechoic chambers is presented by W. H. Emerson.[95] An early study of measurements on absorbers and evaluation of chambers was made by Hiatt, Knott and Senior.[97] Several aspects related to the design of anechoic chambers have been outlined by Galagan.[96]

8.10.2 Rectangular chambers

In the early 1950s, the interest in controlling reflections on both outdoor and indoor antenna ranges had increased to such a level that there was a basis for manufacturing special absorbers for such purposes. In the case of indoor measurements, it was natural in the first attempts simply to line walls, floor and ceiling of rectangular rooms with absorbing material. Hence, the first anechoic chambers were rectangular in shape.

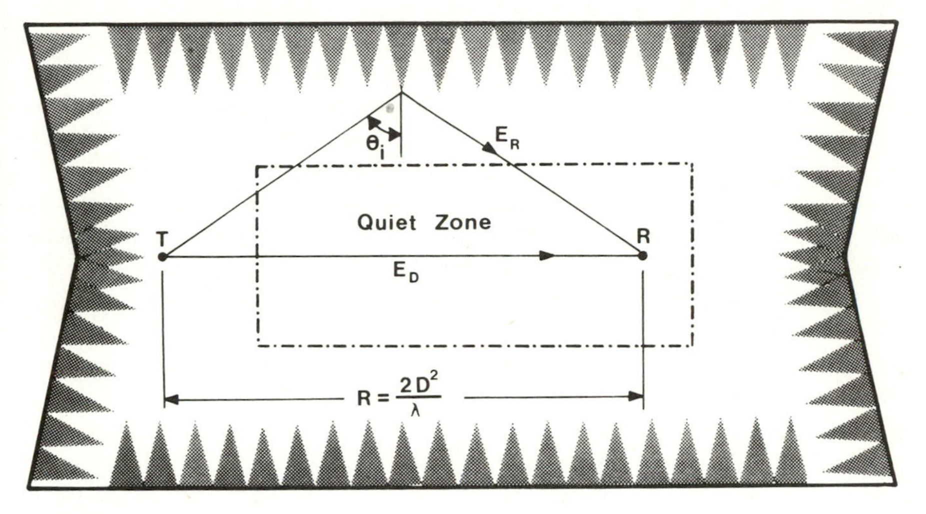

Fig. 8.36 *Rectangular chamber*

In Fig. 8.36 is shown a rectangular chamber. In an antenna pattern or gain measurement a directive illuminating antenna T is usually placed near the middle of one of the end walls and pointing in the direction of the longitudinal axis of the chamber. The test antenna R is placed on the longitudinal axis at a distance R from the illuminating antenna. This sets up a direct field E_D illuminating the test antenna and the surfaces of the chamber. The illumination of the surfaces causes a reflected field E_R to exist in the chamber. Generally, in the design of a rectangular chamber, the planeness of the direct field is required to be in accordance with the conventional far-field criterion and the level of reflections is required to be below a certain specified value in a plane wave test region or a so-called quiet zone placed around the test antenna. The required planeness of the direct field means that the chamber length should allow the criterion $R > 2D^2/\lambda$ to be satisfied. Here, D is the diameter of the test antenna and λ is the wavelength. Since in practice upper limits exist on the length, width and height of chambers, there exists also a limit on the ratio D^2/λ of antennas which can be tested by using conventional techniques. The required low level of reflections also influence the dimensions of the chamber. The chamber width and height should be such that the indicated angle of incidence θ_i on the logitudinal surfaces of the chamber is less than about 60°. A larger angle of incidence will cause large specular reflections from the absorbers. In the assessment of the total length of the chamber, it has to be taken into account that the test antenna usually should be placed at a distance from the tips of the end-wall absorbers equal to half the distance between the tips of the side-wall absorbers. This is done in order to reduce the influence of reflections from the end-wall absorbers including coupling between the absorbers and the test antenna. These considerations lead to a length to width ratio of about 2:1. Often, with the result of larger reflections but in order to satisfy the far-field criterion without having a large width, a length to ratio of 3:1 is accepted.

The major contributions to the inevitable reflections from the absorbers stem from specular reflections. The specular reflections from the longitudinal surfaces of the chamber can be decreased by choosing the directive properties of the illuminating antenna so that the radiation does not reach the longitudinal surfaces through the main lobe. However, some radiation through the side lobes gives rise to reflections. In order to minimise these, baffles or wedges of absorbers were popular in the early days of anechoic chambers, where optical theory was used in design procedures. Quiet zones, for example, were specified as the volume where rays only enter after at least two reflections. Ray tracing resulted in very exotic inventions like the 'coke bottle' and 'throat aperture' designs. However, the simplest wedges were made by letting the entire end walls form wedges, and by introducing longitudinal or transverse baffles along the middle of the remaining surfaces. But it has been found that the dimensions of the baffles have to be optimised carefully. This is due to the fact that baffle constructions are frequency sensitive, edge diffraction may introduce more reflections than are eliminated and absorber surfaces come closer to the centre of the chamber. The result is that rectangular chambers which are constructed at present usually have long absorbers on a plane end wall. Also in regions around specular points of the longitudinal surfaces, absorbers which are longer than on the remaining parts of the surfaces may be installed. It may be advisable to use special fasteners so that absorbers of different length can be tried in order to find the optimum dimension at critical points in the construction.

Absorbing material has been developed to operate in the frequency range from approximately 100 MHz to 100 GHz. At the lower frequency, reflection coefficients of -40 dB have been obtained. At angles of incidence of 60°, reflection coefficients of about -25 dB may have to be accepted. The ideal absorber has an impedance equal to air at the air-absorber interface. In order to convert the electromagnetic energy into heat, the absorbers are manufactured of dielectric and/or magnetic material with loss. Since there should be a smooth impedance variation from the air-absorber interface to the metal plate on which absorbers are usually mounted, most absorbers utilise a resistive taper which is obtained either geometrically (e.g., pyramidal absorbers) or electrically (e.g. absorbers made of alternating layers). High-performance absorbers should have a good impedance match over a wide frequency band for a large range of angles of incidence. The present-day polyurethane foam absorbers loaded with carbon may be up to a few metres in length. Thin ferrite absorbers may be used, but they are costly and may be too frequency sensitive.

By opening the end wall of an anechoic chamber, the indoor test range may be combined with an outdoor range. Thus, a larger test distance can be obtained. In order not to be disturbed by the weather, the end wall can be made as a plastic cover. The roof of the chamber may also be a plastic cover. At low frequencies such a cover may give smaller reflections than a conventional ceiling covered with absorbers. In addition to such arrangements the chamber may be installed on the top of a building so that ground reflections are minimised for the combined indoor and outdoor ranges. Various examples of different indoor test ranges and their design parameters may be found in Appel-Hansen.[94]

Besides frequency ranges, physical size and reflectivity level several other design parameters such as shielding effectiveness, power dissipation capability of absorbers and removability of absorbers have to be specified. In particular, removable floor absorbers may be needed in order to install a ground plane which can take heavy loads. Working floor space in general should be specified together with support capability of special floor absorbers. The test object support should be designed taking the required mobility of the test object into account. To facilitate this, it may be convenient to install special lifts, cranes and motor driven carts. Access ports should also be sufficiently large to bring antennas and equipment to and from the chamber. Anechoic chambers can be used for many other purposes than antenna measurements, e.g. scattering measurements, propagation studies and noise measurements.

8.10.3 Tapered anechoic chambers

At lower frequencies, i.e. below 1000 MHz, it becomes increasingly difficult to obtain accurate measurements in the rectangular chamber. This is due to the fact that the reflections from the absorbers increase with decreasing frequency. In particular, specular reflections from the longitudinal surfaces become large. It becomes difficult to control the illumination of the longitudinal surfaces by using directive antennas. The problem was solved with the construction of the first tapered chambers in the 1960s. As shown in Fig. 8.37 the tapered chamber consists of a funnel or a tapered section and a rectangular test section containing the quiet zone. By placing a transmitting antenna at the apex of the funnel, the test section is illuminated. The funnel length is often about twice as long as the test section. The width of the test section is often close to its length.

A comparison of the rectangular chamber and the tapered chamber will now be made. The different modes of operation may be described with reference to Fig. 8.38. For the sake of illustration the perfect free-space range is also shown. It is seen that introducing surfaces in the form of a rectangular chamber causes reflections. In the Figure only major specular reflections are shown. Wavefronts from the images of the illuminating antenna are indicated. The result is an interference pattern describing the constructive and destructive interference between the direct signal and the reflections. In the tapered chamber the reflections from the longitudinal surfaces are reduced due to the type of illumination. In fact, if the illuminating antenna can be considered as a point source placed at the apex of the funnel no images of the illuminating antenna in the longitudinal surfaces exist. Alternatively, since the funnel may be considered as a horn antenna, although lossy walled, it may be said that the illuminating antenna is an aperture in the form of an entire end wall of the test region. Then the angle of incidence θ_i for specular reflections in the longitudinal walls of rays originating from the center of the aperture is usually less than 45°. Thus, smaller specular reflections can be expected from the longitudinal surfaces in the tapered chamber than in the rectangular chamber, where θ_i equal to 70° may be accepted.

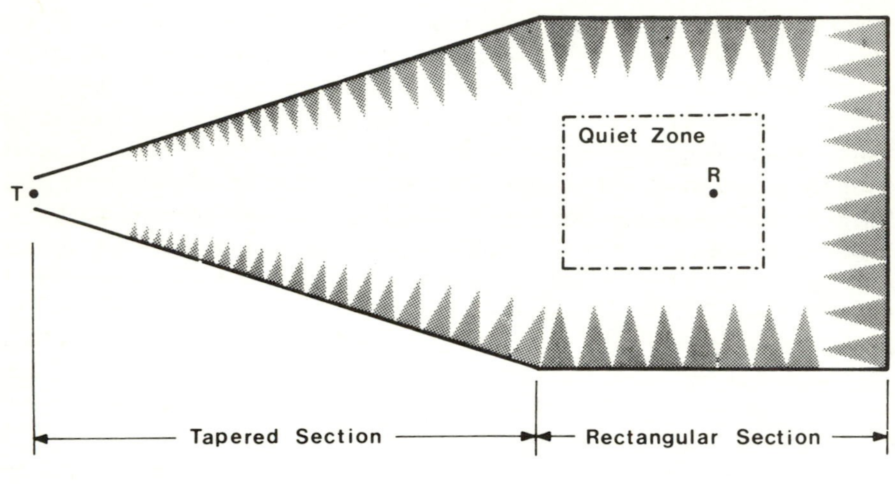

Fig. 8.37 *Tapered chamber*

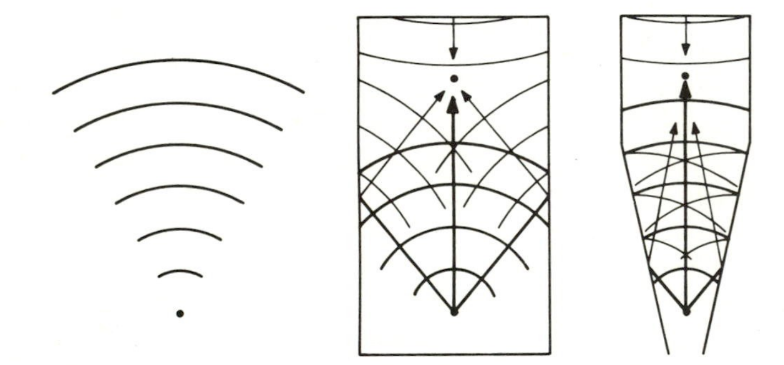

Fig. 8.38 *Comparison of test ranges with respect to major reflections*

In order to place the actually illuminating antenna as close as possible to the apex, the funnel part near the apex may have several access doors. Interchangeable tips for the funnel may be constructed so that the chamber can be illuminated by different small antennas, e.g. open-ended waveguides.

This means that the tapered chamber may not be so critical with respect to directive illuminating antennas as the rectangular chamber. However, no detailed analysis seems to exist of the illumination problem. But, experience has shown that, in comparison with the rectangular chamber, the tapered chamber has a superior performance at low frequencies. Measurements indicate that the wavefront is more plane and the level of interfering reflections is lower. This means that, for the same chamber width, the quiet zone is wider. At mid-frequencies the coupling of the wave to the funnel walls may not be sufficient to obtain better performance. At higher frequencies, it may be easier to illuminate the tapered chamber with a directive antenna placed at some distance from the apex. This means that the operation of the tapered chamber becomes similar to that of the rectangular chamber.

For specific measurement problems it may be advisable to determine the proper location of the illuminating antenna. This can be done experimentally by probing the test region for different positions of the illuminating antenna.

For polarisation measurements, the illumination of the tapered chamber may be critical. Small asymmetries in the funnel and positioning of absorbing material may create cross-polarisation components and differences in propagation of vertically and horizontally polarised waves. In order to reduce such phenomena, it may be advantageous to have a conical funnel part with a circular cross section and a transition to a pyramidal funnel part. Due to the path-loss phenomena, it should be noted that normal field strength and absolute gain measurements based on the transmission formula cannot be made accurately, King, Shimabukuro and Wong.[99]

A tapered chamber may not be so expensive as a rectangular one because a smaller investment in absorbing material is required. This is partly due to the fact that, in the funnel, material with high reflection coefficient is chosen. Near the apex even bare metal walls may improve the illumination of the test region, Hollmann.[98] The funnel cannot be used for anything but launching the wave. This means that in a tapered chamber bistatic scattering measurements cannot be carried out using the total length of the chamber. As in the rectangular chamber, high-quality absorbers are mounted on the end wall.

8.11 Evaluation of test ranges

8.11.1 Introduction

The purpose of a conventional far-field test range is to generate a plane-wave field illuminating the test antenna. Two major sources of error cause this ideal situation not to be obtained. The first source of error stems from the fact that the direct field from the illuminating antenna is not a plane-wave field as discussed in the section on design of test ranges. The second source of error stems from reflected or extraneous signals from the test range. It is mainly this source of error which is studied in the evaluation of test ranges.

On outdoor test ranges, there are usually a finite number of obstacles which give reflections, e.g. the ground, a nearby building and the supports for the transmitting and receiving antennas. Because of the finite number of reflections, there is often no particular difficulty in their evaluation. In the case of indoor test ranges in the form of anechoic chambers, the evaluation of the error due to reflections is more complicated. This is due to the fact that reflections from all directions are incident on the test antenna. Therefore, with the advent of anechoic chambers, the interest in accurate assessment of measurement errors has increased. The evaluation of test ranges will be described mainly with reference to anechoic chambers. However, the procedures can be used for test ranges in general. For a discussion of procedures mainly with reference to outdoor ranges, the reader is referred to the IEEE standard.[1]

It is generally accepted that for different types of measurements, it is necessary to use different evaluation procedures. Two distinct types of measurements exist, i.e. measurements involving one- and two-way transmission. Within each of these types of measurements, the procedure for evaluating the measurement error depends upon the particular measurement procedure. Usually, in the case of one-way transmission measurement, the range is evaluated for antenna pattern measurements and in the case of two-way transmission, the range is evaluated for radar cross-section measurements. In the following only evaluation for antenna pattern measurements will be described. For a discussion of other evaluation procedures the reader is referred to Buckley[105] and Appel-Hansen[100] and their references. Additional information on evaluation of ranges for antenna pattern measurements may be found in Hiatt *et al.*,[97] Hollmann[107] and Appel-Hansen[101,103] and the referenced literature. Also an attempt to find a figure of merit in the case of polarisation measurements can be found in Appel-Hansen and Flemming Jensen[102] and Flemming Jensen.[108]

8.11.2 Basic considerations

The measurement of the reflectivity level is described with reference to the anechoic chamber shown in Fig. 8.39. To demonstrate the meaning of the reflectivity level some basic considerations will be made. A transmitting antenna T is placed near one end wall of the chamber and its main lobe is directed along the main axis of the chamber. A receiving antenna R is placed on this axis at a proper distance from the transmitting antenna. As in conventional pattern measurements the receiving antenna can be rotated in different manners. In the evaluation of ranges, the antennas are often standard gain antennas and the source and detector may be interchanged. In order to measure the reflectivity level the support carrying the receiving antenna can be moved about. In the present case this is arranged by having a rolling cart C in a pit below the base level of the floor absorbers. Longitudinal and transverse movements can be performed. Installation of a lifting gear may provide the possibility of vertical movement. The antenna can be rotated by means of the positioner P on the cart.

When the transmitting antenna illuminates the chamber, there exists at every point a signal which is composed of the direct signal E_D from the transmitting antenna and a reflected signal E_R from the surfaces of the chamber. The reflected signal includes reflections from any obstacles such as frames of instruments placed within the chamber. The direct signal depends in the usual manner upon the radiation characteristic of the transmitting antenna and the distance, whereas the magnitude of the reflected signal varies in a complex manner with the position of the point under consideration. This complex variation is due to the fact that every point of the lining and other objects in the chamber scatter an incident signal. For the sake of illustration three reflections E_{R1}, E_{R2}, and E_{R3} are shown in the Figure.

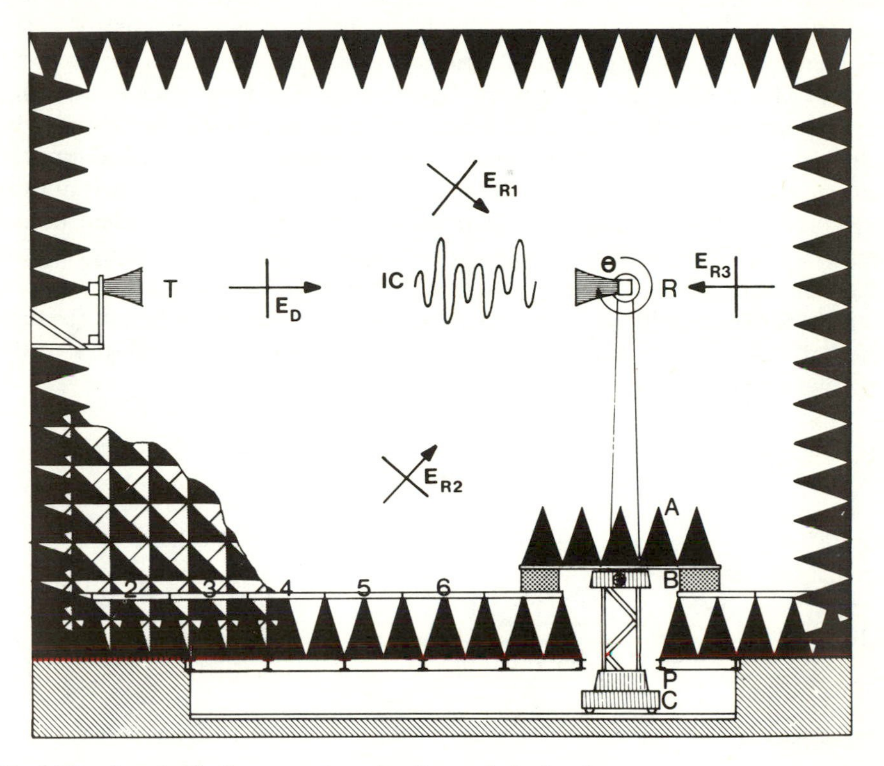

Fig. 8.39 *Longitudinal cross-section of radio anechoic chamber*

The reflected signal E_R will cause errors in measurements made in the chamber. A theoretical evaluation of the errors is prevented due to the very complicated diffraction problem of finding the total interference pattern between E_R and E_D. For determining the errors experimentally, much time is required, since measurements of the reflected field should be made at a sufficient number of points, so that it is possible to derive the destructive and constructive interference between E_D and E_R at every point within the quiet zone of the chamber.

As a measure of the reflected field intensity at a point in the quiet zone, one may use the ratio of reflected field intensity to direct field intensity. This ratio will vary from point to point due to the complex variation of the reflected field. To obtain a complete picture of the variation of the ratio would require much time. Even if this should be possible, an exact evaluation of the measurement error seems to be complicated if not impossible.

The magnitude of the reflected field cannot be determined by a general set up for recording antenna patterns. This is because the reflected field is a vector sum of many components arriving from different parts of the chamber. In order to understand this complication, which influences the measurement of the reflectivity level, let us consider Fig. 8.40. Here, for simplicity, it is assumed that we only have one reflected signal with amplitude E_R and a direct signal with amplitude E_D. The direction of propagation of the reflected signal is along the axis of the receiving antenna which makes an angle ϕ with respect to the direction to the transmitting antenna. By moving the receiving antenna, the ratio between E_R and E_D may be determined from the constructive and destructive interference recorded during the movement. However, because we never have only one reflected field, it is not the magnitude of the total reflected field which is determined. This is evident from a consideration of Fig. 8.40 in the case where E_R is not propagating along the axis of the receiving antenna. The complexity of the problem is also understood by noting that the contribution of the direct signal and of each reflection to the total detected field depends upon the pattern and polarisation characteristics of the receiving antenna.

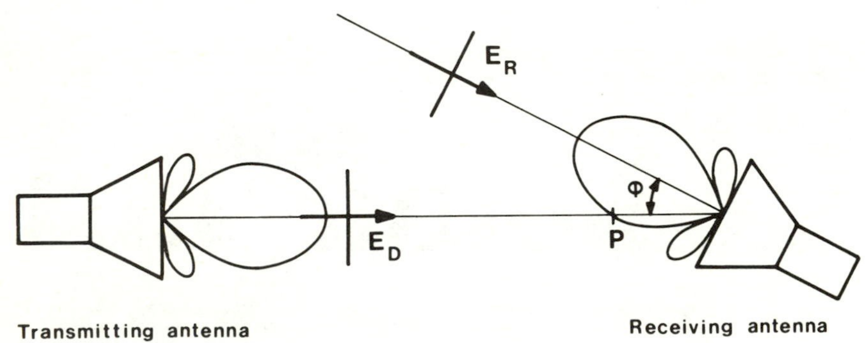

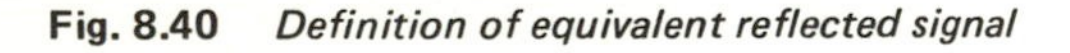
Fig. 8.40 *Definition of equivalent reflected signal*

8.11.3 Analysis of interference curves

The reflectivity level R may be defined in dB, as the ratio between E_R and E_D (Fig. 8.40),

$$R = 20 \log \frac{E_R}{E_D} \tag{8.162}$$

If the test antenna is moved, an in- and out-of-phase interference curve between E_D and E_R can be recorded similar to normal standing-wave curves. In the analysis of the interference curve, it has to be taken into account that the direct signal is detected at a pattern level P corresponding to the aspect angle ϕ. In order to facilitate analysis, we write

$$E_R = E_D\,10^{R/20} \tag{8.163}$$

This means that the total detected field can be considered as the sum of

$$E_D\,10^{P/20} \quad \text{and} \quad E_D\,10^{R/20}$$

In this summation the phase difference between E_D and E_R has to be taken into account. Beforehand, it is not known which one of the two signal levels P and R is the larger, i.e. whether the interference curve varies about the level P or the level R. This is illustrated in Fig. 8.41. It is convenient to introduce the interference distance I, defined as the distance between R and P, i.e. $I = R - P$.

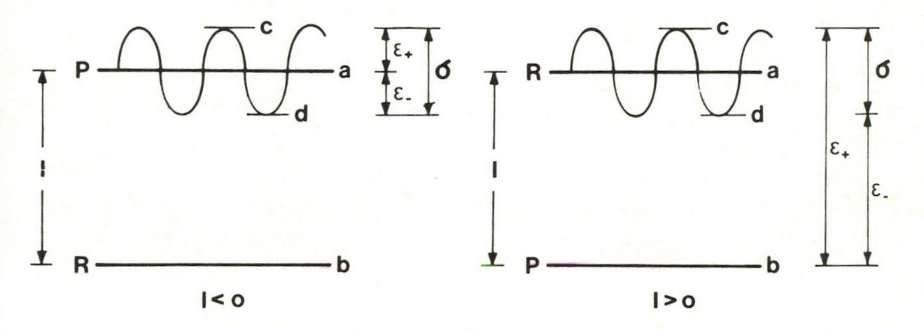

Fig. 8.41 *Interference curves and definition of symbols for negative and positive interference distance $I = R - P$*

Let a denote the larger of P and R and let b denote the smaller. In some cases a will be referred to as the 'level' and b will be referred to as the 'override level'. In addition to this, let c (dB) and d (dB) be the detected signal when E_D and E_R are received in- and out-of-phase, respectively.

As shown in Fig. 8.41, this means that c and d correspond to the maxima and minima of the interference curve, respectively. Then, for conventional normalisation of the main lobe level to 0 dB,

$$\begin{aligned} c &= 20 \log (10^{a/20} + 10^{b/20}) \\ d &= 20 \log (10^{a/20} - 10^{b/20}) \end{aligned} \tag{8.164}$$

Since P is the correct pattern level, let the in-phase error $c - P$ and the out-of-phase error $d - P$ be denoted ϵ_+ and ϵ_-, respectively. In Fig. 8.42a, the errors ϵ_+ and ϵ_- are shown as functions of the interference distance I. It is seen that the out-of-phase error ϵ_- is the largest one for $I < 0$ dB; in particular this is the case for $0 > I > -20$ dB. For $I \leqslant -20$ dB, this worst case error ϵ_- is shown in Fig. 8.42b. However, as seen from Fig. 8.42a, ϵ_+ is nearly equal to $|\epsilon_-|$ for $I \leqslant -20$ dB. As an example, at $I = -20$ dB, $\epsilon_+ = 0.83$ dB and $\epsilon_- = -0.92$ dB. For $0 < I < 3$ dB, $|\epsilon_-| > \epsilon_+$ and for $I > 3$ dB, $|\epsilon_-| < \epsilon_+$. At $I = 6$ dB, $\epsilon_- = 0$ dB.

The range between the curves for ϵ_+ and ϵ_- is the range of variation for the interference curve. The peak-to-peak variation $\sigma = \epsilon_+ - \epsilon_-$ is given by

$$\sigma = 20\log\frac{1+\text{antilog}\,(I/20)}{\pm 1 \mp \text{antilog}\,(I/20)} \tag{8.165}$$

where the upper and lower signs in the denominator are chosen for I negative and positive, respectively. In Fig. 8.42c, σ is shown as a function of I. It is noted that σ is an even function of I. This means that in Fig. 8.42a, the distance between the curves for ϵ_+ and ϵ_-, or the length $c-d$ of the range of variation, is the same for the same numerical value of I.

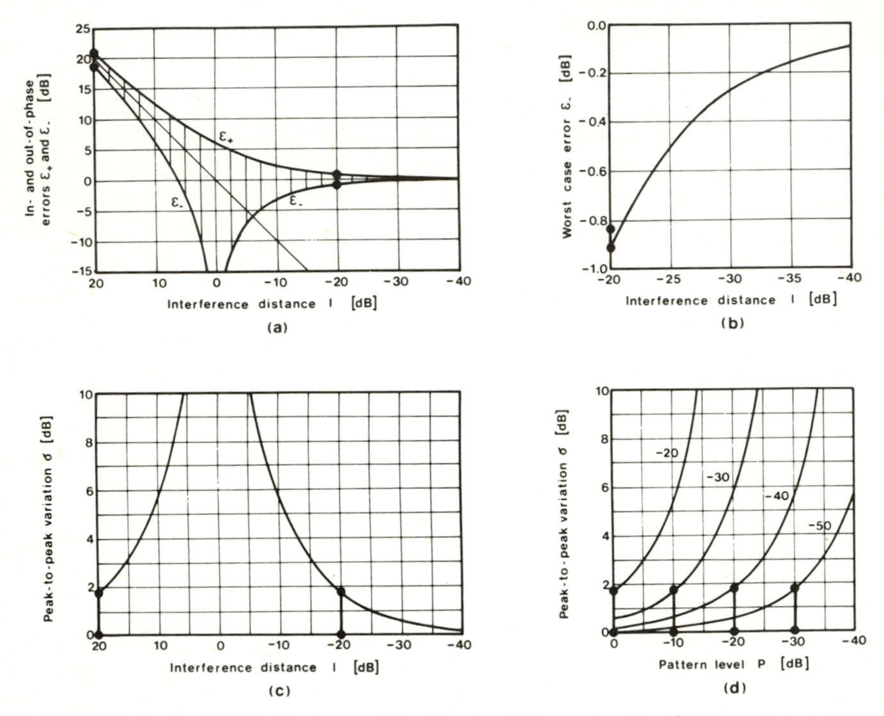

Fig. 8.42 *Interrelationships between interference distance $I = R - P$, in- and out-of-phase errors ϵ_+ and ϵ_-, respectively, peak-to-peak variation σ, reflectivity level R and pattern level P*

In Fig. 8.42d, the peak-to-peak variation σ as a function of the pattern level P is shown for the reflectivity level R equal to -20, -30, -40, and -50 dB. Only curves for the interference distance $I < 0$ dB are shown. Since all curves illustrate the same in- and out-of-phase interference phenomena they are interrelated. As an example, suppose that it is desired to estimate the accuracy at $P = -25$ dB measured on a test range with $R = -50$ dB. Then, the interference distance is $I = -25$ dB and Fig. 8.42a gives $\epsilon_+ \simeq \epsilon_- \simeq 0.5$ dB, Fig. 8.42b gives the worst-case error $\epsilon_- \simeq 0.50$ dB, Fig. 8.42c gives $\sigma \simeq 1.0$ dB, and Fig. 8.42d gives this value directly from knowledge of P and R. As another example, the points marked by dots indicate values for $I = -20$ dB. Thus, it is seen that it is possible from Fig. 8.42

to predict the possible range in which a pattern level P may be recorded in an actual measurement situation, i.e. c and d in Fig. 8.41 can be predicted. Since the analysis given above was based on the interference curves, the VSWR technique for obtaining these curves will be described in the next Section. In this connection it should be noted that σ corresponds to the VSWR in slotted line measurements.

8.11.4 Free-space VSWR technique

The experimental set up for the free-space VSWR technique is described with reference to Fig. 8.43. The transmitting antenna is connected to an oscillator through an attenuator which may be varied for calibration purposes. A counter and power meter monitor the output. The signal from the receiving antenna is guided through a mixer and receiver. By means of the turntable, the antenna may be rotated. The interference curves are recorded when the antenna is moved by rolling the motor-driven cart. Synchros connected to the turntable and cart give readouts of aspect angle and position of the receiving antenna. Signals from the synchros are also fed into the recorder so that the paper chart can follow the movements of the receiving antenna.

It is practice to record interference curves for several aspect angles; e.g. as the aspect angle is varied in steps of 10° from 0° to 360° an interference curve is recorded for each angle. For the sake of brevity we say that interference curves are recorded as the antenna is moved along a traverse line while the main lobe axis of the antenna is scanned in a test plane. An interference curve is shown in Fig. 8.44. This curve shows the interference between the direct and reflected signals. It is seen that oscillations due to reflections are superimposed on the direct signal. Two major types of variations are noticed. The amplitude of the oscillations varies because of the complicated variation of the reflected field with position. The average interference curve level a varies rapidly. In the present case this is due to the fact that the receiving antenna is moved transverse to the transmitting antenna; i.e. the rails in Fig. 8.43 are turned 90°. Thus, the variation in the level a is mainly due to a scanning of the two antenna patterns. If the receiving antenna is moved as shown in Fig. 8.43, the interference curves will often have a sloping character corresponding to the inverse distance variation of the direct signal. This will usually be the case when a corresponds to a pattern level which is several dB over the reflectivity level. However, in case a corresponds to the reflectivity level, the averaged level of the interference curve varies irregularly due to the complicated variation of the reflections.

The analysis of the interference curve in Fig. 8.44 is based on the following arguments. The pattern level P is assumed to correspond to a; i.e. the pattern level is larger than the reflectivity level. If a pattern were recorded, the pattern level P, corresponding to the angle ϕ at which the interference curve is recorded, would be recorded within the limits of the range of variation of the interference curve. In order to indicate this range, the curve is enveloped; i.e. two curves, one connecting the maxima and another connecting the minima, are loosely drawn as shown in the Figure. To determine the worst case errors, the curve is analysed where the maximum peak-to-peak variation is recorded. This means that the maximum value of R is obtained. It is seen that $P = -36.3$ dB, and that the peak-to-peak variation is about 6 dB, which, according to Fig. 8.42*c* corresponds to an interference distance of about -9.5 dB, i.e. a reflectivity level of about -46 dB. In the analysis it has to be taken into account that, for the same value of the peak-to-peak variation $c-d$, the reflectivity level increases with the pattern level.

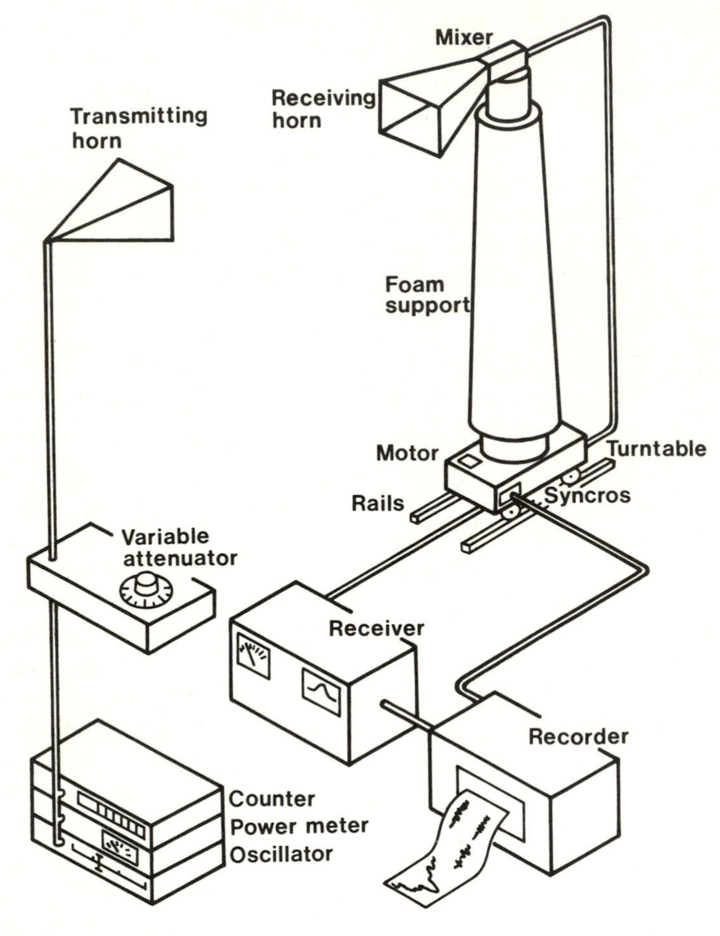

Fig. 8.43 *Experimental set up*

Before the analysis described above can be accomplished, it is necessary to identify the interference curve as one for which the pattern level P is larger than the reflectivity level. Of course, this identification is easy, if P is known beforehand. If this is not the case, it will in general not be erroneous to identify the pattern level with a for interference curves with a regular sloping character, in particular, if this character is according to the inverse distance variation. If for some aspect angles the interference curves vary irregularly, one may renounce information for the considered aspect angles. Alternatively, the ambiguity whether P is a or b may be resolved in different manners, such as analysing the interference curves at different levels a, make continuity and symmetry considerations for the variation in P and R, or change the reflectivity level in the test range, e.g. introducing matallic plates, Appel-Hansen.[101]

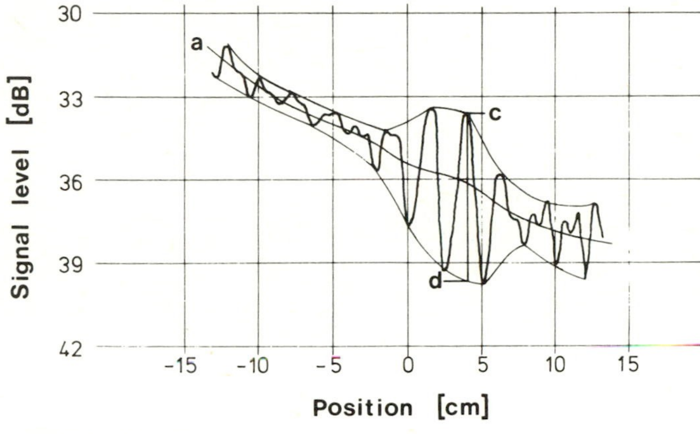

Fig. 8.44 *Analysis of interference curve*

It should be pointed out that identifying R as a or b, at the same time results in identifying P as b or a, respectively. Thus, the VSWR technique can be used for accurate determination of pattern levels. Furthermore, it may even be used to identify pattern levels below the reflectivity level. This is impossible from conventional pattern recordings.

The complete measurement procedure may be described with reference to Fig. 8.45. In order to check that the set up is stable during the recordings of the interference curves, a radiation pattern is recorded before and after the curves. Acceptance of the curves requires that the two patterns agree within a few tenths of a dB. For the present patterns, the test distance corresponds to the middle of the interference curves (0 cm). On top of the Figure one of the patterns is shown. In the middle of the Figure interference curves for ϕ equal to 150° and 180° are shown as representatives for all the curves. The discrepancy between the radiation pattern levels on top of the Figure and the levels of the interference curves at 0 cm is mainly due to imperfections in the mechanical arrangements of the set up. The sloping character of the curves corresponds to the inverse distance variation of the direct signal. Below each curve are given tables with the analysed results. At the bottom of the Figure, the reflectivity level is shown as a function of aspect angle. The maximum value of R equal to -48.4 dB may be used to characterise the test range at 10 GHz for directivity of 16.1 dB. The relative maximum value of R at 0° is due to detectable multiple scattering between the test range and the antennas when they are pointing towards each other. This value indicates inaccuracy in pattern measurements. However, it may be argued that it is irrelevant in the evaluation of the chamber itself.

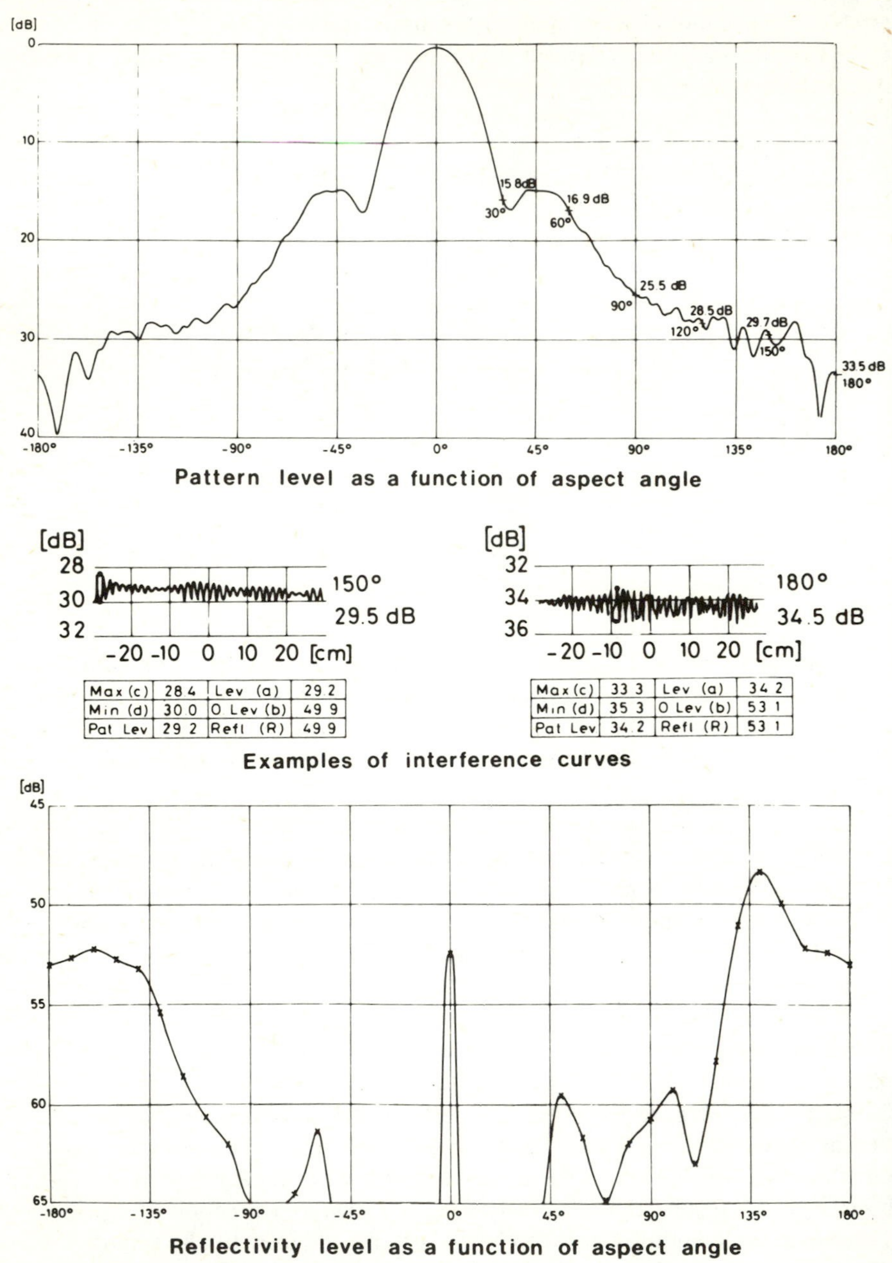

Fig. 8.45 *Analysis and presentation of results for evaluation of chamber at 10 GHz with 16.1 dB directivity antennas*

The multiple scattering mentioned above is a result of reradiation from the receiving antenna. This may also influence measurements at 90° aspect angle for which the reradiation propagating toward the longitudinal surfaces is reflected directly back into the receiving antenna.

8.11.5 Antenna pattern comparison technique

The antenna pattern comparison technique can be carried out with the same type of experimental set up as used for the VSWR technique. The antenna pattern comparison technique is referred to as the APC technique. It is developed from the basic observation that when the pattern of the receiving antenna is recorded at, for example, discrete positions on a traverse line in the quiet zone, small variations occur in the patterns. From the magnitude of these variations the accuracy of the recorded patterns is determined.

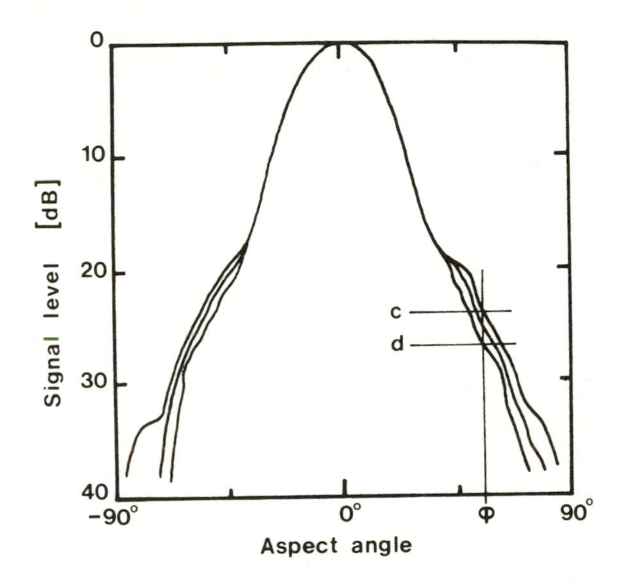

Fig. 8.46 *Antenna pattern-comparison technique*

In order to evaluate the accuracy, all the recorded patterns are superimposed on one another so that their main lobe peak levels coincide, see Fig. 8.46. Then, the maximum variation from d to c at an aspect angle ϕ can be interpreted as the in- and out-of-phase interference between the direct signal and an equivalent reflected signal. Thus, R can be determined as in the case of the VSWR technique. In fact, basically the two techniques are identical. In the VSWR technique, a continuous translation of the receiving antenna is made for discrete values of the aspect angle, whereas in the APC technique, a continuous rotation of the receiving antenna is made at discrete positions. Therefore, from the interference curves discrete points of the antenna patterns can be found and vice versa.

From the APC technique discrete points of the interference curves are often plotted in order to determine R. Thus, by recording a sufficient number of curves and patterns, proper analysis should provide agreement between the two techniques.

However, several comparison studies have demonstrated that in practice it is difficult to obtain agreement between the two techniques. This is mainly due to the fact that, in the APC technique, patterns are usually recorded at discrete positions with an interspacing of a quarter of a wavelength or larger. This interspacing is not sufficiently small to detect all the details of the interference curves. Furthermore, for exact comparison no adjustment of the peak level in the APC technique should be made. However, such an adjustment is common practice in the APC technique. But this should be avoided in case the reflections influence the peak level more than a tenth of a dB. Due to the differences in the final procedures for analysing the patterns and curves, the APC technique often indicates a better reflectivity level than the VSWR technique. A fact influencing this is that usually in the APC technique the reflectivity levels determined at different aspect angles are averaged, whereas in the VSWR technique the maximum value of the reflectivity level is used to characterise the test range. This means that the worst case error is usually indicated when a test range is evaluated by the VSWR technique.

Although the APC technique does not in general give a precise evaluation of a test range, it should be noted that a few patterns recorded and superimposed may give a preliminary insight in the accuracy. If the antenna has some symmetries in its pattern, a consideration of these will give an indication of when the reflections begin to influence the accuracy. This is due to the fact that reflections can be expected to have an asymmetrical effect around pattern maxima, minima and sidelobe levels. Also, a comparison of two azimuthal patterns about the same centre of rotation with the test antenna rotated 180° about its beam axis between cuts may be helpful, IEEE Standard.[1]

8.11.6 Diagnosis of extraneous reflections

During the evaluation of test ranges it is often possible to diagnose the extraneous reflections to such an extent that the major sources of these reflections can be identified. The diagnosis is facilitated by choosing proper traverse lines along which the receiving antenna is moved as well as proper test planes in which the main beam axis of the receiving antenna is scanned. In order to choose these lines and planes, it is convenient to refer to Fig. 8.47. There are three major traverse lines, viz. horizontal longitudinal (or range axis), horizontal transverse and vertical transverse lines which for the sake of brevity are denoted longitudinal, transverse and vertical TL, respectively. Similarly, there are three major test planes, viz. horizontal longitudinal (or azimuthal), vertical longitudinal (or elevation) and vertical transverse (or test aperture) test planes which are denoted horizontal, vertical and transverse TP, respectively.

To illustrate the possible identification of sources of reflections, consider the interference of two plane waves E_D and E_R as illustrated in Fig. 8.48. The longitudinal TL is in the direction of propagation of E_D which may be the direct field of an illuminating antenna. The wave E_R may be a reflection existing in the test range. In the case of transverse movement a spatial period P_t between two adjacent constructive interference points may be observed. From a consideration of wavefronts of E_D and E_R separated by a wavelength λ, P_t is given by

$$P_t = \frac{\lambda}{\sin \phi} \tag{8.166}$$

where ϕ is the angle between the directions of propagation of E_D and E_R. It should be noted that this expression is valid only in the case where the transverse TL is in the plane containing the direction of propagation of E_D and E_R. If the transverse TL makes an angle α with this plane the period is $\lambda/(\sin \phi \cos \alpha)$. Thus, in order to determine the direction to the source of E_R, probing along two traverse lines has to be made.

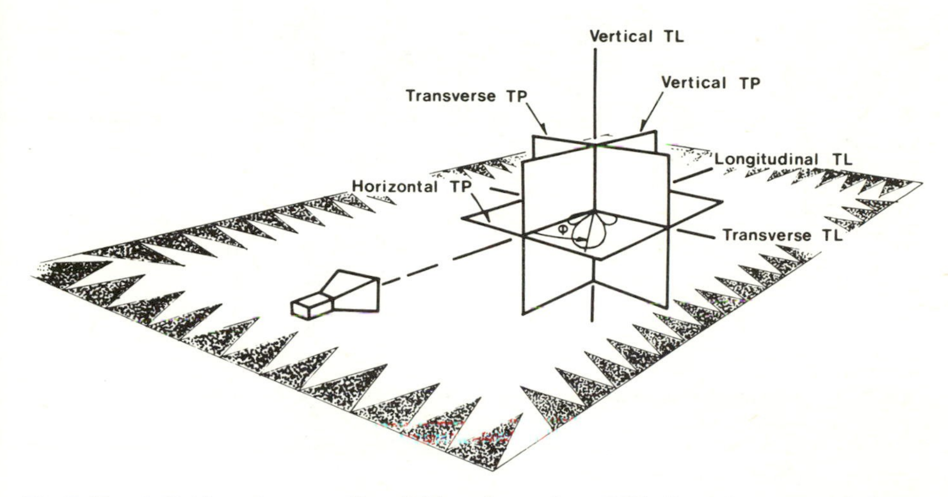

Fig. 8.47 *Definition of traverse lines (TL) and test planes (TP). Receiving antenna is shown with its main direction scanning in the horizontal test plane and moving along a longitudinal traverse line*

In some cases it is convenient to evaluate test ranges using existing positioners which can be moved longitudinally. From a consideration of the lines of constructive interference, it can be shown that the period P_l of the longitudinal interference curve is given by

$$P_l = \frac{\lambda}{2 \sin^2(\phi/2)} \tag{8.167}$$

It is understood that for longitudinal TL, the end wall reflection will cause a period of $\lambda/2$ as in slotted line measurements.

It should be pointed out that the discussion related to Fig. 8.48 is most relevant for tests carried out at high frequencies using a directive receiving antenna pointing towards a specular reflection point, e.g. a specular reflection point of a longitudinal surface of the test range. Furthermore, in actual measurements only approximate agreement with the expressions for P_l and P_t can be observed. This is because the interfering signals are not plane waves and the angle between their directions of propagation changes due to the proximity of the internal surfaces. However, when the difference in path lengths between E_R and E_D is many wavelengths the expressions for P_l and P_t can be considered as good approximations.

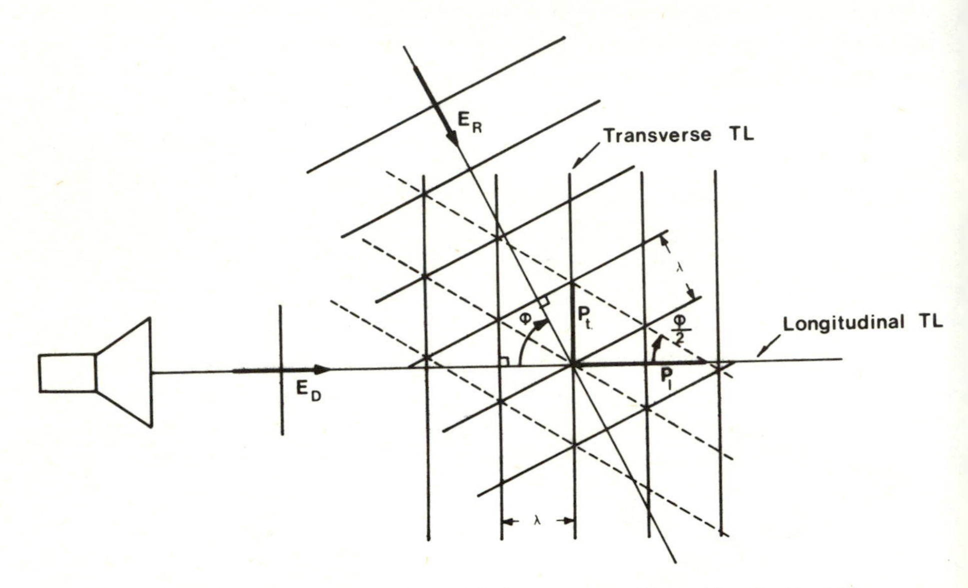

Fig. 8.48 *Interference curve period P_t and P_l for transverse TL and longitudinal TL, respectively*
– – – – Lines of constructive interference (parallel to bisector of angle between directions of propagation)

For tests carried out at low frequencies with low gain antennas in indoor ranges, the discussion related to Fig. 8.48 is not appropriate. This is because low gain antennas are sensitive to reflections from a wide range of directions. An analysis of the interference of specular reflections from the longitudinal surfaces shows that movements within the quiet zone may not be large enough at low frequencies to detect the correct maximum peak-to-peak variation between constructive and destructive interference, Bowman.[104] In order to get an indication of the worst case, the test range may be evaluated at closely spaced frequencies. This will often reveal certain critical frequencies where high values of R are measured. However, the critical frequencies may in general depend upon the position of the illuminating antenna. Therefore, an optimum position of the illuminating antenna may be found in a manner similar to that described below in relation to outdoor ranges.

On outdoor ranges the situation illustrated in Fig. 8.48 may be pronounced because generally only one major reflection in the form of the ground reflection exists. Due to this fact, it is convenient to separate the total illuminating field into two parts: the near-axis incident field and the wide-angle incident field, IEEE Standard.[1] The near-axis incident field consists of the direct field, the ground reflection and other signals coming from directions near to the range axis. The wide-angle incident field consists of reflections from the mounting structure for the receiving antenna and other signals coming from directions having wide angles with respect to the direction of the illuminating antenna. Since the ground reflection is usually the predominant reflection, special attention is paid to the measurement of the near-axis incident field. This is done by using a receiving antenna which discriminates against the wide-angle incident field but which has a directivity such that the direct signal and the ground reflection are detected at nearly the same pattern level. The receiving antenna is moved in the transverse test aperture plane while pointing in the direction of the range axis. Several horizontal and vertical movements may be performed for both horizontal and vertical polarisation. These types of measurements are sometimes referred to as field probe measurements. Interference curves are recorded as in the VSWR technique, but usually only for the zero degree aspect angle.

The field probe measurements can be used to identify the sources of reflections and their magnitudes. Also, they can be used to find the optimum positions of absorbers and diffracting fences used to reduce the reflections. In particular, based on several field probe measurements, the position and alignment of the illuminating antenna can be adjusted to give an adequate symmetric illumination in the test aperture with respect to the range axis. After such field probe measurements, the outdoor range can be evaluated by using the VSWR- or APC-techniques as described in relation to indoor ranges.

As an example, Fig. 8.49 shows the variation of R with aspect angle for tests using a longitudinal TL and a vertical TP for the chamber in Fig. 8.39. Horizontally polarised 20 dB standard gain horns are used at 3 GHz. The reflectivity level is shown for two conditions depending upon whether the tower base absorbers A and some working floor panels are in place or removed. At $\theta = 270°$, it is demonstrated that reflections from the tower base B (Fig. 8.39) may be reduced by making use of the tower base absorbers A. Further information may be found in Appel-Hansen.[101]

If the diagnosis of the sources of reflections is not at a premium, a test probe consisting of three separately detecting small, orthogonal dipole-diode antennas may be used to provide information on the total field amplitude along the traverse line, Crawford.[106] This procedure is valuable in measuring the 'planeness' of the field distributions.

8.11.7 Dependence of reflectivity level on test parameters

Evaluation of a test range can be very time consuming. This is due to the fact that the reflectivity level R depends upon several test parameters, such as frequency, directivity and polarisation of the antennas used in the evaluation. Furthermore, R depends upon the distances and angles which are required to describe the positions and orientations of the illuminating and receiving antennas with respect to the internal surfaces of the chamber. The angles and positions may be described with reference to the major traverse lines and test planes as described in Section 8.11.6. In addition to this, the value obtained for R depends upon the available equipment and the interpretation of the interference errors. This last factor includes the final presentation of the results obtained by analysing the test data.

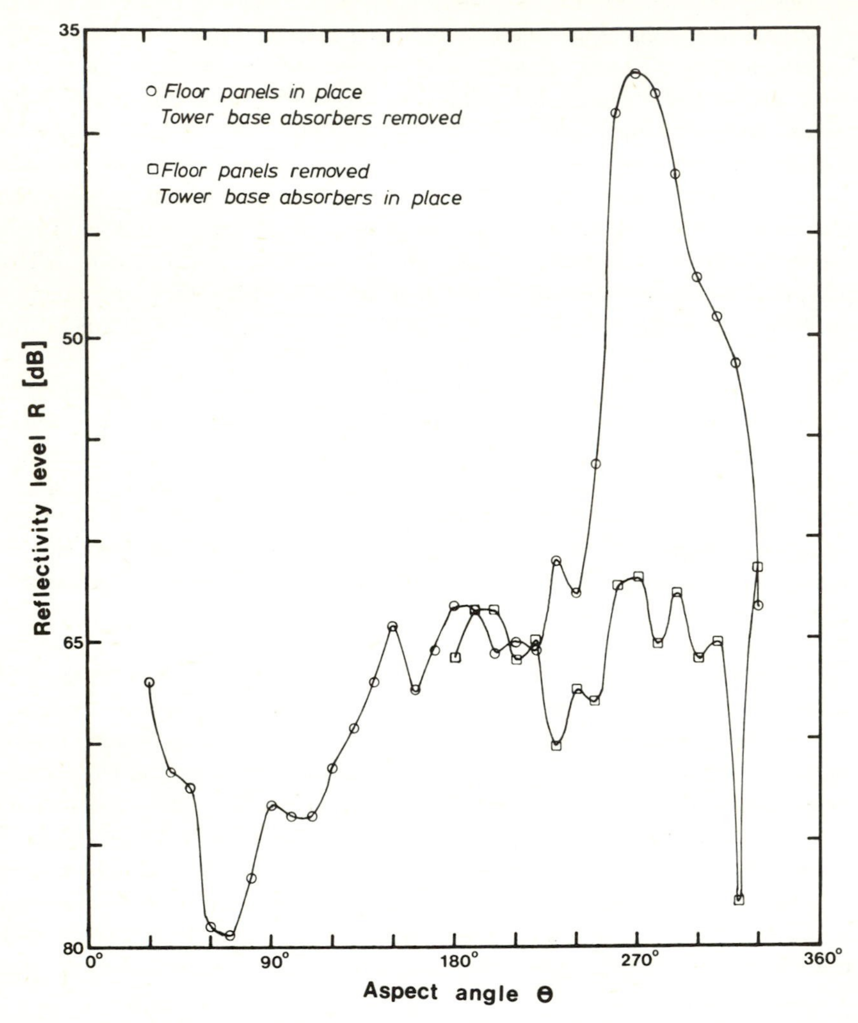

Fig. 8.49 *Reflectivity level for longitudinal TL and vertical TP of chamber shown in Fig. 8.39*

Although detailed information on the dependence of R on all test parameters will probably never be available, some general dependences have been observed. Usually, the reflectivity level decreases with increase of frequency. This is in agreement with the improved behaviour of absorbers as the frequency is increased. Another general observation is a reduction in the reflectivity level as the directivity of the antenna is increased. This is attributed to the fact that there is a reduction in the level of detection of some reflections as the directivity is increased. Also, R can be expected to decrease as the distance from the transverse line to the internal surfaces of the test range is increased. If flat absorbers or a special working floor exist, particular dependence on polarisation may be observed.

From the above it is understood that, depending upon the required information on the test range, it has to be evaluated at several frequencies by using a number of antennas with different directivities and polarisations. Furthermore, several traverse lines and test planes should be used. From Section 8.11.6 it is seen that there are nine combinations of the three major traverse lines and the three major test planes. In addition to this, for each type of traverse line, several lines may be chosen parallel to each other, e.g. 4 or 8 lines properly distributed in the quiet zone and around a center major line. To indicate worst case errors the traverse lines are placed at maximum test distance.

The final selection of traverse lines and test planes depends upon the geometry of the test range and those parts of the test range from where the largest or most critical reflections can be expected. In anechoic chambers, tests carried out using a longitudinal TL combined with both a horizontal TP and a vertical TP will often give information on reflections from all the internal surfaces. However, in order to detect properly reflections from the longitudinal surfaces of the chamber, movements may be required along transverse TL and vertical TL using horizontal TP and vertical TP, respectively. In particular, in order to investigate the reflections from the floor and ceiling of a chamber having a small height length ratio, it may be necessary to use a vertical TL at low frequencies. This is due to the fact that at low frequencies and for small values of ϕ the longitudinal interference curve period P_l may be too long.

It is concluded that for the evaluation of anechoic chambers, most effort should be made at the lower-frequency end of operation of the chamber. It should be evaluated for horizontal and vertical polarisation using low-gain as well as high-gain antennas, in order to indicate measurement accuracy for both types of antennas. It may be sufficient to carry out tests along longitudinal TL using both horizontal and vertical TP. However, if the dimensions of the cross-section are small relative to the length, it may be advisable also to use vertical and transverse TL. It should be emphasised that, if precise information on measurement accuracy is required in an actual measurement situation, the test range has to be evaluated with the actual set up.

8.12 References

1 'IEEE Standard Test Procedures for Antennas', The Institute of Electrical and Electronics Engineers, Inc., ANSI IEEE Std. 149-1979, 1979

2 APPEL-HANSEN, J.: 'The antenna concept in relation to antenna measurements.' Memo, to be published, PC2-SL, 18 Sept. 1980

3 APPEL-HANSEN, J.: 'Centres of structures in electromagnetics – a critical analysis.' Electromagnetics Institute, Technical University of Denmark, Denmark, R 232, Sept. 1980

4 KEEN, K. M.: 'Satellite-antenna measurement techniques,' *IEE Proc.* **127,** PtA, Sept. 1980
5 DYSON, J. D.: 'Measurement of near fields of antennas and scatterers,' *IEEE Trans.* **AP-21,** July 1973, pp. 446–460
6 KING, R. W. P.: 'The theory of linear antennas with charts and tables for practical applications.' Harward University Press, Cambridge, Massachusetts, 1956, pp. 127–141
7 KING, R. J.: 'Microwave homodyne systems,' IEE Electromagnetic Wave Series 3, Peter Peregrimes Ltd., England, 1978, pp. 121–159
8 KNOTT, E. F., LIEPA, V. V., and SENIOR, T. B. A.: 'A surface field measurement facility,' *Proc. IEEE*, **53,** Aug. 1965, pp. 1105–1107
9 KRAUS, J. D.: 'Antennas.' McGraw-Hill Book Company, Inc., 1950, pp. 461–464
10 ABRAMOWITZ, M., and STEGUN, I. A.: 'Handbook of mathematical functions.' Dover Publications, Inc., New York, 1964, pp. 362–363
11 APPEL-HANSEN, J.: 'A note on the rotation coefficient $D^{(j)}_{m'm}(\alpha, \beta, \gamma)$ in addition theorems.' Electromagnetics Institute, Technical University of Denmark, R215, June 1979
12 APPEL-HANSEN, J.: 'On cylindrical near-field scanning techniques,' *IEEE Trans.* **AP-28,** Mar. 1980, pp. 231–234
13 BACH, H., LINTZ CHRISTENSEN, E., HANSEN, J. E., JENSEN, Fr., and HOLM LARSEN, F.: 'Experimental spherical near-field antenna test facility.' Final Report, Vol. I, Electromagnetics Institute, Technical University of Denmark, R 208, Apr. 1979
14 BENNETT, J. C., and SCHOESSOW, E. P.: 'Antenna near-field/far-field transformation using a plane-wave-synthesis technique,' *Proc. IEE*, **125,** Mar. 1978, pp. 179–184
15 BOOKER, H. G., and CLEMMOW, P. C.: 'The concept of an angular spectrum of plane waves, and its relation to that of polar diagram and aperture distribution,' *Journal IEE*, **97,** Pt. III, Jan. 1950, pp. 11–17
16 BORGIOTTI, G. V.: 'Integral equation formulation for probe corrected far-field reconstruction from measurements on a cylinder,' *IEEE Trans.* **AP-26,** Jul. 1978, pp. 572–578
17 BROWN, J., and JULL, E. V.: 'The prediction of aerial radiation patterns from near-field measurements,' *Proc. IEE*, **108B,** 1961, p. 635
18 BRUNING, J. H., and LO, Y. T.: 'Multiple scattering of EM waves by spheres. Part I – Multipole expansion and ray-optical solutions,' *IEEE Trans.* **AP-19,** May 1971, pp. 378–390
19 CARVER, K. R., and NEWELL, A. C.: 'A comparison of independent far-field and near-field measurements of large spaceborne planar arrays with controlled surface deformations.' IEEE Antennas and Propagation Society, International Symposium Digest, Seattle, Washington, Vol. 2, Jun. 1979, pp. 494–497
20 CLAYTON, L., Jr., and HOLLIS, J. S.: 'Calculation of microwave antenna radiation patterns by the Fourier integral method.' Scientific-Atlanta, Inc., The ESSAY, Mar. 1960
21 COREY, L. E., and JOY, E. B.: 'Far-field antenna pattern calculation from near-field measurements including compensation for probe position errors.' IEEE Antennas and Propagation Society, International Symposium Digest, Seattle, Washington, Vol. 2, Jul. 1979, pp. 736–739
22 EDMONDS, A. R.: 'Angular momentum in quantum mechanics.' Princeton University Press, Princeton, N.J., USA, Third printing, with corrections, 1974
23 HANSEN, J. E., and HOLM LARSEN, F.: 'A spherical near-field antenna test facility.' IEEE/AP-S International Symposium, University Laval, Cité Universitaire, Quebec, Canada, IEEE 1980 Symposium Digest, 2–6 June 1980, pp. 264–267
24 HARMENING, W. A.: 'Implementing a near-field antenna test facility,' *Microwave J.*, **22,** Sept. 1979, pp. 44–55
25 FR. JENSEN: 'Electromagnetic near-field–far-field correlations.' Electromagnetics Institute, Technical University of Denmark, LD 15, Jul. 1970
26 JOHNSON, R. C., ECKER, H. A., and HOLLIS, J. S.: 'Determination of far-field antenna patterns from near-field measurements,' *Proc. IEEE*, **61,** Dec. 1973 pp. 1668–1694

27 JOY, E. B., and PARIS, D. T.: 'Spatial sampling and filtering in near-field measurements,' *IEEE Trans.* **AP-20,** May 1972, pp. 253–261

28 KERNS, D. M.: 'Plane-wave scattering-matrix theory of antennas and antenna–antenna interactions.' National Bureau of Standards, NBSIR 78-890, June 1978, pp. 1–277

29 KERNS, D. M., and DAYHOFF, E. S.: 'Theory of diffraction in microwave interferometry,' *J. Res. Nat. Bur. Standards – B.* Mathematics and Mathematical Physics, **64B,** Jan.–Mar. 1960, pp. 1–13

30 HOLM LARSEN, F.: 'Probe correction of spherical near-field measurements,' *Electron. Lett.*, **13,** July 1977, pp. 393–395

31 HOLM LARSEN, F.: 'SNIFTC – Spherical near-field transformation program with probe correction – Manual.' Electromagnetics Institute, Technical University of Denmark, R 201, Oct. 1978

32 HOLM LARSEN, F.: 'Improved algorithm for probe-corrected spherical near-field/far-field transformation,' *Electron. Lett.*, **15,** 13 Sept. 1979, pp. 588–589

33 MARSHALL LEACH, Jr., W., and PARIS, D. T.: 'Probe compensated near-field measurements on a cylinder,' *IEEE Trans.* **AP-21,** July 1973, pp. 435–445

34 NEWELL, A. C., BAIRD, R. C., and WACKER, P. F.: 'Accurate measurement of antenna gain and polarisation at reduced distances by an extrapolation technique,' *IEEE Trans.* **AP-21,** July 1973, pp. 418–431

35 NEWELL, A. C.: 'Upper bound errors in far-field antenna parameters determined from planar near-field measurements. Pt. 2: Analysis and computer simulation.' Lecture notes, National Bureau of Standards, Boulder, Col., July 1975

36 RAHMAT-SAMII, Y., GALINDO-ISRAEL, V., and MITTRA, R.: 'A plane-polar approach for far-field construction from near-field measurements,' *IEEE Trans.* **AP-28,** Mar. 1980, pp. 216–230

37 RODRIGUE, G. P. *et al.*: 'A study of phased array antenna patterns determined by measurements on a near-field range.' Advanced Sensor Directorate, US Army Missile Command, Redstone Arsenal, Alabama, USA, Mar. 1975

38 STEIN, S.: 'Addition theorems for spherical wave functions,' *Quart. J. Appl. Math.*, **19,** 1961, pp. 15–24

39 STRATTON, J. A.: 'Electromagnetic theory.' McGraw-Hill Book Company, Inc., New York, 1941

40 WACKER, P. F.: 'Non-planar near-field measurements: Spherical scanning.' Electromagnetics Division, National Bureau of Standards, Boulder, Col., USA, NBSIR 75–809, June 1975

41 WACKER, P. F., and NEWELL, A. C.: 'Advantages and disadvantages of planar, circular cylindrical, and spherical scanning and description of the NBS antenna scanning facilities.' Antenna Testing Techniques, Workshop at ESTEC, European Space Agency, ESA SP127, Preprint, June 1977, pp. 115–121

42 WACKER, P. F.: 'Plane-radial scanning techniques with probe correction; natural orthogonalities with respect to summation on planar measurement lattices.' IEEE Antennas and Propagation Society, International Symposium Digest, Vol. 2, June 1979, pp. 561–564

43 YAGHJIAN, A. D.: 'Upper-bound errors in far-field antenna parameters determined from planar near-field measurements.' National Bureau of Standards, Technical Note 667, pp. 1–113, Part 1 (Analysis), Oct. 1975

44 YAGHJIAN, A. D.: 'Near-field antenna measurements on a cylindrical surface: A source scattering matrix formulation.' National Bureau of Standards, NBS Technical Note 696, Sept. 1977, pp. 1–32

45 APPEL-HANSEN, J.: 'Antenna measurements using near-field scanning techniques.' Electromagnetics Institute, Technical University of Denmark, Denmark, NB 129, June 1980

46 BICKMORE, R. W.: 'Frauenhofer pattern measurement in the Fresnel region,' *Can. J. Phys.*, The National Research Council, Ottawa, Canada, **35,** July 1957, pp. 1299–1308

47 BROWN, J.: 'A theoretical analysis of some errors in aerial measurements,' *Proc. IEE* Pt. C, **105,** Feb. 1958, pp. 343–351
48 CHENG, D. K., and MOSELEY, S. T.: 'On-axis defocus characteristics of the paraboloidal reflector,' *IEEE Trans.* **AP-3,** Oct. 1955, pp. 214–216
49 JOHNSON, R. C., ECKER, H. A., and MOORE, R. A.: 'Compact range techniques and measurements,' *IEEE Trans.* **AP-17,** Sept. 1969, pp. 568–576
50 JULL, E. V.: 'An investigation of near-field radiation patterns measured with large antennas,' *IRE Trans.* 1962, pp. 363–369
51 KINBER, B. Ye., and TSEYTLIN, V. B.: 'Measurement error of the directive gain and of the radiation pattern of antennas at short range,' *Radio Eng. Electron. Phys.*, 1964, pp. 1304–1314
52 LUDWIG, A. C., and LARSEN, F. H.: 'Spherical near-field measurements from a "compact-range" viewpoint.' Electromagnetics Institute, Technical University of Denmark, Denmark, R 213, Apr. 1979
53 MARTSAFEY, V. V.: 'Measurement of electrodynamic antenna parameters by the method of synthesised apertures,' *Radio Eng. Electron. Phys.*, **13,** 1968, pp. 1869–1873
54 MOSELEY, S. T.: 'On-axis defocus characteristics of the paraboloidal reflector.' Elect. Eng. Dep., Syracuse Univ., Final Rep. USAF Contract AF30(602)-925, 1 Aug. 1954
55 OLVER, A. D., and SALEEB, A. A.: 'Lens-type compact antenna range', *Electron. Lett.* **15,** 5 July 1979, pp. 409–410
56 WACKER, P. F.: 'Theory and numerical techniques for accurate extrapolation of near-zone antenna and scattering measurements.' National Bureau of Standards, Boulder, Col., Unpublished Rep., Apr. 1972
57 HOLLIS, J. S., LYON, T. J., and CLAYTON, Jr., L. (Eds): 'Microwave antenna measurements.' Atlanta, GA, Scientific-Atlanta, Inc., 1970, pp. 7.1–7.9
58 HANSEN, R. C. (Ed): 'Microwave scanning antennas, Vol. 11.' New York, Academic Press, 1966, pp. 38–45
59 STEGEN, R. J.: 'The gain-beamwidth product of an antenna,' *IEEE Trans.* **AP-12,** July 1966, pp. 505–506
60 KUMMER, W. H., and GILLESPIE, E. S.: 'Antenna measurements – 1978.' *Proc. IEEE*, **66,** April 1978, pp. 483–507
61 BEATTY, R. W.: 'Discussion of errors in gain measurements of standard electromagnetic horns.' National Bureau of Standards, Technical Note 351, 1967
62 BOWMAN, R. R.: 'Field strength above 1 GHz; Measurement procedures for standard antennas,' *Proc. IEEE*, **55,** June 1967, pp. 981–990
63 CHU, T. S., and SEMPLAK, R. A.: 'Gain of electromagnetic horns,' *Bell System Tech. J.* **44,** June–Aug. 1965, pp. 527–537
64 JAKES, Jr., W. C.: 'Gain of electromagnetic horns,' *Proc. IRE*, **39,** Feb. 1951, pp. 160–162
65 JULL, E. V., and DELOLI, E. P.: 'An accurate absolute gain calibration of an antenna for radio astronomy,' *IEEE Trans.* **AP-12,** July 1964, pp. 439–447
66 SLAYTON, W. T.: 'Design and calibration of microwave antenna gain standards.' US Naval Research Laboratory, Washington, DC, Rep. 4433, Nov. 1954
67 WRIXON, G. T., and WELCH, W. J.: 'Gain measurements of standard electromagnetic horns in the K and K_a bands,' *IEEE Trans.* **AP-20,** Mar. 1972, pp. 136–142
68 SMITH, P. G.: 'Measurement of the complete far-field pattern of large antennas by radio-star sources,' *IEEE Trans.* **AP-14,** Jan. 1966, pp. 6–16
69 BAARS, J. W. M.: 'The measurement of large antennas with cosmic radio sources,' *IEEE Trans.* **AP-21,** July 1973, pp. 461–474
70 GUIDICE, D. A., and CASTELLI, J. O.: 'The use of extraterrestrial radio sources in the measurement of antenna parameters,' *IEEE Trans.* **AES-7,** Mar. 1971, pp. 226–234
71 KREUTEL, R. W., Jr., and PACHOLDER, A. O.: 'The measurement of gain and noise temperature of a satellite communications earth station,' *Microwave J.*, **12,** Oct. 1969, pp. 61–66

72 WAIT, D. F., DAYWITT, W. C., KANDA, M., and MILLER, C. K. S.: 'A study of the measurement of G/T using Cassiopeia A.' National Bureau of Standards, Boulder, Col., Rep NBSIR 74-382, June 1974

73 BAARS, J. W. M., MEZGER, P. G., and WENDKER, H.: 'The spectra of the strongest nonthermal radio sources in the centimeter wavelength region,' *Astrophys. J.*, **142,** 1965, pp. 122–134

74 BAARS, J. W. M., and HARTSUIJKER, A. P.: 'The decrease of flux density of Cassiopeia A and the absolute spectra of Cassiopeia A, Cygnus A and Taurus A,' *Astron. and Astrophys.*, **17,** 1972, p. 172

75 KUZ'MIN, A. D., and SALOMONOVICH, A. E.: 'Radio-astronomical methods of antenna measurement.' New York, Academic Press, 1966

76 KANDA, M.: 'Study of errors in absolute flux density measurements of Cassiopeia A.' National Bureau of Standards, Boulder, Col., Rep NBIR 75-822, Oct. 1975

77 RUZE, J.: 'Phase centers and phase patterns.' Proceedings Northeast Electronics Research and Engineering Meeting, Boston, Mass., USA, 1963

78 WHEELER, M. S.: 'Phase characteristics of spiral antennas for interferometer applications,' *IEEE International Convention Record*, **12,** Part II, 1964, p. 143

79 VOL'PERT, A. R.: 'On the phase center of antennas,' *Radio Eng.*, (*USSR*), **16,** 1961, (Translation).

80 DYSON, J. D.: 'Determination of the phase center and phase patterns of antennas,' *in* 'Radio antennas for aircraft and aerospace vehicles' W. T. Blackband (Ed). AGARD Conference Proceedings No. 15. Technivision Services, Slough, England, 1967

81 HU, Y. Y.: 'A method of determining phase centers and its application to electromagnetic horns,' *J. of the Franklin Inst.*, **271,** 1961, p. 31

82 BAUER, K.: 'The phase centers of aperture radiators,' *Arch. Elekt. Übertragung.* **9,** 1955, p. 541

83 CARTER, D.: 'Phase centers of microwave antennas,' *IRE Trans.* **AP-4,** 1956, p. 597

84 NAGELBERG, E. R.: 'Fresnel region phase centers of circular aperture antennas,' *IEEE Trans.* **AP-13,** 1965, p. 497

85 GREVING, G.: 'A numerical technique for phase-center calculations using the Poynting-vector approach.' Summaries of Papers 1978 International Symposium on Antennas and Propagation, IECE, Japan, 1978, p. 355

86 DYSON, J. D.: 'The measurement of phase at UHF and microwave frequencies,' *IEEE Trans.* **MTT-14,** 1966, p. 410

87 ELLERBRACH, D. A.: 'UHF and microwave phase-shift measurements,' *Proc. IEEE*, **55,** 1967, p. 960

88 SCHAFER, G. E.: 'Mismatch errors in microwave phase shift measurements,' *IRE Trans.* **MTT-8,** p. 617

89 OH, L. L.: 'Accurate boresight measurements of large antennas and radomes,' *IEEE Trans.* **AP-21,** July, 1970, pp. 567–569

90 LYON, T. J. in WALTON, Jr., J. D. (Ed): 'Radome engineering handbook.' New York, Marcel Dekker, Chap. 7, 1970

91 APPEL-HANSEN, J.: 'Accurate determination of gain and radiation patterns by radar cross-section measurements,' *IEEE Trans.* **AP-27,** Sept. 1979, pp. 640–646

92 GARBACZ, R. J.: 'Determination of antenna parameters by scattering cross-section measurements,' *Proc. IEE*, **111,** Oct. 1964, pp. 1679–1686

93 CHASTAIN, J. B. *et al.*: 'Investigation of precision pattern recording and display techniques.' RADC-TDR-63-247, AD415912, 1963

94 APPEL-HANSEN, J.: 'Radio anechoic chambers.' Electromagnetics Institute, Technical University of Denmark, NB 92, April 1975

95 EMERSON, W. H.: 'Electromagnetic wave absorbers and anechoic chambers through the years,' *IEEE Trans.* **AP-21,** July 1973, pp. 484–490

96 GALAGAN, S.: 'Understanding microwave absorbing materials and anechoic chambers,' *Microwaves*, Pt. I, Dec. 1969; Pt. II, Jan. 1970; Pt. III, Apr. 1970; Pt. IV, May 1970
97 HIATT, R. E., KNOTT, E. F., and SENIOR, T. B. A.: 'A study of VHF absorbers and anechoic rooms.' The University of Michigan, Radiation Laboratory, Report 5391-1-F, February 1963
98 HOLLMANN, H.: 'Reflexionsarme Trichter- und Rechteckkammern für Antennenmessungen,' *NTZ*, **12,** 1972
99 KING, H. E., SHIMABUKURO, F. I., and WONG, J. L.: 'Characteristics of a tapered anechoic chamber,' *IEEE Trans.* **AP-15,** May 1967, pp. 488–490
100 APPEL-HANSEN, J. (Ed): 'Measurements in anechoic chambers.' Lab. of Electromagnetic Theory, Technical University of Denmark, Part I and II, NB 24, February 1968, p. 448
101 APPEL-HANSEN, J.: 'Reflectivity level of radio anechoic chambers.' Special Issue on Antenna Measurements. *IEEE Trans.* **AP-21,** July 1973, pp. 490–498
102 APPEL-HANSEN, J., and JENSEN, Fl.: 'Polarisation measurements and reflectivity level of radio anechoic chambers.' XX Internales Wissenschaftliches Kolloquium, Technische Hochschule Ilmenau, Vol. 3, October 1975, pp. 67–70
103 APPEL-HANSEN, J.: 'The radio anechoic chamber of the Technical University of Denmark and its evaluation.' Workshop on Antenna Test Techniques, ESA, ESTEC, Noordwijk, Holland, 6–8 June 1977
104 BOWMAN, R. R.: 'Prevalent methods for evaluating anechoic chambers: Some basic limitations.' Presented at 1966 Measurement Seminar, Session III, Lecture 3 of High Frequency and Microwave Field Strength Precision, NBS Rep. 9229, 1966.
105 BUCKLEY, E. F.: 'Design, evaluation and performance of modern microwave anechoic chambers for antenna measurements,' *Electron Components*, December 1965, pp. 1119–1126
106 CRAWFORD, M. L.: 'Evaluation of reflectivity level of anechoic chambers using isotropic, three-dimensional probing.' 1974 International IEEE/AP-S Symposium, Georgia Institute of Technology, Atlanta, pp. 28–34
107 HOLLMANN, H.: 'Design and function of anechoic funnel and rectangular chambers for antenna tests.' Deutsche Bundespost, A454 TBr 13, Nov. 1971
108 JENSEN, Fl.: 'On microwave antenna polarisation.' Electromagnetics Institute, Technical University of Denmark, Denmark, LD 29, July 1976
109 OLVER, A. D.: 'The practical performance of compact antenna ranges.' European Space Agency Workshop on Antenna Test Techniques, June 1977, ESA Preprint SP127, pp. 129–134
110 OLVER, A. D., and TONG, G.: 'Compact antenna range at 35 GHz,' *Electron. Lett.* **13,** 14 April 1977, p. 223

Index

TK
7871.6
H34
1982
Vol.1

850634

The Handbook of
antenna design